1993 LECTURES IN COMPLEX SYSTEMS

1993 LECTURES IN COMPLEX SYSTEMS

Editors

Lynn Nadel
Department of Psychology
University of Arizona

Daniel L. Stein
Department of Phyics
University of Arizona

Lectures Volume VI

Santa Fe Institute
Studies in the Sciences of Complexity

Addison-Wesley Publishing Company
The Advanced Book Program

Reading, Massachusetts   Menlo Park, California   New York
Don Mills, Ontario   Wokingham, England   Amsterdam   Bonn
Sydney   Singapore   Tokyo   Madrid   San Juan
Paris   Seoul   Milan   Mexico City   Taipei

Publisher: *David Goehring*
Executive Editor: *Jeff Robbins*
Production Manager: *Michael Cirone*
Production Supervisor: *Lynne Reed*

Director of Publications, Santa Fe Institute: *Ronda K. Butler-Villa*
Production, Santa Fe Institute: *Della L. Ulibarri*

This volume was typeset using TEXtures on a Macintosh IIsi computer. Camera-ready output from a Hewlett Packard Laser Jet 4M Printer.

ISBN 0-201-48368-8

1 2 3 4 5 6 7 8 9 10-MA-9998979695
First printing, July 1995

About the Santa Fe Institute

The *Santa Fe Institute* (SFI) is a multidisciplinary graduate research and teaching institution formed to nurture research on complex systems and their simpler elements. A private, independent institution, SFI was founded in 1984. Its primary concern is to focus the tools of traditional scientific disciplines and emerging new computer resources on the problems and opportunities that are involved in the multidisciplinary study of complex systems—those fundamental processes that shape almost every aspect of human life. Understanding complex systems is critical to realizing the full potential of science, and may be expected to yield enormous intellectual and practical benefits.

All titles from the *Santa Fe Institute Studies in the Sciences of Complexity* series will carry this imprint which is based on a Mimbres pottery design (circa A.D. 950–1150), drawn by Betsy Jones. The design was selected because the radiating feathers are evocative of the outreach of the Santa Fe Institute Program to many disciplines and institutions.

Santa Fe Institute
Studies in the Sciences of Complexity

Proceedings Volumes

Vol.	Editor	Title
I	D. Pines	Emerging Syntheses in Science, 1987
II	A. S. Perelson	Theoretical Immunology, Part One, 1988
III	A. S. Perelson	Theoretical Immunology, Part Two, 1988
IV	G. D. Doolen et al.	Lattice Gas Methods for Partial Differential Equations, 1989
V	P. W. Anderson, K. Arrow, & D. Pines	The Economy as an Evolving Complex System, 1988
VI	C. G. Langton	Artificial Life: Proceedings of an Interdisciplinary Workshop on the Synthesis and Simulation of Living Systems, 1988
VII	G. I. Bell & T. G. Marr	Computers and DNA, 1989
VIII	W. H. Zurek	Complexity, Entropy, and the Physics of Information, 1990
IX	A. S. Perelson & S. A. Kauffman	Molecular Evolution on Rugged Landscapes: Proteins, RNA and the Immune System, 1990
X	C. G. Langton et al.	Artificial Life II, 1991
XI	J. A. Hawkins & M. Gell-Mann	The Evolution of Human Languages, 1992
XII	M. Casdagli & S. Eubank	Nonlinear Modeling and Forecasting, 1992
XIII	J. E. Mittenthal & A. B. Baskin	Principles of Organization in Organisms, 1992
XIV	D. Friedman & J. Rust	The Double Auction Market: Institutions, Theories, and Evidence, 1993
XV	A. S. Weigend & N. A. Gershenfeld	Time Series Prediction: Forecasting the Future and Understanding the Past
XVI	G. Gumerman & M. Gell-Mann	Understanding Complexity in the Prehistoric Southwest
XVII	C. G. Langton	Artificial Life III
XVIII	G. Kramer	Auditory Display
XIX	G. Cowan, D. Pines, & D. Meltzer	Complexity: Metaphors, Models, and Reality
XX	David H. Wolpert	The Mathematics of Generalization
XXI	P. E. Cladis & P. Palffy-Muhoray	Spatio-Temporal Patterns in Nonequilibrium Complex Systems
XXII	Harold Morowitz & Jerome L. Singer	The Mind, The Brain, and Complex Adaptive Systems

Lectures Volumes

Vol.	Editor	Title
I	D. L. Stein	Lectures in the Sciences of Complexity, 1989
II	E. Jen	1989 Lectures in Complex Systems, 1990
III	L. Nadel & D. L. Stein	1990 Lectures in Complex Systems, 1991
IV	L. Nadel & D. L. Stein	1991 Lectures in Complex Systems, 1992
V	L. Nadel & D. L. Stein	1992 Lectures in Complex Systems, 1993
VI	L. Nadel & D. L. Stein	1993 Lectures in Complex Systems, 1995

Lecture Notes Volumes

Vol.	Author	Title
I	J. Hertz, A. Krogh, & R. Palmer	Introduction to the Theory of Neural Computation, 1990
II	G. Weisbuch	Complex Systems Dynamics, 1990
III	W. D. Stein & F. J. Varela	Thinking About Biology, 1993

Reference Volumes

Vol.	Author	Title
I	A. Wuensche & M. Lesser	The Global Dynamics of Cellular Automata: Attraction Fields of One-Dimensional Cellular Automata, 1992

Contributors to This Volume

Eric J. Anderson, Duke University
Philip E. Auerswald, University of Washington
Ann M. Bell, University of Wisconsin
S. N. Coppersmith, AT&T Bell Laboratorties
Rajarshi Das, University of Colorado at Boulder and Santa Fe Institute
William Fortin, Florida Atlantic University
Gyöngyi Gaál, University of California, Berkeley
Liane M. Gabora, University of California, Los Angeles
Maureane Hoffman, Duke University
Alfred Hübler, University of Illinois, Urbana
Jan Tai Tsung Kim, University of Washington
Jan T. Kim, Max-Planck-Institut für Züchtungsforschung
James E. Kittock, Stanford University
José Lobo, Cornell University
Catherine A. Macken, Los Alamos National Laboratory and Santa Fe Institute
Brett McDonnel, Stanford University
Bartlett W. Mel, University of Southern California
Dougald M. Monroe, University of North Carolina
Lynn Nadel, University of Arizona
Arjendu Pattanayak, University of Texas, Austin
S. L. Pepke, University of California, Santa Barbara
B. Kean Sawhill, Santa Fe Institute
Olaf Sporns, The Neuroscience Institute
Peter F. Stadler, Institut für Theoretische Chemie, Austria and Santa Fe Institute
Daniel L. Stein, University of Arizona
William Sulis, McMaster University
Manfred Tacker, Institut für Theoretische Chemie, Austria
Randall Tagg, University of Colorado
Kurt Thearling, Thinking Machines
Patrick Tufts, Brandeis University
Ken Umeno, University of Tokyo
Andreas Wagner, Yale University

Contents

Participant Contributions 395

Preface

The 1993 Complex Systems Summer School once again provided an exciting atmosphere for research, learning, and discussion in a wide variety of fields and topics. As in previous volumes, the contents of this book reflect the topics discussed in the 1993 Summer School, although a few do not appear within. We are also pleased to include a number of contributions from the participants themselves. These are the result of research by individuals or working groups set up during the school. The results are quite impressive. Special thanks to Ann Bell for her efforts on this part of the volume, and to Gyöngi Gaál and Ken Ray for arranging the student seminar series during the Summer School itself.

This will be the last of the summer school lectures volumes in their present form. We plan to begin a new series of volumes, each of which collects both new and previously published Summer School lectures revolving around a common theme. A strong effort will be made in each to find the connections among a wide variety of complex systems with each other and with the central theme. The first of these will be on Pattern Formation, and we hope it's the first of many.

ACKNOWLEDGMENTS

Many people contributed to the success of the summer school. The planning for the school, its day-by-day functioning, and the follow-up after the school finishes are all a reflection of the efforts of a number of people at the Santa Fe Institute. Ed Knapp and Mike Simmons gave much of their time and effort to the Summer School; Ginger Richardson and Andi Sutherland were indispensable from start to finish, as usual; Ronda Butler-Villa and Della Ulibarri played a major role in getting this volume together; Marcella Austin handled the rather complex financial side; and David Mathews got the computational laboratory up and running, and kept it that way. Stuart Kauffman of the Santa Fe Institute committed much of his time to the students, and could often be seen in heated discussion with groups of them in the courtyard of the Institute. Most critical of all, Erin Copeland performed myriad tasks which kept the school running smoothly.

We thank our advisory board, and several institutions that provided computers and associated peripherals. We also thank the University of Arizona, and its Center for the Study of Complex Systems, for permitting the two of us to spend time on this rewarding but time-consuming enterprise. Finally, we must thank those agencies that contributed the funds needed to make the school a reality: financial support was supported by the National Science Foundation, the Department of Energy, Office of Naval Research, the National Institute of Mental Health, Sandia National Laboratories, the Center for Nonlinear Studies at Los Alamos National Laboratory, Institutional Collaborative Research Program of the University of California (INCOR), the Los Alamos Graduate Center of the University of New Mexico, the University of Arizona, and Professor Marcus Feldman; student support was provided by the University of Florida, University of Michigan, and Stanford University; and computer equipment was provided by the University of New Mexico, Digital, Sun Microsystems, Silicon Graphics, and Computerland of Santa Fe.

Lynn Nadel
University of Arizona
Tucson, AZ 85721

Daniel Stein
University of Arizona
Tucson, AZ 85721

January, 1995

Lectures

S. N. Coppersmith
AT&T Bell Laboratories, Murray Hill, NJ 07974

Complex Structures and Dynamics in Condensed Matter Systems

This chapter discusses how systems that condensed matter physicists study exhibit complexity. The behavior of a model system of balls connected by springs in a periodic potential is discussed in detail.

1. INTRODUCTION

Condensed matter physicists study the behavior of condensed phases such as liquids and solids (solid state physics, the study of solids, became overly confining, so a newer, broader field has emerged). Because the systems studied have many interacting degrees of freedom, it is not surprising that they can display complex behavior.

This chapter attempts to communicate a few of the ideas that have emerged from condensed matter physics that could have relevance to complex systems in general. The first section will discuss renormalization group ideas that have proven useful in describing and understanding not only condensed matter systems but

1993 Lectures in Complex Systems, Eds. L. Nadel and D. L. Stein,
SFI Studies in the Sciences of Complexity, Lect. Vol. VI, Addison-Wesley, 1995

also dynamical properties such as the transition to chaos. This section stresses the importance of scale invariance in understanding the properties of some condensed matter systems, which has led to enormous interest in fractals[47,48] and the renormalization group.[76,77,78]

The remaining sections will consider one model system in detail. They will focus on the statics and dynamics of a model system consisting of balls connected by springs in a periodic potential. It will be found that some of the behavior that this system exhibits can be understood using renormalization group ideas, but some of the dynamical properties do not fit simply into a picture based on the ideas in the first section. I will discuss areas where our understanding is lacking and try to indicate areas for future work, including connections of the ball-and-spring system to ideas of self-organized criticality.

2. SCALE INVARIANCE AND THE RENORMALIZATION GROUP

The concept of fractals[47,48] is very important when people discuss complex systems. A fractal object is invariant under a change of scale. There are trivial fractals, such as the unit interval, which looks like a line whether you examine it in the range $[0, 1]$ or $[0, 1/3]$. However, it is possible to have more interesting fractals, one example being the Cantor set, which is constructed by taking the unit interval and removing the middle third, taking the remaining intervals of $[0, 1/3]$ and $[2/3, 1]$ and removing their middle thirds, and so on ad infinitum. The resulting object looks the same whenever the structure is magnified by a factor of three, though it is clearly a lot different than the unit interval itself.

One way to diagnose whether a system is displaying scale invariance is to measure correlation functions. An example of a correlation function that one can define is the local magnetization in a magnet. One can measure the direction of the magnetization at two points x and x' and then ask how the degree of correlation between the two measurements depends on the separation between x and x'. One does this by measuring between all points separated by a given distance, so one is obtaining statistical information about the spin configuration. It is possible to show that if the system is scale invariant, then the correlations must decay as power laws of the separation between x and x'. Thus, power laws are intimately associated with scale invariance.

Many systems in nature display scale invariance, and one of the important questions in the study of complexity is to understand why this is so. In this section two examples where the emergence of scale invariance is understood are discussed. The first arises in the study of second-order phase transitions in statistical mechanics, at a critical point separating an ordered and disordered phase. The second occurs at the transition to chaos in a dynamical system.

2.1 PHASE TRANSITIONS

One example of a physical system which exhibits interesting is a material which undergoes a phase transition into a magnetic state as the temperature is lowered.[1] A simple model that can be used to describe a magnetic transition is the Ising model, where the lth spin can take on the values $S_l = +1$ and $S_l = -1$, and nearest neighbor spins have lower energy if they are identical rather than different. Thus, the system is described using a Hamiltonian:

$$H = -J \sum_{\langle lm \rangle} S_l S_m \,, \tag{1}$$

where the coupling constant is J and the sum is over nearest neighbors on the lattice. The energy of the ith configuration is $E[i] = H$, evaluated for the spin values $S_l\{i\}$. The equilibrium properties of the system can be determined using statistical mechanics. One calculates the partition function Z of the system, which is defined via:

$$Z = \sum_{\text{states i}} e^{-E[i]/k_B T} \,, \tag{2}$$

where $E[i]$ is the energy of state i, k_B is Boltzmann's constant, and T is the temperature. Physical properties of the system, such as the magnetization and spin-spin correlation function, can be expressed as statistical averages also; for instance, the spin-spin correlation function $g(r)$ can be written:

$$g(r) = \frac{1}{Z} \sum_{\text{states i}} S(0)S(r)e^{-E[i]/k_B T} \,. \tag{3}$$

The magnetization M is nonzero for temperatures below the transition temperature T_c and zero above T_c. The region near T_c is known as the critical region. Just below T_c, M obeys a power law:

$$M \propto (T - T_c)^{\delta} \,, \tag{4}$$

and just at T_c $g(r)$ obeys:

$$g(r) \propto (r)^{-(d-2+\eta)} \,. \tag{5}$$

Here, d is the number of dimensions, and δ and η are critical exponents.

The properties near the critical point depend only on the correlation length ξ. Since ξ can be changed by altering either the temperature or the magnetic field, the effects of these two different perturbations enter in a special way. For example, one can write a scaling form that describes the variation of the magnetization M on both temperature T and magnetic field H:

$$M \propto (T_c - T)^{\beta} \mathcal{F}\left(\frac{T_c - T}{H^{\delta}}\right) \,, \tag{6}$$

[1] For a reference, see, e.g., Ma.[44]

where β and δ are "universal" critical exponents.

The observation of power laws as well as the scaling behavior is an indication that this system is exhibiting scale invariance at the critical point. This notion can be quantified using the renormalization group.[76,77,78] The process is shown schematically in Figure 1; it involves a partial summation of the partition function. One sums over (hence eliminates) half the spins in the system, and then rescales lengths and spin magnitudes to write the result as a partition function over renormalized spins.

As indicated in the figure, in general the renormalization group transformation changes the Hamiltonian when it is implemented. However, as one continues iterating the transformation, the Hamiltonian eventually stops changing. The resulting *fixed point* Hamiltonian describes a scale-invariant state. Universality arises because many different Hamiltonia yield the same fixed point Hamiltonian under repeated application of the renormalization group.

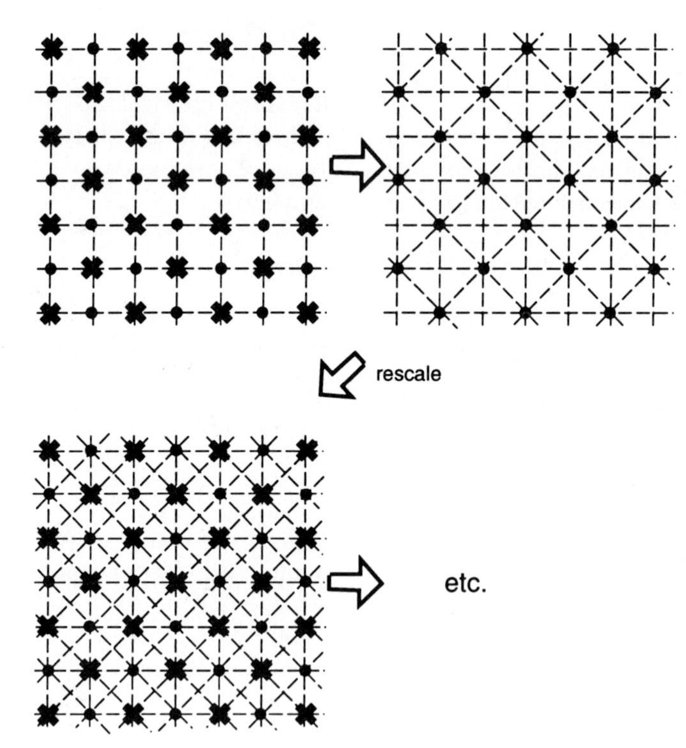

FIGURE 1 Schematic of renormalization group transformation for a two-dimensional spin system.

The Ising model with $J > 0$ in three dimensions has three fixed points. Two so-called trivial fixed points are those that correspond to zero temperature and infinite temperature. At zero temperature all the spins are aligned; if one eliminates every other spin, then the remaining spins are still all aligned, so that the configuration is scale invariant. At infinite temperature, neighboring spins are totally uncorrelated; if one eliminates every other spin then the resulting configuration still has totally uncorrelated neighboring spins. A third fixed point corresponds to the critical temperature T_c; at this fixed point there are nontrivial correlations between the spins, yet the configuration is scale invariant. The properties of this fixed point determine the critical exponents.

2.2 ROUTES TO CHAOS

Another situation where the renormalization group has proven useful occurs when one considers the behavior of a nonlinear dynamical system. For example, one can consider the logistic map, defined by the equation:

$$x_{n+1} = \lambda x_n (1 - x_n) \,, \tag{7}$$

where x is taken to be between 0 and 1. One can examine the behavior of this system as the control parameter λ is increased.

The value $x = 0$ is a fixed point of Eq. (7) since if $x_n = 0$, then $x_{n+1} = 0$ also. However, the solution will not be observed unless small deviations from the fixed point do not grow as the map is iterated. One finds that small perturbations to x decay so long as $\lambda < 1$. When $\lambda > 1$ the fixed point at $x = 0$ is unstable and for $1 < \lambda < 3$ at long times one expects to reach the fixed point at $x = (1 - 1/\lambda)$.

For $\lambda > 3$, neither $x = 0$ nor $x = 1 - 1/\lambda$ is a stable fixed point. One finds that the period of the orbit doubles, so that at long times one observes a period 2 orbit. This can be demonstrated by using Eq. (7) to obtain x_{n+2} and showing that a stable fixed point of this iterated map exists for $3 < x < 3.44949$. As one continues to increase λ, this period 2 orbit becomes unstable and one finds instead a period 4 orbit, and so on. Thus this system displays an infinite cascade of period-doubling bifurcations.

Feigenbaum[24,25] showed that the bifurcation diagram has scale invariance. For instance, if one defines λ_n as the parameter value where the orbit of period 2^n becomes unstable, one finds the power-law behavior:

$$\frac{\lambda_{n+1} - \lambda_n}{\lambda_n - \lambda_{n-1}} = \frac{1}{\delta} \,, \tag{8}$$

where $\delta = 4.669...$ is a critical exponent. The values of x in the $n \to \infty$ orbit form a set that is invariant under rescaling by a factor $\alpha = 2.5029....$

The transition to chaos can be described using the renormalization group. In contrast to the magnet, where the renormalization transformation involves spatial magnification only, here the renormalization transformation involves a combination of spatial and temporal rescaling by a factor of two. The renormalization group is shown schematically in Figure 2; here one eliminates every other n and writes a recursion relation for a map between every other point.

Scale invariance at the transition to chaos is reflected by the existence of a fixed point map $g(x)$ which satisfies $-\alpha g(g(x/\alpha)) = g(x)$.[25]

Thus, the transition to chaos has an underlying scale invariance, which can be quantified using the renormalization group.

2.3 SUMMARY

In this section we have seen that the concept of scale invariance has proven powerful in helping to understand and relate equilibrium phase transitions in magnets and the transition to chaos in dynamical systems. It is a natural question to ask whether

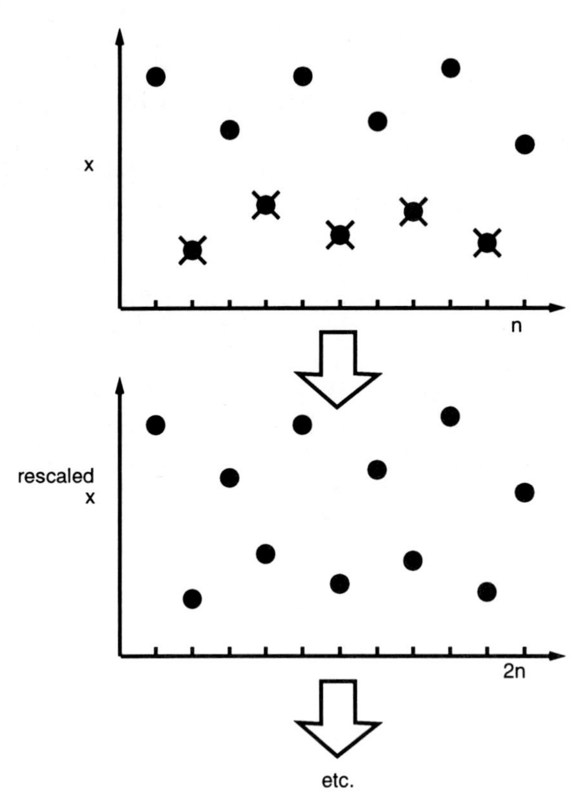

FIGURE 2 Schematic drawing of renormalization group transformation for the transition to chaos of the logistic map.

this concept will prove generally useful in understanding complexity. The remaining sections will ask this question for a specific system of balls and springs in a periodic potential. We will see that some but not all of the properties of this system can be understood within a renormalization group framework.

3. PROPERTIES OF THE FRENKEL-KONTOROVA MODEL

We now turn to examining the behavior of a particular model system consisting of balls connected by Hooke's law springs, each in a periodic potential. This system, which is known as the Frenkel-Kontorova model, is depicted in Figure 3.

We will consider only the case of overdamped dynamics and zero temperature, so that the equation of motion governing the dynamics of the jth particle is:

$$\dot{x}_j = k(x_{j+1} - 2x_j + x_{j-1}) - V\sin(x_j) + F\,, \tag{9}$$

where x_j is the position of the jth particle and the dot denotes a time derivative. The potential has a depth described by V. The dynamics are totally overdamped (one can imagine the whole system being immersed in motor oil), so that the force on each particle is proportional to its velocity (not its acceleration). In this section the external force F (which can be viewed as tilting the potential on which the balls lie) is zero; in the following two sections the case of F nonzero but time-independent will be considered, including a generalization to the case where the springs connecting the particles have different lengths. In the last section the situation where $F(t)$ consists of a series of identical, well-separated, square wave pulses will be discussed.

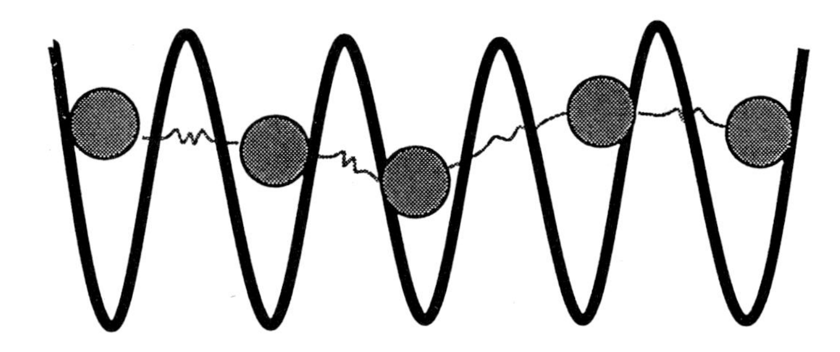

FIGURE 3 Schematic drawing of one-dimensional system of particles connected by Hooke's law springs in a sinusoidal potential $U(x) = V\sin(x)$.

3.1 PINNING TRANSITION OF THE INCOMMENSURATE FRENKEL-KONTOROVA MODEL

We first examine static properties of this system.[2] The time-independent configurations satisfy the equations

$$k(x_{j+1} - 2x_j + x_{j-1}) - V\sin(x_j) = 0\,. \tag{10}$$

The configurations which satisfy Eq. (10) minimize the energy functional:

$$E\{x\} = \frac{1}{2}k(x_{j+1} - x_j - a)^2 - V\cos(x_j)\,. \tag{11}$$

The natural spring length a does not enter into the force equations; different configurations with different mean-particle spacings can exist. We will describe calculation where mean ball spacing is fixed via the boundary conditions, by requiring $x_{j+P} = x_j + 2\pi Q$, where P and Q are integers.[3] The commensurability α, which is a measure of the mean particle spacing, is defined as

$$\alpha = \lim_{N\to\infty} \frac{1}{2\pi N}(x_N - x_0)\,. \tag{12}$$

If α is rational (the ratio of integers), the system is *commensurate*, whereas if α is irrational, then the system is *incommensurate*.

We now ask whether the balls roll down the incline in the presence of an infinitesimal force F. This question is closely (but inversely) related to the question of whether there is a static solution to the force Eq. (10) when F is infinitesimal. We will find that if the system is commensurate, then the balls are always pinned in that for any $V > 0$ the balls do not roll if F is infinitesimal. On the other hand, when the system is incommensurate, then the system undergoes a phase transition as V is increased. If V is small enough and α has the property that $|\alpha n - m|$ is bounded below by a negative power of n for all integers m, n, then the balls are unpinned. On the other hand, if V is larger than a critical value V_c, then the balls are pinned and do not slide when F is infinitesimal. (We will present an argument[14] that yields an upper bound on V_c; its actual value is slightly less than 1.[30]) Therefore, in the incommensurate system there is a phase transition as a function of V at which the balls just become pinned.[2,3,4,5,6,14,60] This phase transition exhibits scale invariance and can be described using the renormalization group.

[2] The discussion here follows that of Coppersmith,[14] to which the reader is referred for further details.

[3] Alternatively, since the natural spring length does determine the relative energies of the different configurations, the ball spacing can be fixed by searching for global minima of the energy. Aubry has shown that every value of the mean ball spacing occurs for the state that minimizes the energy for some value of the natural spring length a.[4]

3.1.1 EXISTENCE OF PHASE TRANSITION. In this section it is shown that this system of balls and springs has both pinned and unpinned phases when α is irrational; this result implies that some value of V must exist where there is a transition between the pinned and unpinned phases.

Whether the balls are pinned is intimately related to the phonon spectrum, which describes the energy cost of small perturbations to the configuration. This spectrum is the set of eigenvalues of the dynamical matrix $\mathbf{M}$, defined as:

$$M_{ij} = (2k + V\cos x_j)\delta_{ij} - k\delta_{ij-1} - k\delta_{ij+1}\,, \qquad (113)$$

where δ_{ij} is 1 if $i = j$ and 0 otherwise. All the eigenvalues of $\mathbf{M}$ must be nonnegative because one is looking at a ground state. If $\mathbf{M}$ has a zero eigenvalue, then the eigenvector corresponds to motion of the chain down the potential well for infinitesimal F. Conversely, if all the eigenvalues of $\mathbf{M}$ are positive, then one can explicitly construct a static solution of the force equations for small enough F, and the balls must be pinned.

The arguments used here all rely on the result of Aubry[2] that the particle positions x_j in the ground states of the system with mean ball spacing $2\pi\alpha$ can be written in terms of a hull function $g(x)$:

$$x_j = 2\pi\alpha j + \Delta_0 + g(2\pi\alpha j + \Delta_0)\,, \qquad (14)$$

where g is odd, bounded by $\pm\pi$, and is periodic with period 2π. A physical interpretation of the hull function can be made by imagining starting off with no potential ($V = 0$) and gradually increasing the potential strength V. The properties of g are such that the particle positions in the ground state can be obtained by taking the particle positions for $V = 0$ ($x_j = 2\pi\alpha j + \Delta_0$) and relaxing each particle towards the bottom of the closest potential well. These properties mean that the particle separations are bounded above by $2\pi\mathrm{int}(\alpha + 1)$, where $\mathrm{int}(x)$ is the largest integer less than x.

The first step is to show that for large enough V/k in the ground state the balls are pinned for any α. We do this by showing that all the eigenvalues of $\mathbf{M}$ are strictly positive. First, since in the ground state the system must be stable with respect to motions of a single particle, one must have

$$0 \le 2k + V\cos x_j\,, \qquad (15)$$

which implies $\cos x_j \ge -2k/V$. Thus the particles cannot be at the tops of wells if V/k is large enough. Since the particle separations are bounded above by $2\pi\,\mathrm{int}(\alpha+1)$, the maximum spring force on the jth ball is $4\pi k\,\mathrm{int}(1+\alpha)$. The magnitude of the force from the potential is $V|\sin x_j|$, so x_j must satisfy

$$|\sin x_j| \le 4\pi k\ \mathrm{int}(\alpha + 1)/V\,. \qquad (16)$$

The two conditions Eqs. (15) and (16) imply that

$$|\cos x_j| \geq [1 - (4\pi k \ \mathrm{int}(\alpha + 1)/V)^2]^{1/2}, \tag{17}$$

which for large enough V/k can only be satisfied if $\cos x_j$ is strictly greater than zero.

Thus, so far we have demonstrated that if V/k is large enough, then each ball is in the lower half of a well. To see that this implies that no zero-frequency mode exists, recall that the balls are pinned if the matrix $\mathbf{M}$ defined above is positive definite. If $V \cos x_j > 0$ for all j, then $\mathbf{M}$ can be written as a sum of two matrices $\mathbf{A}$ and $\mathbf{B}$, with

$$\mathbf{A} = k(2\delta_{ij} - \delta_{ij-1} - \delta_{ij+1}), \tag{18}$$

$$\mathbf{B} = \delta_{ij} V \cos x_j . \tag{19}$$

The eigenvalues of $\mathbf{A}$ are of the form $\lambda = 2k(1 - \cos q)$ and are hence nonnegative, and we have just seen that the eigenvalues of $\mathbf{B}$ are all greater than a bound $\Lambda_0 > 0$ if V is large enough. Since $\vec{v} \cdot \mathbf{M} \cdot \vec{v} = \vec{v} \cdot (\mathbf{A} + \mathbf{B}) \cdot \vec{v} \geq 0 + \Lambda_0$ for any normalized vector $\vec{v}$, it follows that the eigenvalues of $\mathbf{M}$ are bounded below by Λ_0. Thus, if V/k is big enough, then for any α the system is pinned.

We now ask if an unpinned phase exists if V is small but not zero. The natural way to approach this question is to start with a state with $V = 0$ and try to calculate the ball positions when V is small via a perturbative expansion in the quantity V/k. As made clear by Kolmogorov,[37] it is important to do the perturbation expansion keeping α fixed.

The perturbation theory proceeds by writing the position of the jth particle x_j as $x_j = qj + \Delta_0 + u_j$, where $q = 2\pi\alpha$, Δ_0 is a uniform shift, and u_j describes the deviations from the $V = 0$ state. The lowest order contribution to u_j is

$$u_j^{(1)} = -\frac{V/k}{2(1 - \cos q)} \sin(qj + \Delta_0) . \tag{20}$$

This expression is uniformly small unless $\cos q$ is very close to 1; i.e., α is near an integer. If α is exactly an integer, then the divergence can be eliminated by adjusting Δ_0 to be either 0 or π, so that $u_j^{(1)}$ is identically zero. The ground state corresponds to $\Delta_0 = 0$. If α is near to but not exactly an integer [i.e., $(2\pi - q)^2 \lesssim V$], the divergence is not eliminated by adjusting Δ_0 and the perturbation theory cannot be used. Pokrovsky[62] calls this the first "dangerous zone" in q.

The second order correction to u_j can be calculated similarly; one finds:

$$u_j^{(2)} = -\frac{(V/k)^2}{8(1 - \cos q)(1 - \cos 2q)} \sin[2(qj + \Delta_0)] . \tag{21}$$

This expression is troublesome near $\alpha = 1$ (which was already a problem in first order) and also near half-integral values of α. When $\alpha = 1/2$ the ground state corresponds to the values $\Delta_0 = \pi/2,\ 3\pi/2$.

When one continues this process, at nth order a new "dangerous zone" of width (in q) of order $(V/k)^{n/2}$ appears corresponding to a new harmonic of q. If q is not exactly commensurate but in the dangerous zone, the perturbation theory diverges, but if α is exactly an integer, then by adjusting Δ_0 the divergence can be eliminated by locking the balls onto the potential.

One finds that when the balls lock, the value of Δ_0 in the ground state yields a configuration that never has a ball at the top of a potential well and always has the balls symmetrically placed about $x = 0$ or $x = \pi$.

For an incommensurate system with α irrational, whether the perturbation theory converges depends on the relative sizes of the numerators and denominators. Eventually one reaches an order n where for some integer m, $|\alpha n - m| < \epsilon$ for any $\epsilon > 0$, but, intuitively, if n is large enough, then the large denominator will be compensated by the small factor V^n. This situation is precisely the one addressed by the Kolmogorov-Arnol'd-Moser[37,1,54] (KAM) theorem. The theorem implies that if α is a "good" irrational, which means that it has the property that $|\alpha n - m|$ is bounded below by a negative power of n for every integer m, for small enough V the perturbation theory converges.

Aubry[2] has shown that the convergence of the perturbation theory implies that the system has a zero-frequency phonon and hence is unpinned. Thus, for "good" irrationals α the balls are unpinned for small (but nonzero) V and pinned for large V. We now examine the transition between these two regimes.

3.1.2 DISORDER PARAMETER AND SCALING.

In order to make analogies with usual critical phenomena, it is useful to consider one of the two phases as ordered and the other as disordered. The unpinned phase has several features which are characteristic of conventional ordered phases: It has a zero-frequency (Goldstone) mode and has correlations which extend to arbitrarily long distances (in a sense which is described below). However, there is no obvious order parameter, and it is thus simpler to define a *disorder parameter* that is zero only in an unpinned incommensurate unpinned system. So far we have seen that the ground state of a commensurate system is pinned for any V and also does not have balls at well tops, and that when V is large enough the incommensurate system is both pinned and also has no balls at well tops. It is natural to conjecture that the unpinned and pinned phases are distinguished by whether or not there are balls arbitrarily close to the tops of wells in the ground state, and so we define the disorder parameter by

$$\psi = \min_{j,n} \left| x_j - 2\pi \left(n + \frac{1}{2} \right) \right| , \tag{22}$$

which is the minimum distance (measured along the x axis) of any ball from the top of a well. As long as ψ is finite, the balls are pinned, and $\psi = 0$ in the unpinned phase. Note that this disorder parameter is not an average—the average distance from the top can be quite large, even when the solution is unpinned.

To address numerically the properties of an incommensurate system, one can study a sequence of commensurate systems with rational mean particle spacing q_n/p_n, where q_n and p_n are integers. The particular example shown here corresponds to the value $\alpha = \phi = (1+\sqrt{5})/2$, whose optimum rational approximants[58] correspond to ratios of successive Fibonnaci numbers. Thus, we examine the sequence $[p_n, q_n] = [1,2], [2,3], [3,5], [5,8], [8,13], [13,21], \cdots$. One can obtain the ground state for nonzero V by starting with the $V=0$ configuration and then solving the force equations for each V as one slowly increases V.

One calculation that can be performed is to find the disorder parameter ψ (Eq. (22)) in the ground state for different system sizes and values of V. A second calculation involves comparing the ground state to the lowest energy state with a particle at a well top, which we refer to as the "saddle state." One does this by starting with the ball at the top of a well in the saddle state and one of the two balls closest to the top in the ground state and calculating their spatial separation, with the origin chosen to be at the maximum (or the nearest maximum) of the potential. The two solutions are then compared by labeling these balls "0" and then calculating the separation $\Gamma(j)$ between the jth balls in the two solutions. One way to gain insight into the behavior of $\Gamma(j)$ is to consider the $V=0$ and $V=\infty$ limits. When $V=0$, for a commensurate system with p balls in q wells, the two solutions maintain a constant separation of $2\pi/2p$. On the other hand, when $V \to \infty$, the zeroth balls in the two solutions are separated by π, but every succeeding difference is zero. Thus, intuitively one expects that in the small-V unpinned phase the two solutions stay distinct for all j, while in the large-V pinned phase, far from the ball 0 which is locked at the top in the saddle-point solution, the ground state and saddle state are basically indistinguishable.

Since the disorder parameter ψ is nonzero for any finite system, the number of balls per unit cell is analogous to finite size in a spin system. Changing the potential strength V is analogous to changing the temperature in a spin system. This analogy leads one to expect a scaling behavior for ψ as a function of V and p_n for n large and $V \sim V_c$ of the form

$$\psi(V, p_n) \sim |p_n|^{-\sigma/\nu} \tilde{f}(\epsilon {p_n}^{1/\nu}), \tag{23}$$

where $\epsilon = V - V_c$. Figure 4 demonstrates that this scaling form is indeed obeyed.

The scaling behavior in finite-temperature phase transitions is a manifestation of the existence of a diverging correlation length. In this system, we can identify a correlation length which is defined in terms of the differences $\Gamma(j)$ between the ground state and the saddle state. Since the saddle state is obtained by fixing the position of one ball, the large-j behavior of $\Gamma(j)$ measures the extent of the effects of a forced boundary condition. In the pinned phase $V > V_c$, $\Gamma(j)$ falls off

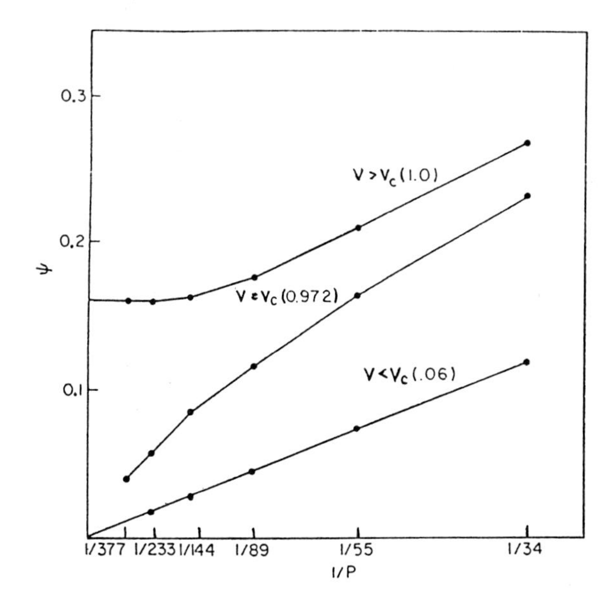

FIGURE 4 Plot of scaled disorder parameter $\psi p_n^{\sigma/\nu}$ as a function of $\epsilon p_n^{1/\nu}$ for various p_n with $\nu = 1$, $\sigma = 0.721$. The plot is of the scaling function $\tilde{f}(y)$ defined in Eq. (23).

exponentially as a function of j; the rate of the falloff can be defined to be the inverse correlation length $\xi^{-1}(V)$. It is found that $\xi(V)$ obeys the relation

$$\xi \propto \epsilon^{-\nu}, \tag{24}$$

with ν the same value (1.0 ± 0.04) as in Eq. (23). As in critical phenomena, the correlation length in the disordered phase measures the falloff of the effects of a boundary condition. In the unpinned phase, the extraction of a correlation length from $\Gamma(j)$ is somewhat more complex because $\Gamma(j)$ has structure that does not decay far from the boundaries. Nonetheless, a correlation length $\xi(V)$ can be defined in terms of the Fourier transform of $\Gamma(j)$.[14] To within the numerical uncertainties, this length also obeys Eq. (24).

Thus, the pinning transition of the discrete Frenkel-Kontorova model exhibits scale invariance in a way quite analogous to that seen in phase transitions in statistical mechanical systems.

3.1.3 RELATION TO STANDARD MAP. The behavior at the pinning transition of the incommensurate Frenkel-Kontorova model is directly related to the transition to chaos of a dynamical system known as the standard map,[29,30,31,68] whose defining equations are:

$$r_{j+1} = r_j + \frac{V}{k} \sin\theta_j \tag{25}$$

$$\theta_{j+1} = \theta_j + r_{j+1}\,. \tag{26}$$

The correspondence between the Frenkel-Kontorova model and the standard map is discussed elsewhere in detail.[14] The renormalization group analysis of this transition has been performed.[45] Therefore, the scale invariance exhibited by the pinning transition of the Frenkel-Kontorova model is intimately connected with the temporal behavior of a nonlinear dynamical system.

3.2 THRESHOLD BEHAVIOR OF A DRIVEN INCOMMENSURATE HARMONIC CHAIN

In the preceding section the behavior of the incommensurate chain in the presence of an infinitesimal force was discussed, and it was found that if the potential well depth V is big enough, the balls remain stationary when a small force is applied. At the transition where the particles are becoming pinned, the system exhibits self-similarity and a diverging length scale.

In this section we examine the pinned regime of large V and ask about the behavior in the presence of a time-independent force F. We choose V large enough so that the balls are pinned for small F, but when F is large enough then one must have the time- and spatially averaged velocity $\bar{v} > 0$. This can be seen by summing Eq. (9) over j; the spring forces sum to zero, and the pinning force per particle is bounded above by V, so the particles must have nonzero velocity if $F > V$. Therefore, there must be some sort of transition between pinned and moving states, which we examine here.[4] The natural order parameter for this transition is the mean velocity of the particles, which is zero in the pinned phase and nonzero in the moving phase.

This depinning transition has some properties that make it quite different than typical phase transitions. One peculiarity is that a true transition exists even for finite systems. This can be seen by examining the equation of motion for a single particle:

$$\dot{x} = -V \sin(x) + F . \tag{27}$$

This equation can be integrated exactly (using the substitution $y = \tan(x/2)$); one finds that a static solution to Eq. (27) exists for $F < V$ and a moving solution with time-averaged velocity $\bar{v} = \sqrt{F^2 - V^2}$ for $F > V$. Thus, even in a zero-dimensional system there is a power law behavior of the velocity versus force relation:

$$\bar{v} \propto f^{\zeta} , \tag{28}$$

where $f = (F - F_T)/F_T$ and $\zeta = 1/2$ for the single particle system.

What happens when one examines the threshold transition for a system with many degrees of freedom? A limit in which considerable analytic progress has been made is that of infinite range couplings, i.e., mean-field theory.[26] In this theory,

[4]The reader is referred to Coppersmith and Fisher[19] for more details of the work outlined in this section.

rather than connecting only nearest neighbors, the springs connect each particle to every other. The velocity characteristic can be obtained analytically; one finds that Eq. (28) holds with $\zeta = 3/2$. Since ζ in mean field theory differs from that for a single particle, it is natural to expect ζ to depend on the dimensionality of the system.

The one-dimensional incommensurate Frenkel-Kontorova is a model with short-ranged interactions with infinitely many degrees of freedom, and here we ask whether it exhibits nontrivial behavior near its depinning threshold, and if so, whether this behavior fits into the renormalization group framework that we have introduced. We will find that although for $\alpha = (\sqrt{5}+1)/2$ the velocity near threshold obeys Eq. (28) with $\zeta \sim 2/3$, several features of the transition are not scale invariant.

Figure 5 shows the mean velocity $\bar{v}$ as a function of the reduced force $f = (F - F_T)/F_T$ for different system sizes. The numerical data are consistent with Eq. (28), with $\zeta \sim 2/3$.

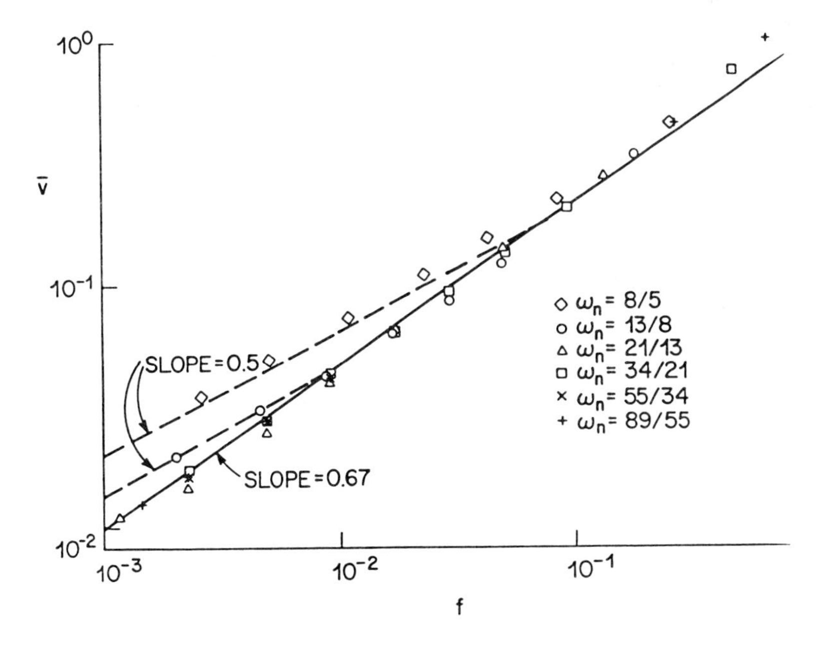

FIGURE 5 Log-log plot of mean velocity $\bar{v}$ as a function of the reduced force $f = (F - F_T)/F_T$ for different chain lengths, demonstrating that $\bar{v} \propto f^\zeta$, with $\zeta = 0.68 \pm 0.05$. The crossover to the one-particle behavior $\zeta = 1/2$ is visible for the smaller systems very close to threshold. The dashed lines have slope 0.5 and the solid line has slope 0.67.

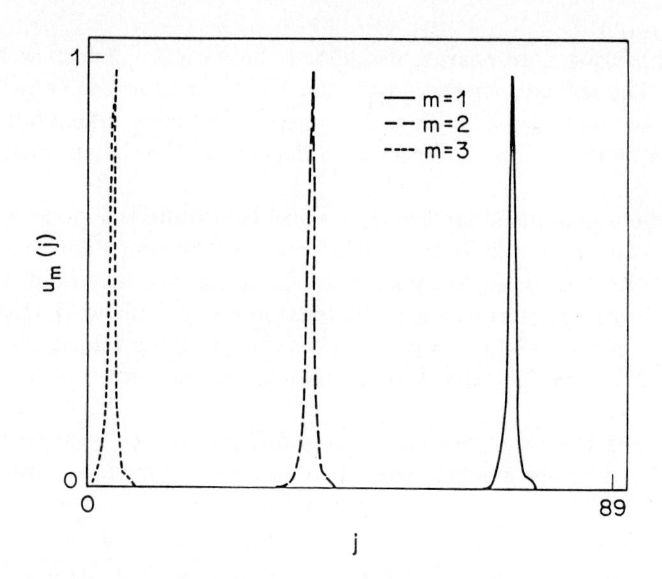

FIGURE 6 Eigenvectors of the three lowest phonon modes $u_m(j)$ of an 89-particle system for $F \approx F_T$. The modes appear highly localized even at threshold.

One can once again address the properties of the incommensurate limit by examining a series of commensurate systems with p_n balls in q_n wells, where p_n and q_n are succeeding Fibonnaci numbers. The dynamical Eq. (9) can be solved numerically for forces F both greater and less than the threshold force F_T. Below threshold, the phonon spectrum is calculated via Eq. (13); the eigenvectors describe the effects on the system of a small perturbation. Above threshold, the steady-state properties of the system can be calculated by numerically integrating the equations of motion until the transients have decayed away.

Thus, the mean velocity appears to behave in a way analogous to an order parameter for a second-order phase transition.

However, when we look in detail near the threshold the system does not appear scale invariant. One symptom that something funny is going on is the phonon spectrum—usually one expects the scaling to indicate that all the properties are determined by a single diverging correlation length, so that the effects of a perturbation will extend for long distances. However, at the threshold the eigenvectors of the lowest phonon modes appear to be localized, as shown in Figure 6.

As F approaches F_T, the lowest eigenvalue Δ_1 appears to vanish as

$$\Lambda_1 \sim |f|^{\mu} , \tag{29}$$

with $\mu = 0.5 \pm 0.005$. This behavior is the same as that exhibited by the one-particle system.

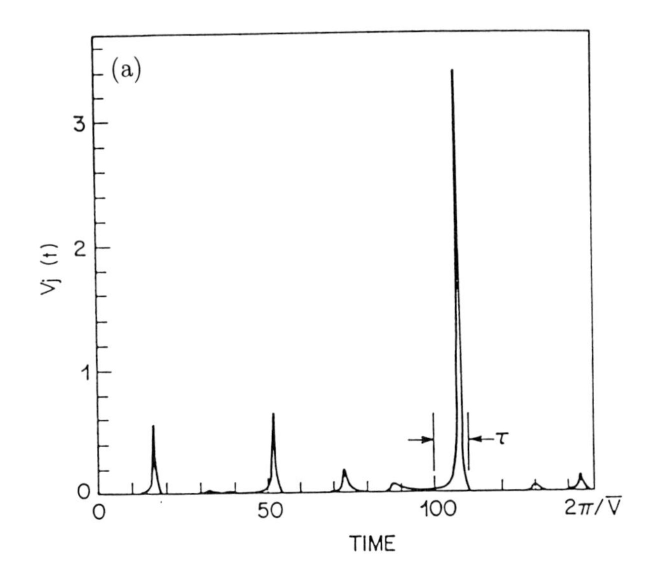

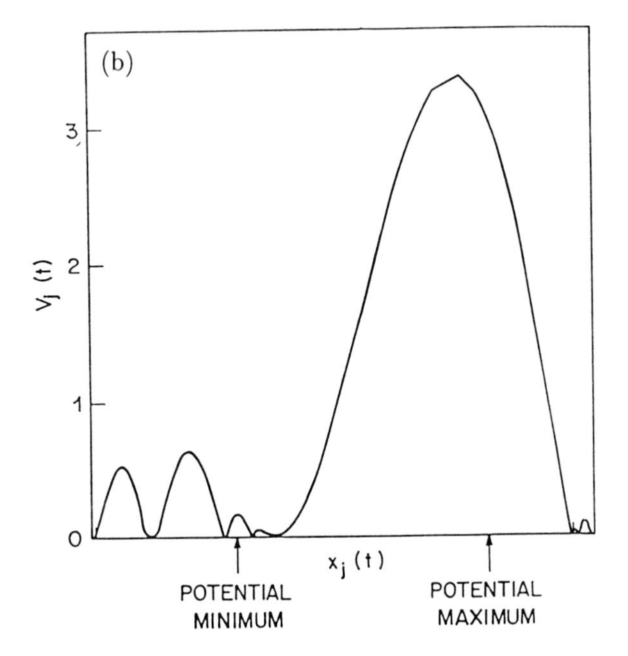

FIGURE 7 (a) Velocity $v_j(t)$ of one particle of a 21-particle system for one period with $V = 4$ and $F = 1.2$. The particle attains a velocity of order unity while hopping over a maximum of the potential, but most of the time it moves very slowly. (b) Velocity of the same particle plotted as a function of its location x_j over one period. This graph demonstrates that almost all the distance it travels is covered during "jumps."

Further evidence that the behavior is not simply described by a single diverging length is the temporal dependence of the velocity of each particle in the moving state. Figure 7 shows the velocity over one period for one particle of a 21-particle system with $F = 1.2$ ($f = 0.009$).

The motion consists almost entirely of nearly discrete jumps. The largest velocity v_1^* is attained when the particle is hopping over a potential maximum, the next highest velocities v_2^* and $v_{2'}^*$ occur when an adjacent particle is hopping over a potential maximum, and so on. The v^* are each of order unity and independent of f as $f \to 0$, but they appear to decay exponentially with distance from the particle hopping over the maximum. Although the jumping motion appears to be localized, the critical behavior is nontrivial because there are many widely separated regions jumping simultaneously.

The dominant feature of the behavior near threshold appears to be well-separated localized "active" regions that occur on either side of the threshold. For $F < F_T$ these regions correspond to the low-frequency eigenmodes, and for $F > F_T$ they correspond, at a given time, to the jumping regions. The localization length (which appears to be finite) corresponds to the size of each active region, but another length scale is needed to describe the distance between active regions. This separation between jumping regions diverges as f tends to zero.

Thus we see that although the behavior of the incommensurate chain near threshold has some properties similar to those displayed by a statistical-mechanical critical point, there are some fundamental differences. In particular, there are two diverging time scales, one corresponding to the time it takes the entire chain to move by 2π, which is of order ($f^{-\zeta}$), and the other to the time it takes for one particle to hop over a well top, which is of order ($f^{-\mu}$). Therefore, it is patently wrong to describe everything in terms of a single diverging correlation length and a single diverging correlation time. Therefore, the scale invariance needed so that a renormalization group description of the transition can be made does not seem to be exhibited at the threshold.

One may ask whether this complication is a result of some peculiarity of the model. One way of answering this question is to look at other models and see whether their properties are similar. It is more interesting to examine models that one expects to describe the behavior of real physical systems, so now we turn to a particular experimental system whose properties are described by a generalization of the Frenkel-Kontorova model that we have been studying.

4. NONLINEAR DYNAMICS OF SLIDING CHARGE DENSITY WAVES

Some materials undergo a phase transition into a state with a charge density wave (CDW), in which the electron charge density is modulated with a periodicity which need not be simply related to the lattice constant of the material. Several CDW materials have nonlinear current-voltage characteristics, which are ohmic at very small drives, but which display a lower differential resistance for voltages above a reasonably well-defined (and small) threshold voltage. The nonlinear conduction is attributed to motion of the CDW, which contributes substantially to the conductivity at fields above the threshold voltage. Thus, the electrical response can be used to probe the CDW dynamics. A significant advantage to CDWs is the wealth of experiments which have contributed to our knowledge of these systems.[5]

The CDW order parameter is described by an amplitude and a phase; the charge density $\rho(\mathbf{r}, \mathbf{t})$ is $\rho(\mathbf{r}, \mathbf{t}) = \bar{\rho} + \delta\rho(\mathbf{r}, \mathbf{t})$, where the modulation of the charge density $\delta\rho(\mathbf{r}, \mathbf{t})$ can be written

$$\delta\rho(\mathbf{r}, \mathbf{t}) \;=\; \rho_0(\mathbf{r}, \mathbf{t}) \cos(\mathbf{Q} \cdot \mathbf{r} \;+\; \phi(\mathbf{r}, \mathbf{t}))\,. \tag{30}$$

The period of the CDW modulation is $2\pi/\mathbf{Q}$; the direction and amplitude of $\mathbf{Q}$ is determined by microscopic properties of the material. The amplitude of the order parameter is described by ρ_0 and the phase by $\phi(\mathbf{r}, \mathbf{t})$.

In the absence of impurities, the CDW would slide whenever an electric field is applied. However, this behavior is not observed because all experimental samples have impurities. In the presence of impurities, it is energetically favorable for the CDW to distort, and both the amplitude $\rho_0(\mathbf{r})$ and phase $\phi(\mathbf{r})$ of the CDW depend on the position $\mathbf{r}$. When an electric field F is applied to a CDW, the field must overcome this pinning energy before the CDW can move. Thus, these deformations caused by the impurities lead to a nonlinear current-voltage characteristic, where CDW motion appears to set in at a nonzero threshold field F_T.

We will focus on the question of how to describe the CDW dynamics in the presence of an applied electric field. The focus is on driven dynamics at zero temperature; thermally induced relaxations in CDWs are slow enough that this is a very good approximation for the situations addressed. First, we will discuss the behavior of the CDW in the behavior of a dc voltage. Because the threshold voltage for CDW motion is so low, indicating that the impurity potential is weak compared to the (appropriately normalized) CDW stiffness, one might expect that it is only necessary to consider low-energy long wavelength deformations in order to describe the behavior. If one does this, then the equations of motion that one examines are a straightforward generalization of those describing the overdamped Frenkel-Kontorova model.

We will examine the behavior of this model and show that the model of balls and springs is actually *unphysical* because it has regions of infinite energy density. Thus, one must explicitly include both the phase and amplitude degrees of freedom in order to describe real CDWs. This result shows that driven dynamics are fundamentally different from thermal equilibrium because the application of an electric field causes the defect-free state to actually become unstable.

Despite this problem with the model, it does yield valuable insight into how a CDW responds to voltage pulses. Experimentally it is observed that the CDW

[5] For a review of CDWs, see, e.g., Grüner.[33]

velocity synchronizes with the *end* of the voltage pulse. It is shown here that this synchronization arises because the CDW selects out a small subset of all possible configurations, and that this subset consists of states which are in some sense the least stable. The CDW exhibits self-organizing properties as a result of this selection. Thus, the remaining two sections will provide two illustrations that the states that result from deterministic driving of a system of many particles differ fundamentally from those that arise in thermal equilibrium.

4.1 INSTABILITY OF THE PHASE DEFORMATION MODEL

4.1.1 DEMONSTRATION THAT PHASE SLIPS MUST OCCUR. The CDW has both phase and amplitude degrees of freedom. Long-wavelength phase deformations can cost arbitrarily low energy whereas amplitude variations must cost an energy of order the CDW gap energy (of order 1000 K). Since CDW threshold fields are usually less than a volt per centimeter, it is natural to assume that the amplitude degrees of freedom are unimportant and consider a model with phase degrees of freedom only.[28,38]

This section will focus on a discretized version of the equations of motion for a model including only the phase degrees of freedom of the CDW. These equations can be interpreted in terms of a simple mechanical analog consisting of a system of overdamped particles connected by springs that obey Hooke's law, each in a sinusoidal potential. In addition, a spatially uniform external force (which describes the effects of an applied electric field on the CDW) can be applied. The equations of motion are:

$$\dot{x}_j \;=\; k \sum_{\delta\ (nn)} (x_{j+\delta} - x_j) \;-\; V_j \sin(x_j - \beta_j) \;+\; F\,. \tag{31}$$

The variables x_j describe the positions of the particles, the V_j and β_j describe the pinning potential, k is the spring constant, and F is an external force, which can be viewed as arising from tilting the corrugated surface on which the particles lie. The sum is over nearest neighbors; the model can be defined in any number of dimensions d. Eq. (9) correspond to the special case of Eq. (31) in one dimension with $V_j = V =$ constant and $\beta_j = 2\pi\alpha j$. For the CDW case, we wish to examine the case of random V_j's and β_j's. In one dimension this model is obtained from the continuum equations of motion for the CDW[71] by considering the case of a pinning potential consisting of δ-function impurities, and interpreting the x_j as the phase at the jth impurity.[61,43,26] The derivation of the model makes explicit use of the assumption that the phase at the jth impurity is well defined, which follows if the amplitude degrees of freedom can be neglected.

Now consider the dynamical behavior of the model described by Eq. (31) for a time-independent F. When the force F is zero, the velocity of each particle for the model described by Eq. (31) is zero. The system has a number of metastable states that grows exponentially with its size. This fact is most easily seen in the limit $V \gg k$, where it is obvious that moving any given particle to the right or to the left by about 2π yields another metastable state. On the other hand, when $F \gg V$, by summing Eq. (31) over all the particles, one finds that the spatially averaged velocity v is nonzero.

As F is increased, the number of metastable states of Eq. (31) decreases. At a well-defined threshold field F_T there is a transition to a unique moving state, and the behavior near threshold is a dynamic critical phenomenon.[26,19,70,42,72,51,57] The steady state time-averaged velocity, which is the same for each particle, is strictly zero for $F < F_T$ and then rises as $(F - F_T)^\zeta$ for $F > F_T$, where ζ is a dimensionality-dependent critical exponent. A diverging correlation length can be defined. Other quantities with singular behavior at threshold can be found, and the exponents describing these singularities appear to obey scaling relations.

Here it is demonstrated that this picture of the nonlinear conduction, although elegant, has a fundamental flaw. It is shown that for an infinite system, Eq. (31) breaks down when $F > 0$ is constant in time. The argument that shows this is simple and may apply to other systems. In addition to demonstrating that the critical behavior predicted by Eq. (31) cannot be observed *in principle* in a physical system, the number density of the defects that are generated in practice is estimated[22] and it is shown that it is quite likely that experimental samples are in a regime where the dynamics of the amplitude of the order parameter play an important role. The dynamics are significantly more complicated than the ones described by equations describing phase degrees of freedom only.

Equation (31) contain the implicit assumption that the separation between every nearest-neighbor pair of particles is bounded, so that

$$|x_{j+\delta} - x_j| \leq S_{\max}, \tag{32}$$

for some fixed $S_{\max}$ and every j and δ. If this condition is violated, the springs no longer obey Hooke's law and Eq. (31) are no longer adequate to describe the system's dynamics. (If Eq. (32) is not satisfied, the model described by Eq. (31) has unbounded energy density.) When one traces through the derivation of Eq. (31) from the original CDW equations of motion, it turns out that the condition of bounded particle separations in Eq. (32) corresponds to the condition that the CDW amplitude $\rho_0(\mathbf{r})$ is nonzero everywhere. Here it is shown that Eq. (32) is violated using a scaling argument as well as by examining a simpler model where exact calculations are straightforward.

To show $|x_{j+\delta} - x_j|$ is unbounded for the model described by Eq. (31), examine the threshold state of an infinite system (the state when $F = F_T(\infty)$). Although there are regions that have different local pinning strengths, the configuration adjusts so that the spring forces hold back the regions with weaker pinning. Thus, just at the threshold no particles are moving.

Now examine the spring forces in the threshold state. Since for Eq. (31) to apply the force exerted by every single spring must satisfy Eq. (32), one is free to choose a region R with less than the typical density of impurities, so that the threshold field of the region $F_T(R)$ is less than $F_T(\infty)$. Choose the region R to have size L, so that its volume $\sim L^d$ and its surface $\sim L^{d-1}$. Define $F_T(R)$ as the threshold field of region R if the springs connecting R to the rest of the system are removed (and perhaps replaced with periodic boundary conditions imposed by rigid bars). Since $F_T(R) < F_T(\infty)$, in the absence of boundary springs R would move with nonzero velocity. Since in the actual system R is not moving, the springs along the boundary of R must exert a force on the region. The magnitude of this spring force scales as $(F_T(\infty) - F_T(R))L^d$. This force is exerted by the L^{d-1} springs on the boundary, so for at least one spring along the boundary one must have

$$\frac{\text{force}}{\text{spring}} \geq (F_T(\infty) - F_T(R))L\,. \tag{33}$$

This scaling is easily seen for the special case of an impurity-free region by summing Eq. (31) over region R. It does not rely on the assumption of linear spring forces; Newton's third law ensures that the spring forces in the interior of R cancel in pairs for any short-ranged interactions between particles.

Using Eq. (33), already one can see that a sufficiently large region with no impurities will have springs stretched more than is admissible ($|x_{j+\delta} - x_j| > S_{\max}$ for some j, δ along the boundary). These very rare regions are discussed below. However, first consider typical fluctuations in the threshold field of regions of size L. One expects the fluctuations in $F_T(L)$ to scale as the fluctuations in the number density of impurities in the region. One expects the total number of impurities in the region, $N_I(R)$, to fluctuate by an amount of order $\sqrt{N_I(R)}$, which implies that the number density fluctuations scale as $L^{-d/2}$. This scaling implies that $F_T(\infty) - F_T(L) \sim L^{-d/2}$, so (using Eq. (33)) the force per spring scales as $L^{1-d/2}$.

This argument implies that in one dimension the typical strains scale as $L^{1/2}$, and in two dimensions it is reasonable to expect them to scale as $\log(L)$.

The scaling argument can be shown to apply rigorously for a simpler model first introduced by Mihaly et al.[53] The main simplification of their model is that the nonlinear pinning force in Eq. (31) is replaced by a random coefficient of friction. The equations of motion are:

$$\begin{aligned} f_j &= k \sum_{\delta\ (nn)} (x_{j+\delta} - x_j) + F - d_j \\ \dot{x}_j &= f_j \quad \text{if } f_j > 0, \\ \dot{x}_j &= f_j + 2d_j \quad \text{if } f_j < -2d_j, \\ \dot{x}_j &= 0 \quad \text{otherwise.} \end{aligned} \tag{34}$$

The x_j are the positions of the particles, F is the uniform force, and the d_j describe the friction coefficients. The sum is over nearest neighbors, and k is the spring constant.

For this model, just at threshold $f_j = 0$ for every j, and the threshold force obeys $F_T = \langle d_j \rangle$, where the brackets denote a spatial average.[21] Since at threshold Eq. (34) are linear in the x's, it is straightforward to Fourier transform them and evaluate $\langle (x_{j+\delta} - x_j)^2 \rangle$ for a system of size L. One indeed finds for this model $\langle (x_{j+\delta} - x_j)^2 \rangle^{1/2} \sim L^{1/2}$ in one dimension and $\langle (x_{j+\delta} - x_j)^2 \rangle^{1/2} \sim \log(L)$ in two dimensions.[6]

In three dimensions, the strain at the boundaries of large regions does not typically diverge as $L \to \infty$, but there will always be a small but nonzero chance that there is a region of size L with no impurities at all.[7] The threshold force $F_T(R)$ for these regions is zero, so that Eq. (33) implies that if $L > kS_{\max}/F_T(\infty)$, the bound Eq. (32) is violated.

4.1.2 ESTIMATING THE PHASE-SLIP DENSITY. So far we have only shown that in principle defects must be present in systems of infinite size. However, experiments can be done only on finite-size samples, and one must address the issue of whether the phase-slip density is large enough to be experimentally relevant. This question is particularly important to address in the CDW case because CDWs are three-dimensional systems with threshold voltages for conduction that are very small on the scale of the gap energy, which is the energy required to drive the CDW amplitude to zero and hence describes the energy to create a phase slip. Therefore, if the density of defects were determined by the relative sizes of the threshold voltage and the defect energy, then the defect density would be unobservably small. However, this naive expectation is not correct because CDWs are described by the limit where the pinning involves the collective action of many impurities. It turns out that the dominant defects are stationary regions present in the sliding state of the CDW, and that the density of defects in this regime is of order unity for experimentally relevant parameters.

The estimation of the density of phase slips must explicitly account for the collective nature of CDW pinning.[8] One must calculate the probability of finding a region of a given size with a given impurity concentration, and comparing the impurity pinning forces with the spring forces both in the interior and boundary of the region. Since there is a greater probability of observing a given variation in impurity concentration in small regions, and because as a rule strongly pinned regions are smaller than weakly pinned regions, most phase slips arise at the boundaries of regions that are anomalously strongly pinned (as opposed to more weakly pinned than average). It is shown here that most of the defects arise at the boundaries of regions that remain stationary while the rest of the CDW is moving. The dominant

[6] Similar calculations have been done in the context of oscillator entrainment: see Sakaguchi et al.[65] and Strogatz and Mirollo.[73,74]

[7] The argument involving very rare fluctuations is related to arguments used for random magnets.[32,63]

[8] The methods are similar to those used to calculate the density of states of impurity states in the band tails of disordered semiconductors[34,79]

contribution arises from regions that are basically undistorted, so the density of phase slips is determined by the relative sizes of the bare impurity potential V_{bare} and the phase slip energy E_{ps}. The threshold field itself plays no role in determining the phase slip density.

For these estimates it is adequate to consider the model which includes only phase degrees of freedom. The energy E of the continuum version of the system is given by

$$E = k \int d^d x \, (\nabla\phi(\mathbf{r}))^2 - \int \mathbf{d^d x} \, \mathbf{V}(\mathbf{r}) \cos(\mathbf{Q} \cdot \mathbf{r} + \phi(\mathbf{r})) \,. \qquad (35)$$

These equations describe a d-dimensional system where ϕ is the CDW phase, $V(\mathbf{r})$ describes the impurity potential, and k describes the CDW stiffness. The first term describes the elastic cost of deformations, and the second term describes the effects of $V(\mathbf{r})$, which couples to the CDW charge density.

First recall how one estimates the threshold field F_T of a CDW. One assumes that F_T is proportional to the pinning energy per unit volume that the CDW gains by deforming in the presence of the impurity potential.[38] This pinning energy density is determined by the competition between the CDW elasticity and the impurity potential. In an infinite system an undistorted CDW would have zero energy density because $V(\mathbf{r})$ is random. However, the CDW can adjust to the impurity configuration by distorting, which increases the elastic energy but can lower the impurity contribution. One can estimate the relative sizes of the elastic energy cost and impurity energy gain by considering a region of size L.[36] Because of statistical fluctuations in the impurity potential, an appropriate deformation typically yields an impurity energy gain proportional to the square root of the number of impurities in the region, so that the energy density gain is $V(n_i/L^d)^{1/2}$, where n_i is the impurity density, d is the number of dimensions, and V is the impurity strength. Because the typical distortion will be an amount $1/Q$ accumulating over a distance L, the elastic energy density cost of the distortion scales as $1/L^2$. (Lengths are measured in units of $1/Q$.)

Minimization of the energy as a function of L leads to a pinning energy density of k/ξ_{LR}^2, where the Lee-Rice length $\xi_{LR} \sim (k/V\sqrt{n_i})^{2/(4-d)}$. The energy gain from the impurity potential and the elastic energy cost are the same order of magnitude on the length scale ξ_{LR}.

The preceding argument gives the pinning energy density of a typical region, but one would like to know the distribution of pinning energies. Since the strongly pinned regions are smaller than the weakly pinned ones, the probability of finding a fluctuation in the impurity potential leading to a strongly pinned region is much greater than one would estimate by looking at a region of size the Lee-Rice length.

It is reasonable that on a length scale ξ, the distribution of impurity potential energies is described by a Gaussian with width proportional to the square root of the number of impurities in the region $(n_i\xi^d)^{1/2}$, so that the distribution of impurity energies per unit volume is described by a Gaussian of width $(n_i/\xi^d)^{1/2}V$.[9] The

[9] This result can be demonstrated simply for a simplified model.[22]

elastic energy cost of the distortion once again scales as Ck/ξ^2, where C is a constant of order unity.

The pinning energy per unit volume of a region ϵ_{pin} is the sum of the elastic and impurity terms. The probability of observing a value of ϵ_{pin} in a region of size ξ, $p(\epsilon_{pin}, \xi)$ is

$$p(\epsilon_{\rm pin}, \xi) \sim \exp - \left\{ \frac{(\epsilon_{\rm pin} + Ck/\xi^2)^2 \xi^d}{(n_i V^2)} \right\} . \qquad (36)$$

The total number of regions with pinning energy density $\epsilon_{\rm pin}$, $P(\epsilon_{\rm pin})$, is $\int d\xi \, p(\epsilon_{\rm pin}, \xi)$.[63] The exponential dependence of the integrand enables one to evaluate the integral using steepest descents, leading to the emergence of a dominant length scale $\tilde{\xi}$ which satisfies

$$\epsilon_{\rm pin} = C \frac{4}{d-1} \frac{k}{\tilde{\xi}^2} . \qquad (37)$$

Evaluating Eq. (36) at $\tilde{\xi}$ yields

$$-\log(p(\epsilon_{\rm pin})) \sim \left(\frac{\epsilon_{\rm pin}}{E_{LR}}\right)^{(2-\frac{d}{2})} , \qquad (38)$$

where E_{LR} is the typical (Lee-Rice) pinning energy $E_{LR} \sim (V^4 n_i^2/k^d)^{\frac{1}{4-d}}$. In three dimensions $-\log(p(\epsilon_{\rm pin})) \sim \epsilon_{\rm pin}^{1/2}$; this result means that strongly pinned regions are vastly more probable than in a Gaussian distribution.

The calculation of the phase slip density can be done via a simple generalization of the arguments we just used to obtain the distribution of pinning energies. The new wrinkle on the situation is that phase slips will arise only when the pinning potential is strong enough to overcome not only the elastic forces in the region but also the springs on the boundary.

The calculation proceeds by defining a quantity $\tilde{G}_{\rm pin}$, which is related to the impurity fluctuation energy density ΔV by $\tilde{G}_{\rm pin} = |\Delta V| - k/\xi^2 - kS_{\rm max}/\xi$. Since the fluctuations in ΔV are described by a Gaussian of width proportional to $\xi^{-d/2}$, the distribution of $\tilde{G}_{\rm pin}$ for regions of size ξ can be written:

$$p(\tilde{G}_{\rm pin}, \xi) \sim \exp - \{(\tilde{G}_{\rm pin} + k/\xi^2 + kS_{\rm max}/\xi)^2 \xi^d/(n_i V^2)\} . \qquad (39)$$

In three dimensions the quantity in brackets in Eq. (39) is a monotonically increasing function of ξ, so that the dominant contribution comes from the smallest possible values of ξ.

The regions with phase slips are those where the impurities happen to be arranged so that there is substantial pinning energy even when the regions are undistorted. The number of regions with phase slips is determined by the ratio of the bare impurity potential to the phase slip energy. One needs to have a statistical fluctuation leading to a correlation of L^d impurities, where L is determined by the condition $L \geq kS_{\max}/V$. Thus one expects the density of phase slips n_{ps} to obey $n_{ps} \sim \exp[-(kS_{\max}/V)^d]$, where d is the number of dimensions.

The estimates presented in this section show that the phase slip density is determined by the ratio of the bare pinning potential to the phase slip energy. Theoretical estimates of both these energies are on the order of the CDW gap energy.[33] In addition, it is possible to obtain information about the size of the bare pinning potential experimentally. One can determine the bare pinning potential if one knows both the threshold field and elastic constant. The threshold field is easily measured, and recently the elastic constant in $NbSe_3$ has been measured by X-ray scattering determination of the position-dependent change in CDW wavevector caused by the effects of the strong pinning centers at the contacts of a sample.[23] This experiment yields a value of the elastic constant (and hence bare impurity potential strength) which is consistent with the theoretical estimates. Since the ratio of the bare impurity potential to the phase slip energy is not small, the density of phase slips in experimental samples could easily be nonnegligible. However, one cannot make a definite theoretical statement because the density of phase slips depends exponentially on this ratio.

4.1.3 EFFECTS OF PHASE SLIPS ON CDW DYNAMICS. The arguments presented above show that a sufficiently large sample of CDW must exhibit phase slips when it is driven by a uniform, time-independent electric field. However, they do not address at all the issue of what happens when phase slips are present. To make progress on this question, one must understand the details of the dynamics at the defects that are generated when the strains get so large that the phase deformation model breaks down. This problem is complex and depends on features of the physical system which did not need to be considered in the phase deformation model.

In this section we discuss dynamics in the presence of defects for CDWs. CDWs differ from other systems such as flux lattices in type II superconductors and Wigner crystals because CDW wavelengths are not conserved. One can have a moving region completely surrounded by stationary CDW; one creates and removes CDW wavelengths at the edges of the region. This is possible because it is possible to interconvert between normal carriers and electrons in the condensate. (A mechanical analogy to this situation is riding an escalator: both the lower and upper stories are stationary, and the escalator creates stairs at the lower story and destroys them at the upper story.) This feature of the CDW system makes it significantly simpler to analyze than other systems, because one expects the phase-deformation model to apply so long as one restricts consideration to a defect-free region.

It is natural to generalize the equations of motion (31) by replacing the Hooke's law springs by springs described by a force law where the maximum force is bounded:

$$\dot{x}_i = k \sum_{\delta\ (nn)} f_{\text{spring}}(x_{i+\delta} - x_i) - V_i \sin(x_i - \beta_i) + F\,, \qquad (40)$$

where the spring forces f_{spring} are now bounded, $|f_{\text{spring}}(x_{i+\delta} - x_i)| < k/S_{\max}$ for all i. Arguments similar to those that demonstrate that the model defined by Eq. (31) has unbounded strains can be used to prove that the velocity of any model of this type must be inhomogeneous for any nonzero force F.[21] One finds two regions with substantially different pinning strengths and averages Eq. (40) over the interiors of the two regions. Since the interaction term is smaller by a factor of $1/L$ than the other terms in the equation, the difference in the pinning strength leads directly to a difference in the velocities of the two regions.

We now address the question of how the presence of defects affects the behavior of the model near the threshold. In one and two dimensions, one expects the CDW to break up into disconnected pieces moving at different velocities, implying that the threshold behavior present in Eq. (31) is completely destroyed. For $d > 2$ a typical region does not have diverging strain at its boundary, so one expects a connected region to start moving all together at a well-defined value of the force. However, when this region starts to move, a nonzero density of regions with stronger than typical pinning will remain stationary; the arguments showing this are exactly analogous to those presented above for the anomalously weakly pinned regions. The implications of these defects for the dynamical behavior depends on their microscopic dynamics.

The simplest case to analyze is a "rubber band" model where the spring force is linear in the particle separations up to a critical separation $S_{\max}$, after which the spring force is zero. For this case it is clear that the spring force exerted at the boundary of a moving region and a stationary region is zero, since the separation between neighboring particles is unbounded at long times. Therefore, when a region breaks free, the spring force changes discontinuously as a function of driving force, leading to a jump in the velocity of the region. In three dimensions, since at the threshold a nonzero fraction of springs break at the same value of the force, the velocity must jump. However, a different model where the spring force is linear in the particle separations up to a critical separation $S_{\max}$, after which the spring force remains at the value $kS_{\max}$ most likely leads to a continuous velocity characteristic with a nonanalyticity at a well-defined threshold. Thus, there is no symmetry argument requiring the transition to a moving state to be continuous; one must consider the dynamics of the depinning process.

In real CDWs the dynamics at the phase slips are complicated. One must understand the interconversion between normal carriers and CDW as well as the interactions between the slow phase deformation modes and the fast amplitude modes. Understanding these dynamics remains a challenging problem.

4.1.4 ROLE OF PHASE SLIPS IN EXPERIMENTS. Most of our knowledge of dynamics in the presence of defects is indirect. Nonetheless, there is good evidence that phase slips play an important role in the dynamics of almost all samples. Although at this stage it is unclear whether the phase slips arise from macroscopic sample inhomogeneities or from the mechanism discussed in this paper, there is evidence that no samples can be described by the Fukuyama-Lee-Rice model in the regime where the phase correlation (or Lee-Rice) length is much smaller than the sample size.

One clearcut feature is that a sample which has two or more regions with different time-averaged velocity must have accompanying phase slips. Therefore, experimental demonstration of inhomogeneous velocity implies the presence of phase slips. Experimentally, inhomogeneous current (which implies the presence of phase slips) has been demonstrated unambiguously in $K_{0.3}Mo\ O_3$.[67] The only material where the current flow is homogeneous enough that the possibility of transport without phase slips can be contemplated (ignoring phase slips that must occur at the electrical contacts) is $Nb\ Se_3$. Although almost all samples exhibit substantial rounding of the current-voltage characteristic,[11] there are a small fraction of samples (about 1 in 10^4) where the threshold is very sharp and where the low-frequency temporal fluctuations in the spatially-averaged CDW velocity are small.[49] The presence of a sharp threshold field is correlated with uniform cross-sectional area, which indicates that there are significant finite size effects which are not included in the model. Even ignoring this complication, when combined ac and dc fields are applied to these samples, the response is significantly more complex than that predicted using the phase deformation model.[35]

A second, more indirect, experimental probe into the the relevance of phase slips is broad-band noise, which is observed above threshold in almost all samples.[9,64,69] The usefulness of broad-band noise as a probe arises because it is known that it does not occur in the phase-deformation model of CDWs.[41,50,52] In samples with a thickness step that leads to two regions with different velocities, the broad-band noise has been shown to be correlated with the presence of the step.[46] However, the question remains whether there are sources of broad-band noise other than macroscopic velocity inhomogeneity. It has been shown that in many $NbSe_3$ samples the instantaneous CDW velocity appears substantially more homogeneous than the time-averaged value,[10,40] indicating that noise sources other than spatial velocity inhomogeneity may be present. Once again, these noise sources are not included in the phase deformation model, so it is reasonable to speculate that they may be identified with phase slips.

In real CDW materials phase slips can arise not only from the mechanism described here but also from sample inhomogeneities (e.g., clumping of impurities) and from thickness variations, as well as because of nonuniform electric fields. However, this issue is semantic, because the arguments here show that an amplification of the variations of the pinning potential occurs (based on surface to volume ratios), so that one is merely asking whether the observed strong-pinning behavior is "intrinsic" or an amplification of weak pinning behavior.

4.1.5 DEFECTS IN OTHER SYSTEMS. Two other systems for which the considerations in this paper are relevant are magnetic flux lattices in type II superconductors and Wigner crystals. If one considers only long-wavelength acoustic modes, both these systems are described by equations of motion that are very similar to the CDW phase-deformation model described above.[66] Therefore, one must again ask whether including only long-wavelength deformations is justified, and if it is not, how the dynamics are affected by the presence of short-wavelength defects (grain boundaries).

The arguments in the preceding sections which show that a model including only long wavelength deformations is unphysical apply to any system where the interactions are short ranged (or screened).[10] However, there is an essential difference between the flux lattice and Wigner crystal on one hand and the CDW on the other hand because CDW wavelengths are not conserved whereas flux lines and electrons are. This difference means that in the latter two systems it is not possible to have an isolated moving region—particles must flow into and out of the region. Therefore, even though a CDW of infinite size has nonzero velocity for any $F > 0$, a flux lattice or a Wigner can have strictly zero velocity at a nonzero value of the force. However, this does not mean that the phase deformation model can be used to describe the behavior near threshold. In two dimensions one expects the system to break up into disconnected pieces, and even in three dimensions, since the dominant defects are stationary regions inside a moving medium, and these are allowed in all the systems, one must account explicitly for the plastic flow around the stationary regions if one hopes to understand the response near the onset of motion.

The nonlinear response of flux lines in type II superconductors is a particularly interesting system to investigate the importance of defects, because the energy cost of creating a defect can be varied by varying the magnetic field. At fields very close to the upper critical field H_{c2}, the nonlinear current-voltage characteristic of the flux lattice in $NbSe_2$ undergoes a systematic evolution that might be indicative of an increasing role of defects as their energy cost is reduced.[12]

The ideas discussed here may also be important when one considers an insulating state of a very low-density two-dimensional electron system, which is most commonly interpreted as a pinned Wigner crystal. The Wigner crystal is composed of electrons which couple to an applied electric field, and therefore one can probe its nonlinear dynamics in ways analogous to those used in the CDW system. The two-dimensionality of the system makes it likely that grain boundaries should be more important than in a three-dimensional system. Estimates of the Lee-Rice length in

[10]The arguments in this paper do not imply an instability of a model including long-wavelength deformations only in systems with unscreened long-ranged interactions if the impurity potential is weak enough. However, this situation is significantly more complicated than that described by the phase deformation model because in such a system shear and rotation cost much less energy than compression, and one must account for this when considering the dynamics of the long-wavelength modes.

this system[59] are of order a lattice constant, which means the energy cost of creating a grain boundary is of order the impurity pinning potential. Therefore it is reasonable to expect substantial creation of defects (grain boundaries) and significant velocity inhomogeneity and rounding of the threshold in this system. Experimentally measured nonlinear current-voltage characteristics in these systems appear quite rounded, and the onset of nonlinear conduction appears correlated with a large increase in broad-band noise.[39]

4.2 SUMMARY

To summarize, in this section it was shown that a model of balls connected by Hooke's-law springs with randomness as strains that diverge in the limit of infinite size in the presence of a time-independent force (which implies the model is unphysical and must be modified). Therefore, the dynamics of a real system such as a CDW in the infinite volume limit cannot be described using a model that includes only long wavelength deformations. Including the effects of the short-wavelength amplitude degrees of freedom in the model fundamentally changes the dynamical behavior of the model. There is strong evidence that phase slips play an important role in many experiments.

5. THE PULSE DURATION MEMORY EFFECT

This section describes some joint work with P. B. Littlewood,[16,75] which concerns a peculiar CDW memory effect that was first seen experimentally.[27,13] Figure 8(a) shows experimentally measured current traces that are induced by repeated voltage pulses.

For each pulse length, the velocity is rising just as the pulse ends. When the length of the pulse is changed, the velocity characteristic undergoes substantial rearrangement, but after a few pulses of the new duration, the velocity rise at the end of the pulse is again observed.

First, note that this experiment does not violate causality. The CDW "knows" when the end of a pulse is coming because it assumes it has the same duration as the preceding pulse. Therefore, we are left only with the question of why the CDW "cares" about the length of the pulse—why the current oscillations are synchronized with it.

Now one expects the velocity of the chain to exhibit a transient oscillation at the start of a pulse. When one particle moves in a periodic potential, its velocity oscillates as it goes down the washboard. When many particles are coupled together, each one has an oscillating velocity. In steady state the oscillations are out of phase and the spatially averaged velocity of the infinite system is independent of time.

However, at the beginning of a pulse each particle starts off at the bottom of a potential well, and hence initially the oscillations are in phase.

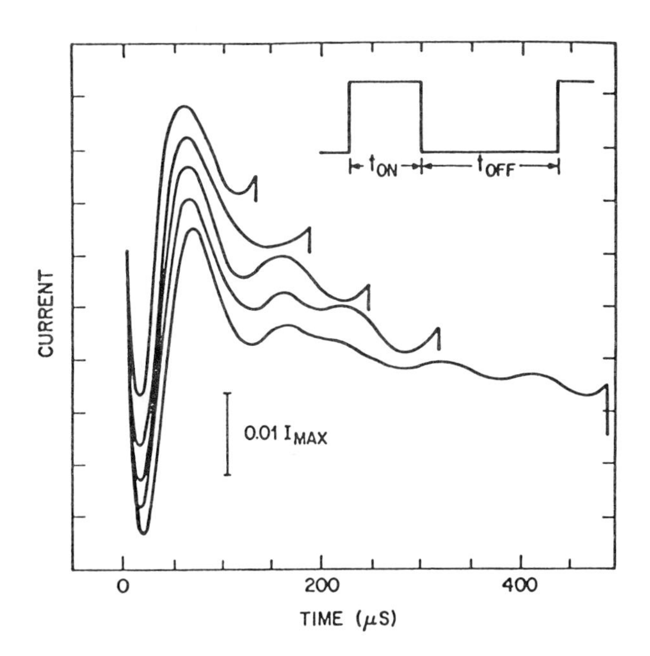

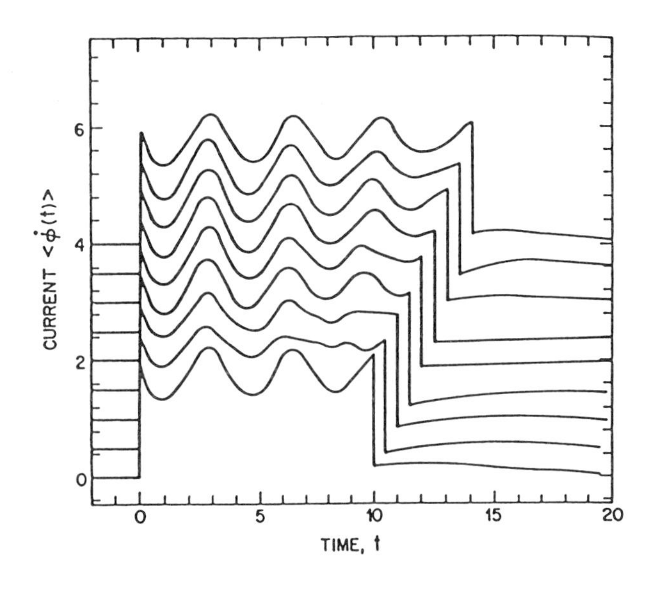

FIGURE 8 (a) Current oscillations in response to a square wave driving field of about 10 times the threshold (see inset) in $K_{0.3}MoO_3$ at 45 K, from Fleming and Schneemeyer.[27] (b) Current oscillations observed in numerical simulations of a modification of Eqs. (31) (described in Coppersmith.[16]) for a one-dimensional system with 50 particles. In both experiment and simulation the pulse length is varied for fixed pulse height; different curves have their vertical axes offset for clarity. Both experiment and simulation show a tendency to have velocity increasing at the end of a pulse.

However, at the end of the pulse this type of argument does not apply. In fact, any deterministic equation describing a single degree of freedom in a periodic potential cannot exhibit this effect. The velocity characteristic of the particle is entirely determined by the equation of motion and the initial conditions. Since the initial conditions are known ($x = \dot{x} = \ddot{x} = \ldots = 0$), there is no freedom left to adjust the velocity at the end of the pulse.

Peter Littlewood and I demonstrated that Eq. (31) can be used to understand this effect.[11] [16] Although Eq. (31) are deterministic, there is freedom to adjust the velocity characteristic because there are many metastable states available to provide different initial conditions. Just at the beginning of a pulse, the CDW is in a metastable state, so the pulse application can be viewed as a mapping from one metastable state to another. Thus, the velocity synchronization arises from the CDW choosing some subset of possible initial conditions.

Our first step was to do a simulation of a modification of Eq. (31) (the modification is unimportant, but I want to be historically accurate), the results of which are shown in Figure 1(b). One can see that although the synchronization is not perfect, there is a tendency for the velocity to be rising at the end of a pulse. Thus it seems that Eq. (31) have a feature that leads to the pulse-duration memory effect. So the problem was now to understand what features in Eq. (31) lead to the effect.

The understanding we reached involves the following logic. By taking a spatial average of the time derivative of Eq. (31), one sees that increasing velocity implies that $\langle V_j \cos(x_j - \beta_j)\rangle < 0$ at the end of a pulse. Thus, the velocity increase implies that a preponderance of particles are at local maxima of their potentials. We claim that the potential maxima are selected because they separate the basins of attraction for different metastable states.

One then needs to understand why the system ends up at the boundary of a basin of attraction. It turns out that the mechanism is closely related to why one always finds a lost item at the last place one looks—once it is found the search ends. Similarly, the CDW is searching for a fixed point—a configuration that is invariant when a pulse is applied. (This evolution occurs because faster-moving regions move ahead and then feel spring forces that make them move more slowly.) As soon as a fixed point is reached, the system stops evolving. Hence the time evolution results in the observation of fixed points that are the least stable in the sense that they are the closest to configurations that are not invariant under application of a pulse.

Thus, the essential features leading to the organization are:

- Many inequivalent metastable states, which leads to many fixed points of the mapping between the metastable configurations. This fixed point must lie in a fairly compact region in the configuration space.
- A source of feedback that causes the system to evolve slowly towards its set of fixed points as the map is iterated.

[11] It can be shown that Eq. (31) need not have unbounded strains when $F(t)$ takes the form of repeated pulses. In addition the synchronization occurs even if the CDW has broken up into several pieces. Thus, studying Eq. (31) is adequate to understand the effect.

When the system starts out, generically it is far from a fixed point, but the iteration of the mapping causes the system to evolve towards its region of fixed points. The system eventually reaches a configuration that is on the boundary of the region of fixed points. This state need not be a typical fixed point, but by definition, no further evolution occurs. Thus, the dynamics cause the system to select out a particular subset of the permissible configurations.

The simplest model that exhibits this effect is one particle in a sinusoidal potential connected by a spring to the point $x = 0$.[18] The equation of motion is:

$$\dot{x} = -kx - V \sin x + F(t) . \tag{41}$$

The particle position is x, the potential strength is V, k is the spring constant (which is assumed to be much less than V), and $F(t)$ once again describes repeated square wave force pulses of magnitude F and duration T. The spring breaks the symmetry so that the system has many inequivalent metastable states.

The effect of applying repeated pulses to this one particle system is pictured in Figure 9, where the evolution is shown for parameter values $F = 20$, $V = 5$, $k = 0.05$, and $T = 0.25$.

The first pulse causes the particle to jump over into the next well. The spring force retarding the motion then increases by $\sim 2\pi k$. After many pulses the spring force increases so much that the particle just barely fails to jump over a well top. This configuration yields an increasing velocity at $t = T$. It is invariant under application of a pulse and hence repeats indefinitely.

We still needed to show that this mechanism still applies when many particles are coupled together. Heuristically, one can view the one-particle equation as representing the motion of each particle relative to the mean field motion of the system as a whole. However, we were not able to construct an analytically solvable model with many degrees of freedom which displays the effect. Instead, we constructed a system of coupled maps which has the ingredients described above and studied its evolution numerically.

The mapping we used can be visualized as consisting of two steps, one analogous to the change in configuration caused by application of a field, and the second analogous to the relaxation to a metastable state that occurs when the field is off. The defining equations are

$$y_j^n = t(k(x_{j+1}^n - 2x_j^n + x_{j-1}^n) + F - d_j) + x_j^n , \tag{42}$$

$$x_j^{n+1} = \mathrm{int}(y_j^n + 1/2) , \tag{43}$$

with periodic boundary conditions $x_1 = x_{N+1}$. The variable x_j^n describes the position of the jth particle at the beginning of the nth pulse, whereas y_j^n is the position of the jth particle just as the nth pulse is turned off. The variables d_j are reminiscent of the pinning potential in Eq. (31), and they are taken to random

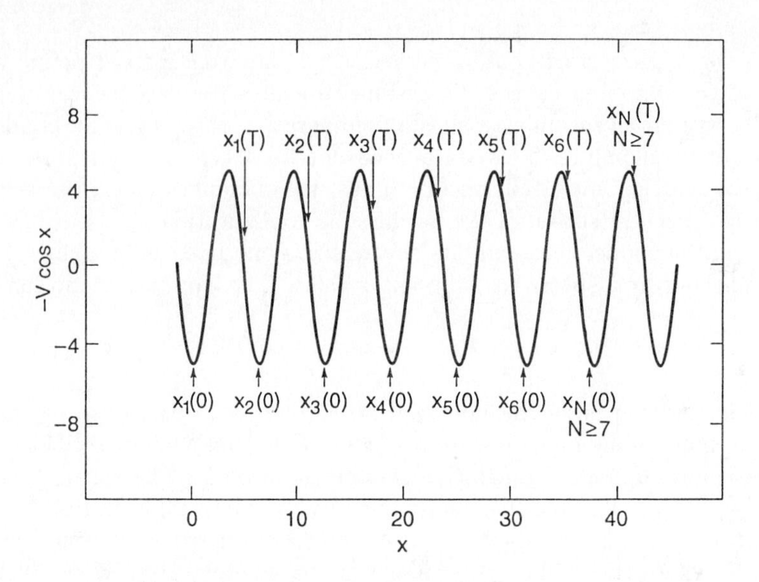

FIGURE 9 Dynamical evolution for the one particle model described by the equation of motion Eq. (41) with $F = 20$, $V = 5$, $k = 0.05$, and $T = 0.25$. When the particle starts out at $x = 0$, a pulse causes it to jump over into the next well. The spring force retarding the motion then increases by $\sim 2\pi k$. After many pulses the spring force increases enough that the particle just barely fails to jump over a well top. This configuration yields an increasing velocity at $t = T$. It is invariant under application of a pulse and hence repeats indefinitely.

numbers uniformly distributed between 0 and 1. The mapping thus depends on the three parameters t, k, and F, with the notation chosen to suggest analogy with the coupled differential Eq. (31).

This system of maps is similar to those that can be derived directly from the differential Eq. (31) in the limit $F \gg V \gg k$.[17] The first step of the map mimics the behavior of the system while the pulse is on, and the second step approximates the effects of turning the pulse off. However, the arguments leading to organization do not depend on the details of the equations of motion (31), but rather on the presence of many metastable states and weak feedback. Therefore, it is perhaps more useful to view the mapping as an example of a dynamical system with these features rather than as a model of charge density waves.

Since the second step of the mapping is discontinuous when frac(y_j), the fractional part of $y_j = 1/2$, this point marks the boundary of a basin of attraction of a fixed point. Thus, the organization manifests itself for the maps when a preponderance of y_j's have fractional part close to 1/2.

Figure 10 shows the the distribution of frac(y_j) for a system of 1000 particles in bins of size 0.001 at the end of different numbers of iterations.

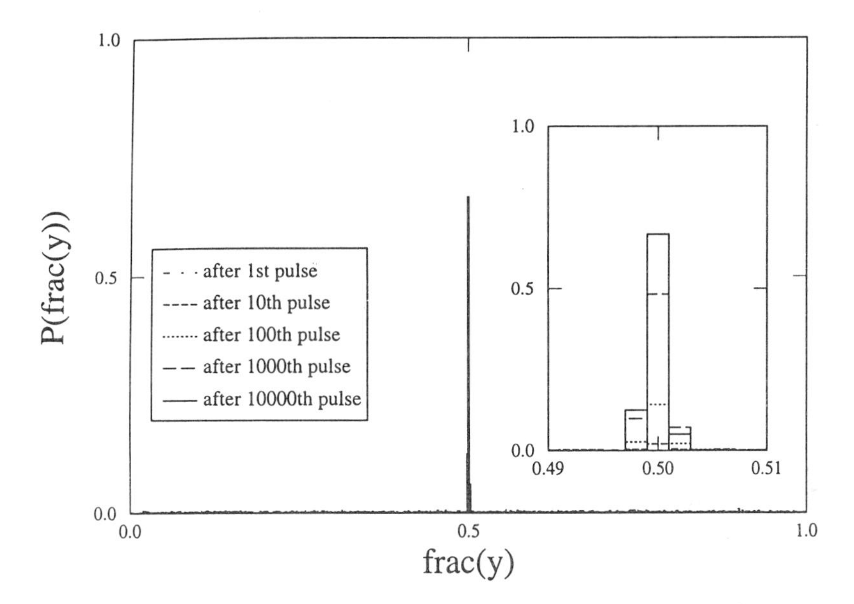

FIGURE 10 Histogram of the fractional part of y_j after 1 (dash-dotted line), 10 (dashed line), 100 (dotted line), 1000 (long-dashed line), and 10000 (solid line) iterations of the map [Eqs. (42),(43)] with 1000 degrees of freedom with parameter values $t = 1$, $F = 2$, and $k = 0.001$. The quantity $P(\mathrm{frac}(y))$ is the fraction of the y's with fractional parts whose values are within a bin of size 0.001. The boundary of every basin of attraction is at $y = 1/2$. The accumulation of y's at half-integer values is clearly visible.

The map defined by Eqs. (42) and (43) was iterated with parameters $t = 1$, $F = 2$, and $k = 0.001$. Initially the $\mathrm{frac}(y_j)$'s are uniformly distributed in the unit interval, but as the map is iterated a pronounced peak in the distribution grows at $\mathrm{frac}(y) = 1/2$.

The peak shows no signs of disappearing as the number of particles is made larger. Thus, increasing the number of degrees of freedom does not destroy the tendency to organization. The peak emerges well before the system reaches a fixed point, for even after 5000 iterations the system is still evolving for the conditions shown. One can envision a process where longer and longer length scales have homogenized their velocities, with piling up at boundaries occurring at each stage. One might hope to describe this process with a mode-coupling theory.

Thus, a simple yet puzzling experimental result can be explained as a self-organization process in a dynamical system with many degrees of freedom. We suspect that the mechanism applies to a broad class of multivariate dynamical systems, but investigating the generality of the phenomenon remains a challenging problem for the future.

To summarize, in this section it was shown that when large force pulses are applied to a CDW, it displays organized behavior which can be understood in terms of a selection mechanism for fixed points of a multivariate dynamical system.

5.1 RELATION TO SANDPILE MODELS.

The maps defined in Eqs. (42), (43) have a direct relation to "sandpile" models that have been introduced in the context of self-organized criticality.[7,8] The correspondence can be seen by performing a series of mathematical transformations.

First one defines a variable α_j so that

$$k\sum_{\delta}(\alpha_{j+\delta}-\alpha_j)=d_j\,, \tag{44}$$

where the sum is over nearest neighbors. Defining $z_j(n)\equiv x_j(n)-\alpha_j$, and "curvature" variables $c_j(n)=\sum_\delta(z_{j+\delta}(n)-z_j(n))$, one obtains

$$z_j(n+1)-z_j(n)=\mathrm{nint}[kc_j(n)+F]\,. \tag{45}$$

Now imagine iterating this map to a fixed point at a given value of F. This means that $\mathrm{nint}[kc_j(n)+F]$ has the same value for all j, which without loss of generality we can take to be zero.

Now, if one increases F slightly, then $\mathrm{nint}[kc_j(n)+F]$ remains zero unless $kc_j(n)+F>1/2$. If F is increased by a tiny amount, typically one expects there to be a very small density of sites where $kc_j(n)+F>1/2$. Let's examine one of those sites, where $z_j(n+1)=z_j(n)+1$. If one updates sequentially, (which shouldn't affect the dynamics at all so long as F is increased by a tiny amount), then the c_j's are updated as:

$$c_j(n+1)=c_j(n)-w \tag{46}$$

$$c_{j+\delta}(n+1)=c_{j+\delta}(n)+1\,, \tag{47}$$

where w is the number of nearest neighbors. These equations define the original BTW sandpile,[7] where the physical interpretation is that sand is gradually added to various sites, and when the "slope" exceeds a critical value, then the sand "topples" onto the nearest neighbors.

The close relation between CDW dynamics and sandpile avalanches reveals an ambiguity with regard to whether the system exhibits "self-organizing" properties. In the CDW context, one explicitly tunes the system by adjusting the applied electric voltage to be near the depinning threshold, while in the sandpile context, there is no obvious tuning parameter.[12] There is a hidden tuning parameter because

[12] We have seen that the CDW critical behavior does not really occur. However, it should be clear that a parameter must be tuned to make a CDW move slowly.

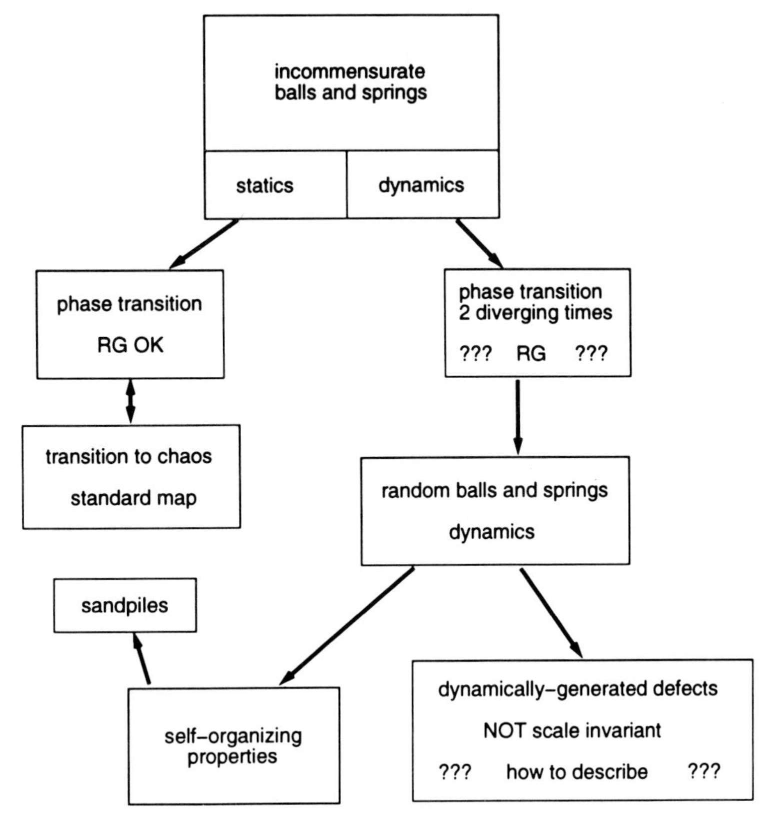

FIGURE 11 Schematic diagram of topics covered in this chapter.

one is adding sand arbitrarily slowly, which is analogous to looking at a CDW moving at a fixed, very slow velocity.

SUMMARY

Figure 11 shows a schematic diagram of the topics covered in this chapter.

First, the importance of scale invariance and the renormalization group was stressed in the context of finite-temperature phase transitions as well as routes to chaos. We then examined the dynamical behavior of a system of balls and springs and showed that some aspects exhibited scale invariance, but other features did not fit simply into the renormalization group framework. We then generalized the model so that it could be used to study the physical problem of sliding CDWs and found that the exotic experimental phenomena that CDWs exhibit yield considerable insight into the complexity that balls and springs can display. In the future we hope to understand whether the phenomena exhibited by CDWs (including those that do not fit into the standard scale-invariant picture) can be understood within a general framework that can be applied to other nonlinear dynamical systems with many degrees of freedom.

REFEREMCES

1. Arnold, V. I. *Usp. Mat. Nauk.* **18** (1963): 91.
2. Aubry, S. "The New Concept of Transition by Breaking of Analyticity in a Crystallographic Model." In *Solitons and Condensed Matter*, edited by A. Bishop and T. Schneider, 264–278. Berlin: Springer-Verlag, 1978.
3. Aubry, S. "Some Nonlinear Physics in Crystallographic Structure." *Lecture Notes in Physics* **93** (1979): 2101.
4. Aubry, S. "Defectibility and Frustration in Incommensurate Structures: The Devil's Staircase Transformation." *Ferroelectrics* **24** (1980): 53–60.
5. Aubry, S. "Violation of the Goldstone Symmetry in Incommensurate Systems." In *Symmetries and Broken Symmetries*, edited by N. Boccara, 313–322. Paris, 1981.
6. Aubry, S. "On the Bifurcation of Certain KAM Tori in the Standard Mapping." In *Numerical Methods in the Study of Critical Phenomena*, edited by J. Della Dora, J. Demongeot, and B. Lacolle, 78–79. Springer, 1981.
7. Bak, P., C. Tang, and K. Wiesenfeld. *Phys. Rev. Lett.* **59** (1987): 381.
8. Bak, P., C. Tang, and K. Wiesenfeld. *Phys. Rev. A* **38** (1988): 36.
9. Bhattacharya, S., J. P. Stokes, M. O. Robbins, and R. A. Klemm. *Phys. Rev. Lett.* **54** (1985): 2453.
10. Bhattacharya, S. et al. *Phys. Rev. Lett.* **59** (1987): 1849.
11. Bhattacharya, S., M. J. Higgins, and J. P. Stokes. *Phys. Rev. Lett.* **63** (1989): 1508.
12. Bhattacharya, S., and M. J. Higgins. "Dynamics of a Disordered Flux Line Lattice." *Phys. Rev. Lett.* **70** (1993): 2617.
13. Brown, S. E., G. Grüner, and L. Mihaly. *Solid State Comm.* **57** (1986): 165.
14. Coppersmith, S. N., and D. S. Fisher. "Pinning Transition of the Discrete Sine-Gordon Equation." *Phys. Rev. B* **28** (1983): 2566.
15. Coppersmith, S. N. "Dynamics of an Incommensurate Harmonic Chain." *Phys. Rev. B* **30** (1984): 410.
16. Coppersmith, S. N., and P. B. Littlewood "Pulse Duration Memory Effect and Deformable Charge Density Waves." *Phys. Rev. B* **36** (1987): 311.
17. Coppersmith, S. N. "Overdamped Frenkel-Kontorova Model with Randomness as a Dynamical System: Mode Locking and Derivation of Discrete Maps." *Phys. Rev. A* **36** (1987): 3375.

18. Coppersmith, S. N. "A Simple Illustration of Phase Organization." *Phys. Lett. A.* **125** (1987): 473.
19. Coppersmith, S. N., and Daniel S. Fisher. "Threshold Behavior of a Driven Incommensurate Harmonic Chain." *Phys. Rev. A* **38** (1988): 6338.
20. Coppersmith, S. N. "Phase Slips and the Instability of the Fukuyama-Lee-Rice Model of Charge Density Waves." *Phys. Rev. Lett.* **65** (1990): 1044.
21. Coppersmith, S. N., and A. J. Millis. "Diverging Strains in the Phase Deformation Model of Sliding Charge Density Waves." *Phys. Rev. B* **44** (1991): 7799.
22. Coppersmith, S. N. "Pinning Energies and Phase Slips in Weakly Pinned Charge Density Waves." *Phys. Rev. B* **44** (1991): 2887.
23. DiCarlo, D. et al. *Phys. Rev. Lett.* **70** (1993): 845.
24. Feigenbaum, M. J. *J. Stat. Phys.* **19** (1978): 25.
25. Feigenbaum, M. J. *J. Stat. Phys.* **21** (1979): 669.
26. Fisher, D. S. *Phys. Rev. B* **31** (1985): 1396.
27. Fleming, R. M., and L. F. Schneemeyer. *Phys. Rev. B* **33** (1986): 2930.
28. Fukuyama, H., and P. A. Lee. *Phys. Rev. B* **17** (1978): 535.
29. Greene, J. M. *J. Math. Phys.* **9** (1968): 760.
30. Greene, J. M. *J. Math. Phys.* **20** (1979): 1183.
31. Greene, J. M., R. S. Mackay, F. Vivaldi, and M. J. Feigenbaum. *Physica D* **3** (1981): 468.
32. Griffiths, R. B. *Phys. Rev. Lett.* **23** (1969): 17.
33. Grüner, G. *Rev. Mod. Phys.* **60** (1988): 1129.
34. Halperin, B. I., and M. Lax. *Phys. Rev.* **148** (1966): 722.
35. Higgins, M. J., and S. Bhattacharya. "Scaling Near Mode Locking in a Charge Density Wave Conductor." *Phys. Rev. Lett.* **70** (1993): 3784.
36. Imry, Y., and S.-k. Ma. *Phys. Rev. Lett.* **35** (1975): 1399.
37. Kolmogorov, A. N. *Dokl. Akad. Nauk. SSSR.* **98** (1954): 527.
38. Lee, P. A., and T. M. Rice. *Phys. Rev. B* **19** (1979): 3970.
39. Li, Y. P., T. Sajoto, L. W. Engel, D. C. Tsui, and M. Shayegan. *Phys. Rev. Lett.* **67** (1991): 1630.
40. Link, G. L., and G. Mozurkewich. *Solid State Comm.* **65** (1988): 5.
41. Littlewood, P. B. "Sliding Charge-Density Waves: A Numerical Study." *Phys. Rev. B* **33** (1986): 6694.
42. Littlewood, P. B., and C. M. Varma. *Phys. Rev. B* **36** (1987): 480.
43. Littlewood, P. B. *Charge-Density Waves in Solids*, edited by L. P. Gor'kov and G. Grüner. Amsterdam: Elsevier, 1989.
44. Ma, Shang-keng. *Modern Theory of Critical Phenomena.* Reading, MA: Benjamin/Cummings, 1976.
45. MacKay, R. S. Princeton University, 1982.
46. Maher, M. P. et al. *Phys. Rev. B* **43** (1991): 9968.
47. Mandelbrot, B. *Fractals: Form, Chance, and Dimension.* San Francisco, CA: W. H. Freeman, 1977.

48. Mandelbrot, B. "The Fractal Geometry of Nature." New York: W. H. Freeman, 1983.
49. McCarten, J. et al. *Phys. Rev. B* **46** (1992): 4456.
50. Middleton, A. A. Princeton University, 1990.
51. Middleton, A. A., and D. S. Fisher. *Phys. Rev. Lett.* **66** (1991): 92.
52. Middleton, A. A. *Phys. Rev. Lett.* **68** (1992): 670.
53. Mihaly, L., M. Crommie, and G. Grüner. *Europhys. Lett.* **4** (1987): 103.
54. Moser, J. "Stable and Random Motions in Dynamical Systems." In *Annals of Mathematical Studies,* Vol. 77. Princeton: Princeton University Press, 1973.
55. Myers, Christopher R., and James P. Sethna. "Collective Dynamics in a Model of Sliding Charge-Density Waves. I. Critical Behavior." *Phys. Rev. B* **47** (1993): 11171.
56. Myers, Christopher R., and James P. Sethna. "Collective Dynamics in a Model of Sliding Charge-Density Waves. II. Finite-size Effects." *Phys. Rev. B* **47** (1993): 11194.
57. Narayan, O., and D. S. Fisher. *Phys. Rev. B* **46** (1992): 11520.
58. Niven, I. "Irrational Numbers." In *Mathematical Association of America.* Wisconsin: Menasha, 1956.
59. Normand, B. G. A., P. B. Littlewood, and A. J. Millis. *Phys. Rev. Lett.* (1992): 3920.
60. Peyrard, M., and S. Aubry. *J. Phys. C* **16** (1982): 1593.
61. Pietronero, L., and S. Strassler. *Phys. Rev. B* **28** (1983): 5683.
62. Pokrovsky, V. L. *J. Physique* **42** (1981): 761.
63. Randeria, M., J. Sethna, and R. Palmer. *Phys. Rev. Lett.* **54** (1985): 1321.
64. Robbins, M. O., J. P. Stokes, and S. Bhattacharya. *Phys. Rev. Lett.* **55** (1985): 2822.
65. Sakaguchi, H. et al. *Prog. Theor. Phys.* **77** (1987): 1005.
66. Schmid, A., and W. Hauger. *J. Low Temp. Phys.* (1973): 667.
67. Segransan, P. et al. *Phys. Rev. Lett.* **56** (1986): 1854.
68. Shenker, S. J., and L. P. Kadanoff. *J. Stat. Phys.* **27** (1982): 631.
69. Sherwin, M. S., and A. Zettl. *Phys. Rev. B* **32** (1985): 5536.
70. Sibani, P., and P. B. Littlewood. *Phys. Rev. Lett.* **64** (1990): 1305.
71. Sneddon, L., M. C. Cross, and D. S. Fisher. *Phys. Rev. Lett.* **49** (1982): 292.
72. Sokoloff, J. B. *Phys. Rev. B* **31** (1985): 2270.
73. Strogatz, S. H., and R. E. Mirollo. *J. Phys. A* **21** (1988): L699.
74. Strogatz, S. H., and R. E. Mirollo. *Physica D* **31** (1988): 143.
75. Tang, Chao, Kurt Wiesenfeld, Per Bak, Susan Coppersmith, and Peter Littlewood. "Phase Organization." *Phys. Rev. Lett.* **58** (1987): 1161.
76. Wilson, K. G. *Phys. Rev. B* **4** (1971): 3174, 3184.
77. Wilson, K. G., and J. Kogut. *Physics Reports* **12C** (1974): 75.
78. Wilson, Kenneth G. "The Renormalization Group: Critical Phenomena and the Kondo Problem." *Rev. Mod. Phys.* **47** (1975): 773.
79. Zittartz, J., and J. S. Langer. *Phys. Rev.* **148** (1966): 741.

Catherine A. Macken,†‡ and Peter F. Stadler†*
†Santa Fe Institute, 1399 Hyde Park Road, Santa Fe, NM 87501
‡Theoretical Division, Los Alamos National Laboratory, Los Alamos, NM 87545
*Institut für Theoretische Chemie, Universität Wien, Währingerstr. 17, A-1090 Vienna, Austria

Evolution on Fitness Landscapes

I. LANDSCAPES AND HIGH-DIMENSIONAL OPTIMIZATION

The image of movement on a landscape has been invoked in many contexts, to lend an intuition about the process of maximization or minimization in multidimensional spaces. For example, one can find landscapes in physics (in particular, with regard to spin glasses), protein chemistry, and molecular biology. Each instance of a landscape is an essentially simple composition of three appropriately defined components: a configuration space, a fitness function, and a move rule. The simplicity of the landscape structure has led to inaccurate uses of the image, and false expectations for the potential of landscapes to make predictions about the underlying physical processes. To address these deficiencies, we lay the foundations for a rigorous analytical treatment of evolution on a fitness landscape. We will exclude from consideration genetic algorithms and other algorithmic approaches for which theoretical results do not exist as yet.

It is instructive to examine some applications of landscapes in order to observe the quantities and processes upon which the landscape is imposed, and the natural way in which questions about optimization can be phrased in terms of movement on a landscape. Therefore, in the first section of this chapter, we describe briefly a sampling of biological landscapes. We adopt a loosely historical perspective in our summary. Then in the second section, we describe in some detail a biological process called affinity maturation of the immune response, and experimental observations that suggest climbing on a landscape as an appropriate view of this process. The particular landscape invoked is a *random* landscape, and represents one end of the spectrum of landscapes from rugged to smooth. The random landscape is the only example for which extensive analytical results are available at present. Other landscapes have required simulations or simplifying arguments, such as a normal distribution for fitness values, in order for progress to be made. Although randomness is an unrealistically simple model for the situation at hand, it has merits as a model of affinity maturation, as will be discussed below. Hence, we shall present some of the available analytical results describing the landscape and movement thereon.

Interest in optimization typically focuses on the probability of reaching the global optimum, the fitness of local optima, and the length of time needed to reach an optimum. Hence, a general characterization of landscapes will include the calculation of the number of local optima, sizes of basins of attraction of optima, distribution of fitness values of local optima and, most importantly, a method of comparison of walks on landscapes having different properties. In this way, it may be possible to predict the outcome of optimization on a novel landscape by the average behavior of a class of landscapes into which it falls. In our final section, then, we define a rigorous mathematical framework for the basic components of a landscape, and introduce a landscape classification measure called the correlation distance. To date, there are limited results available that describe movement on a landscape as a function of this classification variable; these results are presented.

I.1 SOME EXAMPLES OF BIOLOGICAL LANDSCAPES

(A) ADAPTIVE LANDSCAPES: WRIGHT. Sewall Wright can probably be credited with introducing an adaptive landscape for the study of evolution.[111] In some back-of-the-envelope type calculations, he noted that there were an extremely large number of possible combinations of the alleles of all genes in an organism, leading to an extremely large number of different individuals possible. He concluded from these observations that the probable genetic uniqueness of human beings and other organisms that are the product of biparental reproduction is not hard to explain.

In these calculations, Wright referred to a "field" of genetic possibilities. In fact, he had defined what we would now call a *sequence space*, or *configuration space*. Wright then supposed that the entire field was graded with respect to something called "adaptive value under a particular set of conditions." Although he did not elaborate on the meaning of adaptive value, it may be inferred that the value was intrinsic to the individual, since the effects of selection and mutational forces on evolution were later studied for a given gradient. In Figure 2 of his seminal paper, Wright represented the field of genetic possibilities in two dimensions with a continuous scale on each dimension. The adaptive values were represented by contours of constant value on this two-dimensional plot. Thus, in Wright's own words, he had inadequately represented a field that was properly of very high dimension (9000 in his simple calculations, which would be the value of D in the calculations of Section II below); an extra dimension is needed to represent adaptive value. Furthermore, by taking a continuous scale on the two axes, he incorrectly represented variables that were in fact unordered categorical variables (being the alleles of a gene). This representation invited talk like "move up the steepest gradient toward the peak" and other such notions born of our familiarity with the continuous appearance of geographical landscapes. However, because of the unordered categorical nature of the actual variables, the true movement on this landscape differs greatly from that suggested by Wright's 1932 paper.

Nevertheless, Wright did foresee many of the essential features of adaptive landscapes. He talked of the ruggedness of a landscape. He recognized the existence of multiple optima, and therefore the likelihood of being trapped at a local optimum when moving in the field of possibilities. He noted the need for a trial-and-error mechanism to allow exploration of the region surrounding the place in the field of possibilities that a species occupies, in order that evolution may occur. In fact, Wright concluded that the fastest means of evolution is by cross-fertilization among relatively isolated local groups. On the way to reaching this conclusion, he explored the effects of increasing or decreasing the mutational and selection forces. Wright's work quite naturally involved a notion that was later to be called "quasi-species" by Eigen and others (see (c) in this Section below) when he imagines individuals clustering about a peak combination of alleles.

In later work, Wright acknowledged that his pictorial representation of the field of genetic possibilites and their adaptive value had been misleading, and introduced correct descriptions. Even so, succeeding researchers persist in referring to landcapes in a way that is not rigorously correct; fortunately, the intuitive notions suggested by them can often be useful in the correct landscape.

(B) PROTEIN SPACE: MAYNARD SMITH. Maynard Smith[66] suggested that protein evolution generally occurs by means of single mutations leading to a higher fitness. Proteins differing by two mutations with an intermediate configuration having a lower fitness than both of the end configurations cannot occur frequently since the intermediate protein will be expressed at such a low rate in the population that the next mutation is unlikely to take place.

(C) IN VITRO EVOLUTION: EIGEN AND SPIEGELMAN. Spiegelman and coworkers[54,92] showed in the test tube that Darwinian evolution is not restricted to living organisms: RNA molecules can evolve in cell-free medium containing nucleoside triphosphates and the RNA-replicase from the *E. coli* phage $Q\beta$. Eigen[25] devised a mathematical theory decribing evolution of this most simple *in vitro* system based on the following chemical reaction kinetics for the RNA sequences I_k:

$$I_j \overset{Q_{kj} A_j}{\longrightarrow} I_j + I_k. \tag{1}$$

This scheme means that a template I_j is read with a rate constant A_j and the outcome of the copy process is a sequence of type I_k with a probability Q_{kj}. Hence, Q_{jj} is the probability of replicating sequence I_j correctly. Degradation of the molecules can be taken into account by a degradation rate constant D_j.

The main results from this theory can be summarized as follows: Selection acts not on a single molecular type but on an entire mutant distribution called the quasispecies; for low error rates, i.e., accurate replication, these mutant distributions are centered around the sequence with highest fitness, $W_{\max}$; for moderate error rates the outcome of selection depends not only on the fitness W_k of the dominating sequence but also on the fitness values of its neighbors[101]; there is an "error-threshold" separating the localized quasispecies in the low-error regime from an unordered regime of random replication where genetic information is unstable.

The quasispecies model has been explored in detail for uncorrelated landscapes.[67] Leuthäusser[57] systematically explored the relation of this model to spin glasses for a variety of simple model landscapes. This line of investigation has been continued by the computer simulations of Amitrano et al.[2] and Bonhoeffer and Stadler[18] using a long-range spin glass Hamiltonian, and in a recent paper by Tarazona.[104]

The effects of population size have been investigated by Nowak and Schuster.[72] Recently, Derrida and Peliti studied the evolution of finite populations on a flat landscape.[24]

(D) TUNABLY RUGGED LANDSCAPES: KAUFFMAN AND COWORKERS. In 1987, Kauffman and Levin proposed the analogy of walking on a rugged landscape for the process of affinity maturation in an immune response. They suggested that affinity maturation is a selection-driven process that results in higher fitness antibodies being selected until no further point mutations lead to any further improvement in fitness. Thus, the process becomes trapped at a local optimum. Macken and Perelson[60] followed up on this idea. They proposed that selection may result from decreasing antigen concentration due to clearance by the immune system, which in turn necessitates higher antigen-binding affinity in order that the B-lymphocytes continue to attain threshold receptor binding to stimulate proliferation. The model of this process is described in detail in Section II below. Flyvbjerg and Lautrup[31] studied the random model using an approach almost identical to Macken et al.[61]

Kauffman and Weinberger[48] then introduced an NK model, that induced correlation among neighbors in a "tunable," regular fashion. The model of fitness required an additive contribution from each one of the N components of a sequence, where the fitness of an individual component changed with the configurations of a particular K out of the remaining $(N-1)$ components. Using simulation for modest N, Kauffman and coworkers[47,48] explored walking on this landscape. Decreasing K increased the correlation between neighbors.

The use of correlation functions as a characteristic of landscapes was proposed independently by Eigen et al.[27] who applied correlation to RNA landscapes, and by Weinberger[106] who applied correlation to the NK model landscape. We remark that Sorkin[91] used a closely related approach for the landscapes of combinatorial optimization problems. Roughly speaking, the correlation length measures how far neighbors must be separated in sequence space in order that they be essentially uncorrelated. Random landscapes have a correlation length of 0, since nearest neighbors have independent fitnesses. In the NK model, correlation length increases with decreasing K. Correlation length has potential use in comparison of landscapes among model classes, although it does have important limitations on the situations for which it is valid. A recent addition to this research is the work of Bak et al.,[4] who study coevolution of species on a rugged fitness landscape.

(E) RNA LANDSCAPES: THE VIENNA GROUP. The first extensive computational study of a fitness landscape is probably Walter Fontana's thesis research[32] carried out in Peter Schuster's group in Vienna. Instead of assuming a particular structure of the landscape, Fontana constructed the landscapes by folding RNA sequences of chain length 70 into their secondary structures. Secondary structures were then assigned fitness values by combining the effects of structural elements. These early simulations revealed the details of the dynamics of the quasispecies of RNA sequences and confirmed the existence of an error-threshold even on complicated and somewhat correlated landscapes.[33,34] For a recent comprehensive review of the quasispecies model see Eigen et al.[27] Wolfgang Schnabl's thesis[80] provides an extension of the RNA computer simulations to variable chain lengths.

The computer simulations of the Vienna group soon lead to an interest in the fine structure of the fitness landscapes themselves. Applications of the correlation function to RNA folding landscapes (and to combinatorial optimization problems) have been published in series of papers over the last few years: Fontana et al.,[35,36,37] Bonhoeffer et al.,[17] Schuster et al.,[83] Tacker et al.[102] Stadler and Schnabl,[97] Stadler and Happel,[96] Stadler.[93] The landscapes of a variety of quantities have been studied, including: free energies; a caricature-model of replication and hydrolytic constants; and activation energies for folding and unfolding of the RNA secondary structure. Recent work treats the RNA structures themselves instead of numbers assigned to the structures. (See the contribution by Tacker and Stadler in this volume.)

I.2 FITNESS

We began this section with descriptions of some landscape models in biological settings, several of which associated with each sequence or biological entity a quantity called *fitness*. The definition of fitness is the crux of a landscape model. Now, in classical population biology, fitness is often an empirical parameter, identified with the number of offspring reaching adulthood (cf, Crow[21]). Defining fitness thus confounds two effects, one being an intrinsic property of an individual, such as body size or shape, and the other being a measure of that individual's success in a given environment, where success is measured by reproductive capacity. But this confounding of effects leads to problems, namely, that fitness is a function of selection. Survival of the fittest reduces to the tautology "survival of the survivor."

The resolution of this dilemma requires a definition of fitness grounded in *a priori* measureable (physical or chemical) properties of the evolving system. Ideally, we would like to deduce the fitness of the components of a system (such as RNA molecules or cells) from knowledge of the environment and of the basic units that define the component (such as RNA or genomic DAN sequence). This goal is beyond our current capabilities. However, in more restricted contexts, a physicochemical description of fitness has been successful. Some examples follow.

(A) MATURATION OF THE IMMUNE RESPONSE. In Section II, we describe the fitness of an antibody by its affinity for an immunizing antigen. The affinity is a function of association and disociation constants of the antigen and the binding sites of the antibody or cell surface receptor. If the affinity is high enough, there will be proliferation of B-lymphocytes and subsequent production of more B-lymphocytes and antibody. Thus, there is a clear separation of affinity and selection for affinity.

(B) THE MOLECULAR QUASISPECIES. Eigen's[25] quasispecies model was described in Section I.1(c). In Eq. (1), fitness is defined by $W_j = A_j Q_{jj} - D_j$. The concentrations of nucleoside triphosphates needed to build the RNA molecules as well as the concentration of the enzyme $Q\beta$ replicase are considered to be environmental factors and are therefore incorporated in the replication rate constants A_j. Later, the overall replication rate constants A_j were modeled as functions of a number of more-elementary rate constants describing the binding of RNA to the protein, the elongation of the polymer chain, and the dissociation of the replica from the enzyme-template complex.[10] In general, however, it is not sufficient to identify fitness with the effective growth rate W_j. Bauer et al.[7] studied the evolution of RNA molecules in capillaries filled with the *in vitro* RNA replication medium. The RNA molecules form travelling waves in this reaction-diffusion system. Molecules generating faster waves outgrow the slower ones. Fitness is therefore related to the velocity of these travelling waves, which depends now on both the W_j and the diffusion coefficients of the RNA molecules.

(C) HYPERCYCLES. Hypercycles were introduced by Eigen and Schuster[26] as a model for cooperation in a prebiotic world. The key assumption is that each species replicates with the aid of its predecessor in the cycle. Recently, simulations of such a kinetic model in a two-dimensional reaction-diffusion system have revealed that spiral structures arising from cycles with at least five members can undergo selection if locally some of the reaction rate constants are altered.[14,15] In particular, it was found that spirals with higher rotation velocity ω survive, suggesting that, in this particular (computational) model, we can measure fitness by ω, which is a complicated—and at present not analytically known—function of the elementary reaction rate parameters of the hypercyclic reaction kinetics.

(D) STRONGLY INTERACTING SYSTEMS. A definition of fitness becomes more difficult (and may be useless) if the growth of a species is not determined by its own properties and environmental conditions (that remain constant or change at most "very slowly"). As an example, consider predator-prey systems. The survival of a species depends not only on itself but also on the existence and frequencies x_j of the other players in the game. The growth rate of a species I_k is then some function $f_k(x_1, x_2, \ldots, x_n)$. In the simplest cases one can model such an ecosystem by a non-linear dynamical system know as a replicator equation

$$\dot{x}_k = x_k \left(f_k(\vec{x}) - \phi \right) \tag{2}$$

where ϕ is a "dilution" term limiting the total population.[44,81] These systems can show arbitrarily complicated dynamics if a sufficient number of species is involved.[90] In general there is nothing that is optimized; the outcome of selection depends usually on the initial frequency of all involved species. What then is fitness?

One should keep in mind that walks on a landscape give an over-simplified picture of evolution for many biological systems. As an illustration of this fact, we note that adaptive walks and a quasispecies have very different dynamics. A quasispecies, by definition, involves a finite (though usually large) population, and while local optima slow down the evolution of a quasispecies, they cannot trap it forever. This behavior is in contrast to the eventual trapping at a local optimum of an adaptive walk. Thus, even these two very simple models of evolution do not agree in their qualitative predictions. And of course both models ignore sex, recombination, and all the subtleties of the biochemistry of reproduction and mutation in real organisms. However, in somewhat constrained settings, such as that described in Section II, an adaptive walk model of evolution may lead to useful insights into a biological process.

II. AFFINITY MATURATION IN THE IMMUNE RESPONSE

One of the fundamental modes of response of the body to invading antigen is the secretion of antibody by B-lymphocytes. Prior to antigenic challenge, B-lymphocytes express antibody as cell surface receptors. When antigen binds and cross-links these surface receptors, the cell can be stimulated to proliferate and differentiate into an antibody-secreting state, thus leading to a quantitative increase in antigen-specific antibodies.

Recently, it has become apparent that there is complex fine-tuning involved in antibody-mediated immunity. When antibody-secreting cells are examined at various times during primary and secondary immune responses, it is observed that the genetically encoded antibody structures are progressively and extensively altered as a result of somatic mutation in the genes coding for variable regions of the antibody. In conjunction with these accumulating mutations, an increase in the average affinity, or equilibrium binding constant of antibody for the stimulating antigen is found. This phenomenon is called *affinity maturation.*

Increases in affinity of order ten- to fifty-fold are common, while larger increases are much rarer. After an initial improvement in affinity of an order of magnitude or so, further point mutations tend not to lead to additional substantial improvements. It is typical to observe 6 to 8 point mutations leading to amino acid replacements, although the number varies greatly from antibody to antibody. The effects on affinity of particular mutations can be studied by site-directed mutagenesis. It has been suggested that each selected mutation leads to an increase in affinity; thus affinity may be viewed as increasing in a step-wise fashion.[30,53,87] In the following, we describe the formalization of our model in terms of the affinity maturation process.

We begin by describing an antibody variable (V) region by a sequence of N symbols, each symbol being chosen from an alphabet of a letters. If antibody V regions are viewed at the protein level, then $a = 20$ and N is approximately 230, the number of amino acids in both the heavy and light chain V regions. If we view antibody V regions at the level of DNA, then N is approximately 700 and $a = 4$. Our theory is independent of the mode of describing antibodies and we shall discuss only DNA sequences here. The set of all a^N possible configurations of sequences of length N constitutes *sequence space.* Two sequences that differ in one position only are called *one-mutant neighbors.* Two-mutant and j-mutant neighbors can be defined analogously. Since a single point mutation changes one letter in a sequence, evolution by point mutation can be viewed as a connected walk among one-mutant neighbors in sequence space. The number of one-mutant neighbors of a sequence is defined to be D. Because of restrictions in the genetic code, not all point mutations lead to a change in protein composition. On average, only 75% of DNA point mutations are expressed as amino acid changes. Thus for an antibody V region composed of 700 base pairs, the number of one-mutant neighbors $D \simeq 0.75 \times 3 \times 700 = 1575$. To indicate that this value of D is only approximate, we shall assume $D = 1500$ for antibodies. Most of the quantitative predictions of our theory will turn out to depend only on the logarithm of D, so it is somewhat immaterial if D is 1500 or 1600.

Next, we assign a fitness to each sequence in sequence space. A natural measure of fitness, which we adopt here, is the antibody affinity for the immunizing antigen. Although the affinity of each antibody sequence can be determined experimentally, this is clearly not feasible given the large number of possible sequences. Thus one would like a means of predicting affinity from sequence. To do this, one would probably need to predict the three-dimensional structure of both the antibody combining site and the antigen, and then solve what is known as the "docking" problem to determine the interaction of the two molecules. At the moment this is also not feasible. In the absence of a structurally based calculation of fitness, we simply state that affinity can not be predicted from sequence with certainty, and hence assign each antibody independently of all other antibodies a fitness selected randomly from a specified probability distribution (such as normal, lognormal, uniform, etc). The sequences and their fitnesses can now be viewed as a landscape with the fitness being the height of the landscape, i.e., the $(N+1)^{\text{st}}$ dimension above an N-dimensional sequence space.

We do not believe that fitnesses are random functions of sequence. However, this choice of relationship provides information about one extreme possibility for fitness as a function of sequence. If fitnesses are random, neighboring sequences can have very different fitnesses, and the landscape will be *rugged*. Even if the fitness function is not random the landscape can be rugged. Thus a study of random fitness functions provides qualitative information about the class of rugged landscapes. A similar philosophy has proven valuable in physics, where the study of systems with random energy functions has lead to insights into the properties of spin glasses.[74,100]

By adopting Maynard Smith's suggestion that selection operates to retain only mutations of higher fitness, we obtain a rule by which we move in sequence space; i.e., we move in single steps only in a direction of increasing fitness. However, recent experimental evidence suggests that one or two mutations in the antibody V regions may occur per cell cycle.[65] We need theoretical work allowing for the possibility of movement to two-mutant neighbors.

Before formally discussing the predictions of our model, we consider the relationship between the model and the biological process we wish to study. In order to do this, let us consider in greater detail the way in which we model evolution as a walk on a landscape.

The germline sequence defines a starting point on the landscape. At any stage on a walk, single-mutant variants of the current antibody are tested in random order until the first neighbor having a higher fitness is attained. The walk then moves to this new point in sequence space (i.e., new antibody sequence), and the testing process starts anew. If no fitter variant is found among the D one-mutant neighbors, the process stops, since the walk has reached a local optimum.

We assume that expression of improved mutations causes B cell clonal expansion that can be experimentally observed during an immune response.

One way in which this model can be related to biological reality is to invoke a mechanism for allowing the process to stay in place until the first fitter mutant is found, and a mechanism for ensuring expression of the fittest mutant only. For example, we may assume that at each cell division, at least one of the daughter cells is an unmutated copy of the parent cell. (Manser[64] suggests an alternative scheme.) Further, we may assume that expression of improved mutations causes B cell clonal expansion that can be experimentally observed during an immune response and then appeal to selection to ensure that once a fitter mutant is tested, it will be expressed at a higher rate than the parent antibody configuration, effectively "swamping" the expression of the mutant with lower fitness. Clearly, it is unrealistic to imagine an instantaneous replacement of an antibody by its fitter one-mutant variant. Indeed, the relationship between B-lymphocyte proliferation rate and receptor affinity for stimulating antigen is not known and is a major stumbling block to developing a more realistic population-level model for B-lymphocyte proliferation and antibody secretion. However, as we show below, this simplistic model leads to insights into the somatic evolutionary process.

II.1 MODEL PREDICTIONS

Many of the key predictions have been proven to be independent of G, the distribution of fitness values.[61] Whenever this distribution affects the model's predictions, we assume that affinities have a lognormal distribution with parameters μ and σ^2, truncated to lie between 10^2 and 10^{10}. This range is somewhat larger than is usually measured for serum antibodies. In fact, B-cells with affinities less than 10^4 or 5×10^4 are probably not triggered by antigen during an immune response. In our model, antibodies with affinity $u < 10^4$ will be rare but are included for reasons of symmetry in the underlying free energy distribution. We have chosen $\mu = 6$ to ensure that the mean antibody affinity in the model is close to 10^6, i.e., approximately that of the unmutated precursor in the experiments described by Manser.[63] Affinities higher than 10^8 are rarely seen during affinity maturation experiments (e.g. Manser[63]). We chose $\sigma = 2/3$, to assign a probability of less than 1% of attaining an affinity of 10^8 or greater purely by chance selection of an antibody configuration. The truncated lognormal density function is thus

$$g(u) = \frac{K}{u} \exp\left\{ -\frac{1}{2} \left(\frac{\log_{10} u - \mu}{\sigma} \right)^2 \right\}, \quad 10^2 \leq u \leq 10^{10} \tag{3}$$

where K is a constant chosen to ensure that $g(u)$ integrates to 1.

The assumption of a lognormal distribution for U is equivalent to assuming that the free energy of binding has a normal distribution. Besides the lognormal, immunologists have also used the Sips distribution[22,28,89] to describe antibody affinities. However, the Sips distribution is not well behaved—all of its moments are infinite[41]—and thus we will not use it here.

II.1.1 CHARACTERIZATION OF THE LANDSCAPE. Intuitively, we expect a random landscape to have many local optima. If we denote the number of local optima by S_N, then, *for any distribution* G, as N becomes large, the distribution of S_N approaches a normal distribution with mean $a^N/(D+1)$ and variance $a^N[D-(a-1)]/2(D+1)^2$. (cf. Baldi and Rinott,[5] Macken and Perelson.[60])

It is straightforward to extend the calculation of the asymptotic distribution for S_N to allow either one or two mutations to occur per cell generation. In this case we say a sequence is a local optimum if it has a fitness higher than that of all 1- and 2-mutant neighbors. Such optima are rarer than in the case considered above. In fact, the mean of the limiting distribution of S_N, which is still normal, is $a^N/[D+1+D(D-1)/2]$ (Kauffman and Levin[46]). Because there are fewer optima, walks can be expected to be rather longer than in situations in which only one mutation can be performed per generation.

II.1.2 CHARACTERIZATION OF WALKS. There are two essential features of walks on a rugged landscape that we will quantify using our model. One is the length of walks to a local optimum; the other is the fitness attained at positions along a walk. In the following definitions and subsequent analyses, the term *step* will be used exclusively to describe mutations leading to an increase in fitness. Steps should be distinguished from *trials* of mutational variants, which may or may not lead to an increase in affinity.

To describe the lengths of walks to an optimum, we derive the distributions and/or moments of three quantities. For walks beginning at a fitness u_0, we consider $W(u_0)$, the total number of steps to reach a local optimum. Since in our model, steps correspond to mutations leading to a higher fitness, the length of a walk may be compared with experimentally observed numbers of expressed mutations in antibodies. Then we study $M(u_0)$ and $T(u_0)$, the total number of distinct and not necessarily distinct, respectively, mutations tested along a mutational walk. By comparing the number of distinct with the total number of mutations tested, we gain some insight into the efficiency of the mutational process. An efficient process would be expected not to retry a mutation leading to an inferior antibody. We will also use $M(u_0)$ and $T(u_0)$ to relate the model predictions to real time.

To describe the fitness attained along a path, we derive three density functions: $f_k(u; u_0)$, the density of fitnesses attained on the kth step of a walk of length at least k; $f_{\text{evol}}(u; u_0)$, the density of fitnesses of optima attained by an evolutionary walk; and $f_{\text{rand}}(u)$, the density of fitnesses attained by a random sampling of local optima. Using $f_k(u; u_0)$, we can examine predictions of the model about, for example, long walks. We might ask, does the model predict that long walks are more likely to reach unusually high fitnesses, or to simply muddle around at inferior fitnesses? A comparison of the densities f_{evol} and f_{rand} will give another type of insight into the efficiency of the mutational process. One would hope that the extra biological machinery necessary to control an evolutionary process would result in reaching higher affinities than simply choosing antibody sequences or even optima at random, an approach that would probably require little sophistication other than selection.

In the following, the starting point for walks, u_0, was chosen to be 10^6, for which $G(u_0) = 0.5$. The results that follow are not particularly sensitive to choices of the starting fitness below the median. Also, the condition $D(1 - G(u)) \gg 1$ will appear in many of our approximate results. The condition defines a "boundary layer" near the global optimum $G(u) = 1$, where the mathematical behavior of the model changes character.[61]

(A) $W(u_0)$, NUMBER OF STEPS. The distribution of $W(u_0)$ is independent of G other than through $G(u_0)$.[61] The probability $p_W(k; u_0)$ that a walk to a local optimum, starting at fitness u_0, is k steps long is

$$p_W(k; u_0) = G^D(u_0)\,, \quad k = 0, \tag{4a}$$

and

$$p_W(k; u_0) = \frac{1 - G^D(u_0)}{1 - G(u_0)} \frac{1}{(k-1)!} \int_{G(u_0)}^{1} [V(x) - V(G(u_0))]^{k-1} x^{D-1} dx\,, \quad k \geq 1\,, \tag{4b}$$

where

$$V(x) = \int_0^x \frac{1 - z^{D-1}}{1 - z}\, dz\,. \tag{5}$$

For D large and $D(1 - G(u_0)) \gg 1$, $p_W(k; u_0)$ can be approximated by

$$p_W(k; u_0) \sim \frac{1}{(k-1)!} \frac{1}{D(1 - G(u_0))} \int_0^{\infty} \left(\ln[D(1 - G(u_0))/x] - E_1(x)\right)^{k-1} e^{-x} dx\,, \quad k \geq 1\,, \tag{6}$$

where E_1 is the exponential integral.[1] For antibodies, $D = 1500$ is sufficiently large that the condition $D(1 - G(u_0)) \gg 1$ holds for all reasonable starting fitnesses.

From Eq. (3) we obtain, for large D, the mean, $\mathcal{E}[W(u_0)] \sim 1.0991 + \ln[D(1 - G(u_0))]$ and variance $\mathrm{var}[W(u_0)] \sim 0.26 + \ln[D(1 - G(u_0))]$. For the parameter values $\mu = 6$, $\sigma = 2/3$, $u_0 = 10^6$, the mean walk length is 7.7 steps with a standard deviation of 2.6 steps. Less than 9% of the walks will be shorter than 4 steps or longer than 12 steps.

(B) $M(u_0)$, TOTAL NUMBER OF DISTINCT MUTATIONS. $M(u_0)$ has a distribution that is independent of G. For large values of D, the mean $\mathcal{E}[M(u_0)] \sim 0.781D$, with standard deviation $\sim 0.624D$, provided $D(1 - G(u_0)) \gg 1$. Thus, for starting fitnesses outside the boundary layer, and with $D = 1500$, the average number of distinct mutations tested on a path to a local optimum is 1172, with a corresponding standard deviation of 936. We predict that on average, 7.7/1172 or approximately 0.66% of distinct mutations will be improvements.

By focusing on distinct mutations, we assess the proportion of sequence space explored by a mutational process proceeding according to our rules. Sequence space contains $4^{500} \simeq 10^{300}$ antibody V region sequences, of which only 1172 are examined. Thus, the percentage of possibilities explored by the mutational process is extremely small, around 10^{-295}.

(C) $T(u_0)$, TOTAL NUMBER OF NOT NECESSARILY DISTINCT MUTATIONS. The distribution of $T(u_0)$ does not depend on G. The random variable $T(u_0)$ roughly reflects the duration of an evolutionary walk from a starting fitness u_0 until it reaches a local optimum. The strength of correspondence of $T(u_0)$ with time depends on the relationship between the occurrence of mutations and cell divisions, and the effect on cell cycle time of changes in receptor affinity for antigen. Manser[63] calls into question the hypothesis that mutation and cell division are coupled, so that associating $T(u_0)$ with time may not be straightforward.

For large D, $\mathcal{E}[T(u_0)] \sim 1.224D$, for any u_0 such that $D(1 - G(u_0)) \gg 1$.[61] For $D = 1500$, $\mathcal{E}[T(u_0)] \sim 1836$. Of the total number of trials taken to reach an optimum, on average $\{T(u_0) - M(u_0)\} \sim .44D$, or approximately 36% of all trials, will be repeats. Most of these repeats occur when the fitness is close to a local optimum and many trials are needed to find a fitter variant.

(D) $f_k(u; u_0)$, DENSITY OF FITNESSES ATTAINED ON THE KTH STEP OF A WALK HAVING AT LEAST K STEPS. Paths which continue for many steps without reaching a local optimum could be exhibiting one of two quite different behaviors: They continue for a long time because they take many steps at low fitnesses, far from a local optimum; or they are one of the rare paths which attains a high fitness without first becoming trapped at a local optimum. We gain some insight into which of these two scenarios is more likely by examining $U_k(u_0)$, the fitness attained on the kth step of paths having at least k steps for $k = 1, 2 \ldots$. The fitness attained has the density function

$$f_k(u; u_0) = \frac{1 - G^D(u_0)}{1 - G(u_0)} \frac{g(u)}{(k-1)!} \left[V(G(u)) - V(G(u_0))\right]^{k-1} \ , \quad k \geq 1 \ , \qquad (7)$$

where $V(G)$ is given in Eq. (4) and $g(u)$ is the lognormal density function from Eq. (2).

The graphs of f_k in Figure 1 show the modal and mean fitnesses shifting progressively toward higher affinities as the number of steps, k, increases. At the same time, the area under the curves, i.e., the probability that a path continues for at least k steps, decreases. We deduce that it takes rather few steps to improve substantially in fitness and if paths continue for many steps, they do so by virtue of muddling around at high fitnesses where little further progress is made.

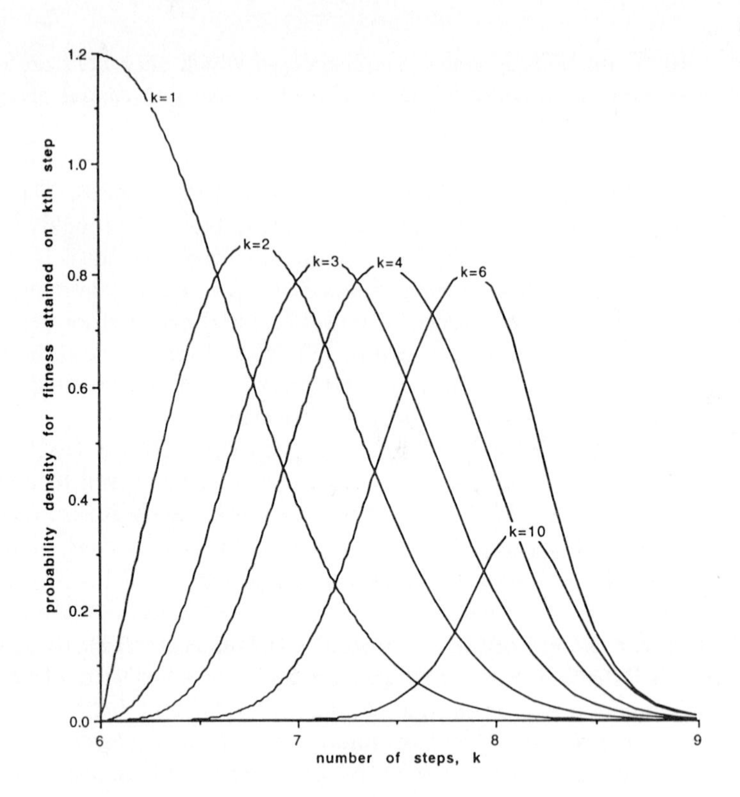

FIGURE 1 The probability that fitness u is attained on the kth step of a walk of at least k steps starting at affinity $u_0 = 10^6$.

(E) $f_{evol}(u; u_0)$, DENSITY OF FITNESSES ATTAINED AT THE END OF A WALK. We denote the fitness attained at the end of an evolutionary walk by $U_{evol}(u_0)$, with a probability density function $f_{evol}(u; u_0)$. For walks starting outside the boundary layer,

$$f_{\text{evol}}(u; u_0) \sim \frac{g(u)}{1 - G(u)} \exp\left\{-D(1 - G(u)) - E_1[D(1 - G(u))]\right\} , \qquad (8)$$

where E_1 is the exponential integral.[61] We can use f_{evol} to study the efficiency of the mutational process. A simple, relevant measure of the success of a mutational walk is the rank of the fitness finally attained. After normalizing so that the rank lies between 0 and 1, the average rank attained is $1 - .6243/D$, or 99.958%, with a standard deviation of $0.68/D$ or At $D = 1500$, the average rank corresponds to a fitness of 1.7×10^8. (See Macken et al.[61])

(F) $f_{\text{rand}}(u; u_0)$, DENSITY ATTAINED BY RANDOM SAMPLING FROM LOCAL OPTIMA. If a local optimum is chosen at random from the sequence space, then its fitness has a density

$$f_{\text{rand}}(u) = (D+1)G^D(u)g(u) \tag{9}$$

with normalized average rank approximately $1 - 1/D$. Thus, a local optimum attained at the end of an evolutionary walk averages 38% closer to the global optimum than does a local optimum selected at random.

II.2 DISCUSSION

Apparently, a simple model in which antibodies are assumed to evolve by point mutations, coupled with selection for higher affinity variants, can explain much of the quantitative and qualitative data on affinity maturation. For example, the observed halting in affinity improvement after only one or two orders of magnitude improvement[8,9] is predicted to occur due to trapping at a local optimum or near a local optimum where improvement mutations become rare. An additional implication results from noting that different germline sequences correspond to different starting positions on the landscape. Therefore, because of the high dimensionality of the space, and the uncorrelated nature of fitnesses, different germline genes should get trapped at different peaks. With enough starting points, the attainment of high affinities may be achieved by rare germline V region gene combinations.

In addition to capturing these qualitative features of the somatic mutation process, the model makes a number of quantitative predictions which could be tested experimentally. For example, the model predicts that the 8 ± 5 (mean $\pm$ 2 s.d.) replacement mutations leading to higher affinities constitute approximately 0.7% of the mutant antibodies tested along an evolutionary walk. The model also predicts that the first few improvement mutations substantially increase the affinity and that later mutations only make small improvements. Experimental work is currently in progress to test this latter prediction.[65] Lastly, in the model, $1 - G(u_0)$ is the fraction of mutations of the unmutated precursor antibody that are improvements. This prediction can be tested by measuring the relative affinities of the one-mutant neighbors of the starting antibody, and may provide a strong test of the random fitness model. By contrast to the random model, in Kauffman's NK model (see Section III.2.1(c)), evolutionary walks start at a fitness that would lie in the boundary layer. Again, Manser's preliminary data will lend insights on the comparison of these two classes of models.

One model prediction that probably deviates from reality in a significant way is the number of mutations to reach a local optimum, $\mathcal{E}[M(u_0)]$ and $\mathcal{E}[T(u_0)]$. At the rate of one mutation per cell cycle, $E[T(u_0)]$ predicts that about 1800 cell cycles, or about 1300 days in our sequential model, is needed to reach a local optimum. Since the primary immune response takes about 9–14 days, there is substantial disagreement between empirical observations and theoretical predictions. Possible explanations are: the process may stop before reaching a local optimum; or, there may be a mechanism for testing variants in parallel; or some combination of both explanations. In support of a parallel testing of variants, we point to recent work of Kepler and Perelson[49,50] who hypothesize (in an optimal control model) that mutation occurs in bursts, perhaps corresponding to the timing of cells migrating through the light, then dark, then outer zones of the germinal center. In the light zone, no selection occurs, but mutations accumulate during clonal expansion. Selection occurs in the dark zone, thus providing parallel lineages for testing for a fitter variant. Clonal expansion ensures that many more than n trials are possible in the time of n cell cycles.

It seems surprising that a model in which fitnesses are assigned at random can lead to reasonable predictions. One means of rationalizing this result is that if affinities are chosen from a lognormal distribution with a small half-width, then the affinities of an antibody and its one-mutant neighbors will most likely be similar. Thus, even though there is no correlation present in the model one will only generally observe small changes in affinity. Because the model is random, these changes can be either an increase or decrease in affinity. Further, on rare occasions, at frequencies determined by the variance in the distribution, a mutant will be chosen that has a much higher or much lower affinity than average. This is typically what is observed. The NK class of models specifies one possible mechanism for inducing correlation among the fitnesses of neighboring sequences. Perelson and Macken[76] introduce an alternative, called a *block model* (see Section III.2.1(d)).

The immune system has been a testing ground for many ideas in molecular biology. Affinity maturation by somatic mutation seems an ideal system in which nature is allowing us to examine in detail the relationships among antibody sequence, structure and function. With appropriate models, one may not only be able to infer information about antibody structure from measurements of affinity and sequence, but also obtain glimpses into the evolutionary driving forces and operating principles of importance in the immune system.

III. A MATHEMATICAL FORMALISM OF LANDSCAPES

The fact that landscapes appear in many contexts is due, in part, to the fact that they are not models, but simply visual representations, and severely limited ones at that, of a relationship between a complex space of entities, such as DNA or protein sequences, and a quality called fitness, which may be measured by, say, a free energy, that is assigned to each entity.

A landscape has no predictive capacity of its own. That is, nothing can be interpolated or extrapolated from the landscape beyond the represented mapping of, say, fitness to sequence. Nevertheless, a landscape provides an attractive means of motivating some of the questions that arise naturally in a combinatorial optimization problem. In landscape terminology, we consider the length of walks to a (local or global) peak or valley (optimum), the size of the basins of attraction of optima, the homogeneity of the surface, the degree of ruggedness of the landscape, etc.

In the following sections, we present a somewhat formal basis for considering evolution on a landscape. To begin, we define a general space of entities to be optimized, called a *configuration space*, and give several important examples of this. Also, we discuss rules for substitution of one configuration by another (*move rules*). The important notion of *isotropy* is defined. If a landscape is anisotropic (loosely speaking, inhomogeneous), little can be said about the behavior of walks on this landscape. For isotropic landscapes, we define a quantity called *correlation length*, being one possible measure of the degree of ruggedness or smoothness of a landscape that enables classification of landscapes. We conclude the section with new results linking the lengths of walks on a landscape with its characteristic correlation length.

III.1. CONFIGURATION SPACES

Although a landscape model is often concerned with the evolution of sequences, such as DNA or protein, the visualization can be applied to problems, such as the traveling salesman problem or a graph bipartitioning, for which the term "sequence" is not an appropriate description of the quantity evolving. In these latter cases, the quantity of interest is an ordering of cities or a list of half of the vertices of a graph. Hence, we adopt the term *configuration* to mean the entity that is to be optimized: The set of all configurations comprises *configuration space.* Associated with each space is a natural notion of derivation of one configuration from another, called a *move rule*, which we discuss in Section III.1.1. We give several examples of configuration spaces and their move rules in Section III.1.2.

III.1.1 MOVE RULE. For any definition of configuration, a particular "move rule" is used to derive a new configuration from an existing one during the optimization procedure. In physical systems like spin glasses, nature defined a move rule. For example, the flipping of a single spin might be a reasonable model for the elementary dynamic step. In molecular biology, the move rule is defined by the elementary mutational events: point mutations, insertions and deletions, and transpositions.

The notion of allowed moves provides us with a canonical (metric) distance between configurations. We define the distance $d(x, y)$ in configuration space, sometimes called the *edit* distance, as the minimum number of moves necessary to convert configuration x into configuration y. This coincides with the usual notion of distance in a graph. See, for example, Buckley and Harary.[19] As we shall see below, the calculation of $d(x, y)$ for two arbitrary configurations can become a formidable problem. In some cases, the calculation is presumably $\mathcal{NP}$ complete.

Further, a move rule induces an arrangement of all configurations in a graph: Each configuration is represented by a vertex, and two vertices are neighbors if the corresponding configurations can be interconverted by a single move. A sensible definition of a move rule yields a connected graph in which there exists a sequence of moves allowing transformation of any configuration into any other. A useful assumption is that each move has an inverse. Relaxing this assumption leads to a number of complicated problems with the interpretation of simulation data (cf, Schnabl[80]).

III.1.2 EXAMPLES OF CONFIGURATION SPACES.

(A) SEQUENCE SPACES. Often a configuration consists of an ordered list of elements; i.e., a sequence. The simplest and best studied example of a sequence space consists of the set of all sequences of common length N taken from an alphabet of a letters. In the case of nucleic acids, the alphabet is given by adenosine, cytosine, guanosine, and uracile or thymidine for RNA or DNA respectively. For proteins, the alphabet consists of the 20 amino acids. In spin glass models the alphabet is given by the possible spin orientations, e.g., + (up) and - (down). The total number of configurations (sequences) is a^n. In the case of a two-letter alphabet, the configuration space is a Boolean hypercube, consisting of the corners and edges of an N-dimensional cube.

The simplest edit operation on a sequence is changing a single letter. For biopolymers, this represents a point mutation; for spin glasses, the flip of a single spin. The distance measure that arises from this edit operation is the Hamming metric[42] which counts the number of positions in two sequences that are occupied by different letters.

Biologically important configuration spaces are not confined to constant sequence length. By allowing single element insertions and deletions, we obtain a highly irregular configuration space. The problem of computing edit distances in different versions of sequence spaces can be solved by dynamic programming techniques (cf, Waterman[105]). Spaces of unequal-lengthed configurations have received little attention to date. Properties that are easily accessible in a space of constant length sequences, such as the consensus sequence, raise major computational problems when insertions and deletions are allowed. However, a consensus sequence can be determined for small numbers of configurations.[29]

Another generalization of sequence space assumes a constant sequence length but different alphabets at different positions. The ensuing graphs have been used recently in the description of RNA landscapes. (See the contribution by Tacker and Stadler in this volume.)

(B) TREE SPACES. In biophysical applications, spaces of trees arise in at least two different contexts: RNA secondary structures can be encoded faithfully by trees,[37,85,86] and the set of binary trees with equal-lengthed branches and N leaves is a configuration space associated with the reconstruction of phylogenies. Spaces of (finite) trees are closely related to sequence spaces.[37,73,103] Edit distances can be calculated by dynamic programming techniques.

(C) PERMUTATION SPACES. A permutation space is composed of all permutations of N objects. Perhaps the best known example of a permutation space is the set of all possible tours of geographically separated cities in the Traveling Salesman Problem.[55] A permutation space may also arise in relation to modeling biological evolution, since rearrangement of the genetic material plays an important role in speciation. As with point mutations at the gene level, a permutation of the genome can be selectively neutral or it may cause specific functional changes, for example by putting a particular gene into a different regulatory context.[79] Therefore, at this board level of definition, a genome may be described by a particular permutation of the genes.

An edit operation in permutation space is a change from one to another of a restricted class of permutations. The distance between two permutations σ and τ ($\sigma \neq \tau$) can be calculated as the distance between $\sigma\tau^{-1}$ and $\varepsilon = (1, 2, 3, \ldots, n)$, where τ^{-1} is the inverse of τ and $\sigma\tau^{-1}$ means applying the two permutations in sequence. Probably the most common edit operator on the symmetric group is a transposition (i, j) for $i < j$, in which the elements in the positions i and j are exchanged. In particular, the operator of the form $(i, i+1)$ defines the set of canonical transpositions. The edit distance is given by the so-called inversion-index of the permutation $\sigma\tau^{-1}$. Another common operator is called, by different authors, inversion, two-opt move, or reversal. In permutation spaces, the calculation of distances between configurations can become a very hard problem, depending on the basic edit operation.

(D) PARTITION SPACES. In the Graph Bipartitioning Problem (GBP) (cf, Fu and Anderson,[39] Liao,[58] Banavar et al.,[6] Wiethege and Sherrington,[110] Fu[38]), the task is to partition an even-sized set of vertices of a graph into two equal-sized subsets, in such a way that a prescribed function of the distance between the two subsets of vertices is minimized. The configuration space $\mathcal{C}$ consists of the set of all such partitions of the vertex set V into equal-sized subsets. A configuration can be encoded as a binary string by labeling a vertex with a "1" if it is contained in subset A and a "0" otherwise. The canonical move set consists of exchanges of single vertices. Two partitions of V, $\{A, B\}$ and $\{C, D\}$, are neighbors of each other if and only if the symmetric differences $A \ominus C$ and $B \ominus D$ both equal the pair of exchanged vertices $\{v_1, v_2\}$, where the symmetric difference of two subsets X and Y is defined as $X \ominus Y = (X \setminus Y) \cup (Y \setminus X) = (X \cup Y) \setminus (X \cap Y)$.

III.1.3 GRAPH THEORETIC PROPERTIES OF CONFIGURATION SPACES. Because a number of rigorous results on the structure of model landscapes require a particular level of symmetry of the underlying configuration space, we introduce some properties of graphs, and tabulate known landscape structures.[36,94,95]

TABLE 1 Some graph-theoretical properties of common configuration spaces.[1]

Set	Metric	Reg.	DDR	VT	DT	C	CC
sequences:							
const. length	Hamming	+	+	+	+	+	+
variable alphabet	Hamming	+	+	+	−	+	+
variable length	edit	−	−	−	−	−	−
trees	edit	−	−	−	−	−	−
permutations	transp.	+	+	+	−	+	−
	can. tr.	+	+	+	−	+	−
	reversals	+	+	+	−	+	−
equipartitions	excange	+	+	+	+	?	?

[1] Abbreviations: Reg.: regular; DDR: distnce degree regular; VT: vertex transitive; DT: distance transitive; C: Cayley graph; CC: Cayley graph of a commutative group; ?: not known.

Let Γ be a graph with N vertices. A graph is *regular* if all vertices have the same number of neighbors. It is called *distance degree regular* (DDR) if each vertex has the same number of neighbors at each distance.

A graph *automorphism* α is a one-to-one mapping of Γ onto itself such that the vertices $\alpha(x)$ and $\alpha(y)$ are connected by an edge if and only if the vertices x and y are connected by an edge. Two vertices x and y of Γ are said to be *equivalent* if there is an automorphism α such that $y = \alpha(x)$. A graph is *vertex transitive* if all vertices are equivalent. A vertex transitive graph is always DDR and hence also regular. A graph is *distance transitive* if, for any two pairs of vertices (x, y) and (u, v) with $d(x, y) = d(u, v)$, there is an automorphism α such that $\alpha(x) = u$ and $\alpha(y) = v$ (cf, Buckley and Harary[19]).

Some configuration spaces can be represented by finite groups $(G, \circ)$.[106,108] Let Φ be a set of generators of G such that the group identity is not contained in Φ and such that each $x \in \Phi$ has its inverse group element x^{-1} also contained in Φ.[1] Let $\Gamma(G, \Phi)$ be the graph with vertex set G and an edge connecting two vertices x and y if and only if $xy^{-1} \in \Phi$. That is, two vertices are connected if and only if there is a $\gamma \in \Phi$ such that $y = \gamma x$. Then $\Gamma(G, \Phi)$ is called a *Cayley* graph (cf Biggs and White[11]). A Cayley graph is vertex transitive. There are distance

[1] "Φ is a set of generators" means that each group element $z \in G$ can be represented as a finite product of elements of Φ, with multiplication defined by the group operation $\circ$.

transitive graphs that are not Cayley graphs and vice versa. Not all Cayley graphs are distance transitive.

It can be shown that spaces of sequences with constant length and permutation spaces can be represented as Cayley graphs. However, in light of the examples above, Cayley graphs cannot encompass all possible configuration spaces, since, for example, tree spaces and spaces of sequences of unequal lengths cannot be represented as a Cayley graph. Also, the group G can be chosen to be commutative for sequence spaces but not for permutation spaces.

III.2. LANDSCAPES

We will use the term "landscape" in a well-defined technical sense.

DEFINITION A *landscape* is a map from a finite metric space of combinatorial complexity into the real numbers.

The mathematical object "landscape" can take on varied forms, some examples of which are given below.

III.2.1. EXAMPLES OF LANDSCAPES.

(A) SPIN GLASS HAMILTONIANS. Spin glasses are magnetic substances in which the interaction among spins is sometimes ferromagnetic, i.e., tending to align the spins, and sometimes antiferromagnetic, i.e., tending to align spins with opposite sign. The classical spin glass materials are noble metals (*Au, Ag, Cu,* or *Pt*) weakly diluted with transition metal ions such as *Fe* or *Mn*. For reviews on spin glasses see Binder and Young[12] and the books edited by Mézard et al.[70] and Stein.[100]

The state of the system is fully described by the orientation of each of the spins, i.e., by a configuration $\sigma = (\sigma_1, \sigma_2, \sigma_3, \ldots, \sigma_n)$ where σ_i is the state of the ith spin and N is the size of the system (i.e., number of spins). Most models assume that the energy of a particular configuration is given by a Hamiltonian of the form

$$\mathcal{H}(\sigma) = \sum_{(i,j)} J_{ij}\sigma_i\sigma_j \tag{10}$$

where σ_i can take the values $+1$ and -1 (so-called Ising spins), and the sum runs over all pairs of interacting spins. In some models there are more than two spin states, e.g., $\sigma_i \in \{-1, 0, +1\}$. The constants J_{ij} describe the spin-spin interactions. These coefficients represent fitness, and are typically drawn from a random distribution. The use of a probability distribution in assigning the fitness contributions can be interpreted either as an admission of ignorance of the true nature of the complex couplings between the bits, or as an attempt to capture the typical statistical properties of a wide class of landscapes.

Sometimes it is assumed that the spins σ_i occupy the nodes of a two- or three-dimensional lattice (Ising models) and that only spins on neighboring lattice sites interact. In long-range models, the spatial arrangement is ignored and one assumes interaction between any two spins. The prototype of the latter case is the Sherrington-Kirkpatrick model.[88] Recently there has been growing interest in multi-spin models of the form

$$\mathcal{H}(\sigma) = \sum_{i_1<i_2<\ldots<i_p} J_{i_1 i_2 \ldots i_p} \sigma_{i_1} \sigma_{i_2} \ldots \sigma_{i_p}, \tag{11}$$

the so-called p-spin models. The Sherrington-Kirkpatrick glass is the special case for $p = 2$.

(B) COMBINATORIAL OPTIMIZATION PROBLEMS. Generally, a combinatorial optimization problem is given by

1. the "domain" of the problem, which defines the family of possible instances;
2. the rules which define a configuration; and
3. a cost function which allows computation of the cost of any configuration.

The spin glass models discussed in the previous section fit this scheme: The domain is given by the possible choice of the spin-coupling coefficients J_{ij}, a configuration is defined by a particular orientation of each of the N spins, and the cost function is the Hamiltonian which the system tries to minimize.

In the following we will give a brief introduction to some classical combinatorial optimization problems which have been studied intensively for decades.

• *Traveling Salesman Problem (TSP)*

The TSP[55] is the most prominent classical example of an $\mathcal{NP}$-complete[40] combinatorial optimization problem. Few mathematical problems have attracted as much attention as the TSP.

Given a collection of N geographically-separated cities, the task is to find the shortest tour visiting each city exactly once and returning to the starting point, with prescribed costs η_{ij} for traveling from i to j. The cost function c is therefore

$$c(\tau) = \sum_{i=1}^{N-1} \eta_{\tau(i)\tau(i+1)} + \eta_{\tau(n)\tau(1)} \tag{12}$$

where τ is the permutation encoding the order of the cities.

The symmetric problem $\eta_{ij} = \eta_{ji}$ has applications in X-ray crystallography,[13] electronics[20] and the study of protein conformations.[16] For these problems, the Lin-Kernighan[59] heuristic has proven to be highly successful. The asymmetric case[71] is much more problematic for heuristic algorithms. Asymmetric costs arise in scheduling chemical processes or from pattern allocation problems in the glass industry.

• *Graph Matching* An example of a matching problem[69] is given by the matrix W of weights of the edges between N vertices of a graph. The task is to find a bipartition of the vertices such that each vertex in the right part is connected with exactly one vertex in the left part and vice versa, and such that the sum of weights along the connections between partitions is minimal (or maximal). It is convenient to encode a particular configuration as an ordered list $\vec{v}$ of the vertices $v(i)$ such that vertices in even positions and odd positions are considered as belonging to the right and left part respectively. A connection is assummed between $v(2k)$ and $v(2k-1)$ for $k = 1, \ldots, N/2$. The cost function is therefore

$$f(\vec{v}) = \sum_{k=1}^{N/2} W_{v(2k),v(2k-1)}. \tag{4}$$

As a canonical move rule, the exchange of two vertices in the list, i.e. a transposition, has been chosen. Different variants of matching problems are obtained with different construction rules for W. For example, W may be a symmetric or asymmetric random matrix with entries drawn independently from a uniform distribution or a distance matrix of N points randomly scattered on the unit square.

• *Graph Bipartitioning* The Graph Bipartitioning Problem (GBP) was introduced in III.1.2(d). Given a graph with an even number N of vertices and an associated $N \times N$ matrix H, the task is to find a partition of the vertex set V into two equal-sized subsets A and B such that

$$f([A;B]) = \sum_{i \in A} \sum_{j \in B} h_{ij} \tag{14}$$

is minimized.

The GBP is closely related to the Sherrington-Kirkpatrick spin glass. In fact, the cost function may be viewed as an SK Hamiltonian with the constraint of vanishing total spin, and the expected value of the global optimum can also be related to the ground state of the SK model.[39]

(C) NK MODELS. The NK model assigns a real-valued fitness to the N-element bit string $\vec{b}$ by first assigning a fitness contribution f_i to the ith bit, b_i, in $\vec{b}$. One such assignment $f(\vec{s}_i)$ is made for each combination of the value of b_i and the value of $0 \leq K < N$ other bits, which we call "neighbors" of the ith bit, these $K+1$ bits comprising the substring, $\vec{s}_i$. $f(\vec{s}_i)$ is assigned by selecting an independent random variable from a prespecified distribution for each of the 2^{K+1} possible values of $\vec{s}_i$, thus generating a "fitness table" for the ith site. There is a different, independently generated table for each of the N sites although the distribution from which the $f(\vec{s}_i)$ are chosen does not change with $\vec{s}_i$. Then the total fitness of the string, F, is defined as the average of the fitness contributions from all sites; that is,

$$F(\vec{b}) = \frac{1}{N} \sum_{i=1}^{N} f(\vec{s}_i). \tag{15}$$

One other aspect of the NK model must be specified; namely, the way in which the substrings, $\vec{s}_i$, are chosen. The simplest—but not the only—way of choosing neighbors, at least for even K, is to use the K sites adjacent to site i; that is, sites $i - K/2$ through $i + K/2$. For odd K we have one more neighbor to the right than to the left. We assume that the sites are arranged in a circle, such that site N is next to site 1. This choice of neighbors gives rise to a class of short-range spin glasses. Alternatively, for each site i, we could assign the neighbors by randomly selecting K other sites. This assignment of neighbors makes the model similar to a long range, dilute spin glass. In a variant called the "purely random" NK model, we choose $K+1$ sites at random.

Predictably, since fitnesses of configurations are chosen independently from identical distributions for any placement of the K interacting sites, features of the landscape, such as the height of local optima and the length of typical uphill walks through a series of fitter neighbors to these local optima, are insensitive to the details of how the $K+1$ bit substrings are chosen.[47,48] However, each selection of fitness tables will lead to a unique set of results. Each selection defines one out of a population of landscapes. (See III.2.2 below for further elaboration on this point.) Therefore, small sample-size effects are seen in some simulation results. Moreover, for realistically large N, the Central Limit Theorem ensures that the fitness of any configuration will have an asymptotically normal distribution. The variance of this distribution will scale as $1/N$ and hence all fitnesses will collapse into a small range of values. The main virtue of the NK model is that it does induce correlation as a function of K, by carrying along $(N - K - 1)$ sites with unchanged fitnesses when a mutation occurs at a single position.

(D) BLOCK MODEL. An alternative approach to generating a "tunably rugged" landscape is introduced by Perelson and Macken,[76] motivated by the observation that molecular sequences often have natural partitions. For example, proteins and RNA can be regarded as compositions of independent structural domains; antibodies contain functionally-distinct framework and complementary determining regions. These domains or regions are called blocks. By treating a sequence as a collection of independent blocks, correlations among the fitnesses of neighboring sequences can be induced and tuned by changing the number of blocks. To be specific, let a sequence of length N be composed of B blocks, having lengths $N_i, i = 1, 2, \ldots, B$ with $\sum_{i=1}^{B} N_i = N$. Each block has random fitness $U_i, i = 1, 2, \ldots, B$. Assume the U_i are distributed independently, with distribution G_i.

Define the total fitness of the sequence to be

$$U = \sum_{i=1}^{B} U_i \,. \tag{16}$$

If $B = 1$, the model reduces to the random landscape model of Section II. If $B = N$, the model is equivalent to an NK model with $K = 0$, and exhibits maximal correlation among neighbors. Intermediate levels of correlation occur for $1 < B < N$. As for the NK model, correlation is induced by carrying along $B - 1$ blocks with unchanged fitness when a mutation occurs at a single position. For large blocks (small B), the analytic results from Section II can be used to great effect in studying these landscapes. Also for small B, the block model is free from the dominance by Central Limit effects under which the NK model suffers.

III.2.2 RANDOM FIELDS AND AVERAGING. In the above examples, the landscapes are defined as an instance of a class of landscapes. The proper formal framework is hence the theory of random fields.

DEFINITION A *landscape* $f : \Gamma \to R$ is an instance of a random field $\mathcal{F}$ on the set of vertices Γ of the configuration space $\mathcal{C}$ defined by the joint distribution function

$$P(y_1, y_2, \ldots, y_n) = \text{Prob}\,\{f(x_i) \leq y_i, \quad 1 \leq i \leq N\} \tag{17}$$

where $x_i \in \Gamma$, $f(x_i)$ is the random fitness at vertex x_i and N is the total number of vertices.

Let $\mathcal{E}[.]$ denote mathematical expectation on this random field. Then the expected value of the product of the random fitness values $f(x_k)$ and $f(x_l)$ at the vertices x_k and x_l is

$$\mathcal{E}[f(x_k)f(x_l)] = \int f(x_k)f(x_l)dP(y_1, y_2, \ldots, y_N) \tag{18}$$

where we use $f(x_k)$ to denote a random fitness and its realization. We will refer to this average as the *ensemble average.*

The sample average $\langle\,.\,\rangle$ is defined for a single realization of the landscape. For instance, the mean fitness of a particular landscape is

$$\bar{f} = \langle f \rangle = \frac{1}{N} \sum_{x_k \in \Gamma} f(x_k) \tag{19}$$

where $f(x_k)$ is now a particular realization of the fitness at vertex x_k. In practice, we estimate $\bar{f}$ using only a sample of values taken from the landscape.

A property of a random field model is called self-averaging if the sample average $\langle\,.\,\rangle$ and the ensemble average $\mathcal{E}[\,.\,]$ coincide. For most statistical landscapes described above, the means and variances are examples of self-averaging properties.

III.3 ISOTROPY

Probably the most basic property of a landscape is its isotropy or anisotropy. By anisotropy we mean that certain properties measured in one part of the landscape differ significantly from these properties in another part of the landscape. It is fairly easy to describe what we mean by isotropy for a random field:

DEFINITION A landscape is *isotropic* if:

1. $\mathcal{E}[f(x)] = \bar{f}$ for all configurations $x \in \Gamma$; and
2. for any two pairs of configurations (x, y) and (u, v) such that there is a graph automorphism α with $\alpha(x) = u$ and $\alpha(y) = v$ then $\mathcal{E}[f(x)f(y)] = \mathcal{E}[f(u)f(v)]$.

Note that we require the same covariance only for geometrically equivalent vertices. Although this definition may not be the most intuitive one, it has proved to be very useful for a rigorous treatment of landscapes. More intuitively, we would like to require that:

(2′).for any two pairs of configurations (x, y) and (u, v) with $d(u, v) = d(x, y)$, then $\mathcal{E}[f(x)f(y)] = \mathcal{E}[f(u)f(v)]$.

We will call a landscape satisfying (2′) *homogeneous*. For distance transitive configuration spaces, homogeneous and isotropic are equivalent. In general, homogeneous implies isotropic but not vice versa.

If we have only a single instance of a landscape, the above definitions become useless. Intuitively, we must base a definition of isotropy on measurements in different parts of the landscape. But what do we mean by "part" of a landscape? It is to be expected that we will find different values of properties of the landscape if we are averaging over sets with different geometry. For example, the mean fitness along a "straight line" in configuration space will generally differ from the mean fitness of a configuration and its neighbors.

In order to study single realizations of landscapes, it has been proposed recently[98] to use a collection $\mathcal{B}$ of test sets having the following properties:

1. $\mathcal{B}$ is a partition of the configuration space, $\mathcal{C}$;
2. $A \in \mathcal{B}$ is a connected subgraph of $\mathcal{C}$;
3. Any two subgraphs $R, S \in \mathcal{B}$ are isomorphic, i.e., they have the same geometry;
4. $1 \ll |A| \ll |\mathcal{C}|$, for any $A \in \mathcal{B}$.

The first three requirements make sure that we take fair samples. The last condition ensures that we have (potentially) enough samples.

We suggest the following:

DEFINITION A realization of a landscape is *empirically isotropic* if, for any collection $\mathcal{B}$ of test sets having the above four properties, the following holds:

$$\langle (f(x) - f(y))^2 \rangle_d^A = \langle (f(x) - f(y))^2 \rangle_d \tag{20}$$

where $\langle\,.\,\rangle_d^A$ means an average over all pairs of configurations (x, y) separated by distance d within a test set $A \in \mathcal{B}$ and $\langle\,.\,\rangle_d$ is the average over all pairs with distance d in the entire configuration space.

An immediate consequence of this definition is that, for empirically isotropic landscapes,

$$\mathrm{var}_{\mathcal{B}}[\langle f \rangle^A] = \sigma^2(1 - \bar{\rho}_A) \tag{21}$$

where var is the variance of the fitness values in the test sets A measured over all $A \in \mathcal{B}$, σ^2 is the variance of the fitness values measured over individual configurations in $\mathcal{C}$ and $\bar{\rho}_A$ is the expected correlation of pairs of fitness values within a test set $A \in \mathcal{B}$, which depends only on the partition $\mathcal{B}$. Since the test sets are by definition "localized" in a fairly small part of the configuration space, we expect $\bar{\rho}_A$ to be non-zero in general. $\bar{\rho}_A$ can be calculated using the autocorrelation function $\rho(d)$ discussed in section III.4.1 by $\bar{\rho}_A = \sum_d p_A(d)\rho(d)$ where $p_A(d)$ is the probability that two randomly chosen configurations in the test set $A \in \mathcal{B}$ are separated by a distance d.

It seems natural to measure anisotropy by the extent to which this relation is violated. We define the dimensionless *coefficient of anisotropy* (with respect to the partition $\mathcal{B}$) as

$$\alpha_{\mathcal{B}} = \frac{\text{var}_{\mathcal{B}}[\langle f \rangle^A]}{\sigma^2} - \bar{\rho}_A \,. \tag{22}$$

We emphasize that our method measures anisotropy with respect to a certain partition of the configuration space. It is possible therefore that landscapes are isotropic with respect to a partition $\mathcal{B}$, while they are anisotropic with respect to an alternative partition $\mathcal{B}'$. From a theoretical point of view it is tempting to define the "true anisotropy" of a landscape as the maximum value of $\alpha_{\mathcal{B}}$ from all partitions $\mathcal{B}$. From a practical point of view, however, one would need prohibitively large computer resources to actually compute $\max_{\mathcal{B}} \alpha_{\mathcal{B}}$.

III.4 RUGGEDNESS

The term *ruggedness* has never been rigorously defined. A landscape is considered rugged if it has many local optima, if the length of adaptive and gradient walks are short compared to the maximal distance in the landscape, and if the correlation between nearest neighbors is small (see, e.g., Kauffman[45]). We suggest *correlation length*, defined below, as one possible measure of ruggedness of a landscape.

III.4.1 CORRELATION LENGTH. Weinberger[106] suggested an investigation of landscapes by means of simple random walks. A random walk on configuration space is called simple if moves are made to nearest neighbors and each nearest neighbor has an equal probability of being chosen. Therefore, a simple random walk visits a series $x_0, x_1, \ldots$ of configurations that can be mapped onto a series $f(x_0)$, $f(x_1)$, ... of fitness values. This series of fitness values can then be analysed in the standard fashion for a time series (cf, Priestley[78]). In particular, for a stationary time series, one is interested in the autocorrelation function

$$r(s) = \frac{\langle f(x_i) f(x_{i+s}) \rangle - \langle f(x_i) \rangle^2}{\sigma^2} \,, \quad s = 0, 1, 2, \ldots \,, \tag{23}$$

where $\sigma^2 = \langle f^2(x_i) \rangle - \langle f(x_i) \rangle^2$.

In Fontana et al.,[36,37] a more direct approach was proposed. The autocorrelation function of the landscape is defined by

$$\rho(d) = \frac{\langle f(x)f(y)\rangle_{d(x,y)=d} - \langle f(x)\rangle^2}{\sigma^2}. \tag{24}$$

The two types of autocorrelation functions in Eqs (23) and (24) are related by the geometric relaxation of the random walk in configuration space. Let φ_{sd} denote the probability that a walk of length s spans a distance d between the initial configuration and the end point. Then

$$r(s) = \sum_d \varphi_{sd}\rho(d), \tag{25}$$

i.e., the two approaches are equivalent.

Since autocorrelation is a useful measure only for stationary processes, $r(s)$ and $\rho(d)$ are only useful for isotropic landscapes. On anisotropic landscapes, we would average away all information on the anisotropy without even noticing that the landscape we investigate is not isotropic. The above ansatz has proved useful, however, because most model landscapes studied so far are isotropic (see Table 2).

Autocorrelation functions have been used for a classification of landscapes.[109] Let us assume for the moment that the distribution of fitness values is normal or reasonably close to normal. Subject to this constraint, all landscapes of practical interest that have been investigated so far fall into one of two classes:

1. "Extremely rugged" landscapes, i.e., those showing a sharp drop-off of the autocorrelation function in the very first step. An example, apart from the random energy model, is a travelling salesman problem with asymmetric cost tables and move rule defined by inversions of parts of tours.
2. "Rugged" landscapes with autocorrelation functions of the form

$$\rho(d) = 1 - \beta d + \dots \tag{26}$$

for small d. Except for the two examples of type (1), all landscapes investigated so far fall into this class.

Smooth landscapes with autocorrelation functions of the form $\rho(d) = 1 - \gamma d^2 + \dots$ have not been found in any "real world" problem. Recently, it has been shown that such landscapes cannot exist on sequence spaces with point mutations as the move rule, but can be constructed over other configuration spaces.[99]

An approximate measure for the ruggedness of a landscape can be extracted from autocorrelation functions by defining an empirical correlation length as the length L at which $\rho(L) = 1/e$, so that $L = \mathcal{O}(1/\beta)$ for class (2) landscapes.

TABLE 2 Combinatorial optimization problems and their autocorrelation functions.[1]

Problem	Metric	λ	L	r(s)	$\rho(d)$
random energy binary	any	n	0	*	$\delta_{0,d}$
	transpositions	$n-1$	$n/4$	$\sim e^{-4s/n}$	
	reversals	$n-1$	$n/2$	$\sim e^{-2s/n}$	$\frac{1-d_H(x,y)}{d_H}$
	canonical tr.	$\frac{n(n-1)}{2}$	$n/2$	$\sim e^{-2s/n}$	
symmetric TSP	transpositions	$n-1$	$n/4$	$\sim e^{-4s/n}$	
	reversals	$n-1$	–	$\sim \frac{1}{2}(\delta_{0,s} + e^{-2s/n})$	$\frac{1-d_H(x,y)}{d_H}$
	canonical tr.	$\frac{n(n-1)}{2}$	$n/3$	$\sim e^{-3s/n}$	
graph matching	transpositions	$n-1$	$n/4$	$\sim e^{-4s/n}$	?
graph bipartitioning	exchanges	$n/2$	$(n-3)/8$	$(1 - \frac{8}{n} + \frac{8}{n^2})^s$	$1 - \frac{n-1}{n-2}[8\frac{d}{n} - 16(\frac{d}{n})^2]$
low aautocorrelated binary strings	Hamming	n	$\approx n/10$	$\sim e^{-10s/n}$	?
adjacent NK binary	Hamming	N	$N/(K+1)$	$\sim (1 - \frac{K+1}{N})^s$	$\frac{K+1}{N} - \frac{1}{\binom{N}{d}} \sum_j (K-j+1)\binom{N-j-1}{d-2}$
random NK binary	Hamming	N	$N/(K+1)$	$\sim (1 - \frac{K+1}{N})^s$	$(1-\frac{d}{N})(1-\frac{K}{N-1})^d$
pure random NK binary	Hamming	N	$N/K+1)$	$\sim (1 - \frac{K+1}{NN})^s$	$1 - (\frac{K+1}{N})^d$
p-spin	Hamming	n	$n/(2p)$	$(1 - \frac{2p}{n})^s$	$\frac{1}{\binom{n}{p}} \sum_j (-1)^j \binom{d}{j}\binom{n-d}{p-j}$
SK-glass	Hamming	n	$n/4$	$(1 - \frac{4}{n})^s$	$1 - \frac{n}{n-1}[4\frac{d}{n} - 4(\frac{d}{n})^2]$

[1] *: The autocorrelation function along a random walk depends on the geometry of the configuration space.

[2] ?: Unknown

[3] λ: The maximum distance between any two configurations.

[4] L: The empirical correlation length of the landscape.

Attributes of the autocorrelation function as a measure for ruggedness are:

i. It is a natural measure, being just the second moment of the value distribution of the landscape.

ii. It is readily computable in numerical simulations (see Table 2).

iii. It can be generalized to so-called *combinatory maps*, for which the fitness value is replaced by a non-numeric quantity, such as spatial structure.

iv. It can be interpreted in term *s* of indepenndence or lack thereof, under appropriate conditions of normality.

An important disadvantage of the autocorrelation function as a measure of ruggedness is that it is not invariant under arbitrary monotonic transformations of the fitness function, as the following example demonstrates. Assume that the activation energies f for, say, the folding of biopolymers, is distributed approximately Gaussian. Suppose that fitness of a biopolymer is measured by the reaction rates, g. By Arrhenius' Law,

$$g(x) = \exp(-\beta f(x)) \tag{27}$$

for configuration x, where $\beta = 1/kT$ is the inverse temperature. Let $\rho(d)$ be the autocorrelation function of the Gaussian landscape of activation energies f. It can be shown,[102] that the landscape of the rate constants g has the autocorrelation function

$$\rho_g(d) = \frac{\alpha^{\rho(d)} - 1}{\alpha - 1} \qquad \text{with} \qquad \alpha = \exp(\sigma^2\beta^2),\ \alpha > 1 \tag{28}$$

where σ^2 is the variance of the activation energies. It is easy to see that $|\rho(d)| \geq |\rho_g(d)|$ with equality holding only if $\rho(d)$ is 0 or 1. If $\rho(d)$ is of the generic type of Eq. (26), with correlation length ℓ, then $\rho_g(d)$ has correlation length

$$\ell_g = \frac{\alpha - 1}{\alpha \log \alpha} \cdot \ell \sim \frac{1}{\sigma^2\beta^2}\ell \tag{29}$$

which can become arbitrarily small for large β. Therefore, a simple exponential transformation can produce an apparently uncorrelated landscape without changing the number of local optima or the length of adaptive walks. The contradiction rests in the application of autocorrelation to these surfaces. In fact, as is well known in statistics, a pairwise correlation of zero implies independence only under the assumption of a bivariate normal distribution for the pairs of observations.

A more detailed picture of the landscape can be obtained from the conditional probability $\wp(f|d)$ that fitnesses differ by f (in absolute value) between two configurations at a distance d in configuration space. The function $\wp(f|d)$ is also called the *fitness density surface.* One can show that the autocorrelation function $\rho(d)$ can be computed from $\wp(f|d)$ by

$$\rho(d) = 1 - \frac{\sum_f f^2 \wp(f|d)}{\sum_d \sum_f p(d) f^2 \wp(f|d)} \tag{30}$$

where $p(d)$ is the probability that two randomly chosen configurations are separated by distance d in configuration space.[37] The density surface contains more information than the autocorrelation function alone. For instance, the chance of finding neutral neighbors is given by $\wp(0|1)$. It turns out that the use of density surfaces is a much more efficient way to compute the autocorrelation function than sampling along a random walk. Furthermore the use of density surfaces is not restricted to landscapes. Any map between two metric spaces can be analyzed this way. For an example see the contribution by Tacker and Stadler in this volume.

III.4.2 LOCAL OPTIMA. One of the most important characteristics of a landscape is the number of local optima. Intuitively, this number will increase with the ruggedness of the landscape. In fact, Palmer[75] used an at-least-exponential increase of the number of local optima with the system size (as measured by, for example, the length of a sequence) as a definition of a rugged landscape.

It is easy to calculate the number of local optima in two extreme cases: the random energy model (see Section II), and a maximally correlated landscape in which the total fitness equals the sum or average of independent fitness contributions from each element in a sequence. In the latter case there is just one optimum. In both cases the expected number of local optima is independent of the distribution of the fitness values. We present below results on the number of local optima for other definitions of landscapes.

(A) LANDSCAPES WITH LOW CORRELATION. For landscapes with low correlation, (i.e., rugged landscapes by the definition in Section III.4.1), we can generalize the formalism developed in section II[93]: Consider a Gaussian landscape with mean μ and variance σ^2. Then, for a random configuration x_0

$$\mathrm{Prob}\{\mathrm{x_0\,loc.opt.}\} = \sqrt{\frac{1+\rho}{2\pi(1-\rho)}}\int_{-\infty}^{+\infty} \exp\left(-\frac{1+\rho}{1-\rho}\frac{\mathrm{y}^2}{2}\right)\Pi^{\mathrm{N}_{(1)}}(\mathrm{y})\mathrm{dy} \tag{31}$$

where $N_{(1)}$ = number of nearest neighbors of x_0, ρ = correlation between neightors, and $\Pi(y)$ is the cumulative normal distribution function evaluated at y. Note that when $\rho = 0$ this expression reduces to $1/(N+1)$, which is the exact result for the random energy model for a binary sequence of length N.

(B) HIGHLY CORRELATED LANDSCAPES. In a maximally correlated landscape, i.e., correlation length $L = \lambda = \max d(x, y)$, one expects to find a single optimum. Now consider a general highly correlated landscape with correlation length $L \approx \mathcal{O}(\lambda)$. Let $N(l)$ denote the number of vertices in a neighborhood of radius l around an arbitrary configuration. We expect $\mathcal{O}(1)$ local optima in a neighborhood with radius L, and hence the probability of finding a local optimum at random can be estimated by

$$p_{lo} = \mathrm{Prob}\{\mathrm{loc.opt}\} \approx \frac{1}{\mathrm{N(L)}}. \tag{32}$$

For sequence spaces we can calculate this more explicitly. For a sequence of length N with alphabet size κ, the number of points within a ball of size l is approximately

$$\begin{aligned} N(1) &= \sum_{j=0}^{\ell} \binom{N}{j} (\kappa-1)^j \approx \binom{N}{1} (\kappa-1)^1 \\ &\approx \frac{(N/e)^N}{[(N-1)/e]^{N-1}(1/e)^1} (\kappa-1)^1 \\ &= \left[(\kappa-1)^{1/N} \left(\frac{N}{1}\right)^{1/N} \left(1-\frac{1}{N}\right)^{1/N-1} \right]^N . \end{aligned} \tag{33}$$

Using the scaled correlation length $\xi = L/N$ we obtain

$$p_{lo} \approx \left[(\kappa-1)^{\xi} (1/\xi)^{\xi} (1-\xi)^{\xi-1} \right]^{-N} . \tag{34}$$

Weinberger[107] gives the following estimate for the probability of finding a local optimum in an NK model for medium and large values of K:

$$p_{lo,NK} \approx [(\kappa-1)(K+1)]^{-\frac{N}{K+1}} . \tag{35}$$

For the NK model in this regime of K, $\xi = 1/(K+1)$ (see Table 2). Hence, Eq. (35) becomes

$$p_{lo,NK} \approx \left[(\kappa-1)^{\xi} (1/\xi)^{\xi} \right]^{-N} . \tag{36}$$

The approximations of Eqs (34) and (36) deviate by a factor $p_{lo,NK}/p_{lo} = g(\xi)^N$. Weinberger's estimate yields good results for small ξ, say $\xi < 0.1$. In this range $g(\xi) = (1-\xi)^{\xi-1}$ is less than 1.1, and it converges to 1 for $\xi \to 0$. Given the crudeness of both approximations, the agreement is encouraging.

For the symmetric TSP with the transposition metric, numerical data on the probability of local optima are published.[93] For this permutation space the following holds:

$$\begin{aligned} N(1) &= \sum_{j=0}^{1} S_j \approx \sum_{j=0}^{1} \frac{(N^2/2)^j}{j!} \approx \frac{(n^2/2)^1}{j1} \\ &\approx \left(\frac{e}{1}\right)^1 \left(\frac{N^2}{2}\right)^1 \approx 1! \left(\frac{e^2}{2\xi^2}\right)^1 , \end{aligned} \tag{37}$$

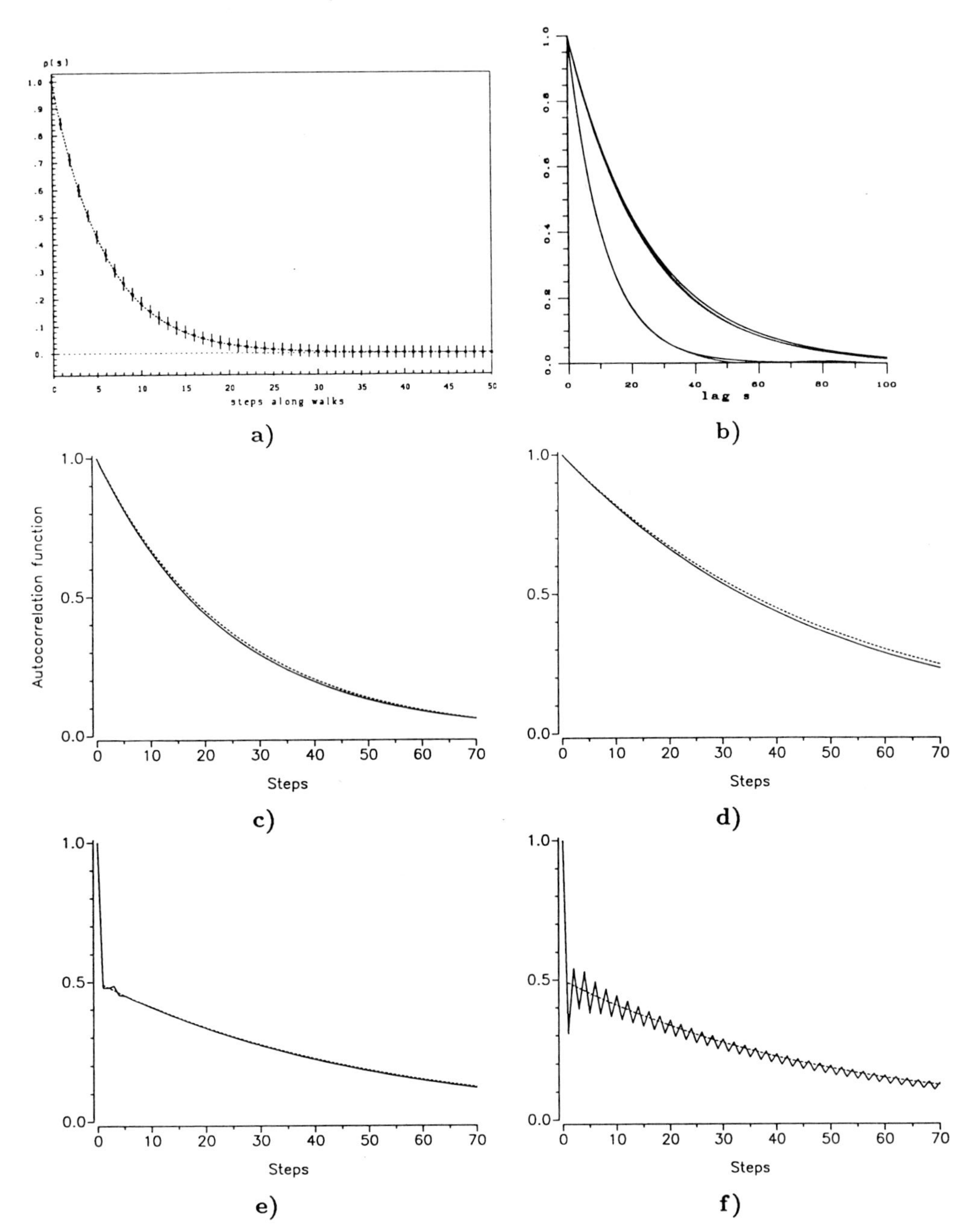

FIGURE 2 Examples of correlation function steps $r(s)$ along random walks.

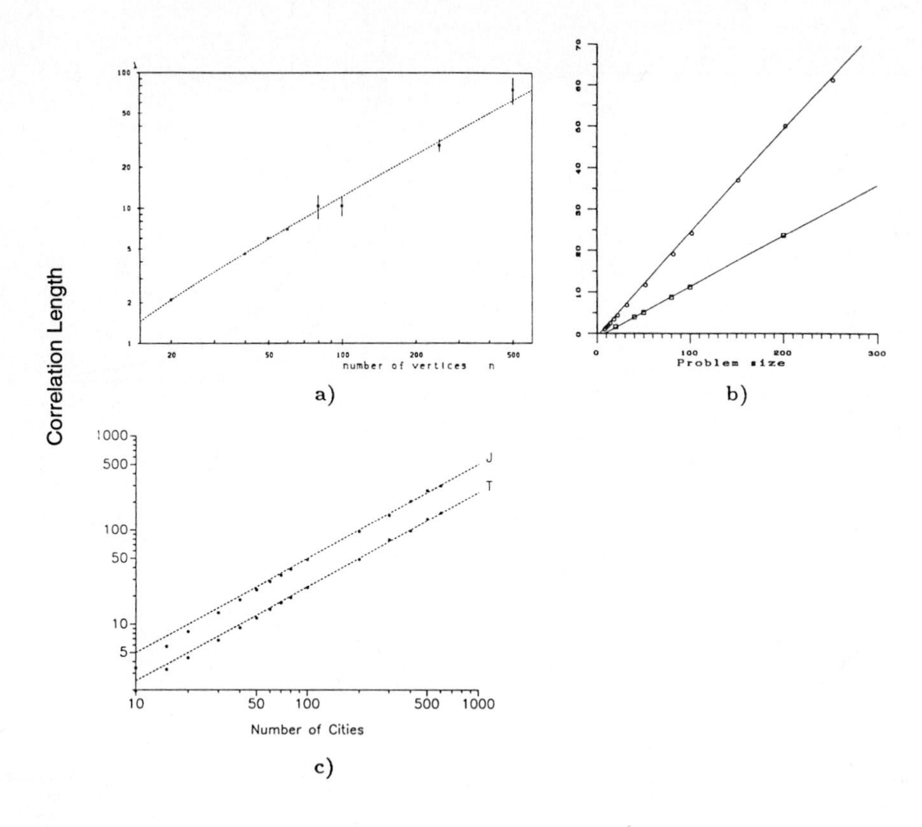

FIGURE 3 Scaling of correlation lengths, for some example landscapes.

where S_j is an unsigned Sterling number of the first kind. Since $\xi = 1/4$ here (see Table 2 and Stadler and Schnabl[93]),

$$p_{lo} \approx \text{const}\frac{2.773^{-N}}{(N/4)!} \tag{38}$$

which fits the data well with a constant of the order 10. $N(l)$ is not known for the inversion metric.

A comparison of the estimate from Eq. (38) with the numerical data of Stadler and Schnabl[93] is shown in Figure 4: the agreement is better than one might expect from such a crude estimate. The number of local optima has also been determined for RNA free energy landscapes.[36] The data are in qualitative agreement with a rate of about one optimum in a neighborhood having a diameter equal to the correlation length. A rigorous theory linking the correlation structure of Gaussian landscapes with the number of local optima is still missing. Ultimately, of course, one would like this link for more general landscapes than Gaussian.

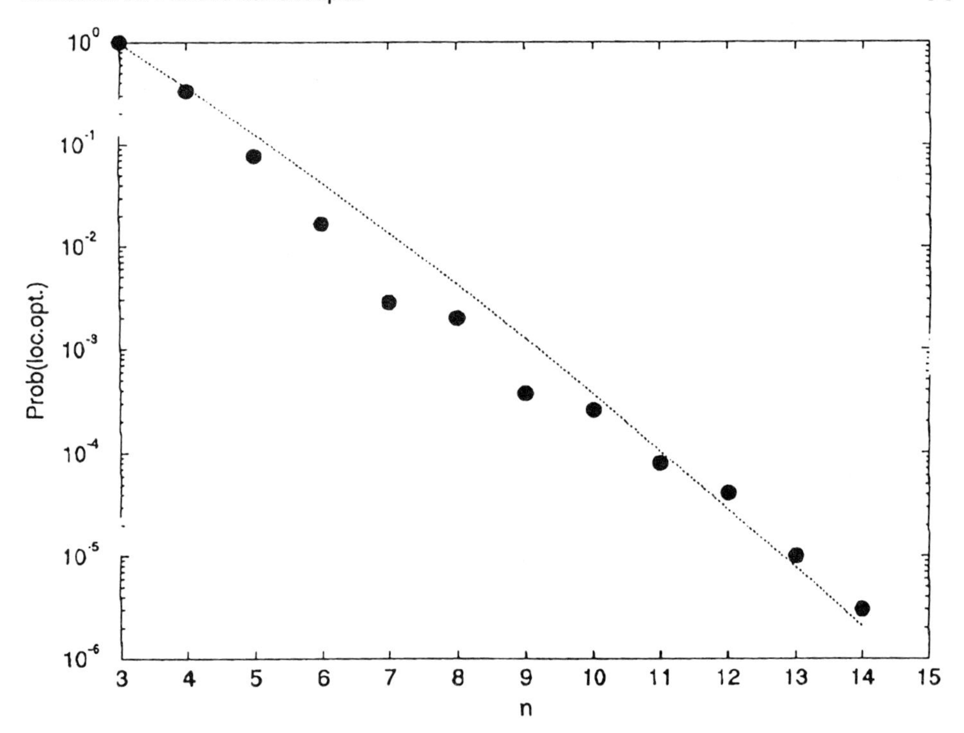

FIGURE 4 Probability of local optima for the TSP with the transposition metric.

III.5 EVOLUTION

Selection acts at the level of fitnesses (or more generally phenotypes). Mutation, on the other hand, takes place at the level of sequences—in fact, modern data used for studying the evolution of life on earth are sequences, and the history of life can be represented as a tree in sequence space. One can therefore view evolution as motion in sequence space and seek the laws of motion in this abstract space. Of course, a single set of laws will not suffice since more than one way exists for passing genetic information from one generation to the next.

The simplest law of motion on a landscape is provided by an uphill (or downhill) walk. Along the walk, a neighboring configuration is accepted if and only if it is more fit (or less fit) than the present one. Such walks have finite length and eventually end at a local optimum. Most studies have been performed with the adaptive walk, which moves to a randomly chosen better neighbor (see Section II). Adaptive walks are a reasonable approximation to real evolutionary dynamics of haploid organisms in the extreme case of low mutation rates and strong selection, since the acceptance of an advantageous mutation occurs in conjunction with all inferior mutants being wiped out. When higher mutation rates lead to the production of multiple one-mutant neighbors per unit of time, gradient or greedy walks can be used. In a gradient walk, the best of these one-mutant neighbors will be accepted provided it is advantageous. When the mutation rate is high enough, non-nearest neighbors will be produced, allowing for a chance of escaping from a local optimum. This last process requires a substantially different model than those described here.

If we consider a bounded population size of haploid organisms and ignore all interactions between organisms except competition for a constant resource, then Eigen's quasispecies model describes the dynamics of evolution (see Section I.1(c)). We emphazise that this model represents an extremely idealized situation, although it can be realized in the lab.[92] If we allow recombination events in addition to mutation, then at present a genetic algorithm is the only technique available to study evolutionary dynamics on a landscape.

III.5.1 ADAPTIVE WALKS AND GRADIENT WALKS IN CORRELATED LANDSCAPES. The length of an adaptive or gradient walk is closely related to the correlation structure of the landscape. Since we have $\mathcal{O}(1)$ local optima in a neighborhood with diameter equal to the correlation length (in a Gaussian landscape), we expect walk lengths to be of the order of the correlation length. More specifically, the length of a gradient walk should be approximately equal to the correlation length since gradient walks move as quickly as possible. On the other hand, adaptive walks should be longer (by a factor that depends only on the structure of the configuration space) since a fitter neighbor is chosen at random. Numerical evidence for the NK model[107] and RNA landscapes[36] is consistent with this statement. For example, in the generalized NK model over an alphabet with κ letters,

$$W_{adap} \approx \ln[(\kappa - 1)(K+1)] \cdot \frac{N}{K+1} \tag{39}$$

for the length of an adaptive walk, and

$$W_{grad} \approx \frac{\kappa - 1}{\kappa} \frac{\ln[(\kappa - 1)(K+1)]}{\ln \kappa} \cdot \frac{N}{K+1} \tag{40}$$

for the length of a gradient walk.

In the block model, the assumption of independence of blocks enables analytical calculations of many of the statistics of interest for an evolutionary walk. It can be shown that, for a sequence of length N and an alphabet size of a,

$$\mathcal{E}(W_{\text{adap}}) \approx BC + \sum_{i=1}^{B} \ln(d_i) + \sum_{i=1}^{B} \ln[1 - G_i(u_{0i})] \tag{41}$$

where $C = 1.0991, d_i = (a-1)N_i$ for N_i the length of the ith block and u_{0i} is the starting fitness of walks in the ith block.[76]

III.5.2 RUGGEDNESS AND THE SPEED OF EVOLUTION. Local optima are obstacles to an optimization algorithm. Therefore, global optimization should be easier when there are fewer local optima. It is conjectured that an equivalent statement is that optimization should be easier when the nearest neighbor correlation is larger. Evidence exists for the validity of this conjecture. For example:

- The Lin-Kernighan[59] optimization heuristic for the TSP, which is based on reversals (two-opt moves), does not perform well in the case of asymmetric costs. Stadler and Schnabl[97] explain this loss of performance by the fact that the nearest neighbor correlation is $\rho < 0.5$ for asymmetric costs while $\rho \approx 1 - 2/N$ in the symmetric case for which the algorithm was designed.
- Fontana[34] finds that a quasipecies simulation using a symmetric TSP landscape yields fitter variants if reversal instead of transposition is used as the move rule. The scaled correlation length ξ for reversals is twice that for transpositions.
- The performance of a simple genetic algorithm has been shown to depend strongly on the correlation length of the underlying landscape (see Table 1 of Manderick et al.[62]).

At present there is no satisfactory theory linking the performance of an algorithm to the structure of the landscape. And how shall we measure and compare the performance of different algorithms in the first place? In the remainder of this section we briefly outline a few partial results and open problems.

(A) QUASISPECIES. The relationship between the critical mutation rate of a quasispecies, p_{crit}, and the structure of a landscape is of utmost importance for evolutionary biotechnology, since evolutionary adaptation is most efficient if operating only slightly below p_{crit} (cf, Eigen et al.[27]). Therefore, predicting the optimal replication accuracy is a central challenge for the theory of landscapes.

McCaskill[67] derived expressions for p_{crit} for uncorrelated landscapes with a variety of fitness distributions. For landscapes with fitness distribution g of the form $g(x) = \exp(-\beta f(x))$, where $f(x)$ is the fitness of a highly correlated Gaussian (mean 0 and variance 1) landscape on a Boolean hypercube, a rough estimate for the error threshold was proposed recently[18]:

$$p_{\text{crit}} \approx d_0\, \beta\, f(opt)\, \xi\, N^{-2}, \tag{42}$$

where, as before, ξ is the scaled correlation length of f, d_0 is a constant which is *a priori* known to be very close to 1 and $f(opt)$ is the fitness of the global optimum. Note, however, that this exponential transformation can produce a landscape with arbitrarily small correlation for β large enough (see Section III.4.1, Eq. (27)). An analogous result for Gaussian (or more general) landscapes is still unknown.

(B) SIMULATED ANNEALING. Simulated annealing[52] is a natural way of looking at the relaxation of spin glasses and related systems. A "mutant" is always accepted if it is favorable, but in contrast to an adaptive walk, it is also accepted with a certain probability u if it is unfavorable, where u depends on the fitness difference Δ between the current state and the "mutant." Usually one assumes $u = \exp(-\Delta/RT)$, where R is constant and T is temperature-like parameter. This algorithm was introduced by Metropolis et al.[68] in order to simulate a thermodynamic system in equilibrium. When the temperature T is gradually reduced during the simulation, one obtains a powerful optimization algortithm,[52] whose performance depends crucially on the choice of the cooling schedule, i.e., the sequence of temperatures, and the structure of the landscapes.

Recent work on the convergence of simulated annealing (cf, Azencott[3] and the references therein) has led to the definition of parameters that are functions of landscape characteristics (configuration space and fitness) and that are closely related to the performance of the algorithm. One parameter, $\mathcal{D}$, describes the *depth* of the landscape. Kern[51] shows that $\mathcal{D}$ is the height of the highest pass that has to be climbed in order to go from one local optimum to an arbitrarily better local optimum. An asymptotically good cooling schedule fulfills $\lim_{k\to\infty} kT_k = \mathcal{D}$. Another parameter, α, which is sometimes called the *difficulty* of the landscape, determines the optimal speed of convergence. The expected values of these parameters for the model landscapes discussed in this chapter and their relevance for other optimization algorithms, including the quasispecies approach, are unknown to date.

REFERENCES

1. Abramowitz, M., and I. A. Stegun. *Handbook of Mathematical Functions.* Washington, DC: National Bureau of Standards, 1966.
2. Amitrano, C., L. Peliti, and M. Saber. "Population Dynamics in a Spin Glass Model of Chemical Evolution." *J. Mol. Evol.* **29** (1989): 513–525.
3. Azencott, R. *Simulated Annealing: Parallelization Techniques.* New York: John Wiley & Sons, 1992.
4. Bak, P., H. Flyvbjerg, and B. Lautrup. "Coevolution in a Rugged Fitness Landscape." *Phys. Rev. A* **46(15)** (1992): 6724–6730.
5. Baldi, P., and Y. Rinott. "Asymptotic Normality of Some Graph Related Statistics." *J. Appl. Prob.* **26** (1989): 171–175.
6. Banavar, J. R., D. Sherrington, and N. Sourlas. "Graph Bipartioning and Statistical Mechanics." *J. Phys. A. Math. Gen.* **20** (1987): L1–L8.
7. Bauer, G. J., J. S. McCaskill, and H. Otten. "Traveling Waves in *in vitro* Evolving RNA." *Proc. Natl. Acad. Sci. USA* **86** (1989): 7937–7941.

8. Berek, C., G. M. Griffiths, and C. Milstein. "Molecular Events During the Maturation of the Immune Response to Oxazolone." *Nature* **316** (1985): 412–418.
9. Berek, C., and C. Milstein. "Mutation Drift and Repertoire Shift in the Maturation of the Immune Response." *Immun. Rev.* **96** (1987): 23–41.
10. Biebricher, C. K., and M. Eigen. "Kinetics of RNA Replication by Qb Replicase." In *RNA Genetics*, edited by E. Domingo, J. J. Holland, and P. Ahlquist, Vol. I. Boca Raton, FL: CRC Press, 1988.
11. Biggs, N. L., and A. T. White. *Permutation Groups and Combinatorial Structures.* Cambridge, MA: Cambridge University Press, 1979.
12. Binder, K., and A. P. Young. "Spin Glasses. Experimental Facts, Theoretical Concepts, and Open Questions." *Rev. Mod. Phys.* **58** (1986): 802–976.
13. Bland, R. G., and D. F. Shallcross. "Large Traveling Salesmen Problems Arising from Experiments in X-ray Crystallography." *Oper. Res. Lett.* **8** (1988): 125–128.
14. Boerlijst, M., and P. Hogeweg. "Spiral Wave Structure in Pre-Biotic Evolution: Hypercycles Stable Against Parasites." *Physica D* **48** (1992): 17–28.
15. Boerlijst, M., and P. Hogeweg. "Self-Structuring and Selection. Spiral Waves as a Substrate for Prebiotic Evolution." In *Artifical Life II*, edited by C. Langton, page nos. Santa Fe Institute Studies in the Sciences of Complexity, Proc. Vol. ??. Reading, MA: Addison-Wesley, 1992.
16. Bohr, H., and S. Brunak. "Travelling Salesman Approach to Protein Conformation." *Complex Systems* **3** (1990): 9–28.
17. Bonhoeffer, S., J. S. McCaskill, P. F. Stadler, and P. Schuster. "RNA Multi-Structure Landscapes. A Study Based on Temperature Dependent Partition Functions." *Eur. Biophys. J.* **22** (1993): 13–24.
18. Bonhoeffer, S., and P. F. Stadler. "Errorthreshold on Complex Fitness Landscapes." *J. Theor. Biol.* **164** (1993): 359–372.
19. Buckley, F., and F. Harary. *Distance in Graphs.* Reading, MA: Addison-Wesley, 1990.
20. Chan, D., and D. Mercier. "IC Insertion: An Application of the TSP." *Int. J. Prod. Res.* **27** (1989): 1837–1842.
21. Crow, J. F. *Basic Concepts in Population, Quantitative, and Evolutionary Genetics.* New York: Freeman, 1986.
22. DeLisi, C. *Antigen Antibody Interactions.* Lecture Notes in Biomath., Vol. 8. New York: Springer-Verlag, 1976.
23. Derrida, B. "Random Energy Model. An Exactly Solvable Model of Disordered System." *Phys. Rev. B* **24** (1981): 2613–2626.
24. Derrida, B., and L. Peliti. "Evolution in a Flat Fitness Landscape." *Bull. Math. Biol.* **53** (1991): 355–382.
25. Eigen, M. "Self-Organization of Matter and the Evolution of Biological Macromolecules." *Naturwissenschaften* **58** (1971): 465–523.
26. Eigen, M., and P. Schuster. *The Hypercycle.* New York: Springer-Verlag, 1979.

27. Eigen, M., J. McCaskill, and P. Schuster. "The Molecular Quasispecies." *Adv. Chem. Phys.* **75** (1989): 149–263.
28. Eisen, H. *Immunology*, 2nd ed. Hagerstown, MD: Harper and Row, 1980.
29. Faulkner, D. V., and A. Jurka. "Multiple Aligned Sequence Editor." *Trends in Biochem. Sci.* (1988): 321–322.
30. Fish, S., M. Fleming, J. Sharon, and T. Manser. "Different Epitope Structures Select Distinct Mutant Forms of an Antibody Variable Region for Expression During the Immune Response." *J. Exp. Med.* **173** (1991): 665–672.
31. Flyvbjerg, H., and B. Lautrup. "Evolution in a Rugged Fitness Landscape." *Phys. Rev. A* **46(15)** (1992): 6714–6723.
32. Fontana, W. "Ein Computer Modell der Evolutionaren Optimierung." Ph.D. Thesis, University of Vienna, 1986 (in German).
33. Fontana, W., and P. Schuster. "A Computer Model of Evolutionary Optimization." *Biophysical Chem.* **26** (1987): 123–147.
34. Fontana, W., W. Schnabl, and P. Schuster. "Physical Aspects of Evolutionary Optimization and Adaptation." *Phys. Rev. A* **40** (1989): 3301–3321.
35. Fontana, W., T. Griesmacher, W. Schnabl, P. F. Stadler, and P. Schuster. "Statistics of Landscapes Based on Free Energies, Replication and Degradation Rate Constants of RNA Secondary Structures." *Mh. Chem.* **122** (1991): 795–819.
36. Fontana, W., P. F. Stadler, E. G. Bornberg-Bauer, T. Griesmacher, I. L. Hofacker, P. Tarazona, E. D. Weinberger, and P. Schuster. "RNA Folding Landscapes and Combinatory Landscapes." *Phys. Rev. E* **47** (1993): 2083–2099.
37. Fontana, W., D. A. M. Konings, P. F. Stadler, and P. Schuster. "Statistics of RNA Secondary Structures." *Biopolymers* **33** (1993): 1389–1404.
38. Fu, Y. "The Use and Abuse of Statistical Mechanics in Computational Complexity." In *Lectures in the Science of Complexity*, edited by D. Stein. Santa Fe Institute Studies in Science of Complexity, Lect. Vol. I. Redwood City, CA: Addison-Wesley, 1989.
39. Fu, Y., and P. W. Anderson. "Application of Statistical Mechanics to NP-Complete Problems in Combinatorial Optimization." *J. Phys. A* **19** (1986): 1605–1620.
40. Garey, M., and D. Johnson. *Computers and Intractability: A Guide to the Theory of NP-Completeness.* San Francisco: Freeman, 1979.
41. Goldstein, B. "Theory of Hapten Binding to IgM. The Question of Repulsive Interactions Between Binding Sites." *Biophysical Chem.* **3** (1975): 363–367.
42. Hamming, R. W. "Error Detecting and Error Correcting Codes." *Bell Syst. Tech. J.* **29** (1950): 147–160.
43. Hofacker, I. L., W. Fontana, P. F. Stadler, L. S. Bonhoeffer, M. Tacker, and P. Schuster. "Fast Folding and Comparison of RNA Secondary Structures (The Vienna RNA Package)." *Mh. Chem.* **125** (1994): 167–188.
44. Hofbauer, J., and K. Sigmund. *Dynamical Systems and the Theory of Evolution.* Cambridge, UK: Cambridge University Press, 1988.

45. Kauffman, S. A. *The Origins of Order.* Oxford: Oxford University Press, 1993.
46. Kauffman, S. A., and S. Levin. "'Towards a General Theory of Adaptive Walks on Rugged Landscapes." *J. Theor. Biol.* **128** (1987): 11–45.
47. Kauffman, S. A., E. D. Weinberger, and A. S. Perelson. "Maturation of the Immune Response via Adaptive Walks on Affinity Landscapes." In *Theoretical Immunology,* edited by A. S. Perelson, Part I. Santa Fe Institute Studies in the Sciences of Complexity, Proc. Vol. II. Reading, MA: Addison-Wesley, 1988.
48. Kauffman, S. A., and E. D. Weinberger. "The N-K Model of Rugged Fitness Landscapes and its Application to Maturation of the Immune Response." *J. Theor. Biol.* **141** (1989): 211–245.
49. Kepler, T. B., and A. S. Perelson. "Somatic Hypermutation in B Cells. An Optimal Control Treatment." *J. Theor. Biol.* **164** (1993): 37–64.
50. Kepler, T. B., and A. S. Perelson. "Cyclic Reentry of Germinal Center B Cells and the Efficiency of Affinity Maturation." *Immunol. Today* **14** (1993): 412–415.
51. Kern, W. "On the Depth of Combinatorial Optimization Problems." *Discr. Appl. Math.* **43** (1993): 115–129.
52. Kirkpatrick, S., C. D. Gelatt, and M. P. Vecchi. "Optimization by Simulated Annealing." *Science* **220** (1983): 671–680.
53. Kocks, C., and K. Rajewsky. "Stepwise Intraclonal Maturation of Antibody Affinity Through Somatic Hypermutation." *Proc. Natl. Acad. Sci. USA* **85** (1988): 8206–8210.
54. Kramer, F. R., D. R. Mills, P. E. Cole, T. Nishihara, and S. Spiegelman. "Evolution *in vitro*: Sequence and Phenotype of a Mutant RNA Resistant to Ethidium Bromide." *J. Mol. Biol.* **89** (1974): 719–736.
55. Lawler, E. L., J. K. Lenstra, A. H. G. Rinnoy Kan, and D. B. Shmoys. *The Traveling Salesman Problem: A Guided Tour of Combinatorial Optimization.* New York: John Wiley & Sons, 1985.
56. Leuthausser, I. "An Exact Correspondence Between Eigen's Evolution Model and a Two-Dimensional Ising Model." *J. Chem. Phys.* **84** (1986): 1884–1885.
57. Leuthausser, I. "Statistical Mechanics of Eigen's Evolution Model." *J. Stat. Phys.* **48** (1987): 343–360.
58. Liao, W. "Graph Bipartitioning Problem." *Phys. Rev. Lett.* **59** (1987): 1625–1628.
59. Lin, S., and B. W. Kernighan. "An Effective Heuristic Algorithm for the Traveling Salesman Problem." *Oper. Res.* **21** (1973): 498–516.
60. Macken, C. A., and A. S. Perelson. "Protein Evolution on Rugged Landscapes." *Proc. Natl. Acad. Sci. USA* **86** (1989): 6191–6195.
61. Macken, C. A., P. S. Hagan and A. S. Perelson. "Evolutionary Walks on Rugged Landscapes." *SIAM J. Appl. Math.* **51** (1991): 799–827.
62. Manderick, B., M. deWeger, and P. Spiessen. "The Genetic Algorithm and the Structure of the Fitness Landscape." In *Proc. of the 4th Internatl. Conf. on Genetic Algorithms,* edited by R. K. Belew and L. B. Booker. San Mateo CA: Morgan Kaufmann, 1991.

63. Manser, T. "Evolution of Antibody Structure During the Immune Response. The Differentiative Potential of a Single B Lymphocyte." *J. Exp. Med.* **170** (1989): 1211–1230.
64. Manser, T. "Maturation of the Humoral Immune Response. A Neo-Darwinian Process?" In *Molecular Evolution on Rugged Landscapes*, edited by A. Perelson and S. Kauffman. Santa Fe Institute Studies in the Sciences of Complexity, Proc. Vol. IX. Redwood City, CA: Addison-Wesley, 1991.
65. Manser, T. Personal communication, 1993.
66. Maynard-Smith, J. "Natural Selection and the Concept of a Protein Space." *Nature* **225** (1970): 563.
67. McCaskill, J. S. "A Localization Threshold for Molecular Quasispecies from Continuously Distributed Replication Rates." *J. Chem. Phys.* **80** (1984): 5194–5202.
68. Metropolis, N., A. W. Rosenbluth, M. N. Rosenbluth, A. E. Teller, and E. Teller. "Equation of State Calculation by Fast Computing Maschines." *J. Chem. Phys.* **21** (1953): 1087–1092.
69. Mézard, M., and G. Parisi. "Mean-Field Equations for the Matching and Travelling Salesman Problems." *Europhys. Lett.* **2** (1986): 913–918.
70. Mézard, M., G. Parisi, and M. A. Virasoro. *Spin Glass Theory and Beyond.* Singapore: World Scientific, 1987.
71. Miller, D. L., and J. F. Pekny. "Exact Solution of Large Asymmetric Traveling Salesman Problems." *Science* **251** (1991): 754–761.
72. Nowak, M. A., and P. Schuster. "Error Thresholds for Replication in Finite Populations. Mutation Frequencies and the Onset of Muller's Ratchet." *J. Theor. Biol.* **137** (1989): 375–395.
73. Ohmori, K., and E. Tanaka. "A Unified View on Tree Metrics." In *Syntactic and Structural Pattern Recognition*, edited by G. Ferrate. Berlin: Springer-Verlag, 1988.
74. Palmer, R. "Statistical Mechanics Approaches to Complex Optimization Problems." In *The Economy as a Complex Evolving System*, edited by P. W. Anderson, K. J. Arrow, A. S. Perelson, and S. A. Kauffman. Santa Fe Institute Studies in the Sciences of Complexity, Proc. Vol. V. Redwood City, CA: Addison-Wesley, 1988.
75. Palmer, R. "Optimization on Rugged Landscapes." In *Molecular Evolution on Rugged Landscapes: Proteins, RNA, and the Immune System*, edited by A. S. Perelson and S. A. Kauffman. Santa Fe Institute Studies in the Sciences of Complexity, Proc. Vol. IX. Redwood City, CA: Addison-Wesley, 1991.
76. Perelson, A. S., and C. A. Macken. "Protein Evolution on Partially Correlated Landscapes." 1994: submitted.
77. Priestley, M. B. *Spectral Analysis and Time Series*, Volumes 1 and 2. London: Academic Press, 1989.
78. Priestley, M. B. *Nonlinear and Non-stationary Time Series Analysis.* London: Academic Press, 1991.

79. Sankoff, D., G. Leduc, N. Antoine, B. Paquin, B. F. Lang, and R. Cedergren. "Gene Comparisons for Phylogenetic Inference. Evolution of the Mitochondrial Genome." *Proc. Natl. Acad. Sci. USA* **89** (1992): 6575–6579.
80. Schnabl, W. "Adaptive Optimization of Polynucleotides with Variable Chain Length." Ph.D. Thesis, University of Vienna, 1990.
81. Schuster, P., and K. Sigmund. "Replicator Dynamics." *J. Theor. Biol.* **100** (1983): 533–538.
82. Schuster, P., and J. Swetina. "Stationary Mutant Distributions and Evolutionary Optimization." *Bull. Math. Biol.* **50** (1988): 636–660.
83. Schuster, P., W. Fontana, P. F. Stadler, and I. L. Hofacker. "From Sequences to Shapes and Back: A Case Study in RNA Secondary Structures." *Proc. Roy. Soc. B* **255** (1994): 279–284.
84. Schuster, P., and P. F. Stadler. "Landscapes, Complex Optimization Problems and Biopolymer Structure." *Comp. Chem.* (1994): in press.
85. Shapiro, B. A. "An Algorithm for Comparing Multiple RNA Secondary Structures." *CABIOS* **4** (1988): 387–397.
86. Shapiro, B. A., and K. Zhang. "Computing Multiple RNA Secondary Structures using Tree Comparisons." *CABIOS* **6** (1990): 309–318.
87. Sharon, J., M. L. Gefter, L. J. Wysoki, and M. N. Margolies. "Recurrent Somatic Mutations in Mouse Antibodies to P-Azophenylarsonate Increase Affinity for Hapten." *J. Immunol.* **142** (1989): 596–601.
88. Sherrington, D., and S. Kirkpatrick. "Solvable Model of a Spin Glass." *Phys. Rev. Lett.* **35** (1975): 1792–1796.
89. Sips, R. "On the Structure of a Catalyst Surface." *J. Chem. Phys.* **16** (1948): 490–495.
90. Smale, S. "On the Differential Equations of Species Competition." *J. Math. Biol.* **3** (1976): 5–7.
91. Sorkin, G. "Combinatorial Optimization, Simulated Annealing, and Fractals." IBM Research Report RC13674 (No. 61253) (1988).
92. Spiegelman, S. "An Approach to the Experimental Analysis of Precellular Evolution." *Qtr. Rev. Biophys.* **4** (1971): 36.
93. Stadler, P. F. "Correlation in Landscapes of Combinatorial Optimization Problems." *Europhys. Lett.* **20** (1992): 479-482.
94. Stadler, P. F. "Random Walks and Orthogonal Functions Associated with Highly Symmetric Graphs." *Disc. Math.* (1994): in press.
95. Stadler, P. F. "Linear Operators on Correlated Landscapes." *J. Physique I* 1994 (in press).
96. Stadler, P. F., and R. Happel. "Correlation Structure of the Landscape of the Graph-Bipartitioning-Problem." *J. Phys. A. Math. Gen.* **25** (1992): 3103–3110.
97. Stadler, P. F., and W. Schnabl. "The Landscape of the Travelling Salesman Problem." *Phys. Lett. A* **161** (1992): 337–344.
98. Stadler, P. F., and W. Gruner. "Anisotropy in Fitness Landscapes." *J. Theor. Biol.* **165** (1993): 373–388.

99. Stadler, P. F., and R. Happel. "Canonical Approximation of Landscapes." (1994): in preparation.
100. Stein, D. L., ed. *Spin Glasses and Biology.* Singapore: World Scientific, 1992.
101. Swetina, J., and P. Schuster. "Self-Replication with Errors. A Model for Polynucleotide Replication." *Biophys. Chem.* **16** (1982): 329–353.
102. Tacker, M., W. Fontana, P. F. Stadler, and P. Schuster. "Statistics of RNA Melting Kinetics." *Eur. Biophys. J.* (1994): in press.
103. Tai, K. "The Tree-to-Tree Correction Problem." *J. ACM* **26** (1979): 422–433.
104. Tarazona, P. "Error Thresholds for Molecular Quasispecies as Phase Transitions. From Simple Landscapes to Spin-Glass Models." *Phys. Rev. A* **45(15)** (1992): 6038–6050.
105. Waterman, M. S. "Sequence Alignments." In *Mathematical Methods for DNA Sequences*, edited by M. S. Waterman. Boca Raton, FL: CRC Press, 1989.
106. Weinberger, E. D. "Correlated and Uncorrelated Landscapes and How to Tell the Difference." *Biol. Cybern.* **63** (1990): 325–336.
107. Weinberger, E. D. "Local Properties of the N-K Model, a Tunably Rugged Landscape." *Phys. Rev. A* **44** (1991): 6399–6413.
108. Weinberger, E. D. "Fourier Series and Taylor Series on Landscapes." *Biol. Cybern.* **65** (1991): 321–330.
109. Weinberger, E. D., and P. F. Stadler. "Why Some Fitness Landscapes are Fractal." *J. Theor. Biol.* **163** (1993): 255–275.
110. Wiethege, W., and D. Sherrington. "Bipartioning of Random Graphs with Fixed Extensive Valence." *J. Phys. A. Math. Gen.* **20** (1987): L9–L11.
111. Wright, S. "The Roles of Mutation, Inbreeding, Crossbreeding, and Selection in Evolution." *Proc. 6th Intl. Congress of Genetics* (1932): 356–366.

Bartlett W. Mel
Department of Biomedical Engineering, University of Southern California MC-1451, Los Angeles, CA 90089; e-mail: mel@quake.usc.edu

Information Processing in Dendritic Trees

Reprinted with permission from *Neural Computation* **6** (1994): 1031–1085.

This review considers the input-output behavior of neurons with dendritic trees, with an emphasis on questions of information processing. The parts of this review are: (1) a brief history of ideas about dendritic trees, (2) a review of the complex electrophysiology of dendritic neurons, (3) an overview of conceptual tools used in dendritic modeling studies, including the cable equation and compartmental modeling techniques, and (4) a review of modeling studies that have addressed various issues relevant to dendritic information processing.

INTRODUCTION

Dendritic trees come in many shapes and sizes, and are among the most beautiful structures in nature (Figure 1). They account for more than 99% of the surface area of some neurons,[41] are studded with up to 200,000 synaptic inputs,[87] are the largest volumetric component of neural tissue,[154] and consume more than 60% of the brain's energy.[177] Most importantly, though, they are the computing workhorses of the brain.

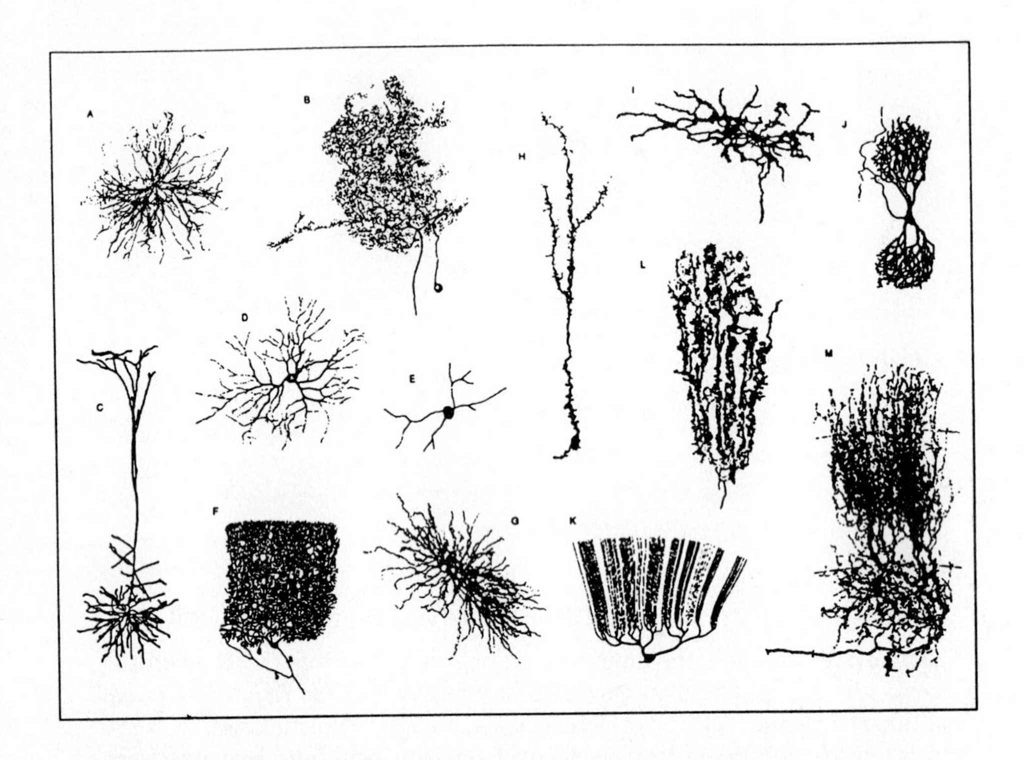

FIGURE 1 Dendrites and other neural trees structures, reproduced with permission. Lengths given are approximate and correspond to direction of maximal extent. (a) Alpha motorneuron in spinal cord of cat (2.6 mm); from Cullheim et al.[32] (b) Spiking interneuron in mesothoracic ganglion of locust (540 μm); courtesy G. Laurent. (c) Layer 5 neocortical pyramidal cell in rat (1030 μm); from Amitai et al.[6] (d) Retinal ganglion cell in postnatal cat (390 μm); from Maslim et al.[94] (e) Amacrine cell in retina of larval tiger salamander (160 μm); from Yang and Yazulla.[181] (f) Cerebellar Purkinje cell in human; from Ramón y Cajal,[133] v. 1, p. 61. (g) Relay neuron in rat ventrobasal thalamus (350 μm); from Harris.[50] (h) Granule cell from olfactory bulb of mouse (260 μm); from Greer.[47] (i) Spiny projection neuron in rat striatum (370 μm); from Penny et al.[113] (j) Nerve cell in the Nucleus of Burdach in human fetus; from Ramon y Cajal,[133] v. 2, p. 902. (k) Purkinje cell in mormyrid fish (420 μm); from Meek and Nieuwenhuys.[99] (l) Golgi epithelial (glial) cell in cerebellum of normal-reeler mutant mouse chimera (150 μm); from Terasshima et al.[165] (m) Axonal arborization of isthmotectal neurons in turtle (460 μm); from Sereno and Ulinski.[145]

This review concerns the question as to how information is processed in dendritic trees. In order to achieve focus, the scope of this review is limited in several ways. First, we bias our discussion toward studies that bear directly on questions of information processing. Less attention is devoted to work that is primarily biophysical in nature, such as studies aimed at matching dendritic neuron models to experimental data—for a thorough recent review see Rall et al.[130] Second, we focus on questions that pertain specifically to *dendritic* computation, i.e., for which the spatially extended structure of the dendritic tree is essential. Questions of relevance to general neuronal computation are not emphasized here, such as the nature of information coding in spike trains,[9,15,97,155,166] or the mechanisms underlying complex neuronal spiking behavior.[18,90,167,169,168,170] Third, we consider computation on fast time scales only. Thus, the central topic of this review concerns how dendritic trees transduce complex patterns of synaptic input into a stream of action potentials at the cell body, over periods of tens of milliseconds. Computations on longer time scales are not considered, such as those involving second-messenger systems,[174] ion channel migration or regulation,[11,80,121] ultrastructural changes,[46] or any consequence of gene-expression.[38] Fourth, the emphasis of this review is primarily on dendritic function in vertebrate neurons. This is a simple consequence of the fact that the bulk of dendritic modeling studies to date have dealt with vertebrate neurons. Finally, specific discussion of the computational significance of dendritic spines is left to a number of excellent reports and reviews available elsewhere.[69,70,75,107,115,132,144,151,182]

The remainder of this section is devoted to a few historical notes regarding the role of dendrites in neuronal function. In Section 2, experimental evidence is reviewed that proves dendrites to be physiologically highly complex objects. In Section 2 the main conceptual and computational tools that have been applied in the study of dendritic function are introduced, including the cable equation, compartmental modeling, and several useful rules of thumb regarding the flow of current in dendritic trees. Finally, modeling studies relating directly to dendritic information processing are reviewed in Section 4.

HISTORY OF IDEAS ABOUT DENDRITES

Historical notes that trace the progression of ideas about dendritic physiology and information processing are available elsewhere and should be consulted by the interested reader.[21,89,123,60,158,128,65,152,142] A few historically significant conceptual landmarks have been collected below.

Dendrites trees, like terrestrial trees, root systems, and vascular systems are a class of branched structures well suited for the penetration of a volume, such as for the extraction or delivery of nutrients. Before the advent of electrophysiolgy, it was reasonable to conclude, as was done by the great neuroanatomist Golgi,[44] that dendrites played a strictly nutritive role in neuronal function. The modern idea, that a dendritic tree exists to extract *information* from the volume in which it sits, has led some to propose that the need to increase neuronal surface area to make room for more synapses has been the main driving force in the evolution of the vertebrate brain.[123] Beyond simply increasing a cell's connectional "fan-in," though, it has also been frequently observed that spatially extended dendritic arborizations make possible the physical segregation of functionally distinct input pathways to a cell, a pattern seen in many areas of the vertebrate brain including the olfactory bulb, hippocampus, and cerebellum.[148]

What of the physiological functions of dendritic trees? The "neuron doctrine" of Cajal,[133] and in particular the "principle of dynamic polarizaton," entailed that signals flowed from the input surfaces of the cell (dendrites and soma) toward the axon, where a wave of nervous energy was then propagated to other neurons through synaptic contacts.[1] How the effects of distinct synaptic inputs were to be combined within the dendrites was at that time a matter beyond the resolution of either experiment or theory. When the all-or-none nature of nervous impulses was first discovered,[3,45] the idea that dendrites, too, might conduct non-decremental impulses prevailed for decades. In a dissenting opinion, Lorente de Nó[88] considered the idea of all-or-none impulse conduction in dendrites to be problematic, pointing out that dendritic integration would in this case degenerate to the propagation of any strong synaptic event directly to the cell's output; the possibility of combining the effects of widely separated synaptic inputs would thus be eliminated.[89] In the 30s and 40s, a "classical" picture emerged based in large part on the study of spinal motorneurons, which held that the exclusive spike trigger zone in a neuron was the axon initial segment, while the dendrites simply collected and summated synaptic inputs—in essence reducing the single neuron to a McCulloch-Pitts-type "morpho-less" unit.[142] In a refinement of this idea, it was commonly believed that inputs onto *distal* dendrites, which were thought to be both weak and slow, acted primarily to modulate the responsiveness of a cell to more powerful, more proximal soma-dendritic inputs.[4,16,28,36,48,122]

In the intervening years, a considerable body of data has accumulated showing that the dendrites of many types of neurons in the vertebrate CNS are replete with complex, interacting voltage-dependent membrane conductances, often capable of generating full-blown action potentials and other highly nonlinear behaviors. These data are reviewed in the next section.

ELECTROPHYSIOLOGY OF DENDRITES

As this review is centrally concerned with the question of dendritic information processing, especially the possibility that dendritic trees are intrinsically sophisticated information processing devices, the experimental studies cited in the following

[1]Cajal came to this conclusion based solely on careful inspection of tissue sections under the microscope.

sections have been chosen to emphasize interesting nonlinear voltage behavior in dendritic trees. For a review of experimental work relevant to the passive cable properties of dendritic trees, see Rall et al.[130]

MOTORNEURONS

The earliest experiments suggesting active behavior in dendrites were carried out in "chromatolyzed" motorneurons, i.e., spinal motorneurons whose axons to the peripheral musculature had been cut 2-3 weeks prior to recording.[37] These authors recorded small, spikelike, excitatory potentials several millivolts in height, which they termed *partial responses* (Figure 2). Partial responses were distinguished from conventional synaptic EPSP's in three ways: (1) they were brief and spikelike in character, (2) they occured with a greater range of latencies in response to synaptic stimulation, and (3) they were blocked in all-or-none fashion by hyperpolarizing current injection at the soma; EPSP's would normally *increase* in size under these conditions. A distal dendritic origin was inferred since the partial responses could

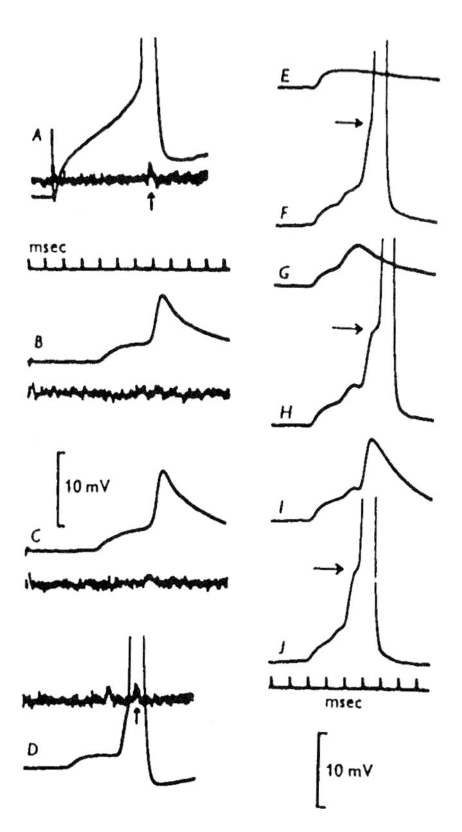

FIGURE 2 Partial responses in chromatolyzed motor neurons.[37] (a) Response to depolarizing current pulse that generated an action potential. (b)-(j). Responses to afferent synaptic volleys. Traces in (e)-(j) were selected to show the wide range of variability of partial responses that are superimposed on the EPSP shown in (e). Arrows indicate initiation of full spikes.

only be blocked with large hyperpolarizing currents at the cell body. Kuno and Llinas[76] provided further evidence for a distal dendritic origin of partial responses in chromatolyzed motorneurons by showing that they were more easily blocked by synaptic inhibition delivered to the distal dendrites than by hyperpolarizing currents at the soma. They emphasized that somatic action potentials could arise from partial responses of different sizes, suggesting multiple sites of spike origin in the dendritic tree. In spite of the profoundly abnormal anatomical and physiological condition of these axotomized cells (see Eccles et al.[37]), the idea that dendrites could contain discrete excitable "hot spots" gained first popularity in these early results. The methodological problems associated with the study of chromatolyzed motorneurons were later highlighted,[146] where it was suggested that dendritic spikes in these cells may result from a pathological redistribution of excitable Na^+ channels into the dendrites that might otherwise have been destined for the severed axon.

CEREBELLAR PURKINJE CELLS

Cerebellar purkinje cells have yielded one of the most clearcut examples of dendritic spiking among vertebrate neurons (Figure 3). In early work in cerebellar slices, Hild and Tasaki[52] recorded spikelike blips from an extracellular electrode pressed up against a dendritic branch of a Purkinje cell. Intradendritic recordings from Purkinje cells both *in vitro* and *in vivo* have since confirmed the existence of active dendritic spike generation in mammalian,[43,85,138,162] avian,[82] reptilian,[83,84] and amphibian[58] cerebella. The probability that dendritic spikes in Purkinje cells were due to Ca^{++} and not Na^+ currents was first demonstrated by Llinas and Hess.[82] Unlike fast somatic sodium spikes that peak in 1-2 ms, dendritic calcium spikes were slow rising, typically reaching a peak of 30-60 mV at 5-10 ms, and could be evoked by a 10 mV depolarizing intradendritic current injection (Llinas and Nicholson[83]; Llinas and Sugimori[85]; Figure 3). A subsequent detailed study of the electrophysiological properties of dendritic membrane in these cells also revealed voltage-dependent calcium "plateau potentials" of between 10 and 30 mV, i.e., non-inactivating excitatory potentials that could outlast a small depolarizing current stimulus by hundreds of milliseconds.[85] In these experiments, calcium spiking and plateau potentials were ubiquitous in Purkinje cell dendrites; more recent optical recording experiments have established that calcium influx occurs across essentially the entire dendritic tree.[138,162,164] In contrast, Na^+-mediated spiking and plateau channels were found to be localized near the soma.[58,85]

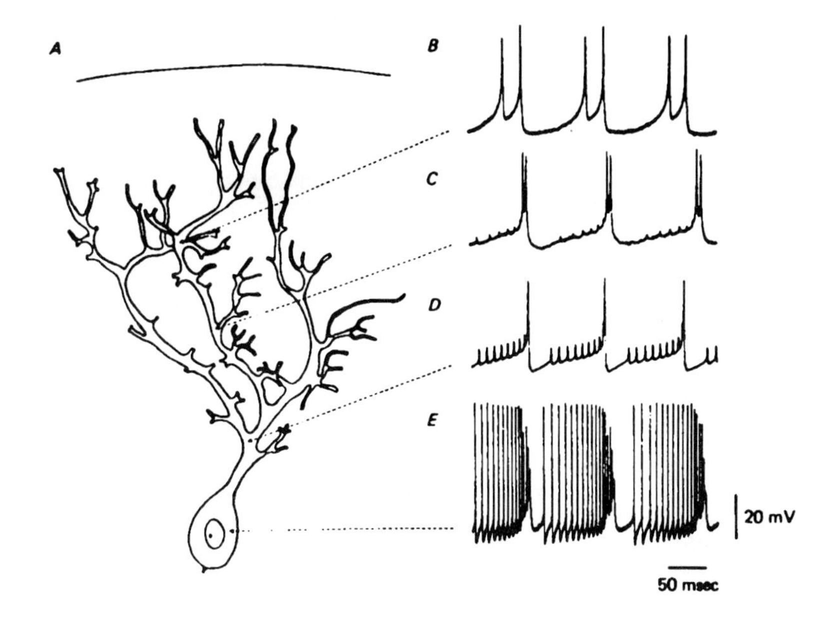

FIGURE 3 Composite picture showing the relationship between somatic and dendritic action potentials following DC current injection through the recording electrode. A clear shift in amplitude of the somatic spike against the dendritic Ca^{++}-dependent potentials is seen when comparing the more superficial recording in B with the somatic recording in (e). Note that at increasing distances from the soma the fast spikes are reduced in amplitude and are barely noticeable in the more peripheral recordings. Reprinted with permission from Llinas and Sugimori.[85]

HIPPOCAMPAL PYRAMIDAL CELLS

Spencer and Kandel[159] first observed "fast prepotentials" (FPPs) in hippocampal pyramidal cells, citing similarities to the partial responses described by Eccles et al.[37] Somatic or axonal origins of the spikes were again ruled out since they were not blocked by hyperpolarizing current injection. Following Eccles[36] the authors conjectured that the spikelike prepotentials may have originated at the main bifurcation of the apical dendritic tree, where a patch of excitable membrane could act as a booster to otherwise ineffective distal synaptic inputs (Figure 4). Subsequent studies using a variety of techniques have provided extensive further evidence for dendritic spiking mechanisms in hippocampal pyramidal cells.[7,12,51,95,120,124,141,140,171,178,179] In the first direct intradendritic recording in CA1 and CA3 hippocampal cells *in vitro*, Wong et al.[178] showed that bursts consisting first of fast Na^{+}-dependent spikes followed by slow (presumably Ca^{++}-dependent) spikes, could occur either spontaneously or in response to depolarizing current pulses. A subsequent study on CA1 dendrites that had been surgically isolated from their cell bodies[12] showed similar mixtures of fast Na^+ and slow Ca^{++} spikes (as well as spikes of intermediate duration) in response to intradendritic current injections (Figure 5(a)). The depedence of fast and slow dendritic spikes on sodium and calcium channels, respectively, was demonstrated in CA1 pyramidal cells by Poolos and Kocsis,[120] as illustrated in Figure 5(b). The occurence of multiple points of inflection, and mixtures of spikes of differing amplitudes in these studies was interpreted as evidence of multiple dendritic sites of spike generation; other workers have reached similar conclusions.[64,95,171,179] The relative independence of dendritic and somatic action potentials was explicitly demonstrated by impaling the same neuron with two electrodes,[179] as illustrated in Figure 5(c). Finally, in a recent optical recording study, Jaffe et al.[61] reported that when potassium currents were blocked with TEA,

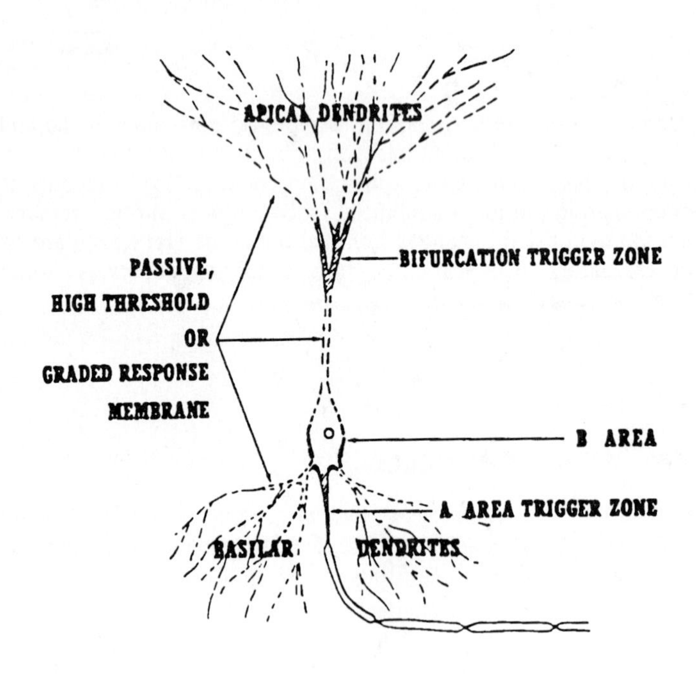

FIGURE 4 The idea that fast prepotentials recorded at the cell body arise from a patch of excitable membrane at the main apical bifurcation of a hippocampal pyramidal cell was illustrated by Spencer and Kandel[159]; reprinted with permission.

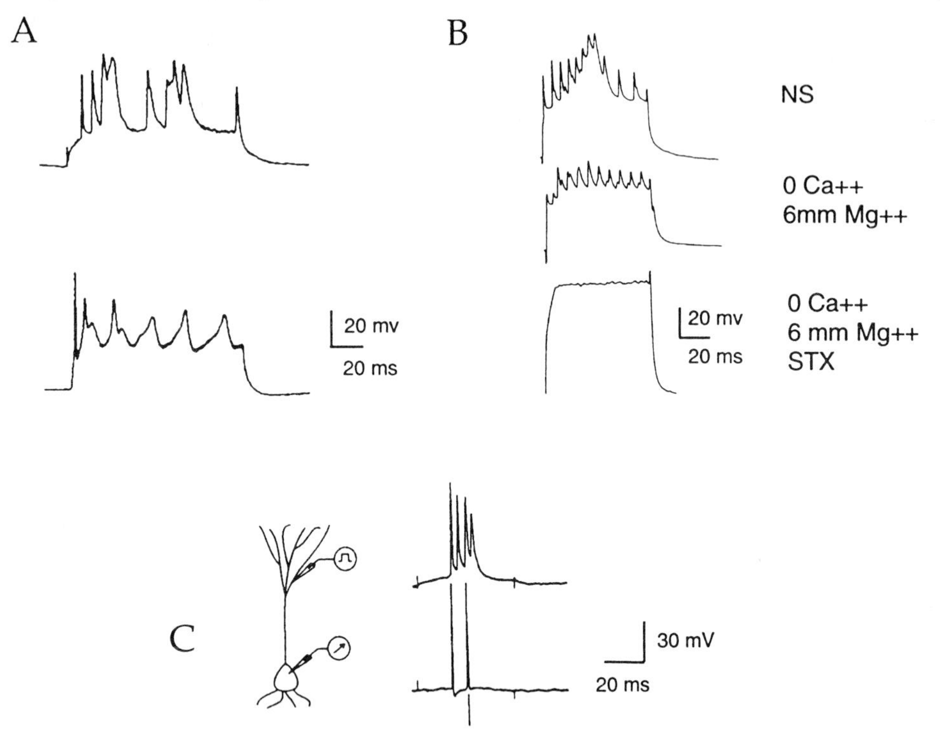

FIGURE 5 (a) Recordings from intact (upper trace) and surgically isolated (lower trace) dendrites of CA1 pyramidal cells reveal complex spiking traces in response to current injections[12]; recordings were made between 100 and 350μm from the cell body (reprinted with permission). (b) Intradendritic recordings at more than 200 μm from the cell body in CA1 pyramidal cells in response to 0.9 nA current injections are shown for the same cell under three conditions: normal saline (upper), 0-calcium (middle), and STX (lower). Note disappearance of underlying slow spike when calcium was removed from the bath, and disappearance of fast spikes in the presence of a sodium channel blocker (reprinted with permission from Poolos and Kocsis[120]). (c) Simultanous recording in apical dendrites and soma in CA1 pyramidal cell in response to direct dendritic depolarization. Somatic response (lower trace) to dendritic burst (like that in upper trace) reveals independence of dendritic and somatic spiking activity. Reprinted with permission from Wang and Stewart.[179]

voltage-dependent calcium entry could be recorded across the entire apical and basilar dendritic tree of a CA1 cell, whereas rapid sodium entry indicative of active penetration of sodium spikes was present only in proximal dendrites up to a distance of 200 μm from the cell body.

In addition to dendritic spikes, subthreshold voltage-dependent conductances have been observed in intact and isolated CA1 dendrites that give rise to larger responses to depolarizing (excitatory) than hyperpolarizing (inhibitory) current pulses.[12] This anomalous rectification was considered to be due either to voltage-dependent Ca^{++} channels or to background activation of NMDA channels, whose negative slope conductance near resting potential[96] can result in an apparent increase in input resistance with depolarization.[163] In hippocampal dentate granule cells depolarized to −50 mV, Keller et al.[66] have shown that when synaptic inhibition has been blocked, the NMDA (vs. non-NMDA) component of the EPSP accounts for half the peak current and three quarters of the total injected synaptic charge. This experiment argues that the voltage-dependence of the NMDA channels is likely to be an important nonlinearity influencing subthreshold synaptic integration (see also Salt[139]). An NMDA-dependent potentiation in the dendritic spiking response in CA1 cells following a tetanus has also been shown,[120] revealing an intricate interplay between at least two kinds of excitatory, voltage-dependent nonlinearites in these cells. Interestingly, neither fast nor slow spikes were generated in the dendrites of dentate granule cells in response to large depolarizing current pulses,[12] highlighting the fact that significant differences are found in the distribution of membrane nonlinearities in different types of neurons within the same brain area, even among closely related neuron types.

NEOCORTEX

Unlike their hippocampal and cerebellar counterparts, the dendritic trees of neocortical pyramidal cells are not confined to specific neural laminae that are essentially free of cell bodies (such as *stratum radiatum* in hippocampus or *stratum moleculare* in cerebellum). It is therefore technically difficult to systematically record from neocortical dendrites in order to study their electrophysiological properties directly. Early evidence for active dendritic responses in neocortical pyramidal cells was mostly indirect.[8,28,31,125,157,158] Deschenes[34] found fast prepotentials in 40% of the fast-conducting pyramidal tract neurons of the primary motor cortex in anaesthetized cats, of a form very similar to those previously described in the hippocampus[159] and elsewhere (see Spencer[158]). However, the possibility that the FPPs were electrotonically attenuated spikes from other neurons transmitted through gap junctions was not entirely ruled out.[34] In cultured neonatal pyramidal cells from rat sensorimotor cortex, Huguenard et al.[59] demonstrated the existence of voltage-dependent Na^+ channels on the proximal apical dendrites up to at least 80 um from the cell body, using cell-attached patch and whole cell recordings. Evidence for distal dendritic calcium spikes has also been provided in mature neocortical neurons *in vitro*.[160] In these experiments, depolarizing clamp voltages at the cell body were frequently seen to initiate large uncontrolled Ca^{++} spike currents, suggesting that the Ca^{++} spikes were occuring in an electrotonically remote dendritic location. Also in somatic recordings, Reuveni et al.[136] blocked Na^+ spikes using TTX and K^+ currents using TEA to reveal prolonged Ca^{++} spike-plateaus.

Since repolarization of the calcium spike was often seen to occur in several discrete steps (Figure 6(a)), and based on the results of modeling studies, the calcium spikes were concluded to originate from several discrete dendritic Ca^{++} hot spots separated by passive membrane. In the most direct evidence for dendritic excitability in these cells, several labs have in the past three years recorded action potentials in intradendritic recording from layer 5 pyramidal cells,[6,117] in some cases including complex superpositions of spikes of varying widths and amplitudes in response to constant current injections (Pockberger et al.[117]; Amitai et al.,[6] see Figure 6(b)). In another vein, Cauller and Connors[27] have shown *in vitro* that stimulation of layer 1 afferents alone is sufficient to drive strong responses at the cell bodies of layer 5 pyramidal cells. In an attempt to understand this result using compartmental modeling techniques, they subsequently demonstrated that in a cell

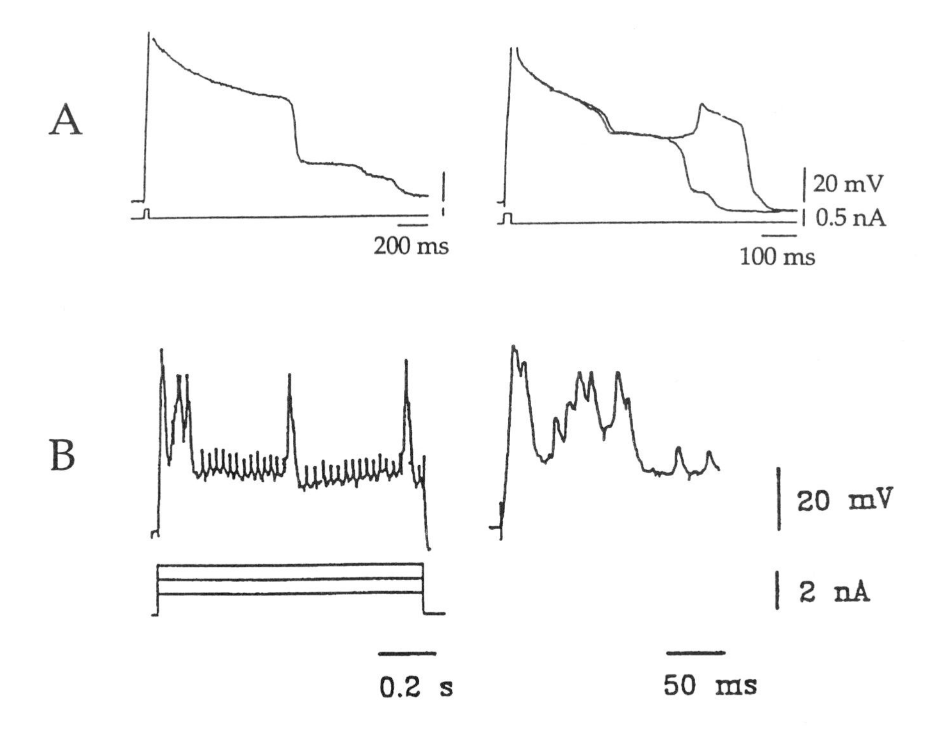

FIGURE 6 (a) Stepwise repolarization in neocortical pyramidal cells of calcium plateaus induced by small intracellular current pulses in TTX-TEA medium. Duration of plateaus was varible from trial to trial, but breakpoint voltages remained relatively constant. Two plots are from two different neurons. Reprinted with permission from Figure 1CD.[136] (b) Intradendritic recording near main apical branchpoint (200–300 μm from the cell body) also in a neocortical pyramidal cell (reprinted with permission from Amitai et al.[6]). Complex superposition of fast and slow spikes were seen in response to a 2 nA current injection. Two plots are same trace at different time scales.

with a purely passive dendritic tree, stimulation of distal apical synapses was never sufficient to generate cell responses of a magnitude observed in their experiments. Their conclusion: active dendritic conductances are involved in the amplification of distal dendritic input in these neurons.

As in hippocampus, another type of voltage-dependent nonlinear membrane mechanism known to play a important role in dendritic integration in neocortex is the NMDA channel.[96] In pharmacological blocking studies, NMDA channels have been shown to account for a major proportion of the excitatory synaptic drive onto neocortical pyramids,[42,108] consistent with histochemical labelling studies that reveal relatively high concentrations of NMDA receptor binding sites in the superficial, synapse-rich layers of cerebral cortex where the dendrites of these cells receive much of their synaptic input.[30] The voltage-dependence of the NMDA channel[96] has in modeling studies proven to be capable of significantly influencing the integrative behavior of a pyramidal cell's dendritic tree.[102,105]

OTHER CELLS

Evidence for voltage-dependent dendritic membrane mechanisms has been acquired in other types of vertebrate neurons, including cells in the thalamus,[62,91] inferior olive,[86] and substantia nigra.[81] Maekawa and Purpura[91] reported fast depolarizing potentials in response to synaptic stimulation, distinct from slow EPSP's, which they interpreted as possible partial spikes of dendritic origin that could account for the "extraordinary responsiveness" of these cells to synaptic input. In another study of thalamic neurons, Jahnsen and Llinas[62] reported high-threshold Ca^{++} spikes lasting 18-22ms in presumed intradendritic recordings that were similar to those observed in Purkinje cell dendrites. In the inferior olive, Llinas and Yarom[86] have described a number of voltage-dependent conductances, including multiple all-or-none high-threshold Ca^{++} spikes of presumed dendritic origin. Finally, in the substantia nigra, Llinas et al.[81] provided evidence for two types of dendritic Ca^{++} spikes, one low the other high-threshold with respect to somatic current injection. In these cells, the Ca^{++} spikes are thought to be involved in the dendritic release of dopamine.

SUMMARY

While it is difficult to achieve a meaningful summary of the experimental work discussed above, certain general tendencies in the experimental data may be identified. For several important classes of vertebrate neurons, the conception of the dendritic tree as an essentially passive collector of synaptic inputs has not been borne out in the results of 30 years of electrophysiological work. First, in several major types of output neurons in the cerebellum and mammalian forebrain, mechanisms capable of generating dendritic spikes of one or more ionic varieties have been either positively demonstrated to exist through intradendritic recordings, or are strongly suggested in intrasomatic recordings or through a variety of other techniques. Dendritic calcium spikes are a particularly widespread phenomenon, having been observed in cerebellar Purkinje cells, hippocampal and neocortical pyramidal cells, and cells in the thalamus and inferior olive.

Second, in many of these same neuron types, especially hippocampal and neocortical pyramidal cells, but also neurons in the pyriform cortex, thalamus, basal ganglia, midbrain, spinal cord and other areas[30,96,109]), a component of the excitatory synaptic input to the dendrites is carried by voltage-dependent NMDA-type channels; in some cases a the NMDA component may predominate, such as in response to high-frequency stimulation and/or "natural" sensory input.[42,63,66,108,139]

In the next section, we review the mathematical and conceptual tools that have been used to model the electrical behavior of dendritic trees.

CONCEPTUAL AND COMPUTATIONAL TOOLS

In order to understand computation in dendritic trees, it is necessary to first understand the principles that govern the flow of electric current in dendrites. For example, when an excitatory synapse is activated on a dendritic spine, where does the injected current flow, how long does it take, and what is its effect on the membrane potential both locally and elsewhere in the dendritic tree? In this section we review conceptual and computational tools and basic results relevant to current flow in dendrites. For excellent introductory reviews of the biophysical mechanisms underlying membrane resistance and capacitance, membrane potential, time constants, and basic steady state and transient responses of passive and active membranes.[53,60,65,98,152,128] A working knowledge of these concepts is assumed in the following.

One of the most important early developements in the study of neural information processing was the application of the one-dimensional "cable" equation to problems of current flow in branched, 'passive neuronal structures.[126,127] To begin, we present the cable equation, consider the physical interpretation of its terms, and discuss an important analytical solution. We then enumerate several useful rules of thumb regarding the electrical behavior of passive dendritic trees. Finally, we introduce the enterprise of compartmental modeling, a discretized numerical approach to the solution of the cable equation that allows treatment of arbitrary neuronal geometries and voltage-dependent membrane mechanisms.

THE CABLE EQUATION

Due to their physical construction, neurites (dendrites and axons) have been likened to electrical cables. A cable consists of a long, thin, electrically conducting core surrounded by a thin membrane whose resistance to transmembrane current flow is much greater than that of either the internal core or the surrounding medium. In the case of a dendrite, both the internal cytoplasm and the extracellular space are thought to conduct nearly as well as seawater (see Jack et al.,[60] ch. 1). Because the resistance to current flow through the cytoplasm is relatively low, injected current can travel long distances down the dendritic core before a significant fraction leaks out across the highly resistive membrane. The fundamental equation used to describe current flow in passive dendrites is the "cable equation":

$$\frac{1}{r_i}\frac{\partial^2 V}{\partial x^2} = c_m \frac{\partial V}{\partial t} + \frac{V}{r_m} \tag{1}$$

where $V = V(x, t)$ is the transmembrane voltage at location x and time t, and for a unit length of cable, r_i (in kΩ/cm) is the resistance to internal current flow along the core, r_m (in kΩ cm) is the transmembrane resistance, and c_m (in μF/cm) is the membrane capacitance (see Jack et al.,[60] Shepherd and Koch,[152] Rall et al.[130] for discussion of units). The cable equation was developed in 1855 as a part of a mathematical theory with practical applications for transatlantic telegraph lines (Kelvin[67]; see Rall[128] for historical perspective). Derivations and assumptions of the cable equation can be found in.[60,128,129,152]

The basic cable equation may be easily interpreted as a balance of three kinds of electrical current. Consider a short length of a dendritic branch labelled x (Figure 7). A fundamental law of conservation of electrical current (Kirchhoff's current law) tells us that the net accumulation of "axial" current at x (i.e., *axial current entering—axial current leaving*) must be equalled by the net current flowing out of the cell across the membrane at location x (i.e., *ionic membrane current* + *capacitive membrane current*). This equality is represented directly in Eq. (1). The single term on the left represents the net axial current into node x through the cytoplasm from neighboring "compartments" to the left and right of x. The second-derivative form of the term indicates that the axial current flowing into x is, roughly speaking, related to the "curvature" of the voltage profile along the dendritic branch centered at x. For example, when the voltage at x is less than the average of its neighbors (positive voltage curvature), then there is a net accumulation of axial current at x. The two terms on the right represent the capacitive and ionic currents, respectively, that must flow across the membrane at x in order to balance the net axial current influx. These terms simply state that (i) the capacitive current at x is proportional to the rate of change of the local transmembrane voltage—rapid changes in voltage are associated with large capacitive currents, and (ii) the resistive, or ionic, current at x is, according to Ohm's law, directly proportional to the membrane voltage at x.

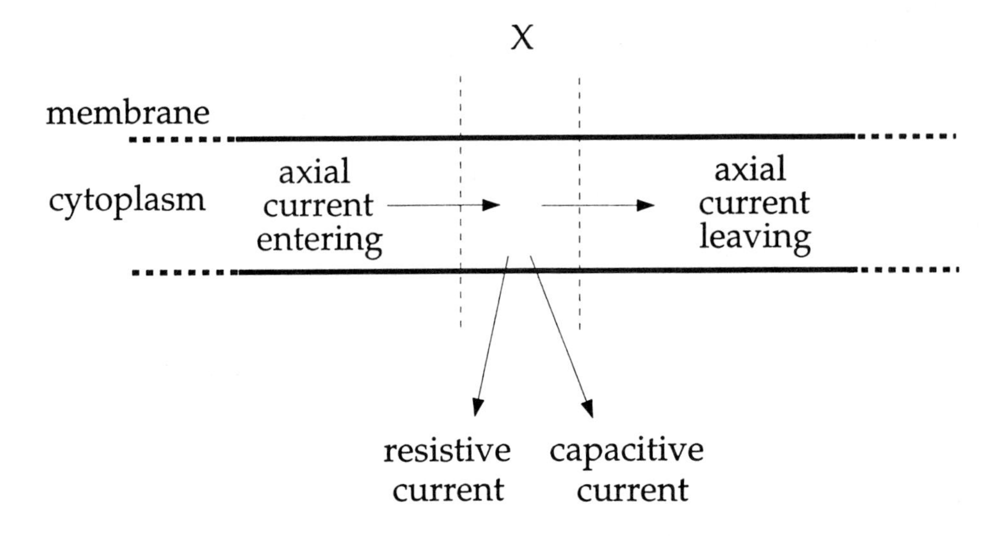

FIGURE 7 Section x of a one-dimensional cable. Arrows show axial current flowing into and out of section x, and the ionic (resistive) and capacitive transmembrane currents. The cable equation (Eq. (1)) specifies that, for all x, these currents must always be in balance.

To summarize, there are three types of current flowing in and around point x in a dendritic branch: (1) net axial current into x through the cytoplasm, related to the local membrane voltage "curvature," (2) capacitive current, proportional to the rate of change of membrane voltage, and (3) ionic current, proportional to the membrane voltage itself in the case of a passive dendrite. The cable equation simply says that these three quantities must always be in balance.

A CLASSICAL SOLUTION TO THE CABLE EQUATION

One historically important solution gives the voltage of a uniform infinite cable in response to a current step I_0 injected at the origin $X = 0$:

$$V = \frac{r_i I_0 \lambda}{4}\{\exp(-X)\mathrm{erfc}(\frac{X}{2\sqrt{T}} - \sqrt{T}) - \exp(X)\mathrm{erfc}(\frac{X}{2\sqrt{T}} + \sqrt{T})\} \qquad (2)$$

where $X = x/\lambda$ measures distance from the origin in space constants in a cylinder with infinite extension, and $T = t/\tau$ measures time in time constants. The space constant $\lambda = (r_m/r_i)^{1/2}$ is the distance at which the voltage has fallen off to V_0/e, i.e., to about one third its value at the site of stimulation; the membrane time constant $\tau = r_m c_m$ is the time required for an isopotential patch of membrane to charge to within $1/e$ of its steady state value. The charging of the inifinite cable about the origin during one time constant is plotted in Figure 8. If attention is restricted to the transient voltage change at $X = 0$,

$$V_0(T) = \frac{r_i I_0 \lambda}{2} \operatorname{erf}(\sqrt{T}), \tag{3}$$

which represents growth that is significantly faster than a single exponential (bold contour at $X = 0$). If attention is restricted to the steady state voltage along the entire cable at long times, we see that the voltage profile at long times is given by

$$V(X) = \frac{r_i I_0 \lambda}{2} \exp(-X), \tag{4}$$

a simple exponential decay with distance (dashed contour at $\mathrm{T} \gg \tau$).

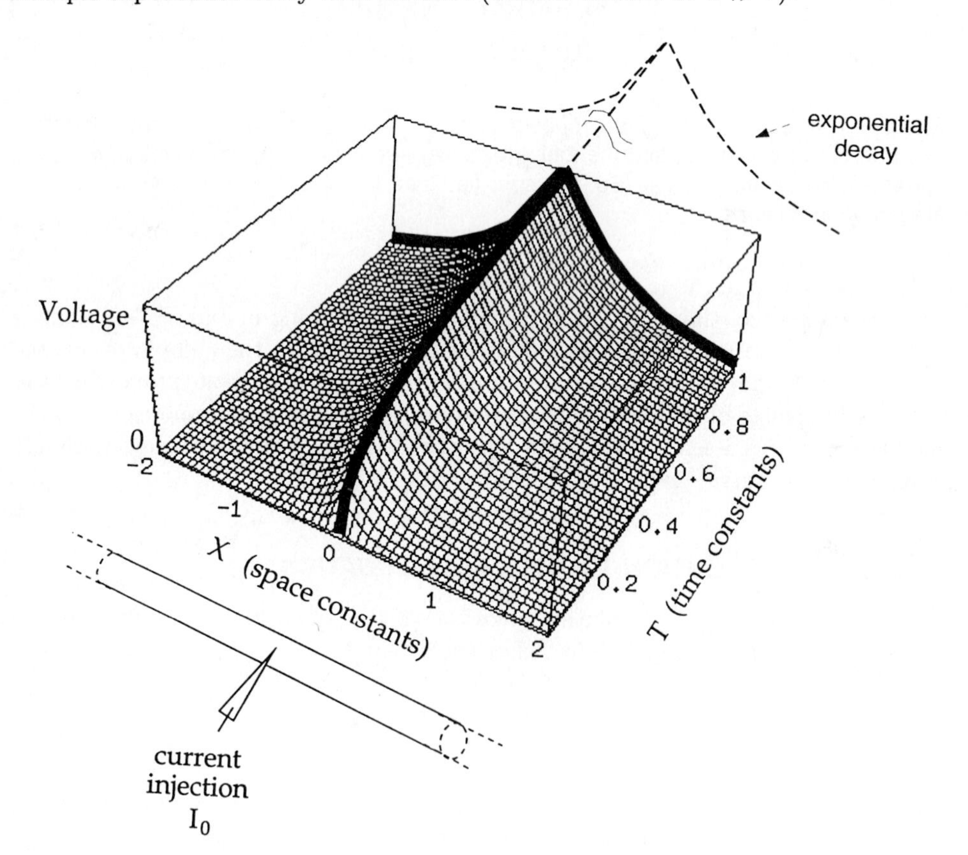

FIGURE 8 Plot of infinite cable charging in response to current step I_0 at $X = 0$. Transient voltage change at $X = 0$ is faster than exponential, given by Eq. (3). At long times, voltage decays symmetrically and exponentially about $X = 0$, as given by Eq. (4).

For derivations and solutions of the cable equation under these and a number of other boundary conditions see (Jack et al.[60]; Rall[128]), such as the response to step currents for finite or semi-infinite cables, with either sealed or open ends, response to voltage steps for finite or infinite cables, response to non-instantaneous voltage changes, and step changes in membrane conductance; a variety of other quantities, such as input resistance, electrotonic length, and velocity of the propagating decremental voltage wave in response to a stimulus, have also been derived explicitly.

For branched passive cables, an early and historically important contribution was the idea of an "equivalent cylinder" representation of a dendritic tree. Rall[126] showed that when (i) all branch points in a dendritic tree obey the $d^{3/2}$ law (see item 4 below), (ii) all terminal tips have identicaly boundary conditions, and (iii) all terminal tips lie at same electrotonic distance from the soma, then the entire dendritic tree may be replaced by a single "equivalent" cylinder for certain restricted input conditions, greatly simplifying the calculation of dendritic responses—see Rall[128] for details.

An algorithm for solving the case of arbitrary passive dendritic trees was also first provided by Rall.[126] Butz and Cowan[26] later developed a graphical method to compute the Laplace transform for arbitrary passive trees,[26] and Horwitz[56,57] extended and actually applied this method for arbitrary trees. Abbott et al. showed how the path integral can be computed for arbitrary dendritic trees, using a method borrowed from statistical physics that is both computationally efficient, and allows the assumtion of time and/or spatially varying membrane resistivities. Most recently, Evans et al.[39] and Major et al.[92] have treated arbitrary passive dendritic trees in the most complete way. Another recent paper describes a method for computing explicit signal delays between arbitrary pairs of stimulus and recording sites,[5] and examine the consequences for a cell's sensitivity to synchronous synaptic inputs. Some interesting analytical extensions of cable theory have also been developed for dendritic trees containing active membrane, including a linearized analysis of active neuronal cables valid for voltage perturbations of a few millivolts,[68] and a continuum-limit analysis of the propagation of signals in dendrites with active spine heads.[10]

One of the most important applications of the cable equation and its extensions to branched structures has been to the estimation of biophysical parameters of real neurons based on experimental data, especially membrane and cytoplasmic resistivity (R_m and R_i), and electrotonic length L. An excellent recent review of this large body of work is available.[130]

DENDRITIC ELECTROTONUS: RULES OF THUMB

One of the most significant legacies of cable theory lies in the rules of thumb it has provided relating to the electrical behavior of passive dendritic trees. A paradigmatic assumption common to many electrophysiologists and modelers of single-neuron function has been that a thorough understanding of the electrical behavior of passive neural processes is desirable, even in cases when active voltage-dependent nonlinearities are known to be present *en force.* We thus close our overview of cable theory by consolidating several rules of thumb regarding the flow of current and distribution of voltage in passive dendrites and dendritic trees.

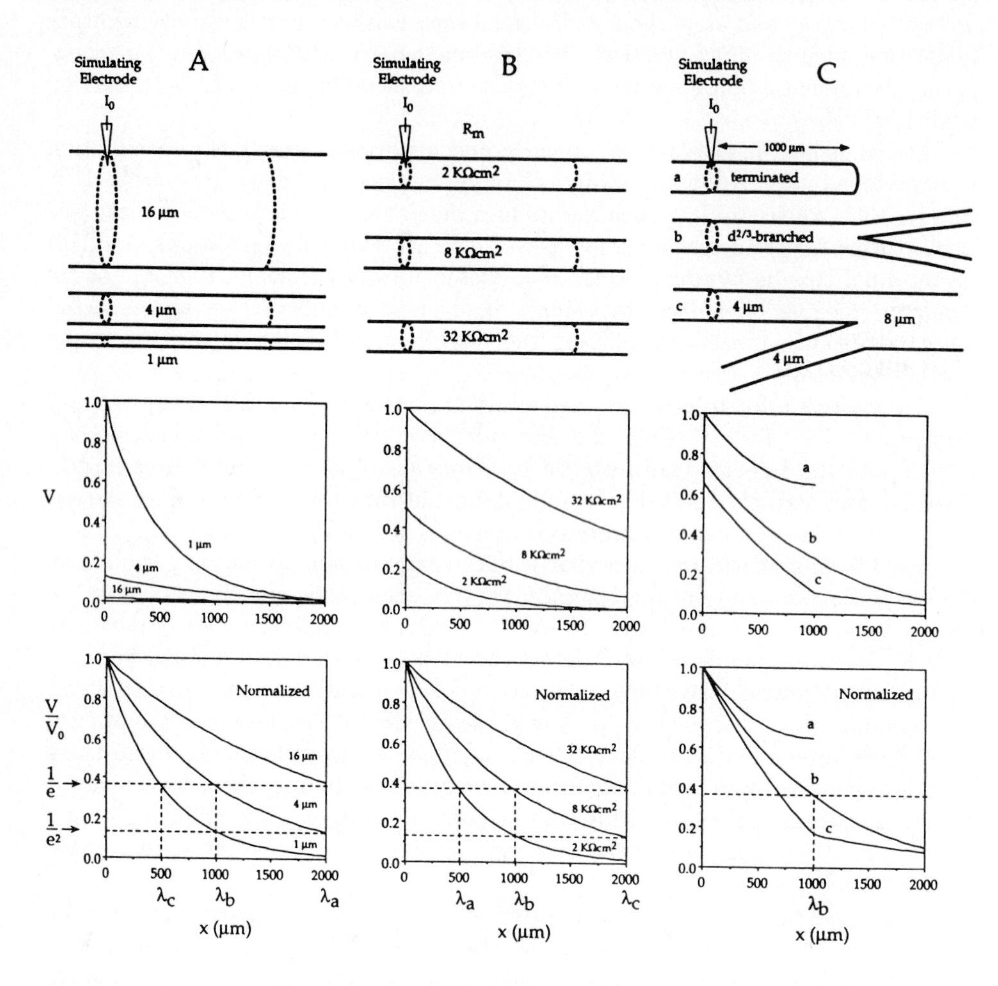

FIGURE 9 Steady state spread of voltage from a stimulus origin (e.g., a voltage clamp at $X = 0$) in an infinite cable, adapted from [152]. Default cable parameters are $R_m = 8000\Omega\text{cm}^2, R_i = 80\Omega\text{cm}$. Larger diameters (a) and higher specific membrane resistance (b) give longer space constants. C. Steady state spread of voltage under 3 branching conditions: (a) terminated, (b) connection to 2 or more daughter branches that satisfy the $d^{3/2}$ law, and (c) connection at a branch point to one equivalent and one thicker branch. Adapted and extended from material in Shephert and Koch.[152]

1. Voltage signals attenuate with distance. In the reference case of a cylinder with infinite extent, a steady state input decays exponentially with distance, with $\lambda = (r_m/r_i)^{1/2} = (d/2)(R_m/R_i)^{1/2}$, where d is the process diameter. Hence, voltage attenuation is more severe for thinner dendrites (Figure 9(a)) and/or leakier membrane (Figure 9(b)). The upper graphs in both A and B show the voltage at $T \gg \tau$ along the length of the cable in response to a current step I_0 at the origin; in the lower graphs the three curves are normalized to allow comparison of electrotonic lengths.
2. Voltage attenuation is more severe for high frequency components of an input waveform. For $\omega \gg 1/\tau$, $\lambda(\omega) \propto 1/\sqrt{\omega}$. Thus, the peaks of signals that are fast relative to τ, such as spikes or fast synaptic inputs, are much more strongly attenuated with distance than are steady state inputs.[137,183]
3. The input resistance R_{in} of a neurite is the magnitude of the voltage response at $T = \infty$ to a unit DC current step. For an infinitely long uniform neural process, $R_{in} = \sqrt{r_m r_i}/2$ grows as $1/d^{3/2}$. As a consequence, inputs to distal dendritic branches, which are often of very small diameter, can result in large local depolarizations in comparison to identical inputs delivered to large branches, which are often found closer to the cell body. The upper graph in Figure 9(a) illustrates the variation in R_{in} that results from changes in branch diameter: the relatively large voltage deflection at $X = 0$ for the 1μm diameter case is indicative of its relatively high input resistances; the same is true for the high R_m case in B.
4. Voltage attenuation depends on boundary conditions, that is, what a branch is connected to (Figure 9(c)). For example, when a stimulus is applied near to a branch end (a), the voltage attenuation is significantly reduced in the direction of the closed end, since the charge that would have flowed past the closed end "piles up" locally and causes a relative increase in membrane potential (see Jack et al.[60]; Shepherd and Koch[152]). When a dendritic "parent" branch connects to a set of k smaller daughter branches (b), where $d_1^{3/2} + \ldots d_k^{3/2} = d_{parent}^{3/2}$, then the voltage attenuation is uninterrupted through the branch point, as if the daughter tree were a simple a continuation of the parent branch (see Rall[128]). When a stimulus is applied to a thin branch that connects to a thick branch (c), then the voltage attenuation is exaggerated in the direction of the thick branch, since some of the charge that would have depolarized the thinner branch near the branch point is "sucked" into the relatively low resistance pathway offered by the thicker branch.
5. Voltage attenuation is directionally asymmetric in a dendritic tree, as illustrated for an idealized neuron in Figure 10(a). If a constant stimulus is applied at distal tip I, then the steady state voltage response is strongly attenuated in the direction of the cell body (upper solid curve), whereas if the same stimulus is applied at the cell body, the voltage attenuation from the cell body to the distal tip is modest (lower dashed curve; figure from Rall and Rinzel[131]). This asymmetry is due to the difference in cable boundary conditions looking toward or away from the cell body, as discussed in item 4, and has been frequently

treated in the literature.[22,72,131] An excellent graphical representation of this asymmetry is provided by the morpho-eletrotonic transform (MET) introduced by Zador.[183] As shown in Figure 10(b) for a hippocampal CA1 pyramidal cell, the length of each section of dendrite is scaled by the log steady state voltage attenuation *from* the soma *to* that section. (The *log-attenutation* is defined as $L_{ij} = log|A_{ij}|$, where $A_{ij} = V_i/V_j$ is the voltage attenuation from i to j for a stimulus at i; see Zador[183]). The entire tree is highly reduced, particularly the basal dendrites and small apical side branches due to their closed-end boundary conditions. By contrast, in Figure 10(c) the distance from every point to the soma is made proportional to the log-attenuation from that point *to* the soma. In this case, small side branches and thin basal dendrites are exaggerated in length, reflecting the strong attenuation of voltage signals in the direction of the cell body.

6. Voltage and current attenuation are reciprocal: voltage attenuation from i to j is exactly equal to *current* attentuation from j to i in a passive dendritic tree (i.e. $A^V_{ij} = A^I_{ji}$), for any locations i and j.[72] Thus, current attenuation from a distal site to the cell body can be modest, implying that a large fraction of the charge injected at a distal dendritic site flows to the cell body—compare responses at cell body due to somatic vs. distal stimulus in Figure 10(a).
7. Speed, delay, and input synchronization. A transient input to a dendritic branch, such as a synaptic current, is reduced in size and smoothed out in time as it propagates away from the site of stimulation. In the case of an infinitely long unbranched passive dendrite, the centroid of the wave propagates at a speed of $2\lambda/\tau$, i.e., two space contstants per time constant. More generally, the total signal delay is symmetric between any two points in a passive dendritic tree, and is independent of the shape of the input signal. Delays from dendrites to soma are on the order of one membrane time constant in morphologically realistic dendritic trees.[5] Importantly, local charging times on thin dendritic branches may be an order of magnitude faster than the membrane time constant τ. Consequently, distal dendritic arbors may function more as coincidence detectors for local synaptic inputs whereas the soma functions more as an integrator.

COMPARTMENTAL MODELING

For all of their considerable conceptual appeal, analytic solutions to the cable equation become increasingly cumbersome to the extent that the case under study diverges from a passive unbranched uniform cable stimulated with a constant current or voltage source. When a cell has a complex irregular branching structure, nonuniform passive membrane properties, contains voltage- or concentration-dependent membrane channel conductances, or is driven by synaptic conductance changes in lieu of current inputs, then "compartmental" modeling is the technique

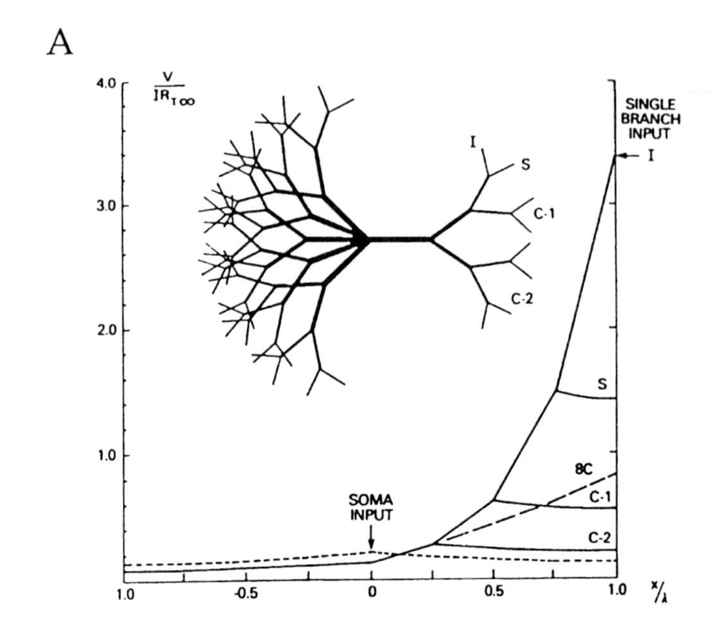

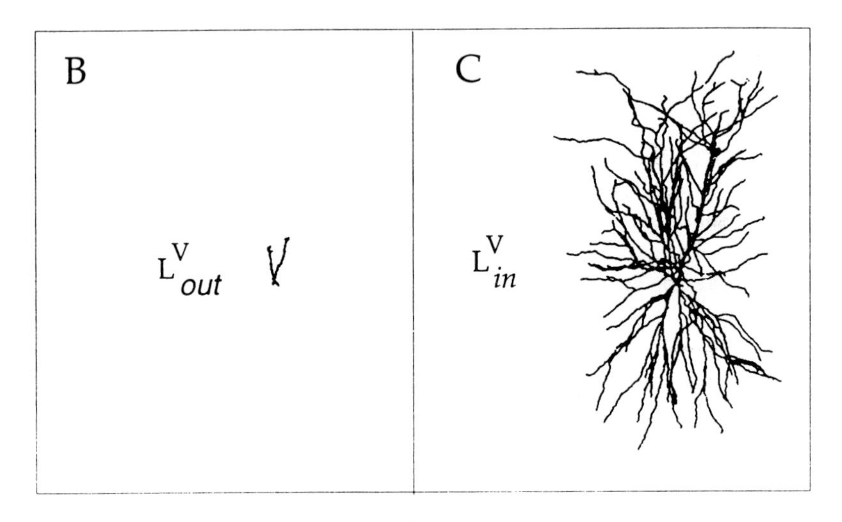

FIGURE 10 (a) Diagram of idealized neuron and plot of steady-state voltage for different stimulus conditions. Solid curve shows voltage profile in dendritic tree for constant current stimulus I at a single distal tip. Note steep voltage attenuation along trajectory to soma vs. gradual attenuation outbound along sister (S) and cousin (C-1,2) branches. Curve with short dashes shows voltage profile when same input I is delivered to soma. Reprinted with permission from Rall and Rinzel.[131] (b) Morphoelectrotonic transform (METs) of hippocampal CA1 pyramidal cell. Distance from cell body (cont'd)

FIGURE 10 (continued) to every dendritic section is proportional to the log DC voltage attenuation L^V_{out} from the soma to that dendritic section. Reprinted with permission from Zador.[183] (c) Different MET of same cell; now distance to cell body from each dendritic section is proportional to L^V_{in}, i.e., the log-attenuation from the dendritic section inward to the soma.

of choice. Originally introduced by Rall,[127] compartmental modeling represents a finite-difference approximation of the linear cable equation, or its nonlinear extensions. Compartmental modeling entails that the dendritic tree, axonal tree, or other cable-based stucture be broken into a branched network of discrete isopotential compartments. Each compartment consists of a set of lumped circuit elements representing the biophysical properties of the corresponding length of neuronal cable, and the compartments are connected together via lumped axial resistances (Figure 11). The time evolution of voltages and other variables within this "equivalent circuit" structure in response to an arbitrary pattern of synaptic or other input is computed using standard numerical integration techniques. The advantage of such a representation is that the biophysical properties of the membrane and the cytoplasm can vary arbitrarily from compartment to compartment if so desired, and the membrane or synaptic conductances within a compartment can be defined to have complex dependencies on voltage, time, and other variables—Hodkin-Huxley channels are one example. The nuts and bolts of compartmental modeling are available elsewhere, for example see Perkel et al.,[114] Segev et al.,[143] Claiborne et al.[29] for various treatments of the method. See also Traub et al.,[170] Borg and Graham,[19] Brown et al.,[24] and Mel[105] for examples inluding modeling of nonlinear membrane conductances, Mascagni[93] for a lucid discussion of numerical issues, Shepherd[147] for an interesting historical perspective, and De Schutter[33] for an overview of currently available software for creating and running compartmental models.

COMPUTATIONAL STUDIES OF DENDRITIC FUNCTION

A variety of modeling studies have been carried out over the past three decades to explore various aspects of dendritic function beyond simple summation of synaptic inputs. In the following, we discuss this work in the context of four main ideas that have dominated the conceptual landscape. These are:

- The spatially extended nature of a dendritic tree permits useful spatio-temporal interactions among active synapses.

A

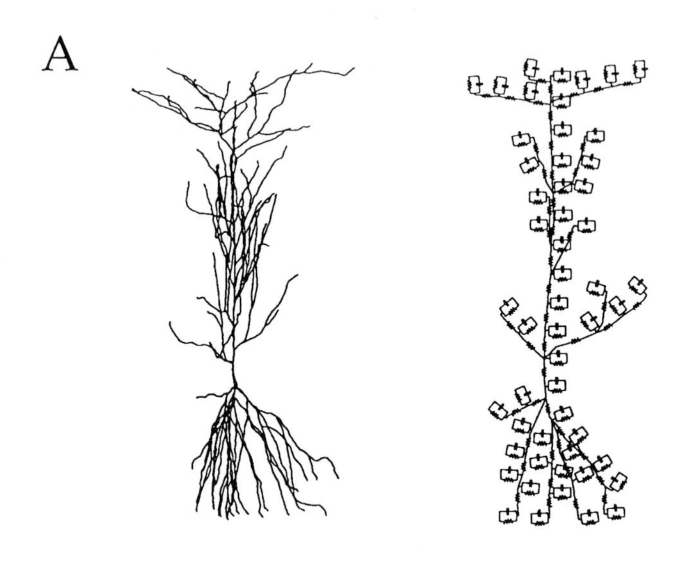

B

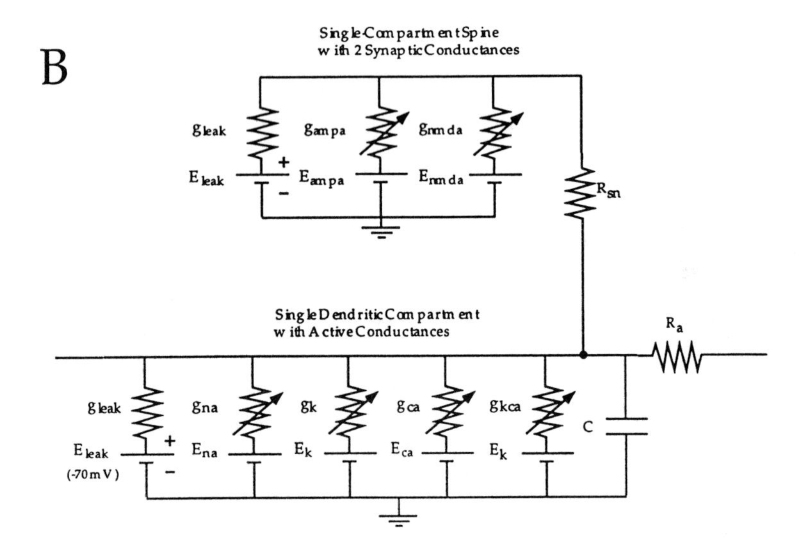

FIGURE 11 (a) Simplified compartmental representation of the dendritic tree of a hippocampal pyramidal cell (courtesy of Tony Zador). (b) Blowup of equivalent circuit for a single dendritic compartment with attached spine. Main dendritic compartment is depicted with voltage-dependent Na^+ and K^+ conductances for fast Hodgkin-Huxley spiking, as well as a slow voltage-dependent Ca^{++} conductance and a Ca^{++}-dependent K^+ channel.

- Dendritic trees can have multiple pseudo-independent processing subunits.

- Passive dendritic structure may be modulated by external influences to alter the input-output behavior of the cell as a whole, or of individual subunits.
- Nonlinear membrane mechanisms appropriately deployed can allow the dendritic tree of a single neuron to act as a powerful multilayer computational (e.g., logical) network.

SPATIO-TEMPORAL INTEGRATION

Wilfred Rall[127] first demonstrated that a passive dendritic branch, by virtue of its spatial extension, can act as a spatio-temporal filter that selects for specific temporal sequences of synaptic inputs. Since time is required for signals to propagate along a dendritic branch, it matters what part of the dendrite gets stimulated at what time. For example, the largest superposition of signals at the cell body occurs when distal synapses are activated before proximal synapses. This principle is illustrated in Figure 12(a): the peak of the voltage waveform at the cell body is twice as large when inputs are activated in a sweep toward the cell body (DCBA) than in a sweep away from the cell body (ABCD).

Poggio and Torre[119] and Koch et al.[72,73] pursued this basic idea further in the effort to explain direction-selective (DS) responses in retinal ganglion cells. They amplified on Rall's basic idea by showing that the relative placement and timing of excitatory and *inhibitory* synapses on the same dendrites could lead to a much more pronounced directional difference than in Rall's study case of excitation alone. Essentially, a large synaptic conductance increase whose reversal potential is close to the resting potential, usually called "silent" or "shunting" inhibition, acts like a hole in the membrane that shunts a large fraction of any passing current directly to the extracellular ground. While a shunting synapse cannot by itself alter the potential at the cell body, it can effectively short out the path to the cell body for any more distal depolarizing or hyperpolarizing influences[72,73,119]; see Shephert and Koch[152] for explanation of shunting inhibition.

In the simplest instance, the Koch-Poggio-Torre model for retinal direction selectivity entails that photoreceptors are topographically mapped onto each ganglion cell dendrite, and each photoreceptor is assumed to activate both an excitatory and an inhibitory synapse at approximately the same dendritic locus (Figure 12(b)). The inhibitory conductances are assumed to activate with slower kinetics than the excitatory conductances, however, such that the excitatory input has time to begin propagating toward the cell body before the co-localized shunting inhibition is sufficiently activated to exert its "veto" effect. Thus, if the photoreceptor-induced stimulus sweeps along the branch *toward* the cell body, the slowly activating inhibitory conductances consistently exert their influence *distal* to the snowballing excitatory wave bound for the cell body, and are therefore ineffective at blocking it. If the photoreceptor stimulus sweeps along the branch *away* from the cell body, then the inhibitory conductances consistently exert their influence on the direct path to the cell body for all subsequently activated more distal excitatory inputs.

This elemental nonlinear synaptic interaction was shown to produce strongly direction selective responses in a modeled retinal ganglion cell (Koch et al.[74]; Figure 12(c)).

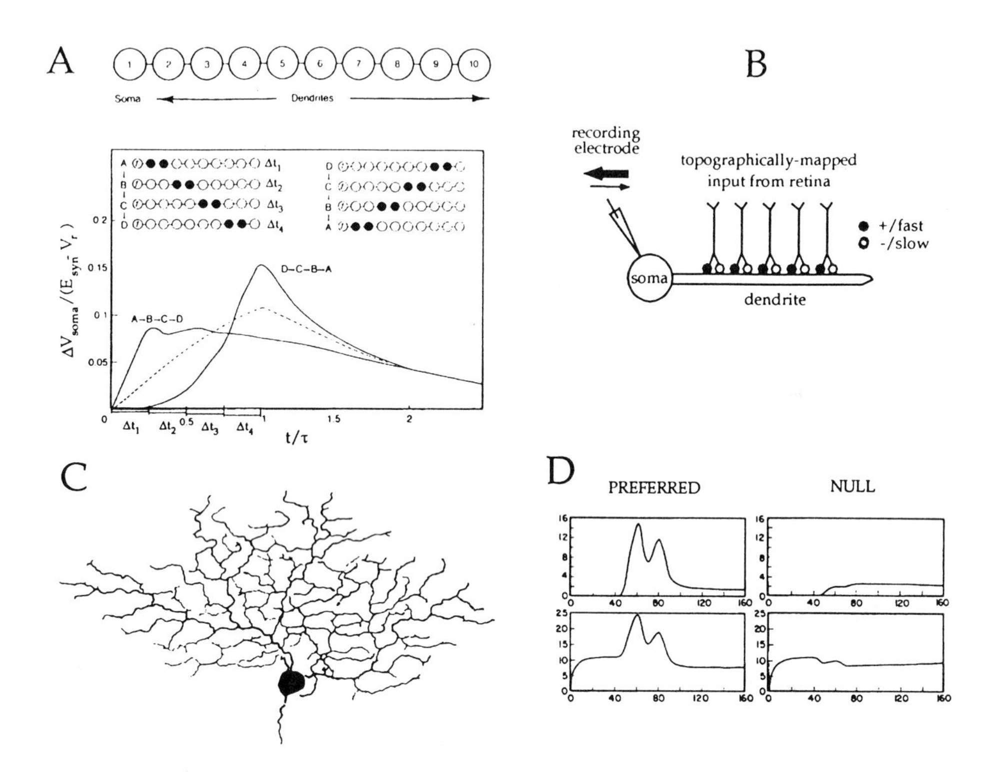

FIGURE 12 (a) Passive 10-compartment model demonstrates directional difference for input sequence ABCD vs. DCBA; reprinted with permission from Rall.[127] (b) Single branch schematic of Koch-Poggio-Torre[73] model of direction selectivity as measured at the cell body. Fast acting excitatory inputs (black circles) are effectively "vetoed" by slow-activating shunting inhibition (open circles) only when the direction of sweep is away from the cell body. (c) Reconstructured direction-selective cell from the rabbit retina (reprinted with permission from Koch et al.[74]). (d) Cell in C was modeled assuming passive dendrites and on-path inhibition as schematized in (b). Responses to preferred-direction stimulus are shown at left, null-direction at right. Lower traces include small DC current injection at cell body. Units are mV above rest (ordinate) vs. msec (abscissa).

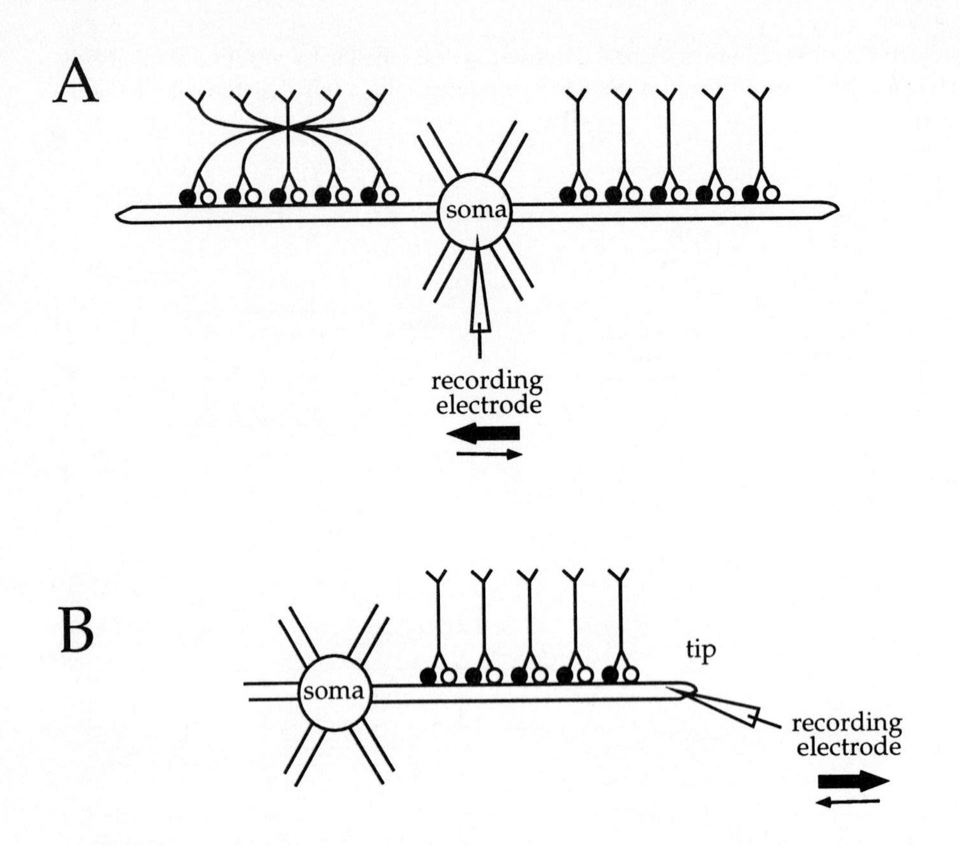

FIGURE 13 (a) One possible asymmetrical system of input connections giving rise to direction selectivity as measured at cell body (adapted from Koch et al.[72]) (b) Tip of long dendritic branch is always direction selective (adapted from Borg-Graham and Grzywacz[20]).

This modeling study depended critically, however, on a neuroanatomical assumption that has remained unsubstantiated—i.e., that inputs to retinal ganglion cell dendrites are precisely "wired" such that each inhibitory synapse is positioned to veto subsequently activated excitatory synapses when the stimulus sweep is in the "null" direction, but *not* when the stimulus sweep is in the preferred direction. An example of such an asymmetric wiring diagrams is illustrated in Figure 13(a). An interesting alternative source of the necessary asymmetry for DS responses is discussed by Borg-Graham and Grzywacz[20]; (see also Vaney et al.[172]), that does not depend on precise asymmetric deployment of excitatory and inhibitory synapses onto the dendrites of retinal ganglion cells. The authors point out that even in the case of entirely symmetric, topographically mapped mixed excitatory and inhibitory input onto a circularly symmetric dendritic tree, the *tip* of each dendritic branch is direction selective when considered as an output, i.e., responds more strongly to photoreceptor sweeps toward the tip (Figure 13(b)). Evidence that amacrine cells in the rabbit retina preferentially make contacts onto retinal ganglion cells via their branch tips has led these authors to the conjecture that the DS of retinal ganglion cells is due to DS inputs from amacrine cell branch tips rather than due to internal processing in retinal ganglion cells themselves. Consistent with this hypothesis, Borg-Graham and Gryzwacz[17] demonstrated that retinal ganglion cells remained direction selective even when their inhibitory inputs were blocked. For an elegant demonstration of the probably passive dendritic basis of direction-selectivity in an invertebrate, the blowfly, see Haag et al.[49]

Another issue relevant to dendritic spatio-temporal integration is the question as to whether synchronously activated synapses scattered about a dendritic tree are more effective at driving a cell than the same number of inputs activated asynchronously—as has been commonly postulated, e.g., Abeles.[2] This question has been investigated in Bernander et al.[14] for passive dendrites (though NMDA synapses were considered in one condition), where it was shown that when the number of active synapses is less than the number of fully synchronized inputs needed to fire a single action potential, then synchronous inputs are always more effective than asynchronous. When the number of activated synapses exceeds this threshold, then the saturating nonlinearity associated with excitatory synaptic action tips the balance gradually in favor of desynchronized inputs. With regard to precise timing of synaptic inputs, Softky and Koch[155] and Softky[156] have discussed the plausibility and consequences of sub-millisecond coincidence of synaptic inputs to dendritic trees that contain very large, very brief synaptic conductances and/or the potential for fast spike generation.

DENDRITIC SUBUNITS

A second important idea relevant to dendritic information processing is that of "dendritic subunits," i.e., the idea that pseudo-independent computations can be carried out simultaneously in different dendritic subregions. An early discussion of dendritic subunits can be found in Llinás and Nicholson[83], where it was proposed that synaptic integration and consequent local spiking activity could occur pseudo-independently in different branches of the Purkinje cell dendritic tree. Koch et al.[72] first formally defined a dendritic subunit as a region within which the voltage attentuation is small between any pair of synapses i and j in the subunit, but for which the voltage attentuation is large between any subunit synapse and the soma s. More precisely, a subunit consisted of any group of synapses such that $A^V_{is}/A^V_{ij} > c, (c > 1)$ for all i and j in the subunit. (The original definition was expressed in terms of transfer resistances instead of attenuations.) For a particular choice of membrane parameters ($R_m = 2500\Omega\text{cm}^2, R_i = 70\Omega\text{cm}, C_m = 2\mu\text{F/cm}^2$)

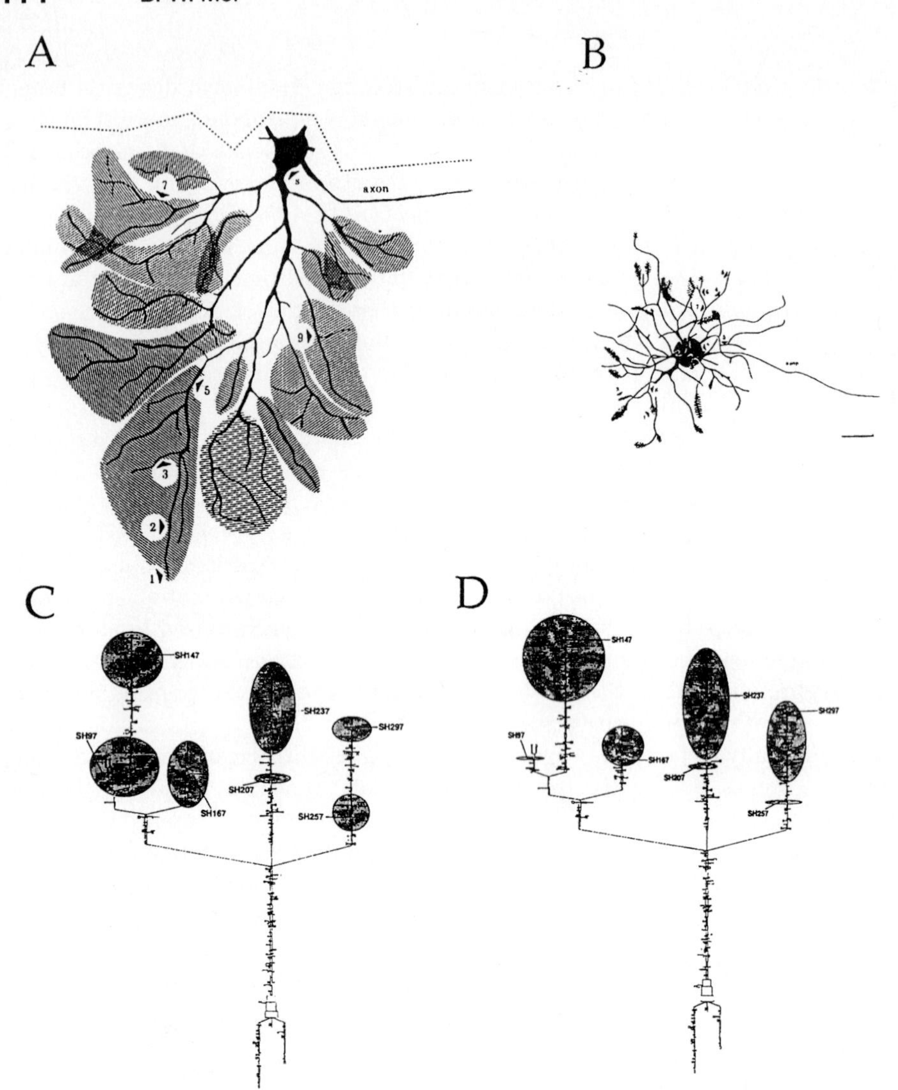

FIGURE 14 (ab) Subunit structure of two types of retinal ganglion cells based on passive cable properties (reproduction of Figures 2, p. 240 and 3, p. 241, with permission from Koch et al.[72]). (a) Large subunits in an α retinal ganglion cell. (b) Much smaller subunits are seen in β ganglion cell in same study. (cd) Study of subunit structure of granule cell of the mouse olfactory bulb (reprinted with permission from Woolf et al.[180]). (c) Subunits around 7 reference spines are illustrated, where subunit is defined as region around input spine with no more than 5% steady state voltage attenuation. (d) Same cell, where subunit criterion is region of greater than 10mV depolarization in response to 4nS input at reference spine.

and the value $c = 4$, and under the assumption that subunits should not overlap, a set of subunits was determined as shown in Figure 14(a) for an α-type retinal ganglion cell. The implication of this result is that an α-type retinal ganglion cell does indeed have a considerable capacity for independent processing within its dendritic tree. A β-type ganglion cell with a much smaller dendritic tree ($\sim 100\mu$m) and relatively thick branches had very small subunits. The subunit boundaries of Figure 14(a) and (b) should not, however, be taken too literally, as they depend on c, which was arbitrarily chosen, on the cell morphology, on the algorithm used to grow subunits beginning at the dendritic tips, and on membrane paramters—they disappear almost completely when $R_m > 8000\Omega\text{cm}^2$. It is also important to emphasize that the definition of dendritic subunits from Koch et al.[72] stresses the electrotonic independence of subunits from the cell body, but not electrotonic independence of subunits from each other.

Woolf et al.[180] carried out a somewhat different analysis of subunit structure in granule cells of the olfactory bulb. In this case a subunit was defined to be a neighborhood in the dendrites about an arbitrarily chosen reference spine, for example consisting of all spines at which the steady-state voltage attenuation was less than 5% relative to the stimulated reference spine (Figure 14(b)). When the subunit criterion was was changed to include all neighboring spines that were depolarized by more than 10 mV in response to a 4 nS transient synaptic conductance input at the reference spine, some subunits remained essentially unchanged, others grew dramatically, and still others essentially disappeared (Figure 14(c)). When the EPSP rise time was slowed from 0.2 ms to 1 ms, subunits grew so large at to encompass the entire dendritic tree.

In these two subunit studies, the observed sensitivity of subunit structure to biophysical parameter assumptions and to subunit definitions has double-edged significance. Though the concept of dendritic subunits is heuristically useful, and has guided important questions as to the passive integrative properties of dendritic trees, the marked sensitivity of subunit size to changes in modeling assumptions makes their explicit graphical enumeration less informative than it might be hoped. Any attempt to characterize a dendritic tree in terms of the locations of a fixed number of discrete subunits is thus necessarily misleading. On the other hand, the sensitivity of subunit size to parameters and assumptions in these studies is informative in and of itself, as it makes explicit the notion that the effective electrotonic structure of a dendritic tree depends strongly on both biophysical membrane parameters (see next section), and on the specific type of intradendritic voltage communication under consideration. In this latter case, for example, synaptic interactions that have sharp voltage thresholds may be expected to operate within a radically different virtual subunit structure than those that depend smoothly on voltage (Figures 14(cd)).

Given the virtual impossibility of counting discrete subunits or assigning their boundaries in a meaningful way, a complementary statistical approach may be used to quantify the relative electrotonic independence of dendritic synapses from each other. In a recent study of the input-output behavior of neocortical pyramidal cell dendrites, a histogram was generated to quantify the steady-state voltage and current attentuation between randomly chosen pairs of synapses in the passive dendritic tree (Figure 15); from Mel.[103] The histogram shows that the average steady state attenuation factor for voltage or current between randomly chosen synapse pairs is nearly 70; for about half the pairs of synapses in the dendritic tree, the attenuation factor is greater than 25. The histogram representation of a dendritic tree is weak in that it quantifies the electrotonic independence of each dendritic locus from each other in only a probabilistic sense, but it is immune to the dramatic parameter sensitivity characteristic of efforts to define and discretely label dendritic subregions.

MODULATION OF PASSIVE MEMBRANE PROPERTIES

The fact that dendritic subunit structure is sensitive to biophysical membrane parameters leads directly to the suggestion that intradendritic information processing could be modulated by any outside influence acting on passive membrane properties. A third idea relevant to dendritic integration thus entails that outside modulating influences can act to alter the cable properties of part of all of a dendritic tree, thereby changing its integrative behavior in response to patterns of synaptic input.

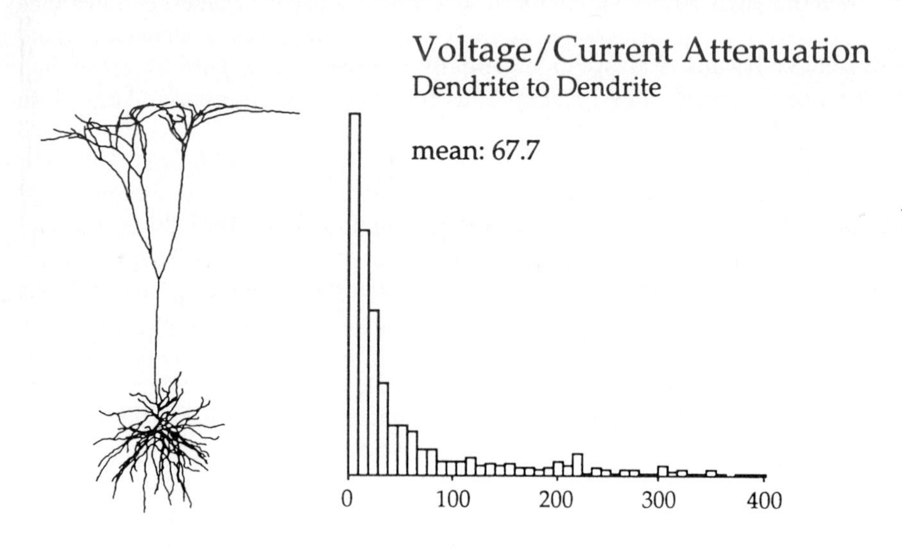

FIGURE 15 Voltage or current attenuation histogram for DC current inputs to electrically passive pyramidal cell dendritic tree, with $R_m = 10,000\Omega\text{cm}^2, R_i = 200\Omega\text{cm}$, $C_m = 1\mu F/\text{cm}^2$. Pairs of input and recording locations were chosen at random, uniformly in dendritic length. Average steady-state attenuation factor was 67.7. Pyramidal cell morphology courtesy of Rodney Douglas and Kevan Martin.

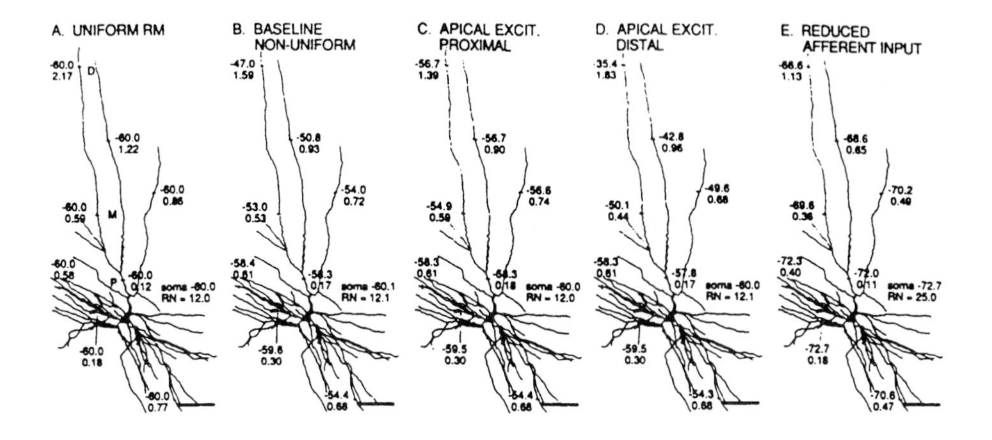

FIGURE 16 Steady-state response of passive neuron for different synaptic activity distributions. The resting potential (upper number) and electrotonic distance (lower number) to selected dendritic locations in the cortical pyramidal cell were computed for each of the three conditions (with permission from Holmes and Woody[55]).

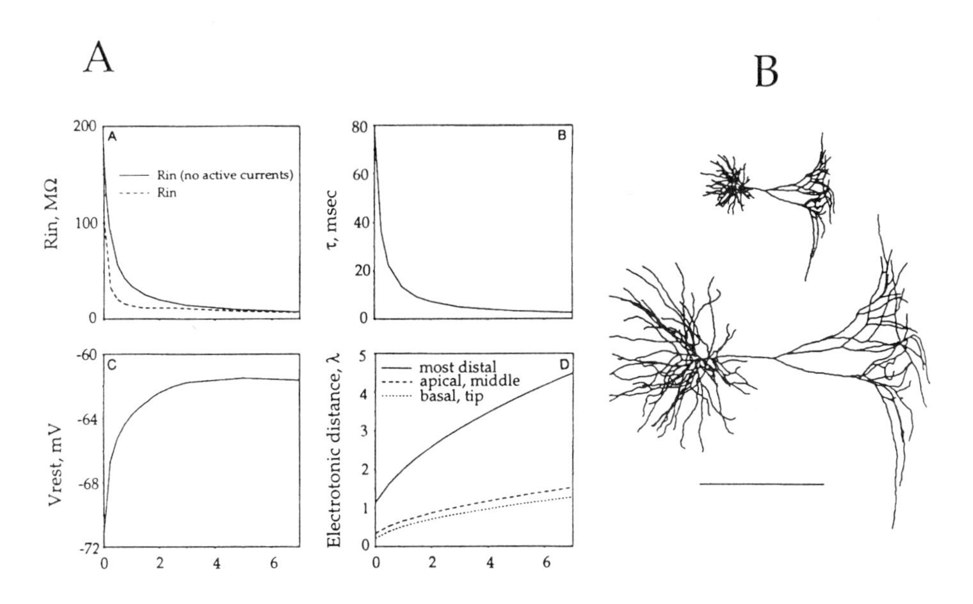

FIGURE 17 (a) Impact of synaptic background firing frequency from 0 to 7 Hz on cell parameters, including input resistance (upper left), time constant (upper right), resting potential (lower left), and electrotonic length (lower right) (b) Electronic size of cell at two levels of synaptic background activity: 0 Hz (upper cell), representative of conditions in a slice, and 2 Hz (lower cell), representative of condition of low-level background activity. Reprinted with permission from Bernander et al.[13]

Holmes and Woody[55] first demonstrated that different spatially non-uniform patterns of background synaptic activity impinging onto a modeled cortical pyramidal dendritic tree give rise to different non-uniform resting membrane potential distributions, various distortions in the effective length constants of dendritic branches, location-dependent changes in the ability of distal synapses to influence the cell body, and pronounced changes in membrane time constants (see Figure 17). Essentially, the spontaneous low-frequency openings of 10,000–20,000 synapses in the dendritic tree of the modeled cell yielded, in the aggregate, a significant change in effective membrane resistivity which in turn induced the observed changes in electrotonic structure and time constants of the cell. Though these authors considered only the passive membrane case, the straightforward inference could be made from this work that induced inhomogeneities in the dendritic voltage environment under variable patterns of background synaptic activity could lead to variable operating regimes for any voltage-dependent membrane mechanisms residing in the dendritic tree.

Bernander et al.[13] further explored the effects of background synaptic activity on the passive cable structure of a layer 5 neocortical pyramidal cell. A fixed spatial distribution of 4000 excitatory and 1000 inhibitory synapses was modeled, while frequency of background activity was varied from 0 to 7 Hz. Over this range, the time constant and input resistance of the cell measured at the cell body were reduced by a factor of 10, while the electrotonic length of the cell grew by a factor of 3 (Figure 17). The authors further demonstrated that the reduction in membrane time constant associated with more vigorous background activity could lead to an increased selectivity for synchronous vs. asynchronous activation of other synaptic inputs; see also Bernander et al.[14] and Rapp et al.[134] This study thus demonstrates that the activity of the intrinsic cortical network is likely to exert a powerful influence on the integrative behavior of individual neurons.

In a different vein, Laurent and Burrows[78] and Laurent[77] have proposed that non-spiking interneurons in the metathoracic ganglion of the locust may have independently modulable input-output regions within their dendritic trees. In these invertebrate neurons, input and output synapses intermingle along the same dendritic branches (Figure18(a)), in contrast to most vertebrate neurons for which dendrites are input structures exclusively (see Shepherd[148] for rules and exceptions). A biophysical mechanisms was proposed whereby a system of intersegmental control axons could modulate a sensory-motor reflex arc. In the reflex circuit of Figure 18(b), afferent inputs from mechanosensory receptors on the legs are spatially intermingled with outputs to motor neurons in the dendrites of a non-spiking interneuron. Intersegmental control inputs make additional synaptic contacts onto these same branches. While their precise function is unknown, these inputs seem capable of locally modulating membrane properties in such a way as to enhance or supress the afferent-to-motor reflex connection within a restricted region of the dendritic tree. One putative mechanism involves large shunting synaptic conductances activated by intersegmental inputs that simply lower the local input resistance, reducing

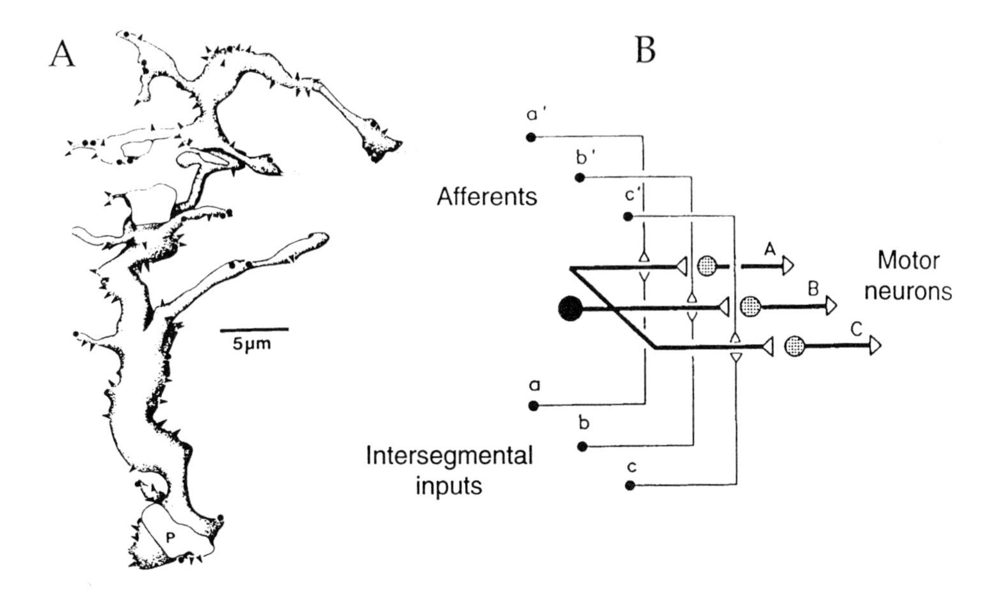

FIGURE 18 Possible input-output function of intersegmental interneuron dendritic tree in locust. (a) Reconstruction of portion of a non-spiking local interneuron in thoracic nervous system of locust. 22 input (circles) and 64 output (triangles) synapses are spatially intermingled. Reprinted with permission from Watson and Bkurrows.[175]
(b) Model of a non-spiking interneuron that receives input from 3 local afferents (a',b',c') and from 3 intersegmental interneurons (a,b,c); output connections (A,B,C) project to three motor neurons. In this model, the 3 local circuits (a',a,A), (b', b, B), and (c', c, C) can be modulated separately. Reprinted with permission from Laurent and Burrows.[78]

the size, spread, and effectiveness of afferent EPSP's.[78] The subunit structure of the dendritic tree has also been shown in modeling studies to permit the input-resistance to be pseudo-independently modulated in different dendritic regions,[79] consistent with the idea that different reflex arcs may indeed be controllable by different intersegmental inputs.

NONLINEAR PROCESSING IN DENDRITES

One of the questions of greatest intellectual interest in the study of neuronal information processing regards the limits of computational power of the single neuron. In this vein, the fourth idea we consider here is that nonlinear membrane mechanisms, if appropriately deployed in a dendritic tree, can allow the single neuron to act as a powerful multilayer computational network. The most common instantiation of this idea has been the proposal that individual neurons may implement a hierarchy of

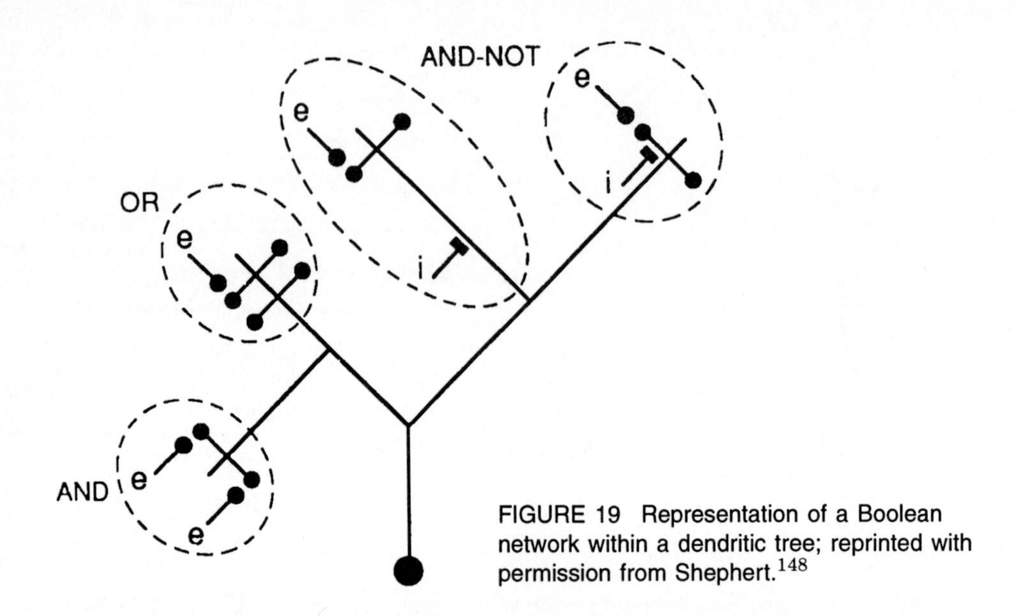

FIGURE 19 Representation of a Boolean network within a dendritic tree; reprinted with permission from Shephert.[148]

logical operations within their dendritic trees, consisting of AND, OR, AND-NOT, and XOR operations[72,73,83,89,119,132,149,150,153,184]; see Figure 19).

SYNAPTIC NONLINEARITIES. One version of this idea was based on the mathematical observation that synaptic conductance changes interact in a nonlinear way on a dendritic branch[119,127]; the emphasis of this latter study was on the second-order multiplicative interaction between excitatory and shunting inhibitory synapses underlying the so-called AND-NOT operation[72,73,74]; see Koch and Poggio[70,71]) for discussions of multiplicative and other nonlinear mechanisms in neuronal computation.

VOLTAGE-DEPENDENT MEMBRANE AND LOGIC OPERATIONS. A number of other modeling studies have further pursued the neuron-as-logic-network metaphor, primarily by demonstrating that dendrites appropriately configured with voltage-dependent membrane can approximatively implement two-input logical operations, such as AND, OR, and XOR. For example, Shepherd and Brayton[149] showed that simultaneously synaptic input to two neighboring spines with excitable Hodgkin-Huxley spine heads could, once both spines fired action potentials, result in sufficient depolarization in the underlying dendritic branch as to fire off two additional nearby spines; a single synaptic input was presumably insufficient (Figure 19(a)). They argued that this behavior was AND-like in that the output of the dendritic region, signalled by the activation of the entire cluster of four spines, depended on the simultaneous activation of two inputs. By increasing the synaptic conductance onto each spine head, a single presynaptic event could be made to trigger suprathreshold activity in all four spines; this was termed an OR-gate, since only one of many possible inputs was needed to generate an output for the region (Figure 20(b)). While the existence of Hodgkin-Huxley membrane in spine heads has not been demonstrated experimentally, results of this kind generalize well when the excitable membrane resides instead on dendritic shafts,[153] or when altogether other excitatory voltage-dependent mechanisms are assumed.[102,105]

In analogy to a logical XOR function, Zador et al.[184] have recently demonstrated that a combination of two voltage-dependent membane mechanisms could produce a non-monotonic output from a dendritic region in response to monotonically increasing synaptic input (Figure 21). Thus, low levels of synaptic input to dendritic sites A and B (corresponding to a logical 0/0) were insufficient to cause a somatic spike. At intermediate levels of net synaptic input delivered to the two sites

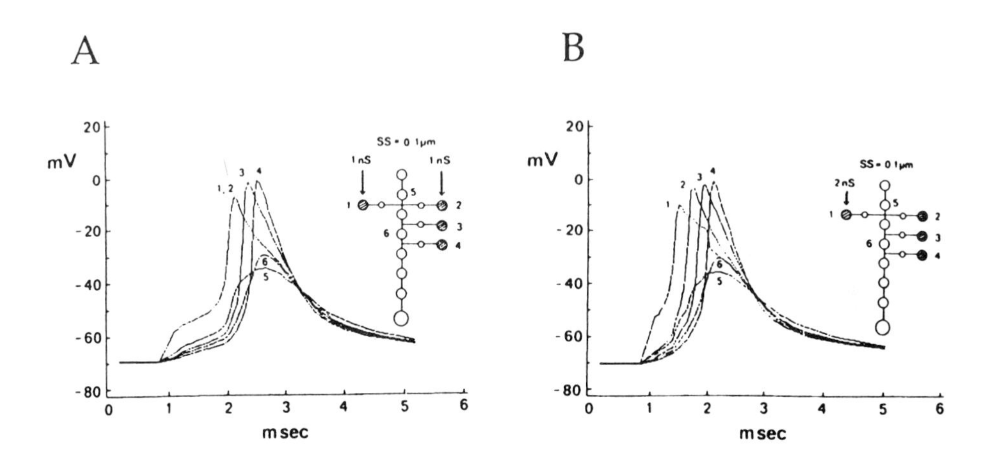

FIGURE 20 Dendritic implementation of two input logic functions. (a) A group of 4 neighboring spine heads contained active (Hodgkin-Huxley) membrane. Synaptic conductances were chosen such that a single synaptic input was insufficient to fire any of the spines, whereas simultaneous activation of spines 1 and 2 led to firing of all 4 spines heads (voltage traces for each spine head are numbered). This thresholding behavior was likened to a two-input logical AND (this and part B reprinted with permission from Shepherd and Brayton[149]). (b) When synaptic conductances were doubled, input to spine 1 alone led to firing of all 4 spine heads. This condition was likened to a logical OR.

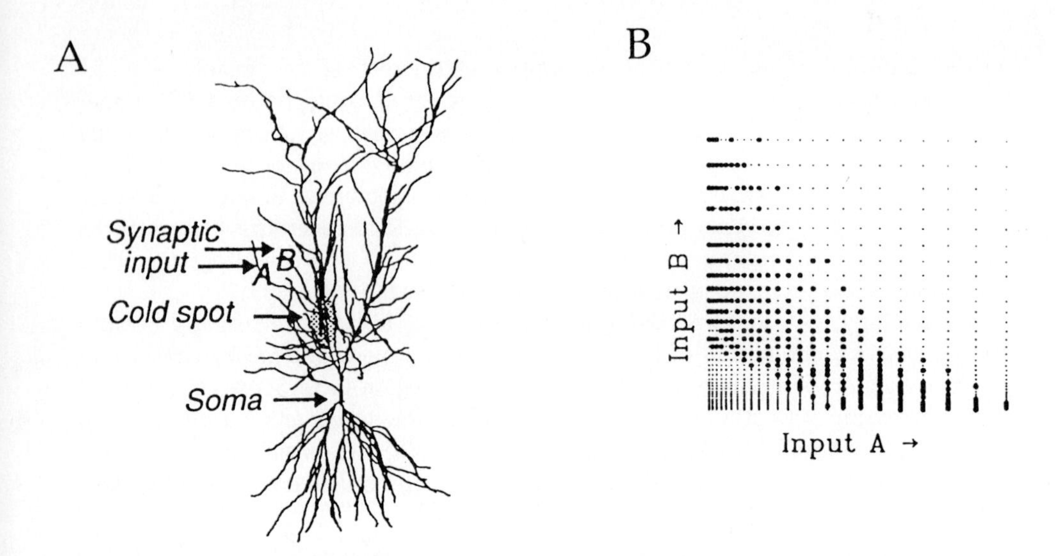

FIGURE 21 (a) In a reconstructed hippocampal pyramidal cell, a "cold spot" was placed in the dendritic membrane, consisting of a high density of Ca^{++}-dependent K channels, and a low-density of voltage-dependent Ca^{++} channels. Synapses were activated on dendritic branches A and B. Non-monotonic responses at the cell body with increasing input at branches A and B are shown in plots B (bold dot for single spike); the non-monotonicity was likened to that of a logical XOR. Reprinted with permission from Zador et al.[184]

(correponding to logical 1/0 or 0/1), action potentials could be elicited. However, at higher levels of combined synaptic input to A and B (corresponding to logical 1/1), membrane depolarization in the dendritic tree was sufficient to activate voltage-dependent calcium channels, leading to an influx of calcium ions, followed by activation of a calcium-dependent potassium channels that gave rise to a strong inhibitory (outward) current. In this case, action potentials at the cell body were suppressed (Figure 21(b)).

In an attempt to explore the complexity of interactions among many synaptic inputs in a dendritic tree, Rall and Segev[132] tested the input-output behavior of dendritic branches containing passive and excitable dendritic spines (Figure 22). Several synaptic input conditions are shown at left, where black spines are excitable and white spines are passive. The corresponding output condition shows all those spines that fired action potentials as a result of the given synaptic input. The resulting somatic depolarization is also given for each case. Based on the complexity of interactions seen in these demonstrations, the authors concluded that dendritic trees with excitable spine clusters (and presumably other varieties of excitable membrane nonlinearities) afford rich possibilities for pseudo-independent logiclike computations.[132]

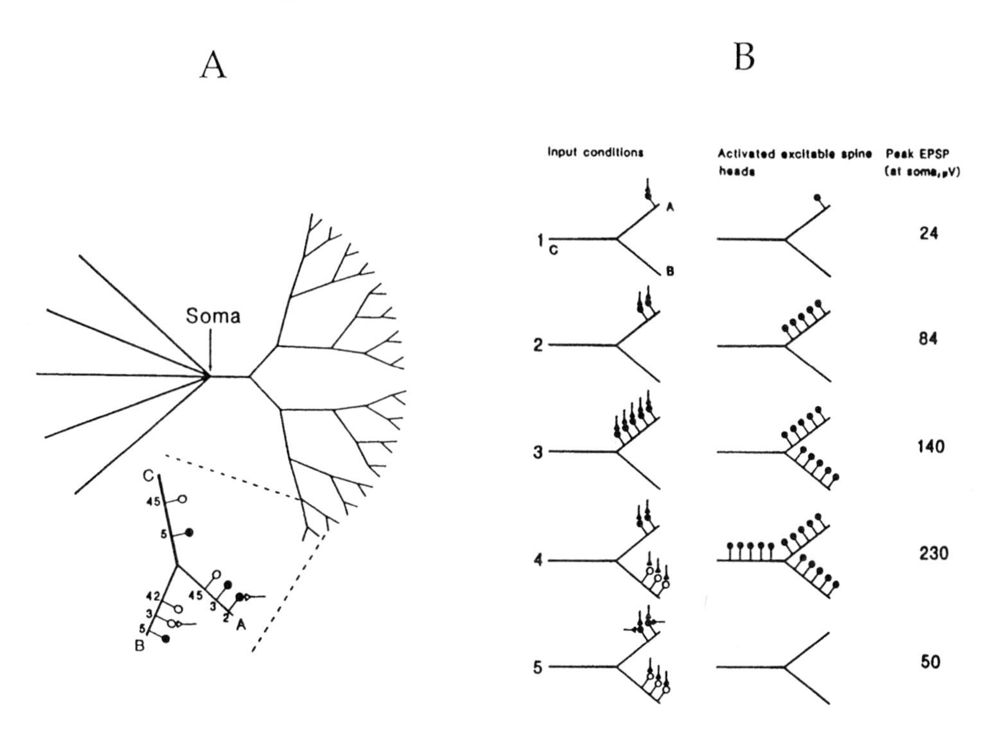

FIGURE 22 Summary of 5 different cases with active spine heads on distal dendritic branches; reprinted with permission from Rall and Segev.[132] (a) Location of subtree relative to soma of passive cell; each branch had 50 spines, 45 passive (open), 5 active (filled). (b) Left column shows spines receiving synchronous excitatory (open synapses) or inhibitory (filled) synaptic input. Center column shows active spine heads that fired action potentials. Right column shows peak of passively propagated wave as measured at cell body.

PROBLEMS WITH THE LOGIC-NETWORK METAPHOR. The dendritic-tree-as-logic-network metaphor has been a useful concept that has motivated a variety of analyses of the nonlinear computational properties of single neurons. However, two issues may be raised that suggest a potential misfit between logical computation and dendritic computation. First, logical computation is inherently "ill behaved"—changes in any and every input variable must in general be capable of altering or not altering the output of the logic function depending on the values of all other inputs. Second, inputs to a logic function are in general unrelated to each other in terms of their effects on the output. Both of these aspects of logical computation are at odds with the fact that the electrotonic structure of a dendritic tree gives rise to extensive voltage sharing and smooth neighborhood relations among synaptic inputs (see Figures 9 and 10(a)). Furthermore, logic functions require precise wiring diagrams, where a single erroneous input connection or malfunctioning computational element can corrupt the input-output behavior of an entire circuit. A crucial accompaniment of any logic-network theory of dendritic processing, therefore, is a mechanism, such as some form of learning, that can drive the appropriate micro-organization of individual dendrites, synapses, and membrane channels. Such a process would, for example, need to explain the development of the precise spatial juxtaposition of each afferent synapse both with other specific afferent synapses and with the appropriate type of nonlinear membrane as suggested by Figure 19. Such a theory must also cope with accumulating evidence that significant dendritic remodeling occurs the mammalian brain, even during adulthood,[46] implying that a dendritic logic circuit must continuously adapt to "on line" changes in its basic computing architecture.

A literal interpretation of the dendritic-tree-as-logic-network metaphor would be significantly bolstered by a demonstration in which a biologically relevant nontrivial (e.g., multiple input) logic function is mapped onto a realistically modeled dendritic tree, such that the output of the cell follows the specified truth table. The case would be particularly strong if accompanied, as suggested above, by a biologically plausible account for the establishment of the dendritic logic circuit. No such demonstration has yet appeared in the published literature.

LOW-ORDER POLYNOMIAL FUNCTIONS. An alternative metaphor for nonlinear dendritic computation, which may be viewed as a smooth, analog version of the "logical dendrites" hypothesis, is the idea that a dendritic tree acts as an approximative low-order polynomial function with many terms—in short, a big sum of little products. A number of authors have fielded conjectures along these lines[35,40,100,106,118,119] where the requisite multiplicative nonlinearity has typically been assumed to derive from some nonlinear membrane mechanisms of an excitatory nature. The Hodgkin-Huxley thresholding mechanism previously used in demonstrations of AND-like synaptic interactions is one example[149,153]; NMDA channels and a variety of other mechanisms have also been considered good candidates (see Koch and Poggio[71]; Wilson[176]; Mel.[101,102,104,105] As a nonlinear approximator, the low-order polynomial representation is highly constrained in that (1) only two levels of computation are involved—a sum of products, (2) only the simplest nonlinear interaction is allowed—multiplication, and (3) the number of terms in each product is small—e.g., 2 or 3. An example of a biologically relevant computation of this order of complexity is a correlation between two high-dimensional input patterns.

An abstract model neuron called a "clusteron" has recently been introduced that maps low-order polynomial functionality onto a dendritic tree in a way that is directly testable within a detailed biophysical model.[101,104,105] The clusteron consists of a "cell body" where the global output of the unit is computed, and a dendritic tree, which for present purposes is visualized as a single long branch

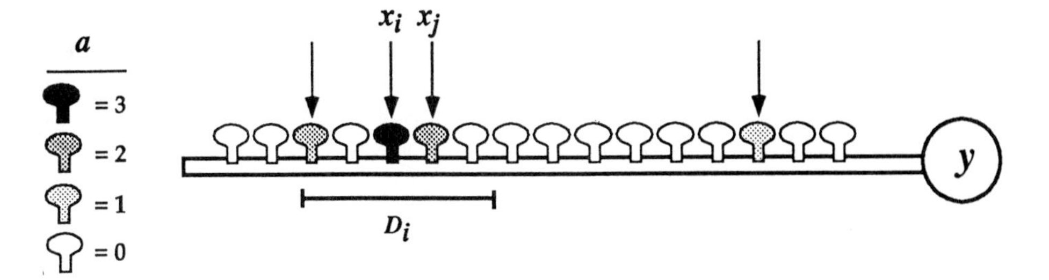

FIGURE 23 The clusteron is a limited second-order generalization of a thresholded linear unit, in which the excitatory effect of each synaptic input depends on the activity of other synapses in the neighborhood (active inputs designated by arrows). A cluster-sensitive Hebb-type learning rule says that synapses that are frequently coactivated with their neighbors should be stabilized; those that tend to act alone are destabilized and allowed to re-establish connections at new dendritic loci.

attached to the cell body (Figure 23). The dendritic tree receives a set of N excitatory weighted synaptic contacts from a set of afferent "axons." The output of the clusteron is given by

$$y = g\left(\sum_{i=1}^{N} w_i a_i\right), \tag{4}$$

where a_i is the net excitatory activity due to synapse i, w_i is its weight, and g is an optional thresholding nonlinearity. Unlike the thresholded linear unit, in which the net input due to the ith synapse is $w_i x_i$, the net input at the ith clusteron synapse is its weight w_i times its activity a_i, where

$$a_i = x_i\left(\sum_{j\in\mathcal{D}_i} w_j x_j\right), \tag{5}$$

x_i is the direct input stimulus intensity at synapse i, and $\mathcal{D}_i = \{i-r, \ldots i, \ldots, i+r\}$ represents the neighborhood of radius r around synapse i. The clusteron as defined includes only pairwise interactions among synapses in the same neighborhood, denoted by the $x_i x_j$ terms in Eq. (5). Thus, the underlying "biophysical" assumption implicit in clusteron "physiology" is that the output of a dendritic neighborhood grows quadratically, i.e., expansively, with increasing input.

We note that as a fixed number of synaptic inputs are delivered to a clusteron, first in a diffuse spatial pattern, and then in progressively more clustered spatial patterns, the response of the clusteron steadily increases. Experiments of exactly this kind were carried out in a detailed biophysical model of a layer 5 neocortical pyramidal cell containing various complements of excitatory voltage-dependent mechanisms in its dendrites (Mel[102,105]; Figure 24). Under a variety of conditions in which either NMDA synapses, slow calcium spikes, fast sodium spikes, or combinations of the three were placed in the dendrites in differing spatial distributions, an initial positive-slope regime in response to increasingly clustered synaptic input was indeed observed (Figure 24); results for a passive dendritic tree as control

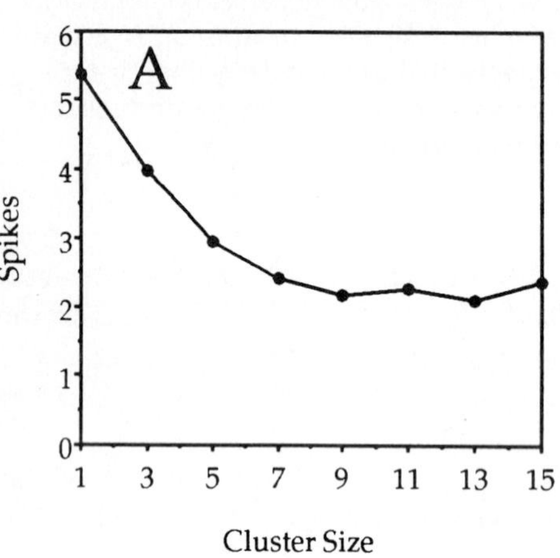

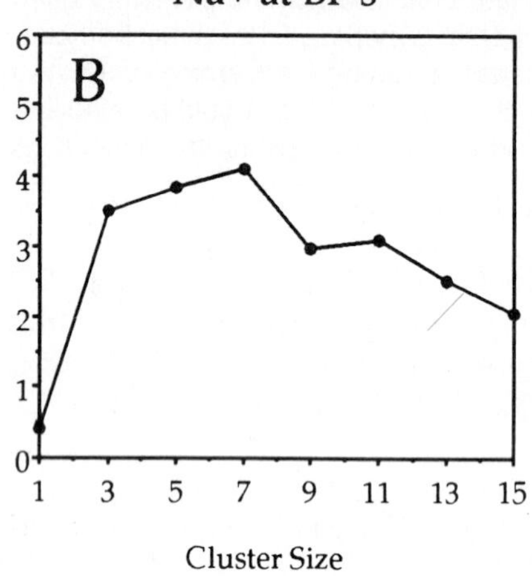

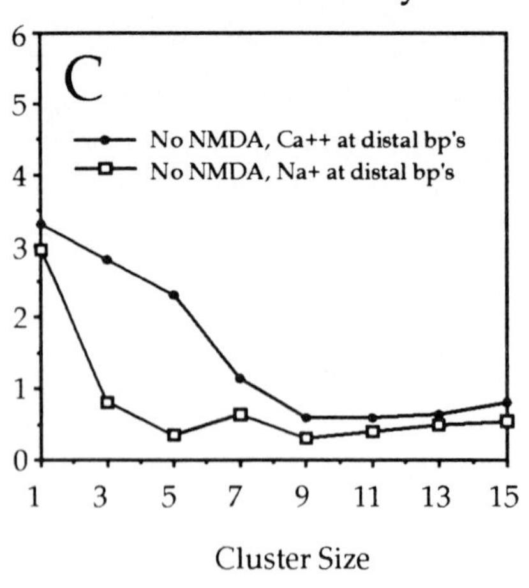

FIGURE 24 Plots of average cell response vs. cluster size under three biophysical conditions.[105] 100 synapses were activated at 100 Hz for 100 ms and resulting somatic spikes were counted A. Response of cell with passive dendrites and *ampa* synapses falls off monotonically due to classical synaptic (saturation) nonlinearity as cluster size is increased from 1 to 15. (b) Combination of high-NMDA, slow-spiking membrane distributed distally, and fast-spiking membrane distributed at branch points gave strong cluster sensitivity. (c) In two cases when no *nmda* was present, and dendritic spiking conductances were sparse (at branch-points only), cluster-sensitivity was abolished. Reprinted with permission from Mel.[105]

condition are shown in (a). Since the biophysically modeled cell was subject to saturation effects, unlike its abstract clusteron counterpart, responses to highly clustered synaptic input patterns were gradually diminished. The positive-slope "cluster-sensitivity" regime was observed to be present as long as the dendrites contained a sufficiently powerful and widely distributed complement of expansive nonlinear membrane mechanisms. No significant dependence on the kinetics, voltage dependence, or localization of the voltage-dependent membrane channels was observed in these steady-state stimulus-response experiments. Figure 24(c) shows two conditions in which a sparse, patchy distribution of either sodium or calcium spikes was insufficient to yield a cluster-sensitive regime.

FUNCTIONAL SIGNIFICANCE OF DENDRITIC CLUSTER SENSITIVITY. The functional significance of dendritic cluster sensitivity was demonstrated in Mel,[102] where it was shown that the nonlinear input-ouput behavior of an NMDA-rich dendritic tree could provide a capacity for nonlinear pattern discrimination (Figure 25(a)). In a subsequent study, it was estimated that a 5×5 mm slab of neocortex containing cluster-sensitive cells has the capacity to represent on the order of 100,000 sparse input-output pattern associations with high accuracy.[104] Another recent study has shown that a cluster-sensitive neuron can implement an approximative *correlation* operation entirely internal to its dendritic tree, with possible relevance to the establishment of nonlinear disparity tuning in binocular visual cells (Figure 25(b); see Ohzawa et al.[111]). Interestingly, a sum-of-products computation has been proposed in various forms as a crucial nonlinear operation in other types of visual cell responses, including responses to illusory contours,[116] responses to periodic gratings,[173] and velocity-tuned cell responses.[110]

The underlying idea in both examples of Figure 25 is that groups of frequently coactivated afferent axons represent prominent higher-order "features" in an input stream. These features may be encoded as groups of neighboring synaptic contacts onto a cluster-sensitive dendritic tree (see Brown et al.[23] for discussion of related ideas).

In the context of pattern memory, a "prominent" higher-order feature is any group of frequently coactivated input lines corresponding to a frequently observed conjunction of elemental sensory features in the input stream. If the prominent higher-order features accumulated from a set of training patterns are dendritically encoded in this way, then training patterns, which contain relatively many of these higher-order features, will (1) activate relatively many clustered pairs of synapses in the dendritic tree, and hence, (2) produce stronger cell responses than unfamiliar control patterns. The memory capacity of a cluster-sensitive dendritic tree is studied empirically in Mel.[104]

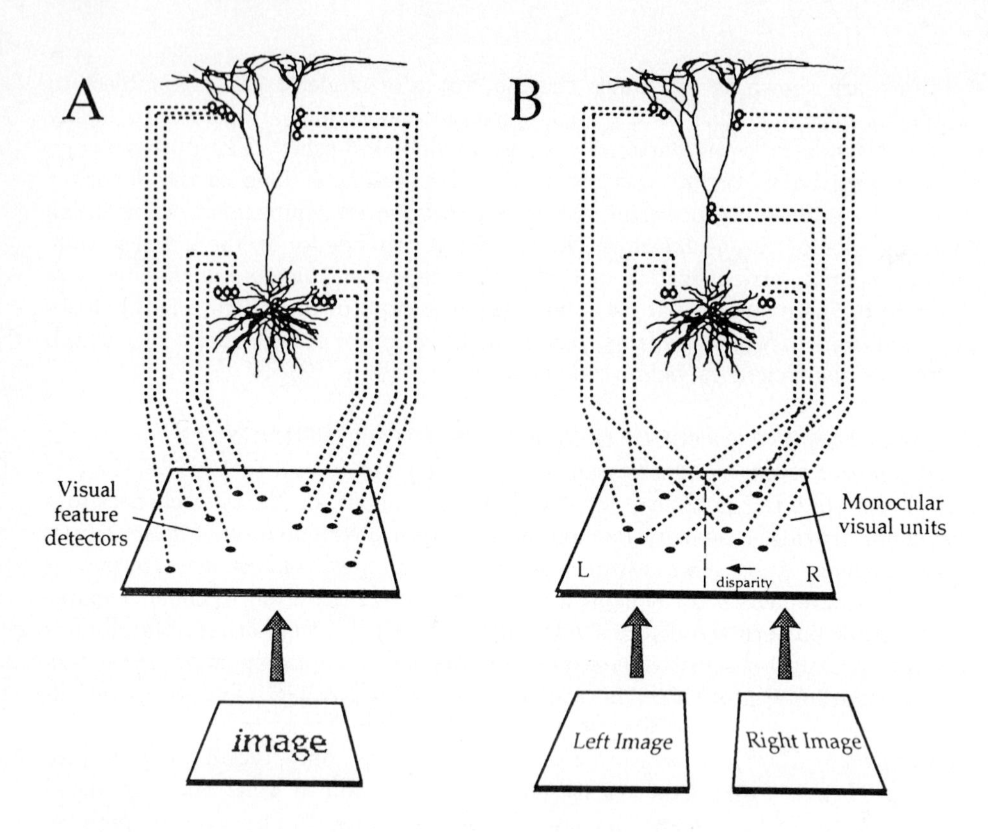

FIGURE 25 (a) Discriminating familiar from unfamiliar patterns within a cluster-sensitive dendritic tree. Visual input patterns drive a layer of visual feature-selective cells whose axons terminate on the cluster-sensitive dendrites. Patterns that drive synapses in clusters elicit relatively strong cell responses. Reprinted with permission from Mel.[102] The spatial ordering of afferent synaptic connections onto the dendritic tree may thus be of crucial importance for information storage. (b) Implementation of a "tuned excitatory" disparity-selective binocular cell in a schematized visual cortex. Axons from corresponding left and right monocular units, which are strongly correlated during normal vision, make synaptic connections onto neighboring patches of the binocular dendritic tree. Zero-disparity binocular images thus tend to activate synapses in pairs, a relatively effective (i.e., clustered) stimulus condition. Non-matching input to the left and right receptive fields activate synapses diffusely, and are thus relatively ineffective stimuli. Reprinted with permission from Mel.[105]

In the case of stereopsis, a prominent higher-order feature is any pair of afferents from corresponding locations in the left and right-eye receptive fields, which have a high probability of being coactivated during normal visual behavior. As illustrated in Figure 25(b), if these features are dendritically encoded, a zero-disparity visual stimulus contains relatively many higher-order "correspondence" features, activates many clustered pairs of synapses in the dendritic tree, and produces a stronger cell response than a non-corresponding stereo stimulus. Such a preference for stimuli at fixed disparity across the receptive field is a characteristic of many binocularly drivable complex cells in the primate visual system.[111]

Both of the scenarios of Figure 25 entail that the ordering of synaptic connections onto a dendritic tree be manipulated by some form of learning process, such that frequently coactivated input lines tend ultimately to form neighboring synaptic connections. An abstract learning rule with these properties is discussed in detail elsewhere.[101,104] Biophysically detailed modeling studies of Hebbian learning mechanisms in dendritic trees have also been carried out (Holmes and Levy[54]; Brown et al.[23]; Pearlmutter[112]).

CONCLUSIONS

We are at present in an awkward period, where many of the new ideas relating to dendritic function are the products of modeling studies, but where limited experimental access to dendritic trees means that modeling work proceeds mostly without direct experimental support. Recent progress in the development of optical recording methods portends an unprecedented period in which modeling techniques and experimental hypothesis testing can proceed in concert. The secrets of dendritic information processing may then be fully told.

ACKNOWLEDGMENTS

Thanks to Idan Segev and Tony Zador and for many helpful comments on the MS. This review would not have been possible without the generous working environment provided by Christof Koch at Caltech. This work was supported by the McDonnell-Pew Foundation and the Office of Naval Research.

REFERENCES

1. Abeles, M. "Role of the Cortical Neuron: Integrator or Coincidence Detector?" *Israel J. Med. Sci.* **18** (1982): 83–92.
2. Adrian, E. D. "The 'All-or-None' Principle in Nerve." *J. Physiol.* **47** (1914): 460.

3. Adrian, E. D. "The Spread of Activity in the Cerebral Cortex." *J. Physiol. London* **88** (1937): 127–161.
4. Agmon-Snir, H., and I. Segev. "Signal Delay and Input Synchronization in Passive Dendritic Structures." *J. Neurophysiol.* **70** (1993): 2066–2085.
5. Amitai, Y., A. Friedman, B. W. Connors, and M. J. Gutnick. "Regenerative Electrical Activity in Apical Dendrites of Pyramidal Cells in Neocortex. *Cerebral Cortex* **3** (1993): 26–38.
6. Andersen, P., and T. Lomo. "Mode of Activation of Hippocampal Pyramidal Cells by Excitatory Synapses on Dendrites." *Exp. Brain Res.* **2** (1966): 247–260.
7. Arikuni, T., and S. Ochs. "Slow Depolarizing Potentials and Spike Generation in Pyramidal Tract Cells." *J. Neurophysiol.* **36** (1973): 1–12.
8. Attick, J. J., and A. N. Redlich. "Toward a Theory of Early Visual Processing." *Neural Comp.* **2** (1990): 308–320.
9. Baer, S. M., and J. Rinzel. "Propagation of Dendritic Spikes Mediated by Excitable Spines: A Continuum Theory." *J. Neurophysiol.* **65(4)** (1991): 874–890.
10. Bell, A. J. "Self-Organization in Real Neurons: Anti-Hebb in 'Channel Space'." In *Advances in Neural Information Processing Systems*, edited by J. E. Moody, S. J. Hanson, and S. P. Lippmann, Vol. 4, 59–66. San Mateo, CA: Morgan Kaufmann, 1992.
11. Benardo, L. S., L. M. Masukawa, and D. A. Prince. "Electrophysiology of Isolated Hippocampal Pyramidal Dendrites." *J. Neurosci.* **2** (1982): 1614–1622.
12. Bernander, O., R. Douglas, K. Martin, and C. Koch. "Synaptic Background Activity Influences Spatiotemporal Integration in Single Pyramidal Cells." *Proc. Natl. Acad. Sci. USA* **88** (1991): 11569–11573.
13. Bernander, O., C. Koch, and M. Usher. "The Effect of Synchronized Inputs at the Single Neuron Level." (1993): Submitted.
14. Bialek, W., F. Rieke, R. R. de Ruyter van Steveninck, and D. Warland. "Reading a Neural Code." *Science* **252** (1991): 1854–1857.
15. Bishop, G. H. "The Dendrite; Receptive Pole of the Neurone." *Clin. Neurophysiol. Suppl.* **10** (1958): 12–21.
16. Borg-Graham, L., and N. M. Grzywacz. "An Isolated Turtle Retina Preparation Allowing Direct Approach to Ganglion Cells and Photoreceptors, and Transmitted-Light Microscopy." *Investigative Opthalmology & Visual Science* **31** (1990): 1039.
17. Borg-Graham, L. J. "Modelling the Somatic Electrical Response of Hippocampal Pyramidal Neurons." Master's thesis, MIT, 1987.
18. Borg-Graham, L. J. "Modelling the Non-Linear Conductances of Excitable Membranes." In *Cellular and Molecular Biology: A Practical Approach*, edited by H. Wheal and J. Chad. Oxford: Oxford University/IRL Press.
19. Borg-Graham, L. J., and N. Grzywacz. "A Model of the Direction Selectivity Circuit in Retina: Transformations by Neurons Singly and in Concert."

In *Single Neuron Computation*, edited by T. McKenna, J. Davis, and S. F. Zornetzer. Cambridge, MA: Academic Press, 1992.

20. Brazier, M. A. B. The Historical Development of Neurophysiology." In *Handbook of Physiology*, edited by J. Field, H. W. Magoun, and V. E. Hall, Sec. 1 *Neurophysiology*, Vol. 1, 1–58. Washington, DC: American Physiological Society, 1959.
21. Brown, T. H., V. C. Chang, A. H. Ganong, C. L. Keenan, and S. R. Kelso. "Biophysical Properties of Dendrites and Spines That May Control the Induction and Expression of Long-Term Synaptic Potentiation." In *Long-Term Potentiation: From Biophysics to Behavior*, 201–264. New York: Alan R. Liss, 1988.
22. Brown, T. H., Z. F. Mainen, A. M. Zador, and B. J. Claiborne. "Self-Organization of Hebbian Synapses in Hippocampal Neurons." In *Advances in Neural Information Processing Systems*, edited by R. Lippmann, J. Moody, and D. Touretzky, Vol. 3, 39–45. Palo Alto, CA: Morgan Kauffman, 1991.
23. Brown, T. H., A. M. Zador, Z. F. Mainen, and B. J. Claiborne. "Hebbian Modifications in Hippocampal Neurons." In *Long-Term Potentiation: A Debate of Current Issues*, edited by J. Davis and M. Baudry, 357–389. Cambridge, MA: MIT Press, 1991.
24. Brown, T. H., A. M. Zador, Z. F. Mainen, and B. J. Claiborne. "Hebbian Computations in Hippocampal Dendrites and Spines." In *Single Neuron Computation*, edited by T. McKenna, J. Davis, and S. F. Zornetzer, 88–116. Boston, MA: Academic Press, 1992.
25. Butz, E. G., and J. D. Cowan. "Transient Potentials in Dendritic Systems of Arbitrary Geometry." *Biophys. J.* **14** (1974): 661–689.
26. Cauller, L. J., and B. W. Connors. "Functions of Very Distal Dendrites: Experimental and Computational Studies of Layer I Synapses on Neocortical Pyramidal Cells." In *Single Neuron Computation*, edited by T. McKenna, J. Davis, and S. F. Zornetzer, 199–229. Boston, MA: Academic Press, 1992.
27. Chang, H. T. "Cortical Neurons and Spinal Neurons. Cortical Neurons with Particular Reference to Apical Dendrites." *Cold Spring Harbor Symp. Quant. Biol.* **17** (1952): 189–202.
28. Claiborne, B. J., A. M. Zador, Z. F. Mainen, and T. H. Brown. "Computational Models of Hippocampal Neurons." In *Single Neuron Computation* edited by T. McKenna, J. Davis, and S. F. Zornetzer, 61–80. Boston, MA: Academic Press, 1992.
29. Cotman, C. W., D. T. Monaghan, O. P. Ottersen, and J. Storm-Mathisen. "Anatomical Organization of Excitatory Amino Acid Receptors and Their Pathways." *TINS* **10(7)** (1987): 273–280.
30. Cragg, B. G., and L. H. Hamlyn. "Action Potentials of the Pyramidal Neurones in the Hippocampus of the Rabbit." *J. Physiol. London* **130** (1955): 326–373.

31. Cullheim, S., J. W. Fleshman, and R. E. Burke. "Three-Dimensional Architecture of Dendritic Trees in Type-Identified Alpha Motoneurons." *J. Comp. Neurol.* **255** (1987): 82–96.
32. De Schutter, E. "A Consumer Guide to Neuronal Modeling Software." *TINS* **15(11)** (1992): 462–464.
33. Deschenes, M. "Dendritic Spikes Induced in Fast Pyramidal Tract Neurons by Thalamic Stimulation." *Exp. Brain Res.* **43** (1981): 304–308.
34. Durbin, R., and D. E. Rumelhart. "Product Units: A Computationally Powerful and Biologically Plausible Extension to Backpropagation Networks." *Neural Comp.* **1** (1990): 133–142.
35. Eccles, J. C. *The Physiology of Nerve Cells.* Baltimore: The Johns Hopkins Press, 1957.
36. Eccles, J. C., B. Libet, and R. R. Young. "The Behavior of Chromatolysed Motorneurones Studied by Intracellular Recording." *J. Physiol.* **143** (1958): 11–40.
37. Esterle, T. M., and E. Sandersbush. "From Neurotransmitter to Gene—Identifying the Missing Links." *Trends Phar.* **12(10)** (1991): 375–379.
38. Evans, J. D., G. C. Kember, and G. Major. "Techniques for Obtaining Analytical Solutions to the Multicylinder Somatic Shunt Cable Model for Passive Neurons." *Biophys. J.* **63** (1992): 350–365.
39. Feldman, J. A, and D. H. Ballard. "Connectionist Models and Their Properties." *Cog. Sci.* **6** (1982): 205–254.
40. Fox, C. A., and J. W. Barnard. "A Quantitative Study of the Purkinje Cell Dendritic Branchlets and Their Relationship to Afferent Fibers." *J. Anat. (London)* **91** (1957): 299–313.
41. Fox, K., H. Sato, and N. Daw. "The Effect of Varying Stimulus Intensity on NMDA-Receptor Activity in Cat Visual Cortex." *J. Neurophysiol.* **64** (1990): 1413–1428.
42. Fujita, Y. "Morphological and Physiological Properties of Neurons and Glial Cells in Tissue Culture." *J. Neurophysiol.* **31** (1968): 131–141.
43. Golgi, C. *Sulla Fina Anatomia degli Organi Centrali del Sistema Nervoso.* Milan: Hoepli, 1886.
44. Gotch, F. "The Sub-Maximal Electrical Response of Nerve to a Single Stimulus." *J. Physiol.* **28** (1902): 395.
45. Greenough, W. T., and C. H. Bailey. "The Anatomy of A Memory: Convergence of Results Across a Diversity of Tests." *TINS* **11** (1988): 142–147.
46. Greer, C. A. "Golgi Analyses of Dendritic Organization Among Denervated Olfactory Bulb Granule Cells." *J. Comp. Neurol.* **257** (1987): 442–452.
47. Grundfest, H. "Electrical Inexcitability of Synapses and Some Consequences in the Central Nervous System." *Physiol. Rev.* **37** (1957): 337–361.
48. Haag, J., M. Egelhaaf, and A. Borst. "Dendritic Integration of Motion Information in Visual Interneurons of the Blowfly." *Neurosci. Letters* **140** (1992): 173–176.

49. Harris, R. M. "Morphology of Physiologically Identified Thalamocortical Relay Neurons in the Rat Ventrobasal Thalamus." *J. Comp. Neurol.* **254** (1986): 382–402.
50. Herreras, O. "Propagating Dendritic Action Potential Mediates Synaptic Transmission in CA1 Pyramidal Cells *in situ.*" *J. Neurophysiol.* **64** (1990): 1429–1441.
51. Hild, W., and I. Tasaki. "Morphological and Physiological Properties of Neurons and Glial Cells in Tissue Culture." *J. Neurophysiol.* **25** (1962): 277–304.
52. Hille, B. *Ionic Channels of Excitable Membranes*, 1st ed. Sunderland, MA: Sinauer Associates, 1984
53. Holmes, W. R., and W. B. Levy. "Insights into Associative Long-Term Potentiation from Computational Models of NMDA Receptor-Mediated Calcium Influx and Intracellular Calcium Concentration Changes." *J. Neurophysiol.* **63** (1990): 1148–1168.
54. Holmes, W. R., and C. D. Woody. "Effects of Uniform and Non-Uniform Synaptic 'Activation-Distributions' on the Cable Properties of Modeled Cortical Pyramidal Neurons." *Brain Res.* **505** (1989): 12–22.
55. Horwitz, B. "An Analytical Method for Investigating Transient Potentials in Neurons with Branching Dendritic Trees." *Biophys. J.* **36** (1981): 155–192.
56. Horwitz, B. "Unequal Diameters and Their Effects on Time-Varying Voltages in Branched Neurons." *Biophys. J.* **41** (1983): 51–66.
57. Hounsgaard, J., and J. Midtgaard. "Intrinsic Determinants of Firing Pattern in Purkinje Cells of the Turtle Cerebellum *in vitro.*" *J. Physiol.* **402** (1988): 731–749.
58. Huguenard, J. R., O. P. Hamill, and D. A. Prince. "Sodium Channels in Dendrites of Rat Cortical Pyramidal Neurons." *Proc. Natl. Acad. Sci. USA* **86** (1989): 2473–2477.
59. Jack, J. J. B., D. Noble, and R. W. Tsien. *Electric Current Flow in Excitable Cells.* Oxford: Oxford University Press, 1975
60. Jaffe, D. B., D. Johnston, N. Lasser-Ross, J. E. Lisman, H. Miyakawa, and W. N. Ross. "The Spread of Na^+ Spikes Determines the Pattern of Dendritic Ca^{2+} Entry into Hippocampal Neurons." *Nature* **357** (1992): 244–246.
61. Jahnsen, H., and R. Llinás. "Ionic Basis for the Electroresponsiveness and Oscillatory Properties of Guinea-Pig Thalamic Neurons *in vitro.*" *J. Physiol.* **349** (1984): 227–247.
62. Jones, K. A., and R. W. Baughman. "NMDA- and Non-NMDA-Receptor Components of Excitatory Synaptic Potentials Recorded from Cells in Layer V of Rat Visual Cortex." *J. Neurosci.* **8** (1988): 3522–3534.
63. Jones, O. T., D. L. Kunze, and K. J. Angelides. "Localization and Mobility of ω-Conotoxin-Sensitive ca++ Channels in Hippocampal ca1 Neurons." *Science* **244** (1989): 1189–1193.
64. Kandel, E. R., and J. H. Schwartz. *Principles of Neural Science*, 2nd ed. New York: Elsevier, 1985.

65. Keller, B. U., A. Konnerth, and Y. Yaari. "Patch Clamp Analysis of Excitatory Synaptic Currents in Granule Cells of Rat Hippocampus." *J. Physiol. London* **435** (1991): 275–293.
66. Kelvin, W. T. "On the Theory of the Electric Telegraph." *Proc. Roy. Soc.* **7** (1855): 382–399.
67. Koch, C. "Cable Theory in Neurons with Active, Linearized Membranes." *Biol. Cybern.* **50** (1984): 15–33.
68. Koch, C., and T. Poggio. "A Theoretical Analysis of Electrical Properties of Spines." *Proc. R. Soc. Lond. B* **218** (1983): 455–477.
69. Koch, C., and T. Poggio. "Biophysics of Computation: Neurons, Synapses, and Membranes." In *Synaptic Function*, edited by G. E. Edelman, W. F. Gall, and W. M. Cowan, 637–697. New York: Wiley, 1987.
70. Koch, C., and T. Poggio. "Multiplying with Synapses and Neurons." In *Single Neuron Computation*, edited by T. McKenna, J. L. Davis, and S. F. Zornetzer, 315–345. Cambridge, MA: Academic Press, 1992.
71. Koch, C., T. Poggio, and V. Torre. "Retinal Ganglion Cells: A Functional Interpretation of Dendritic Morphology." *Phil. Trans. R. Soc. Lond. B* **298** (1982): 227–264.
72. Koch, C., T. Poggio, and V. Torre. "Nonlinear Interaction in a Dendritic Tree: Localization, Timing and Role of Information Processing." *Proc. Natl. Acad. Sci. USA* **80** (1983): 2799–2802.
73. Koch, C., T. Poggio, and V. Torre. "Computations in the Vertebrate Retina: Gain Enhancement, Differentiation and Motion Discrimination." *TINS* **May** (1986): 204–211.
74. Koch, C., A. Zador, and T. H. Brown. "Dendritic Spines: Convergence of Theory and Experiment." *Science* **156** (1992): 973–974.
75. Kuno, M., and R. Llinás. "Enhancement of Synaptic Transmission by Dendritic Potentials in Chromatolysed Motorneurones of the Cat." *J. Physiol.* **210** (1970): 807–821.
76. Laurent, G. "Voltage-Dependent Nonlinearities in the Membrane of Locust Nonspiking Local Interneurons, and Their Significance for Synaptic Integration." *J. Neurosci.* **10** (1990): 2268–2280.
77. Laurent, G., and M. Burrows. "Intersegmental Interneurons Can Control the Gain of Reflexes in Adjacent Segments of the Locust by Their Action on Nonspiking Local Interneurons." *J. Neurosci.* **9** (1989): 3030–3039.
78. Laurent, G., and Haiyun. "A Modeling Study of Voltage-Dependent Integration of Synaptic Potentials by Locust Non-Spiking Local Neurons." Western Nerve-Net Conference, Seattle, WA, 1993.
79. LeMasson, G., E. Marder, and L. F. Abbott. "Activity-Dependent Regulation of Conductances in Model Neurons." *Science* (1993): in press.
80. Llinás, R., S. A. Greenfield, and H. Jahnsen. "Electrophysiology of *pars compacta* Cells in the *in vitro* Substantia Nigra—A Possible Mechanism for Dendritic Release." *Brain Research* **294** (1984): 127–132.

81. Llinás, R, and R. Hess. "Tetrodotoxin-Resistant Dendritic Spikes in Avian Purkinje Cells." *Proc. Natl. Acad. Sci. USA* **73** (1976): 2520–2523.
82. Llinás, R., and C. Nicholson. "Electrophysiological Properties of Dendrites and Somata in Alligator Purkinje Cells." *J. Neurophysiol.* **34** (1971): 534–551.
83. Llinás, R., C. Nicholson, J. A. Freeman, and D. E. Hillman. "Dendritic Spikes and Their Inhibition in Alligator Purkinje Cells." *Science* **160** (1968): 1132–1135.
84. Llinás, R., and M. Sugimori. "Electrophysiological Properties of *in vitro* Purkinje Cell Dendrites in Mammalian Cerebellar Slices." *J. Physiology (London)* **305** (1980): 197–213.
85. Llinás, R., and Y. Yarom. "Properties and Distribution of Ionic Conductances Generating Electroresponsiveness of Mammalian Inferior Olivary Neurones *in vitro*." *J. Physiol.* **315** (1981): 560–584.
86. Llinás, R. R., and K. D. Walton. "Cerebellum." In *The Synaptic Organization of the Brain*, edited by G. M. Shepherd, 214–245. Oxford: Oxford University Press, 1990.
87. Lorente de Nó, R. "Studies on the Structure of the Cerebral Cortex. II. Continuation of the Study of the Ammonic System." *J. Psychol. Neurol. Leipzig.* **46** (1934): 113–177.
88. Lorente de Nó, R., and G. A. Condouris. "Decremental Conduction in Peripheral Nerve. Integration of Stimuli in the Neuron." *Proc. Natl. Acad. Sci. USA* **45** (1959): 592–617.
89. Lytton, W. W., and T. J. Sejnowski. "Simulations of Cortical Pyramidal Neurons Synchronized by Inhibitory Interneurons." *J. Neurophysiol.* **66** (1991): 1059–1079.
90. Maekawa, K., and D. P. Purpura. "Properties of Spontaneous and Evoked Synaptic Activities of Thalamic Ventrobasal Neurons." *J. Neurophysiol.* **30** (1967): 360–381.
91. Major, G., J. D. Evans, and J. J. B. Jack. "Solutions for Transients in Arbitrarily Branching Cables: I. Voltage Recording with a Somatic Shunt." *Biophys. J.* **65** (1993): 423–449.
92. Mascagni, M. V. "Numerical Methods for Neuronal Modeling." In *Methods in Neuronal Modeling*, edited by C. Koch and I. Segev, 439–484. Cambridge, MA: Bradford, 1989.
93. Maslim, J., M. Webster, and J. Stone. "Stages in the Structural Differentiation of Retinal Ganglion Cells." *J. Comp. Neurol.* **254** (1986): 382–402.
94. Masukawa, L. M., and D. A. Prince. "Synaptic Control of Excitability in Isolated Dendrites of Hippocampal Neurons." *J. Neurosci.* **4** (1984): 217–227.
95. Mayer, M. L., and G. L. Westbrook. "The Physiology of Excitatory Amino Acids in the Vertebrate Central Nervous System." *Prog. Neurobiol.* **28** (1987): 197–276.

96. McClurkin, J. W., L. M. Optican, B. J. Richmond, and T. J. Gawne. "Concurrent Processing and Complexity of Temporally Encoded Neuronal Messages in Visual Perception." *Science* **253** (1991): 675–677.
97. McCormick, D. A. "Membrane Properties and Neurotransmitter Actions." In *The Synaptic Organization of the Brain*, edited by G. M. Shepherd, 32–66. Oxford: Oxford University Press, 1990.
98. Meek, J., and R. Nieuwenhuys. "Palisade Apattern of Mormyrid Purkinje Cells—A Correlated Light and Electron-Microscopic Study." *J. Comp. Neurol.* **306** (1991): 156–192.
99. Mel, B. W. "The Sigma-Pi Column: A Model of Associative Learning in Cerebral Neocortex." *CNS Memo 6*, Computation and Neural Systems Program, Caltech, 1990.
100. Mel, B. W. "The Clusteron: Toward a Simple Abstraction for a Complex Neuron." In *Advances in Neural Information Processing Systems*, edited by J. Moody, S. Hanson, and R. Lippmann, Vol. 4, 35–42. San Mateo, CA: Morgan Kaufmann, 1992.
101. Mel, B. W. "NMDA-Based Pattern Discrimination in a Modeled Cortical Neuron." *Neural Comp.* **4** (1992): 502–516.
102. Mel, B. W. "Information Processing in an Excitable Dendritic Tree." *CNS Memo 17*, Computation and Neural Systems Program, California Institute of Technology, 1–69, 1992.
103. Mel, B. W. "Memory Capacity of an Excitable Dendritic Tree." (1993): In revision.
104. Mel, B. W. "Synaptic Integration in an Excitable Dendritic Tree." *J. Neurophysiol.* **70(3)** (1993): 1086–1101.
105. Mel, B. W., and C. Koch. "Sigma-Pi Learning: On Radial Basis Functions and Cortical Associative Learning." In *Advances in Neural Information Processing Systems*, edited by D. S. Touretzsky, Vol. 2, 474–481. San Mateo, CA: Morgan Kaufmann, 1990.
106. Miller, J. P., W. Rall, and J. Rinzel. "Synaptic Amplification by Active Membrane in Dendritic Spines." *Brain Res.* **325** (1985): 325–330.
107. Miller, K. D., B. Chapman, and M. P. Stryker. "Visual Responses in Adult Cat Visual Cortex Depend on N-Methyl-D-Aspartate Receptors." *Proc. Natl. Acad. Sci. USA* **86** (1989): 5183–5187.
108. Nicoll, R. A., R. C. Malenka, and J. A. Kauer. "Functional Comparison of Neurotransmitter Receptor Subtypes in Mammalian Central-Nervous-System." *Physiol. Rev.* **70** (1990): 513–565.
109. Nowlan, S. J., and T. J. Sejnowski. "Filter Selection Model for Generating Visual Motion Signals." In *Advances in Neural Information Processing Systems*, edited by S. Hanson, J. Cowan, and L. Giles, Vol. 5, 369–376. San Mateo: Morgan Kaufmann, 1993.
110. Ohzawa, I., G. C. DeAngelis, and R. D. Freeman. "Stereoscopic Depth Discrimination in the Visual Cortex: Neurons Ideally Suited as Disparity Detectors." *Science* **279** (1990): 1037–1041.

111. Pearlmutter, B. A. "Hebbian Self-Organization is Jointly Controlled by Passive and Input Structure." *Neural Comp.* (in press).
112. Penny, G. R., C. J. Wilson, and S. T. Kitai. "Relationship of the Axonal and Dendritic Geometry of Spiny Projection Neurons to the Compartmental Organization of the Neostriatum." *J. Comp. Neurol.* **269** (1988): 275–289.
113. Perkel, D. H., B. Mulloney, and R. W. Budelli. "Quantitative Methods for Predicting Neuronal Behavior." *Neurosci.* **6** (1981): 823–837.
114. Perkel, D. H., and D. J. Perkel. "Dendritic Spines: Role of Active Membrane in Modulating Synaptic Efficacy." *Brain Res.* **325** (1985): 331–335.
115. Peterhans, E., and R. von der Heydt. "Mechanisms of Contour Perception in Monkey Visual Cortex. II. Contours Bridging Gaps." *J. Neurosci.* **9** (1989): 1749–1763.
116. Pockberger, H. "Electrophysiological and Morphological Properties of Rat Motor Cortex Neurons *in vivo*." *Brain Research* **539** (1991): 181–190.
117. Poggio, T., and F. Girosi. "Regularization Algorithms for Learning that are Equivalent to Multilayer Networks." *Science* **247** (1990): 978–982.
118. Poggio, T., and V. Torre. "A New Approach to Synaptic Interactions." In *Lecture notes in Biomathematics. Theoretical Approaches to Computer Systems*, edited by H. Heim and G. Palm, Vol. 21, 89–115. Berlin: Springer-Verlag, 1977.
119. Poolos, N. P., and J. D. Kocsis. "Dendritic Action Potentials Activated by NMDA Receptor-Mediated EPSPs in CA1 Hippocampal Pyramidal Cells." *Brain Res.* **524** (1990): 342–346.
120. Popov, S., and M.-M. Poo. "Diffusional Transport of Macromolecules in Developing Nerve Processes." *J. Neurosci.* **12(1)** (1992): 77–85.
121. Purpura, D. P. "Nature of Electrocortical Potentials and Synaptic Organizations in Cerebral and Cerebellar Cortex." *Intern. Rev. Neurobiol.* **1** (1959): 47–163.
122. Purpura, D. P. "Comparative Physiology of Dendrites." In *The Neurosciences. A Study Program*, edited by G. C. Quarton, T. Nelnechuk, and F. O. Schmitt, 373–392. New York: Rockefeller University Press, 1967.
123. Purpura, D. P., J. G. McMurtry, C. F. Leonard, and A. Malliani. "Evidence for Dendritic Origin of Spikes Without Depolarizing Prepotentials in Hippocampal Neurons During and After Seizure." *J. Neurophysiol.* **29** (1966): 954–979.
124. Purpura, D. P., and R. J. Shofer. "Cortical Intracellular Potentials During Augmenting and Recruiting Responses. I. Effects of Injected Hyperpolarizing Currents on Evoked Membrane Potential Changes." *J. Neurophysiol.* **27** (1964): 117–132.
125. Rall, W. "Branching Dendritic Trees and Motoneuron Membrane Resistivity." *Exp. Neurol.* **1** (1959): 491–527.
126. Rall, W. "Theoretical Significance of Dendritic Trees for Neuronal Input-Ouput Relations." In *Neural Theory and Modeling*, edited by R. F. Reiss. Stanford, CA: Stanford University Press, 1964.

127. Rall, W. "Core Conductor Theory and Cable Properties of Neurons." In *Handbook of Physiology: The Nervous System*, edited by E. R. Kandel, J. M. Brookhardt, and V. B. Mountcastle, Vol. 1, 39–98. Baltimore, MD: Williams and Wilkins, 1977.
128. Rall, W. "Cable Theory for Dendritic Neurons." In *Methods in Neuronal Modeling*, edited by C. Koch and I. Segev, ch. 2. Cambridge, MA: MIT Press, 1989.
129. Rall, W., R. E. Burke, W. R. Holmes, J. J. B. Jack, S. J. Redman, and I. Segev. "Matching Dendritic Neuron Models to Experimental Data." *Physiological Rev.* **72(4)** (1992): S159–S186.
130. Rall, W., and J. Rinzel. "Branch Input Resistance and Steady Attenuation for Input to One Branch of a Dendritic Neuron Model." *Biophys. J.* **13** (1973): 648–688.
131. Rall, W., and I. Segev. "Functional Possibilities for Synapses on Dendrites and on Dendritic Spines." In *Synaptic Function*, edited by G. E. Edelman, W. F. Gall, and W. M. Cowan, 605–636. New York: Wiley, 1987.
132. Ramòn y Cajal, S. *Histologie du système nerveux de l'homme et des vertébrés, translated by L. Azoulay.* Paris: Malaine, 1909.
133. Rapp, M., Y. Yarom, and I. Segev. "The Impact of Parallel Fiber Background Activity on the Cable Properties of Cerebellar Purkinje Cells." *Neural Comp.* **4** (1992): 518–533.
134. Reuveni, I., A. Friedman, Y. Amitai, and M. J. Gutnick. "Stepwise Repolarization From ca^{2+} Plateaus in Neocortical Pyramidal Cells: Evidence for Non-Homogeneous Distribution of hva ca^{2+} Channels in Dendrites." *J. Neurosci.* **13** (1993): 4609–4621.
135. Rinzel, J., and W. Rall. "Transient Response in a Dendritic Neuron Model for Current Injected at One Branch." *Biophysics J.* **14** (1974): 759–789.
136. Ross, W. N., N. Lasser-Ross, and R. Werman. "Spatial and Temporal Analysis of Calcium-Dependent Electrical Activity in Guinea Pig Purkinje Cell Dendrites." *Proc. R. Soc. Lond. B* **240** (1990: 173–185.
137. Salt, T. E. "Mediation of Thalamic Sensory Input by Both NMDA Receptors and Non-NMDA Receptors." *Nature* **322** (1986): 263–265.
138. Schwartzkroin, P. A., and D. A. Prince. "Changes in Excitatory and Inhibitory Synaptic Potentials Leading to Epileptogenic Activity." *Brain Res.* **183** (1980): 61–76.
139. Schwartzkroin, P. A., and M. Slawsky. "Probable Calcium Spikes in Hippocampal Neurons." *Brain Res.* **135** (1977): 157–161.
140. Segev, I. "Single Neurone Models: Oversimple, Complex and Reduced." *TINS* **15** (1992): 414–421.
141. Segev, I., J. W. Fleshman, and R. E. Burke. "Compartmental Models of Complex NNeurons." In *Methods in Neuronal Modeling*, edited by C. Koch and I. Segev, 63–96. Cambridge, MA: MIT Press, 1989.
142. Segev, I., and W. Rall. "Computational Study of an Excitable Dendritic Spine." *J. Neurophysiol.* **60** (1988): 499–523.

143. Sereno, M. I., and P. S. Ulinski. "Caudal Topographic Nucleus Isthmi and the Rostral Nontopographic Nucleus Isthmi in the Turtle, *Pseudemys Scripta*." *J. Comp. Neurol.* **261** (1987): 319–346.
144. Sernagor, E., Y. Yarom, and R. Werman. "Sodium-Dependent Regenerative Responses in Dendrites of Axotomized Motorneurons in the Cat." *Proc. Natl. Acad. Sci. USA* **83** (1986): 7966–7970.
145. Shepherd, G. "Canonical Neurons and Their Computational Organization." In *Single Neuron Computation*, edited by T. McKenna, J. Davis, and S. F. Zornetzer, 27–60. Boston, MA: Academic Press, 1992.
146. Shepherd, G. M. *The Synaptic Organization of the Brain.* Oxford: Oxford University Press, 1990.
147. Shepherd, G. M., and R. K. Brayton. "Logic Operations are Properties of Computer-Simulated Interactions Between Excitable Dendritic Spines." *Neurosci.* **21** (1987): 151–166.
148. Shepherd, G. M., R. K. Brayton, J. P. Miller, I. Segev, J. Rinzel, and W. Rall. "Signal Enhancement in Distal Cortical Dendrites by Means of Interactions Between Active Dendritic Spines." *Proc. Natl. Acad. Sci. USA* **82** (1985): 2192–2195.
149. Shepherd, G. M., and C. A. Greer. "The Dendritic Spine: Adaptation of Structure and Function for Different Types of Synaptic Integration." In *Intrinsic Determinants of Neuronal form and Function*, edited by R. Lasek and M. Black, 245–262. New York: Alan R. Liss, 1988.
150. Shepherd, G. M., and C. Koch. "Dendritic Electrotonus and Synaptic Integration." In *The Synaptic Organization of the Brain*, edited by G. M. Shepherd, 439–473. Oxford: Oxford University Press, 1990.
151. Shepherd, G. M., T. B. Woolf, and N. T. Carnevale. "Comparisons Between Active Properties of Distal Dendritic Branches and Spines: Implications for Neuronal Computations." *Cognitive Neurosci.* **1** (1989): 273–286.
152. Sirevaag, A. M., and W. T. Greenough. "Differential Rearing Effects on Rat Visual Cortex Synapses. III. Neuronal and Glial Nuclei, Boutons, Dendrites, and Capillaries." *Brain Research* **424** (1987): 320–332.
153. Softky, W., and C. Koch. "The Highly Irregular Firing of Cortical Cells is Inconsistent With Temporal Integration of Random EPSP's." *J. Neurosci.* **13** (1993): 334–350.
154. Softky, W. R. "Sub-Millisecond Coincidence Detection in Active Dendritic Trees." (1993): In preparation.
155. Spear, P. J. "Evidence for Spike Propagation in Cortical Dendrites." *Exp. Neurol.* **35** (1972): 111–121.
156. Spencer, W. A. "The Physiology of Supraspinal Neurons in Mammals." In *Handbook of Physiology, Sec. 1 The Nervous System*, edited by J. M. Brookhart, V. B. Mountcastle, and E. R. Kandel, Vol. 1, 969–1021. Bethesda, MD: American Physiological Society, 1977.
157. Spencer, W. A., and E. R. Kandel. "Electrophysiology of Hippocampal Neurons. iv. Fast Prepotentials." *J. Neurophysiol.* **24** (1961): 272–285.

158. Stafstrom, C. E., P. V. Schwindt, M. C. Chubb, and W. E. Crill. "Properties of Persistent Sodium Conductance and Calcium Conductance of Layer V Neurons from Cat Sensorimotor Cortex *in vitro*." *J. Neurophysiol.* **53** (1985): 153–170.
159. Sugimori, M., and R. R. Llinás. "Real-Time Imaging of Calcium Influx in Mammalian Cerebellar Purkinje Cells *in vitro*." *Proc. Natl. Acad. Sci. USA* **87** (1990): 5084–5088.
160. Sutor, B., and J. J. Hablitz. "EPSPs in Rat Neocortical Neurons *in vitro*. II. Involvement of N-methyl-D-aspartate Receptors in the Generation of EPSPs." *J. Neurophysiol.* **61(3)** (1989): 621–634.
161. Tank, E. W., M. Sugimori, J. A. Connor, and R. Llinás. "Spatially Resolved Calcium Dynamics of Mammalian Purkinje Cells in Cerebellar Slice." *Science* **242** (1988): 773–777.
162. Terashima, T., K. Inoue, Y. Inoue, M. Yokoyama, and K. Mikoshiba. "Observations on the Cerebellum of Normal-Reeler Mutant Mouse Chimera." *J. Comp. Neurol.* **252** (1986): 264–278.
163. Theunissen, F. E., and J. P. Miller. "Representation of Sensory Information in the Cricket Cercal Sensory System. II. Information Theoretic Calculation of System Accuracy and Optimal Tuning-Curve Widths of Four Primary Interneurons." *J. Neurophysiol.* **66** (1991): 1690–1703.
164. Traub, R. D. "Simulation of Intrinsic Bursting in CA3 Hippocampal Neurons." *Neurosci.* **7** (1982): 1233–1242.
165. Traub, R. D., F. E. Dudek, C. P. Taylor, and W. D. Knowles. "Simulation of Hippocampal Afterdischarges Synchronized by Electrical Interactions." *J. Neurosci.* **4** (1985): 1033–1038.
166. Traub, R. D., and R. Llinás. "Hippocampal Pyramidal Cells: Significance of Dendritic Ionic Conductances for Neuronal Function and Epileptogenesis." *J. Neurophysiol.* **42** (1979): 476–496.
167. Traub, R. D., R. K. S. Wong, R. Miles, and H. Michelson. "A Model of a CA3 Hippocampal Pyramidal Neuron Incorporating Voltage-Clamp Data on Intrinsic Conductances." *J. Neurophysiol.* **66** (1991): 635–650.
168. Turner, R. W., and T. L. Richardson. "Apical Dendritic Depolarizations and Field Interactions Evoked by Stimulation of Afferent Inputs to Rat Hippocampal CA1 Pyramidal Cells." *Neurosci.* **42** (1991): 125–135.
169. Vaney, D. I., S. P. Collin, and H. M. Young. "Dendritic Relationships Between Cholinergic Amacrine Cells and Direction-Selective Retinal Ganglion Cells." In *Neurobiology of the Inner Retina*, edited by R. Weiler and N. N. Osborne , 157–168. Berlin: Springer-Verlag, 1989.
170. von der Heydt, R., E. Peterhans, and M. R. Dursteler. "Grating Cells in Monkey Visual Cortex: Coding Texture?" In *Channels in the Visual Nervous System: Neurophysiology, Psychophysics, and Models*, edited by B. Blum, 53–73. London: Freund, 1991.
171. Waterhouse, B. D, F. M. Sessler, W. Liu, and C. S. Lin. "2nd Messenger-Mediatedactions of Norepinephrine on Target Neurons in Central Circuits—A

New Perspective on Intracellular Mechanisms and Functional Consequences." *Prog. Brain Res.* **88** (1991): 351–362.

172. Watson, A. H. D., and M. Burrows. "Distribution and Morphology of Synapses on Nonspiking Local Iinterneurones in the Thoracid Nervous System of the Locust." *J. Comp. Neurol.* **272** (1988): 605–616.
173. Wilson, C. J. "Dendritic Morphology, Inward Rectification and the Functional Properties of Neostriatal Neurons." In *Single Neuron Computation*, edited by T. McKenna, J. L. Davis, and S. F. Zornetzer. Cambridge, MA: Academic Press, 1992.
174. Wong-Riley, M. T. T. "Cytochrome Oxidase: An Endogenous Metabolic Marker for Neuronal Activity." *Trends in Neurociences* **12(3)** (1989): 94–101.
175. Wong, R. K. S., D. A. Prince, and A. I. Busbaum. "Intradendritic Recordings from Hippocampal Neurons." *Proc. Natl. Acad. Sci. USA* **76** (1979): 986–990.
176. Wong, R. K. S., and M. Stewart. "Different Firing Patterns Generated in Dendrites and Somata of CA1 Pyramidal Neurones in Guinea-Pig Hippocampus." *J. Physiol.* **457** (1992): 675–687.
177. Woolf, T. B., G. M. Shepherd, and C. A. Greer. "Local Information Processing in Dendritic Trees: Subsets of Spines in Granule Cells of the Mammalian Olfactory Bulb." *J. Neurosci.* **11** (1991): 1837–1854.
178. Yang, C.-Y., and S. Yazulla. "Neuropeptide-like Immunoreactive Cells in the Retina of the Larval Tiger Salamander: Attention to the Symmetry of Dendritic Projections." *J. Comp. Neurol.* **248** (1986): 105–118.
179. Zador, A., C. Koch, and T. H. Brown. "Biophysical Model of a Hebbian Synapse." *Proc. Natl. Acad. Sci. USA* **87** (1990): 6718–6721.
180. Zador, A. M. "Biophysics of Computation in Single Hippocampal Neurons." Ph.D. Thesis, Yale University, Interdepartmental Neuroscience Program, 1993
181. Zador, A. M., B. J. Claiborne, and T. J. Brown. "Nonlinear Pattern Separation in Single Hippocampal Neurons with Active Dendritic Membrane." In *Advances in Neural Information Processing Systems*, edited by J. Moody, S. Hanson, and R. Lippmann, Vol. 4, 51–58. San Mateo: Morgan Kaufmann, 1992.

B. Kean Sawhill
Santa Fe Institute, 1399 Hyde Park Road, Santa Fe, NM 87501

Self-Organized Criticality and Complexity Theory

WHAT IS SELF-ORGANIZED CRITICALITY?

INTRODUCTION

Self-organized criticality (abbreviated as SOC from here on) describes a large and varied body of phenomenological data and theoretical work. As is the case with many such catch-all terms, its current meaning has evolved far from any initial precise meaning the term may have had. This chapter will attempt to elucidate the basic concepts behind SOC, describe some of the relevant experimental and theoretical findings, and indicate possible future research directions.

The functional focus that I will choose to adopt is the understanding and description of broad classes of pattern formation in nature. The original motivation for SOC came from recursive mathematics. Iterative maps produced fascinating images—*fractals*—which are now in common usage in popular culture.[6,19,35,29] Some of them produced patterns which looked tantalizingly similar to patterns

1993 Lectures in Complex Systems, Eds. L. Nadel and D. L. Stein,
SFI Studies in the Sciences of Complexity, Lect. Vol. VI, Addison-Wesley, 1995

found in nature, such as river erosion patterns, plant and leaf structure, and geological landscapes. The question then presented itself, *What kind of dynamical mechanism would be required to produce such patterns in nature?* In addition to being able to produce complex patterns, this mechanism would also have to be simple, because if the mechanism were as complex as the pattern produced, very little would be gained by such an "explanation."

In the late 1980s, various researchers created model computational systems that showed SOC behavior, building on the work of Bak, Tang, and Wiesenfeld.[2,3,6,7,16,30,38] These results were particularly compelling because it was possible to generate complex and coherent structures from very simple models. One of the key features of these systems was the capability of producing spatially distributed patterns of a fractal nature without requiring interaction rules which had explicit long-range dependences. One could say that the long-range correlations "emerged" from short-range interaction rules. Another key feature was the fact that these systems converged on configurations which exhibited long-range coherence even if they did not start with such a configuration. To do so did not require the tuning of any external parameter; hence the term "self-organization."

The notion of criticality comes from thermodynamics. In general, thermodynamic systems become more ordered as the temperature is lowered. Cohesion wins out over thermal motion, and structure emerges. Thermodynamic systems can exist in a number of phases ranging from the everyday (solid, liquid, gas, crystal) to the unfamiliar (superfluid, plasma, antiferromagnet, spin glass). A system is said to be critical if it is poised at a phase transition. The most well-understood phase transitions are those that take place under conditions of thermal equilibrium, describing systems in which the temperature is a well-defined quantity and in which the free energy is minimized. These conditions characterize systems that have had enough time to settle into their most probable configurations, thus allowing simplifying statistical assumptions to be made. Many phase transitions have associated with them a *critical point* which separates two or more phases. (More than two phases can meet at a point in a two-dimensional phase diagram whose axes are temperature and pressure.) As a system approaches the critical point from above, it begins to organize itself on a microscopic level. Large structural fluctuations appear, yet they come from local interactions. These fluctuations result in the disappearance of a characteristic scale in the system at the critical point. This behavior is remarkably general and independent of the specific detail of the system's dynamics.[33,34]

The kind of structures which SOC aspires to explain look like equilibrium thermodynamic systems near critical points. Typical SOC systems, however, are not in equilibrium with their surroundings, but instead have nontrivial interactions with them. For instance, water flowing over erodable soil produces erosion fans with a self-similar structure, and once erosion has occurred in one place, it is more likely to continue to occur there rather than somewhere uneroded. This disqualifies conditions of randomization which are essential assumptions for equilibrium thermodynamics, as not every part of the system is interchangeable with every other part. In addition, SOC systems do not require the tuning of an external control parameter such as temperature to exhibit critical behavior, but instead tune themselves to a point at which "critical-type" behavior occurs. This critical behavior is the bridge between fractal structures and thermodynamics. Fractal structures appear the same at all size scales, hence they have no characteristic scale, just as fluctuations near a phase transition have no characteristic scale. It is then logically compelling to seek a dynamical explanation of fractal structures in terms of thermodynamic criticality. A proper understanding of SOC requires an extension of the notion of critical behavior to nonequilibrium thermodynamical systems and hence an extension of the tools required to describe these systems. This chatper will begin to explore a strategy for undertaking this task which involves re-examining the basic assumptions behind statistical thermodynamics and examining the concept of organization by introducing notions from information theory and the theory of computation.

SELF-SIMILARITY IN SPACE AND TIME

A *fractal* is a spatial curve which has no characteristic length scale, and is said to be *self-similar*, in that any subset of the curve magnified by an appropriate amount has a form similar to the whole. A standard example is the Koch, or "snowflake" curve. This curve is constructed by a simple iterative procedure in which an initial triangle is operated up on as shown in Figure 1.

In each succeeding iteration, each perimeter line segment is replaced by a segment in which the middle third consists of an outward pointing equilateral triangle with one of its edges being collinear with the original segment, as shown at the top of the two figures above. In the Koch curve, it can be seen that the number of triangles produced by iteration step k is given by $N(k) = 3 \times 4^{k-1}, k \geq 1$, which is easy to derive by observing that each iteration replaces one edge by four. The triangles produced in each successive iteration are $1/3$ the size of the previous ones. Size and iteration number are then related by $s(k) = (1/3)^k$ and hence the number of triangles are related to their size by

$$N(s) \propto s^{-\left(\frac{\ln 4}{\ln 3}\right)}.$$

Scale invariance occurs for any system in which $N(s)$ has a polynomial or *power-law* dependence on s. If we rescale a system by $s \to as$, where a is a multiplicative constant, we find $N(as) \to CN(s)$ where C is a multiplicative constant and the form of the scaling function is preserved.

Fractals can also exist as a function of time. Consider a time series of data of the form $X(t)$. Mandelbrot defines scaling as follows: A noise $X(t)$ is to be called *scaling* if X itself or its integral or derivative (repeated, if need arises) is *self-affine*. That is, if $X(t)$ is statistically identical to its transform by contraction in time followed by a corresponding change in intensity. Thus, there must exist an exponent $\alpha > 0$ such that for every $h > 0$, $X(t)$ is statistically identical to $h^{-\alpha}X(ht)$. More generally, and especially in case t is discrete, $X(t)$ is to be called asymptotically scaling if there exists a slowly varying function $L(h)$ such that $h^{-\alpha}L^{-1}(h)\,X(ht)$ tends to a limit as $h \to \infty$.

This definition requires that one check every mathematical characteristic of $X(t)$ and $h^{-\alpha}X(ht)$. Thus, scaling can never be proved in empirical science, and in most instances the scaling property is inferred from a single test that is only concerned with one facet of sameness.

The most widely used test of scaling is based on spectra. A noise is spectrally scaling if its measured spectral density at the frequency f is of the form $1/f^{\beta}$ with β a positive exponent. When β is close enough to 1 to justify $1/f^{\beta}$ being abbreviated into $1/f$, one deals with a "$1/f$ noise."[29]

A fractal time series has some special properties. It is not reducible to a periodic signal plus a random noise term of the form $X(t) = A\cos(\omega t) + \eta(t)$. If this were so, the contribution due to the random term would average out as $t \to \infty$ and the signal would acquire a well-defined expectation value for its frequency. Instead, there is a distribution of frequencies $D(f)$. The expectation value of frequency can be calculated:

$$\langle f \rangle = \int D(f)\, f df\,.$$

If $D(f) \propto 1/f$, then there is no characteristic frequency, just as there is no characteristic length scale for fractal geometries.

1/F NOISE AND DYNAMICS

$1/f$ noise is found everywhere in nature. A sampling of the data sets which demonstrate this include:

1. Freeway traffic patterns (fluctuations in the number of vehicles past a certain point per unit time and fluctuations in traffic density (number of vehicles per length of roadway).[31]
2. Fluctuations of river water levels. Large data sets exist for this phenomena. The Nile, for instance, has been accurately documented for over two millenia.[37]
3. Price fluctuations on the stock exchange. This has the advantage of being a well-documented large data set with high resolution.[1,8]
4. Distributions of earthquake magnitudes.[9,4,5,20,30,21] (Phenomenologically fitted by the Gutenberg-Richter Law.)
5. Current fluctuations in a resistor, otherwise known as "shot noise." This is probably the first source of $1/f$ noise for which it was attempted to gain a theoretical understanding. Shot noise becomes significant for very low currents as the discrete nature of electric charge becomes important in relation to its transport properties.[6]
6. Distributions of the size (area and duration) of forest fires.[17,18,19]

It is clear that the physical interaction dynamics of these processes vary so widely that they cannot be ascribed to any single mechanism or force. Instead, there seems to be some underlying "logical dynamics" describing the interrelationship of the degrees of freedom of the system. All of the above data sets appear to be history dependent; i.e., the future evolution of the system depends on more than the current state of the system. Such processes are called non-Markovian.

What is needed is a dynamical mechanism to generate scaling behavior in both space and time in an extended system. By inspection of the above phenomena, these spatially extended dynamical systems seem to have correlated behavior at all size (hence all time and frequency) scales, yet they do not obviously possess mechanisms of interaction at all time and space scales. If anything, the interactions are local: segments of earth in an earthquake zone only react to the earth next to them, electrons in a resistor only significantly interact with neighboring electrons, trees in a forest fire can only catch fire if the ones next to them do, etc. All of the above systems are also dissipative and open and do not obviously conserve a quantity such as energy or a well-defined analogue. In fact, it is easy to imagine in many of the above cases that if the system were to become closed and not have significant interaction with its surroundings, the data set would become uninteresting. (For example, the Nile would dry up.) In addition, none of the systems above have any fine-tuning requirement: There is no externally applied adjustment of the system aimed towards maintaining critical-type behavior.

MODELING SOC BEHAVIOR

BOOLEAN NETS AND COMPUTATION

The "logical dynamics" in the systems described above can best be characterized by distributed information processing. This means a system which is capable of processing, storing, and transmitting information about the states of the elements of which it is composed. The most general computational paradigm which can implement distributed information processing is called a Boolean Net, schematically depicted in Figure 1.

The circles represent logic elements which produce an output (or outputs) as a function of inputs using expressions constructed from Boolean functions (AND, $\overline{OR}$, etc.). The lines are directed communication pathways, and they transmit state information in the form of zeroes and ones. The nodes are capable of memory, as it possible to construct information storage devices (flip flops) using Boolean functions. Hence, the system is capable of information storage, transmission, and processing at an arbitrarily complex level. This is actually more power than we need, as such nets can be constructed with arbitrary topologies, removing the regularity

FIGURE 1

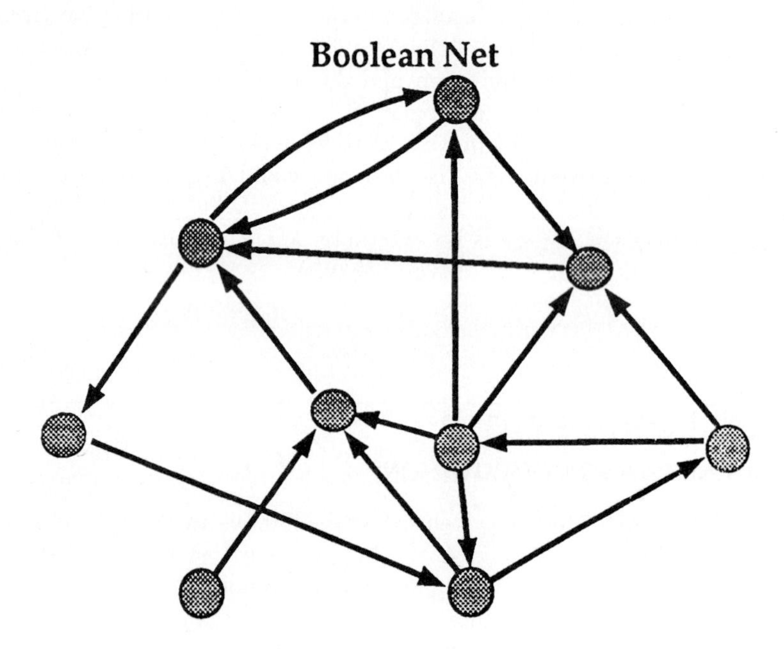

FIGURE 2

imposed by a physically distributed system of the nature that we are interested in. The variety of dynamics that a Boolean Net can generate is so great that it constitutes an over-generalization for our current purposes.

There is an entire hierarchy of computational systems that can be derived from the above general system by imposing a series of nested constraints. Each imposed assumption reduces the generality but also increases the ease of analysis. The challenge is to find the most constrained and regular system which still contains the ingredients necessary to generate SOC. It turns out that no generality is lost by assuming synchronous updating. A significant assumption which affects the set of possible topologies is to require that each logical node have the same number of inputs as every other one. This is called the N-K model, and it is due to Stuart Kauffman, who devised it to study the evolution of fitness landscapes in biological systems.[26] N refers to the number of nodes, and K to the number of inputs per node.

The most important assumption in generating classifiable behavior is to impose the regularity of logical function at each site in addition to imposing uniform connectivity. The system now possesses a *rule* which characterizes the logical dynamics at any node in the system. Uniform connectivity and an overall rule are the essential requirements for a Boolean Net to look like a spatially extended dynamical system. This is because uniform connectivity means that the Boolean Net can be mapped onto a regular physical structure which possesses a well-defined metric. Mapping such a system onto a spatial structure does not always preserve locality, however, hence even stronger constraints are required to allow the imposition of conservation laws.

In sum, we have started with the most general computational paradigm (the Boolean Net) and have imposed the following constraints: synchronicity, uniformity of connectivity, regularity of interaction rules, and locality of connections. The resultant structure looks like a lattice of uniform matter, all of which obeys the same local dynamics. The nodes can refer to a great variety of things, from automobiles on a freeway to trees in a burning forest to landscape elements under rainfall to stock traders in a market. The rules refer to how each node relates to the ones around it, and hence encode information such as when a car should slow down to not hit the one in front of it or when a trader should buy or sell depending on what other traders are doing. The SOC systems described above usually consist of uniform elements, each of which only has information about what is going on nearby; hence it is desirable to attempt to model their behavior using an appropriately constrained model.

This computational paradigm that we have finally arrived at is known as a cellular automaton (CA), and is due to work done in the 1940s at Los Alamos by Stanislaw Ulam and John von Neumann. They were first used in vacuum tube computers to study the behavior of models of coupled masses and springs, resulting in the first computational evidence of chaotic behavior in dynamical systems. Incidentally, this discovery was accidental and unexpected, a characteristic signature of the majority of significant discoveries in science.

CELLULAR AUTOMATA

A cellular automaton consists of a countable array of discrete sites or cells i in d dimensions, operated upon by a local update rule ϕ in parallel on local neighborhoods of a given radius r. At each time step the sites take on values in a finite alphabet A, represented by a set of k possible symbols, conventionally numbered $\{0, \ldots, k-1\}$. The value at each time-space site is denoted by σ_t^i, and the update function is written $\sigma_{t+1}^i = \phi\left(\sigma_t^{i-r}, \ldots, \sigma_t^{i+r}\right)$. The state Σ_t of the CA at time t is the configuration of the entire spatial array $\Sigma_t \in A^N$, where N is the total number of cells in the lattice.

The CA global update rule operates by analogy with the local update rule

$$\Phi : \Sigma_t \rightarrow \Sigma_{t+1}$$

by applying ϕ in parallel to all sites on the lattice. For finite N it is also necessary to specify a boundary condition, and this is of particular importance in SOC models, as we are interested in modeling systems which have open and dissipative boundaries.

A simple example of a CA is a $\{d = 1, r = 1, N = 8\}$ CA. The rule table encodes all of the information necessary to determine the dynamics of the CA. The three binary digits in the left-hand column indicate all of the possible states that the radius 1 neighbors can take, and the right-hand column indicates the value that the central bit takes at the next time step. The rule table and four steps of time evolution are shown below, and periodic boundary conditions are assumed.

In this particular CA, it is evident that the rule table can be filled out 256 ways, as there are 8 slots, each of which can have a value $\sigma \in \{0, 1\}$. In general, a radius r rule with an alphabet of k symbols will have $R = k^{k^{(2\alpha+1)}}$ different possible rule tables. Hence the diversity of possible local interaction rules which can be modeled becomes enormous rapidly.

Cellular Automaton

RULE TABLE	
States(t-1)	State(t)
0 0 0	1
0 0 1	0
0 1 0	0
0 1 1	1
1 0 0	1
1 0 1	0
1 1 0	0
1 1 1	0

t=0
t=1
t=2
t=3
t=4

FIGURE 3

The kinds of systems in which we are interested often possess some sort of local conservation principle, such as conservation of money, of water, of force, or of charge. Local conservation rules can always be expressed as a differential equation (the *equation of continuity*). In general, differential equations can be written as difference equations over a discrete lattice, provided that a continuous limiting process can be demonstrated. These difference equations can be implemented using a CA with the appropriate rule table, with the size of the alphabet corresponding to resolution level and the radius of the rule being a function of the order of the spatial dependence of the original differential equation ($r = \mathcal{O}/\in$). The systems that we would like to model are characterized by having local interactions and transport properties. This defines a class of differential equations known as *reaction-diffusion* equations. These systems are also open-ended, both absorbing and dissipating their salient quantity.[10,4,23,39]

DIFFUSION. A likely candidate for such a system is driven diffusion with open boundaries. This allows for both dissipation and for transport. Diffusion is the macroscopic process resulting from a random walk occurring on a microscopic scale, a relation first shown in Einstein in his work on Brownian motion for which he was awarded the Nobel Prize. The microscopic evolution of probability on a discrete one-dimensional lattice where each site distributes its probability to its two neighboring sites in the next time step is depicted schematically in Figure 4.

The evolution of probability can be written as the following difference equation:

$$P(x_i, t_j) = \frac{1}{2}\left[P(x_{i-1}, t_{j-1}) + P(x_{i+1}, t_{j-1})\right] .$$

If we subtract the quantity $P(x_i, t_{j-1})$ from both sides, we can construct the following difference equation:

$$P(x_i, t_j) - P(x_i, t_{j-1}) = \frac{1}{2}\Big[\left[P(x_{i+1}, t_{j-1}) - P(x_i, t_{j-1})\right] - \left[P(x_i, t_{j-1}) - P(x_{i-1}, t_{j-1})\right]\Big] .$$

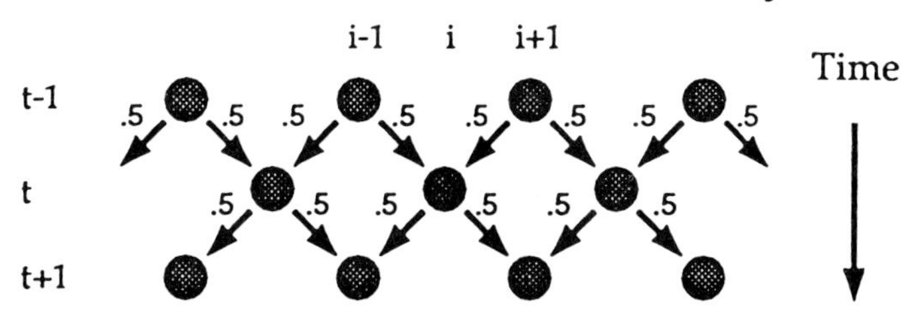

FIGURE 4

If we let $x_{i+1} - x_i = \Delta x$ and $t_i - t_{i-1} = \Delta t$, we now have

$$\frac{\Delta P}{\Delta t} = \frac{1}{2}\frac{\Delta}{\Delta x}\left[\frac{\Delta P}{\Delta x}\right]$$

which, so long as the limiting operation $\Delta \to \partial$ is well defined, results in the homogenous one-dimensional diffusion (heat) equation

$$\frac{\partial P}{\partial t} = D\frac{\partial^2 P}{\partial x^2}$$

where the diffusion constant D has units of L^2T^{-1} and contains the factor $1/2$ in its definition. This is a special case of a general diffusion equation of the form:

$$\frac{\partial u(\vec{x},t)}{\partial t} = \vec{\nabla}\cdot\left[k\vec{\nabla}u(\vec{x},t)\right] + F(\vec{x},t)\ .$$

This equation allows for nonisotropic and inhomogeneous diffusion in an arbitrary number of dimensions, depending on the structure of $k = k(\vec{x},t)$. The F term allows for the presence of sources and sinks.

If k is a constant, the resulting equation is linear and can be solved by the methods of Laplace or Fourier transforms, resulting in behavior which is continuous and stable under perturbations. We are interested in cases where k is not a constant and thus the system will generate nonsuperposable inhomogeneities due to the nonlinearity of the resulting equation. The easiest and least arbitrary way to implement such nonlinearities is to go back to the microscopic roots of the diffusion equation, namely processes which discretely evolve distributions, and postulate a plausible process which could result in nonlinearities. This brings up philosophical questions as to whether nature is fundamentally discrete or continuous, but in this paper we are primarily concerned with networks of well-defined discrete interacting functions.

SANDPILES

One of the most common and obvious models for introducing nonlinearities in a discrete computational model of diffusion is called the "sandpile" model, due to Bak, Tang, and Wiesenfeld.[11,12,15,22,25,27,28,32,38] This makes use of the "stick-slip" phenomenon which is intuitively very familiar. In it, a driving force is applied to a system, which to first approximation, causes no effect until a critical force is reached, whereupon the effect is sudden. Such behavior is manifest in the action of many systems that have friction: Avalanches of snow and sand, critical firing voltages for neurons, earthquakes, even the action of wax on cross-country skis. Yet it is more complex than simple friction: An initial perturbation may cause a large chain reaction, so it is as if the effective friction were reduced once the system was set into motion. A graph of force versus displacement is shown in Figure 5, comparing stick-slip with Hooke's Law, which is the linear model.

This type of model is called a sandpile model because real sand has similar characteristics—it may by piled up until it reaches a critical angle, at which point additional grains of sand may set off large avalanches of sand. To express the force-displacement relation of the stick-slip graph (a Heaviside step function) in terms of continuously differentiable functions suitable for use in a continuum differential equation would result in an infinite series of functions of x of differing orders for the diffusion parameter k, accomplishing our desired nonlinearity in a microscopically motivated fashion.

Sandpile models come in many flavors. In general, they consist of a set of *sites* s_i which may take on a set of values $z \in \{0, 1, \ldots, z_c\}$, where z_c is a *critical value* of the parameter z. A standard interpretation of z_c is that of local slope, i.e., the differential in height or potential required to initiate a displacement. The sites are updated by an update rule similar to that of a CA, in which each site is affected by its neighbors. Computational "sand" is added either at random or at the center of the pile, and if any site attains its critical value, a universal rule (the same for all sites) specifies how the sand gets distributed to its neighbors. This is the "slip" in the stick-slip process. Sandpile models differ from conventional CAs in that they are not closed systems in which the update rule acts on a given initial configuration-they are driven with a constant input. Their update rules can be cast in the form of CA rule tables, though these tables consist of many entries which do not change the central site and hence have lots of "filler."

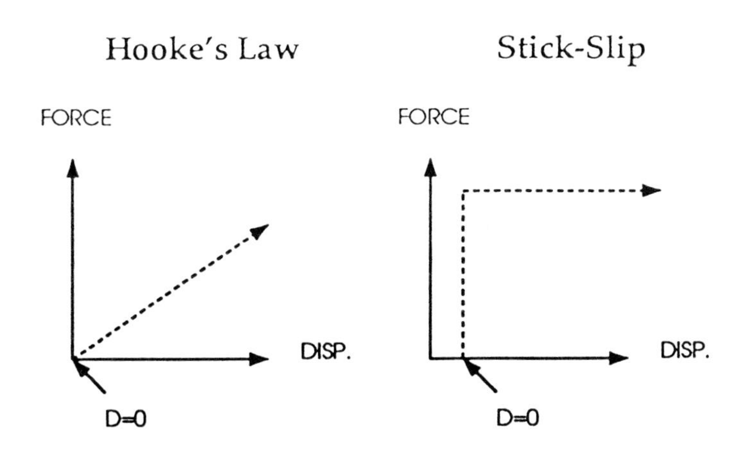

FIGURE 5

Sandpile models provide a limited metaphor for the study of fundamentally nonequilibrium thermodynamical systems. There is no overall conserved quantity such as energy, the systems are both driven and dissipative, and they are only characterized by local interaction laws. They are limited in that the set of possible interactions is somewhat arbitrarily constrained by having fixed lattice sites,

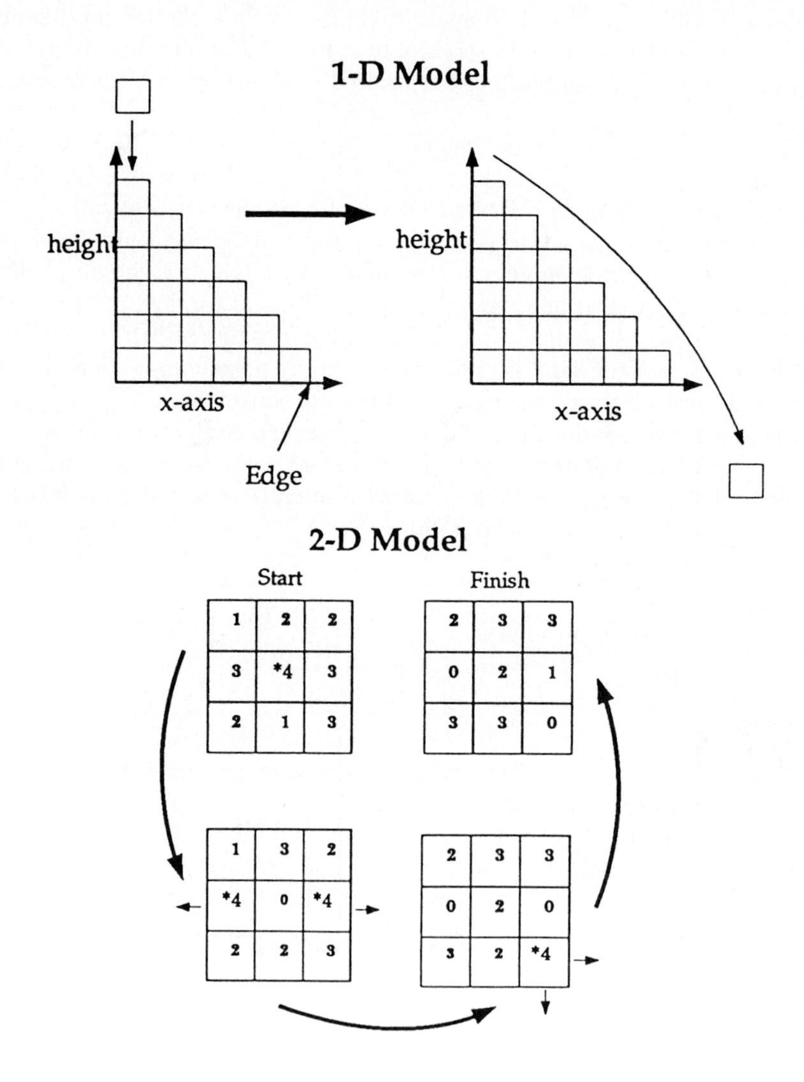

FIGURE 6

both in space and in number. Nonetheless, many different sandpile models are capable of showing $1/f$-type behavior, and there seems to be a considerable amount of independence from the actual structure of the update rule, so long as it is of the "critical value" type. Not all sandpile models produce interesting behavior, however. In Figure 6 we depict two sandpile models in their states before and after a grain of sand is added.

This particular one-dimensional model has $z_c = 1$ and one-neighbor interactions, and after a time $t \approx s^2$, where s is the number of sites, the behavior settles down into the uninteresting behavior shown above, and all avalanches are the same size, namely one particle of sand. The two-dimensional model with $z_c = 4$ and four neighbor uniform interaction is more interesting, as reaching equilibrium from the configuration shown above after having a grain of sand dropped in the middle is a three-step process involving the dissipation of four units of sand (shown by the small arrows). At each time step, the unstable site(s) are indicated by asterisks. After a little contemplation, it is easy to see that avalanches of many different sizes can occur. After a little more contemplation, it can be seen that the system does not approach any single-configuration attractor as does the one-dimensional model. In fact, attractor cycles are approached which can have very long periods.

In general, the models with multiple-neighbor interactions produce a kind of dynamical statistical equilibrium. This means that the input of one grain of sand does not in general result in the dissipation of one grain of sand, but that the time averages of input and output balance. The interesting characteristic of this dynamical balance is that the fluctuations around it are described by power laws. They do not have a characteristic size, except that there is an upper limit, namely that which is imposed by the size of the system. The process is fundamentally a nonequilibrium process, and the genesis of the term "self-organized" lies in the fact that a system will tune itself to this statistical equilibrium, at which point it will show power-law behavior.

The sandpile model is only a convenient computational paradigm for what is quite possibly a much more general phenomenon of non-equilibrium dynamics. Certainly the simplest systems which show power law behavior can be modeled by sandpiles, but what about systems that have more complex internal states, inhomogeneous update rules, adaptive rules, unfixed spatial coordinates, memory, or non-local interactions? Why is power-law behavior interesting, after all, and what interesting phenomena can it point to other than fractal structure? To begin to probe these questions requires an inquiry into the study of the dynamics of structure and patterns, separate from the considerations of a specific model.

INFORMATION THEORY AND ENTROPY

INTRODUCTION

The concept of information theory first found expression in Shannon's 1949 book, "The Mathematical Theory of Communication."[36] At that time, the principal area of application was that of telephony, i.e., sending sequential messages across a one-dimensional wire. Shannon founded the notion of information in the statistical mechanics of Gibbs and Boltzmann, and the two bodies of theory are bound together by their mutual dependence on a notion of a system having discrete *states.*

The thermodynamics of Boltzmann and Gibbs utilize the concept of a "phase space" that is occupied by a system of particles. The interactions between particles are governed by local rules. The phase space has $2ND$ dimensions, where D is the number of dimensions of the embedding space and N is the number of particles. The factor 2 comes from the fact that the classical equations of motion in one dimension require specifying both position and momentum in order to know the trajectory of the particle completely, which is tantamount to knowing the first and zeroth order derivatives of the second-order differential equations of motion. The exact structure of the local rules is not important, merely the assumption that they conserve energy and momentum and that the system is sufficiently random such that if an initial configuration is specified by choosing any point in the phase space, the system trajectory will eventually cover the entire phase space uniformly. This is the concept of *ergodicity.* The size of the state space is determined by the size of the system in question. A given total energy can usually be achieved by a large number of possible configurations of the state space, and this degeneracy is the basis of the concept of *entropy.* The fact that this degeneracy had to be finite and well defined and not divergent for statistical mechanics to make sense is one of the principal motivations behind quantum mechanics—in quantum mechanics, phase space is discretized.

Entropy, in a sense, measures what we don't know about a system—we could call it "missing information" instead. Imagine a system that can be in one of 2^n configurations for a given set of external parameters. We could represent the state space as a sequence of length 2^n in which each individual configuration is specified by putting the number 1 in one of the digits of the sequence, all the others being zero. This is not a very efficient way to represent a state space, but it serves for illustration. We now want to find out what state the system is in, given no information about it initially. We could play a generalization of the game "20 Questions" called "n questions," in which each question is, "Is there a 1 on the (left or right) side of the interval?" Each question will narrow the possible interval in which the 1 might reside by a factor of 2. Hence, after we've asked n questions, we will know the state of the system. In general, the missing information of a system with W states will be given by $MI = \log_2 W$. A simple two-stage construction allows this relation to be extended to systems which have states occurring with differing probabilities, rather than all the same as in the example above. Consider a system with

N equally probable states. The amount of information required to specify the state of the system should be independent of the process by which it is done, as long as no redundancies occur. Hence it should be possible to converge upon the desired state by using different processes of elimination. This is shown diagramatically in Figure 7.

The diagram on the right is subject to the constraint

$$\sum_{i=1}^{j} m_i = N\,.$$

We can associate a probability $P_i = m_i/N$ with the intermediate stage of the tree on the right. Now we can equate the "missing information" of the right side to the "missing information" of the left side. The right side can be expressed as the sum of the amount of information required to specify each stage of the two stage construction. We then have

$$MI\left(N, \frac{1}{N}\right) = MI\,(P_1, \ldots, P_j) + \sum_{i=1}^{j} P_i \log_2 m_i\,.$$

The expression on the left is the MI associated with a N-state system, each state of which has the same probability of occurring, namely $1/N$. We know the quantity on the left, so we can rewrite the above equation as

$$\log N = MI\,(P_1, \ldots, P_j) + \sum_{i=1}^{j} P_i \log m_i\,.$$

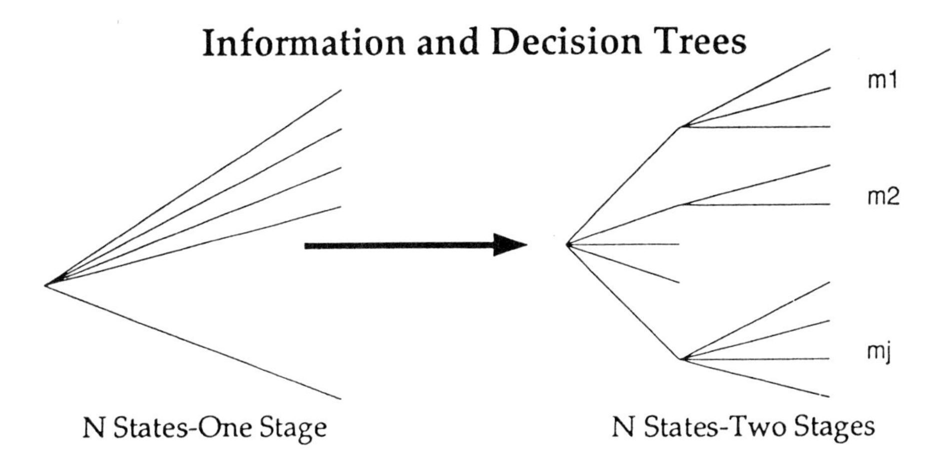

FIGURE 7

Recalling the definition of P_i, we finally arrive at

$$MI\,(P_1, \ldots, P_j) = -\sum_i P_i \log P_i \,.$$

We have now derived the form of entropy arrived at both by Shannon and by Gibbs, although they did not both use the same approach.

MISSING INFORMATION AND EQUILIBRIUM THERMODYNAMICS

As a warm-up, we can consider how information theory can be applied to uniform statistical ensembles. The connection between MI and thermodynamics can be established as follows. Consider a macroscopic system composed of many states. We do not possess detailed information about the state structure, but we know the total energy of the system. We can sum over all internal states j, and the system is subject to the constraints

$$\widetilde{E} = \sum_j E_j P_j$$

where $\sum P_j = 1$. If the observer possess no information about the internal state of the system, the observer is "unbiased," and this maximizes the missing information, as anything s(he) knows about the system decreases this quantity:

$$MI = -\sum_j P_j \log P_j \,.$$

We can now use the method of Lagrange multipliers to maximize the MI subject to the above constraints on the energy and the total probability. Form the quantity

$$\Lambda = -k \sum_j P_j \log P_j + \lambda_0 \left[1 - \sum_j P_j\right] + \lambda_1 \left[\widetilde{E} - \sum_j E_j P_j\right]$$

where $\{\lambda_o, \lambda_1\}$ are Lagrange multipliers and k is a constant of proportionality, determined by experimental means. If we differentiate Λ with respect to $\{P_j, \lambda_0, \lambda_1\}$, and set the resulting equations to zero, after some algebra we eventually find the well-known equilibrium relation between probability and energy:

$$P_j = \frac{e^{-\beta E_j}}{\left(\sum_j e^{-\beta E_j}\right)}$$

where $\lambda_1/k \equiv \beta$. This is interpreted as the "most probable" distribution of population as a function of energy. The identification of β with inverse temperature comes from the ideal gas law, and does not emerge from the above considerations alone. Hence we have seen that very basic assumptions have produced a powerful mechanism for characterizing an equilibrium system. What we would like is to be able to extend the utility of these tools to systems that are far from equilibrium.

INFORMATION, COMPUTATION, AND COMPLEXITY

INTRODUCTION

Consider a system in which the total energy (or its analogy) is unknown, yet something is known about the internal state structure. This characterizes many systems which are considered to be "complex"—such as economic systems, ecological systems, the immune system, social organizations, the origin of life, etc. These systems are characterized by one or more of the following points:

1. Open system (does not possess well defined boundaries).
2. Coherent behavior, correlations over long length and time scales.
3. Dynamics involving interrelated spatial and temporal effects.
4. A combination of randomness and regularity.
5. Strongly coupled degrees of freedom.
6. Noninterchangeable system elements.
7. Adaptive behavior.
8. Self-tuning parameters.

Many such systems generate complex behavior from simple local rules, hence we have some information about the internal structure of these systems, yet we have very little predictively useful information about the system as a whole. This is somewhat the inverse problem of equilibrium statistical mechanics, where the precise structure of the local rules is unimportant as long as they satisfy general constraints, and these general considerations about such rules allow the powerful machinery of statistics to be brought to bear on problems of many degrees of freedom.

QUANTIFYING COMPLEXITY

How would one go about quantifying a notion of complexity in dissipative systems? It would be desirable to do it in such a way that a smooth connection can be made with systems that are not complex, such as an ideal gas, perhaps by varying a parameter such as energy flow into the system. We have many intuitive notions of what constitutes complexity, some more amenable to quantification than others. One of the more concise heuristic definitions of a complex system is one that "transmits, stores, and processes information." Another one is "description length"—How long does the instruction set have to be to describe the behavior of the system under study? Can these definitions be planted on firmer ground?

The key to formalizing these questions lies in work done in the 1930s on the theory of computation by Gödel, Turing, and Church. At that time, there was an imprecise notion of what it meant to "compute" or "prove" a number or theorem, but as of yet no consistent formal definition. The motivation was a challenge by Hilbert as to whether or not there existed undecidable mathematical statements.

Even though it is not precisely relevant to this discussion, Gödel was able to show that no general "theorem proving machine" can be constructed. Turing, building on the work of Gödel, was able to construct an abstract notion of a "computing machine" that was the most general "theorem proving machine" possible, the well-known Turing machine. These advances constitute the beginnings of formal computational theory.

How could these formal advances be related to physical occurrences "out there" in Nature? Somehow our interaction with Nature has to be formalized as well. Scientists often do this without thinking explicitly of this aspect of the process, namely by doing experiments. We consider an "experiment" to be an interaction with Nature, often aided by mechanical or electronic means. Depending on the design of the experiment, we constrain the set of possible outcomes. If one's experiment is set up to find flies, it may not detect worms. "Data" is the result of this action, and usually comes in Boolean "off/on" form—"the protein is synthesized by the DNA," or numerical form—"the positron energy is 1.43 GeV." Science is a language for describing Nature, and as with all languages, some things are easier to describe than others. In this case, we are seeking quantifiable repeatable experiments which will more precisely pin down our intuitive notion of complexity.

If we are trying to understand the properties of a dynamical extended system which produces complex outputs, we need somehow to convert this dynamics into a stream of information which we will then analyze. As it turns out, the notion of "language" referred to above will take on a much more precise meaning. A simple example is provided by the ubiquitous pendulum: We could take a measurement of the position of the pendulum bob every tenth of a second and discretize the measurement space into two regimes—left of center (L) and right of center (R). A typical free unforced pendulum will yield a series of the following form: $RRRLLLRRRLLL\ldots$ Admittedly, this is not the most interesting experiment that can be postulated, but it does produce a data series for use as an example. The connection with formal computation theory comes when we attempt to create a logical process that will generate this same series.

Describing the above series is very easy: "Three R's, Three L's, repeat forever." If we were to relate complexity to the length of algorithm required to construct the sequence (perhaps as a function of the length of the sequence), measurement would corroborate with intuition—the pendulum is not very complex. This notion of "description length" was pursued by Kolmogorov and Chaitin in the 1960s under the rubric of *Algorithmic Complexity.* An interesting consequence of Algorithmic Complexity is that the more random a sequence is, the more "complex" it is. A truly random sequence has no deterministic method of algorithmic compression, no shorter description than just writing down the entire series. This is slightly distressing. It would mean that a very sensitive pressure sensor placed on the wall of a box containing an ideal gas would tell us that the process inside was maximally complex. Neither the pendulum nor the ideal gas seem maximally complex, but rather they seem to be extremes of simplicity, one simple because of regular determinism, the other because of statistical indistinguishability. The notion of algorithmic complexity does not allow for the simplicity associated with statistical averaging. Complexity associated with ensembles of dynamical extended systems is something else again.

A method which captures this intuitive notion that randomness is not any more complex than extreme regularity is called *machine reconstruction*, which also has its origins in the 1960s due to work of Renyi and has subsequently been applied to stochastic systems by Crutchfield.[13] The key notions are that of a *time series* of data and of a *statistical ensemble* of systems. The method of machine reconstruction relates complexity to diversity of time series data patterns that a system can generate. A pendulum is not diverse because of its extreme regularity, and the motion of a particle in an ideal gas is not diverse because of statistical self-similarity—every particle looks like every other particle over any large enough time segment, on the average—particles could only be distinguished by putting numbers on them or painting them different colors. Consider a data sequence in one dimension: **A**: $\{\alpha_1, \alpha_2 \ldots, \alpha_n\}$ where the α run over the alphabet of symbols particular to the dynamical process being studied. No loss of generality is incurred if we translate this alphabet into binary for purposes of clarification. We would like to look for patterns in the data. To do this, imagine a window of size L that fits over the sequence, covering the interval $[\alpha_i, \alpha_{i+L-1}]$. Sliding this window along the sequence will generate a series of "words" of length L. Their patterns can be classified using a *parse tree* of depth L. This is shown below, with the sequence 0101 outlined in a heavy black line.

Pattern Parse Tree

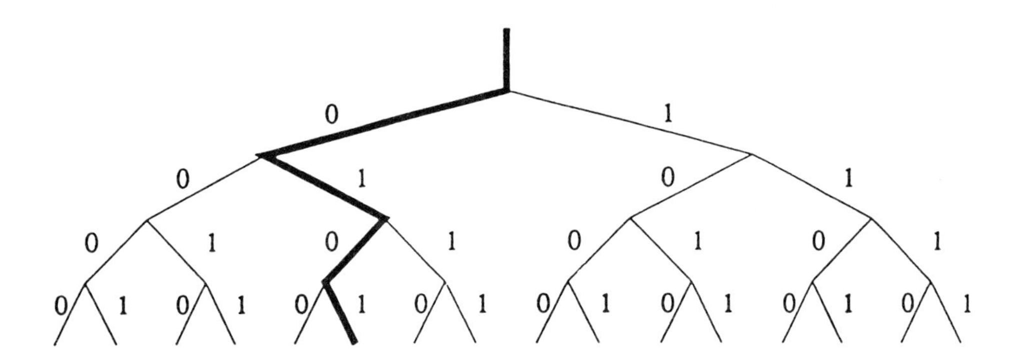

FIGURE 8

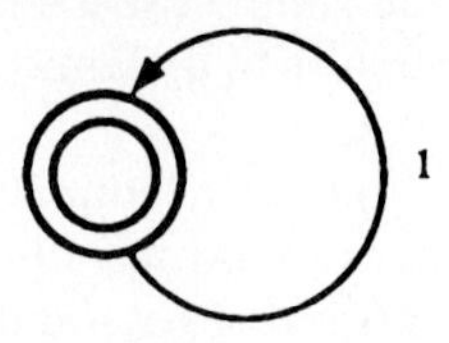

FIGURE 9

As we slide the window over the sequence under analysis, we will generate a sequence of patterns, up to a possible diversity of 2^L. We can view each pattern as a "state" of a computational device. A couple of examples will serve to clarify the utility of this construction. Consider a very simple sequence of all ones or zeroes. The same path through the tree will be filled out at each window position. At the other extreme, consider a random sequence. After a number of window-sliding steps much greater than the diversity of the tree, all possible paths through the tree will be filled out with very close to equal probability.

To see how this method can be used to quantify a notion of complexity, we will reconstruct computational "machines" which reproduce the features of the series under study. The first example above would look like in Figure 9.
This is called a Deterministic Finite Automaton (DFA) because the transition rules are deterministic, and they operate between a finite number of states. The arrows with the numbers next to them signify that if the machine receives the number given, it effects a state transition. A double circle signifies the "start" state. One says that the machine thus "accepts" only certain sequences. The machine above does not accept zeroes, for instance. The second example is slightly more interesting, as it makes use of the notion of statistical ensemble. If the transition rules are not deterministic, we can construct a slightly different kind of machine called a Non-deterministic Finite Automaton (NFA). Here we consider a statistical interpretation of "state," and construct a machine which will utilize this. The totally random system has only one state (that which fills out the tree completely with equal probability), and since it accepts zeros or ones with equal probability, there are two transition arrows, each weighted with probability 0.5. Such a machine is depicted in Figure 10.

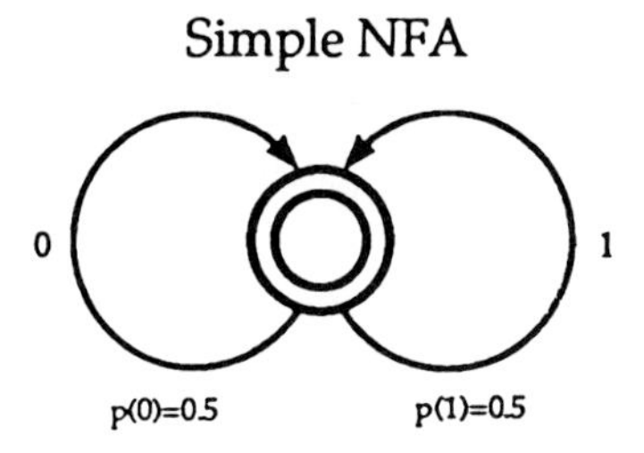

FIGURE 10

For a general machine, there are two quantities which characterize the complexity of the series produced which are called *topological entropy* and *dynamical entropy*. Both are needed to characterize the amount of information required to specify a series produced by such a machine. If one wanted to specify a sequence produced by a machine, one would have to specify the number of nodes or states of the machine (the topology) and then would also have to specify the action taken at the various decision bifurcations of the machine. The procedure for specifying what state the machine is in at a given time requires $C = \log_2 N$ bits of information. In addition, knowing the sequence produced requires specifying one out of all of the possible trajectories the machine may have produced since it started. The machines shown above are very simple-there is only one node in each of them, so $C = 0$ for both of these examples. We can graph the information required to specify the sequence produced by an arbitrary machine as a function of the length of the sequence it produces; see Figure 11.

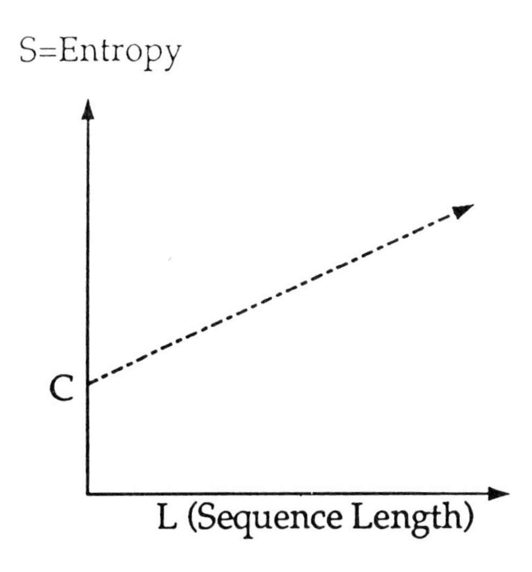

FIGURE 11

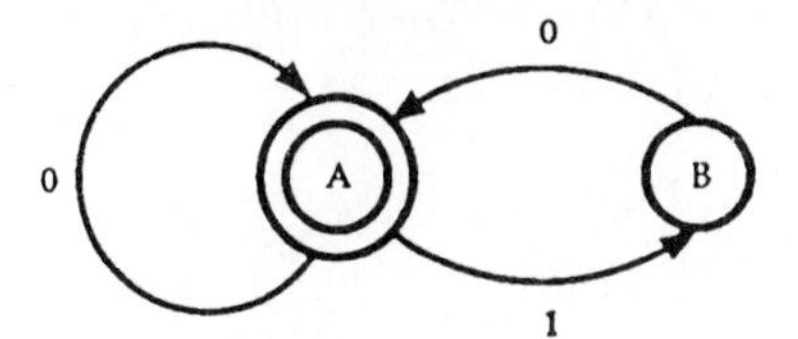

FIGURE 12

Note that the plot does not start at the origin, but rather has a certain amount of offset. This intrinsic amount of required information that is independent of sequence length is called "Complexity" by Crutchfield and is the y−intercept of the entropy versus sequence length plot above. For the DFA example, the plot would start at $S = 0$, and in addition, its slope would be zero, as there is only one choice (hence no choice) per unit of sequence length, meaning no new information is needed at later times to specify the sequence. The NFA would also start at $S = 0$, but a specific sequence would have two choices per time step (unbiased random coin flip) and hence would generate entropy at the rate of one bit per unit of sequence length, as $\Delta S = \Delta MI = -\sum_{i=1,2} p_i \ln_2 p_i = 1$. Hence information theory applied to computational theory can produce a quantitative measure of complexity.[14]

More sophisticated machines than the two examples used above, such as the "Golden Mean Machine" DFA pictured in Figure 12, are possible. This machine produces information at a per bit rate equal to the log of the Golden Mean in the limit of infinite series length.

The pattern parse tree for the GM machine gives a diversity at the nth level equal to $\mathcal{F}_{\rm n}$, where $\mathcal{F} \in \{1, 1, 2, 3, 5, 8, \ldots\}$ is the Fibonacci series. The pattern parse tree is shown below, with the machine state superimposed at each node and diversity shown on the right-hand side for each layer of the tree. The graphic reconstruction of the sequences shows a striking visual self-similarity. In fact, it looks strikingly like an avalanche or a cascade. This suggests an application to SOC systems.

There are limitations to this type of analysis, however. Non-Markovian processes cannot be analyzed in this fashion. Consider the following problem: How does one enumerate all of the "proper" ways of nesting $2n$ parentheses? There is no finite set of local rules which will enumerate all of the possible ways to nest an arbitrary number of parentheses, because this particular process is history dependent—a running tab must be kept of the number of parentheses of each kind, hence the system requires memory. This type of process is called *context-free*, as the constraints are global, not local (contextual). Constant local rules can be applied anywhere as long as a global constraint is satisfied, but no augmented set of local rules can implement this global constraint.

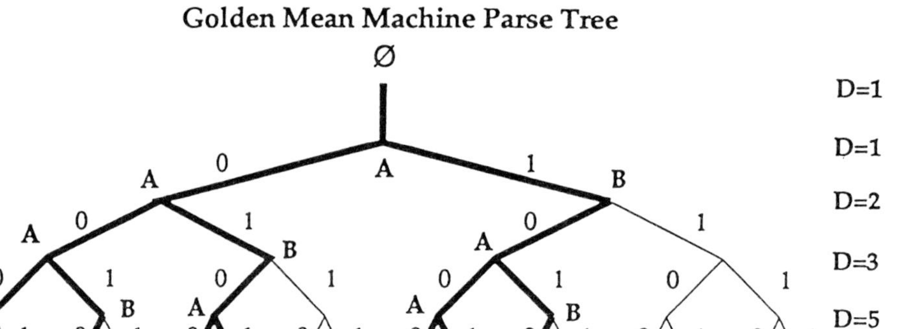

FIGURE 13

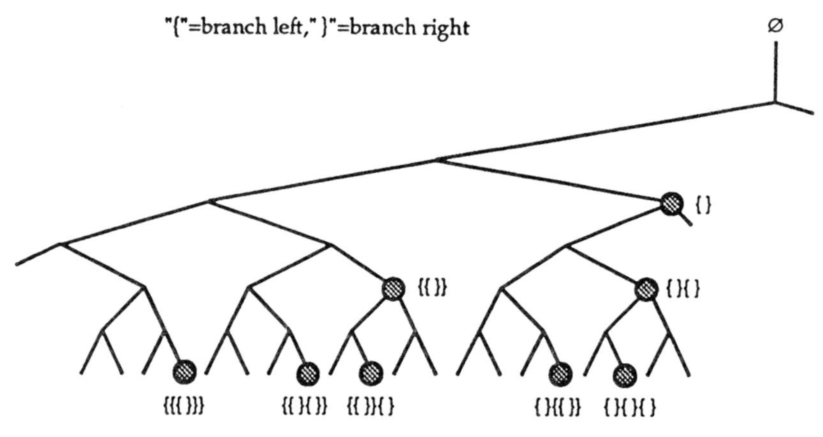

FIGURE 14

The Golden Mean machine is a *regular language*, characterized by a finite number of types of vertices in a parse tree and no global constraint. A brief inspection of the graph in Figure 13 reveals that there are two kinds of vertices in the GM

Machine, and they can occur anywhere in a graph—one comes in from the right and forks, the other comes in from the left and cuts right. Their iterated application produces all of the series acceptable by the machine. In contrast, a parse tree is applied to the parentheses problem in Figure 14.

If one starts out at the top and traces out the path to the accepted states, the intermediate states are not also acceptable states, unlike the GM Machine parse tree construction. Hence we do not have the type of self-similarity that allows us to construct a finite automaton. The problem of nesting parentheses is intrinsically more complex. Unlike the GM Machine, the parentheses problem has a global constraint, not just a local rule. The global rule is that the summed number of left parentheses can never exceed the summed number of right parentheses in the course of the construction of an acceptable series, and in addition must be equal at the termination of the construction process.

In fact, this complexity can be formalized using what is known as the "Chomsky hierarchy," an arrangement of formal languages in order of their computational power. Chomsky divided formal languages into four categories, each a subset of the next one higher up.

We have only discussed the first category in a great deal of detail. In addition, we have provided an example from the second category, namely that of the parentheses nesting problem. The third category has branching rules which are only activated within a specified context. The fourth category is the most powerful known, and it differs from the third in that not only are there global rules and memory, but that these rules can change as a function of the state of the machine+input, in other words, the machine is "programmable" in the sense that we are familiar with. Chomsky calls these languages *recursively enumerable.* In principle, all postulatable

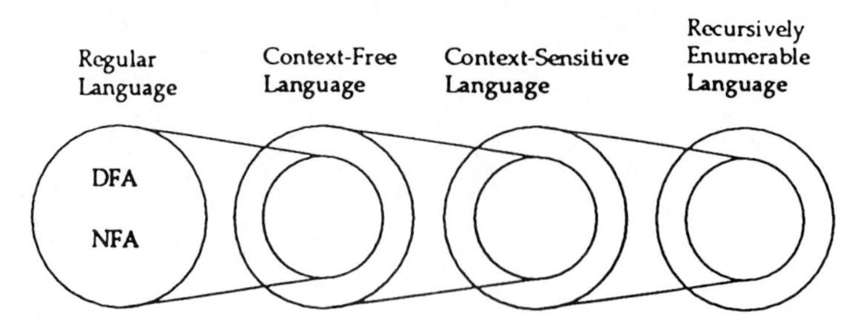

FIGURE 15

problems can be expressed as Turing machines; hence, we say that a Turing machine is capable of *universal computation*—that which can be computed can be computed by a Turing machine. A computation may have an undecidable result and the process of computation may not halt, but the full logical content of any computation can nonetheless be embedded in a such a machine.[24]

TYING IT TOGETHER: COMPUTATION, DYNAMICS, AND SOC

The last two sections have discussed approaches to quantifying the concepts of information and computation. These are essential to understanding the ideas of pattern, complexity, and organization in a formal sense. What we would like to do is to be able to apply these techniques to the dynamics of physical systems to understand the genesis of complex systems in Nature. Statistical thermodynamics provides the simplest application of information theory to physical processes, but it makes assumptions which must be retracted in order to study complex systems that are not in equilibrium.

The first chapter discussed physical systems in Nature and what sort of computational paradigms are best suited for modeling them. This is in itself a context-dependent process, as our understanding of nature has been deeply structured by the intellectual discipline of the scientific method as cultivated over the last 350 years. Our understanding of nature has been traditionally characterized by symmetry principles, global and local conservation laws, and the separation of systems of interest from the surrounding environment. The study of complexity constitutes a significant turning point in this centuries long history, as the traditional boundaries of mathematizable science are being extended into unexplored regimes. Indeed, by exploring the relation of language to our description of Nature, it is just becoming possible to establish boundaries which demarcate that which is describable by scientific method and that which is not, due to the application of results of Gödel and Turing to physical systems.

The formal power of the theory of computation allows us to make abstractions about the languages we use to describe nature and to understand their descriptive limits. We have seen above how to establish a connection between physical processes and language, namely by measuring the system of interest. Therefore it is essential to incorporate the theory of computation into a description of a physical system if we are to attempt to quantify the notion of complexity in a consistent and useful manner.

So how is this useful for understanding SOC systems? SOC systems are an entire class of nonequilibrium systems which seem to be capable of generating complexity of their own accord. The simplest such models, such as computational sandpiles, are characterized by either regular or context-free languages, as they may or may not have global constraints. (conservation of "sand," etc.) Attempting to understand such simple systems may point the way to general principles of complex system dynamics. Such SOC indicators as $1/f$ noise may prove to be very coarse measures of a broad class of systems, and may be necessary but not sufficient requirements for interesting manifestations of complexity, depending on what class of complexity (what kind of computational capability) one is looking for. A sandpile, after all, is not nearly so interesting as a cell, though understanding the dynamics of a sandpile may point the way to understanding some of the processes which take place in a cell. We are currently working on defining general classes of regular language rules which can produce $1/f$ noise.

It is unlikely that "complexity" will ever have such a well defined interpretation as "energy" now has. Complexity is dependent on the descriptive power of the language being used, and cannot cross categories of the Chomsky hierarchy; hence, it only makes sense to discuss complexity within the context of a descriptive paradigm. In a sense this is a natural evolution in the history of science—quantum mechanics began this process by incorporating the notion of observer into the process of observation. The theory of complexity takes this a step further; however, the situation is serious but not hopeless.

We define a system by what sort of measurements we make. We used the constraint of global energy measurements to make powerful statements about equilibrium statistical mechanics using a minimum-information principle. This is an extremely coarse-grained approach to the structure of systems with many degrees of freedom, namely summing over all of them. Understanding noise in an open system is only slightly less coarse-grained, and is a well defined enough problem that it can be approached computationally. Perhaps there is something like an additional Law of Thermodynamics which maximizes a specific measure of complexity for certain classes of dissipative systems, a result that indicates that "nature tends toward the complex." It may be true that a complex system cannot be characterized by a single number, but that is hardly surprising, as it may be precisely that property that causes us to intuitively classify such a system as complex. We are working on it.

ACKNOWLEDGMENTS

Many thanks for useful discussions to John Casti, Jim Crutchfield, Cris Moore, Harald Atmanspacher, Daniel Shapiro, Mats Nordahl, and Erica Jen. Thanks also to Patrisia Brunello for the thinly veiled threat of death which caused this manuscript to finally get produced.

REFERENCES

1. Arthur, B. A. "Positive Feedbacks in the Economy." *Sci. Am.* **Feb.** (1990): 92.
2. Bak, P., and K. Chen. "Mean Field Theory of Self-Organized Critical Phenomena." *J. Stat. Phys.* **51** (1988): 797.
3. Bak, P., and K. Chen. "The Physics of Fractals." *Physica D* **38** (1989): 5.
4. Bak, P., and K. Chen. "Dynamics of Earthquakes." Technical report BNL-43452, Brookhaven National Laboratory, 1989.
5. Chen, K., P. Bak, and S. P. Obukhov. "Self-Organized Criticality in a Crack-Propagation Model of Earthquakes." *Phys. Rev. A* **43** (1991): 625.
6. Chen, K., P. Bak, and K. Wiesenfeld. "Self-Organized Criticality: An Explanation of $1/f$ Noise." *Phys. Rev. Lett.* **59** (1987): 381.
7. Chen, K., P. Bak, and K. Wiesenfeld. "Self-Organized Criticality." *Phys. Rev. A* **38** (1988): 364.
8. Chen, K., J. Scheinkman, P. Bak, and M. Woodford. "Self-Organized Criticality and Fluctuations in Economics." Working paper 92-04-018, Santa Fe Institute, Santa Fe, NM, 1992.
9. Carlson, J. M., and J. S. Langer. "Properties of Earthquakes Generated by Fault Dynamics." *Phys. Rev. Lett.* **62** (1989): 2632.
10. Hugues, C., and . Manneville. "Criticality in Cellular Automata." *Physica D* **999** (1990): 1.
11. Chau, H. F., and K. S. Cheng. "Generalized Sandpile Model and the Characterization of the Existence of Self-Organized Criticality." *Phys. Rev. A* **44** (1991): 6233.
12. Creutz, M. "Abelian Sandpiles." *Computers In Physics* **5** (1991): 198.
13. Crutchfield, J. P. "Semantics and Thermodynamics." Working paper 91-09-033, Santa Fe Institute, Santa Fe, NM, 1991.
14. Crutchfield, J. P. "Critical Computation, Phase Transitions, and Hierarchical Learning." Working paper 93-10-061, Santa Fe Institute, Santa Fe, NM, 1993.
15. Dhar, D. "Self-Organized Critical State of Sandpile Automaton Models." *Phys. Rev. Lett.* **64** (1990): 1613.
16. Dhar, D., and R. Ramaswamy. "Exactly Solved Model of Self-Organized Critical Phenomena." *Phys. Rev. Lett.* **63** (1989): 1659.
17. Drossel, B. W. K. Mossner, and F. Schwabl. "Computer Simulations of the Forest-Fire Model." *Physica A* **190** (1992): 295.
18. Drossel, B., and F. Schwabl. "Self-Organized Critical Forest-Fire Model." *Phys. Rev. Lett.* **69** (1992): 1629.
19. Drossel, B., and F. Schwabl. "Formation of Space-Time Structure in a Forest-Fire Model." Preprint. Inst. für Theor. Physik, Tech. Univ. München, Garching, August, 1993
20. Feder, H. J. S., and J. Feder. "Self-Organized Criticality in a Stick-Slip Process." *Phys. Rev. Lett.* **66** (1991): 2669.

21. Feder, H. J. S., Z. Olami, and K. Christensen. "Self-Organized Criticality in a Continuous, Nonconservative Cellular Automaton Modeling Earthquakes." *Phys. Rev. Lett.* **68(8)** (1992): 1245.
22. Grassberger, P., and S. S. Manna. "Some More Sandpiles." *J. Phys. France* **51** (1990): 1077.
23. Hanson, J. E., and J. P. Crutchfield. "The Attractor-Basin Portrait of a Cellular Automaton." *J. Stat. Phys.* **66** (1992): 1415.
24. Hopcroft, J. E., and J. D. Ullman. *Introduction to Automata Theory, Languages, and Computation.* Reading, MA: Addison-Wesley, 1979.
25. Hwa, T., and M. Kardar. "Avalanches, Hydrodynamics, and Discharge Events in Models of Sandpiles." *Phys. Rev. Lett.* **62** (1989): 1813.
26. Kauffman, S. A. *The Origins of Order.* Oxford University Press, 1993.
27. Krug, J. "Landslides on Sandpiles: Some Moment Relations in One Dimension." *J. Stat. Phys.* **66** (1992): 1635.
28. Liang, N. Y., S.-C. Lee, and W.-J. Tzeng. "Exact Solution of a Deterministic Sandpile Model in One Dimension." *Phys. Rev. Lett.* **67** (1991): 1479.
29. Mandelbrot, B. *The Fractal Geometry of Nature.* San Francisco, CA: W. H. Freeman, 1982.
30. Nagel, S. R., L. Wu, L. P. Kadanoff, and S.-M. Zhou. "Scaling and Universality in Avalanches." *Phys. Rev. A* **39** (1989): 6524.
31. Nagel, K. "A Discrete Automaton for Simulated Freeway Traffic Showing a (Self-Organizing) Critical Transition." Preliminary draft of thesis, 1992.
32. Nagel, S. R. "Instabilities in a Sandpile." *Rev. Mod. Phys.* **64** (1992): 321.
33. Prigogine, I. *Non-Equilibrium Statistical Mechanics.* New York: John Wiley & Sons, 1962.
34. Ruelle, D. *Statistical Mechanics.* New York: W.A. Benjamin, 1969.
35. Schroeder, M. *Chaos, Fractals, Power Laws.* San Francisco, CA: W. H. Freeman, 1991.
36. Shannon, C. E., and W. Weaver. *The Mathematical Theory of Communication.* University of Illinois Press, 1963.
37. Takayasu, H., and H. Inaoka. "New Type of Self-Organized Criticality in a Model of Erosion." *Phys. Rev. Lett.* **68(7)** (1992): 966.
38. Tang, C., K. Wiesenfeld, and P. Bak. "A Physicist's Sandbox." In *Los Alamos Workshop on Noise and Spatially Extended Dynamical Systems.* 1988.
39. Theiler, J., K. Wiesenfeld, and B. McNamara. "Self-Organized Criticality in a Deterministic Automaton." *Phys. Rev. Lett.* **65** (1990): 949.

Olaf Sporns
The Neurosciences Institute, 3377 North Torrey Pines Court, La Jolla, California 92037
email: sporns@nsi.edu

Neural Models of Perception and Behavior

INTRODUCTION

This chapter is about a complex system, the brain, and how computer modeling can help to understand the mechanisms by which neural activity gives rise to mental phenomena, such as perception. The first part will try to provide a brief and necessarily sketchy overview of brain function, including the brain's evolutionary origins, its embryonic development, its anatomic structure, and the physiology of its constituent cells, the neurons. I will limit myself to topics that are important for computational modeling of neuronal networks. Then, I will sketch out some fundamental concepts in what has come to be called "computational neuroscience" and give an overview of a biological theory of brain function based on principles of selection. The main part of this chapter will be devoted to work carried out over the last six years by my colleagues and myself at The Neurosciences Institute. I will focus on realistic models of the visual cortex and discuss their relationship to problems of visual perception. Then I will deal with some issues related to the design of whole automata, simulated or real, that possess a nervous system as well as a body structure, and that "live" in various environments. I will demonstrate how these types of models can help to understand the relationship between neuronal function and behavior.

THE BRAIN, A COMPLEX SYSTEM

The human brain is perhaps the most complex object in the known universe. This fact alone should be sufficient to convince all those interested in complex systems that the brain is an object well worth studying. Not only is the brain very complex, it is also the only known material object that is able to produce "awareness" or consciousness.

The number of neurons and connections in the human brain is simply staggering: the cerebral cortex alone contains about 10 billion neurons (estimates for the entire nervous system are about 10^{12} neurons). There are 1 million billion connections between these neurons, and 1 mm^3 of cortex alone contains about 1 billion connections.

These numbers are huge, although perhaps not *that* huge given that 1 cm^3 of air contains 2.7×10^{19} molecules. However, while the physico-chemical properties of gases can be described by relatively simple sets of equations (for example, in terms of the statistical mechanics of their constituent molecules), no such theoretical framework exists to relate the behavior of the brain's individual neurons to its behavior as a whole. This is because the brain has an elaborate macroscopic and microscopic anatomic structure (or architecture), which is the result of millions of years of evolution, and its constituent elements (neurons) come in many different types and interact in very characteristic ways. A main theme running through this chapter is that in order to understand the brain we must pay deliberate and painstaking attention to its architecture. In the course of this chapter it will become evident that, for example, many phenomena of visual perception depend crucially on the underlying neuronal architecture (the way neurons are connected and function together as populations).

The student interested in learning more about neuroscience can choose among a number of excellent textbooks.[16,24,63,85,99,104,105] Obviously, each of these books gives a more complete and detailed overview of brain function than can be provided in this chapter. In the next section we will focus on topics that are of special interest to computational modeling.

BASIC FUNCTIONS OF THE NERVOUS SYSTEM

Nervous systems subserve a variety of functions. Many of these functions are of evolutionary advantage to the whole organism. Nervous systems enable an organism to sense its environment; all sense organs contain nerve cells that conduct their signals to other, more central parts of the brain. They also are responsible for the initiation and control of all motor activity, such as locomotion and speech. In addition to neurons devoted to sensory and motor function, most nervous systems, especially those of higher vertebrates contain very elaborate network-like circuitry that "processes" incoming sensory inputs. Ultimately, the activity of neurons gives rise to a variety of mental phenomena, e.g., sensation, perception, attention, dreaming, memory, and learning. Many, if not all, of these functions contribute to or enhance the survival of the organism and its capacity to adapt to changing circumstances in its environment. Perhaps the most puzzling and, for a neuroscientist, most challenging "function" of higher nervous systems is consciousness. Consciousness has recently become the focus of several books by physicists, cognitive scientists, and philosophers.[23,73,89,102] A biological theory aimed at explaining how consciousness arises from the activity of the brain has been proposed by Gerald Edelman.[32,33] Some elementary aspects of this theory are discussed in the next main section, although it is beyond the scope of this chapter to deal with the issue of consciousness itself.

SOME EVOLUTIONARY ASPECTS

The architecture of a nervous system (like any other part of an organism's morphology) is the result of evolution. During evolution there is selective pressure for enhanced sensory and motor capabilities as well as adaptive behavior. This will, in many cases, result in the enlargement and structural elaboration of neural circuitry. "Lower" vertebrates, such as fish and reptiles, have smaller brains (relative to body mass) than "higher" vertebrates, such as birds and mammals.[60] Over the last two million years in the evolution of primates, their cranial capacity (i.e., brain size) almost quadrupled, while body mass stayed fairly constant.[113] This last observation is an example of an important global evolutionary trend, called encephalization (or corticalization), i.e., the tendency for relative enlargement of the brain (or cortex). Current evolutionary theory suggests that this progressive enlargement does not reflect the accumulation over time of "new" circuitry solving "new" problems. Rather, evolution tends to be conservative, that is, if a given problem is solved by existing circuitry (i.e., in the midbrain), evolution will try to elaborate this circuitry rather than evolve a new system (i.e., cortical) to solve the problem better. Thus, if a particular function is performed by a particular set of circuits in one species, it is likely to be performed by homologous circuits in descendent (more highly evolved) species. Consequently, the number of types of circuitry that the brain uses to deal with different problems might be rather small. Another important evolutionary point is that the ecological requirements facing a given species will determine to a large extent what neural mechanisms evolve to solve certain problems. For example, the visual systems of rats, cats, monkeys, and humans will certainly reflect the different ecological requirements posed by these animal's environments. There is no single "solution" to a given visual problem, for example the recognition of visual forms; each species will tend to evolve its own solution.[60]

Much can be learned about the evolution of nervous systems by comparing existing species in an evolutionary framework. Even the simplest invertebrate species possess nervous systems. The first specialized nerve cells, sensory, and/or motor, probably arose rather early in the evolution of multicellular organisms, together with other specialized cells such as muscle or secretory cells. Very primitive multicellular organisms (for example coelenterates, which arose some 700 million years ago) contain neurons organized in two-dimensional nerve nets enabling some species (such as jellyfish) to show coordinated swimming movements. These nerve nets show some specialization into sensory and motor cells, and there is cell-cell communication (signal transfer) between them. At this stage, nerve cells do not aggregate to form a separate organ (central nervous system).

Among the simplest central nervous systems (CNS) are those found in flatworms, with a simple "brain" localized in the animal's head region, surrounded by a dense nerve plexus. In these animals, sensory cells are beginning to be concentrated in certain regions of the body and form sense organs, such as eyes. The brain is located at one end of a longitudinally arranged and segmented nerve cord. At this stage the major evolutionary trends are: concentration of nerve cells in the head region, segmentation, specialization among sensory neurons (into photoreceptors, mechanoreceptors, and chemoreceptors), and specialization of different subregions of the brain. Consequences of these trends are: an increasing repertoire of motor behaviors, the evolution and elaboration of sensorimotor reflexes, and the development of major principles of neuronal networks, among them the distinction between inhibitory and excitatory cells, and reciprocal excitation and inhibition.

Most of the over one million animal species in existence today are arthropods (mostly insects). The variety of arthropod nervous systems reflects their enormous evolutionary success in adapting to many different environments. Insects have developed many specialized sense organs such as compound eyes, chemoreceptors, and several types of mechanoreceptors connected to the jointed legs. Their brains have complicated internal structure and receive inputs from the various sense organs located in the head segments; neural pathways associated with the visual system are particularly highly developed. Motor control functions are performed by nerve ganglia associated with the wings and legs. Specialized nerve cells are involved in neurosecretory and hormonal functions, including sexual functions and metamorphosis.

Many important neural functions, from sensorimotor activity to learning and memory, can be studied at the level of invertebrate (or "miniature") nervous systems. Higher nervous systems (e.g., those of primates), however, cannot be regarded merely as aggregations of numerous "miniature" nervous systems; they have their own structure and functional properties, to which we now turn.

DEVELOPMENT AND STRUCTURE OF VERTEBRATE NERVOUS SYSTEMS

Vertebrate nervous systems are characterized by a vast increase in the number of neurons and in the complexity of their interactions. Encephalization continues and most neurons in vertebrate brains become "interneurons"; that is, they are neither purely sensory nor motor, but dedicated to the elaboration (or "processing") of sensory inputs and motor outputs. Vertebrate nervous systems are divided into central and peripheral nervous systems (CNS and PNS). The dorsally situated CNS forms a tube-shaped nerve cord that develops through invagination from a single layer of ectodermal epithelium. The anterior portion of this tube forms enlarged and thickened vesicles, the cerebral ganglia or "brain." The posterior portion forms the spinal cord, surrounded by the bones of the vertebral column.

The major anatomical subdivisions of the brain are: the rhombencephalon (with the cerebellum, a hemispheric structure at the back of the brain playing an important role in motor function), the mesencephalon (containing the inferior and superior colliculi, structures participating in auditory and visual perception), the diencephalon (containing the thalamus, whose various nuclei serve to transmit incoming sensory inputs to the cerebral cortex, and containing important nuclei involved in the control of visceral functions, eating, drinking, sex, and emotional states), and the telencephalon (with the cerebral cortex).

Originally much of the telencephalon was dedicated to olfaction. The evolutionarily oldest part of the cortex is the olfactory cortex (or paleocortex). The newest part is the neocortex, a structure that has been very much enlarged and expanded in the evolution of higher vertebrates. The neocortex forms a much convoluted sheet of neurons, organized into six distinct layers. Another part of the telencephalon is called the basal ganglia (located below the cortex in the interior part of the cortical hemispheres); these structures are involved in a variety of sensorimotor functions.

The elementary subunits of the cerebral cortex are vertically oriented columns of cells, spanning all six cortical layers, containing neurons that are strongly interconnected. Horizontally, the cortex is organized in functionally specialized cortical fields (also called areas or maps). These maps can be sensory or motor, and often contain ordered or topographic representations. For example, many visual cortical maps contain ordered representations of the visual field, somatosensory maps contain representations of the body surface, etc. Often, these maps are distorted in characteristic ways. For example, visual maps contain enlarged representations of the central or foveal part of the visual field, and somatosensory and motor maps contain enlarged representations of certain parts of the body that are particularly well innervated and are capable of very refined movements (e.g., the hand or tongue). Over the extent of the cortical surface, sensory or motor maps are multiply represented. The part of the monkey cortex dedicated to the "processing" of visual inputs alone contains more than 30 distinct visual maps, many of which are topographic maps of the visual field. This mosaic organization of the cortex reflects the functional specialization of cortical areas. Each area contains neurons that are specialized to respond to a particular aspect of the visual world, e.g., motion, color, or texture.

NEURONS

There is an enormous diversity of neuronal forms. Common to most neurons is that they possess a cell body (or soma) containing a nucleus and most of the machinery needed for protein synthesis, and neurites. Neurites are either called dendrites or axons. Dendrites are mainly receptive (i.e., they receive inputs coming from other cells); often there is more than one dendrite extending from the cell body and dendrites tend to branch extensively in relatively large volumes of space (typically 0.001 to 0.1 mm^3). Axons are transmitting the output of a cell to other cells. These cells can be located nearby, or some distance (millimeters or centimeters) away. In general, only one axon leaves the cell body, but the axon may branch and may connect to a large number of target neurons. Axons can be very long; the axons of some cells in the human motor cortex extend over 1 meter in the spinal cord.

When neurons make contact with one another, they form synaptic junctions, or synapses. Synapses are specialized cellular structures that allow the transfer of a signal (a change in electrical potential) from one cell to another. For the vast majority of synapses, signal transfer can only occur in one direction, from the pre- to the postsynaptic cell. Most synapses are formed between axon terminals (presynaptic) and dendrites (postsynaptic). The two sides are separated by the synaptic cleft. Synapses are formed during embryonic development and, while the overall pattern of synaptic connectivity tends to remain unchanged, they are formed, broken down, and reorganized throughout the adult lifetime of the organism. Synaptic transmission can be electrical or chemical. Direct electrical communication between neurons is relatively rare. In chemical synapses a change in the electrical potential of the presynaptic cell causes the release of a chemical neurotransmitter into the synaptic cleft. Receptors in the membrane of the postsynaptic cell bind neurotransmitter molecules and thus cause a change in the electrical potential of the postsynaptic cell.

Neurons generate electrical potentials across their cell membrane, a capacity they share with all other types of cells (including many plant cells). The potential difference across the cell membrane when the cell is not excited is called the resting potential. The resting potential results from the selective permeability of the cell membrane for different types of ions which produces different ionic concentrations on the two sides of the membrane. The concentration gradients for different ions across membranes are generated and maintained by ion pumps, protein channels (extending through the cell membrane) that depend for their action on metabolic energy. These channels are selective for particular kinds of ions (e.g., Na^+, K^+, Ca^{++}), depending on their electric charge and size.

Neuronal signal transmission depends on disturbances of the resting potential that reflect ionic currents across the cell membrane (ions entering or leaving the cell). An electric potential generated by inputs across synaptic junctions can spread laterally in the neuronal membrane. This spread is passive and no amplification of the signal takes place; in fact, the amplitude of the potential is attenuated as it spreads across the membrane. Passive electrotonic conductance takes place, for example, in dendrites and has specific spatial and temporal properties. These define two very important principles of neural integration, i.e., the integration of several inputs arriving at a nerve cell. This integration is important because in the cerebral cortex each neuron receives about 10,000 synaptic inputs while producing only one output. Repetitive activation of a presynaptic neuron can lead to a temporal overlap in the elicited postsynaptic potentials. This leads to a temporal summation effect and can result in higher efficacy (i.e., ability to trigger a postsynaptic response) of a train of presynaptic action potentials delivered at high temporal frequency. Spatial summation, on the other hand, involves the summation of inputs that are spatially separated on a dendritic tree. The time it takes for each of the inputs to travel down the dendrite towards the cell body is critical for spatial summation to occur. Effective spatial summation can lead to a higher probability that a postsynaptic response is elicited.

Transmission of neuronal signals over long distances requires an active process, the conversion of an electrical potential into a self-regenerating, all-or-nothing electrical pulse, called an action potential. Action potentials do not decrement with distance, can be conducted at relatively high speed, and are of relatively fixed temporal duration (a few milliseconds). Action potentials are the main means by which vertebrate neurons communicate with each other.

Repetitive activation of a synaptic connection between two neurons can result in an enhancement of the strength of this connection. Such synaptic plasticity is thought to be the cellular basis for learning and memory. Plasticity can occur over several different time scales, from changes that last for seconds to others that are virtually permanent. Long-term changes are observed, for example, in the hippocampus, a structure that is important in the formation and consolidation of new memories.

Over 70 percent of the neurons use glutamate as a neurotransmitter, project axons to relatively distant cortical or subcortical targets, and are excitatory, i.e., they tend to depolarize postsynaptic target neurons, increasing the probability that these neurons will generate an action potential. Essentially all other cortical neurons use gamma-aminobutyric acid (GABA) as their neurotransmitter, send out short axons that terminate on neurons nearby in the same cortical or subcortical area, and are inhibitory; i.e., they act to hyperpolarize target neurons and decrease their likelihood of firing. Some structural themes reappear over and over again in different neuronal networks. Networks often incorporate divergence and convergence of neural connections (often occurring in conjunction); this is a result of the simple fact that most neurons receive large numbers of inputs and in turn (by axonal branching) connect to many other neurons. Signals can travel in sequential chains of neurons (as in a simple reflex arc). Often, neurons are reciprocally interconnected, either by excitatory or inhibitory connections. Lateral inhibition refers to the mutual inhibition of (neighboring) cells and is used in many sensory systems to sharpen responses to an incoming stimulus. Furthermore, a complex nervous system often allows multiple pathways between neurons. These types of connectivity can be used in many variations and combinations and can thus account for much of the brain's architecture.

THE VISUAL SYSTEM

Perhaps the most highly evolved sensory system in primates is the visual system (for an introduction and overview see Hubel[56]). A large part of the primate cerebral cortex is devoted to visual processing. Primates can sense several visual submodalities, for example movement, form, depth, and color. Photoreceptors in the retina detect incoming light signals and convert them into electrical signals. These signals leave the retina via the optic nerve, and by way of a part of the thalamus (the lateral geniculate bodies) and the optic radiation, these signals reach the primary visual cortex.

The primary visual cortex forms a topographic map of the visual field, with each visual hemifield mapped in the opposite cortical hemisphere. Each neuron in this map is responsive only to stimuli located in a small part of the visual field (the receptive field of that neuron); no single neuron in the primary visual cortex "sees" the entire visual field. Neurons are tuned to respond strongly only when certain preferred stimuli appear. For example, orientation-selective neurons respond when an appropriately oriented line appears in their receptive field. Neurons with similar response properties are often found close together, for example grouped together in orientation columns. Such columns form an ordered (not always continuous) map of orientations across the surface of the cortex.

"Higher" visual areas contain neurons that respond specifically to colored, textured, or moving visual stimuli.[76] They often have more extended receptive fields and form separate maps of the visual world. All these areas are interconnected by numerous cortico-cortical pathways. Eventually visual areas connect to so-called polysensory areas of the brain, i.e., areas that receive inputs from vision as well as other sensory modalities. The neural bases for most visual perceptual phenomena can be found in the visual cortex, in both "lower" and "higher" areas.[54]

COMPUTATIONAL MODELING OF THE BRAIN AND BRAIN THEORY

THE BRAIN AS A COMPUTER

In this century, the most powerful metaphor to understand brain function, and the functions and properties of (human) minds, has certainly been the computer. The computer metaphor basically implies that the brain is a piece of specialized hardware, and the mind a kind of software implemented to run on this hardware.[43,61,91] The use of the computer as an explanatory model for brain function has been subject to severe criticism from neuroscientists and philosphers of mind.[33,90,102] The work presented here is based on a different view of the nervous system, one that takes into account its evolutionary origins and embryonic development and offers biologically based explanations for a wide variety of brain functions, including perception.

But computers are useful in another sense. They increasingly serve as tools to investigate the brain as a complex system by modeling and simulating neurons and neuronal networks. Recently, the advent of powerful computers has made it possible to recreate realistic neuronal structures inside a computer. "Computational neuroscience"[103] refers to the study of the algorithms and computational strategies employed by the brain. Rather than subscribing to the view that the brain actually implements algorithms to solve problems, we will use the term "computational neuroscience" loosely to refer to the use of computers in studying principles of brain function. A recent introductory textbook[20] provides a useful, if somewhat biased, overview of computational neuroscience. More in-depth reading material is contained in several volumes with collected reprints of important papers in the field.[5,6]

MODELS OF NEURONS AND NETWORKS

There is a number of very important questions that concern the function of single nerve cells and can be addressed by modeling single neurons in great detail. Classical work in this area has been performed by Rall,[92] who studied the cable properties of dendrites and axons and their characteristic spatial and temporal characteristics. In this year's Summer School, single neuron modeling is dealt with by Bartlett Mel. Here, I will deal only with relatively crude approximations of single neurons and concentrate on the emergent properties of large populations and networks of neurons.

Binary threshold units as conceptual models of neurons in a network were introduced by McCulloch and Pitts.[77] These units are simple logical switches very much analogous to the on-off units of digital computers. Interconnected, they carry out symbolic logical operations. McCulloch and Pitts proved that for every computable function there exists an equivalent logical circuit consisting of binary units. A neural net of this kind could therefore serve as an implementation of a Turing machine.[117] Even though these ideas seem rather improbable in view of modern neuroscientific evidence, they effectively launched the ongoing efforts of researchers to build computer models of brain circuits.

Neuronal connections are not fixed but their strength can vary with experience. This has been proposed as the mechanistic basis for learning and memory.[52,53] This modification of synaptic connections follows certain rules. The most important set of rules states that a connection from neuron A to neuron B will be strengthened by temporally correlated activity in these neurons, as would be the case if neuron A succeeds (repeatedly) in firing neuron B. The original rule (see Hebb,[53] p. 62) reads: "When an axon of cell A is near enough to excite a cell B and repeatedly or persistently takes part in firing it, some growth process or metabolic change takes place in one or both cells such that A's efficiency, as one of the cells firing B, is increased." Hebb's formulation reflects one of the fundamental laws of association, the law of close temporal contiguity. The application of associationist thinking to the nervous system was anticipated by various other authors before Hebb.

Artificial neural networks often use "learning algorithms" such as back-propagation[101] or other methods based on optimizing a cost function by gradient descent. Such algorithms are biologically unrealistic in that they require global comparison of synaptic weights controlled by a teacher external to the model. Hebbian rules are more realistic because they are local (i.e., they do not require inputs or error signals from outside the locale of the synapse). There is much neurobiological evidence in support of such local rules.[65]

The dynamics of many neuronal networks are characterized by the activation (or response) function $s_i(t)$ of its constituent neuronal units i, and by the learning rule used to update the connection strengths c_{ij} between them. Typically, the response of a neuronal unit depends on extrinsic factors (most importantly inputs arriving from other neurons) and some intrinsic factors such as noise (spontaneous fluctuations of the membrane potential) or depression (the fact that neurons become less responsive after a certain number of action potentials have been fired). The activity of a neuron s_i can be interpreted as the frequency of action potentials in a certain time interval, as a firing probability, or as a membrane potential. The strength of a synaptic connection c_{ij} between cells i and j is usually given a value between 0 and 1 (or 0 and -1 for inhibitory connections). A typical activation function might read:

$$s_i = \phi(\sigma_j c_{ij} \times s_j + N) \qquad (1)$$

with $\sigma_j c_{ij} \times s_j$ denoting the sum of all inputs from units j connected to unit i and N denoting noise. The nature of the function ϕ defines if the neural unit is binary or continuous, linear or nonlinear.

Hebbian learning rules take the general form:

$$\delta c_{ij} = \delta \times A(s_j) \times B(s_i) \tag{2}$$

with δ defining the overall learning rate and $A(s_j)$ and $B(s_i)$ defining pre- and postsynaptic activity (as functions of s_j and s_i), respectively. Variations of this rule take into account that synaptic strengths can decrease if pre- and postsynaptic activity are uncorrelated (see below); i.e., neuron B fires while A does not, or vice versa. Most versions of Hebb's rule assume that no modification takes place if both neurons are inactive.

Synaptic rules may deviate from simple Hebb rules, most significantly by the incorporation of temporal and heterosynaptic effects. Some very detailed synaptic rules incorporating such effects have been proposed by Finkel and Edelman.[40,41] Heterosynaptic modification at a given site depends in direction or extent on events occurring at nearby synapses on the same cell, potentially mediated through second messengers. This differs from a simple Hebb rule, which specifies synaptic change without taking such "contextual" influences into account. Diffusible substances, such as nitric oxide, that are produced in an activity-dependent fashion, may play a special role in mediating widespread heterosynaptic or heterocellular effects.[44,82] In the work discussed below (see the last two main sections) we have incorporated heterosynaptic effects (although without explicit treatment of the kinetics of the chemical substances involved or of spatial dimensions) to model the influence of internally registered "value" states on synaptic modification related to behavior.

These relatively simple activation and learning rules can be used in a variety of contexts. As a broad theoretical basis for applying such rules in models of brain functions, we now turn to a discussion of the theory of neuronal group selection, proposed by Gerald Edelman.[28,30,32] To understand this theory and its implications we need to first examine the issue of selection and the application of selectionist thinking to the brain.

THE BRAIN AS A SELECTIVE SYSTEM

Newborn animals are confronted with a world full of sensory stimuli about which very little is known *a priori*. The partitioning of this stimulus world into meaningful categories will depend on the individual experience of that animal, and in particular on the adaptive value that this partitioning has for the organism. Different organisms belonging to the same species can and will often differ in their ways of responding to identical sensory stimuli. The evolution of different animal species living in the same habitat will often result in divergent modes of behavior and different adaptive strategies. For an organism lacking extensive experience or prewiring, the world is an "unlabeled place."[30,95] For instance, the category membership of even simple objects may be unknown and not intrinsically determined. Categorization of such objects requires exploration and choice, both depending extensively on various criteria of adaptive value, which themselves may be evolutionarily fixed. If the relations between the objects of the world on the one side and the organism with its complicated brain structure on the other side are not predetermined (in fact, not predeterminable), a brain theory has to address this issue and present a solution that is consistent with known ontogenetic mechanisms and neural structures.

The indeterminacy of the informational content of world objects and events is matched by the structural variability of animal nervous systems, at many levels of organization. As Karl Lashley pointed out in 1947: "The brain is extremely variable in every character that has been subjected to measurement. [...] individuals start life with brains differing enormously in structure" (Lashley,[71] p. 333). Lashley does not offer a theoretical explanation for this effect but ventures to speculate that "such variations [...] cannot be disregarded in any consideration of the causes of individual differences in mental traits" (ibid., p. 333). Lashley anticipated the potential influence of structural variability on individual behavior, although he did not see the implications for the potential role of selection in the brain. The structural variability of the brain does not usually affect gross anatomical features which are characteristic for the animal species. But the size and organization of cortical areas,[78] the thickness of fiber tracts, the number of neurons constituting a nucleus, and particularly the microanatomy of neuronal circuits can vary significantly from specimen to specimen (examples are presented by Edelman[30] and Reeke[98]).

The precise axonal and dendritic arborization patterns of neurons in any given neuroanatomical area show enormous variation both within and between individuals, to a degree which makes it very likely that this structural diversity cannot be genetically coded but must be the result of epigenetic regulatory processes acting during development. Such regulatory processes, for example, involve the adhesion, movement, differentiation, growth, division, and death of cells. It was proposed that the generation of diversity in growing neuronal structures is a necessary outcome of the action of morphoregulatory controls on cellular driving forces that are sensitive to local influence and context during development.[29,31] The enormous degree of variability within the nervous system is irreconcilable with theories of the nervous system based on the manipulation of information; machines specialized for such functions usually require precise wiring at all levels and scales. Indeed, most information processing theories of the brain perpetuate the erroneous conclusion that complex behaviors must be carried out by precise circuitry.

It is the basic tenet of the theory of neuronal group selection[28,30,32] that selection can resolve the apparent contrast between highly variable neuronal circuits facing an "unlabeled world" and the reproducible and adaptive behavior that they produce. The theory proposes that selectional mechanisms govern the formation, adaptation, and interactions of local collections of hundreds to thousands of strongly interconnected neurons called neuronal groups. These mechanisms can be roughly divided into developmental selection, experiential selection, and reentrant mapping.

DEVELOPMENTAL SELECTION. Developmental processes generate not only the well-defined macroscopic anatomical order of the nervous system that is characteristic for each animal species, but also lead to the formation of highly variant local neuronal circuitry. Nonselective systems usually need to be protected against variance in their structural components; such variance tends to increase the amount of noise and requires redundancy to restore reliable function. However, in a selective system, a certain degree of variability is mandatory and thus might itself constitute an adaptive trait in evolution. Developmental mechanisms that give rise to variable structures during epigenesis may thus be actively preserved and refined during evolution.

EXPERIENTIAL SELECTION. Postnatally, after much of the meshwork of anatomical connections has been laid down, synaptic mechanisms become the main agents of plasticity and adaptation. Synaptic selection guides the formation and the dynamics of the functional circuitry of neurons as they respond during experience. These experiential selectional processes are also responsible for the formation of neuronal groups as distinct entities, under competition for incoming and outgoing signaling pathways. Some of these synaptic mechanisms also act in the developmental formation of the network, and activity-dependent processes continue to shape the morphology of neurons postnatally, such that a clear temporal border between the developmental and experiential phases cannot be drawn.

In order for somatic selection to operate in the nervous system, its variable functional units must be exposed (either directly or through intermediate units) to a sufficient sample of the afferent sensory signals to permit it to respond differentially to various objects and events in the environment. Units undergoing selection must be able to contribute, through the output signals they emit, to some aspect of the behavior of the organism. Finally, the units must have the capacity to change their responses according to the relative success of the behaviors to which they contribute. (The term "success" can only be broadly defined and is not always directly related to the survival of the organism.)

The functional units of the nervous system are not single nerve cells but rather groups of hundreds to thousands of strongly interconnected cooperating neurons. Neuronal groups may sometimes correspond loosely to regions of strong anatomical connectivity. More often, however, their borders will not coincide with any anatomically distinguishable entity. Neuronal groups in the cerebral cortex most likely extend through all the cortical layers. The cells in a group will tend to share their response properties and receptive field structure. Furthermore, they will respond to incoming excitatory signals in a temporally correlated fashion. Such correlations both serve to strengthen the interactions of neurons within a group via synaptic mechanisms and contribute to their response characteristics. The selectivity of neurons within a group thus results from the specificity of their input fibers as well as from their dynamic interactions. The interactions of neurons both within and between groups continually increase or decrease as a result of selective changes occurring in response to changes in the patterns of their inputs. Thus, the set of neurons that constitutes a particular group can vary over time, but at any one time, neighboring groups will be nonoverlapping, spatially distinct, and separated by borders.

Group organization has several advantages for the nervous system: It reduces the need for specific point-to-point wiring in map formation by providing spatially extended targets for large afferent arborizations; it permits units with fixed anatomy to undergo functional reorganization as required by the changing needs and growth of the organism (this point has been demonstrated in a model by Pearson et al.[88] based on experiments of cortical reorganization in the somatosensory cortex by Kaas et al.[62]); it permits essential reciprocal mappings between distant cortical areas to be maintained during such reorganization; it permits signals reflecting sensory context to have consistent effects on units with related function by placing those units in close spatial proximity; it permits selective changes in synaptic efficacy to be coordinated across collections of neurons with related function; and it fosters the long-term stability of connections receiving common patterns of correlated input, reducing the danger that useful outputs will be disrupted by uncoordinated synaptic changes induced by any unusual strong inputs. The general importance of groups in the nervous system is attested to by the widespread occurrence of grouplike local structures such as ocular dominance columns, blobs, slabs, barrels, fractured somatotopies, etc. In any event, the dense interconnections throughout the nervous system make it most unlikely that cells could ever function as individuals. The opposite point of view argues for the autonomous function of single cells at many or all levels of the nervous system. Essential for these theories is that local cooperative effects between neurons tend to be ignored or considered to be of negative impact.

Evidence for the existence of neuronal groups is provided, for example, by studies of map organization and plasticity in the somatosensory cortex[79] and by observations on the dynamical interactions of visual cortical cells, which will be dealt with in great detail later. The latter experiments also provide direct evidence for reentry as a fundamental mode of interaction between neuronal groups in cortical areas.

REENTRANT MAPPING. In order for comparison and association of neuronal responses registered in separate cortical maps to take place, these neuronal responses must become correlated with each other. This correlation or conjunction is brought about by reentry, the ongoing and often recursive exchange of signals between maps along parallel anatomical connections. The anatomical and physiological basis for reentry will be discussed more extensively in the subsequent sections. Reentry can take many forms, including connections running backwards from a "higher" to a "lower" area, as well as laterally, between areas at the same hierarchical level but in different pathways. Reentry is typically reciprocal, involving the exchange of signals in both directions between two areas. Often, these areas are located in different sensory pathways (modalities), and reentry provides a mechanism for the correlation of responses to different aspects (i.e., form and color) of the same object in the environment. These different forms of reentry are of course associated with different functional roles; several of them occur in the various models described in detail in this chapter.

Feedback (negative and positive) is the central mechanism operating in the control of cybernetic systems.[121,122] Reentry and feedback share some properties but differ in others. Reentry is inherently parallel and involves populations of interconnected units, whereas feedback usually involves the recursion of a single scalar variable. Reentry is distributed; i.e., each area simultaneously reenters to many other areas (note that reentry can occur between areas at the same hierarchical level as well as between higher and lower levels in a system). Reentry has a statistical nature inasmuch as not all connections comprising a reentrant structure are used at all times. Finally, as we will show in the following main section, reentry can establish *correlations* between distant events in the cortical sheet[109] and give rise to the *construction* of novel operations.[42] It is generally used more for correlation than for error correction or gain control. A striking example of a densely interconnected distributed system in which reentry may be of critical importance is provided by the multiple functionally segregated areas of the visual system. In this system, most cortico-cortical and many thalamo-cortical connections are in fact reciprocal; this structural fact points to the importance and generality of reentrant mappings between different areas of the brain.

Reentry can also provide a structural and functional basis for associative memory, in that the responses of reentrantly activated groups often reflect the existence of stimulus correlations having potential significance[34] (see also Darwin III's categorization system described in the last main section). The modification of synapses involved in these responses, biased according to the strength of the reentrant response, provides the physical substrate for this form of memory. Functionally, such memory appears as the facilitation of categorical responses that have previously been selected in response to similar stimuli in the past; such facilitated responses are of course modified according to the current context. In this view, memory is a process of recategorization rather than a replicative storage of discrete data. Associations are developed across appropriate reentrant signaling pathways through the same mechanisms of synaptic modification that are used elsewhere to stabilize initial categorical responses. Frequently, these associations are formed between neuronal maps and reflect different aspects of a stimulus complex. Reentrant mapping on the basis of developmental and experiential selection is a major component in the performance of higher brain function.

MODELS OF PERCEPTUAL GROUPING AND FIGURE-GROUND SEGREGATION

What is the relationship between visual perception and underlying neural mechanisms? We know for sure that all visual perceptual phenomena (and all mental phenomena in general) are caused by the activity of populations of neurons in our brains. As an example of how it is possible to explain a perceptual function in terms of patterns of neural activity we will deal in detail with perceptual grouping and figure-ground segregation. In order to understand these phenomena we have to expand on our discussion of cortical anatomy (in the first part of this chapter) and deal with some general principles of cortical anatomy.

THE PROBLEM OF CORTICAL INTEGRATION

The mammalian cerebral cortex is partitioned into several areas, each specialized to carry out a relatively specific function.[123,124,125] For example, there are areas specialized for color vision, motion detection, texture, visual form, and location. In view of this high degree of *functional segregation* the question arises how neural activity in the various areas is integrated to form a unified percept of the world and to allow coherent behavioral responses. This problem of *cortical integration* has been of central concern in theoretical accounts of brain function.[22,30,32,38,42,106,115,116,125] We have designed a series of realistic computer models of cortical networks that demonstrate how cortical integration might be achieved.

Most earlier attempts to explain integration have started from the notion that the visual cortex is organized hierarchically with a progressive increase in the specificity of its neurons from the sensory periphery to more central areas. According to one version of the model, perceptual integration is achieved by the confluence of diverse processing streams at a very high hierarchical level or "master area." However, while there are areas of the brain that receive convergent inputs from multiple modalities or submodalities,[22] no single area has been identified that receives a sufficiently rich set of afferents to allow it to carry out a universal integrative function.

An alternative view of integration takes into account the cooperative interactions between different areas of the brain. This view is consistent with a characteristic anatomic feature of cortical organization: the abundance of reciprocal pathways linking functionally segregated areas. In the visual system, such reciprocal pathways exist between areas at the same or different hierarchical levels.[39] In addition neurons within each area are often densely and reciprocally interconnected (see e.g., Kisvarday and Eysel[67]). Some return projections are anatomically diffuse[124] and may serve a modulatory function, others can directly drive neurons in lower areas.[81] Edelman[28] proposed that such reciprocal pathways form the substrate of cortical integration by allowing the dynamic, bidirectional exchange of neural signals between areas (reentry, see previous section). Computer modeling of reentrant projections has shown that there are at least two ways in which reentry can act. It can directly influence the response properties of neurons (*constructive* function; e.g., Finkel and Edelman[42]), or it can give rise to patterns of temporal correlations (*correlative* function; e.g., Sporns et al.[109]).

Reentry can occur between neuronal groups located in the same cortical area (intraareal reentry) and between different maps (interareal reentry). First we will deal with a perceptual phenomenon which involves the activity of neurons in one visual area (the primary visual cortex) alone. Thus, we will address the problem of cortical integration both within as well as between cortical areas, focusing on the reentrant interactions between multiple segregated visual areas.

PERCEPTUAL GROUPING AND FIGURE-GROUND SEGREGATION

A classical problem in visual perception is that of perceptual grouping and figure-ground segregation. These two processes, both of fundamental importance in perceptual organization, refer to the ability to group together elementary features into discrete objects and to segregate these objects from each other and from the background. At the beginning of this century, Gestalt psychologists thoroughly investigated the factors influencing grouping and the distinction between figure and ground.[64,68,69] They described a number of perceptual laws ("*Gestalt laws*"), such as those of similarity, proximity, continuity, and common motion (see Figure 1).

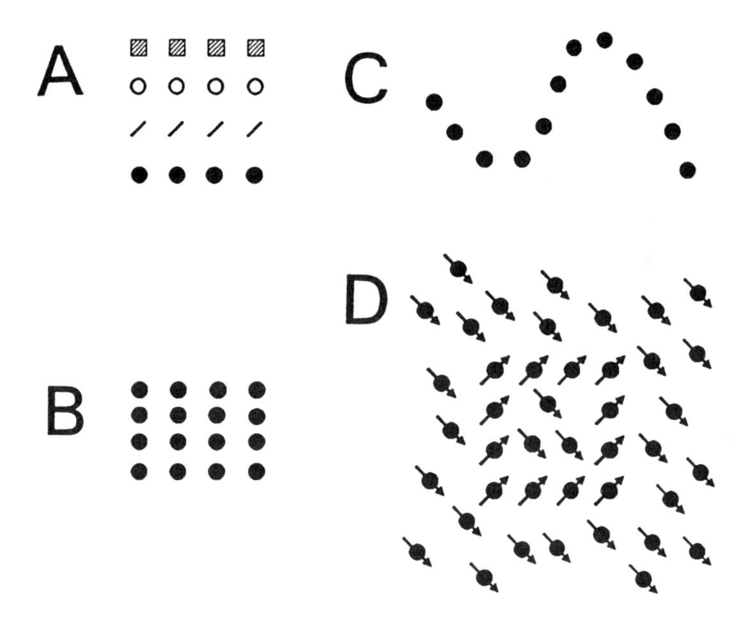

FIGURE 1 Examples of Gestalt laws. The panels show grouping and segregation of elementary stimulus features according to the laws of (a) similarity, (b) proximity, (d) continuity, and (d) common motion. In (d), arrows indicate the movement of two separate populations of dots.

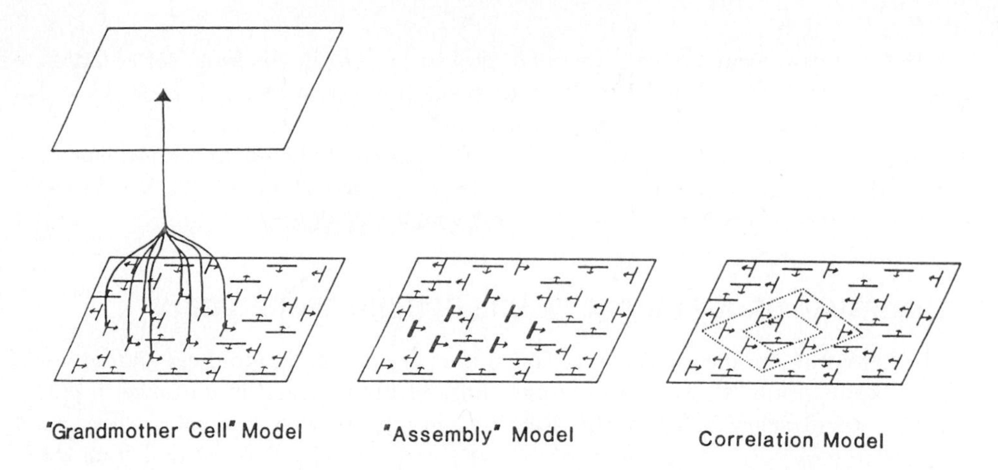

FIGURE 2 Different possible models of perceptual grouping and figure-ground segregation. In all three cases an identical display of moving short light bars is presented to a primary layer of direction-selective neurons. Eight of these bars move coherently to the right (the "figure," shaped like a diamond), the other bars move in random directions ("ground"). (Left) A specialized layer of "grandmother cells" that are appropriately connected to the primary layer detects the coherently moving object. The firing of a cell within this layer "codes" for the presence of the figure. (Middle) Neurons within the primary layer responding to the coherently moving figure enhance their firing frequency (indicated by thick bars). Other neurons responding to the noncoherent background remain at a lower level of firing. The mean activity level of an assembly of cells "codes" for the figure. (Right) Neurons within the primary layer responding to the coherently moving figure correlate their firing which becomes synchronous. This temporally correlated collection of neurons is called a cohort, the spatial extent of which is marked by the stippled line. Other neurons responding to the background are uncorrelated to neurons within the cohort and can therefore be segregated.

For example, looking at Figure 1(a), we tend to perceive a horizontal organization of the display by grouping similar stimulus elements together. In Figure 1(d), the randomly moving dots and the dots moving together forming the outline of a figure (a square in this case) segregate and the human observer vividly perceives a square moving in front of a background. While Gestalt psychology was very successful in identifying basic laws in human perception, the neural explanations offered by some Gestalt psychologists were less convincing.

In principle, perceptual grouping and figure-ground segregation could be accomplished by a number of neural architectures (Figure 2). In a hierarchical model based on *specialized detector units* (so-called "grandmother" cells; Figure 2, left) an appropriately connected detector "reassembles" the elementary features and signals the presence of the figure (a kind of hierarchical strategy has been proposed e.g., by Barlow[7]). However, the number of detectors needed to guarantee a response for each possible object in varying positions and contexts would be prohibitively large. More recently, several modelers and theoreticians have argued that cooperative processes may play a role in figure-ground segregation.[19,51,66] Expressed more directly in neural terms, a cooperative "assembly" model proposes that all those neurons that respond to a coherent figure enhance their *mean activity*[107] (Figure 2, middle). Ambiguity arises, however, if several "assemblies" responding to different objects or to an object and a coherent background have to be distinguished.[49] This would not be the case in a model based on *temporal correlations* (Figure 2, right). This model achieves grouping and segmentation by linking features through correlated activity among neuronal groups, made possible by intraareal reentry.

Recently, microelectrode recordings in the cat visual cortex have provided evidence that temporal correlations might play a role in visual perception, especially perceptual grouping and segregation. This experimental evidence suggests a role for correlative reentry in setting up stimulus-dependent patterns of correlated neural activity. Gray and Singer[47] reported that orientation-selective neurons show oscillatory discharges at around 40 Hz when presented with an optimally oriented stimulus. These oscillations can be observed in single cells,[59] as well as local populations of cells.[47] As discussed above, such populations, characterized by shared receptive field properties and temporally correlated discharge patterns, are examples of *neuronal groups.*[28] Oscillatory activity of cortical neurons has also been reported in motor cortex,[83] inferotemporal cortex[84] and visual cortex of the monkey.[27,70] Electroencephalographic[15] (EEG) and magnetoencephalographic[72,100] (MEG) recordings in human subjects have also revealed 40-Hz rhythmic cortical activity in a variety of brain regions and during various behavioral and cognitive states.

Further experiments revealed widespread patterns of correlations in the cat visual system and provided support for the notion that these patterns depend on reentrant connectivity. When a single long light bar is moved across the receptive fields of spatially separated neurons with similar orientation specificity, cross-correlations reveal that their oscillatory responses are synchronized.[25,48] The synchronization becomes weaker if a gap is inserted into the stimulus contour and it disappears completely if two parts of the contour are moved separately and in opposite directions. Synchrony is established rapidly, often within 100 msec, and single episodes of coherency last for 50 to 500 msec. Frequency and phase of the oscillations vary within the range of 40–60 Hz and $+/-3$ msec, respectively.[50] Synchronized neural activity was observed between cortical areas V1 and V2[25,86] and between striate and extrastriate cortical areas.[36] Engel et al.[37] have shown stimulus-dependent correlations between neurons located in the two hemispheres of cat visual cortex in response to two simultaneously presented stimuli, one on each side of the visual midline. These correlations disappear when the corpus callosum is transected, an indication that reentrant cortico-cortical pathways are responsible for their generation. After transection, the hemispheres continue to show neuronal activity at normal levels. This supports the view that mean activity rates alone are an insufficient indicator of neural integration, and that the temporal characteristics of neuronal firing play an important role in this process.

DYNAMIC BEHAVIOR OF SINGLE NEURONAL GROUPS

We have modeled the dynamic behavior of neuronal groups in several computer simulations.[109,110,111] In these models, neuronal groups contain both excitatory

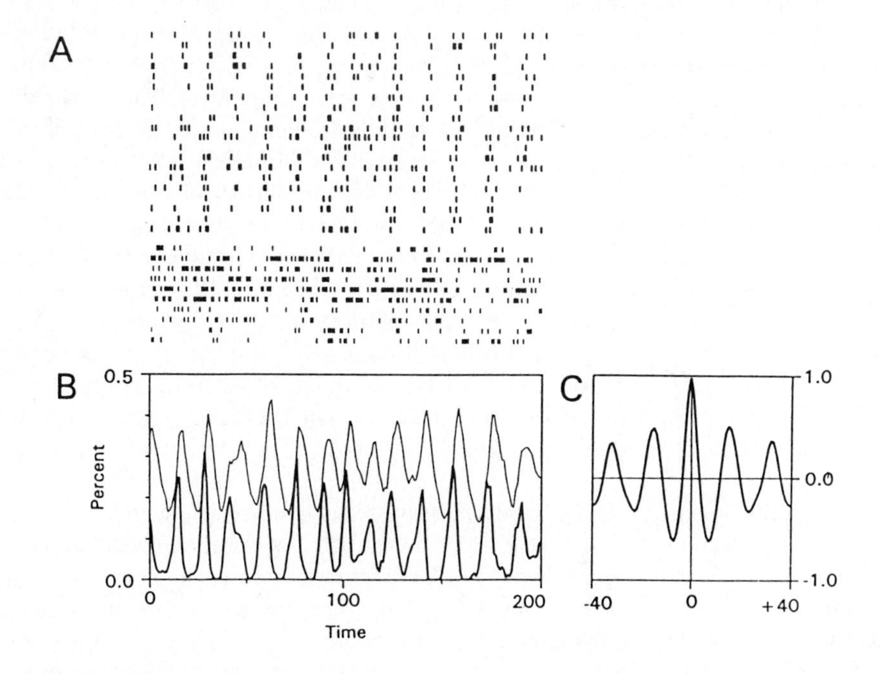

FIGURE 3 Rhythmic discharges in a simulated neuronal group. The group responds to an optimally oriented moving light bar and contains 160 excitatory and 80 inhibitory cells.[109] (a) Firing patterns of 20 randomly selected excitatory (top) and 10 randomly selected inhibitory cells (bottom) for a time period of 200 msec. Over this short time period, individual excitatory cells show a fairly irregular discharge pattern. It is not immediately obvious that these cells have oscillatory characteristics. In the model, inhibitory neurons have a higher spontaneous firing rate which further obscures their oscillatory discharge pattern. (b) Population activity (analogous to a local field potential) of the same neuronal group recorded simultaneously with the single neurons shown in (a). The activity of both excitatory and inhibitory populations (thick and thin traces, respectively) shows a pronounced oscillatory time course. The inhibitory population trails the excitatory population by about 5 msec (about a quarter of the oscillation period). Note that the frequency shows considerable variations. (c) Autocorrelation of the excitatory cell population for the same time period shown in (b).

and inhibitory neurons; there are connections within the excitatory cell population and reciprocal connections linking the excitatory and inhibitory populations. The connectivity pattern is sparse and random; for example, a given excitatory cell connects to 10 percent (chosen at random) of all other excitatory cells within the group. This local anatomical pattern is consistent with statistical data obtained from real cortical circuits.[114]

The discharge characteristics of a neuronal group are largely determined by cooperative interactions. In the simulations, locally correlated oscillatory activity is generated by the interactions between excitatory and inhibitory cells within a group (Figure 3). The mean oscillation frequency depends critically on the temporal delay introduced by the recurrent inhibitory connections. In accordance with experimental results,[50] the frequency of the oscillations when estimated for a short time-interval of 200 msec ("instantaneous frequency") varies significantly (Figure 3). This is due to the fact that each neuronal group acts as a population oscillator, composed of many sparsely connected and partly independent neurons. Over short periods of time the coherent activity of such neuronal groups is statistically more reliable than the activity of each of its constituent neurons. Thus, neuronal groups represent a first, elementary step for establishing functionally significant correlations.

COUPLED NEURONAL GROUPS

When two neuronal groups are reentrantly coupled (with fixed coupling strength and in the absence of any conduction delays), a variety of new dynamic phenomena emerge. Given a constant input and sufficient coupling the groups will synchronize their rhythmic discharge patterns and become temporally correlated. In the simulations, the instantaneous frequencies of the two groups become more similar, and the frequencies as well as their dispersion decrease with increasing coupling strength.[111] Thus, even when coupled to other groups, a given neuronal group will display dynamic variability in its oscillatory discharge pattern, although this variability is somewhat reduced. As observed by Gray et al.,[50] the phase shifts of the cross-correlations of two coupled groups are normally distributed with a mean phase shift of zero. At very low coupling strength all phase shifts are equally probable and the distribution is flat. Stronger coupling is accompanied by a reduced dispersion of phase shifts and a pronounced phase locking with zero phase lag (Figure 4(a)).

Another important observation is that neuronal groups with sparse and random intrinsic connectivity, display dynamic behavior that is highly variable across a population of individual examples of such groups. We must assume that as a result of stochastic processes in neural development, cortical neuronal groups are generated according to the same gross anatomic plan but differ in the precise pattern of termination of connections. In simulations, such variable groups show different mean frequencies and frequency dispersions in the absence of mutual coupling. In short,

A

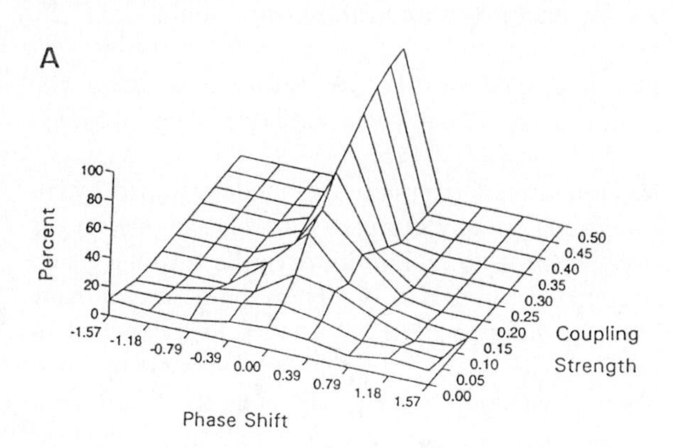

B

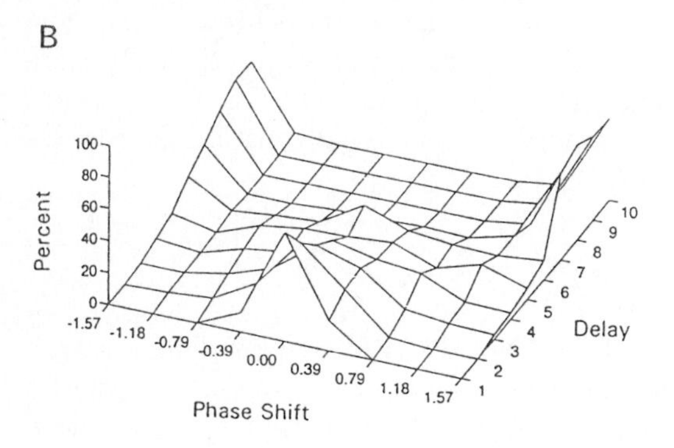

FIGURE 4 (a) Distribution of phase shifts for two reentrantly coupled neuronal groups as a function of coupling strength. Phase shifts are obtained using fitted Gabor functions and plotted from $-\pi/2$ to $+\pi/2$ (corresponding to approximately -10 and $+10$ milliseconds). Distributions are unimodal, normal, and centered around 0. At high coupling strengths, the two groups are strongly phase-locked with a phase shift of 0. There are no conduction delays in the reentrant connections. (b) Effect of allowing conduction delays in reentrant connections. Shown is the distribution of phase shifts for two reentrantly coupled neuronal groups as a function of delay (in msec) for a constant coupling strength of 0.25.

dynamic variability is the result of underlying structural variability. Such dynamic variability is of functional importance because it prevents accidental phase locking of groups that are not interacting or responding to common input.

Conduction delays in the reentrant connections linking the two groups influence the distribution of phase shifts significantly (Figure 4(b)). For small delays (1 to 4 msec) the distribution remains centered around a phase shift of zero. At around 5 msec (corresponding to about one quarter of the 20 msec oscillation period) a transition occurs and for delays above 5 msec most phase shifts are around $+/-(\pi/2)$. According to these simulations phase shifts are not equiprobable even with arbitrary conduction delays; phase shifts of 0 or $\pi/2$ are more likely to occur than $\pi/4$. In the brain, correlations (even if they exist over long distances, e.g., cross-callosally[37]) most often have a mean phase shift of 0 msec. On the other hand, the conductance velocity of many cortical fibers and pathways seems to be in the critical range of 5 msec or more.[18] Zero phase shifts could be achieved by the presence of at least some proportion of fast axons, or by other mechanisms including common input or short-term synaptic modification.

FIGURE-GROUND SEGREGATION IN A NETWORK MODEL OF THE VISUAL CORTEX

An interconnected network of many neuronal groups can produce a very large number of patterns of temporal correlations that depend on its anatomical connectivity and its inputs. The role of correlations in sensory segmentation has been explored in a number of models.[26,57,87,108,119,120] We have investigated the potential role of temporal correlations in perceptual grouping and figure-ground segregation in a model[110] which consisted of an array of orientation- and direction-selective neuronal groups forming a map of visual space. These groups are linked by a pattern of reentrant connections similar to that existing in visual cortex. There are preferential connections between neuronal groups with similar orientation and direction selectivity, and connection density falls off with distance. The response function of the modeled neuronal units is very similar to the one given by Eq. (4) (see below). An important feature of the model is the distinction between synaptic strength c_{ij} and synaptic efficacy e_{ij} (Figure 5). Changes in synaptic strength are generally considered to represent long-term synaptic plasticity; in the present model all strengths c_{ij} remain fixed. However, the synaptic efficacies e_{ij} of reentrant (and local excitatory-excitatory) connections can change on a short time scale (within tens of milliseconds). This change depends on pre- and postsynaptic activity, according to

$$e_{ij}(t+1) = (1-\gamma)\times e_{ij}(t)+\delta\times\phi(e_{ij}\times e_{ij}(t))+\delta\times\phi(e_{ij})\times(\bar{s}_i-\theta_l)\times(\bar{s}_j-\theta_j)\times R \quad (3)$$

with: δ = amplification factor, a parameter which adjusts the overall rate of synaptic change; γ = decay constant for synaptic efficacy; $\bar{s}_i$, $\bar{s}_j$ = time-averaged activity of cells i, j; θ_l, θ_j = amplification thresholds relating to post- and presynaptic activity, respectively; $R = 0$ if $\bar{s}_i < \theta_l$ and $\bar{s}_i < \theta_j$ and $R = 1$ otherwise; $\phi(x)$ = decreasing sigmoidal function. In the model, we assumed that $0 < c_{ij} + e_{ij} < 1$ and used the sum of strength and efficacy to compute the inputs $\sigma(c_{ij} + e_{ij}) \times s_j$ to a given cell i.

In general, the efficacy e_{ij} among correlated groups increases (with the effect of rapidly amplifying and stabilizing the correlations) and that among uncorrelated groups decreases. These changes are transient; in the absence of activity, the efficacy rapidly (within 100–200 msec) returns to its resting value. Evidence for voltage-dependent short-term increases in the efficacy of horizontal connections has been found in cat visual cortex.[45,55] Fast modulation of synaptic efficacy has also been observed in other parts of the nervous system.[2,3,4,46] Aertsen has pointed out[14] that short-term changes in the efficacy of connections can be achieved either by an actual biophysical change at the synapse or by dynamic influences exerted by other groups of neurons. From our perspective this underscores that, irrespective of actual biophysical mechanisms at the synaptic level, short-term modifications should be a general and widespread phenomenon in the cerebral cortex.

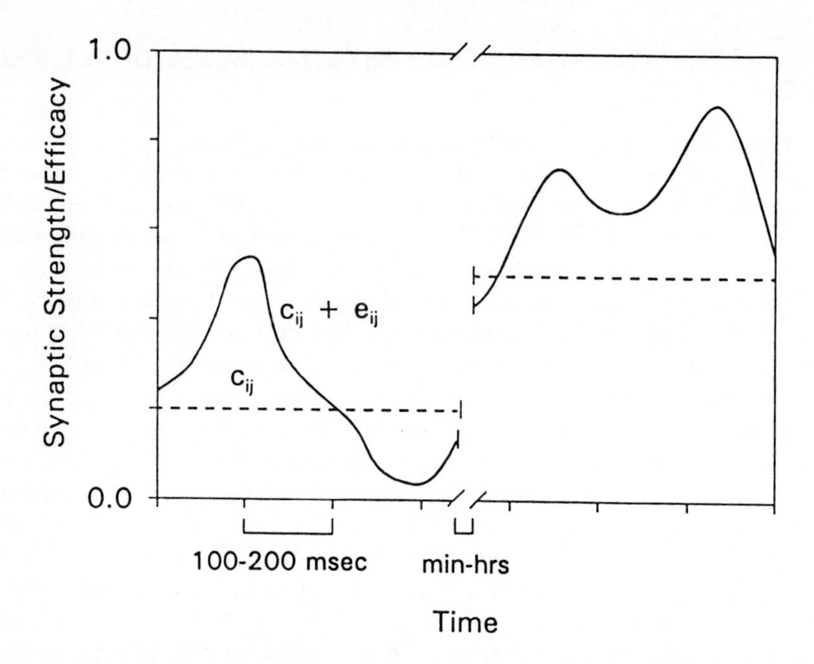

FIGURE 5 Schematic diagram of the time courses of synaptic strength (c_{ij}) and efficacy (e_{ij}). Efficacy and strength change on time scales of hundreds of milliseconds and of minutes to hours, respectively.

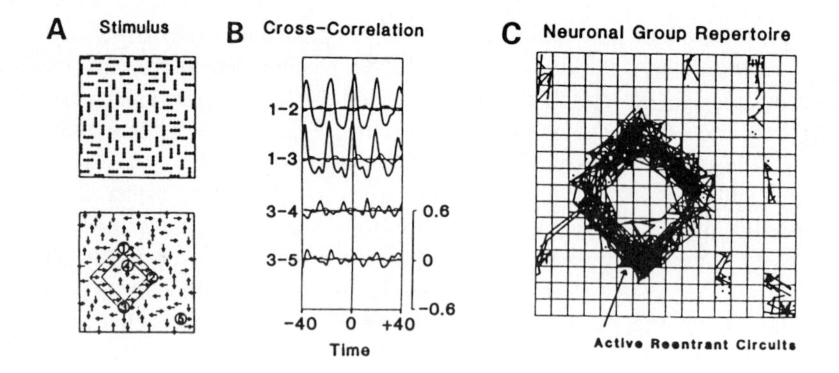

FIGURE 6 An example of grouping and segmentation in the model of Sporns et al.[110] (a) Stimulus presented to the model, consisting of a diamond-shaped figure, composed of vertically oriented bars, and a set of randomly oriented bars forming a background. In the top panel the bars are shown at their starting positions, the bottom panel shows their corresponding directions of movement indicated by arrows. (cont'd.)

FIGURE 6 (cont'd.) Encircled numbers with arrows in the bottom panel refer to the locations of recorded neuronal activity. The corresponding cross-correlations are displayed in (b). "Electrodes" 1 and 2 recorded from neurons responding to the figure, "electrodes" 3, 4, and 5 from neurons responding to the background. (b) Cross-correlograms of neuronal responses to the stimulus shown in (a). Cross-correlograms are computed over a 100 msec sample period and subsequently averaged over 10 trials. Numbers refer to the locations of neuronal groups within the direction-selective networks (see a). Four correlograms are shown, computed between msec 201 and 300 after stimulus onset. The correlograms are scaled and shift-predictors (thin lines, averaged over 9 shifts) are displayed for comparison. (c) Establishment of active reentrant circuits among neuronal groups selective for vertically oriented bars moving to the right, 250 msec after the onset of the stimulus shown in (a). Black lines indicate strongly enhanced functional connections.

Figure 6 shows an example of the responses of the model. The groups responding to the bars which composed the object are rapidly linked by coherent oscillations and are segregated from those groups responding to elements of the background. We found that the ability to establish specific linking (grouping) is directly related to the ability of achieving segmentation. Accordingly, there is no coherency among groups responding to elements of the figure and others responding to elements of the background; the latter include elements moving in the same direction as the figure, but placed some distance away. In the model, synchronization after stimulus onset is rapid and—in accordance with perceptual data—occurs usually within 100–200 msec. Multiple coherent episodes of varying length may occur at different times in different trials. Furthermore, synchrony is transient (between 100–500 msec), and its offset is fast, as would be clearly required by the fact that the visual scene continuously changes due to eye movements. In the model, episodes of correlated activity coincide with transient enhancement of reentrant connectivity due to short-term changes in synaptic efficacy.[110] Thus, the perceptual time scale of several hundred milliseconds may in part be determined by voltage-dependent effects at cortical synapses.

DISCUSSION

This computer model shows that, at least in principle, the neural basis for the integration and segregation of elementary features into objects and background might be represented by the pattern of temporal correlations among neuronal groups mediated by reentry. In addition, since the resulting grouping and segregation are consistent with the Gestalt laws, of continuity, proximity, similarity, common orientation, and common motion, it suggests that the neural basis for these laws is to be found implicitly in the specific pattern of reentrant connectivity incorporated into the architecture.

This proposed mechanism for the neural basis of perceptual grouping and figure-ground segregation creates a definite link between specific features of neural architectures (in this case the pattern of reentrant interconnections between cells in the visual cortex) and perception. The perception of visual Gestalt qualities is constrained, if not determined, by the specific connection pattern present in the visual cortex. Interestingly, to a large extent this pattern is not genetically predetermined but arises in postnatal development during exposure of the animal to its visual world. Thus, early experience gives rise to neural structures which in turn determine or at least severely constrain the function of the visual system in the adult. First, spatio-temporal correlations in the world (such as continuity of visual edges) lead to correlated neural activity in the developing nervous system selectively stabilizing those circuits that best match these correlations.[74] Second, in adult function, these correlations are, in a sense, "recalled" after having been embodied in the neural architecture. The world looks the way it does because reentrant connectivity reflecting significant patterns (correlations) in the world has been selected during development. It is an intriguing hypothesis that these intertwined processes of experiential selection and reentrant correlation play key roles in numerous other brain areas as well.

A NEURAL MODEL OF THE PRIMATE VISUAL CORTEX AND A SOLUTION TO THE BINDING PROBLEM

COOPERATIVE INTERACTIONS AMONG MULTIPLE CORTICAL AREAS

In the previous sections, we have dealt with integration and temporal correlations at the level of neuronal groups and within a single cortical area (linking). Another important problem concerns the "binding" of different visual attributes belonging to an object. Specific neural responses to these attributes occur in the various functionally segregated areas of the visual system. Coherent perception or behavior requires the integration of this distributed neural activity. An earlier model,[109] not dissimilar to the one discussed in the previous section, presented an initial example of how temporal correlations generated through reentry can solve the binding problem. Neural responses to an extended visual contour were integrated by local linking within V1 and reentrant binding from pattern-motion detectors in V5. Recently, in order to explain the unity of visual perception across submodalities, we generalized these results and simultaneously implemented linking and binding as well as constructive and correlative functions of reentry in a model of multiple areas of the visual system.[116]

We introduced a computational scheme that deals explicitly and efficiently with short-term correlations among large numbers of units. In the model, cortical maps are made up of basic units that stand for populations of neurons organized in groups. The activity of a unit thus reflects the temporal discharge pattern of an entire group rather than that of an individual cell. At each computer iteration (corresponding to several tens of milliseconds) the state of each unit i is characterized by an activation variable $s_i(t)(0 \leq s_i(t) < 1)$, and a phase variable $p_i(t)$ (circular, $0 \leq p_i(t)2\pi$). While s_i indicates the *activity level*, p_i indicates *when* the unit fires within the given iteration. Thus, we approximate the activity profile of a group of neurons for a given iteration by using an amplitude and a phase variable. In our

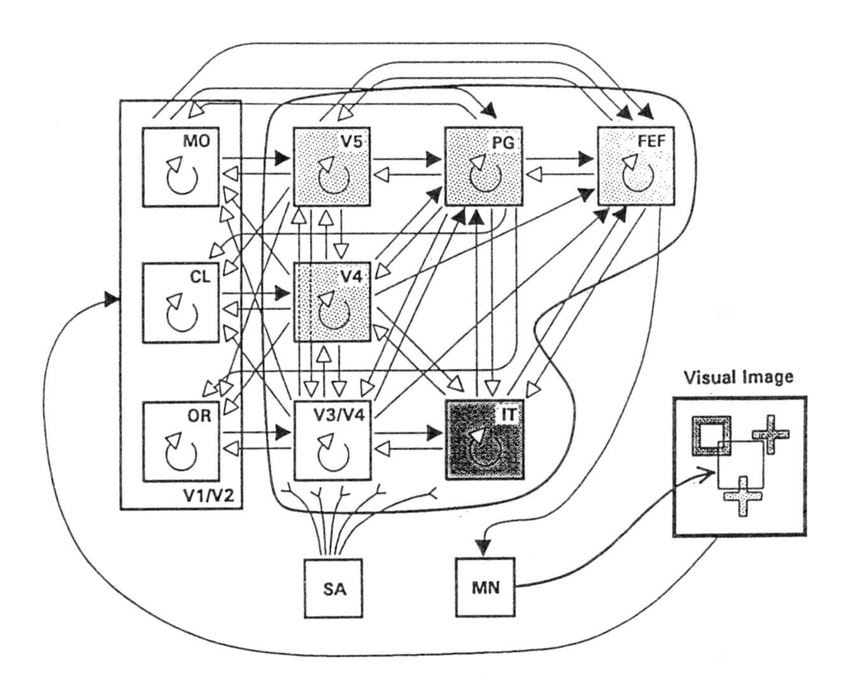

FIGURE 7 Architecture of a model of the visual system.[116] Segregated visual maps are indicated as boxes, pathways (composed of many thousands of individual connections) are indicated as arrows. The model comprises three parallel streams involved in the analysis of visual motion (top row), color (middle row), and form (bottom row). Areas are finely (no shading) or coarsely topographic (light shading), or nontopographic (heavy shading). The visual image (sampled by a color CCD camera) is indicated at the extreme right. The output of the system (simulated foveation movements under the control of eye motoneurons MN) is indicated at the bottom. Filled arrows indicate voltage-independent pathways, unfilled arrows indicate voltage-dependent pathways (incorporating short-term synaptic plasticity). Curved arrows within boxes indicate intra-areal connections. Box labeled SA refers to the diffusely projecting saliency system used in the behavioral paradigm; the general area of projection is outlined. V1/V2, V3/V4, V4 and V5 = corresponding visual cortical areas in the macaque monkey; MO, CL, OR = motion-, color-, and orientation-selective parts of V1/V2; PG = areas of parietal cortex; FEF = frontal eye fields; IT = inferotemporal cortex; SA = saliency system; MN = eye motor neurons.

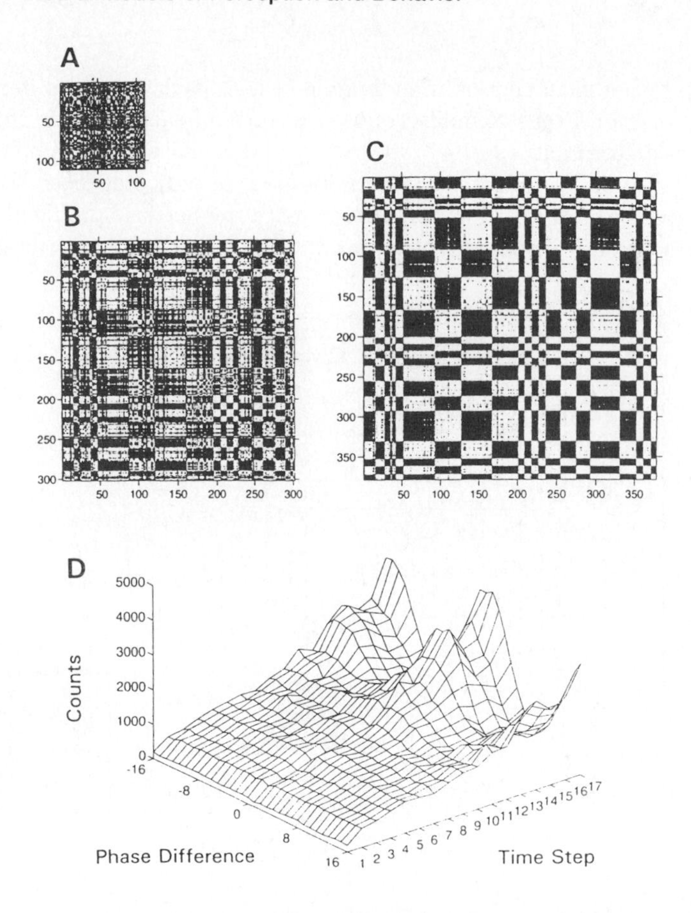

FIGURE 8 (a)-(c) Matrices of phase differences obtained from the model of the visual cortex.[116] All active units ($s_i > 0.2$) of the system are tabulated on the vertical and horizontal axes and their phase difference ($p_i - p_j$) is displayed (coded in shades of gray, white = 0, black = $+/-$ 16; note that p_i is circular and runs from $0 \leq p_i < 2\pi$, divided in 32 bins). The model responded to a set of two objects, a cross and a square (same as in Figure 9 in Tononi et al.[116]), displayed are time steps 1 (a), 10 (b), and 17 (c). In (a), not many units are responding and phase differences are randomly and fairly uniformly distributed. In (b), more units have become active and coherent domains of units with phase difference 0 begin to form. In (c), a checkerboard pattern indicates that the active units have split into two coherent, but mutually uncorrelated, populations. (d) Distribution of phase differences over time from the onset of stimulation (time steps 1 to 17, same run as in (a)–(c). The distribution is flat at the beginning of the run; at the end most phase differences are either 0 or $+/-$ 16.

scheme, correlations between two or more units can be segregated into two components, long-term (hundreds of milliseconds or seconds) and short-term (tens of milliseconds). While the long-term component is reflected in the covariance between activity variables over many iterations, the short-term component is reflected by phase differences at each iteration. As in the model of figure-ground segregation we incorporated a mechanism of short-term plasticity (voltage-dependency) for many of the visual pathways.

The model receives visual input from a color camera and contains nine functionally segregated areas divided into three parallel anatomical streams for form, color, and motion. The areas are connected by mostly reciprocal pathways. Altogether, 10,000 units are linked by about 1,000,000 connections between areas at different levels (forward and backward), areas at the same level (lateral), and within an area (intrinsic). A detailed description of the model is given by Tononi et al.[116] (for a schematic diagram see Figure 7).

We used the model to investigate a number of problems all of which are related to the overall theme of how integrated cortical function is achieved. We studied two psychophysical phenomena (the perception of form from motion and motion capture) which, respectively, illustrate the constructive and correlative functions of reentry. We showed that the reentrant interactions between the motion and form streams can be used to construct responses to oriented lines from moving random dot fields (form from motion). In the model, the activity of a "higher" area of the motion stream (V5) is reentered into a "lower" area of the form stream (V1) and modifies its responses to an incoming stimulus. Motion capture involves the illusory attribution of motion signals from a moving object to another object (defined by chromatic or color boundaries) that is actually stationary. Based on the model, we propose that the basis for the perceptual effect of motion capture is the emergence of short-term correlations between units in the motion and color streams as a result of reentrant connections linking these streams.

The correlative properties of the model are explored further in simulations involving all three streams. When presented with a single object, the model solves the so-called "binding problem" and displays coherent unit activity both within and between different areas, including a nontopographic one. Two or more objects can be simultaneously differentiated. We show that coherent unit activity depends on the presence of reentrant pathways giving rise to widespread cooperative interactions among areas. In order to efficiently evaluate patterns of correlations we simultaneously display short-term correlations in a correlation matrix. Consecutive displays of the correlation matrix reflect the functional connectivity of areas and units within areas over time (Figure 8(a)-(c)).

A key characteristic of this model is that successful integration is linked to an observable output, a simulated foveation response (for another example of how an integrated state can affect behavior, see Tononi et al.[115]). This eliminates the problem of deducing potential outputs by interpreting specific patterns of neural activity and correlations from the point of view of a privileged observer; it also shows that temporal correlations are not merely epiphenomena of neural activity. The foveation response is also used as a basis for conditioning. Reward for a correct discrimination response is mediated by activation of a saliency system that resembles diffuse projection systems in the brain, such as the monoaminergic and cholinergic systems. The simulated diffuse release of a modulatory substance from the saliency system regulates synaptic changes in multiple cortical pathways. After conditioning, the model performs a behavioral discrimination of objects that requires the integration through reentry of distributed information regarding shape, color, and location. This behavior depends critically upon the presence of short-term temporal correlations brought about by reentry and it does not require integration by a hierarchically superordinate area.

SUMMARY AND CONCLUSION

An anatomically and physiologically detailed model of the visual cortex can provide new insight into a fundamental problem of brain function, the problem of cortical integration. Central to the solution of this problem are the patterns of temporal correlations that result from the cooperative reentrant interactions within and between multiple brain areas.

The models illustrate that cortical integration takes place at *multiple levels of organization*: locally within neuronal groups, as well as within and between cortical areas (for a schematic diagram see Figure 9). These various levels are mutually interdependent. The cooperative interactions of neurons within groups are important because in order to express the linking and binding of, for example, object attributes, temporal correlations need to be statistically significant and be able to affect other sets of neurons. Significant correlations between small sets of single neurons spanning multiple cortical areas could not be formed or maintained over time. All our models have shown that dynamic integration is strongly dependent on underlying anatomy. The dynamic distinction of linking and binding follows closely the anatomic distinction of intra-map and inter-map reentry. Other levels that we have not taken into account involve the binding of neural signals relating different sensory modalities, or the binding of neuronal activity related to memory and actual sensory input.

Cortical integration has to meet important *temporal constraints*. Speaking somewhat loosely, the "perceptual time scale" is in the range of hundreds of milliseconds. For instance, in vision, the minimum duration of a perceptual experience appears to be around 100 msec.[12] Simple visual stimuli as well as complex visual scenes can be perceived within this time span.[11] According to these and other psychophysical results, any mechanism accounting for perceptual integration must be fast, on the order of a few hundreds of milliseconds. As both experimental and modeling studies have shown, short-term correlations provide a means to achieve integration within the limits of the perceptual time scale. We have suggested that the transient and dynamic character of perceptual integration is the result of changing patterns of short-term correlations and short-term modification of synaptic efficacies.

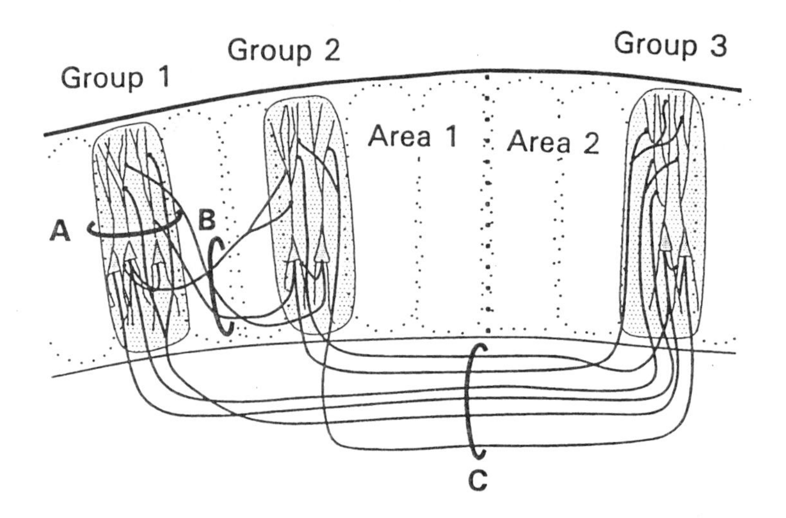

FIGURE 9 Schematic diagram illustrating different levels of neural integration (intragroup, intraareal, interareal). The diagram shows a cross-section through the cortex, oriented such that the outer layers are on top and the cortical white matter is at the bottom. The section shows the border region between two different areas (heavy dots). Three neuronal groups are outlined in detail, other groups are located in between (light dots); Groups 1 and 2 are in Area 1, Group 3 is in Area 2. The groups are comprised of pyramidal cells (small triangles); their dendrites are indicated as thin lines, their axons as thicker lines emerging from the bases of the triangular cell bodies. Pyramidal axons form three different types of connections: Cells connect to other cells within each neuronal group; cells in Group 1 connect to cells in Group 2 and vice versa; cells in Groups 1 and 2 (Area 1) connect to cells in Group 3 (Area 2); Group 3 in turn connects to cells in both Group 1 and 2. The three types of connections form the anatomical basis for integrated activity within and between neuronal groups (A = intragroup, B = intraareal and C = interareal). Note that the diagram omits connections to and from other neuronal groups and does not show multiple cell layers or dendritic termination patterns.

While it is generally accepted that changes in the firing frequency of neurons can influence and control behavior, it is less obvious that changing patterns of temporal correlations (even in the absence of any changes in firing frequency) can do the same. Experiments as well as modeling studies have suggested that individual cortical neurons are sensitive to temporal correlations in their inputs[1,9]; thus, the timing of activity in one set of neurons can have a well-defined effect on the activity of other neurons and can ultimately influence behavior. In some of our models[115,116] patterns of correlations were crucial in determining behavioral responses to complex stimuli. Some evidence[118] has pointed to a role for correlated neuronal activity in the control of behavior.

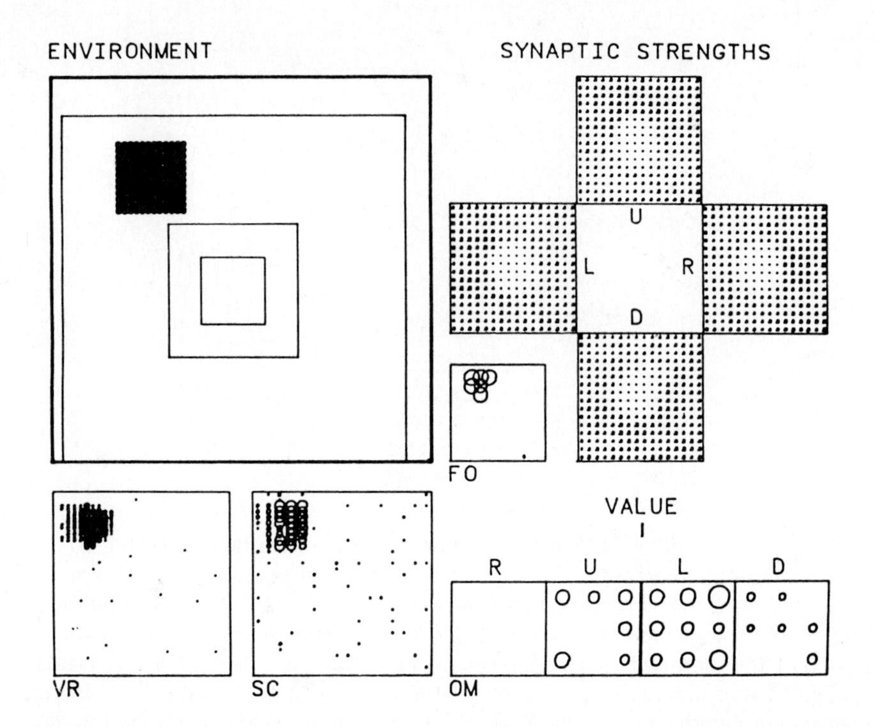

FIGURE 10 Display of Darwin III's environment and neuronal networks of the oculomotor system. The object is represented by the dark square on the upper left, Darwin III's visual field is indicated by the three squares centered on each other within the environment; the outer and the two inner squares delineate the outer boundaries of the visual field, and the foveal and perifoveal regions, respectively. Circles within the boxes labelled VR, SC, FO, and OM correspond to responses of neuronal units in that network. VR receives topographically mapped connections from the moving visual field. Its excitatory units map onto another network, SC, which in turn connects to four collections of neurons that move the eye (OM). A network of value-sensitive units (FO) receives dense connections from the center of the foveal region of the visual field and responds vigorously to a bright stimulus in that position, less vigorously (as shown) to a stimulus in the periphery. The map of the excitatory connections from SC to OM is displayed for the motor neurons that move the eye to the right, up, left, and down, respectively (R,U,L,D). A unit in OM receives connections from all positions in SC, and the squares in the connection strength plot are proportional to the strength of the connections that arrive from the corresponding position in SC. Initially these connections have random strength, and consequently movements elicited by sensory stimulation appear to be uncoordinated. After selective amplification under the influence of the global heterosynaptic input from FO, the maps show a characteristic pattern, with a central hole and a shallow gradient across the array. The asymmetric distribution of connections across SC will guide the eye towards the stimulus, and once centered, only small corrective movements will be made. Notice that in OM motor units for left- and upward motion are active, corresponding to a movement towards the target.

Our model of the visual cortex[116] shows that integration involves cooperative interactions of neurons both within and between cortical areas. We proposed that in integrating multiple cortical areas reentry can operate in two fairly distinct modes, *constructive and correlative reentry.* Reentrant inputs can modify the response properties of neurons directly and thus help to construct new response properties. An example of this was the generation of responses to form-from-motion boundaries in V1/V2 by reentrant inputs from V5. Reentrant interactions can also give rise to temporal correlations both within and between cortical areas; several examples have been given in this chapter. These two modes of reentry are not mutually exclusive; correlations may give rise to subtle changes in neuronal response properties, which may in turn affect patterns of correlations.

MODELS OF BEHAVING AUTOMATA: SIMULATIONS AND REAL-WORLD DEVICES

INTRODUCTION

The behavior of organisms is controlled by neural activity. Behavioral adaptation is the result of synaptic plasticity in the neuronal circuits giving rise to the behavior. This last part of the chapter will deal with behavior in simulated systems and real-world devices. The models we designed use principles and mechanisms of selection to achieve behavioral plasticity and adaptation as well as a rudimentary form of learning. The modeling strategy we have employed is called "Synthetic Neural Modeling" (SNM). Synthetic neural modeling aims at an understanding of complex neural phenomena by taking into account processes and interactions at all relevant levels, from the synaptic to the behavioral.[96] It attempts to model behaving neural systems in three important domains: the environment (containing objects and events, resulting in stimulus patterns), the body structure or phenotype of the system (its configuration of, for example, eyes and arms), and the nervous system itself (with its neuronal networks, patterns of anatomical interconnectivity and neuronal physiology). These domains cannot be viewed in isolation from one another; in fact, one main goal of SNM is to model and analyze their various interactions. In this section, we give some examples of how these interaction can be understood using a synthetic neural model.

DARWIN III

Darwin III is an example of a synthetic neural model consisting of a simple but realistic nervous system that is embedded in a specific "phenotype" acting in a specific "environment," all of which are simulated in a computer. It constitutes a complete selective system that can be studied at many different levels, separately or conjunctively. It has, in a crude form, some of the behavioral capabilities of an animal, without the use of *a priori* definitions of categories, codes, or information-processing algorithms. Darwin III allows us to examine sensory processes involving recognition and classification, motor acts, such as visual saccades, reaching, and touch-exploration, and the combination of both in a behaving automaton in which the effects of selection on output can influence the perceptual categorizations of the system through the rearrangement of the environment produced by the creature's own actions. The pattern of connections, and rules for cell responses and synaptic modifications are specified by the experimenter (using the CNS program) for each network. However, no instructions are given concerning the nature of particular stimuli, and no algorithms or other procedures are specified to govern the dynamics of its neuronal circuits above the single-cell level.

We will only give a brief summary of Darwin III's structure and performance; the interested reader may find a more detailed account elsewhere.[96,97] Darwin III consists of some 50 interconnected networks, with over 50,000 cells and over 620,000 connections between them and receives sensory input from three modalities: vision, touch, and kinesthesia. On the motor side, there is an eye and an arm with multiple joints, each controlled by collections of motor neurons.

Every cell in Darwin III has a scalar activity state determined by a response function. This function has terms corresponding to synaptic inputs, Gaussian noise, decay of previous activity, depression and refractory periods, and long-term potentiation (LTP). The relative magnitudes of these terms can be varied parametrically. The response function used in Darwin III is given in simplified form:

$$s_i(t) = [\sigma_j c_{ij} \times s_j] \times \phi(D) + N + W \tag{4}$$

where $s_i(t)$ = state of cell i at time t; $\sigma_j c_{ij} \times s_j$ = total synaptic input; c_{ij} = strength of connection from input j to cell i ($1 > c_{ij} > 0$, inhibitory); j = index over individual connections; $D = u_D s_i(t-1) + \omega_D D(t-1), u_D$ = growth coefficient for depression, ω_D = decay coefficient for depression (when $D > \theta_D$, where θ_D is a refractory threshold, then $\phi(D)$ is set to 0 for a specified number of cycles, after which D is set to 0 and $\phi(D)$ returns to 1.0); N = Gaussian noise; W = decay term $= \omega s_i(t-1)$; and $\phi(x)$ = decreasing sigmoidal function, approximated as $\phi(x) = 1 - 2x^2 + x^4$. The term $[\sigma_j c_{ij} \times s_j] \times \phi(D)$ must exceed a given firing threshold or it is ignored.

There are two major departures from simple Hebb rules (see previous section on computational modeling) in Darwin III, both of which are aimed at implementing the principle of *value-dependent modification*: (1) the changes in c_{ij} values of a given type may be made to depend heterosynaptically on the activity of cells in some network (*value system*) reflecting the organism's evaluation of its recent behavior; and (2) the dependence on pre- and post-synaptic activity may be based on moving-window time averages of these quantities rather than instantaneous values ("slow synaptic modification"). The coupling of synaptic change to value is essential if the automaton is to modify its behavior in a truly adaptive manner, as opposed to merely tuning chance responses, and the application of this principle will be discussed in detail for several different types of behavior in the following sections. Darwin III's synaptic modification rule is:

$$c_{ij}(t+1) = c_{ij}(t) + \delta \times \phi(c_{ij}) \times (\bar{s}_i - \theta_I) \times m_{ij} - \theta_J) \times (v - \theta_v) \times R \qquad (5)$$

where δ = amplification factor (learning rate), a parameter which adjusts the overall rate of synaptic change; $\bar{s}_i = \theta_I$ time-averaged activity of cell i, calculated according to $\bar{s}_i(t) = \lambda s_i(t) + (1 - \lambda)s_i(t-1)$, where λ = damping constant for averaged activity; θ_I = amplification threshold relating to postsynaptic activity; m_{ij} = average concentration of hypothetical postsynaptic "modifying substance" produced at a synapse made on cell i by cell j according to $m_{ij}(t) = m_{ij}(t-1) + u_M s_j - \text{Min}(T_M m_{ij}(t-1), T_M^\circ)$, where u_M = production rate for m_{ij}, T_M = decay constant for m_{ij}, T_M° = maximum decay rate for m_{ij} (m_{ij} may be replaced simply by $\bar{s}_j$ if desired; see similar synaptic rule for changes in efficiency e_{ij}, Eq. (3)); θ_j = amplification threshold relating to presynaptic activity; v = magnitude of heterosynaptic input from relevant value system neurons; θ_v = amplification threshold relating to value; and R = rule selector, in most cases $R = 0$ if $\bar{s}_j < \theta_l$ and $m_{ij} < \theta_J$ and $R = 1$ otherwise.

Darwin III consists of four neuronal subsystems: (a) a saccade and fine-tracking oculomotor system, (b) a reaching system using a single multijointed arm, (c) a touch-exploration system using a different set of "muscles" in the same arm, and (d) a reentrant categorizing system.

Darwin III's oculomotor system (Figure 10), by selection from a set of initially random eye motions, acquires the ability to move the eye toward objects in its visual field and to track moving objects. Visual signals are mapped topographically to a network containing excitatory and inhibitory neurons that loosely represents an area of the brain called the superior colliculus. Excitatory cells in this network are densely connected in the model to oculomotor neurons, whose activation causes eye motion. The initial connection strengths of these connections are assigned at random, so that eye motions are initially uncorrelated to visual stimulation.

An innate value scheme imposes a global constraint on the selection of eye movements: only those motions are selected which bring visual stimuli into the foveal part of the retina. All that is needed to accomplish this is a value system containing cells connected densely to the fovea and more sparsely to the periphery. The responses of these cells provide a heterosynaptic component which modulates the modification of connections from the "colliculus" to the motor neurons. Connections which are active in a short time interval before foveation occurs are thus selected and strengthened. Thus, selection affects synaptic and cellular populations *after* they have contributed to a motor act. What is selected are populations of connections that favor the reoccurrence of motions that happen to give foveation (resulting in "good" value). The selection process, however, is "blind" with respect to the individual contribution that each one of these connections makes (unlike "back-propagation"[101]). There is no *a priori* analysis indicating which connections may have adaptive value, and no error signal is computed and "back-propagated" into the network. The system only repeats those behaviors that result in a "positive" response of its value system. Our simulations showed that the oculomotor system improves considerably after only 150 presentations of an object at different locations (Figure 11); after training the eye quickly centers on a stimulus anywhere in the visual field. This simulation exemplifies how selection based on value can shape the development of a sensorimotor system.

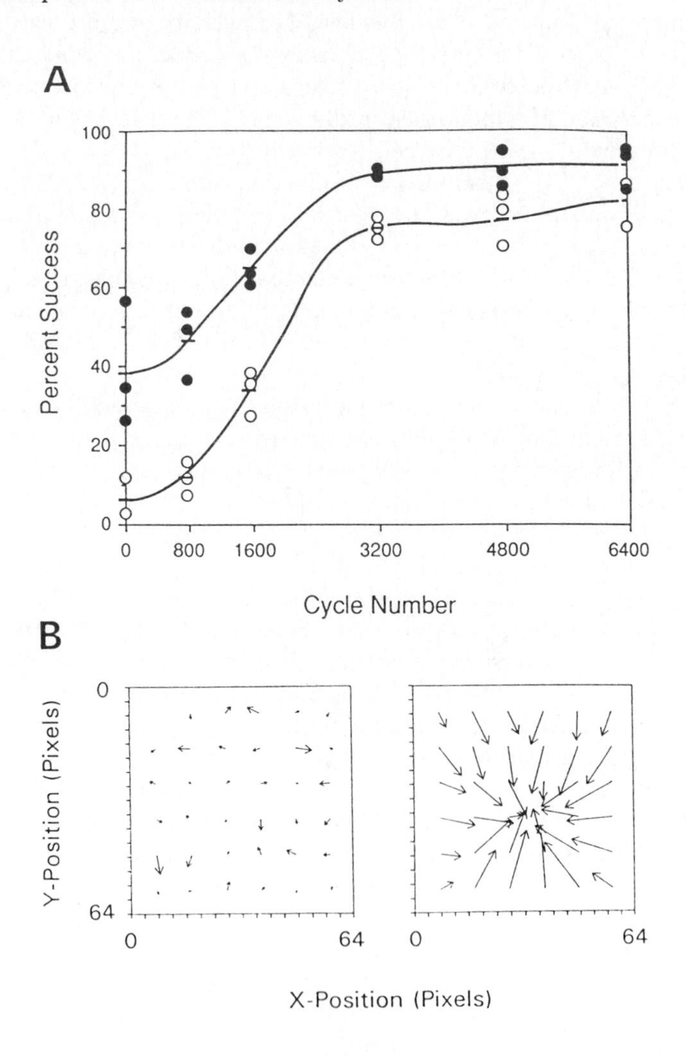

FIGURE 11 Ways to evaluate Darwin III's improvement in foveation. (a) Training curves for Darwin III's oculomotor system. Three versions of the oculomotor system were run, which were identical with respect to all system parameters but contained (cont'd.)

FIGURE 11 (cont'd.) different random number seeds determining the initial connection strengths and addresses for the connections from networks SC to OM and the sequence of object locations during training. After 0, 800, 1600, 3200, 4800, and 6400 cycles the performance of each of the three versions was evaluated as follows: connection strengths were kept fixed for 1600 cycles during which the object location was varied systematically to cover the input array. one-hundred trials of 16 cycles each were recorded and the percent success rate of the run was computed as the number of trials that were successful according to a given criterion. This criterion was foveation for one cycle any time during the trial (solid circles) and foveation for at least 6 consecutive cycles (open circles). The three values entered at each time point correspond to the three randomized versions tested. The data suggest that randomization of connections and of object locations during training does not alter significantly the progress towards optimal behavior. (b) Movement vectors recorded from simulations of Darwin III's oculomotor system before (left) and after training (right). The eye was set in turn at each of 36 defined positions (origins of arrows) in the input array (square enclosure, 64×64 pixels), while the object was stationary at the center of the array throughout the simulation. The eye was allowed to move freely, starting 20 times from the same position before being set to a new starting point. The maximal time allowed to reach the target was 16 cycles. For each of the 36 starting positions the movement of the eye was recorded and the average displacement of the eye on the array was calculated. An arrow is drawn for each of the starting positions indicating the average movement vector between cycles 1 and 3. Before training (left), average movement vectors are short, point in arbitrary directions, and show no correlation between neighboring positions. After training (right), the vectors are longer and point towards the target at the center; neighboring vectors show a high degree of correlation indicative of a continuous mapped representation of movement direction.

The problem of reaching with a multi-jointed arm is more difficult and complex than the visual tracking problem, because there are multiple degenerate solutions due to the mechanical redundancy of arm motions in three-dimensional space.[10] The nervous system needs to master such situations routinely by reducing the number of degrees of freedom to reach a controllable state. We propose that the mechanism by which the nervous system achieves this is essentially selective and leads to the amplification of gestural motions with adaptive couplings of relevant movement parameters to form motor synergies.[112] As an example of how this is done we examined Darwin III's multi-jointed arm (Figure 12) which develops smooth reaching movements by selection of such movements from a prior repertoire of spontaneous gestural motions (see Figure 13). In the model, movements are initiated by a simple "motor cortex" (MC), and the neural activity is transmitted via an intermediate network (IN), representing brainstem nuclei or an additional motor cortical layer, to the motor neurons (SG) that move arm extensors and flexors. Gestures emanating from these systems are filtered by a cerebellum-like structure, here simplified to just "granule cells" (GR) and "Purkinje cells" (PK). "Granule cell" inputs come from both kinesthesia (which senses the angular positions of the various joints in the arm) and target vision. As a result, the firing of "granule" cells corresponds to combinations of joint positions, i.e., particular conformations of the arm, and positions of the target. The system has the task of associating these arm positions with appropriate and inappropriate gestures through modification of synaptic connections, primarily those between "granule cells" and "Purkinje cells." "Purkinje cells" then act directly to inhibit inappropriate gestures at the motor cortical level (deep cerebellar nuclei are omitted here).

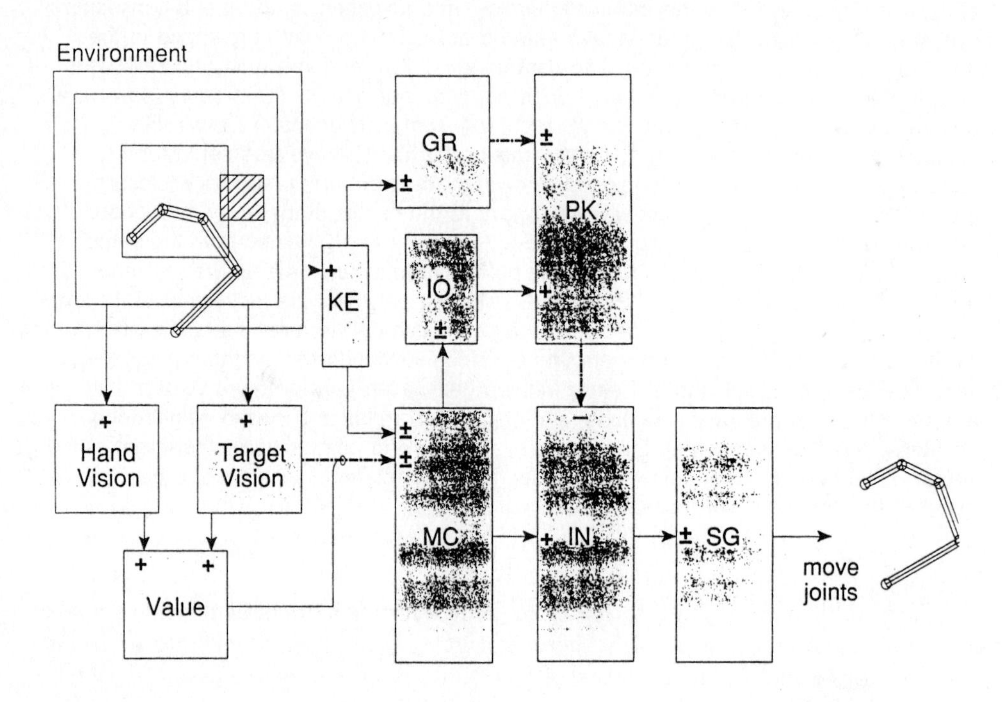

FIGURE 12 Schematic diagram of one version of the arrangement mediating arm reaching in Darwin III. The two large squares in the environment represent peripheral and foveal visual fields; the smaller hatched square is a stimulus object. The bent structure with circular links is the arm. The main main motor pathways across the bottom of the diagram are responsive to both visual signals (indicating the location of the target) and kinesthetic signals (indicating the conformation of the arm). "+," "−," and "+/−" indicate excitatory, inhibitory, and mixed connections, respectively. Heavy dashed arrows indicate modifiable connections. The figure also indicates a biased connection between "value" and the modifiable connections efferent from the target vision network. The model "cerebellum" (top) acts to inhibit inappropriate gestures and thus improves the directional specificity and smoothness of reaching.

Nothing in these circuits prejudices the arm to move in a coordinated way, or even in a particular direction, before training. Instead, motions are selected when they are successful in getting the arm closer to the object. Success is evaluated by a simple value system in which signals from a visual area responsive to the creature's own hand and signals from another visual area responsive to stimulus objects are combined in a common map such that its activity increases when the two visual responses map near a common location. Activity in this value system is carried to the "motor cortex," where it influences the extent of synaptic modification as described earlier, as well as to the "inferior olive" (IO), where it gates activity emanating from the motor cortical network to the Purkinje cell layer. Connections whose activation leads to an increase in the value response, representing motion of the arm closer to the object, are selectively favored.

With training and selective amplification of synaptic connections, Darwin III's reaching system improves significantly. A systematic assessment of its behavior shows how motor synergies emerge during the training process restricting the initially very broad envelope of paths to a narrow bundle, with most of the paths intersecting the target object (Figure 13). Plotting movement variables such as joint angles[94,112] shows that these variables are no longer independent of each other. The system has made choices and the resulting movements represent gestural "wholes," a movement Gestalt; this is a strategy that is taken without external instruction and is of great adaptive value to the organism because it considerably reduces the number of independently controlled variables. The four-jointed arm model described here acts as a gestural module. Because of the lack of appropriate neuronal mappings, it is unable to perform evenly over a larger region of space. The design of such mappings was investigated in a related but nonredundant reaching model.[96]

The arm of Darwin III, once it has established contact with the surface of an object, starts moving along the edges of objects guided by tactile signals. During such exploration it assumes a straightened posture by "reflex" to facilitate the tracing of objects. Tracing object contours provides Darwin III with sensory signals concerning the surface (or outline) properties of objects and these signals are used as a second sensory modality in subsequent object categorization.

In order to deal with nontrivial categorizations, a recognition system must be responsive to stimulus features of more than one kind. Subsystems responsive to different sensory modalities (or submodalities) form "classification couples,"[28] which correlate their respective neuronal signals by reentry. This arrangement reflects the need for an animal in a natural environment to combine apparently arbitrary combinations of features. Again, the issue of value is important; most likely, animal classification does not exist as an abstract faculty of "animal cognition," but subserves very definite behavioral functions. Indeed, its relevance for the organism becomes manifest only when it is coupled to appropriate behaviors having survival value for the organism. There will be, in general, neither a simple computable function dividing the stimulus space into distinct categories, nor a deterministic mapping relating perceptual categories to behavior.[13] This poses the difficult problem of how associations between sensory and motor signals can be formed.

These and other considerations, including simplicity, led to the design of the coupled categorization-response system in Darwin III. Since category formation in some form must exist prior to learning, we have made use of relatively impoverished low-level sensory networks capable of distinguishing just two categories to construct the first version of this system. The categories are "rough-striped" objects (bad for this species) and all others (good). The result of a "rough-striped" categorization is one of the simplest possible behaviors, a reflex-like arm motion (a "swat") that frequently removes the stimulus from the automaton's vicinity.

We will omit a detailed description of the connectivity and dynamic properties of the system and instead discuss its overall behavior (see Reeke et al.[96] for more details). As a result of activity in reentrant connections between visual and tactile areas the firing thresholds of neuronal units that are activated in a correlated fashion are lowered. When no novel features have been found for a time, a pattern is triggered in the reentrant categorization area that is characteristic of the particular class to which the object belongs (although different examples of the same class will give different patterns). This pattern, if it is one that activates the ethological value system, evokes the swat reflex to remove the stimulus. Stimuli not recognized as noxious are left undisturbed. Figure 14 shows a distribution of stimulus objects ordered by the frequency of occurrence of the rejection response. Although the

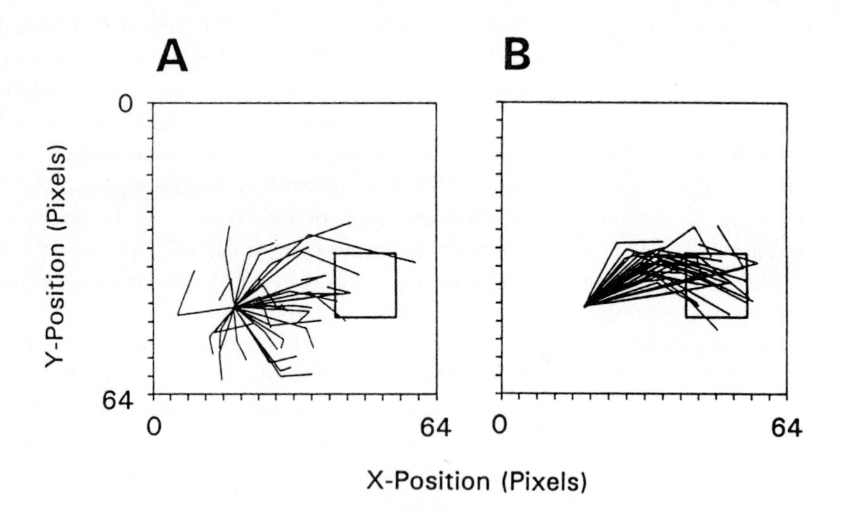

FIGURE 13 Traces of paths taken by the distal tip of the four-jointed arm before (a) and after (b) 180 training cycles. Training proceeded as follows: for each of 30 trials, the arm was placed in a standard position with the tip at the point where the trajectories diverge (lower left). It was then allowed to move for 6 cycles, and synaptic changes occured depending upon the success of the movements relative to a target object whose position is shown by the square at the right. After training (b), movements which reached the object on a direct path have been selected.

design of the categorization system was kept simple, the resulting behavior of the system is fairly complex. Classification of objects seems to result in graded category boundaries and is probabilistic rather than deterministic in character. Class

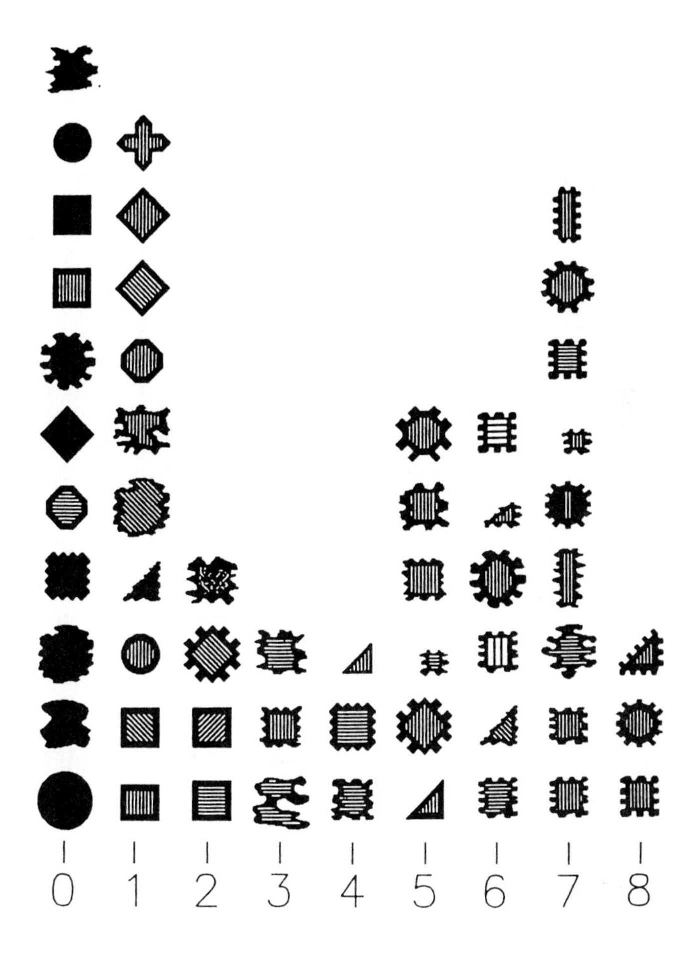

FIGURE 14 Objects grouped according to the responses of one version of Darwin III's categorization system. Each object was presented eight times and a maximal tracing time of 50-cycles per object was allowed. Tracing started at different positions on the object (usually close to the center) and the activity of the reflex output was constantly recorded. In the case of a rejection response the automaton's arm hit the object and removed it from the environment. An activation of the rejection response at any time within the 50-cycle time interval was counted as a "rejection" response. If no response occurred within the 50 cycle time limit, the trial was ended and a new object was entered. Objects are arranged in nine columns depending on the frequency with which they met with a response. In this version of Darwin III, an intrinsic negative ethological value is attached to "bumpy and striped" objects.

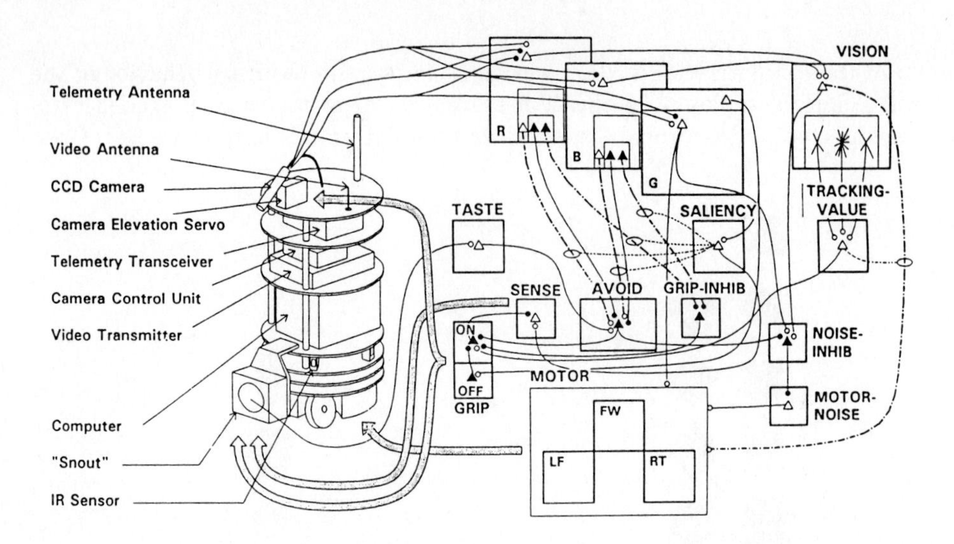

FIGURE 15 Schematic diagram of Darwin IV showing the major components of the mobile artifact (left) and the simulated nervous system used for the block sorting task (right). Boxes represent neuronal networks. Open and filled triangles denote excitatory and inhibitory cells, respectively. Lines connecting cells or networks indicate neuronal pathways (only a few representative cells and connections are shown); heavy arrows indicate efferent motor pathways leading to effectors. Open and filled circles indicate excitatory and inhibitory synapses, respectively. Synaptic pathways subject to value-dependent selective amplification are indicated by ·-·-·. Dashed lines originating at the value systems involved end in loops around the affected pathways. Beginning at the upper left, the red, blue, and green channels from the CCD video camera provide input to color-opposition cells in areas R, B, and G, respectively. Areas R and B jointly provide input to a VISION network (top right), which directly excites MOTOR areas FW (for forward motion), LF (for left motion), and RT (for right motion), bottom center (see also Figure 17, insets). These areas in turn activate NOMAD's wheels to produce locomotion. MOTOR neurons also receive noise input from MOTOR-NOISE area, which generates spontaneous (exploratory) motions of NOMAD. MOTOR-NOISE is inhibited by activity in NOISE-INHIB, which is excited by VISION, thereby reducing spontaneous locomotion when a potential tracking target is in sight (other inputs to NOISE-INHIB are described below). Connections from VISION to MOTOR are amplified under the selective influence of the TRACKING-VALUE network. TRACKING-VALUE responds most strongly to light falling directly in front of NOMAD, which excites the rectangular area near bottom of VISION network, corresponding to the proximal region of the field of view. Cells in the foveal and perifoveal regions of R and B (smaller rectangles at bottoms of R and B networks) are also connected via selectively modified synapses (see below) to excite area AVOID and to inhibit area GRIP-INHIB when an object is sighted at a moderate distance. After training, these areas are responsible, respectively, for initiating the avoidance and preventing the gripping reflexes discussed in the text. (cont'd.)

FIGURE 15 (cont'd.) GRIP-INHIB is normally active, leading to inhibition of GRIP. When GRIP-INHIB is itself inhibited, GRIP is released from inhibition and can activate the snout magnet and camera elevation effectors, but only when excitatory input is received at the same time from TRACKING-VALUE, causing NOMAD to pick up objects that have been foveated. AVOID acts by inhibiting both NOISE-INHIB and GRIP, causing NOMAD to move randomly without activating its gripping magnet. The green (G) visual network is responsible for recognition of the green home area. When the camera has been elevated and the green target is seen from a distance (when cells at bottom of G are active), MOTOR and NOISE-INHIB are excited, causing NOMAD to move towards the green target. When the target is reached, other cells (top of G) become active, exciting GRIP-OFF cells and providing the SALIENCY signal. SALIENCY immediately excites SENSE neurons, activating the conductivity sense in the snout. "Bad taste" activates TASTE cells, which excite reflex avoidance via a direct pathway to AVOID. Activation of AVOID by this mechanism while SALIENCY is still active leads to strengthening of connections from R or B (whichever was active when the object was recently viewed) to AVOID; as a consequence, after training, avoidance can be activated directly by visual signals without need for the object to be tasted.

membership is a matter of degree (as determined by the frequency of the associated behavioral act) rather than all-or-none. These results are in agreement with many experimental studies on human and animal categorization. In this version of Darwin III the selection of rough-striped objects for rejection is built in as an evolutionary imperative; a more general solution to the problem of associating behavioral acts with categories must involve the action of "ethological" value systems, analogous to the ones used in the sensorimotor systems. Such value systems imply the ability to evaluate consequences of behavioral acts; they permit the formation of stimulus-response associations based on experience rather than innate connectivity.

Darwin III is an example of an adaptively behaving creature synthesized from units defined at the neuronal level. The aim of synthetic neural modeling is to help us bridge the gap between the neuronal and psychological levels of description of mental functions, and to test by simulation the consequences of structural and dynamical variations at all levels. Selection of neuronal responses contributing to adaptive behavior seems to be all that is required to "bootstrap" an artificial organism into states in which it can make categorical discriminations.

DARWIN IV AND NOMAD

The simplified and predesigned nature of the objects in Darwin III's environment could have led to inadvertent bias in the simulations, and the resulting limited behavior could not readily be compared with that of animals. For example, in most simulations, the visual input was presented at rather low resolution with carefully controlled noise content and a featureless background. The environment was two-dimensional, and the arm and the objects it manipulated were not subject to inertia or friction.

To avoid these limitations, we have extended synthetic neural modeling to incorporate a real-world artifact in a new automaton, Darwin IV.[35] This restricts the computer simulation solely to the organization and dynamics of the nervous system; the environment and the artifact are real. NOMAD (Neurally Organized Multiply Adaptive Device), the artifact itself, moves about in the environment and provides visual and other sensory inputs to a simulated nervous system in a supercomputer. NOMAD in turn receives telemetered signals from the neural portions of Darwin IV that govern its behavior. Darwin IV can execute built-in reflexes as well as several modes of behavior that are subject to selective amplification. These behavioral modes and reflexes are combined during experience to perform a number of exemplary tasks, some of which are described here.

THE MOBILE DEVICE. The mobile artifact called NOMAD (Neurally Organized, Multiply Adaptive Device) is small enough to move about effectively in a single experimental room, but large enough for easy construction and adjustment. Its design (mostly using commercially available components) is modular, permitting new sensor and effector elements affecting its "phenotype" to be added without major redesign.

NOMAD is based on a battery-powered mobile platform with three steerable wheels that permit independent translational and rotational motion (Figure 15, left). Modules stacked on the platform provide effectors, sensors, and a telemetry interface to the nervous system, which is simulated on an nCUBE/10 parallel supercomputer. A rigid "snout" fitted with an electromagnet permits NOMAD to grip small metal objects. NOMAD's primary visual input is provided by a miniature color CCD video camera mounted near the top of the device. Camera azimuth can be varied by rotation of the platform as a whole; elevation can be directly controlled by NOMAD's nervous system. Additional "senses" are provided by electrical contacts in the snout (Figure 15, left), which can detect the conductivity of objects gripped by the electromagnet ("taste"), as well as by infrared (IR) proximity sensors mounted around the periphery of the base. In current versions of NOMAD, the primary function of the IR sensors is to redirect the motion of the base when a collision with a large fixed object, such as a wall, is imminent.

An on-board computer provides interfaces linking the various sensors and effectors to the simulated nervous system. The principle was followed that all signals exchanged between the nCUBE and this on-board computer should be in the form of encoded neuronal responses. The on-board computer initiates no behavior by itself (other than manual override and collision avoidance) and is strictly under control of the nervous system simulated in the nCUBE.

THE ENVIRONMENT. Experiments with NOMAD are carried out in a specially arranged room (Figure 16), equipped with a frame containing an 8' x 10' raised floor (covered with either opaque or translucent plexiglass) and surrounded by "walls" consisting of projection screens. Depending on the studies to be carried out, the experimental area may be configured to include a collection of objects and can contain real or projected landmarks upon the floor or walls. The area under the raised floor contains a second mobile platform controlled by a conventional robotics program. This platform carries a flashlight which projects a stimulus upward through the translucent floor for tracking by NOMAD. A television camera mounted on the ceiling is used to record the behavior of NOMAD for later evaluation.

THE NERVOUS SYSTEM. The same program (CNS, see above) used for simulating Darwin III was used to simulate and display the nervous system of Darwin IV (Figure 15, right) and to collect performance data. The previous network specifications for Darwin III[96,97] served as a prototype for the development of the new networks that were required to extend the simulation to real-world sensors and effectors. A topographically mapped visual network receives input from the three color channels of the video camera and projects to three groups of motoneurons, one each for translation, for rotation to the left, and for rotation to the right (Figure 15). Other networks handle sensory and motor pathways pertaining to the snout.

Neuronal responses are updated in discrete time steps using rules for cellular activity and synaptic modification based on those employed in Darwin III (Eqs. (4) and (5)). Visual units cease to respond to stationary visual images as a result of simulated cellular depression. Synaptic modification depends not only on pre- and post-synaptic activity but also on the activity of a set of neuronal units (a value system) reflecting the adaptive value of NOMAD's recent behavior. Such value systems embed structural features of particular nervous systems and phenotypes selected during evolution. They provide broad constraints for possible adaptive behaviors but, in general, do not fully determine the specific behaviors or categorizations of the individual organism. Darwin IV contains two distinct value systems. The first is implemented by visual units that respond more strongly when a target appears in the region of the visual field adjacent to NOMAD's snout. Its activity influences the probability of changes in synaptic strength between visual and motor networks leading to tracking (Figure 15, right). The second value system, referred to as "saliency," is triggered whenever Darwin IV activates its snout sensor to assess surface conductivity. "Saliency" modulates changes in the strength of the connections linking visual networks with reflex centers (Figure 15, right). The actions of neuronal networks controlling Darwin IV's built-in reflexes are described in detail in the next section.

TRACKING A MOVING LIGHT. The behavioral tasks investigated using NOMAD are relatively simple and serve as examples of how adaptive behavior can be achieved in a real-world device using selective principles. First, we investigated the ability of Darwin IV to approach and track a moving light mounted on the mobile base below the translucent floor of the environment. NOMAD's video camera was installed at the front of the device pointing at an angle of roughly 45° down from the horizontal. NOMAD's camera covers about 14% of the entire environment. Note that the oblique angle of the camera gives rise to keystone distortion of the visual image and that distant stimuli appear to grow larger when approached. This poses additional problems in constructing an appropriate neural map transforming sensory inputs to motor outputs.

Darwin IV's motor system includes separate sets of units which are spontaneously active. In the absence of a visual stimulus, NOMAD rotates and translates at random, driven by this spontaneous activity (search mode). When a stimulus appears and motoneurons are directly activated by the visual network (tracking mode), spontaneous activity is inhibited by separate connections from the visual area. After a movement has occurred, the synaptic populations giving rise to this movement are probabilistically strengthened or weakened by selection (see insets in Figure 17) depending on whether or not the movement resulted in an increase in value. After some time, those movements that facilitate close approaches and tracking occur more frequently than others. During training, selective modification of synaptic strengths automatically accommodates to any nonlinearities resulting from the distorted visual image and from the mechanics of NOMAD's motor apparatus.

During early trials, NOMAD rarely approaches the light (Figure 17). Even if NOMAD approaches the light by chance, it is unable to track the target's movements because there is no innate adaptation to carry out this task. After experience and the concomitant selectional events in the nervous system, NOMAD consistently tracks the light along complicated trajectories and loses contact only occasionally. When contact is lost, NOMAD briefly reverts to search mode, but it resumes tracking after encountering the light again.

SORTING COLORED BLOCKS. Following successful training for the tracking task, we used the networks developed by Darwin IV while tracking a moving light for a different but related task: locating and approaching stationary objects. Following exposure to the tracking task, NOMAD was placed on the environmental platform after an opaque floor under roomlight illumination was installed. The environment contained colored blocks (hollow 4" cubes) made of 0.008" gauge sheet steel (Figure 16). Surfaces of the blocks were coated with translucent plastic sheets of two kinds, one of which was electrically conductive while the other was not. The conductivity of each block was associated arbitrarily with its color—in the experiments described here, conductive blocks were blue and nonconductive blocks were red. We distributed the blocks in various patterns without providing any information to Darwin IV on their order or position or on the nature of the correlation of conductivity with color. In one part of the environment (see Figure 16), a raised green landmark signaled the "home position," a potential collection point for colored blocks.

To permit the device to distinguish between different colored blocks, the camera input was transmitted to three visual maps, each sensitive to a different primary color (red, green, or blue) and topographically mapped to the tracking vision network (Figure 15). Four motor reflex networks were also used. One, the "gripping reflex," turns the snout magnet on whenever an object appears in its immediate vicinity. Prior to training, this network is activated with a probability of around 50% and, as a result, the device picks up blocks randomly with this frequency. Value-dependent modification of connections from the red and blue color networks can modify this probability. A second reflex network controls the elevation of the camera. Two elevations were used in the sorting experiment: one pointing 45° down from the horizon for searching for blocks, the other horizontal for searching for the "home position." Normally the camera is positioned to search for blocks, but it is automatically elevated when the snout magnet is turned on and released when it is turned off. A third reflex is triggered whenever the green landmark activates the upper portion of the visual field (signaling proximity to home); NOMAD stops, the camera returns to face downward, and conductivity receptors in the snout are activated. The fourth reflex network strongly activates the set of neuronal units generating spontaneous locomotion if a conductive block ("bad taste") is sensed via the snout, resulting in an avoidance response. This network also receives modifiable connections from the red and blue color networks.

To test the global behavior of Darwin IV in searching for, locating, and approaching objects, we devised a simple block sorting task using conductive blue blocks and nonconductive red blocks. The aim of this task was to train Darwin IV, through experience with "good" and "bad" tasting blocks, to avoid the blocks whose color is associated with "bad taste" without the need to approach and grip them. Before training, NOMAD approached all blocks in its vicinity under the guidance of its tracking system, and, in approximately 50% of all approaches, established physical contact with the block's surface. In these cases, the block was gripped with the magnetic snout and the camera was raised by reflex into a horizontal position. The resulting change in Darwin IV's field of view allowed it to use the tracking system to search for the designated home position while pushing the block with its snout across the floor. As soon as the home position was reached, the camera turned downward again towards the block and the conductive properties of the block were sensed with the snout, leading to a strong signal in the "saliency" value system (Figure 15). This signal increases the probability of concurrent synaptic changes within Darwin IV's nervous system regardless of whether the block is in fact conductive or nonconductive. In general, the result is a strengthening of the association between the active vision networks and the gripping reflex. As a consequence, the probability of gripping an object in the future increases. If high conductivity ("bad taste") is sensed, however, the avoidance reflex is activated; this causes NOMAD to turn away from the direction of the object. Value thus allows an association to be formed between the color of the object in the foveal region and an aversive response. Because of resulting changes in synaptic strengths, in future encounters the appearance in the perifoveal region of an object of the color previously sensed as having bad taste is sufficient to trigger an avoidance response. After having delivered an object of either color to the home position, NOMAD resumes its search for new objects after habituation in the visual networks sets in.

If snout sensing and the corresponding value network are disabled, NOMAD collects red and blue blocks but does not behaviorally discriminate according to their color. If the value network is enabled, after a few encounters with red and blue blocks NOMAD sorts them according to color: only red blocks are taken to the home position, and blue blocks are actively avoided.

DISCUSSION. These first experiments using NOMAD were deliberately limited to very simple tasks. Our main objective was to demonstrate the feasibility of extending synthetic neural modeling to an artifact capable of performance in the real world. We developed a methodology that involves long-lasting runs and the simultaneous recording of neural and behavioral data. These data can be used for the

FIGURE 16 Photograph of NOMAD in its environment as arranged for the block sorting experiments (showing NOMAD, randomly distributed blocks, and the landmark for "home position"). Blue blocks are labeled with an asterisk.

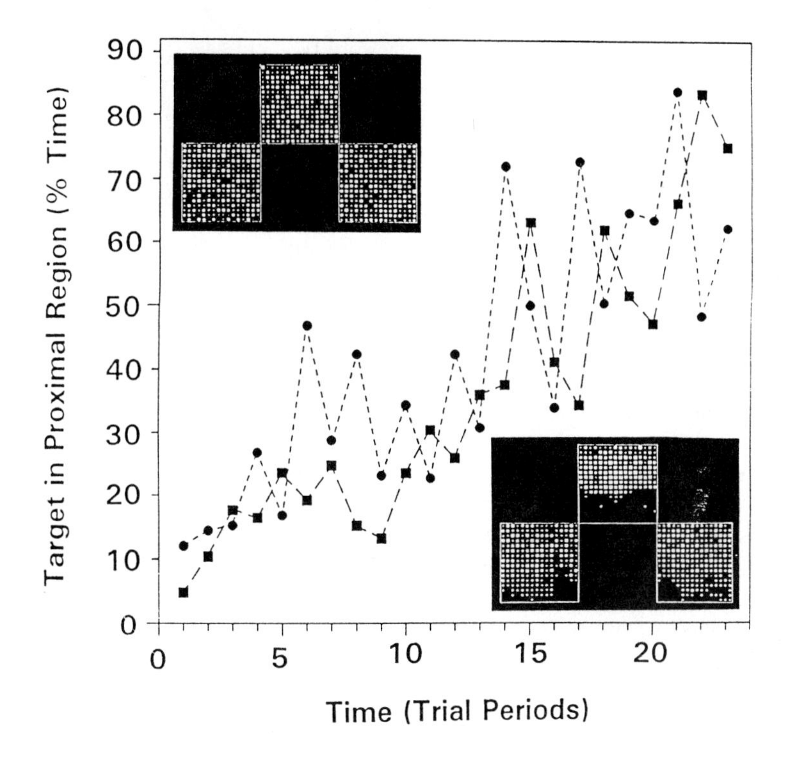

FIGURE 17 Training curve for two individual runs of Darwin IV's tracking system. Time is given in trial periods, with each trial period lasting for 250 time steps. The success rate is given as the percent of total time that the target was in the proximal part of NOMAD's visual field. Insets show strengths of connections between VISION and MOTOR networks before (upper left) and after (lower right) training. Stronger connections are indicated by larger and lighter squares.

analysis of cross-level interactions among synaptic changes, environmental stimuli, phenotypic variations, and behavior. This methodology is flexible and expandable; an experimenter studying NOMAD may devise different behavioral tasks, as well as neural and phenotypic structures.

Darwin IV's behavioral repertoire includes several reflex responses (gripping, camera elevation, snout sensing, and avoidance) as well as adaptive behaviors resulting from sensorimotor interactions (such as random search, tracking, approaching, and homing). In general, adaptive behaviors depend upon experience and are not predictable in detail, although they follow constraints posed by value schemes. Some of Darwin IV's tasks, for example block sorting, require the successful combination, based on experience, of sequences of reflex and adaptive behaviors. Within bounds, the device is able to transfer previously "learned" behaviors (such as approaching and tracking) to different environmental situations. For example, after training it was possible to use the same networks for tracking a moving light and for approaching stationary colored blocks or a colored home position on the basis of the differential reflectance of all three stimuli relative to the background.

Others working more directly in the field of robotics have attempted to design mobile devices operating in a real-world environment (for a review of some significant earlier efforts see Eyengar and Elfes[58]). Despite much effort, however, "classical" AI programs have generally been unable to deal effectively with autonomous behavior within a rich environment.[93] More recent work[8,17,21,75,80] on real or simulated behaving robots has emphasized how systems composed of independent modules can, in fact, give rise to composite behaviors in the absence of centralized control. Unlike Darwin IV, the design used for many of these systems is based on invertebrates (particularly insects), is hard-wired and nonplastic, and minimizes direct interactions between the constituent modules (for a detailed and critical discussion of other approaches see Reeke and Sporns[98]).

While the simple behaviors we have modeled so far using Darwin IV bear certain resemblances to these systems, our approach differs in several fundamental respects. Unlike that of other systems, Darwin IV's behavioral repertoire consists of a combination of several built-in reflexes with various modes of ongoing adaptive behaviors emerging from the encounters of a selectionally based nervous system with an environment. Most of the elementary behaviors are subject to value-dependent modification of synapses. Sensor-driven modes depend directly on selective synaptic processes to mold the resulting behavioral patterns. As a consequence of selective synaptic change, the elementary behaviors combine with each other and with reflexes during exploration to yield associative sequences that allow sorting of objects in the absence of a fixed sequence of programmed instructions. Each behavior is controlled by multiple networks in the nervous system; a strict one-to-one mapping does not exist between neural centers and the behaviors elicited by their activation.

The problem of understanding how adaptive action in a rich environment is initiated and controlled by vertebrate nervous systems requires the development and testing of global models of brain function. SNM is aimed at realistic modeling of such nervous systems and as such it uses biologically based synaptic rules to modify ongoing behavior. NOMAD, as embedded within the Darwin IV simulation, provides a valuable testing ground for neural models in real-world environments. It allows the investigator to explore many possible combinations of phenotypic structure and neuronal architecture. Specifically, it allows modeling of both modular ("invertebrate-like") and cortically integrated ("higher") nervous systems. So far, our studies provide a basis for incorporating increasingly complex neuronal structures into the simulation and for evaluating the impact of their function on behavior.

SUMMARY AND GENERAL CONCLUSIONS

This chapter has covered a wide terrain of models and problem areas in neuroscience. We have started from the assumption that to understand the brain, especially the complex interactions of processes across several levels of organization, it is necessary to design biologically based simulations. The examples presented in this chapter were intended to show that such simulations can indeed help to elucidate neural mechanisms of perception and behavior. Although making extensive use of computational tools and methods, the basic underlying principles that are implemented in these simulations are taken from biology and are not based on algorithmic conceptions of the brain or mind. Specifically, we have taken an approach based on a general theory of brain function that emphasizes principles of selection. Many of the examples in this chapter show how selectionist thinking can be applied to the analysis and simulation of brain function.

There are many challenges left. In particular, biologically realistic models of higher brain functions are still in their infancy, and the construction of whole automata (robots or autonomous systems) is still aiming more towards the construction of insects than towards instantiations of complex "higher" organisms. However, even at this early stage, computer modeling has become a valuable adjunct to experimental neuroscience. Undoubtedly, it will continue to contribute to our understanding of that very special complex system, the brain.

ACKNOWLEDGMENTS

All of the original work reviewed in this chapter has been done in collaboration with my colleagues at The Neurosciences Institute, most notably Gerald Edelman, George Reeke, and Giulio Tononi. The work was carried out as part of the Institute Fellows in Theoretical Neurobiology program at The Neurosciences Institute, which is supported through the Neurosciences Research Foundation. The Foundation received major support from the J. D. & C. T. MacArthur Foundation, the Lucille P. Markey Charitable Trust, and the van Ameringen Foundation. O. S. is a W. M. Keck Foundation Fellow.

REFERENCES

1. Abeles, M. *Corticonics.* Cambridge, MA: Cambridge University Press, 1991.
2. Aertsen, A. M. H. J., G. L. Gerstein, M. K. Habib, and G. Palm. "Dynamics of Neuronal Firing Correlations: Modulation of 'Effective Connectivity.'" *J. Neurophysiol.* **61** (1989): 900–917.
3. Aertsen, A., E. Vaadia, M. Abeles, E. Ahissar, H. Bergman, B. Karmon, Y. Lavner, E. Margalit, I. Nelken, and S. Rotter. "Neural Interactions in the Frontal Cortex of a Behaving Monkey: Signs of Dependence on Stimulus Context and Behavioral State." *J. für Hirnforschung* **32** (1991): 735–743.
4. Ahissar, M., E. Ahissar, H. Bergman, and E. Vaadia. "Encoding of Sound-Source Location and Movement: Activity of Single Neurons and Interactions between Adjacent Neurons in the Monkey Auditory Cortex." *J. Neurophysiol.* **67** (1992): 203–215.
5. Anderson, J. A., and E. Rosenfeld, eds. *Neurocomputing: Foundations of Research.* Cambridge, MA: MIT Press, 1988.
6. Anderson, J. A., A. Pellionisz, and E. Rosenfeld, eds. *Neurocomputing 2: Directions for Research.* Cambridge, MA: MIT Press, 1990.
7. Barlow, H. B. "Critical Limit Factors in the Design of the Eye and Visual Cortex." *Proc. R. Soc. Lond. B* **212** (1981): 1–34.
8. Beer, R. D. *Intelligence as Adaptive Behavior. An Experiment in Computational Neuroethology.* Boston, MA: Academic Press, 1990.
9. Bernander, O., R. J. Douglas, K. A. C. Martin, and C. Koch. "Synaptic Background Activity Influences Spatiotemporal Integration in Single Pyramidal Cells." *Proc. Natl. Acad. Sci. USA* **88** (1991): 11569–11573.
10. Bernstein, N. A. *The Coordination and Regulation of Movements.* Oxford: Pergamon, 1967.
11. Biederman, I., R. J. Mezzanotte, and J. C. Rabinowitz. "Scene Perception: Detecting and Judging Objects Undergoing Relational Violations." *Cog. Psychol.* **14** (1982): 143–177.
12. Blumenthal, A. L. *The Process of Cognition.* Englewood Cliffs, NJ: Prentice Hall, 1977.
13. Bongard, M. *Pattern Recognition.* New York: Spartan Books, 1970.
14. Boven, K.-H., and A. Aertsen. "Dynamics of Activity in Neuronal Networks give rise to Fast Modulations of Functional Connectivity." In *Parallel Processing in Neural Systems and Computers*, edited by R. Eckmiller, G. Hartmann, and G. Hauske, 53–56. Amsterdam: Elsevier, 1990.
15. Bressler, S. L., and R. Nakamura. "Inter-Area Synchronization in Macaque Neocortex During a Visual Discrimination Task." In *Computation and Neural Systems*, edited by F. Eeckmann and J. Bower, 515–522. Norwell, MA: Kluwer, 1993.
16. Brodal, P. *The Central Nervous System.* New York: Oxford University Press, 1992.

17. Brooks, R. A. "New Approaches to Robotics." *Science* **253** (1991): 1227–1232.
18. Bullier, J., M. E. McCourt, and G. H. Henry. "Physiological Studies on the Feedback Connection to the Striate Cortex from Cortical Areas 18 and 19 of the Cat." *Exp. Brain Res.* **70** (1988): 90–98.
19. Caelli, T. "Three Processing Characteristics of Visual Texture Segmentation." *Spatial Vision* **1** (1985): 19–30
20. Churchland, P. S., and T. J. Sejnowski. *The Computational Brain.* Cambridge, MA: MIT Press, 1992.
21. Connell, J. H. *Minimalist Mobile Robotics. A Colony-Style Architecture for an Artificial Creature.* Boston, MA: Academic Press, 1990.
22. Damasio, A. R. "The Brain Binds Entities and Events by Multiregional Activation from Convergence Zones." *Neural Comp.* **1** (1989): 123– 132.
23. Dennett, D. C. *Consciousness Explained.* Boston, MA: Little, Brown, and Company, 1991.
24. Dowling, J. E. *Neurons and Networks.* Cambridge, MA: Harvard University Press, 1992.
25. Eckhorn, R., R. Bauer, W. Jordan, M. Brosch, W. Kruse, M. Munk, and H. J. Reitboeck. "Coherent Oscillations: A Mechanism of Feature Linking in the Visual Cortex? Multiple Electrode and Correlation Analyses in the Cat." *Biol. Cybern.* **60** (1988): 121–130.
26. Eckhorn, R., H. J. Reitboeck, M. Arndt, and P. Dicke. "Feature Linking via Synchronization Among Distributed Assemblies: Simulations of Results from Cat Visual Cortex." *Neural Comp.* **2** (1990): 293–307.
27. Eckhorn, R., A. Frien, R. Bauer, T. Woelbern, and H. Kehr. "High Frequency (60-90 Hz) Oscillations in Primary Visual Cortex of Awake Monkey." *NeuroReport* **4** (1993): 243–246.
28. Edelman, G. M. "Group Selection and Phasic Re-entrant Signalling: A Theory of Higher Brain Function." In *The Mindful Brain*, edited by G. M. Edelman and V. B. Mountcastle, 51–100. Cambridge, MA: MIT Press, 1978.
29. Edelman, G. M. "Cell Adhesion and Morphogenesis: The Regulator Hypothesis." *Proc. Natl. Acad. Sci. USA* **81** (1984): 1460–1464.
30. Edelman, G. M. *Neural Darwinism.* New York: Basic Books, 1987.
31. Edelman, G. M. *Topobiology.* New York: Basic Books, 1988.
32. Edelman, G. M. *The Remembered Present.* New York: Basic Books, 1989.
33. Edelman, G. M. *Bright Air, Brilliant Fire.* New York: Basic Books, 1992.
34. Edelman, G. M., and G.N. Reeke, Jr. "Selective Networks Capable of Representative Transformations, Limited Generalizations, and Associative Memory." *Proc. Natl. Acad. Sci. USA* **79** (1982): 2091–2095.
35. Edelman, G. M., G. N. Reeke, Jr., W. E. Gall, G. Tononi, D. Williams, and O. Sporns. "Synthetic Neural Modeling Applied to a Real-World Artifact." *Proc. Natl. Acad. Sci. USA* **89** (1992): 7267–7271.

36. Engel, A. K., A. K. Kreiter, P. König, P., and W. Singer. "Synchronization of Oscillatory Neuronal Responses Between Striate and Extrastriate Visual Cortical Areas of the Cat." *Proc. Natl. Acad. Sci. USA* **88** (1991): 6048–6052.
37. Engel, A. K., P. König, A. K. Kreiter, and W. Singer. "Interhemispheric Synchronization of Oscillatory Neuronal Responses in Cat Visual Cortex." *Science* **252** (1991): 1177–1179.
38. Engel, A. K., P. König, A. K. Kreiter, T. B. Schillen, and W. Singer. "Temporal Coding in the Visual Cortex: New Vistas on Integration in the Nervous System." *Trends Neurosci.* **15** (1992): 218–226.
39. Felleman, D. J., and D. C. Van Essen. "Distributed Hierarchical Processing in the Primate Cerebral Cortex." *Cerebral Cortex* **1** (1991): 1–47.
40. Finkel, L. H., and G. M. Edelman. "Interaction of Synaptic Modification Rules within Populations of Neurons." *Proc. Natl. Acad. Sci. USA* **82** (1985): 1291–1295.
41. Finkel, L. H., and G. M. Edelman. "Population Rules for Synapses in Networks." In *Synaptic Function*, edited by G. M. Edelman, W. E. Gall, and W. M. Cowan, 711–757. New York: Wiley, 1987.
42. Finkel, L. H., and G. M. Edelman. "The Integration of Distributed Cortical Systems by Reentry: A Computer Simulation of Interactive Functionally Segregated Visual Areas." *J. Neuroscience* **9** (1989): 3188–3208.
43. Fodor, J. A. *The Modularity of Mind.* Cambridge, MA: MIT Press, 1983.
44. Gally, J. A., P. R. Montague, G. N. Reeke, Jr., and G. M. Edelman. "The NO Hypothesis: Possible Effects of a Short-Lived, Rapidly Diffusible Signal in the Development and Function of the Nervous System." *Proc. Natl. Acad. Sci. USA* **87** (1990): 3547–3551.
45. Gilbert, C. D. "Horizontal Integration and Cortical Dynamics." *Neuron* **9** (1992): 1–13.
46. Gochin, P. M., E. K. Miller, C. G. Gross, and G. L. Gerstein. "Functional Interactions Among Neurons in Inferior Temporal Cortex of the Awake Macaque." *Exp. Brain Res.* **84** (1991): 505–516.
47. Gray, C. M., and W. Singer. "Stimulus-Specific Neuronal Oscillations in Orientation Columns of Cat Visual Cortex." *Proc. Natl. Acad. Sci. USA* **86** (1989): 1698–1702.
48. Gray, C. M., P. König, A. K. Engel, and W. Singer. "Oscillatory Responses in Cat Visual Cortex Exhibit Inter-Columnar Synchronization Which Reflects Global Stimulus Properties." *Nature* **338** (1989): 334–337.
49. Gray, C. M., P. König, A. K. Engel, and W. Singer. "Synchronization of Oscillatory Responses in Visual Cortex: A Plausible Mechanism for Scene Segmentation." In *Synergetics of Cognition*, edited by H. Haken, 82–98. Berlin: Springer-Verlag, 1990.
50. Gray, C. M., A. K. Engel, P. König, and W. Singer. "Synchronization of Oscillatory Neuronal Responses in Cat Striate Cortex: Temporal Properties." *Visual Neurosci.* **8** (1992): 337–347.

51. Grossberg, S., and E. Mingolla. "Neural Dynamics of Perceptual Grouping: Textures, Boundaries, and Emergent Segmentations." *Perception & Psychophys.* **38** (1985): 141–171
52. Hayek, F. A. *The Sensory Order.* Chicago: Chicago University Press, 1952.
53. Hebb, D. O. *The Organization of Behavior: A Neuropsychological Theory.* New York: Wiley, 1949.
54. Held, R. "Perception and Its Neuronal Mechanisms." *Cognition* **33** (1989): 139–154.
55. Hirsch, J. A., and C. D. Gilbert. "Synaptic Physiology of Horizontal Connections in the Cat's Visual Cortex." *J. Neuroscience* **11** (1991): 1800–1809.
56. Hubel, D. H. *Eye, Brain, and Vision.* New York: Scientific American Library, 1988.
57. Hummel, J. E., and I. Biederman. "Dynamic Binding in a Neural Network for Shape-Recognition." *Psychol. Rev.* **99** (1992): 480–517.
58. Iyengar, S. S., and A. Elfes, eds. *Autonomous Mobile Robots: Control, Planning, and Architecture.* Los Alamitos, CA: IEEE Computer Society Press, 1991.
59. Jagadeesh, B., C. M. Gray, and D. Ferster. "Visually Evoked Oscillations of Membrane Potential in Cells of Cat Visual Cortex." *Science* **257** (1992): 552–554.
60. Jerison, H. J. *Evolution of the Brain and Intelligence.* New York: Academic Press, 1973.
61. Johnson-Laird, P. N. *The Computer and the Mind.* Cambridge: MIT Press, 1988.
62. Kaas, J. H., M. M. Merzenich, and H. P. Killackey. "The Reorganization of Somatosensory Cortex Following Peripheral-Nerve Damage in Adult and Developing Mammals." *Ann. Rev. Neurosci.* **6** (1983): 325–356.
63. Kandel, E. R., J. H. Schwartz, and T. M. Jessell. *Principles of Neural Science,* 3rd ed. New York: Elsevier, 1991.
64. Kanizsa, G. *Organization in Vision: Essays on Gestalt Perception.* New York: Praeger, 1979.
65. Kelso, S. R., A. H. Ganong, and T. H. Brown. "Hebbian Synapses in the Hippocampus." *Proc. Natl. Acad. Sci. USA* **83** (1986): 5326–5330.
66. Kienker, P. K., T. J. Sejnowski, G. E. Hinton, and L. E. Schumacher. "Separating Figure from Ground with a Parallel Network." *Perception* **15** (1986): 197–216
67. Kisvarday, Z. F., and U. T. Eysel. "Cellular Organization of Reciprocal Patchy Networks in Layer III of Cat Visual Cortex (Area 17)." *Neuroscience* **46** (1992): 275–286.
68. Koffka, K. *Principles of Gestalt Psychology.* New York: Harcourt, 1935.
69. Köhler, W. *Gestalt Psychology.* New York: Liveright, 1947.
70. Kreiter, A. K., and W. Singer. "Oscillatory Neuronal Responses in the Visual Cortex of the Awake Macaque Monkey." *Eur. J. Neurosci.* **4** (1992): 369–375.

71. Lashley, K. S. "Structural Variation in the Nervous System in Relation to Behavior." *Psychol. Rev.* **54** (1947): 325–334.
72. Llinas, R., and U. Ribary. "Coherent 40-Hz Oscillation Characterizes Dream State in Humans." *Proc. Natl. Acad. Sci. USA* **90** (1993): 2078–2081.
73. Lockwood, M. *Mind, Brain, and the Quantum.* Cambridge: Basil Blackwell, 1989.
74. Löwel, S., and W. Singer. "Selection of Intrinsic Horizontal Connections in the Visual Cortex by Correlated Activity." *Science* **255** (1992): 209–212.
75. Maes, P., ed. *Designing Autonomous Agents: Theory and Practice from Biology to Engineering and Back.* Cambridge: MIT Press, 1991.
76. Maunsell, J. H. R., and W. T. Newsome. "Visual Processing in Monkey Extrastriate Cortex." *Ann. Rev. Neurosci.* **10** (1987): 363–401.
77. McCulloch, W. S., and W. Pitts. "A Logical Calculus of the Ideas Immanent in Nervous Activity." *Bull. Math. Biophys.* **5** (1943): 115–133.
78. Merzenich, M. M., R. J. Nelson, J. H. Kaas, M. P. Stryker, W. M. Jenkins, J. H. Zook, M. S. Cynader, and A. Schoppmann. "Variability in Hand Surface Representations in Areas 3b and 1 in Adult Owl and Squirrel Monkeys." *J. Comp. Neurol.* **258** (1987): 281–296.
79. Merzenich, M. M., G. Recanzone, W. M. Jenkins, T. T. Allard, and R. J. Nudo. "Cortical Representational Plasticity." In *Neurobiology of Neocortex*, edited by P. Rakic and W. Singer, 41–67. Chichester: Wiley, 1988.
80. Meyer, J.-A., and S. W. Wilson, eds. *From Animals to Animats. Proceedings of the First International Conference on Simulation of Adaptive Behavior.* Cambridge: MIT Press, 1991.
81. Mignard, M., and J. G. Malpeli. "Paths of Information Flow Through Visual Cortex." *Science* **251** (1991): 1249–1251.
82. Montague, P. R., J. A. Gally, and G. M. Edelman. "Spatial Signaling in the Development and Function of Neural Connections." *Cerebral Cortex* **1** (1991): 199–220.
83. Murthy, V. N., and E. E. Fetz. "Coherent 25- to 35-Hz Oscillations in the Sensorimotor Cortex of Awake Behaving Monkeys." *Proc. Natl. Acad. Sci. USA* **89** (1992): 5670–5674.
84. Nakamura, K., A. Mikami, and K. Kubota. "Oscillatory Neuronal Activity Related to Visual Short-Term Memory in Monkey Temporal Pole." *NeuroReport* **3** (1992): 117–120.
85. Nauta, W. J. H., and M. Feirtag. *Fundamental Neuroanatomy.* New York: Freeman, 1986.
86. Nelson, J. I., P. A. Salin, M. H.-J. Munk, M. Arzi, J. and Bullier. "Spatial and Temporal Coherence in Cortico-Cortical Connections: A Cross-Correlation Study in Areas 17 and 18 in the Cat." *Visual Neurosci.* **9** (1992): 21–37.
87. Neven, H., and A. Aertsen. "Rate Coherence and Event Coherence in the Visual Cortex: A Neuronal Model of Object Recognition." *Biol. Cybern.* **67** (1992): 309–322.

88. Pearson, J. C., L. H. Finkel, and G. M. Edelman. "Plasticity in the Organization of Adult Cortical Maps: A Computer Model Based on Neuronal Group Selection." *J. Neuroscience* **7** (1987): 4209–4223.
89. Penrose, R. *The Emperor's New Mind.* Oxford: Oxford University Press, 1989.
90. Putnam, H. *Representation and Reality.* Cambridge: MIT Press, 1988.
91. Pylyshyn, Z. W. *Computation and Cognition: Toward a Foundation for Cognitive Science.* Cambridge: MIT Press, 1984.
92. Rall, W. "Cable Theory for Dendritic Neurons." In *Methods in Neuronal Modeling*, edited by C. Koch and I. Segev, 9–62. Cambridge, MA: MIT Press, 1989.
93. Reeke, G. N., Jr., and G. M. Edelman. "Real Brains and Artificial Intelligence." *Proc. Am. Acad. Arts and Sciences (Daedalus)* **117** (1988): 143–173.
94. Reeke, G. N., Jr., and O. Sporns. "Selectionist Models of Perceptual and Motor Systems and Implications for Functionalist Theories of Brain Function." *Physica D* **42** (1990): 347–364.
95. Reeke, G. N., Jr., L. H. Finkel, and G. M. Edelman,. "Selective Recognition Automata." In *An Introduction to Neural and Electronic Networks*, edited by S. F. Zornetzer, J. L. Davis and C. Lau, 203–226. New York: Academic Press, 1990.
96. Reeke, G. N., Jr., L. H. Finkel, O. Sporns, and G. M. Edelman. "Synthetic Neural Modelling: A Multi-Level Approach to Brain Complexity." In *Signal and Sense: Local and Global Order in Perceptual Maps*, edited by G. M. Edelman, W. E. Gall, and W. M. Cowan, 607–707. New York: Wiley, 1990.
97. Reeke, G. N., Jr., O. Sporns, and G. M. Edelman. "Synthetic Neural Modeling: The 'Darwin' Series of Automata." *Proc. IEEE* **78** (1990): 1498–1530.
98. Reeke, G. N., Jr., and O. Sporns. "Behaviorally Based Modeling and Computational Approaches to Neuroscience." *Ann. Rev. Neurosci.* **16** (1993): 597–623.
99. Reichert, H. *Introduction to Neurobiology. Stuttgart, FRG: Thieme.* New York: Oxford University Press, 1992.
100. Ribary, U., A. A. Ioannides, K. D. Singh, R. Hasson, J. P. R. Bolton, F. Lado, A. Mogilner, and R. Llinas. "Magnetic Field Tomography of Coherent Thalamocortical 40-Hz Oscillations in Humans." *Proc. Natl. Acad. Sci. USA* **88** (1991): 11037–11041.
101. Rumelhart, D. E., G. E. Hinton, and R. J. Williams. "Learning Representations by Back-Propagating Errors." *Nature* **323** (1986): 533–536.
102. Searle, J. R. *The Rediscovery of the Mind.* Cambridge: MIT Press, 1992.
103. Sejnowski, T. J., C. Koch, and P. S. Churchland. "Computational Neuroscience." *Science* **241** (1988): 1299–1306.
104. Shepherd, G. M. *Neurobiology*, 2nd ed. Oxford: Oxford University Press, 1988.
105. Shepherd, G. M., ed. *The Synaptic Organization of the Brain*, 3rd ed. Oxford: Oxford University Press, 1990.
106. Sherrington, C. *The Integrative Action of the Nervous System.* New Haven: Yale University Press, 1947.

107. Singer, W. "Activity-Dependent Self-Organization of the Mammalian Visual Cortex." In *Models of the Visual Cortex*, edited by D. Rose and V. G. Dobson, 123–136. London: Wiley, 1985.
108. Sompolinsky, H., D. Golomb, and D. Kleinfeld. "Global Processing of Visual Stimuli in a Network of Coupled Oscillators." *Proc. Natl. Acad. Sci. USA* **87** (1990): 7200–7204.
109. Sporns, O., J. A. Gally, G. N. Reeke, Jr., and G. M. Edelman. "Reentrant Signaling among Simulated Neuronal Groups Leads to Coherency in their Oscillatory Activity." *Proc. Natl. Acad. Sci. USA* **86** (1989): 7265–7269.
110. Sporns, O., G. Tononi, and G. M. Edelman. "Modeling Perceptual Grouping and Figure-Ground Segregation by Means of Active Reentrant Connections." *Proc. Natl. Acad. Sci. USA* **88** (1991): 129–133.
111. Sporns, O., G. Tononi, and G. M. Edelman. "Dynamic Interactions of Neuronal Groups and the Problem of Cortical Integration." In *Nonlinear Dynamics and Neural Networks*, edited by H. G. Schuster, 205–240, Weinheim, FRG: Verlag Chemie, 1991.
112. Sporns, O., and G. M. Edelman. "Solving Bernstein's Problem: A Proposal for the Development of Coordinated Movement by Selection." *Child Devel.* **64** (1993): 960–981.
113. Stebbins, G. L. *Darwin to DNA, Molecules to Humanity.* New York: W. H. Freeman, 1982.
114. Stevens, C. "How Cortical Interconnectedness Varies with Network Size." *Neural Comp.* **1** (1989): 473–479.
115. Tononi, G., O. Sporns, and G. M. Edelman. "The Problem of Neural Integration: Induced Rhythms and Short-Term Correlations." In *Induced Rhythms in the Brain*, edited by E. Basar and T. H. Bullock, 367–395. Boston, MA: Birkhäuser, 1992.
116. Tononi, G., O. Sporns, and G. M. Edelman. "Reentry and the Problem of Integrating Multiple Cortical Areas: Simulation of Dynamic Integration in the Visual System." *Cerebral Cortex* **2** (1992): 310–335.
117. Turing, A. M. "On Computable Numbers, with an Application to the Entscheidungsproblem." *Proc. Lond. Math. Soc.* **42** (1937): 230–265.
118. Vaadia, E., E. Ahissar, H. Bergman, and Y. Lavner. "Correlated Activity of Neurons: A Neural Code for Higher Brain Functions?" In *Neuronal Cooperativity*, edited by J. Krüger, 249–279. Berlin: Springer-Verlag, 1991.
119. von der Malsburg, C., and J. Buhmann. "Sensory Segmentation with Coupled Neural Oscillators." *Biol. Cybern.* **67** (1992): 233–242.
120. von der Malsburg, C., and W. Schneider. "A Neural Cocktail-Party Processor." *Biol. Cybern.* **54** (1986): 29–40.
121. Wiener, N. *Cybernetics.* Cambridge: MIT Press, 1948.
122. Wisdom, J. O. "The Hypothesis of Cybernetics." *Brit. J. Phil. Sci.* **2** (1951): 1-24.
123. Zeki, S. "Functional Specialization in the Visual Cortex of the Rhesus Monkey." *Nature* **274** (1978): 423–428.

124. Zeki, S., and S. Shipp. "The Functional Logic of Cortical Connections." *Nature* **335** (1988): 311–317.
125. Zeki, S. "Parallelism and Functional Specialization in Human Visual Cortex." *Cold Spring Harbor Symp. Quant. Biol.* **55** (1990): 651–661.

Randall Tagg
Department of Physics, University of Colorado at Denver, Denver, CO 80217-3364
e-mail: rtagg@cudnvr.denver.colorado.edu

Instabilities and the Origin of Complexity in Fluid Flows

1. INTRODUCTION

Legend has it that boatmen traveling down the Rhine River near Sankt Goarshausen would hear the enchanting songs of a beautiful maiden whose spirit inhabited a rock abutting the river. The enchantment drew the boatmen towards the so-called Lorelei rock (Figure 1), where they wrecked their boats and were swallowed up by the river. Once the inspiration of romantic poems by Heine and Eichendorf, the Lorelei in today's more prosaic age has a different sort of enchantment. The swirls and eddies in the river flowing past could draw the curious scientist to a similar fate. Even an earlier investigator, Leonardo DaVinci, was lured into fascination with such fluid motions (Figure 2), but fortunately he did not drown in the subject of his interest. To avoid such disaster, today's scientist has learned to capture and to some extent control the phenomena in the laboratory. Still, the patterns are no less enchanting (Figures 3 and 4) and might even produce the haunting song of aeolian tones from the wind blowing through power lines. Danger still exists, however, for the unwary: those who attempt to model and understand complex, evolving flows can soon be swallowed up by the mathematics!

1993 Lectures in Complex Systems, Eds. L. Nadel and D. L. Stein,
SFI Studies in the Sciences of Complexity, Lect. Vol. VI, Addison-Wesley, 1995

FIGURE 1 An illustration of one version of the Lorelei myth, with the Rhine surging up the famous cliff to carry the enchantress into the depths, away from avenging soldiers. Like the Lorelei's song, fluid instabilities lure unwary students into some tricky mathematics—but the phenomena are beautiful. [Illustration from W. Ruland, Legends of the Rhine (Köln: Verlag von Hoursch & Bechstedt).]

The purpose of this chapter is to navigate the phenomena associated with fluid instabilities and to begin to understand how these are connected to the complexity that is the sine qua non of turbulent motion. As much as possible, we will try to develop a heuristic understanding of the wide variety of mechanisms and dynamics that are encountered in different flow problems. However, we cannot avoid working through some mathematics along the way. There are several reasons for this. First, in order to make each problem manageable, we try to use a combination of heuristics and mathematics to identify a small number of modes of motion that dominate the behavior of the flow. Second, we aim to test our ideas through a quantitative comparison with experiments. Finally and most importantly, the universal features of the flows—the qualities that might be exhibited in many problems—are best expressed mathematically, for example through amplitude equations. The mechanistic interpretations are often specific to particular flows, but there are overriding considerations such as symmetry and dimensionality that are shared by many problems. For people who do not specialize in studying fluid motions, this understanding—mediated by some mathematics—is the principal means by which similarities to other topics will be found.

FIGURE 2 Leonard da Vinci was an astute observer of fluid motions, with many sketches of swirls and eddies in rivers and clouds. In this sketch, the note compares water flow to the form of human hair. [Her Majesty the Queen, Royal Library, Windsor Castle; Windsor Leoni volume (12579 r).]

FIGURE 3 Wake flow

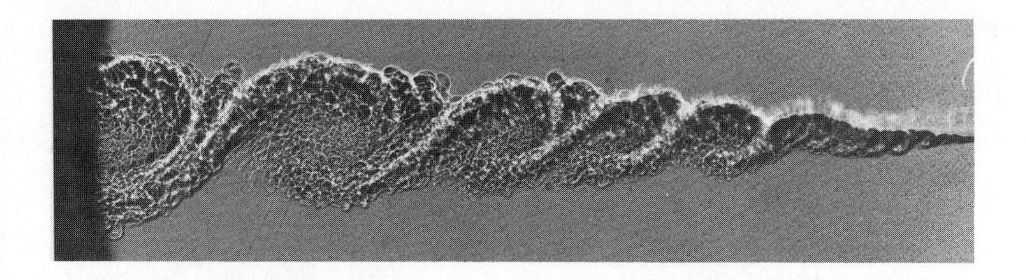

FIGURE 4 Turbulent mixing layer

I emphasize that it is surprisingly easy to find or generate instability phenomena.[1] This led me in early experience with these problems to expect that making systematic measurements and connecting them to underlying models would be straightforward. It is not. Nature, if not luring us into peril like the Lorelei, at least might be accused of the salesperson's tactic of "bait and switch." We are enticed to examine flows like those shown in Figure 4 only to be forced to shift our attention to somewhat simpler problems—and still to bear the expense of a nontrivial mathematical description. Fortunately, most of the details may be deferred to the references and we can concentrate on a few results and their graphical representation.

I will start with a brief mention of the first instability problem I encountered—the shape of free, rotating drops—because it accentuates the connection between symmetry and the beginnings of complexity in fluid motion. Then I will use the problem I have studied most recently—flow between rotating cylinders (the Couette-Taylor problem)—as a vehicle to explain most of the formalism we will need for the rest of the chapter. The Couette-Taylor experiment is unique for the large number of distinct flow patterns that can be obtained, providing many pictures and

[1] An instability "scavenger hunt" was organized as part of this chapter. Prizes were awarded to the participants who documented the widest variety of hydrodynamic instabilities that could be identified in weather patterns, coffee cups, fish ponds, etc. at the Santa Fe Institute Summer School.

plots with which to illustrate more general concepts. Since we will aim to use these concepts again later on, this is the most detailed section of the chapter.

Next I will visit, one by one, a wide range of instabilities that have been chosen because of the variety of physical mechanisms that seem to drive them. This is the bulk of the chapter, during which I will use graphs and pictures that have become the icons of their subjects. Often they have amusing names, like the "Busse balloon" or the "stability nosecone." Like any good icon, each of these figures symbolizes an important way of thinking. Overall, the subject has been broken down into the following main themes:

- flows subject to centrifugal forces;
- flows subject to buoyancy forces;
- shear flows and wakes; and
- motion of interfaces between two fluids.

Another important theme—rotating flows that model the motion of the oceans and the atmosphere—will have to wait for a future lecturer. For further reading on these topics, the reader may consult the books by Craik,[60] Drazin and Reid,[79] Manneville,[167] Sherman,[238] and Swinney and Gollub[251]; also strongly recommended are the fluid dynamics "picture books" by van Dyke[273] and Nakayama et al.[183] Some "classic" works on hydrodynamic stability are Lin[165] and Chandrasekhar.[38]

I will conclude with two epilogues. The first will be a purposefully irreverent view of the subject in connection to the ideas of complexity that are the domain of other work connected with the Santa Fe Institute. I must admit that I was caught by surprise at the difficulty in making a link between the onset of turbulence in fluid motion and the fascinating area of adaptive complexity studied at the Santa Fe Institute. The second epilogue is a slightly more serious set of speculations on where I hope some such connections can be made.

2. ROTATING DROPS

It is appropriate to begin with the example of rotating drops held together by surface tension because this problem is closely related to a fundamental problem in classical mechanics: the shape of rotating self-gravitating masses. This problem has occupied some of the best thinkers in mechanics, from Newton to Poincaré to Chandrasekhar (see Chandrasekhar[39]). In both the surface-tension and self-gravitating cases, when the fluid is at rest a sphere is the minimum-energy shape. Intuition tells us that when the drop or fluid mass starts to rotate, it bulges at the equator and flattens at the poles but retains an axisymmetric shape. What absolutely caught me by surprise when I was an undergraduate starting to do experiments on rotating drops was the symmetry-breaking that occurred when the drop was spun fast enough. Nonaxisymmetric shapes were, of course, by then a well-established fact. . .but there's nothing like first-hand experience in the laboratory to sharpen one's intuition.

Figure 5 shows the prototype for the apparatus I used—a system designed by J. Plateau more than one hundred years ago[206] in which he, with the aid of an assistant (Plateau was blind!), observed drops to deform into oblate spheres, two-lobed shapes, and tori. The drop is formed around a disk attached to a shaft and is supported by an immiscible and much less viscous fluid of equal density. Rotating the shaft spins up the drop and observations are made of the shapes formed before too much motion is induced in the surrounding fluid. Figure 6 shows a sketch taken from 16-mm movie film of a typical three-lobed shape that was obtained as a transient by spinning the drop up from rest. Other shapes, including a two-lobes, four-lobes, and a torus were obtainable—although the drop ultimately either broke up into free-floating droplets or became a rather distorted one- or two-lobed drop attached to one side of the disk.[257]

Computations[28] of the shapes for the ideal case of free-rotating drops showed that there are indeed several families of drop shapes branching from the axisymmetric family, but only a segment of the two-lobed branch is stable.

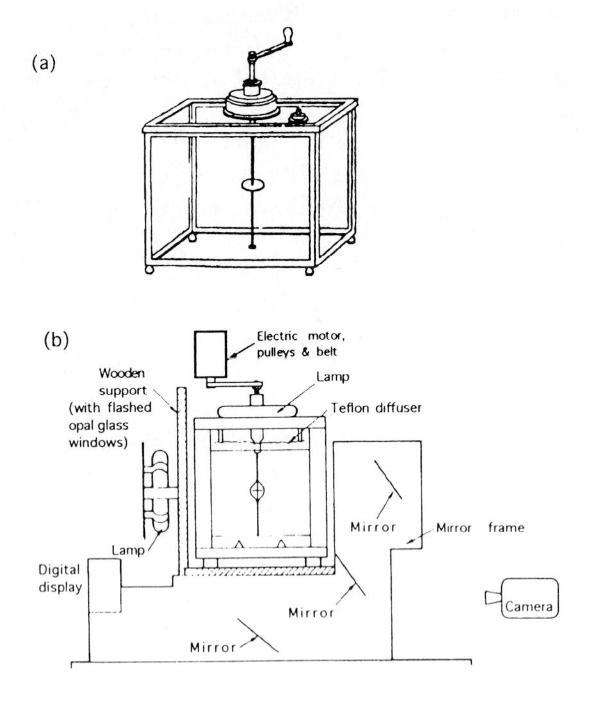

FIGURE 5 (a) Plateau's apparatus[206] for studying the shape of rotating drops. A drop of one fluid (for example, oil) is formed around the disk in the center of the shaft. Another less viscous fluid (for example, a mixture of water and alcohol) that has the same density buoyantly supports the drop. The crank is turned and the drop deforms into a variety of shapes. (b) Arrangement used to repeat Plateau's experiments while recording drop shapes from two orthogonal views using a high-speed 16-mm motion picture camera.[257]

The problem is a special case in the subject of bifurcations of fluid motions because the equilibrium shapes and their stability can be established by computing the energy of the system and various perturbations. More typically, such energy methods can only set lower bounds on the amount of stress needed to induce instability in a pattern of fluid motion: beneath such bounds, the pattern is guaranteed to be stable, but an experiment may need to go well beyond the established limits to actually produce an instability. However, the energy methods can be very useful as a guide to when instabilities might occur, because very little is assumed about the specific nature of the perturbations.[126,245]

In addition to the idea of symmetry-breaking, this example shows that multiple solutions to the equations of motion exist for given constraints, for example, for a given angular momentum. The understanding of what solutions may exist and how families of such solutions are interconnected as one or more parameters are varied is the subject of bifurcation theory. Indeed, it was Poincaré who enlisted the term "bifurcation" to denote the branching of axisymmetric and non-axisymmetric shapes in the case of spinning self-gravitating masses. The behavior of spinning drops is a far cry from complexity—nothing like the milieu of neuronal activity that we summarize in the phrase "my head is spinning." However, it is one of the simplest examples of a symmetry-breaking bifurcations and spontaneous evolution of new patterns in nonlinear systems. Also, it would be a fun experiment to try the next time you are on board the space shuttle.[276]

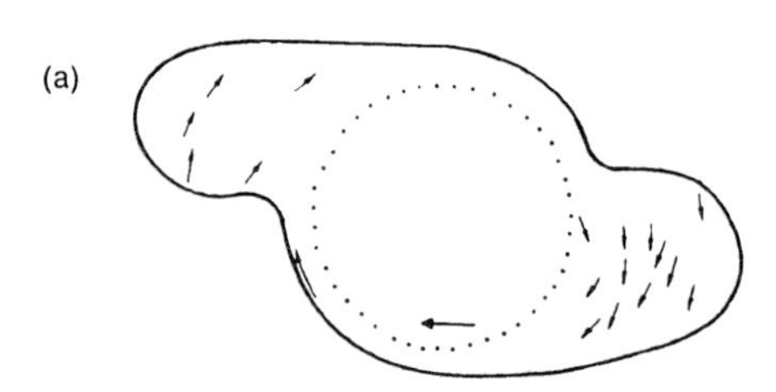

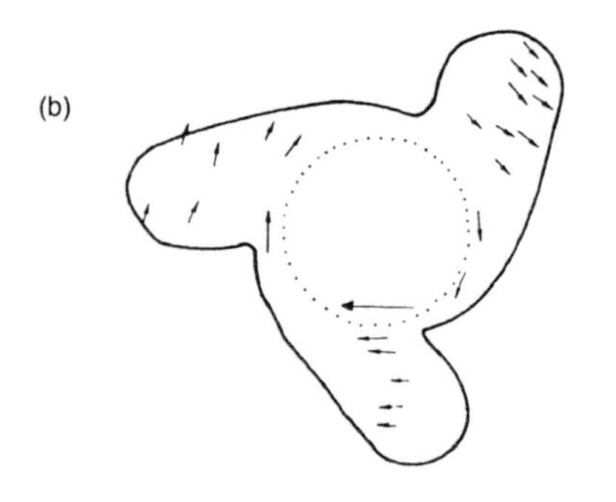

FIGURE 6 View of rotating drops along the shaft axis. The arrows mark the motion of trace particles inside the drop. The dotted circle is the outline of the supporting disk. (a) Two-lobed shape. (b) Three-lobed shape. In both cases, the shapes were obtained as transients after the drop was spun up by the disk but before too much surrounding fluid was entrained into the motion.

3. CENTRIFUGAL INSTABILITY—THE COUETTE-TAYLOR EXPERIMENT

3.1 LAMINAR COUETTE FLOW—THE PRIMARY FLOW STATE

In the Couette-Taylor experiment (Figure 7) fluid fills the annulus between two coaxial cylinders and is set into motion by rotating one or both cylinders. The inner cylinder of radius r_1 rotates with angular velocity ω_1 and the outer cylinder of radius r_2 rotates with angular velocity Ω_2, with Ω_2 taken to be negative if the outer cylinder rotates in the opposite direction from the inner cylinder. Unlike most natural and engineering flows, the fluid never leaves the system, making it a so-called "closed system." This might seem to be very limiting in terms of the range of dynamics that could be obtained, but in fact the experiment is a lucky compromise between confinement and a large number of degrees of freedom. As a consequence, many distinct types of flow can be stabilized, with levels of dynamical complexity ranging in steps from featureless laminar flow all the way up to fine-scaled turbulence. Moreover, control and measurement of very high precision have been achieved, at a level comparable to the other favorite of experimental physicists: convection in horizontal layers (see Section 4). As a prototype for testing basic theories of the nonlinear dynamics of fluid motion, the Taylor-Couette experiment might be considered as the "hydrogen atom" of fluid physics, with roughly 2,000 papers connected to the problem or its variations.[255] The following discussion will

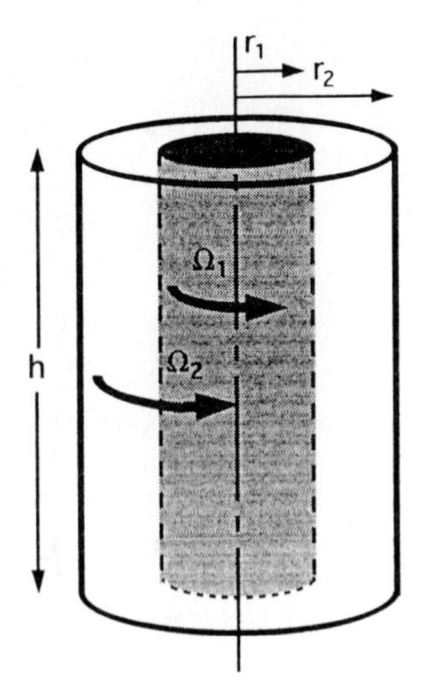

FIGURE 7 Geometry of the Couette-Taylor apparatus. Fluid fills the annulus between two independently rotating, coaxial cylinders.

necessarily be selective in citations to this vast literature; for access to the many important papers that are omitted, see reviews by DiPrima and Swinney[72] and Tagg,[256] as well as the recent books by Koschmieder[145] and Chossat and Iooss.[46]

When there is no applied stress (both cylinders are at rest) the fluid is in thermodynamic equilibrium and adopts the full cylindrical symmetry of the apparatus. We suppose that the cylinders have been machined precisely enough that there is nearly perfect rotational symmetry about the axis. We also imagine an ideal system that is infinitely long, so that there are no ends to break the translational symmetry parallel the axis. For sufficiently small amounts of stress, the flow still has the symmetry of the imposed conditions: the rotational symmetry, the translational symmetry, but now also a direction set by the rotation of the cylinders. The fluid travels in simple circular paths around the inner cylinder and is called circular Couette flow. The torque transmitted through the fluid from one cylinder to the other in this flow regime is proportional to the difference in angular velocities of the cylinders, which is the operating principle of the Couette viscometer. For fluid density ρ, kinematic viscosity ν, and radius ration $\eta \equiv r_2/r_1$, the torque per unit length G is:

$$G = -4\pi\rho\nu\frac{\eta^2}{1-\eta^2}r_2^2(\Omega_1 - \Omega_2). \tag{1}$$

For low enough rotation rates, circular Couette flow is the only possible type of motion. This is a quite general idea: when the stresses on a fluid system are small enough, the state of motion that results is unique, has the symmetry of the imposed conditions, and is a continuous extension of the rest state when the fluid was in thermodynamic equilibrium. Thus we call the family of solutions parametrized by

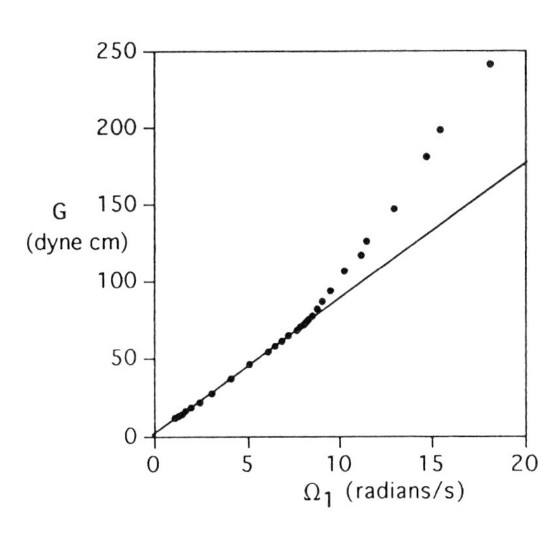

FIGURE 8 Torque measurements of Donnelly,[74] based on Table 2 of Donnelly and Simon[75]: the torque G transmitted to a stationary outer cylinder is plotted against the angular velocity Ω_1 of the inner cylinder. Notice the break away from linear dependence; this was found at an angular velocity of 8.330 radians/sec, corresponding to a critical Reynolds number (see text) of 67.94, compared to a theoretical value of 68.2. The system consisted of cylinders with inner radius 1.0 cm and outer radius 2.0 cm, with a working fluid (oil) of kinematic viscosity 0.1226 cm^2/s and density 0.8404 g/cm^3. The height h was 5.0 cm.

increasing but small values of the stress the "thermodynamic branch." We refer to the state of fluid motion along this branch as the "primary flow." Experience shows that following this branch soon leads to instability and a break in the slope of the plot of torque versus angular velocity (Figure 8). The next step, then, is to predict when and how such instabilities occur. Indeed, the first successful quantitative prediction of a hydrodynamic instability was accomplished by G. I. Taylor[263] with this system.

3.2 EQUATIONS OF MOTION

In order to proceed, we must now introduce the equations which govern the fluid motion in this and other systems we will be discussing. Note first that with one or two important exceptions, we consider the fluid density ρ and temperature T as constants. The independent variables are spatial position $\mathbf{r} = (x, y, z)$ (in a reference frame fixed to our laboratory) and time t. We seek to find, as functions of these variables, the velocity vector field $\mathbf{u}(\mathbf{r}, t) = [u(\mathbf{r}, t), v(\mathbf{r}, t), w(\mathbf{r}, t)]$ and the pressure field $p(\mathbf{r}, t)$. Here, u, v, and w are usually the Cartesian components of the velocity field; however, we will also find it useful to use cylindrical polar coordinates and specify the radial, azimuthal and axial velocity components as u, v, and w, respectively. The choice will be made clear in the context of each problem.

There is an important complication arising from specifying fluid properties as functions of the laboratory coordinates $\mathbf{r} = (x, y, z)$. The problem is alluded to in the famous quote of Heraclitus (c. 500 B.C.): "Upon those who step into the same rivers, different and different waters flow." In order to describe dynamical changes, we must follow individual fluid elements. In describing the change in any quantity c that moves with the fluid, we must take into account the fact that in time Δt, the fluid element has moved to a new position $\mathbf{r} + \mathbf{u}\Delta t$. That is,

$$\begin{aligned} \Delta c &= c(\mathbf{r} + \mathbf{u}\Delta t, t + \Delta t) - c(\mathbf{r}, t) \\ &= f(\partial c, \partial x)u\Delta t + f(\partial c, \partial y)v\Delta t + f(\partial c, \partial z)w\Delta t + f(\partial c, \partial t)\Delta t\ . \\ &= [\mathbf{u} \bullet \nabla c + f(\partial c, \partial t)]\Delta t \end{aligned} \tag{2}$$

Here ∇c is shorthand for the gradient vector: $(\partial c/\partial x, \partial c/\partial y, \partial c/\partial z)$. We define the limit of $\Delta c/\Delta t$ as $\Delta t \to 0$ to be the "convective derivative"

$$\frac{Dc}{Dt} \equiv \frac{\partial c}{\partial t} + \mathbf{u} \bullet \nabla c. \tag{3}$$

In this discussion, c could be concentration of dye. It could also be a component of the velocity of the fluid element. Indeed, we find that the rate of change in momentum of a fluid element of mass Δm is given by:

$$\Delta m \frac{D\mathbf{u}}{Dt} = \rho \Delta V \frac{D\mathbf{u}}{Dt} \equiv \rho \delta V \left[\frac{\partial \mathbf{u}}{\partial t} + (\mathbf{u} \cdot \nabla)\mathbf{u} \right], \tag{4}$$

where we have identified the mass δm with the product $\rho\delta V$ of density and volume of the fluid element and where

$$(\mathbf{u}\cdot\nabla)\mathbf{u} \equiv \begin{pmatrix} \mathbf{u}\bullet\nabla u \\ \mathbf{u}\bullet\nabla v \\ \mathbf{u}\bullet\nabla w \end{pmatrix}. \tag{5}$$

Newton tells us that this rate of change in momentum must equal the forces acting on the fluid element. These include gravitational force $\rho\delta V\mathbf{g}$, the resultant of a gradient in pressure forces acting on the fluid element $-\nabla p\delta V$, and the resultant of viscous stresses applying friction to the boundaries of the fluid element $\rho\nu\nabla^2\mathbf{u}\delta V$. For details of how these expressions are determined, see texts such as Tritton[266] or Acheson.[1] Thus Newton's second law, or the momentum equation, is translated into what are called the Navier-Stokes equations:

$$\frac{\partial\mathbf{u}}{\partial t} + (\mathbf{u}\bullet\nabla)\mathbf{u} = \mathbf{g} - \frac{1}{\rho}\nabla p + \nu\nabla^2\mathbf{u}. \tag{6}$$

Note that this is actually a set of three equations, because there are three components to the velocity vector $\mathbf{u}$.

Since gravity is a conservative force, it can be expressed as the gradient of a potential: $\mathbf{g} = -\nabla\phi$. If we also take advantage of our assumption that the density ρ is a constant, we can rewrite the Navier-Stokes equation as:

$$\frac{\partial\mathbf{u}}{\partial t} + (\mathbf{u}\bullet\nabla)\mathbf{u} = -\frac{1}{\rho)\nabla\tilde{\mathbf{p}}} + \nu\nabla^2\mathbf{u}\,, \tag{7}$$

where $\tilde{p} \equiv p + \rho\phi$ is an effective pressure. For example, if there is a constant gravitational field in the downward z-direction, $\phi = gz$ and $\tilde{p} \equiv p + \rho gz$. By dealing with the dynamics in terms of $\tilde{p}$, we are effectively able to ignore the contribution of gravity and other conservative forces. This is possible so long as we don't deal with fluid interfaces or other situations where the density is not constant. Thus for the Couette-Taylor system, we should be able to see the same kinds of flows if the cylinders are upright or lying on their sides and, indeed, experiments are done both ways.[2]

Since we have four variables, u, v, w, and p, we need another equation. This is obtained by requiring that mass be conserved. By integrating the rate at which mass flows through an arbitrary closed surface in the fluid and equating it to the rate of change of mass inside the surface, one can show[266] that the general equation for mass conservation is:

$$\frac{\partial\rho}{\partial t} + \nabla\cdot(\rho\mathbf{u}) = 0. \tag{8}$$

[2] Differences can arise between horizontal and vertical orientation if the particles used for flow visualization are not exactly neutrally buoyant within the fluid; see Dominguez-Lerma et al.[73]

Since we are assuming that the density is constant, this reduces to:

$$\nabla \bullet \mathbf{u} = 0. \tag{9}$$

Either form is known as the "continuity equation."

Accompanying these equations must be a combination of boundary conditions and initial conditions. We specify that the fluid in contact with any solid surface (here the walls of the cylinders) must have the same velocity as that of the solid surface. This is called the "no slip" boundary condition and is satisfied for all but extremely rarefied fluids. Indeed, one by-product of the successful prediction of instability in the Couette-Taylor experiment was confirmation of the validity of the no-slip condition and the assumed model for viscous forces. In other problems, we may need to specify the velocity of the fluid "at infinity" but here this is not necessary. Since there is also a time derivative, we must in principle also specify an initial condition. Sometimes, instead, we look for solutions that are independent of time or have a simple, say periodic, time-dependence. Later, of course, in dealing with truly complex flows, we cannot make such restrictive assumptions.

3.3 SCALING AND DIMENSIONLESS PARAMETERS

Having written down the equations, another important step in studying fluid motions is to scale all the variables so that experiments in different laboratories with cylinders of different sizes might be compared. In this way, the results of a miniature Couette system designed to fit within the poles of a high-field magnet for nuclear magnetic resonance[146] can be directly compared to results of a one-meter sized system designed for turbulence studies.[157,158] Also, experiments with different fluids (water, silicone oil, mixtures of water and glycerol, etc.) can be compared.

First, we chose to scale distance in terms of some natural length over which the velocities, pressures, etc. vary appreciably in the system, which in this case is the gap $d \equiv r_2 - r_1$ between the cylinders. Time can be expressed with respect to a couple of possible scales. One, an inertial time scale, measures the typical time it takes fluid elements to travel the characteristic distance of the system. In this case, we divide the characteristic distance d by the velocity $\Omega_1 r_1$ at the surface of the inner cylinder to get $\tau_i = d/\Omega_1 r_1$. In other circumstances, we might be interested in the rate at which viscous effects are propagated through the fluid—a viscous time scale. By dimensional reasoning, we find this to be $\tau_v = d2/\nu$. Let us for now arbitrarily chose the first (τ_i) to be our time scale. The scale for the pressure is set by the rate at which fluid with the typical velocity $\Omega_1 r_1$ carries momentum crosses an imaginary unit surface: $\rho(\Omega_1 r_1)^2$. We thus make the following scalings in the equations of motion, where all the primed variables denote dimensionless quantities:

$$\begin{array}{ll} \mathbf{r} = \mathbf{r}' d & \nabla = \nabla' \frac{1}{d} \\ \mathbf{u} = \mathbf{u}' \Omega_1 r_1 & \frac{\partial}{\partial t} = \frac{\partial}{\partial t'} \frac{\Omega_1 r_1}{d} \\ p = p' \rho \Omega_1^2 r_1 2 & \end{array} \tag{10}$$

In terms of the dimensionless variables, the equations of motion become:

$$\frac{\partial \mathbf{u}'}{\partial t'} + (\mathbf{u}' \bullet \nabla')\mathbf{u}' = -\nabla p' + \frac{1}{\text{Re}}\nabla'^2\mathbf{u}', \tag{11a}$$

$$\nabla' \bullet \mathbf{u}' = 0, \tag{11b}$$

where $\text{Re} \equiv (\Omega_1 r_1 d/\nu)$ is called the Reynolds number, named for O. Reynolds, who first explained the significance of this combination of variables in describing flow transitions. In most of the following, we use the dimensionless form of the equations and will drop the primes from the variables.

Sometimes the Reynolds number is described as the ratio of "inertial forces" to "viscous forces." Another useful point of view is to recognize that $\text{Re} = \tau_v/\tau_i$, which compares the viscous time scale to the inertial time scale. A low Reynolds number implies that viscosity can transport momentum on a time scale comparable to or shorter than the rate of motion of the fluid. A high Reynolds number implies that velocity changes occur quickly compared to the viscous diffusion time. Caution must be exercised in ignoring viscosity altogether in high-Reynolds-number flows: near solid boundaries, at so-called "critical layers" in shear flows, and at the finest level of detail in fully developed turbulence, there can be a redefinition of appropriate length scales such that the time-scales are comparable and the modified Reynolds number is near unity.

We must try to describe all other relevant features of the system in dimensionless terms, so that different experiments can be compared. For the case of the Taylor-Couette experiment, other important parameters are the radius ratio $\eta \equiv r_1/r_2$ and the angular-velocity ratio $\mu \equiv \Omega_2/\Omega_1$. Note that μ can take both positive and negative values, corresponding to co-rotating and counter-rotating cylinders, respectively. The case $\mu = 0$ (outer cylinder at rest) has by far been the most studied. An alternative to the parameter μ is another Reynolds number, $\text{Re}_2 \equiv \Omega_2 r_2 d/\nu$, which likewise measures the outer cylinder angular velocity in non-dimensional form. Later, we will find through a heuristic discussion of the instability mechanism, that there is even a substitute for Re_1, called the Taylor number Ta, that in some sense more naturally measures how close the system is to its first point of instability. Finally, since real experiments must be done with cylinders of finite height h, we define the aspect ratio $\Gamma \equiv h/d$. For many purposes, it appears from experiment that for $\Gamma \gtrsim 25$, the flow is a good approximation to that obtained for infinitely long cylinders. There can, however, be subtle exceptions, some of which will be discussed below.

3.4 TOUR DE COUETTE

Before immersing ourselves any further in the mathematics (do you hear the song of the Lorelei?), let's vicariously do an experiment by looking at what happened when someone systematically varied the parameters Re_1 and Re_2 for fixed values of η and

Γ. Figure 9 shows David Andereck's richly detailed exploration of the flow states obtained for a radius-ratio $\eta = 0.883$ and aspect ratio $\Gamma = 30$. I have added the label "Here there be dragons" to a region that has not yet been successfully divided into distinct flow regimes; typically, the flow in this region is a mess of undulating vortices and turbulence. Space does not allow inclusion of all of Andereck's photos of the well-defined flow states, so Figure 10 shows a selection of some of the "prettier" flows. The patterns were made visible by mixing into the fluid a suspension of tiny flakes called "Kalliroscope."[173] These flakes align themselves on the average with the local patterns of shear in the flow, and act optically like little mirrors that reflect light differently depending on their orientation. Indeed, one remarkably simple quantitative approach to experiments of this sort is to point a laser at the flow and measure the intensity of reflected laser light as a function of time. We will see later that this purely temporal information gives us important clues about the complexity of the flow.

Even the compendium reported by Andereck is incomplete, because the observed flow patterns were obtained by a very specific protocol: fix Ω_2 and very slowly increase Ω_1 to see what states arise. Also, care was taken to ensure that when the Taylor-vortex flow was obtained, there were always 30 vortices filling the finite-length annulus from top to bottom. Variation of the protocol or manipulation of the system to get a different number of vortices can produce a much different diagram. Indeed, in a landmark work on this experiment, Coles[52] showed how a large variety of wavy-vortex states could be obtained for identical settings of the parameters, purely by taking different protocols in arriving at the final parameter values. He has a three-dimensional model of the diagram shown in Figure 11 which

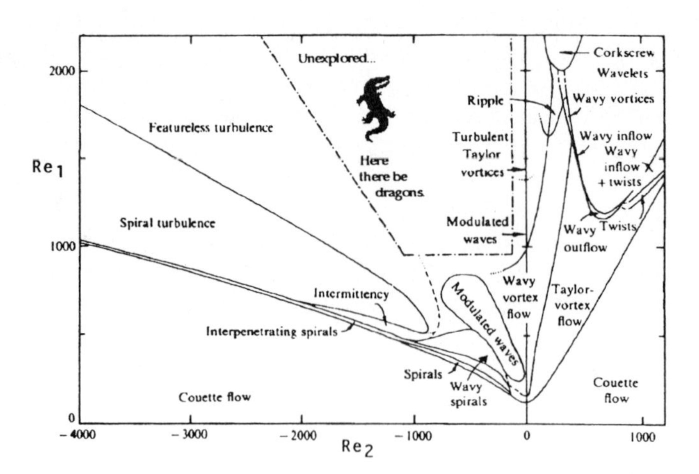

FIGURE 9 Diagram of flow states obtained between rotating cylinders of radius ration 0.883. The protocol was to first set the outer cylinder Reynolds number Re_2 and then slowly increase the inner cylinder Reynolds number Re_1. After Andereck et al.[6]

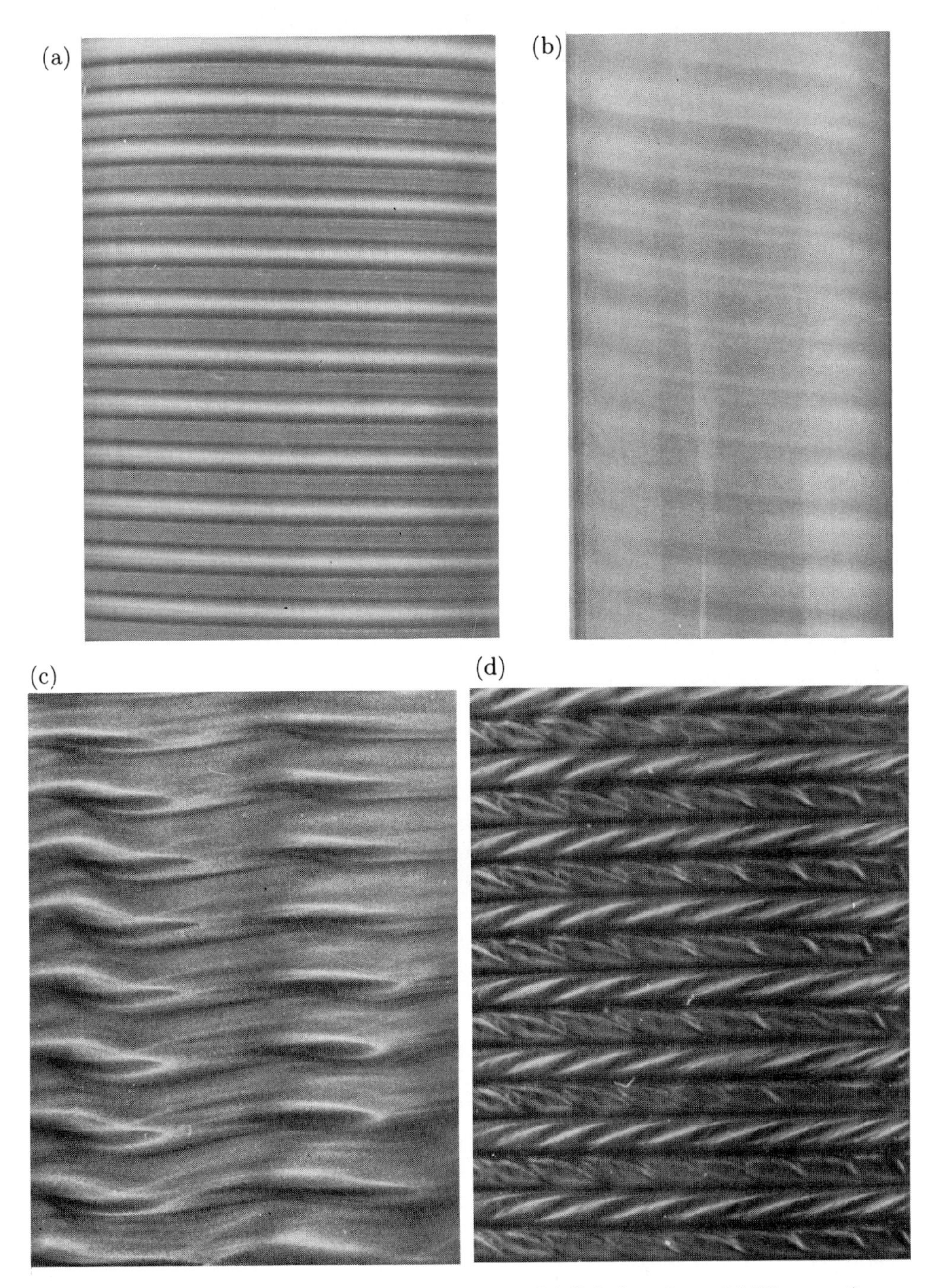

FIGURE 10 Andereck photos. (a) Taylor vortices; (b) Spiral vortices; (c) Wavy vortices; and (d) Twist vortices.

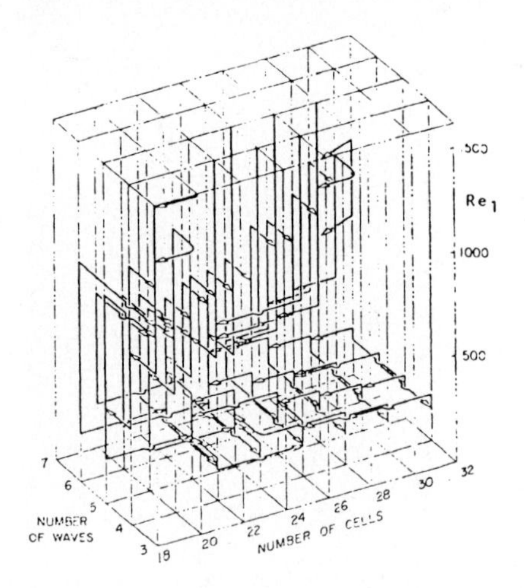

FIGURE 11 Multiplicity of states and transitions between states in the Couette-Taylor system. A state is characterized by the number of vortices ("cells") in the axial direction and number of waves in the azimuthal direction. States exist over a range of Reynolcs number (vertical direction), as indicated by bolder vertical segments. At the end of these segments, transitions were observed to occur to a different state, indicated by an arrow. From Coles.[52]

he has offered as a prize to anyone who can explain this intricate connection of states to his satisfaction. The prize has been unclaimed for nearly 30 years. (This diagram will be explained further in Section 3.10.) In addition to selecting a path or protocol through parameter space, it is important in conducting experiments with such systems to proceed slowly across transition boundaries like those shown in Figure 9. As we will see later, the process of transition can become infinitely slow at the boundaries—a phenomenon called "critical slowing down"—while, in other cases, there may be hysteresis: the transition occurs for one value of the parameter as it is increased, but for a lower value as it is decreased. Fortunately, the painstaking work involved in such studies can now be automated by computers and time-lapse video. As an example, I prepared a computer-guided tour to many of these states; a similar tour, with illustration of the time dependence recorded from scattered laser light, is given in my videotape "Tour de Couette" shown with this chapter.

3.5 PRIMARY INSTABILITY—HEURISTICS

One of the great advantages of Taylor-Couette experiments idealized to have infinitely long cylinders is that one can write down a solution to these equations in closed form. Using cylindrical polar coordinates, the only nonzero velocity is the azimuthal component v, which is found to be

$$v = Ar + \frac{B}{r}, \tag{12}$$

where A and B are determined by the boundary conditions $v(r_1) = \Omega_1 r_1$ and $v(r_2) = \Omega_2 r_2$:

$$A = \frac{\Omega_2 r_2^2 - \Omega_1 r_1^2}{r_2^2 - r_1^2} \qquad \text{and} \qquad B = (\Omega_1 - \Omega_2) r_1^2 r_2^2, r_2^2 - r_1^2. \tag{13}$$

The pressure is obtained by integrating $dp/dr = \rho v^2/r$, which we will discuss further below. This solution for velocity and pressure is the circular Couette flow state.

We consider this as the "base flow" and try to find out if any sort of disturbance is likely to lead to instability and a new flow pattern. Let us try to understand heuristically how instability can arise. Of course, we realize that heuristic reasoning may be right in spirit but wrong in detail. Indeed, sometimes it is even wrong in spirit, letting us get the right answer for the wrong reasons. *Sic transit scientia mundi.*[3] For different discussions of this instability, see Stuart,[249] Coles,[53] and Manneville.[167]

Think of a centrifuge: if you spin an object, it tends to be ejected from its path of rotation unless some force acts to keep it constantly changing the direction of its velocity to stay on the circular path. For fluid traveling around a curved path, this force must arise from a difference in pressure acting on the fluid from the outside and from the inside. This pressure gradient provides the necessary centripetal force per unit volume:

$$\frac{dp}{dr} = \rho \frac{v^2}{r}. \tag{14}$$

Just such a pressure gradient is set up in circular Couette flow to maintain the system in equilibrium. However, is this situation stable?

Consider a set of curved streamlines as shown in Figure 12(a). In addition to flow between rotating cylinders, there are other situations where such curved streamlines can occur. These include flow through curved channels (the Taylor-Dean problem), flow in boundary layers over curved surfaces (forming Görtler vortices), and flow out of a pipe where the fluid is diverted by a facing wall.

Imagine fluid traveling along one of the streamlines being bumped by some noise from its radius r to a new radius $r + \xi$, where ξ is much smaller than the representative scaled of the problem (in this case, d is the gap between the cylinders). For convenience, let us think of a parcel of fluid of volume ξ^3 moving along this streamline. In a reference frame rotating with the fluid, there is an opposition of forces which may or may not restore the fluid to its original radius. First, there is the "centrifugal force" that seems, in this rotating reference frame, to pull the fluid radially outwards. At radius r, the magnitude of this "force" is $F_c = \rho \xi^3 v^2/r$. What is the change ΔF_c in this force when the parcel of fluid is moved to the new radius $r+\xi$?

[3]The actual words are *Sic transit gloria mundi*, from Book I, Chapter 3 of the *The Imitation of Christ* by Thomas A Kempis. Fittingly, the chapter, entitled "Of the knowledge of truth," is an essay on the futility of learning in the absence of divine guidance!

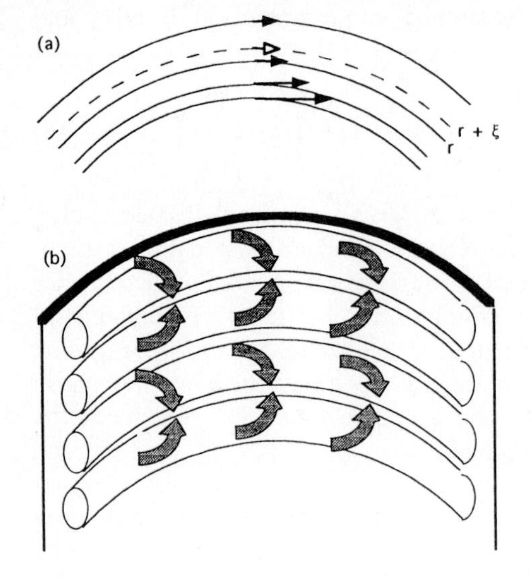

FIGURE 12 Origins of centrifugal instability. (a) A generic set of curved streamlines. We imagine one streamline, with radius of curvature r, being displaced to a new radius $t + \xi$ and ask if the pressure field will act to restore the streamline to its original radius. If not, and we have a wall containing the flow (b), then the enusing instability leads to an overturning of the fluid as it tends to be ejected outwards, producing a stack of curved vortices.

Here we must be careful: we assume that when the fluid was bumped, no torque was applied; indeed, in the absence of viscosity or any external body forces this would be rigorously true. Thus the angular momentum $L = \rho\xi^3 vr$ of the fluid will have been preserved. The centrifugal force may be written as $F_c = (\rho\xi^3)^{-1}L^2/r^3$. The change in centrifugal force under the constraint that L remains constant is then

$$\Delta F_c \approx \frac{dF_c}{dr}\xi = [-3(\rho\xi^3)^{-1}L^2/r^4]\xi = -3\rho\xi^4 v^2/r_2. \tag{15}$$

Opposing the centrifugal force is a radially inward force F_r due to the ambient pressure gradient: $F_r = -\rho\xi^3 dp/dr$ which is related to the undisturbed velocity field by $Fr = -\rho\xi^3 v^2/r$. At the new radius, this force is different by an amount:

$$\Delta F_r \approx \frac{dF_r}{dr}\xi = -\rho\xi^4\left(\frac{2v}{r}\frac{dv}{dr} - \frac{v^2}{r_2}\right). \tag{16}$$

The net force acting on the fluid parcel is thus:

$$\begin{aligned} F_{\text{net}} =& F_c + \Delta F_c + F_r + \Delta F_r \\ =& \Delta F_c + \Delta F_r \quad \text{(because the forces } F_c \text{ and } F_r \text{ were originally in balance)} \\ \approx& -2\rho\xi^4\left(\frac{v}{r}\frac{dv}{dr} + \frac{v^2}{r_2}\right) \\ =& -\rho\xi^4\frac{1}{r^3}\frac{d(vr)^2}{dr}. \end{aligned} \tag{17}$$

We can now reach our first conclusion: the net force will be radially inwards and will restore the fluid to its original radius if $d(vr)^2/dr > 0$. This criterion for stability was first established by Rayleigh.[212,213,214] Notice that we have not yet specified the dependence of v on r, so this criterion could apply to a variety of curved flows. When applied to Couette flow, the criterion for stability becomes: $\Omega_1\Omega_2 \geq 0$ (cylinders are co-rotating) and $\Omega_1 r_1^2 < \Omega_2 r_2^2$.

Suppose instead that $d(vr)^2/dr < 0$, so the net force is radially outwards. If we take viscosity into account at this point, we will discover that dissipative processes may still prevent instability. Instability will occur only if the imbalance in forces has time to act before the momentum difference that produces the imbalance is diffused away by viscosity. Suppose the fluid starts to accelerate outwards due to F_{net}. If it has a radial velocity u, its motion will be opposed by a viscous drag whose magnitude is estimated by:

$$\begin{aligned} F_{\text{drag}} &= -\text{ dynamic viscosity } x \text{ velocity gradient } x \text{ area} \\ &\approx -\rho\nu(u/\xi)\xi^2 \\ &= -\rho\nu u\xi. \end{aligned} \tag{18}$$

A velocity is reached where this drag balances the force F_{net}:

$$F_{\text{drag}} + F_{\text{net}} = 0. \tag{19a}$$

$$-\rho\nu u\xi - \rho\xi^4 \frac{1}{r^3}\frac{d(vr)^2}{dr} = 0. \tag{19b}$$

Solving for u:

$$u = -\xi^3, \nu r^3 \frac{d(vr)^2}{dr}. \tag{20}$$

The time taken for the fluid element to move a distance ξ is

$$\tau_1 = \xi/u = [-(\xi^2/\nu r^3)d(vr)^2/dr]^{-1}. \tag{21}$$

Meanwhile, momentum diffuses to fluid at nearby radii on a time scale

$$\tau_2 = \xi^2/\nu. \tag{22}$$

Thus we might only expect instability to occur if $\tau_2/\tau_1 > 1$, or

$$\frac{\xi^4}{\nu^2}\frac{1}{r^3}\frac{d(vr)^2}{dr} > 1. \tag{23}$$

Let $\xi = \varepsilon d$ where $\varepsilon \ll 1$, and define the Taylor number

$$Ta \equiv \max\left[-\frac{d^4}{\nu^2}\frac{1}{r^3}\frac{d(vr)^2}{dr}\right] \tag{24}$$

where the expression in brackets is maximized over the radii r for which $d(vr)^2/dr < 0$. Then, when dissipation is present, instability is delayed until $Ta > 1/\varepsilon^4$. This threshold is a numerical value which we call the critical Taylor number $\mathrm{Ta_c}$, which we expect is much larger than unity.

So far this argument has been applied to any curved flow. Now consider circular Couette flow, $v = Ar + B/r$, and evaluate the Taylor number:

$$
\begin{aligned}
\mathrm{Ta} &\equiv \max\left[-\frac{d^4}{\nu^2}\frac{1}{r^3}\frac{d(vr)^2}{dr}\right] \\
&= \max\left[-4\frac{d^4}{\nu^2}A\left(A+\frac{B}{r_2}\right)\right] \\
&= -4\frac{d^4}{\nu^2}A\left(A+\frac{B}{r_1^2}\right) \quad \text{(the maximum occurs for minimum } r, \text{ which is } r_1) \\
&= -4\frac{d^4}{\nu^2}A\Omega_1 \\
&= \frac{4\Omega_1 d^4}{\nu^2}\frac{\Omega_1 r_1^2 - \Omega_2 r_2^2}{r_2^2 - r_1^2}.
\end{aligned}
\tag{25}
$$

For small gap $d \ll r_1$ and $\Omega_2 = 0$, the critical Taylor number is $\mathrm{Ta_c} = 3{,}416$—a number much larger than unity, as expected.

In summary, we have established that a curved flow with azimuthal velocity profile given by a function $v(r)$:

a. is stable provided $d(vr)^2/dr > 0$ throughout the profile;

b. does not become unstable in the presence of viscous dissipation until $\mathrm{Ta} > \mathrm{Ta_c}$, where Ta is the Taylor number defined by

$$\mathrm{Ta} \equiv \max\left[-\frac{d^4}{\nu^2}\frac{1}{r^3}\frac{d(vr)^2}{dr}\right],$$

and the critical value $\mathrm{Ta_c}$ is typically a number much larger than unity.

3.6 LINEAR ANALYSIS OF THE ONSET OF INSTABILITY

What happens when instability does occur? If the fluid tends to be ejected outwards, it eventually comes into contact with a bounding wall and is deflected in the axial direction. Thus we might expect a picture like Figure 12(b) where the fluid is constantly being overturned and follows helical streamlines, forming layers of vortices. Indeed, when G. I. Taylor performed experiments with this system, he observed such motion by injecting dye into the flow. The vortices are now called Taylor vortices, and are more clearly visualized by the Kalliroscope suspension (Figure 10(a)). When the cylinders are rotating in the same direction or the outer cylinder is only weakly rotating in the opposite direction, these vortices lie flat on top of one another and the flow remains independent of time. A key feature is the fact that the pattern is spatially periodic in the axial direction with a wavelength λ.

If the cylinders are counter-rotating at a sufficiently strong rate, then a pattern known as spiral vortex flow emerges (Figure 10(b)): the vortices wind from one end of the cylinder to the other in a helical pattern similar to the American barber pole. Furthermore, the flow is time-dependent, with the velocity at any given location cycling periodically with a fixed frequency. The flow is still spatially periodic in the axial direction, but is now also periodic in the azimuthal direction, repeating itself an integer (m) number of times as one observes the full 2π radians of angle around the cylinder. The pattern appears to be moving as a wave, traveling towards either one end of the cylinder or the other. Either of these flows—Taylor vortices or spiral vortices—are called cellular flows because the original translational symmetry along the axis of the cylinder has been broken into a pattern of flow "cells." When such flows establish themselves after the first instability has occurred, we also refer to them as "secondary flows." Much of the recent work on the emergence of complexity in fluid systems has concentrated on understanding the details by which certain secondary flow patterns are developed and their subsequent evolution as parameters are varied further.

At this point, we can introduce an important experimental diagnostic of flows that we wish to study. We can record a sequence of measurements of flow velocity at a single point in the flow as a function of time. If the flow is time-independent, the sequence is flat (except for noise). If the flow varies periodically, a periodic function of time is produced (Figure 13(a)) An important further step in the analysis is to compute a Fourier transform of the time-series data in order to produce a power spectrum (Figure 13(b)). This new function tells how much of the signal is composed of a sinusoidal variation of a given frequency ω. Sharp peaks at some nonzero fundamental frequency ω_1 and its harmonics $n\omega_1$ (n is an integer) indicate a periodic flow, as is the case with the spiral vortices. Sometimes the peaks are broadened (Figure 13(c)), indicating an irregular large-scale modulation of the spiral pattern.

Can we predict the critical Taylor number, the resulting wavelength, and the resulting frequency ω_1 (which may be zero)? Yes, and the first step is the method called linear stability analysis. We consider infinitesimal disturbances, the idea being that noise or thermal fluctuations in an experimental system can provide tiny disturbances which, if they are able to amplify, could destabilize the base flow. If the base flow is found to be stable to all conceivable small disturbances we must, of course, also consider the possibility that sufficiently large disturbances are needed to upset the flow. By analogy, a pencil balanced on its tip requires the slightest push to fall over, but a wine glass (fortunately) requires a considerable shove to be tipped.

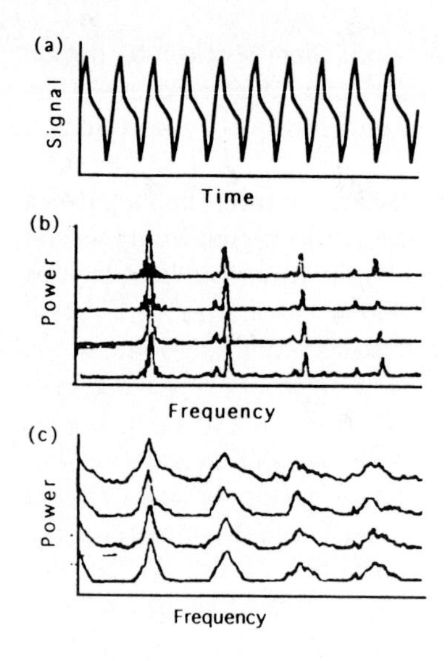

FIGURE 13 Flow measurements in time and frequency domains. (a) A traveling wave passing a fixed observation point would generate a periodic signal that is recorded as a time series. (b) The signal may also be represented by a power spectrum, which shows how much power is present in the signal at various frequencies. A periodic signal will have a spectrum with sharp peaks; peaks at larger frequencies are integer multiples (harmonics) of the lowest-frequency peak. [Author's actual data for spiral vortex flow.] (c) If the signal amplitude undergoes irregular modulation on time scales long compared to the signal period, the peaks in the power spectrum are broadened. [Author's actual data for alternating spiral vortex flow.]

In order to examine the effect of small disturbances, we arbitrarily express the velocity and pressure fields as sums of the original flow fields (denoted by capital letters $\mathbf{U}$ and P) and the disturbances (denoted by lower-case letters $\mathbf{u}$ and p):

$$\mathbf{u}_{\text{total}} = \mathbf{U} + \mathbf{u} \tag{26a}$$

$$p_{\text{total}} = P + p. \tag{26b}$$

We substitute this into the equations of motion (Eqs. 11(a) and 11(b)) and eliminate the parts that are already satisfied by the original flow fields, remembering that $(\mathbf{U}, p)$ are by themselves valid solutions to the equations of motion. We obtain equations for the disturbances:

$$\frac{\partial \mathbf{u}}{\partial t} + (\mathbf{U} \bullet \nabla)\mathbf{u} + (\mathbf{u} \bullet \nabla)\mathbf{U} + (\mathbf{u} \bullet \nabla)\mathbf{u} = -\nabla p + \frac{1}{\text{Re}}\nabla^2 \mathbf{u}. \tag{27a}$$

$$\nabla \bullet \mathbf{u} = 0. \tag{27b}$$

Any boundary conditions on $\mathbf{u}_{\text{total}}$ are already satisfied by $\mathbf{U}$, so the disturbance $\mathbf{u}$ is zero at any boundaries. Now a crucial step is taken to simplify the equations: we ignore terms involving products of the disturbance with itself or its gradient: that is, we ignore the underlined expression $(\mathbf{u} \bullet \nabla)\mathbf{u}$. This makes the equations "linear," meaning that if any two sets of fields $(\mathbf{u}_1, p_1)$ and $(\mathbf{u}_2, p_2)$ are solutions to the equations, then so is their sum $(\mathbf{u}_1 + \mathbf{u}_2, p_1 + p_2)$. This "principle of superposition"—so important in other areas of physics like electrodynamics and quantum mechanics—is not generally valid for fluid systems, and only arises here because we are assuming that disturbances remain very small. Later, it will be important for us to try to correct for this assumption and we will try to "bootstrap" our way back to a solution to the full nonlinear equations using what we have learned from the linearized analysis.

We will refer to Eqs. (27(a) and 27(b)) with the nonlinear term $(\mathbf{u}\bullet\nabla)\mathbf{u}$ removed as the linearized equations. There is an extremely powerful body of mathematics for handling such equations. Here again we will concentrate on how the results from this theory arise on physical grounds. The first observation is that we can deduce a lot about the solution $(\mathbf{u}, p)$ from the symmetries of the physical system. The boundary conditions and the base state $\mathbf{U}$ only depend on the radius r and not on time t, axial position z, or angular position θ. Another way of saying this is that the coefficients of Eqs. (27(a) and 27(b)) and their boundary conditions do not depend on the independent variables t, z, or θ. We recall that simple ordinary differential equations with constant coefficients have solutions that are either exponential or complex exponential (sinusoidal). So we try a solution of the form:

$$\mathbf{u} = A\hat{\mathbf{u}}(r)e^{st+ikz+im\theta}, \quad (28a)$$

$$p = A\hat{p}(r)e^{st+ikz+im\theta}. \quad (28b)$$

Here the complex number A represents the amplitude of the solution which will remain arbitrary in the linear problem but will become an essential ingredient in the later nonlinear treatment; its value depends on how we chose to normalize the functions $\hat{\mathbf{u}}$ and $\hat{p}$. The numbers k and m are called wavenumbers. Here k must be real because we insist that an initial disturbance ($t = 0$) not become unbounded at either $z = +\infty$ or $-\infty$; k is related to the length λ over which the disturbance repeats itself by the relation $k = 2\pi/\lambda$. The number m must be a real integer so that changing θ to the identical spatial position $\theta + 2\pi$ gives an identical value for $\mathbf{u}$ and p.

The number s is permitted to be complex: $s = \sigma - i\omega$. If its real part, σ, is positive, the disturbance grows exponentially and we say that the original flow is unstable. If σ is negative, the disturbance dies away and the base flow is *linearly* stable. The conditions that give $\sigma = 0$ are called conditions of "marginal stability" provided a slight change one direction renders σ positive (the flow becomes unstable) while a slight change the other way renders σ negative (the flow remains stable).

In fact, this is not the only game in town: for some flows (such as shear flows, to be discussed below), we will find it more appropriate to think in terms of a purely imaginary $s = -i\omega$ acting to "wiggle" the flow at one location. Then we look for the possibility of spatial growth or decay away from this location described by a complex wavenumber $k = k_r + ik_i$. Thus we have two types of stability analysis: temporal and spatial. For now, we concentrate on the temporal stability, since this tends to be the most appropriate for closed systems like the Couette-Taylor problem.

It is important to remember that an actual fluid flow is represented by the real part of the above complex functions $\mathbf{u}$ and p; one of the advantages of working with linear equations is that we can the simpler complex notation for most of the analysis and obtain the real part by adding the complex conjugate. The complex notation also helps the role of phase in periodic solutions to the physical problem. Of course, when we look at nonlinear equations, it will be crucial that we include these solutions *and* their complex conjugates.

Substitution of the assumed form of solutions into the linearized equations yields a system of ordinary differential equations for the functions $\hat{\mathbf{u}}(r), \hat{p}(r)$ that we can write schematically as:

$$L\Psi = sM\Psi. \tag{29}$$

Here Ψ represents the four functions of radius contained in $\hat{\mathbf{u}}(r), \hat{p}(r)$:

$$\Psi = \begin{pmatrix} \hat{u}(r) \\ \hat{v}(r) \\ \hat{w}(r) \\ \hat{p}(r) \end{pmatrix}. \tag{30}$$

L is a linear operator: it can be viewed as a matrix containing functions of r, derivative operators like d/dr, the wavenumbers k and m, and (importantly) parameters describing the externally imposed conditions like the Reynolds number (Re), the speed ratio μ, and the radius ratio η. To remind ourselves of this dependence, we write

$$L = L\left(r, \frac{d}{dr}, k, m; \mathrm{Re}, \mu, \eta\right). \tag{31}$$

M is a 4-by-4 constant matrix. Details of the form of L and M are given in the appendix.

The key feature of Eq. (29) is that it, combined with boundary conditions that $\hat{\mathbf{u}}(r) = 0$ when $r = r_1$ and $r = r_2$, forms an eigenvalue problem: this means that for given Reynolds number and choice of wavenumbers k and m, only certain values of $s = s_n(k, m; \mathrm{Re}, \mu, \eta)$ will permit a solution to be found. The index $n = 1, 2, 3, \ldots$ refers to the fact that, for the Couette-Taylor problem with radius bounded by $r_1 \leq r \leq r_2$, there is a countably infinite set of eigenvalues s_n for given $(k, m; \mathrm{Re}, \mu, \eta)$. We order the eigenvalues according to the size of their real part: $\sigma_1 \geq \sigma_2 \geq \sigma_3$, etc. The dependence of the eigenvalues s_n on wavenumbers is often referred to as a dispersion relation. The corresponding eigenfunctions $\Psi_n(r)$ give disturbance fields $\mathbf{u}_n = \hat{\mathbf{u}}_n(r)\exp(s_n t + ikz + imq)$ and $p_n = \hat{p}_n(r)\exp(s_n t + ikz + imq)$, which are referred to as "normal modes" of this system. Computation of the eigenvalues and eigenfunctions typically requires numerical solution to the system of ordinary differential equations represented by Eq. (29) and can be done with a personal computer.[156]

We look for the threshold of instability: the marginal stability condition $\sigma_1(k, m; \mathrm{Re}, \mu, \eta) = 0$. This implicitly gives the Reynolds number for marginal instability as a function of the other parameters: $\mathrm{Re} = \mathrm{Re}_{\mathrm{marginal}}(k, m; \mu, \eta)$. Plots of $\mathrm{Re}_{\mathrm{marginal}}$ versus wavenumber k are called marginal stability curves. Figure 14(a)

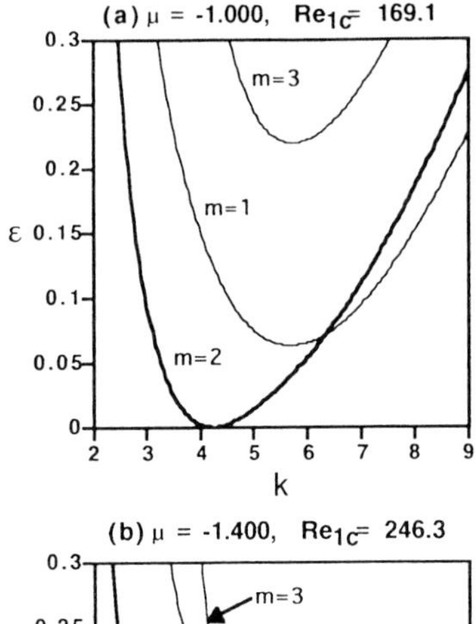

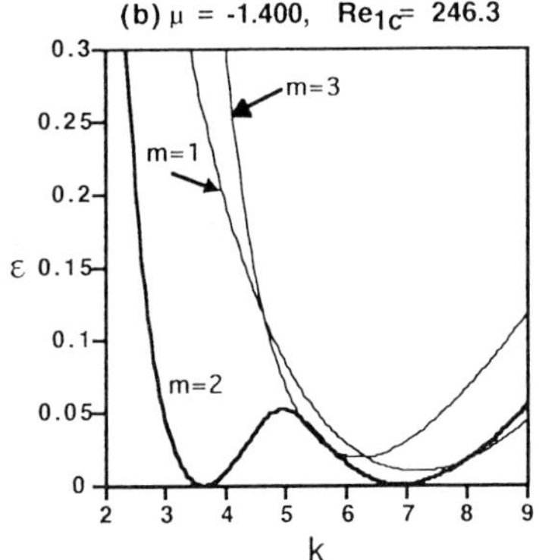

FIGURE 14 Marginal stability curves for a Couette-Taylor system with radius ration 0.640. The abcissa is the axial wavenumber k while the ordinate is the Reynolds number in reduced form $\varepsilon = (\mathrm{Re}_1 - \mathrm{Re}_{1c}/\mathrm{Re}_1)$. Distince curves are drawn for azimuthal wavenumbers $m = 1$, 2, and 3. (a) For speed ration $\mu = -1.000$, the $m = 2$ curve is the deepest, with a single minimum at $k_c = 4.227$. (b) For speed ration $\mu = 1.400$, the $m = 2$ curve is still the deepest, but there are two nearly equal minima at $k_{1c} = 3.628$ and $k_{2c} = 6.896$. [New computations following Tagg et al.[258]]

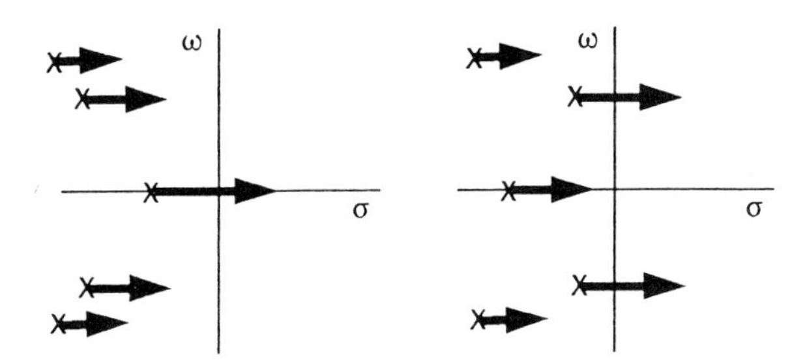

FIGURE 15 Two types of paths taken by eigenvalues of the linear stability problem as a parameter is varied. (a) A real eigenvalue crosses the imaginary axis, producing an "exchange of stabilities" where the resulting secondary flow remains steady in time. (b) A complex-conjugate pair of eigenvalues cross the imaginary axis, producing a Hopf bifurcation with a time-periodic secondary flow.

shows a set of such curves for the Couette-Taylor experiment for a case of counter-rotating cylinders. Note that there is a minimum Reynolds number on the marginal curve, which we will call the critical Reynolds number Re_c; in Figure 14(a), the minimum occurs for $m_c = 2$ at a value $k_c = 4.227$. These values of the wavenumbers giving the absolutely lowest marginal Reynolds number are called the critical wavenumbers. It is commonly assumed that the curves of Re versus k for a given m have single minima. In fact, I have discovered cases in the Couette-Taylor problem where there can be two minima, and even more special situations where the minima are equally deep so that there are two critical axial wavenumbers k_{1c} and k_{2c} (Figure 14(b)). So far, this situation has not been explored experimentally. On the other hand, there have been several experiments where two different values of m have equal minima; these will be discussed further in the Section 3.9.

Note that the imaginary part ω_c of the eigenvalue s is now also determined by k_c, m_c: we can say that when $\mathrm{Re} = \mathrm{Re}_c$, the complex critical eigenvalue $s = 0 - i\omega_c$. Actually, if $m_c = 0$, it turns out that $\omega_c = 0$. This corresponds to the onset of Taylor vortices, and is sometimes referred to as an "exchange of stabilities": one steady flow is replaced with another steady flow. Conversely, if $m \neq 0$, then $\omega_c \neq 0$: a steady flow is replaced by an oscillating flow whose frequency of oscillation is nonzero at onset; this is called a Hopf bifurcation. The disturbance is in fact a traveling wave, with "crests" of constant phase advancing in both axial and azimuthal directions: this is the spiral vortex found for counter-rotating cylinders. A useful point of view of these two cases is obtained by plotting the "trajectories" of the eigenvalue s in the complex-plane as the Reynolds number passes through Rec. In one case (exchange of stabilities—Figure 15(a)), a single trajectory crosses through $\sigma = 0$ along the real axis. In the other case (Hopf bifurcation—Figure 15(b)), two trajectories of complex-conjugate eigenvalues cross through $\sigma = 0$ at points $\pm i\omega_c$ displaced a nonzero distance from the real (σ) axis. In both cases, all other eigenvalues are assumed to remain in the left-hand side of the graph where $\sigma < 0$.

Now we make a very important step. Consider that the Reynolds number has just been slowly nudged above the critical value Re_c. Out of a noisy soup of fluctuations characterized by all possible values of m and k, we expect that the system will amplify just that fluctuation with wavenumbers m_c and k_c. This means that the system will spontaneously select scales of length in the axial and azimuthal directions. Taylor vortices have $m = 0$; our theory predicts that they should repeat themselves with a wavelength $\lambda_c = 2\pi/k_c$ and this is indeed well established by experiment. In fact, $k_c \approx \pi$ and so $\lambda_c \approx 2d$: each vortex has a roughly "square" cross section of size equal to the gap d between the cylinders.

Spontaneously selected numbers like k_c and m_c that describe the new state of order after an instability has occurred are sometimes called "emergent parameters"—adapting (perhaps somewhat dangerously) terminology given in Anderson and Stein.[7] The numbers represent, in part, the macroscopic ordering that can take place in a nonequilibrium system in a manner that would by no means be obvious from a purely microscopic model of the interaction between individual fluid particles. The new order comes at a price: the basic flow before instability looked the same if we made arbitrary shifts in the axial or azimuthal directions. Now, such shifts amount to a shift in phase of the sinusoidal functions describing the cellular flows. The price we have paid is that there is a complete ambiguity in these phases; nature will select them by accident. Associated with such phase variables can be large length and time-scale dynamics which we will discuss in Section 3.8.

Actually, the above conclusion—that the system produces a cellular flow of a single length scale—is too simplistic even within the scope of a linear analysis. When the secondary flow is a traveling wave—or in open systems where fluid can flow in one end and out the other—we must consider the grow of initial fluctuations more carefully. We have implicitly relied on an assumption that the normal modes of our linear analysis form a "complete set" of functions: that is, any arbitrary disturbance to the base flow can be expressed as a superposition of these modes.

$$\begin{pmatrix} u(r,\theta,z,t) \\ v(r,\theta,z,t) \\ w(r,\theta,z,t) \\ p(r,\theta,z,t) \end{pmatrix} = \int_{-\infty}^{\infty} dk \sum_{m=1}^{\infty} \sum_{n-1}^{\infty} A_n(k,m) \begin{pmatrix} P\hat{u}_n(r) \\ \hat{v}_n(r) \\ \hat{w}_n(r) \\ \hat{p}_n(r) \end{pmatrix} e^{i(kz+m\theta)} e^{s_n(k,m)t} + \text{c.c.}, \tag{32}$$

where the "+c.c." means "add the complex conjugate of the preceding expression" and the amplitudes $A_n(k,m)$ are determined by the initial conditions.

Suppose in fact that we contrive an initial disturbance that is highly localized in the z-direction. If the fastest growing mode is a traveling wave (e.g., a spiral vortex), then a computation of the expression (Eq. (32)) for later times reveals that the disturbance spreads out into two packets of spiral vortices that travel in either direction along the cylinder. These packets continue to grow if this system has a Reynolds number greater than Re_c. However, we must distinguish two cases, as shown in Figure 16. The packet might travel away from the initial disturbance sufficiently fast that, even though the waves are growing in amplitude within the traveling packet, the fluid returns to its basic flow state at the origin of the disturbance. We say in this case that we have a "convective instability" because the growing instability convects away from its source. However, if the instability grows sufficiently fast relative to the rate at which the wave packet moves away, the origin will remain in an altered state long after the center of the packet has moved away. We call this case "absolute instability." These ideas were given in the classic fluid dynamics text by Landau and Lifshitz[155] and were well known to the plasma-physics community (see Bers[21]). Only in the last ten years have they been more widely applied to fluid dynamics problems: see Huerre[119] and Monkewitz[178] and references therein.

The distinction between convective and absolute instability is extremely important for open flow systems, because in such cases the instability is typically of the convective type. This type has the ability to amplify noise appearing at the upstream end of an apparatus with the result that the traveling waves that arise from the instability have a randomly varying phase. Consequently, power spectra of downstream velocity measurements for such systems show broadened, rather than sharp, peaks at the frequency ω. The packets that evolve from upstream noise are called "noise-sustained structures."[69] Another consequence of convective instabilities is the possibility that one can miss the onset of instability entirely if the experimental system is not long enough to permit fluctuations to grow to a detectable level. Some very careful experiments with convective and absolute instabilities have been performed by Babcock et al.[8] and Tsameret and Steinberg[267] in a Couette-Taylor experiment modified so that the base flow has a steady axial component (fluid is pumped in one end and out the other). The distinction between convective and absolute instabilities was also found to be important for spiral vortex flow in the closed system,[258] although interpretation is complicated by the need to understand what the waves do at the ends—still an open problem for this system.

It would be interesting to find applications of these ideas outside of fluid dynamics and plasma physics. Anytime there is a propagating, growing disturbance, one might look for convective versus absolute instability. Propagation of neural signals

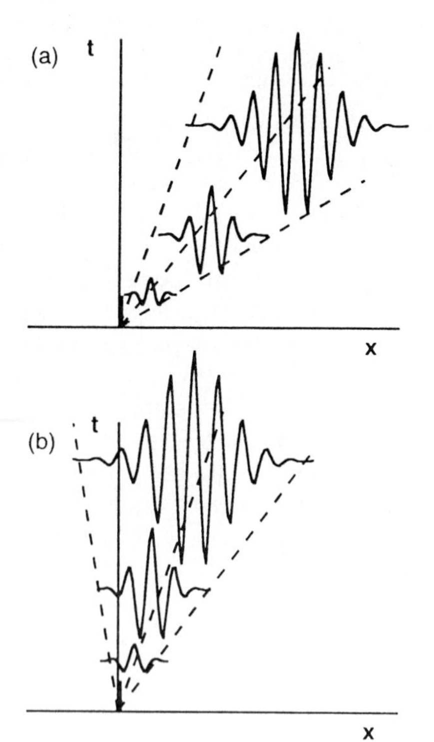

FIGURE 16 Evolution in space and time of local disturbances. (a) Convective instability: a local disturbance at position $x = 0$ at time $t = 0$ propagates away as it spreads and grows, leaving no permanent change behind. (b) Absolute instability: a local disturbance, though its peak may propagate away, still leaves a change in the system at the origin of the disturbance.

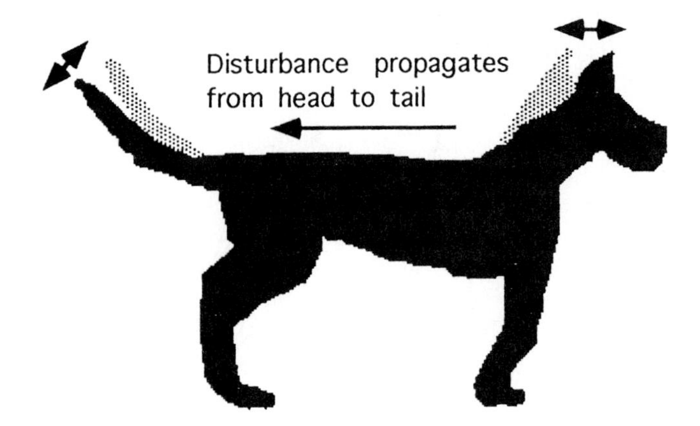

FIGURE 17 Propagation of a disturbance in the author's dog, who was trying to shake a bandage from his ear. Is there a common connection to the idea of convective instabilities, or is this just a shaggy dog story?

is certainly visible within biological systems, but I'm not sure that this fits within the framework I have been discussing. Nonetheless, while I writing this chapter, I noticed (see Figure 17) wavepackets that seemed to travel from head to tail of my dog when he tried to shake a bandage from his ear!

3.7 WEAKLY NONLINEAR THEORY OF SECONDARY FLOWS

We have cheated. We have been discussing about the system undergoing instability and establishing a new cellular pattern of flow or a traveling wave packet. However, our mathematical description of the growing disturbances had them growing exponentially in time. Clearly, this exponential growth will not continue forever; the disturbances soon become so large that we can no longer neglect the nonlinear terms in the equations. How can we determine, once the nonlinearity is felt, what features of our linear theory are preserved in the actual flow? Also, how can we deal with the problem of instability to disturbances of finite, rather than infinitesimal, amplitude? Often, the answers to these questions rely on experiment and numerical computations. However, in several important cases—the Couette-Taylor experiment is one of them—we can proceed systematically from the linear analysis to a so-called weakly nonlinear analysis that allows us to determine the flow states that ultimately result. The requirement for this analysis to proceed is that we remain quite close to the onset of instability. Typically, this means that the Reynolds number is within about 1% of its critical value Re_c. We will find it convenient to refer to a reduced Reynolds number defined by:

$$\varepsilon \equiv \frac{\mathrm{Re} - \mathrm{Re}_c}{(\mathrm{Re}_c)} \tag{33}$$

so that $\varepsilon < 0$ below transition and $\varepsilon > 0$ (and presumed small) above transition.

This sort of analysis and its connection to some very careful experiments was been the subject of earlier lectures at the Santa Fe Institute summer school.[2,184] Related work on the delicate issues in describing the formation and dynamics of patterns in nonequilibrium systems is also the subject of a monumental review by Cross and Hohenberg.[66] I will therefore give an abbreviated discussion, oriented towards topics connected with experiments I have done with the Couette-Taylor system.

Suppose that we expect Taylor vortices to grow. Due to nonlinearity, the growing linear mode will interact with itself and with the base flow to generate higher harmonics (components with integer multiples of the fundamental spatial and temporal frequencies) and even alterations to the base flow. These in turn interact and provide a sort of negative feedback that limits the growth of the linear eigenmode. This idea was first formally developed by Stuart[247,248] and Watson,[277] applied in detail to the Couette-Taylor problem by Davey,[68] and given a rigorous formulation by Kirchgässner and Sorger.[134] However, much earlier, Landau[154] used heuristic reasoning (such as symmetry considerations) to arrive at an equation describing this process in terms of the amplitude A of the linear mode:

$$\tau_0 \frac{dA}{dt} = (1 + ic_0)\varepsilon A - g(1 + ic_2)|A|^2 A\,. \tag{34}$$

This "amplitude equation" can be derived from the more formal or rigorous approaches mentioned above; it is often referred to as the Landau equation or the Landau-Stuart equation.

Several comments are in order. First, the linearized equation

$$\frac{dA}{dt} = \tau_0^{-1}(1 + ic_0)\varepsilon A \tag{35}$$

has eigenvalue $\tau_0^{-1}(1 + ic_0)\varepsilon$, which is just the first-order term in an expansion of the eigenvalue s in the parameter ε:

$$s = i\omega_c + \tau_0^{-1}(1 + ic_0)\varepsilon + o(\varepsilon^2). \tag{36}$$

Considering this as a Taylor series expansion, we identify:

$$\tau_0^{-1}(1 + ic_0) = \begin{cases} \frac{\partial s}{\partial \varepsilon} & \text{at } \varepsilon = 0; \\ \mathrm{Re}_c \frac{\partial s}{\partial \mathrm{Re}} & \text{at } \mathrm{Re} = \mathrm{Re}_c. \end{cases} \tag{37}$$

A second and very important comment is that this Landau equation is the only form up to cubic order in A that is consistent with translational invariance in the z-direction and rotational invariance in the θ-direction. In another coordinate frame with coordinates $(r, \theta', z') = (r, \theta - \theta_0, z - z_0)$, the amplitude is $A' = A\exp(ikz_0 + im\theta_0)$. The physics must be the same, so the same amplitude equation must be obtained for A'. Thus any phase shifts in A must cancel from the amplitude equation; the reader may verify that this is the case for the Landau equation. We say that the amplitude equation is "equivariant" under the symmetry operations of translations in z and rotations in θ.

Finally, the complex Landau constant $g(1 + ic_2)$ describes the nonlinear saturation of the amplitude and the finite-amplitude corrections to the frequency ω. Before showing this, we note that computation of the Landau constant depends on the particular flow and requires a detailed perturbation analysis. The numerical value of g depends on the normalization chosen for the eigenfunction Ψ; it is important not to overlook the requirement to state this normalization when quoting a value for g. Lately, a powerful method called "center manifold reduction"[37] has become widely used for reducing the Navier-Stokes equations to equations like the Landau equation and computing the nonlinear coefficients like g and c_2. A discussion of such methods applied to the Couette-Taylor problem is found in the recent book by Chossat and Iooss[46] and the use of symbolic algebra computer programs to aid in the often formidable algebraic analysis is discussed in Rand and Armbruster.[211]

In order to solve the Landau equation, we write the amplitude in terms of a modulus and phase: $A = \rho e^{i\phi}$. Then two equations result:

$$\tau_0 \frac{d\rho}{dt} = \varepsilon\rho - g\rho^3, \qquad (38a)$$

$$\tau 0 \frac{d\phi}{dt} = c_0\varepsilon - gc_2\rho_2. \qquad (38b)$$

Looking for solutions with ρ steady in time $(d\rho/dt = 0)$, we get

for $g > 0$:

$\varepsilon < 0:$	$\phi = \tau_0^{-1} c_0 \varepsilon t,$	**stable**;	
$\varepsilon > 0:$	$\rho = 0,$	$\phi = \tau_0^{-1} c_0 \varepsilon t,$	unstable;
	$\rho = (\varepsilon/g)^{1/2},$	$\phi = \tau_0^{-1}(c_0 - c_2)\varepsilon t,$	**stable**;

for $g < 0$:

$\varepsilon < 0:$	$\rho = 0,$	$\phi = \tau_0^{-1} c_0 \varepsilon t,$	**stable**;
	$\rho = (-\varepsilon/g)^{1/2},$	$\phi = \tau_0^{-1}(c_0 - c_2)\varepsilon t,$	unstable;
$\varepsilon > 0:$	$\rho = 0,$	$\phi = \tau_0^{-1} c_0 \varepsilon t,$	unstable.

The stability of the solutions is determined, as before, by adding a small perturbation and seeing, within the amplitude equation description, if it grows or decays. Note that Eq. (38(a)) for the magnitude ρ is de-coupled from the phase equation, a result of the symmetry of the problem. However, Eq. (38(b)) for the phase ϕ retains a dependence on ρ through the nonlinear term $-gc_2\rho_2$. The solutions for ϕ may be interpreted as corrections to the frequency: $\omega = \omega_c + (d\phi/dt)$. Thus for $\varepsilon > 0$ in the supercritical case ($g_r > 0$), the frequency is altered according to

$$\omega = \omega_c + \tau_0^{-1}(c_0 - c_2)\varepsilon, \tag{39}$$

where the term $\tau_0^{-1}c_0\varepsilon$ is a linear correction and the term $-\tau_0^{-1}c_2\varepsilon$ is the leading-order nonlinear correction. The solutions for ρ and their stability are shown in the bifurcation diagrams of Figure 18. Note the characteristic $\varepsilon^{1/2}$ dependence of the

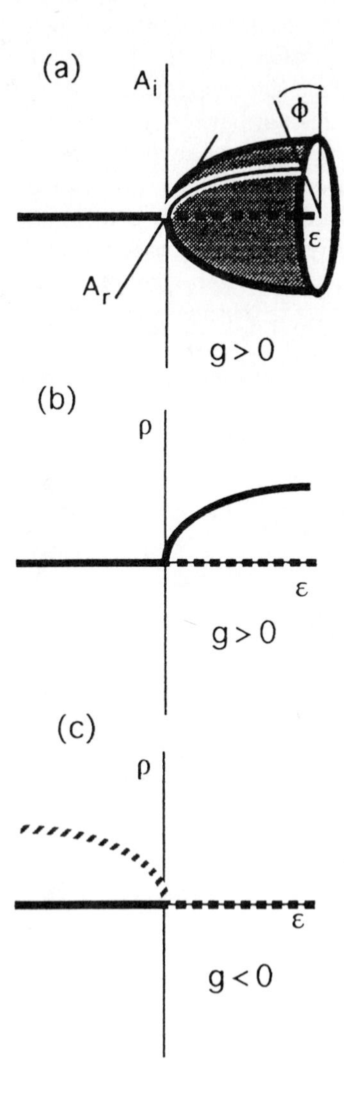

FIGURE 18 Bifurcation of solutions to the Landau equation (Eq. (24)) as the control parameter ε is varied: stable solutions are solid curves and unstable solutions are dashed curves. (a) In the case of a supercritical bifurcation (Landau constant $g > 0$), the complex amplitude of a new solution rises continuously from zero as ε is increased above the bifurcation point $\varepsilon = 0$. However, the new solution is underdetermined up to a phase ϕ. (b) For a supercritical bifurcation, the modulus of the amplitude, $\rho \equiv |A|$ grows as $\varepsilon^{1/2}$ for $\varepsilon > 0$ and the new solution is stable. (c) For a subcritical bifurcation (Landau constant $g > 0$), the modulus of the amplitude for the new mode grows as $(-\varepsilon)^{1/2}$ for $\varepsilon < 0$ but the new solution is unstable in the neighborhood of the bifurcation point $\varepsilon = 0$.

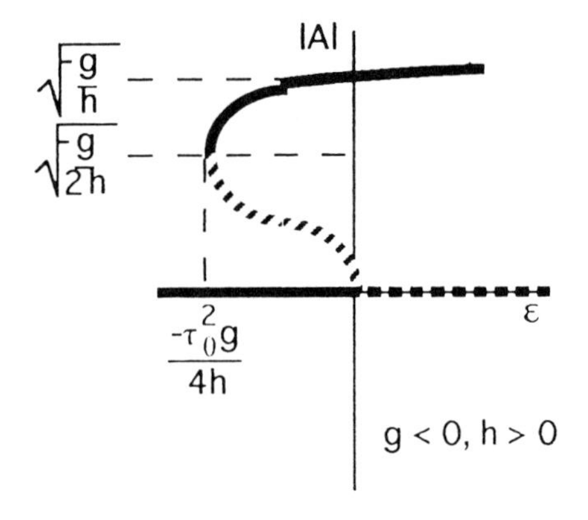

FIGURE 19 Subcritical bifurcation with a turning point: a new stable solution (upper continuous curve) may be reached for sufficiently large initial disturbances away from the base solution (lower solid line for $\varepsilon < 0$). Careful experiments will show hysteresis: as ε is increased, the lower solution persists up to $\varepsilon < 0$. Then the system jumps to the upper branch. If ε is now decreased, the upper branch solution persists down to the turning point, whereupon the solution returns to the original branch. The features of this bifurcation are described by the fifth-order amplitude equation shown.

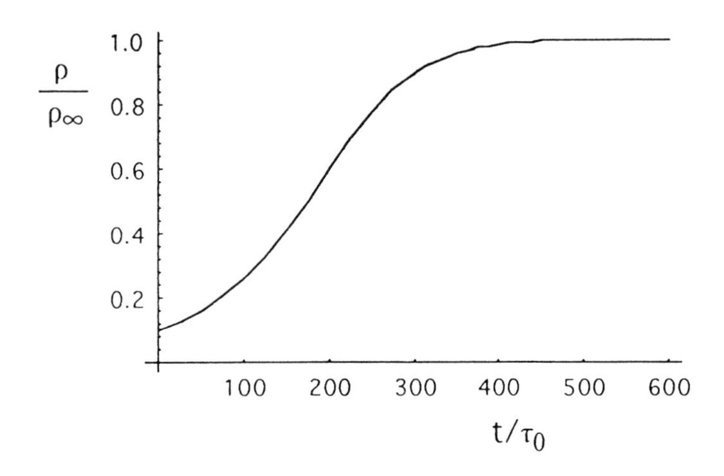

FIGURE 20 Growth of an initial disturbance (Eq. (40)) as predicted by the Landau equation (see Eqs. (34) and (38a)). The modulus of the mode amplitude starts at $\rho(0)/\rho_\infty = 0.1$ and levels off for large times at $\rho(t)/\rho_\infty = 1$, where $\rho_\infty = (\varepsilon/g)^{1/2}$. As a typical value, we set $\varepsilon = 0.01$. As the system is taken closer and closer to transition ($\varepsilon \to 0$) the time to level off diverges at $1/\varepsilon$; this is called "critical slowing down."

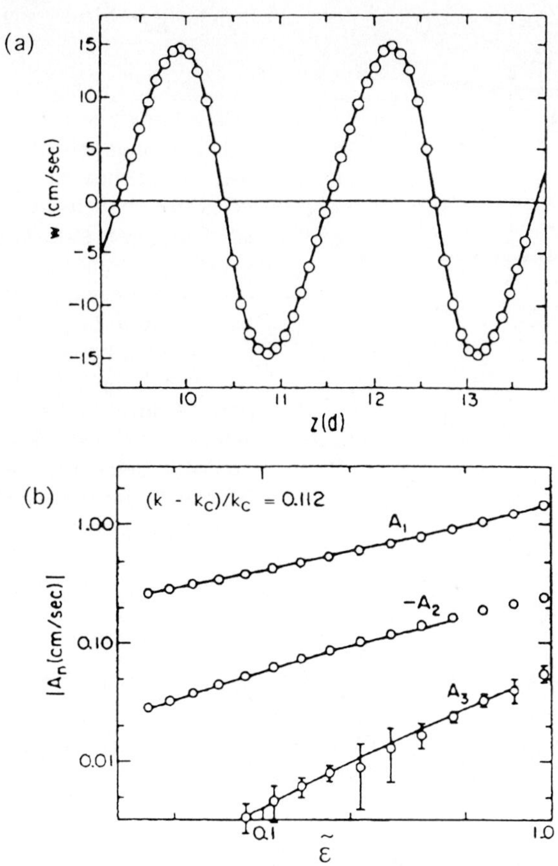

FIGURE 21 Data on growth is mode amplitude for Taylor vortices. (a) The mode amplitudes are obtained by fitting (solid curve) a Fourier expansion (Eq. (41)) to axial velocity data (circles). Such data were obtained by precise velocity measurements using laser doppler velocimetry. (b) Data extracted from the above fits show the variation with $\tilde{\varepsilon}$ of the amplitudes of the fundamental (A_1) and two higher harmonics (A_2 and A_3) of the spatial fluctuation of the axial velocity near transition. The initial slope (on this log-log plot) of the curve for the fundamental A_1 confirms the $\tilde{\varepsilon}^{1/2}$ dependence of the fundamental variation with wavelength $2\pi/k$; the solid curves represent fits to expressions of the form Eq. (41) that take into account higher-order corrections. Here $\tilde{\varepsilon}$ measures the distance away from the marginal curve for the actual wavenumber k of the flow, which was slightly different from the critical wavenumber k_c. From Heinrichs et al.[109]

bifurcating solutions. We see that the sign of the real part of the Landau constant is crucial. For positive g there is a stable supercritical mode bifurcating from the base state. For negative g, there is subcritical bifurcation to a solution that is unstable and higher-order terms in the amplitude equation are needed to resolve what happens for $\varepsilon > 0$; see Figure 19.

In fact, an exact time-dependent solution may be found for Eq. (38(a)) given an initial nonzero disturbance ρ_o:

$$\rho_2 = \frac{\varepsilon}{\tau_0 g} \frac{1}{1 + \left(\frac{\varepsilon}{\tau_0 g \rho_0^2}\right) - 1} e^{-2\varepsilon t/\tau_0} . \tag{40}$$

Experiments can verify this time dependence by making a jump in ε from subcritical to supercritical and observing that the amplitude of the critical eigenmode follows the curve shown in Figure 20; see Section 5.5 below, where such a procedure was used for wake flows. Experiments can also verify dependence $A \sim \varepsilon^{1/2}$, as shown in Figure 21(a) using data from Heinrichs et al.[109] In making such comparisons, it is important to extract only the component of velocity variations that corresponds to the "fundamental," that is the to distinguish the part of a cellular pattern that varies with wavenumber k from the parts that vary with higher harmonics $nk(fon > 1)$. This is usually done, as in Figure 21(b), by fitting any given velocity component u (at fixed r and θ) to a function of form

$$u = u_o + A\cos(kx + \phi) + \sum_{n=2}^{\infty} A_n \cos(nkz + \phi_n). \tag{41}$$

In fact, identifying $A_1 \equiv A$, it is a consequence of Davey's[68] analysis that the leading order dependence of the fundamental and harmonics on ε is given by

$$A_n \sim \varepsilon^{n/2}(1 + \alpha_n \varepsilon + \ldots). \tag{42}$$

3.8 ENVELOPE EQUATIONS, PATTERN SELECTION, AND PHASE DYNAMICS

We are still cheating: we assumed that beyond the point of instability, the only relevant modes were the linear eigenmode and harmonics. In fact, it is evident from the marginal stability curve (Figure 14(a)) that, for $\varepsilon > 0$, there is a continuous band of axial wavenumbers k that have positive growth rate. Important questions, then, concern the mutual interaction of this band of wavenumbers, the selection of a particular wavenumber for supercritical flows, and the stability of secondary flows with given wavenumbers. I will deal with this important topic quite briefly, because it is discussed at length in the substantial articles by Newell,[184] Ahlers,[2] Cross and Hohenberg,[66] and Newell et al.[185] For the moment, we will consider only the transition to time-independent Taylor vortices, so $m = 0$ and $\omega_c = 0$.

Anyone familiar with communication theory will recognize that a signal containing a distribution of frequencies around some carrier frequency will appear as a fast varying oscillation enclosed within a slowly varying envelope. In this present case, we think of k_c as the carrier frequency (in space rather than time) and look for modulations of the fast varying part e^{ikz} by a slowly varying envelope $A(z,t)$. Newell and Whitehead[186] and Segel[236] derived an appropriate extension of the Landau equation to describe both the slow time and space dependence of A:

$$\tau_0 \frac{\partial A}{\partial t} = \varepsilon A + \xi_0^2 \frac{\partial^2 A}{\partial z^2} - g|A|^2 A. \tag{43}$$

The coefficients τ_0 and ξ_0 (ξ_0 is called the correlation length) are determined by

$$\tau_0 = \left(\frac{\partial \sigma}{\partial \varepsilon}\right)^{-1} \quad \text{at } \varepsilon = 0, \tag{44a}$$

$$\xi_0^2 = -\frac{\tau_0}{2}\left(\frac{\partial^2 \sigma}{\partial k^2}\right) \quad \text{at } k = k_c. \tag{44b}$$

Equation (43) is usually referred to as the Ginzburg-Landau equation, after an equation of similar form (but much different physics) used in the Ginzburg-Landau[97] theory of superconductivity; it is a one-dimensional version of the Newell-Whitehead-Segel envelope equations used to describe patterns in buoyancy-driven convection. We now require boundary conditions for A, which for Taylor vortices in finite-length Couette systems are $A = O(1)$; physically, this corresponds to the presence, even for $\varepsilon < 0$, of strong end cells of at each end of the annular column.

One important result that can be derived from the Ginzburg-Landau equation is the fact that cellular flows of given wavenumber k are stable for a vales of k that is narrower than the band contained between the boundaries of the marginal stability curve. At a given value of ε, a steady solution to the GL equation is a state of uniform wavenumber k given by

$$A(k) = A_0 e^{i(k-kc)z}, \tag{45}$$

where

$$A_0 = \left[\frac{\varepsilon - \xi_)0^2(k-k_c)^2}{g}\right]^{1/2}. \tag{46}$$

Suppose this state is perturbed by a modulation:

$$\tilde{A} = [A_0 + a_+(t)e^{iqz} + a_-^*(t)e^{-iqz}]e^{i(k-k_c)z}. \tag{47}$$

The added terms are called "sidebands," because they represent shifts $\pm q$ away from the original wavenumber k. If this is substituted into the GL equation and terms nonlinear in $a_\pm$ are discarded, it is found that $a_\pm(t)$ grow exponentially in time and thus the state $A(k)$ is unstable provided

$$(k-k_c)^2 > 1/3(k_m - k_c)^2. \tag{48}$$

Here k_m is the wavenumber at the marginal stability curve where $A_0 = 0$:

$$(k_m - k_c)^2 = \frac{\varepsilon}{\xi_0^2}. \tag{49}$$

Near the boundary of the unstable region, the fastest growing modulation corresponds to

$$q \sim 0;$$

the instability is due to a long wavelength modulation. This instability is known as the Eckhaus instability.[82] Kogelman and DiPrima[140] first applied these ideas to the Couette-Taylor problem, while Eckhaus' original approach was recast using envelope equations by Stuart and DiPrima[250] and Kramer and Zimmermann.[147] A bifurcation analysis of the instability, separating the competing wavenumbers by restricting the system to finite length, was recently done by Tsiveriotis and Brown[268] and Tuckerman and Barkley.[271] A discussion of some beautiful experiments on the Eckhaus instability for Taylor vortices is found in the Santa Fe lectures of Ahlers.[2]

Another result, that leads to a generalization outside the range of validity of the Ginzburg-Landau equation, is the concept of phase dynamics. Think of suddenly stretching some of the vortex cells at a particular point in the column, for example, by moving one of the end boundaries. How do the vortices in the whole column respond? Are the vortices springy, so that the perturbation propagates as a pulse through the column, or does the system relax by letting the disturbance diffuse? For Taylor vortices, the latter is the case. Suppose we perturb our steady-state solution to the Ginzburg-Landau equation (Eq. (43)) so that it becomes

$$A'(k) = [A_0 + a(z,t)]e^{i(k-kc)z + i\psi(z,t)}. \tag{50}$$

If this is inserted into the Ginzburg-Landau equation and terms nonlinear in the perturbation are discarded, one finds that the amplitude perturbation $a(z,t)$ undergoes an initial transient whose time scale is τ_0/ε. This can be long, if ε is small, but eventually the amplitude adiabatically follows (is "slaved" to) the phase:

$$a = -\frac{\xi_0^2 q}{q A_0}\frac{\partial\psi}{\partial z}. \tag{51}$$

The phase perturbation, on the other hand, relaxes on a time scale that diverges for any value of ε with the length scale of the perturbation. The phase dynamics is described by a diffusion equation:

$$\frac{\partial\psi}{\partial t} = D_\parallel \frac{\partial^2\psi}{\partial z^2}, \tag{52}$$

where $D_\parallel$ is related to q and the parameters of the Ginzburg-Landau equation by

$$D_\parallel = \frac{\xi_0^2}{\tau_0}\frac{\varepsilon - 3\xi_0^2 q^2}{\varepsilon - \xi_0^2 q^2}. \tag{53}$$

This diffusive behavior of the phase, first derived by Pomeau and Manneville[207] in a model of fluid convection, is expected to be valid well beyond the validity of the Ginzburg-Landau equation; however, the diffusion coefficient will no longer be given by Eq. (53). This treatment, called "phase dynamics," describes the large-scale, long-time dynamics of strongly nonlinear cellular flows. Indeed, the form of the phase equation, including extension to include nonlinear terms, is dictated by the dimensionality and symmetry of the problem. For a review, see Brand.[25] Several recent experiments exploring phase dynamics in the Couette-Taylor system are described in Wu and Andereck.[281]

3.9 NONLINEAR TRAVELING WAVES AND MODE INTERACTIONS

The richness of dynamics near the first transition increases a lot when one considers the nonlinear development of spiral vortices (Figure 10(b)) between counterrotating cylinders. First of all, the flow is periodic in time. Second, the reflection symmetry of the system requires that spiral vortices traveling towards positive z and spirals traveling towards negative z are both valid linear solutions. One mode is the reflection of the other through any plane normal to the cylinder axes. These two modes are coupled, so that the appropriate extension of the Landau equation is the system of two equations:

$$\tau_0 \frac{dA}{dt} = (1 + ic_0)\varepsilon A - g1(1 + ic_2)|A|^2 A - g_2(1 + ic_3)|B|^2 A\,, \tag{54a}$$

$$\tau_0 \frac{dB}{dt} = (1 + ic_0)\varepsilon B - g_1(1 + ic_2)|B|^2 B - g_2(1 + ic_3)|A|^2 B\,. \tag{54b}$$

Here A is the amplitude of the vortex traveling towards positive z and B is the amplitude of the mode traveling towards negative z; we will call these right- and left-traveling waves, respectively.[4] These equations admit two types of solutions besides the trivial solution $A = B = 0$:

I. $A \neq 0; B = 0.$ $\quad$ TW_R : right-traveling waves (spiral vortices)
$\quad A = 0; |B| \neq 0.$ $\quad$ TW_L : left-traveling waves (spiral vortices).
II. $|A| = |B| \neq 0.$ $\quad$ SW : Standing waves in the z-direction (but traveling waves in the θ-direction): also known as "ribbons."

The mode of bifurcation and relative stability of the two modes is determined by the values of g_1 and g_2. When both modes branch supercritically, only one is stable.

$$0 < g_1 < g_2 \Longrightarrow \text{stable, supercritical spirals;}$$
$$|A| = (\varepsilon/g_1)^{1/2}, B = 0; \quad \text{or} \quad A = 0, B = (\varepsilon/g_1)^{1/2};$$
$$0 < g_1 \text{ and } -g_1 < g_2 < g_1 \Longrightarrow \text{stable, supercritical ribbons,}$$
$$|A| = |B| = [\varepsilon/(g_1 + g_2)]^{1/2}$$

[4] We use "right" and "left" rather than "up" and "down" to be consistent with the literature on traveling waves in buoyancy-driven convection in binary fluid mixtures.

The dependence of these coefficients g_1 and g_2 on the experimental control parameters (Re_1, μ, and η) is nontrivial, requiring numerical computations.[70] The full set of bifurcation diagrams is shown in Figure 22; this is a generic description for Hopf bifurcations in systems with so-called O2 symmetry (see, for example, Crawford and Knobloch[61]). In this context, O2 symmetry is equivalent to translational symmetry in the z-direction along with reflection symmetry through any plane normal to the z-axis. In fact, spiral vortices are commonly seen in experiments, but the "ribbons" state was presumed not to be stable until Demay and Iooss[70] identified a range of control parameters where they should occur. A ribbons-like state was observed experimentally and computed numerically by Tagg et al.,[259] but the observations are complicated by end-effects that will be discussed below.

Suppose the spiral vortex state is selected. One key question concerns the nonlinear development of the frequency. Figure 23 shows a comparison between experiment, direct numerical simulation, and weakly nonlinear theory. At first, numerical and theoretical results were in bad disagreement with experiment. Then it was realized[84] that the pressure was assumed to be spatially periodic. However, the traveling waves under such conditions will generate a mean flow; since this cannot occur in a closed system, there will be a difference in average pressure from one end of the traveling wave to the other. When this is taken into account, there is much better agreement. Note that it was necessary to go to fifth-order amplitude equations[209] to match the experimental data over the range shown.

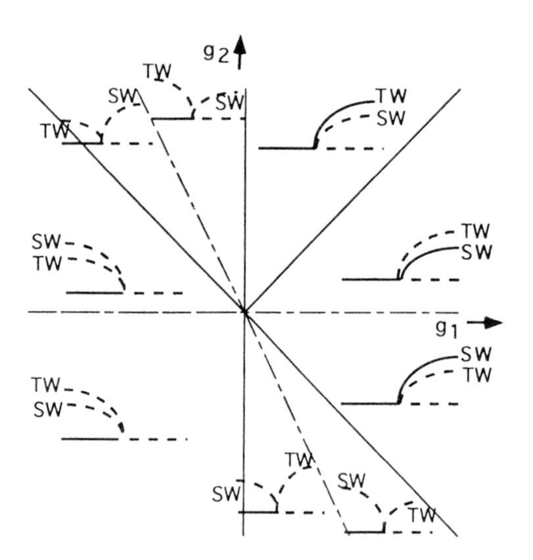

FIGURE 22 Bifurcation diagrams for systems with translational and reflection symmetry (O2 symmetry). The axes are the nonlinear coupling parameters g_1 and g_2 of Eq. (54). Stable branches are shown as heavy solid curves; unstable branches are shown as heavy dashed curves. To cubic order, there are only two regimes of stable bifurcating solutions: traveling waves (TW: either left or right) for $g_2 > g_1 > 0$ and standing waves (SW) for $-g_1 < g_2 < g_1$ and $g_1 > 0$.

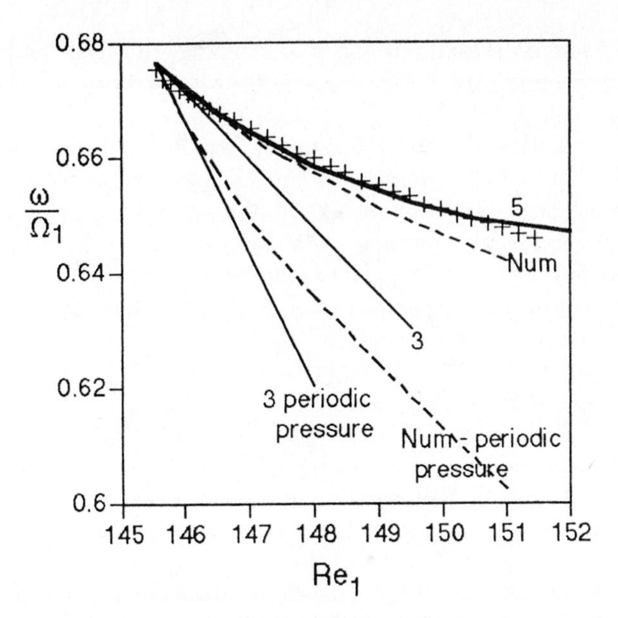

FIGURE 23 Nonlinear development of spiral vortex frequency ω (in units of the inner cylinder frequency Ω_1) as the Reynolds number Re_1 is raised above onset. Experimental data points are shown by crosses $(+)$. The curves labeled with a suffix "periodic pressure" were evaluated using the assumption of periodic pressure. The labels 3 and 5 refer to cubic- and fifth-order amplitude calculations respectively. The label "Num" refers to numerical simulations. The fifth-order amplitude calculation and numerics without periodic pressure assumption agree well with the data, although it is uncertain why there is a discrepancy between the two curves. (All curves were generated assuming the critical axial wave number k_c; the actual wavenumber k was not determined in the experiment.) Experimental and numerical data are from Edwards et al.[83] and corrected amplitude-equation curves are from Raffai and Laure.[209] The system had radius ratio $\eta = 0.7992$ and speed ratio $\mu = -0.74$, giving (from linear stability) $\mathrm{Re}_{1c} = 145.57, k_c = 3.6997, m_c = 2$, and $\omega_c \Omega_1 = 0.67664$.

We have again been ignoring the interaction of modes with a finite band of axial wavenumbers. The appropriate generalization leads to coupled complex Ginzburg-Landau equations:

$$\begin{aligned}\tau_0 \frac{dA}{dt} =& (1+ic_0)\varepsilon A - \tau_0 v_g \frac{\partial A}{\partial z} + \xi_0^2(1+ic_1)\frac{\partial^2 A}{\partial z^2} - g(1+ic_2)|A|^2 A \\ & - h(1+ic_3)|B|^2 A\,, \end{aligned} \tag{55a}$$

$$\begin{aligned}\tau_0 \frac{dB}{dt} =& (1+ic_0)\varepsilon B + \tau_0 v_g \frac{\partial B}{\partial z} + \xi_0^2(1+ic_1)\frac{\partial^2 B}{\partial z^2} - g(1+ic_2)|B|^2 B \\ & - h(1+ic_3)|A|^2 B\,. \end{aligned} \tag{55b}$$

Recalling that $s \equiv \sigma - i\omega$, the linear coefficients are defined by derivatives evaluated at $\varepsilon = 0$ and $k = k_c$:

$$\tau_0^{-1}(1 + ic_0) \equiv \frac{\partial s}{\partial \varepsilon}, \tag{56a}$$

$$\tau_0^{-1}\xi_0^2(1 + ic_1) \equiv \frac{\partial^2 s}{\partial k^2}, \tag{56b}$$

$$v_g \equiv \frac{\partial \omega}{\partial k}. \tag{56c}$$

Here v_g is called the group velocity, corresponding to the speed at which a spatially localized wavepacket of spiral vortices would travel if nonlinear effects were ignored. The consistency with which the terms in the complex Ginzburg-Landau equations are of the same perturbation order is problematic (see, for example, Cross and Hohenberg,[66] p. 881). It is generally assumed that the group velocity is "small." When this is not the case, Knobloch and De Luca[138] have derived non-local amplitude equations, that is, equations that involve integrals over the domain of flow. So far, the consequences have not been explored for spiral vortex flow. For the Couette-Taylor system, control parameters may be chosen so that group velocity is positive, zero, or negative.[258] Indeed, one of the key interests in this regime of flow is the wide range over which the coefficients in the complex Ginzburg-Landau equation may be "tuned."

Zaleski et al.[283] discussed an equation related to Eqs. (55) and boundary conditions for the amplitude A of one of the spiral modes in the case of small gaps and without coupling between left and right traveling waves. Using a boundary condition that $A = 0$ at the ends, they were able to recover many features of amplitude measurements made by Tabeling and Trakas[252] for spiral vortex flow in a conducting fluid placed in a magnetic field. Solving Eq. (55(a)) without nonlinear terms and imposing the boundary condition that A vanish at the ends gives the "whale" shaped amplitude variation shown in Figure 24, which is similar to the form of the data found by Tabeling and Trakas and characteristic of spiral eigenmodes of *finite* cylinders. It is also possible to obtain a superposition of these modes with a "defect" or grain boundary demarking the region where they overlap.

Experimentally, such defect states are often observed: spirals form domains of right traveling waves at one end and left traveling waves at the other, with a grain boundary in between.[6] The grain boundary can drift, in which case power spectra indicating the time-dependence of the flow show a broadened frequency peak centered on ω. However, it is possible[254] to obtain either a dominant left or right spiral or a chevron structure in which left and right spirals occupy opposite halves of the cylinder with a fixed grain boundary in between (Figure 25(a)). In such cases, the power spectrum gives a sharp frequency peak at ω. Conditions were found where, starting with the chevron structure, the grain boundary "broke loose" and oscillated periodically from one end of the cylinder to the other. Initially,

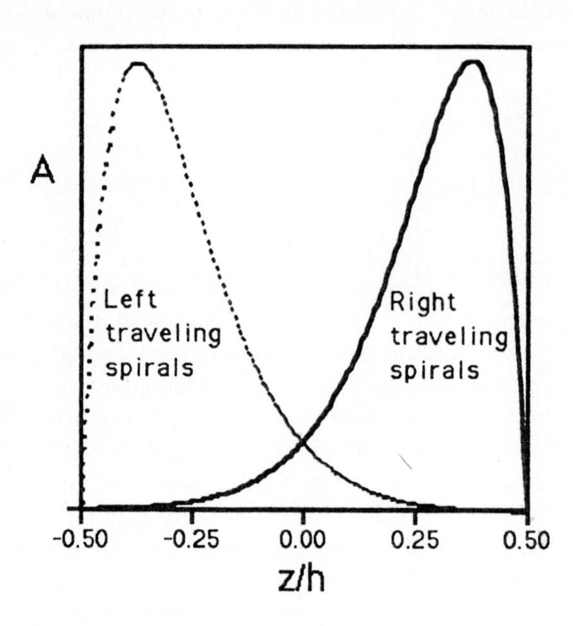

FIGURE 24 Linear eigenmodes for a simplified model of traveling waves in a finite container. The solid curve is for right-traveling waves and the dotted curve is for left-traveling waves. These are solutions to the linear part of the complex Ginzburg-Landau equations (Eqs. 55(a) and 55(b)) with boundary condition that the amplitudes vanish at the ends of the cylinders. (An investigation of the *nonlinear* equations with various boundary conditions is found in Cross.[65]) The linear model solutions have form $\cos(\pi z/h)\exp(\pm\alpha z/h)$ where h is the length of the system and a is related to the parameters in the complex Ginzburg-Landau equations by $\alpha \equiv (\tau_0 v_g h)/[2\xi_0^2(1+c_1^2)]$.

the position of the grain boundary oscillated sinusoidally in time (see Figure 25(b)); for larger ε however, the grain boundary would remain near one end of the cylinder for nearly half a cycle and then suddenly shift to the other end of the cylinder for nearly half a cycle (see Figure 25(c)). The period of such "alternating spiral" dynamics was typically a factor of 50 longer than the spiral wave period.

These phenomena are reminiscent of the blinking states observed in binary fluid convection[141,91] (see also the review of Cross and Hohenberg[66] although they do not always occur in the corresponding range of ε. One important feature found in binary fluid convection experiments was a sensitivity to small variations in the length of the system. Such apparent correspondences in various experimental systems led Knobloch[137] to create an explicit model for the dynamics of traveling wave patterns in finite systems. Amplitude equations are formulated for the linear eigenmodes of the *finite* system; terms are included that result from the broken translational invariance. The finite cylinder eigenmodes have been computed by Edwards[83] with laminar Couette flow boundary conditions. Knobloch's model recovers much of the observed dynamics for selected values of the parameters in the amplitude equations. This includes a strong sensitivity of the dynamics to the length of the system, modulo one spiral wavelength. An intriguing result is the existence of chaotic dynamics in which the system remains close to one of one eigenstate for a while and then suddenly shifts towards another. This seems very similar to the observed alternating spiral dynamics, although further computational and experimental work is needed to pin down the correspondence. In any case, an important lesson from this is that dynamics exists in large aspect-ratio systems involving competition of patterns over large length and time scales. This dynamics can be missed, or even worse, confused with dynamics predicted for strictly temporal equations that ignore spatial variations in the amplitudes of competing modes.

With this caveat that complex spatio-temporal dynamics can result from "putting waves in a box," we will cautiously examine a problem originally posed assuming spatially uniform amplitudes. Suppose that modes with different azimuthal wave numbers m and $m + 1$ simultaneously become unstable. This can be achieved by varying two control parameters, for example the Reynolds number and the angular velocity ratio μ. The locus in parameter space of such conditions is called a set of "co-dimension two points." For example, $m = 0$ and $m + 1 = 1$ corresponds to competition between Taylor vortices and spirals, while $m = 2$ and $m + 1 = 3$ corresponds to competition between spirals of different "pitch." Plots locating curves of such co-dimension two points are shown in Figure 26. By analyzing the dynamics near one of the points, it is possible to explore the variety of dynamics that may occur well into regions far away from the co-dimension two point. In other words, a "local" analysis gives global information.

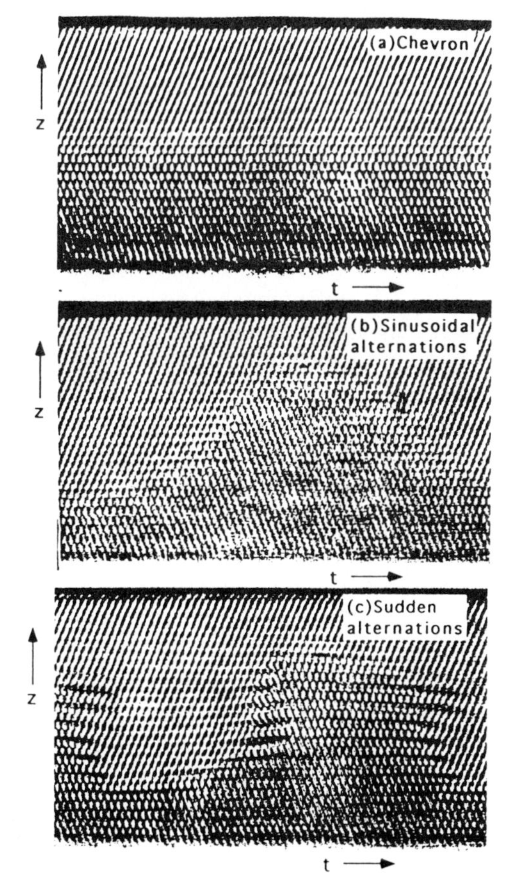

FIGURE 25 Space-time diagrams of spiral-vortex traveling waves in a finite-length Couette-Taylor experiment. Vortex boundaries as a function of axial position (vertical axis) are plotted at successive times (horizontal axis). Up-traveling waves are the stripes at the top of each figure and down-traveling waves are the stripes at the bottom. A "grain-boundary" separates the two types of waves. (a) The grain boundary remains fixed, giving a chevron pattern. (b) After less than 1% increase in Reynolds number, the axial position of the grain boundary moves sinusoidally. (c) After about 2% further increase in Reynolds number, the grain boundary makes sudden shifts from one end of the cylinder to the other.

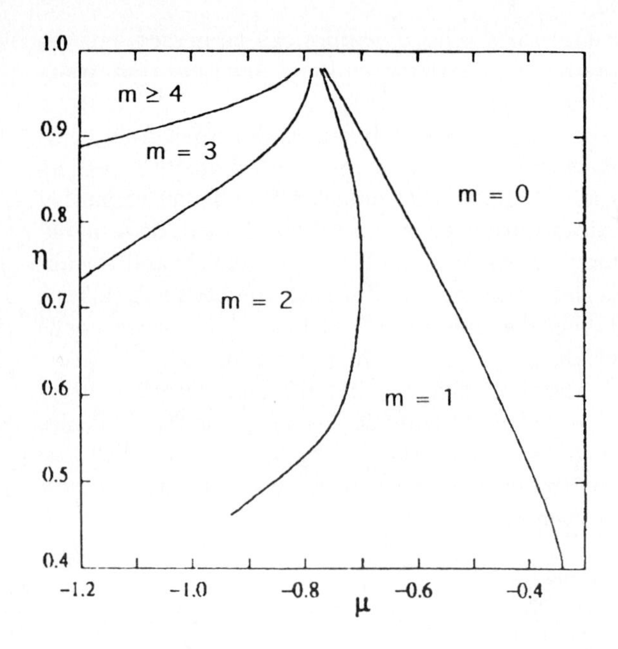

FIGURE 26 Parameter ranges over which the critical azimuthal wavenumber $m = 0, 1, 2, 3,$ and 4 in flow between counter-rotating cylinders. Here μ is the speed ratio and η is the radius ratio. The boundaries between the regions are curves of codimension-two points, near which several modes strongly interact. The convergence of these curves for small gaps (radius ratio $\eta \rightarrow 1$) indicates that the azimuthal wavenumber starts to behave like a quantity that can have a continuum of values. [From Langford et al.[156]]

As an example, Golubitsky and Stewart[102] established coupled amplitude equations and delineated possible bifurcation scenarios near the Taylor vortex / spiral vortex codimension-two point. Several states were identified that matched the symmetry of states reported by Andereck et al.[6]: these included wavy vortices and "twists" as well as Taylor vortices and spirals. The choice of scenario and relative stability of states depends on the numerical value of coefficients in the symmetry-deduced amplitude equations. Golubitsky and Langford[103] computed such coefficients for various radius ratios and predicted that for a range of radius ratios not close to 1, a wavy vortex state should be seen as the codimension-two point is orbited from the Taylor-vortex side to the spiral-vortex side. Moreover, considerable hysteresis should be observed as the orbit is reversed. These findings were subsequently confirmed experimentally at radius ratio 0.800.[260] However, large-scale spatio-temporal dynamics (alternating spirals) was also seen, and the reconciliation of this fact with the mode-interaction picture remains an open problem.

As a concluding note, Chossat et al.[45] have studied the temporal dynamics of modes interacting near the $m = 2/m + 1 = 3$ codimension-two point. A rich variety of flows states is predicted (including "interpenetrating spirals" indicated on the Andereck diagram), Figure 9 and the corresponding coupled amplitude equations admit chaotic solutions for appropriate values of the parameters. Many of these features have been found in the experiments described by Stern et al.[242] However, again, it will be important to sort out what part of the observed motion—especially if it is chaotic—is due instead to large-scale spatio-temporal dynamics.

3.10 HIGHER INSTABILITIES AND CHAOS

We have been concentrating so far on the linear and weakly nonlinear analysis of flows near the first instability, because this is where we expect to make the most detailed connection between the equations of motion (the Navier-Stokes and continuity equations) and the development of complexity in fluid flows. Looking at the Andereck diagram, one would conclude that we are working on the ground floor of the intricate structure of possible flow states, although I would prefer to say we are laying a foundation. Pursuing this analogy further, we would conclude that the "penthouse" corresponds to the fully turbulent states. In this section, we will make a rapid journey up to the penthouse to see what we can learn along the way. Our most direct route will be to hold the outer cylinder stationary and see how the flow evolves as we increase inner cylinder Reynolds number Re_1. We will leave the solid ground of analytical solutions for base flow states, relying on numerical and more indirect methods for characterizing the patterns and their dynamics. The Reynolds number will be expressed as the ratio $\mathrm{Re}_1/\mathrm{Re}_{1c}$, where Re_{1c} is the value where Taylor vortices first appear; the following discussion is summarized, as a function of $\mathrm{Re}_1/\mathrm{Re}_{1c}$, in Table 1.

Most of the studies of the path to turbulence have been done for systems with relatively narrow gap between the cylinders, such as radius ratio near 0.88. At this radius ratio, Taylor vortices (Figure 10(a)) appear for $\mathrm{Re}_{1c} = 120$. These are stable only for a narrow range or Reynolds numbers up to about $\mathrm{Re}_1/\mathrm{Re}_{1c} = 1.2$, beyond which the vortices become wavy (Figure 10(c)): undulations appear on the boundaries of the vortices that travel in the azimuthal direction at about 1/3 of the speed of the inner cylinder. This wavy vortex state was the subject a extensive series of experiments by Coles[52] in which the nonuniqueness of flow states was vividly established. By varying the protocol of changing cylinder speeds, Coles was able to obtain many combinations of the number of vortices N and the number of waves m at the same final value of Re_1. Once a state (N, m) was obtained, Coles could find the limits of its stability by slowly ramping inner cylinder speed up or down. In this way, he constructed the intricate diagram shown in Figure 11 that shows how the states are interconnected in parameter space. This reminds me of the child's board game called "chutes and ladders."

Since then there have been several studies of the wavy vortex state. For example, King and Swinney[132] did detailed experiments on the limits of stability of wavy vortex flow states as a function of Reynolds number and axial wavelength of the underlying vortices. Pockets of parameter space appeared where the motion was irregular, possibly chaotic, when wavy states competed with one another. King et al.[133] also carefully examined the speed of the waves and good agreement was achieved in the numerical work of Marcus,[168,169] who discussed a possible mechanism for the instability of Taylor vortex flow to wavy vortices. If you watch wavy vortex flows, just after transition has occurred to a new wavy vortex state, it is possible

TABLE 1 Flow states obtained between coaxial cylinders of radius ratio $\eta = 0.88$, with only the inner cylinder rotating. The rotation rate is expressed as the ratio of the Reynolds number Re_1 to the critical value Re_{1c} where Taylor vortices first appear.

Re_1/Re_{1c}	Flow state
$0-1$	Base flow (Couette flow).
$1-1.2$	Taylor vortex flow: N vortices in a system of aspect ratio Γ (typically N is the nearest integer to Γ).
$1.2-7.5$	Wavy vortex flow: N vortices deformed into m waves traveling azimuthally around the cylinders. N typically decreases as Reynolds number increases (the wavy vortices become fatter).
$7.5-9.5$	Modulated wavy vortices with small-scale modulation (ZS mode), visible as "blips" traveling along the wavy vortex boundaries.
$9.5-12.5$	Modulated wavy vortices with large-scale modulation (GS mode), visible in some cases as periodic flattening of the waves.
$12.5-20$	Weakly turbulent vortices, still showing modulated waves. Experimental evidence of low-dimensional chaos.
$20-40$	Turbulent vortices, often with fast wave-like oscillations buried inside.
$40-500$	Strongly turbulent vortices.
> 500	Strongly turbulent flow: vortex structure is hard to discern but may still be present!

to see isolated turbulent patches where the vortices are not correctly linked around the cylinder; these can be long-lived[77] and were given the name "turbators."[199] We learn, then, that irregular motion can be induced by such localized disturbances as well as by competition or instability that fills the whole system.

As Reynolds number increases, the wavy state remains—barring mode competition or dislocations—laminar and strictly periodic in time, with frequency f_1 that may change as new states (N, m) are selected. Overall there is a tendency for the number of vortices to decrease as they are "squeezed out" by the fat wavy vortices in the center of the column.[34] At a value of $Re_1/Re_c \sim 7.5$, a new frequency f_2 appears in the flow so that the dynamics is now quasi-periodic. We call this the "modulated wavy vortex" state, and it turns out that there are at least two distinct varieties. The one seen first as Reynolds number is increased (but found later in experiments: see Zhang and Swinney[284]) has a fast, short-length-scale modulation traveling along the wavy vortex boundaries. We call it the ZS mode. At $Re_1/Re_{1c} \sim 9.5$, a much different modulation appears[105,106] which has such a large scale that it often appears as a slow periodic flattening and reemergence of the waviness of the vortices; this is called the GS mode. Power spectra of these two types of modes are shown in Figure 27; numerical and experimental work by Coughlin et al.[56] (see also Coughlin and Marcus[54,55] explored the transition from one mode to another and the possibility that competition between them could generate chaotic dynamics. An interesting feature of the quasi-periodic modes is that symmetry prevents the two frequencies f_1 and f_2 from "locking" together into proportions f_2/f_1 that equal ratios of integers[107,210]; such frequency-locking—sustained over a nonzero range of the control parameter—is otherwise a common feature of systems exhibiting modes of competing frequencies. In the case of modulated wavy vortices, however, the first frequency is "special": it corresponds to a traveling wave, which means that we can bring it down to zero by going into a rotating reference frame that moves with the traveling wave at angular velocity $2\pi f_1$.

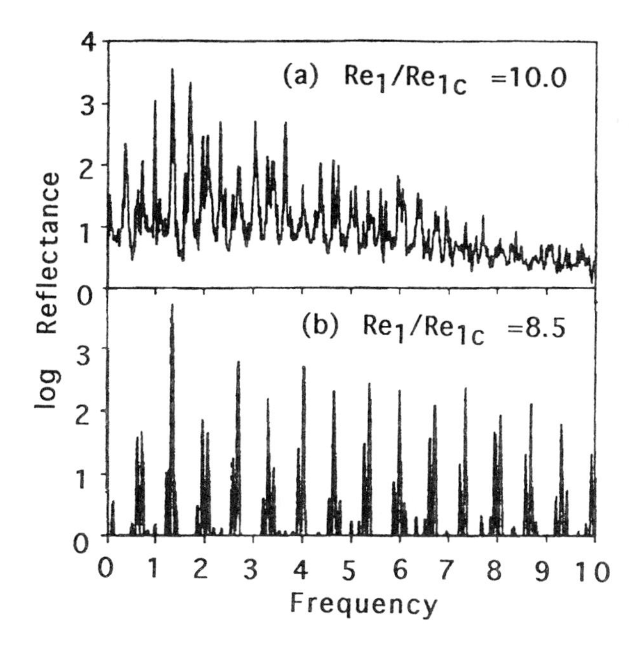

FIGURE 27 Power spectra from two types of modulated wavy vortex flow in the Couette-Taylor experiment (radius ratio 0.876). The abscissa is frequency in units of the inner cylinder frequency, while the ordinate is a measure of the amount of fluctuation, within a small frequency interval, of reflectance of laser light from the visualized flow. In both figures, the large peak at frequency 1.3 is the fundamental frequency of the wavy motion. (a) The modulation of these waves at Reynolds number $\mathrm{Re}_1/\mathrm{Re}_{1c} = 8.5$ has significant fluctuations at higher frequency, corresponding to visible small-scale fluctuations in the wavy vortex boundaries. (b) The modulation at Reynolds number $\mathrm{Re}_1/\mathrm{Re}_{1c} = 10.0$ has more fluctuations at lower frequency, corresponding to visible slow modulation of the wavy vortex boundaries. Neither mode is chaotic, although their interaction has been speculated to produce chaos [see Coughlin et al.[56]].

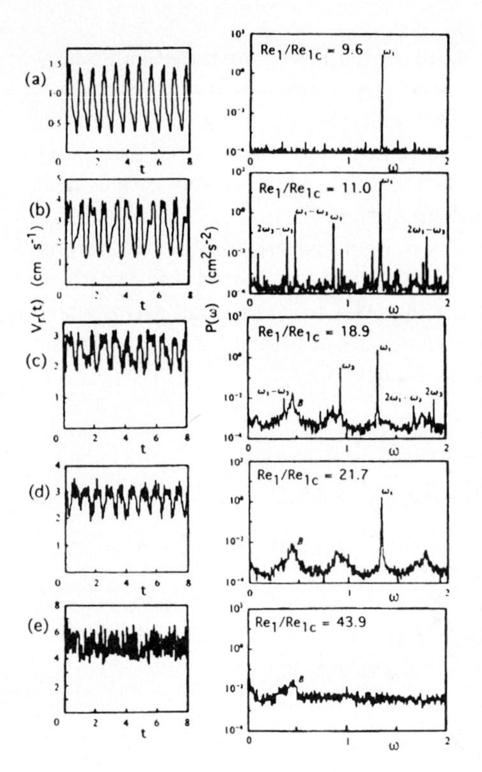

FIGURE 28 Evidence from a Couette-Taylor experiment that turbulence appears in fluid flows after the appearance of only a few independent frequencies and their linear combinations. The cylinder radius ratio was 0.877. Time and frequency scales are based on the inner cylinder period and frequency respectively. Fluid radial velocity was measured at successively higher Reynolds numbers, yielding time series (left) and corresponding power spectra (right). (a) Wavy flow: only a single frequency is evident. (b) Modulated wavy flow. (c) The spectra now contains broad-band features, corresponding to weakly turbulent flow. Waves and their modulations are, however, still present. (d) The modulation frequencies have disappeared. (e) No sharp frequencies remain. From Fenstermacher et al.[89]

As Reynolds number is further increased, beyond a value $\mathrm{Re}_1/\mathrm{Re}_{1c} \sim 12.5$, the result was—when first observed—an experimental confirmation of a major change in point of view of how turbulence arises in fluid flows. Landau[154] had proposed that turbulence arose be the continued appearance of new frequencies as Reynolds number was increased, each with a random phase in relation to the other frequencies. Eventually, the many distinct and incommensurate frequencies would become indistinguishable, and the flow would have the strong irregularity in time that we associate with turbulence. However, Ruelle and Takens,[222] using recent developments in the mathematical theory of dynamical systems, proposed that the broadband spectral characteristics of turbulence could appear immediately after the first two incommensurate frequencies. Experimental confirmation came from velocity measurements at a single point in the flow in the Couette-Taylor system by Gollub and Swinney[100] (see also Fenstermacher et al.[89]). The evolution of power spectra from single frequency to two frequencies to broad band is shown in a close-up view in Figure 28 and in a plot covering the whole range of significant frequency content in Figure 29. The latter figure in particular shows, as a signature (but not conclusive proof) of chaotic dynamics, the steady rise of a broad-band "floor" that appears to decay exponentially with frequency.

From such observations emerged the idea that weakly turbulent flow in Taylor vortices could be described in dynamical systems language by motion along a "strange attractor." This point of view was more firmly established by direct reconstruction from experimental data of such attractors by Brandstater et al.[26] (see also Brandstater and Swinney[27]). This reconstruction is done using the time-delay method,[196,261] where a "phase space" is constructed out of coordinates $u(t)$, $u(t+\tau_1)$, $u(t+\tau_2)$, etc. Here the velocity measurements u are sampled at regular intervals in time t and at delayed times $t+\tau_1$, $t+\tau_2$, etc. A plot of the evolution in time of points constructed from coordinates is called a "phase portrait." The result of such reconstructions for the Couette-Taylor system is shown in Figure 29. When the flow is quasi-periodic, that is, characterized by two frequencies f_1 and f_2, the attractor is a torus (Figure 29(a)), which is a two-dimensional surface. When chaos emerges, the torus becomes wrinkled (Figure 29(b)–(d)) and acquires noninteger dimension. This is more clearly seen by taking a cross section of the torus (called a "Poincaré" section) and observing that the smooth closed curve (Figure 30(a)) from the torus becomes fuzzy (Figure 30(b)–(d)). More quantitatively, it is possible to

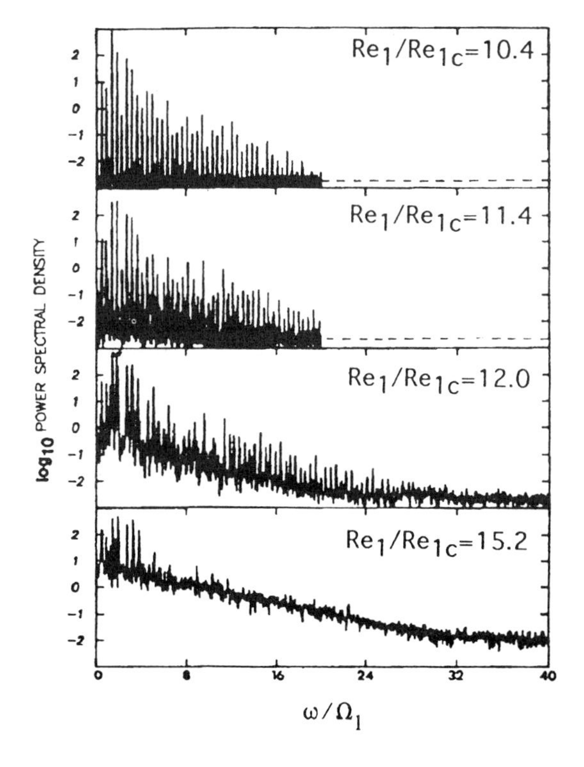

FIGURE 29 Evolution of broad-band frequency content as flow in the Couette-Taylor experiment becomes turbulent. This is similar to the right side of Figure 28, except the power spectra are shown over a much broader frequency range. The top two spectra are for modulated wavy vortex flow. The bottom two spectra are for chaotic or weakly turbulent flow. Note the rise in broad-band background in these two spectra, with an exponential fall-off of power with frequency (which appears as a linear fall-off in this figure because it is a semi-log plot). This is characteristic of broad-band spectra produced from deterministic rather than stochastic systems. [From Brandstater and Swinney.[27]]

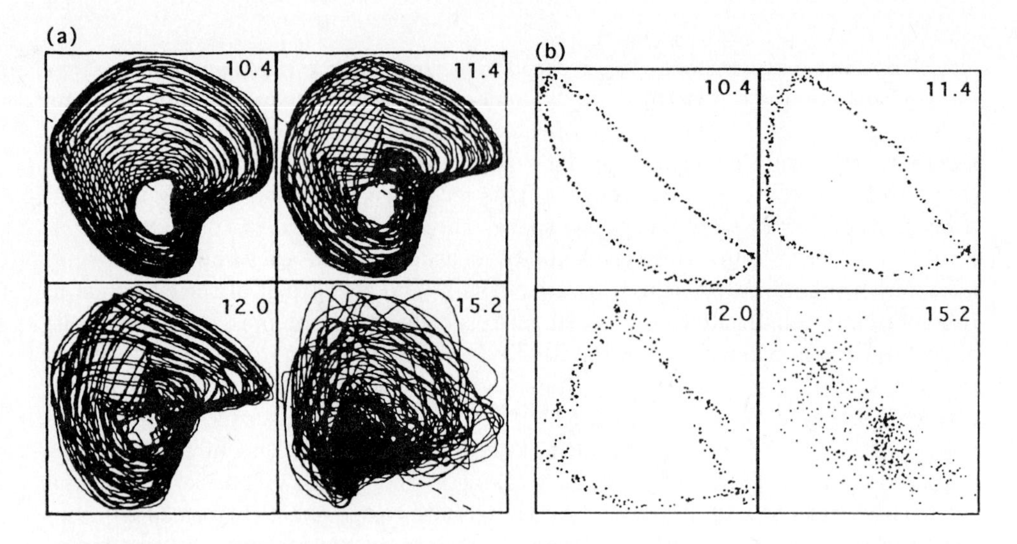

FIGURE 30 (a) Projections of phase portraits onto a plane and (b) Poincaré sections reconstructed from velocity data taken under the same set of conditions as Figure 29. The ratio of Reynolds number to critical Reynolds number is shown in the upper right of each figure. The top figures correspond to modulated wavy vortex flow. This flow, with two independent frequencies, produces a phase portrait with the topology of a torus. The corresponding Poincaré sections (top two figures of (b)) are closed curves. Chaotic flow, as shown in the bottom figures, yields more diffuse phase portraits and Poincaré sections. In fact, the phase portrait becomes fractal, with non-integer dimension. [From Brandstater and Swinney.[27]]

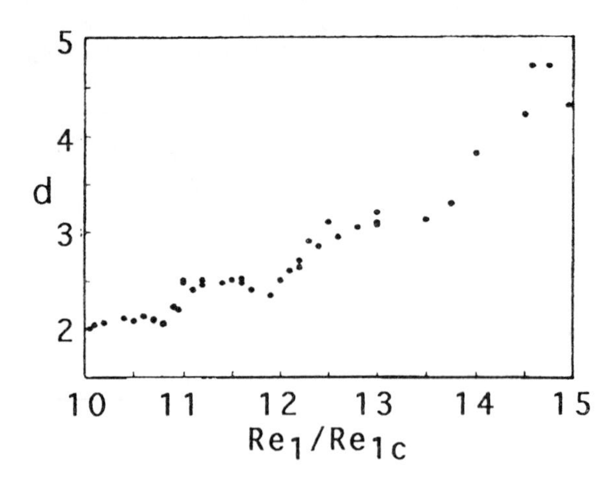

FIGURE 31 Attractor dimension versus Reynolds number. The attractors were reconstructed from Couette-Taylor flow velocity data, as already shown in Figures 29 and 30. Note that the dimension appears to rise from two (for a torus) to noninteger values beyond $\mathrm{Re}_1/\mathrm{Re}_{1c} \approx 11$, corresponding to the onset of chaotic flow. [From Brandstater and Swinney.[27]]

extract attractor dimension and other diagnostics of chaos (such as Lyapunov exponent, measuring the average rate of separation of nearby trajectories in phase space) from this data. Figure 31 shows how the fractal dimension of the attractors increases smoothly as Reynolds number increases. For more detailed discussion of these diagnostics of chaotic motion, see Manneville.[167]

Since all the data supporting this dynamical systems point of view of weak turbulence came from a single-point measurement, an important question concerns how well this describes motion throughout the system. What if, instead, we based our attempt to create a phase portrait on measurements at several locations, in separate vortices? Are the vortices (or vortex pairs) acting as independent, loosely coupled chaotic oscillators or is the whole system synchronized as one chaotic machine? Significant early work on this question can be found in L'vov et al.[151] The thesis of Hirst[116] showed that the attractor dimension did not vary with size of the system. Finally, recent investigations reported in Buzug et al.[35] indicate that the chaos is indeed "global," that is, the chaotic dynamics is a property of the whole system. However, it may be possible to vary the degree of coupling by choosing different radius ratios, and the subject of local versus global chaotic dynamics remains one for further investigation.

3.11 TURBULENCE

Finally, we are approaching conditions for well-developed turbulence. At $\mathrm{Re}_1/\mathrm{Re}_{1c} \sim 20$, the sharp frequency components have diminished and large, well-defined waves have disappeared from the turbulent vortices. However, at somewhat higher Reynolds numbers, new sharp high frequency components appear for a while—this has been referred to as "re-emergent order."[275] An important and surprising fact about turbulent Taylor vortices is that the vortex structure seems to persist to very high Reynolds numbers, even above $\mathrm{Re}_1/\mathrm{Re}_{1c} > 500$. The record for highest Reynolds number turbulence in the Couette-Taylor system seems to be $\mathrm{Re}_1/\mathrm{Re}_{1c} \approx 14,600$ in the studies reported by Lathrop et al.[157,158] This, however, exceeds our province of discussing the "emergence" of complexity in fluid flows.

There are at least four distinct types of turbulence seen in the Couette-Taylor system (see Figure 32). Turbulent vortices occur for stationary outer cylinder. When the outer cylinder is strongly counterrotated, the turbulence appears to be featureless and fine-scaled, with no evidence of vortex structures comparable to the size of the gap. However, in an intermediate regime, there are striking phenomena: spiral vortices and turbulent bursts. Both of these are characterized by regions of fine-scale turbulence interrupted (in space or in time) by regions of nearly laminar flow. The laminar regime may still contain some structure, such as hints of interpenetrating spirals. It will be interesting to see if the emergence of fine-scale turbulence out of this quasi-laminar soup of traveling waves might be comparable to the emergence of turbulence in open flow systems.

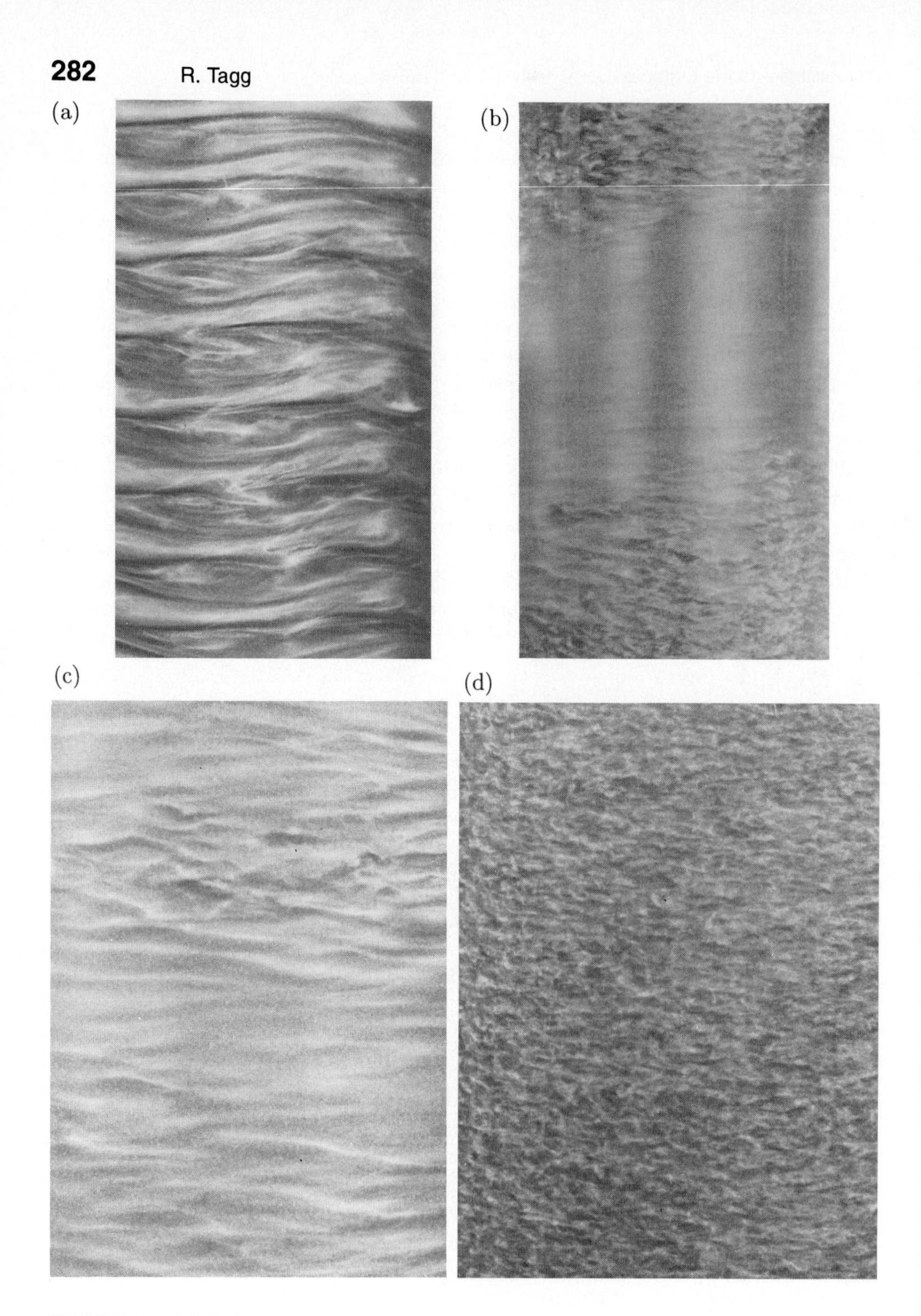

FIGURE 32 (a) Turbulent Taylor vortices, (b) Spiral turbulence, (c) Turbulent bursts, and (d) Featureless turbulence.

3.12 FINITE LENGTH EFFECTS AND EXPERIMENTS AT SHORT ASPECT RATIO

No description of the Couette-Taylor problem is complete without mention of effects of finite length and, in turn, the important theory and experiments that have been developed from studies of short-aspect ratio systems. It is useful to think anew about even moderately long systems: instead of trying to make the theory of Fourier modes work within a finite container, one may think that this system simply defines

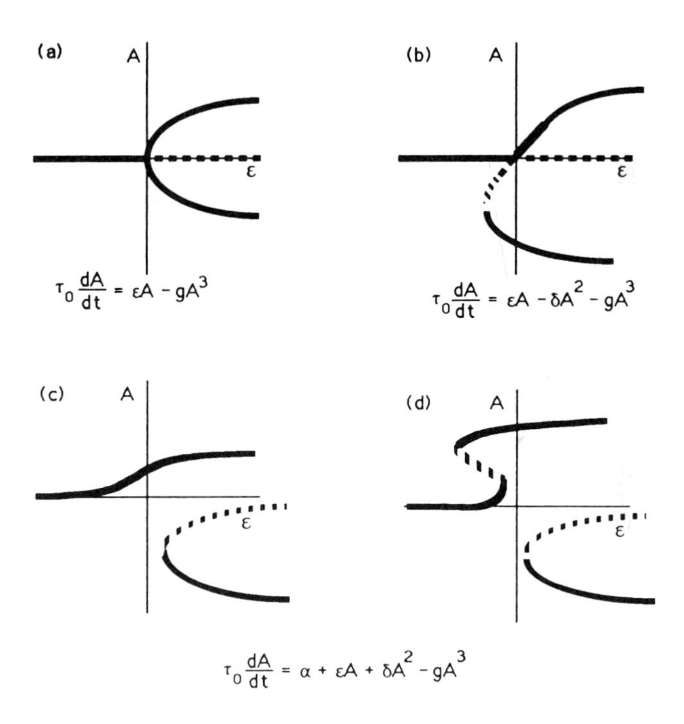

FIGURE 33 Perfect and imperfect bifurcations. (a) A pitchfork bifurcation is shown with the simplest corresponding amplitude equation; this bifurcation is very unlikely because the two branches are not equivalent in real systems. (b) A transcritical bifurcation, obtained by generalizing the amplitude equation by adding a quadratic term; this is also an unlikely scenario for most real systems. (c) Imperfect bifurcation, in which the original state smoothly deforms. Now all terms up to cubic order are present in the amplitude equation. Nearby, a separate state occurs via at a turning point via a "saddle node" bifurcation. (d) Another possibility for the imperfect bifurcation, in which the smoothly deformed solution turns back into an unstable section; there is hysteresis between the lower and upper stable portions of this solution. [After Crawford and Knobloch.[61]]

its own characteristic states of flow. The objective is to delineate, as far as possible, the types and interconnections of states that occur. In the following, we will assume that only the inner cylinder rotates.

A consequence of doing real experiments in finite-length cylinders is the breaking of translational invariance in the axial direction. The flow cannot penetrate the end walls, so we immediately expect that the number of vortices in the system must be quantized. Naively, we might think of the original translational invariance—that gave us a continuous family of solutions distinguished by phase—being reduced to a choice of two possible states at the first instability. One state has an integer number n of vortices with wavelength $\lambda = 2\Gamma/n$ (where Γ is the aspect ratio and wavelength is in units of the gap d); the other state is the first one shifted by half a wavelength. The appropriate bifurcation diagram, with two equivalent states, is called a pitchfork bifurcation (see Figure 33(a)). However, the states are not equivalent and the bifurcation has a much different character, as we will now see.

Near the ends of the system, recirculating cells of flow are generated, even when $\mathrm{Re}_1 < \mathrm{Re}_{1c}$, by the so-called Ekman pumping mechanism: away from the ends, a pressure gradient builds up with increasing radius to provide the centripetal acceleration of the rotating fluid. However, at the ends (which we assume are solid rings attached to a stationary outer cylinder), the fluid cannot move because of the no-slip boundary condition. Viscous friction in the layer near the end walls provides the centripetal force and thus there is, near the walls, less of a pressure build up with radius. Consequently, the pressure further up the outer cylinder wall is higher and tends to drive the fluid down the wall. As the fluid approaches the ends, it must turn inwards, and a recirculating "Ekman cell" is set up with fluid flowing in at the ends. For this reason, it is common "lore" of Taylor vortex experimenters that Taylor vortices occur in pairs, such that the vortices at the ends of the system have their inward flow adjacent to the end boundaries (see Figure 34(a)). Actually, this is not always so: it is possible to squeeze the Ekman cell at one end so that the system has an odd number of fully formed vortices in between one normal Ekman vortex and one squeezed Ekman vortex (see Figure 34(b)). There may also be a squeezed Ekman cell at each end and an even number of vortices in between that circulate the "wrong" way (Figure 34(c)). We can imagine such a short length—say equal to the gap (aspect ratio equal to 1)—that the only possibility is a fat Ekman vortex and a skinny Ekman vortex, the "Laurel and Hardy"[5] of Couette-Taylor experiments (see Figure 34(d)). See Pfister et al.[203] for experiments at such short aspect ratios.

Squeezing one of the Ekman cells is unfavorable, however, so that it requires much stronger driving to produce this mode. The consequence is that, as Reynolds number is increased, one state of vortex flow smoothly develops from the normal, equal-sized pair of Ekman cells; there is no sharp transition with a well-defined

[5] For readers not familiar with American comic culture, Laurel and Hardy were two comedians of the 1920s–1940s, one of whom (Stan Laurel) was tall and skinny and the other heavy and short (Oliver Hardy).

critical Reynolds number. Instead, for moderately large aspect ratios (20–30), a small but nonzero range of Reynolds number is traversed over which most of the development of vortices in the column occurs. The transition is said to be an "imperfect bifurcation" (see Figure 33(b)). To some extent, this can be modeled in the Ginzburg-Landau equation with boundary conditions that the amplitude is of order unity at the ends. By such modeling, experimentalists have been able to infer the value of the critical Reynolds number that would have occurred without the end effects.[109]

Disconnected from the smoothly developing state, in the sense that they cannot be achieved by steady increase of Reynolds number from zero, are the "anomalous mode" states with a squeezed Ekman cell (possibly at both ends), whose neighboring vortex circulates inwards near the ends. Such states can only be achieved by sudden jumps in Reynolds number followed by the chance that the fluctuations in the system lead to the anomalous mode rather than the "usual" mode. In a beautiful paper full of photographs of vortex cells seen in cross section, Benjamin and Mullin[16] demonstrate no less than 20 distinct states for the *same* final Reynolds number 359, aspect ratio 12.61, and radius ratio 0.600. Many of these states were anomalous modes.

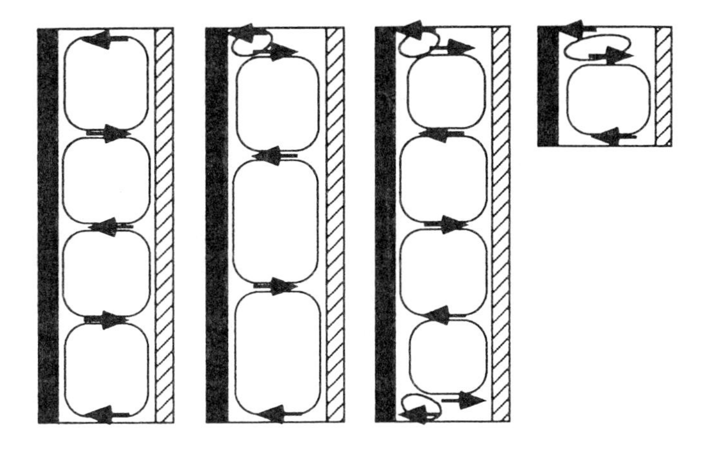

FIGURE 34 Normal and anomalous modes in short Couette-Taylor experiments. (a) A normal four-cell state, with flow inwards at both ends. (The inner cylinder is on the left in all cases.) (b) A three-cell anomalous mode. Note the outward flow near the top, except for a small vortex in the corner near the inner cylinder. (c) A four-cell anomalous mode. (d) A very short system with one large vortex accompanied by a thinner vortex. The author suggests naming this mode after Laurel and Hardy. [Based on Cliffe and Mullin[50] and Pfister et al.[203]]

More basic still is the question of state selection at small aspect ratios. Motivated by theoretical considerations and experiments of Benjamin,[14,15] further careful experiments were conducted by Mullin[180] to study the selection of four- and six-cell flow states when the Reynolds number was varied up and down, both continuously and after sudden starts, for various aspect ratios between 4 and 6. Depending on aspect ratio, different bifurcation scenarios could occur, with ranges of overlap (hysteresis) in Reynolds number where both states could be obtained. The bifurcation scenarios can be shown to alter in a systematic way as aspect ratio is varied, explaining many of the experimental observations. Associated with such work were full numerical simulations of the real problem, that is cylinders of finite length with solid end walls, by Cliffe.[51] By identifying the individual states, both stable and unstable—a denumerable cast of "actors" in the dynamics of short-aspect ratio systems—one hopes to clearly identify a low-dimensional dynamical system whose evolution to chaos may be understood in great detail. For further discussion and extension to the consideration of time-dependent (both periodic and chaotic) modes in short aspect ratio systems, see the reviews by Mullin.[181,182]

4. THERMAL (BUOYANCY-DRIVEN) INSTABILITY

We have spent a major portion of this article laying down, in the context of the Couette-Taylor problem, the machinery for dealing with instabilities and the beginnings of complexity in fluid flows. Now we will turn, more briefly, to other characteristic types of problems. It would be a mistake to think that the points of view established so far will transplant "one-to-one" into these new areas. Indeed, some ideas will be discarded as irrelevant and many new ideas will be learned. However, we will maintain the framework of linear stability analysis followed, where possible, by weakly nonlinear theory and mapping of bifurcations toward chaotic flows.

The first new problem is buoyancy-driven convection. The system is shown in Figure 35: a horizontal layer of fluid is heated from below, with temperature decreasing from T_b at the bottom to T_t at the top. Note the choice of coordinate system, with z being the vertical direction and the plates parallel to the xy-plane. Here the physical origin of the instability is perhaps the most intuitive of all the systems we are considering: warm fluid is less dense than cool fluid, so it will tend rise to the top. The simplicity is, as should now be expected, an illusion: the activity in studying this system remains unabated after nearly one hundred years since the experiments with convection in whale oil (spermaceti) conducted by Bénard[13] and the theoretical explanations offered by Rayleigh.[213,214]

One reason, of course, for considerable interest in convection is the fact that convection processes are extremely important in geophysical flows and in providing heat transport in engineering applications. If anything, the literature is more vast than that for the Couette-Taylor problem. However, we are concerned with a subset of this work that deals, through ultra-precise experiments and often extremely subtle theories, with spontaneous pattern formation and the origins of chaotic dynamics in dissipative systems. I will give only a glimpse of this work, because the convection problem has been the primary example of other general articles, including those of Newell[184] and Cross and Hohenberg[66]—as well the book by Manneville.[167] There are also several reviews more specific to the convection problem, including Palm,[197] Normand et al.,[192] Busse,[30,32] and Behringer[10]; a collection of earlier papers is in Saltzman.[227]

Since convection experiments are among the most precise experiments in fluid mechanics, it is worth mentioning a few details about how they are done (Figure 36). In the ideal case, the layers have infinite horizontal extent while the top and bottom plates have infinite thermal conductivity (and thus completely uniform temperature). In practice, materials are chosen so the top and bottom plates have much larger conductivity than the fluid. The sidewalls, with thermal properties close to those of the fluid, must have excellent thermal contact with the top and bottom plates or else temperature inhomogeneities result that "force" the flow. Transparent top plates (such as sapphire, with relatively good thermal conductivity compared to glass) are often used in order for the fluid motion to be visible, while the bottom plate is often made of polished nickel-plated copper that serves as a mirror. By shining collimated light down through the top plate and letting it reflect off the bottom, one obtains a shadowgraph that visualizes the convecting cells. Density differences between upflowing and downflowing fluid act like lenses to deviate the light into some directions and away from others, showing the patterns as in Figure 37. Other methods, like interferometry and schlieren techniques, are also used. A recent innovation is a high-resolution infrared camera system, costing nearly $100,000, that directly "sees" temperature differences.

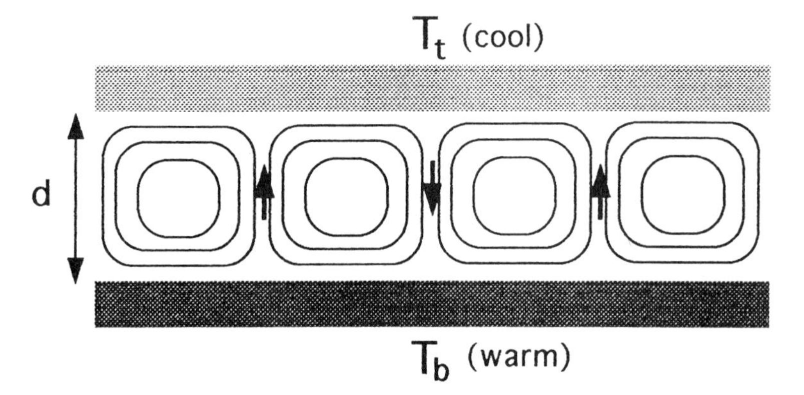

FIGURE 35 Geometry for establishing Rayleigh-Bénard convection.

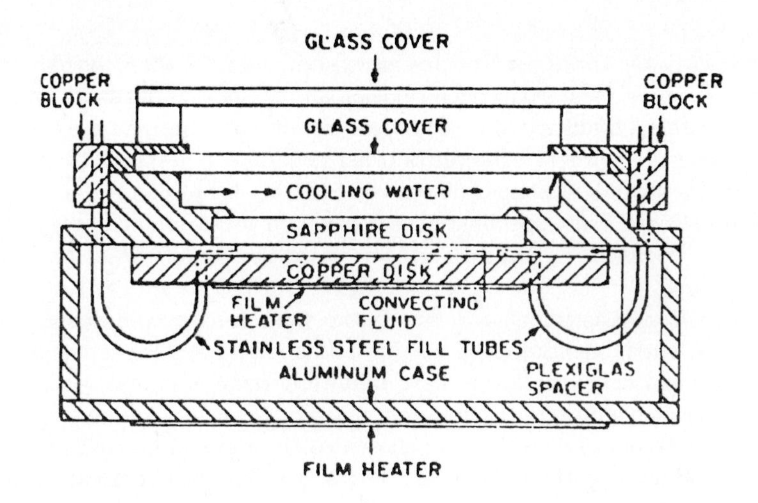

FIGURE 36 Actual apparatus for Rayleigh-Bénard convection experiments, which are among the most precise experiments done in fluid mechanics. Convection occurs in fluid in the thin gap between the sapphire and copper disks. A plexiglass spacer separates the disks. [From Heutmaker and Gollub.[115]]

Heat is usually supplied by heaters embedded in the bottom plate, with servo control to establish either constant temperature difference or constant heat flux. The top-plate temperature may be regulated by carefully directed fluid currents from a temperature controlled bath. The whole system is isolated from room temperature fluctuations by an insulating container. Sometimes (for example, with liquid helium experiments—see Behringer[10]) no visualization is used so both top and bottom surfaces may be made of high thermal conductivity copper. Very sensitive thermometers can measure temperature and temperature gradients in the bounding walls, allowing heat flux measurements to made. This is the analogue to torque measurements in the Couette-Taylor system.

A rectangular cell with horizontal dimensions L_x and L_y and with gap d between the plates has aspect ratios L_x/d and L_y/d; a cylindrical container of diameter D has an aspect ratio D/d. Experiments are often designed with either large-aspect ratio (say greater than 30) or small aspect ratio (1 to 3). Sometimes one aspect ratio (say in the y-direction) is intentionally made smaller than the other in order to established a preferred alignment of the convective rolls, rendering the horizontal variation of the observed patterns nearly one-dimensional. The large aspect ratio systems conform better to the ideal of infinite horizontal extent; however, after instability occurs the situation can be very complex because the (near-) isotropy and translational invariance of the system produces a quasi-continuum of interacting modes. The small-aspect ratio systems pose a substantially different stability problem, but one which may be solvable by numerical methods. In this case, a small number of modes can dominate the behavior of the system and the dynamics is in principle reducible to a system of ordinary differential equations. An interesting variation on the design of convection apparatus has been the construction of annular cells. People who do convection experiments must be quite patient, because the time scales are slow and experimental runs often last for weeks.

4.1 PRIMARY INSTABILITY

As before, we start by attempting to understand heuristically what factors determined the onset of convection. If the bottom plate is cooler than the top, the system is stably stratified and fluid remains at rest. If the temperature gradient is reversed by heating the fluid layer from below, instability sets in when the thermally induced buoyancy is strong enough to overcome the dissipative effects of viscosity ν and thermal diffusivity κ. Suppose the fluid has density ρ_b at the bottom and a thermal expansion coefficient $\alpha \equiv -(1/\rho)(\partial\rho/\partial T)$. Imagine a small parcel of fluid of size ξ that is slightly warmer (by an amount θ) than the surrounding fluid. The buoyancy force is $\rho_b\alpha\theta\xi^3 g$ where g is the acceleration due to gravity. As result of this force, the parcel accelerates until the buoyancy force is opposed by a viscous drag $\sim \rho\nu u\xi$, just as we argued in the case of the Taylor vortex instability. Balancing the forces gives $u \sim g\alpha\theta\xi^2/\nu$. Thus the time for the parcel to move a distance comparable to it size is $\tau_1 \sim \xi/u \sim \nu/(g\alpha\theta\xi)$. However, heat diffuses out of the parcel with a time scale $\tau_2 \sim \xi^2/\kappa$. This time must be longer than the time scale of motion of the parcel; otherwise the cause of the buoyancy (the temperature difference) is removed. Thus, to have instability, we require $\tau_2/\tau_1 > 1$, meaning $\alpha g\theta\xi^3/\nu\kappa > 1$. Now we assume that the temperature fluctuation is a fraction of the difference in temperature between top and bottom plates, $\theta = \varepsilon_1(T_b - T_t)$, while the size of the parcel may be expressed as $\xi = \varepsilon_2 d$. Thus we expect instability to occur when

$$\frac{\alpha g\varepsilon_1(T_2 - T_1)(\varepsilon_2 d)^3}{\nu\kappa} > 1 \qquad (57a)$$

or

$$\mathrm{Ra} \equiv \frac{\alpha g(T_2 - T_1)d^3}{\nu\kappa)} > \frac{1}{\varepsilon_1\varepsilon_2^3} \gg 1. \qquad (57b)$$

Here the parameter Ra is called the Rayleigh number; the critical value Ra_c where convection begins for fluid between two solid plates is calculated to be 1708—a large number, as expected from the above inequality.

Since the fluid is initially at rest, we do not need to solve equations of motion in order to write down the base state; this is also called the conduction state, since heat is only transported by thermal conduction through the stationary fluid. The base state is:

$$\mathbf{U} = 0 \qquad \text{(fluid is at rest)} \tag{58a}$$

$$T = Tb + \frac{T_t - T_b}{d}\left(z + \frac{d}{2}\right) \qquad \text{(linear temperature profile)} \tag{58b}$$

$$\rho = \rho_b(1 - \alpha(T - T_b)) \qquad \left(\text{neglecting } \frac{d\alpha}{dT}\right) \tag{58c}$$

$$P = P_b - \int_0^z \rho(z) g dz) \tag{58d}$$

Note that in this problem, we have an additional field variable, the temperature T.

We now proceed to write down the equations of motion for the conditions when convection occurs. The deviations from the base state are written as $\mathbf{u} = (u, v, w)$, temperature θ, and pressure p. To simplify matters, we adopt the so-called Oberbeck-Boussinesq approximation: the temperature variation is coupled to the dynamics only through the buoyancy force per unit volume $\rho_b \alpha \theta g$; otherwise, the density and other material properties of the fluid are assumed constant. Laboratory systems come close to this approximation provided the temperature differences remain moderate (in practice, of order several degrees Kelvin). The physical quantities can be scaled by d (distance), $\tau_\theta \equiv d^2/\kappa$ (time), κ/d (velocity), and $\kappa\nu/\alpha g d^3$ (temperature). With these scalings the full nonlinear equations (the Boussinesq equations) for the disturbance fields are:

$$\nabla \bullet \mathbf{u}, = 0 \tag{59a}$$

$$\frac{1}{Pr}\left[\frac{\partial \mathbf{u}}{\partial t} + (\mathbf{u} \bullet \nabla)\mathbf{u} + \nabla p\right], = \nabla^2 \mathbf{u} + \theta \hat{\mathbf{z}} \tag{59b}$$

$$\frac{\partial \theta}{\partial t} + (\mathbf{u} \bullet \nabla)\theta - \mathrm{Ra} w = \nabla^2 \theta\,. \tag{59c}$$

Here $Pr \equiv (\nu/\kappa)$ is called the Prandtl number, a quantity solely based on the fluid physical properties; it expresses the relative importance of the two nonlinear terms: $(\mathbf{u} \bullet \nabla)\mathbf{u}$ versus $(\mathbf{u} \bullet \nabla)\theta$ and thus may be expected to strongly affect the nonlinear behavior. Interestingly, it does not enter into the linear stability calculation. Equation (59(a)) is just the continuity equation, expressing conservation of mass. Equation (59(b)) is the momentum equation, which in comparison to Eq. (11(a)), contains the additional term $\theta\hat{\mathbf{z}}$ due to the buoyancy force. Finally, Eq. (59(c)) expresses how heat is transported: the first term $(\mathbf{u} \bullet \nabla)\theta$ represents advective transport of the temperature disturbance, while the term $(\mathrm{Ra} w)$ represents advective transport of thermal energy deposited in the *base* state by the z-component w of the fluid velocity. The term $(\nabla^2\theta)$ on the right-hand side represents diffusive transport.

There are two common boundary conditions at $z = \pm 1/2$ (note that z now refers to the dimensionless vertical coordinate). Free (also called stress-free) boundary conditions are appropriate if the convecting fluid is bounded by another fluid of much lower viscosity and the interface is assumed to remain flat: $\theta = \partial u/\partial z = \partial v/\partial z = w = 0$. Rigid (no-slip) boundary conditions are appropriate if the convecting fluid is bounded by rigid walls, as in Figure 35: $\theta = u = v = w = 0$.

The linearized problem is obtained by neglecting the advective terms $(\mathbf{u} \bullet \nabla)\mathbf{u}$ and $(\mathbf{u} \bullet \nabla)\theta$. The system originally has translational invariance in both x and y directions, assuming a system of infinite horizontal extent. Furthermore, until a convection pattern is established, these two directions are equivalent—we say the system is isotropic in the xy-plane. This is a major difference between this system and the Couette-Taylor problem, for which there was no equivalence between the z and θ directions. These symmetry considerations lead us to expect solutions of the form:

$$(u, v, w, \theta, p) = (\hat{u}(z), \hat{v}(z), \hat{w}(z), \hat{\theta}(z), \hat{p}(z)) \exp[st + i(k_x x + k_y y)]. \quad (60)$$

Substituting this expression into the linearized equations and making some manipulations (see, for example, Kundu[149]) yields two equations coupling $\hat{w}$ and $\hat{\theta}$:

$$Pr[(D^2 - k^2)^2 \hat{w} - k^2 \hat{\theta}] = s(D^2 - k^2)\hat{w} \quad (61a)$$

$$(D^2 - k^2)\hat{\theta} + \mathrm{Ra}\hat{w} = s\hat{\theta} \quad (61b)$$

where $D \equiv d/dz$ and $k^2 \equiv k_x^2 + k_y^2$ The two types of boundary conditions at $z = \pm 1/2$ translate to:

$$\hat{\theta} = \hat{w} = D^2 \hat{w} = 0 (\text{stress-free or "free"}); \quad (62a)$$

$$\hat{\theta} = \hat{w} = D\hat{w} = 0 (\text{no-slip or "rigid"}). \quad (62b)$$

The eigenvalue s for this problem, with either type of boundary condition, is real, so the secondary flow that results from instability is a steady flow—another case of "exchange of stabilities." Although the case of stress-free boundary conditions at both walls is unrealistic, it has the advantage that closed-form solutions for the fields and dispersion relation $s = s(k, \mathrm{Ra}, \mathrm{Pr})$ can be obtained. They are:

$$(\hat{w}, \hat{\theta}) = (W_n, \Theta_n) \sin(n\pi z); \quad (63a)$$

$$s^2 + (1 + \mathrm{Pr})(k^2 + n^2\pi^2)s + \mathrm{Pr}\frac{(k^2 + n^2\pi^2)^3 - k^2\mathrm{Ra}}{k^2 + n^2\pi^2} = 0. \quad (63b)$$

The condition $s = 0$ gives a family of marginal stability curves indexed by the integer n:

$$\mathrm{Ra}(k, n) = \frac{(n^2\pi^2 + k^2)^3}{k^2}. \quad (64)$$

TABLE 2 Critical Rayleigh number (Ra_c) and critical wavenumber (k_c) for various boundary conditions on convection in a horizontal fluid layer.

Boundary condition	Ra_c	k_c
Free-free	$27\pi^4/4 \approx 657.512$	$\pi/\sqrt{2} \approx 2.2214$
Rigid-free	1100.6	2.68
Rigid-rigid	1707.76	3.11632

The minimum Rayleigh number is obtained for $n = 1$; minimization must also be done over k. The resulting critical Rayleigh number Ra_c and critical wavenumber k_c is summarized in Table 2 for all three possible pairs of boundary conditions. Again, note that these critical values do not depend on Prandtl number. The stability problem with rigid boundaries requires numerical solution of a transcendental equation to obtain the marginal curve $\mathrm{Ra}(k)$; a good discussion is found in Kundu.[149]

4.2 WEAKLY NONLINEAR THEORY AND PATTERN SELECTION

A nonlinear analysis must determine if the resulting eigenmodes saturate into a stable secondary flow. Since there is no preferred direction for the wavevector $\mathbf{k}$, it is necessary to investigate the nonlinear selection of a stable pattern out of all possible combinations of modes with wavenumber of the same magnitude k but different direction. Most study has been given to the patterns that can "tile" a plane, such as rolls, squares, triangles, and hexagons. It is also possible to consider sates formed from concentric rolls (see Koschmieder[145]).

A derivation of the amplitude equations for different cellular patterns and computation of the coupling (Landau) coefficients has been given by Cross.[63] Earlier nonlinear treatments (beginning with Gorkov[104]; Malkus and Veronis[166]; see also Schlüter et al.[234]) use a somewhat different formalism that tends to be more common in the literature. However, for continuity with the rest of our discussion, we will adapt some of the features of Cross' treatment. By analogy to the Couette-Taylor problem, we measure the distance above the stability threshold Ra_c by the quantity $\varepsilon \equiv (\mathrm{Ra}/\mathrm{Ra}_c - 1)$. We let $\Psi_0 \exp(ikx)$ represent the two-dimensional "plane-wave" eigenmode given by the field $\hat{u}(z), 0, \hat{w}(z), \hat{\theta}(z), \hat{p}(z)) \exp(ikx)$, oriented so that the y component of velocity is identically zero ($\hat{v}(z) = 0$). Eigenmodes corresponding to wavevectors oriented at angles θ from the x-axis are then

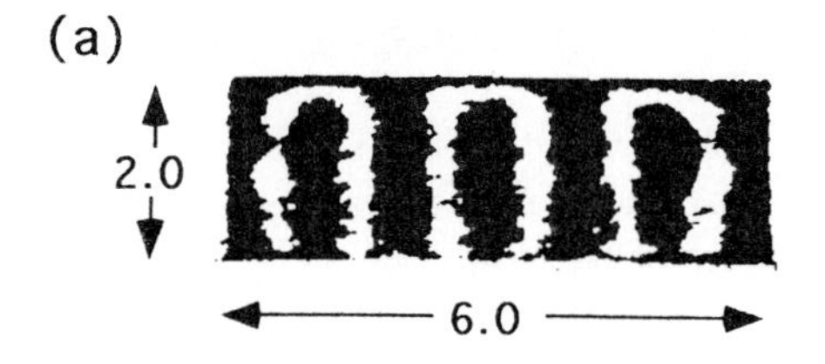

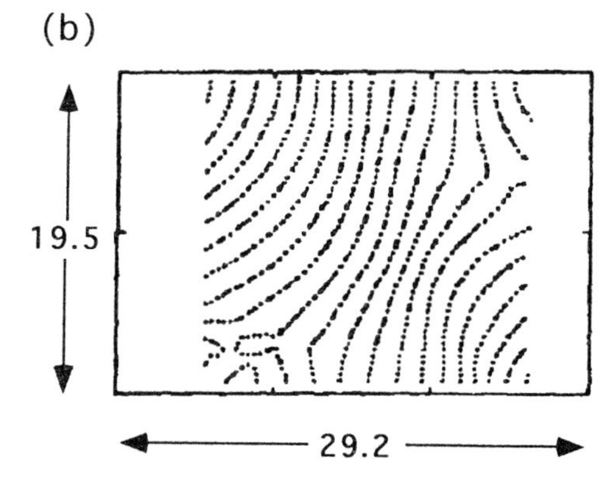

FIGURE 37 Convection in rectangular boxes. (a) Small box with one side significantly longer than the other: the rolls tend to align with the short side.[244] (b) Large box: rolls tend to orient perpendicular to the wall. Note the existence of defects.[101]

denoted by $R_\theta(\Psi_0)\exp(ik\cos\theta x + ik\sin\theta y)$, where $R_\theta(\Psi_0)$ denotes the rotated state $(\hat{u}(z)\cos\theta, \hat{u}(z)\sin\theta, \hat{w}(z), \hat{\theta}(z), \hat{p}(z))$. The following gives an example of the appropriate combination of modes that could lead to hexagonal patterns:

$$\begin{aligned}\Psi =& A_1\Psi_0\exp(ikx) + A_2R_{\pi/3}(\Psi_0)\exp\left[i\left(-\frac{1}{2}kx + \frac{\sqrt{3}}{2}k_cy\right)\right] \\ &+ A_3R_{2\pi/3}(\Psi_0)\exp\left[i\left(-\frac{1}{2}kx - \frac{\sqrt{3}}{2}k_cy\right)\right] + \text{c.c.}\end{aligned} \tag{65}$$

The corresponding cubic-order amplitude equations are:

$$\tau_0\frac{dA_1}{dt} = \varepsilon A_1 - [g_0|A_1|^2 + g_{\pi/3}(|A_2|^2 + |A_3|^2)]A_1\,, \tag{66a}$$

$$\tau_0\frac{dA_2}{dt} = \varepsilon A_2 - [g_0|A_2|^2 + g_{\pi/3}(|A_1|^2 + |A_3|^2)]A_2\,, \tag{66b}$$

$$\tau_0\frac{dA_3}{dt} = \varepsilon A_3 - [g_0|A_3|^2 + g_{\pi/3}(|A_1|^2 + |A_2|^2)]A_3\,. \tag{66c}$$

Hexagons are given by the steady solution $A_1 = A_2 = A_3 = [\varepsilon/(g_0 + 2g_{\pi/3})]^{1/2}$, while parallel rolls would be given by solutions such as $A_1 = (\varepsilon/g_0)^{1/2}$ and $A2 = A3 = 0$. Note that the nonlinear coupling coefficients depend on the angle θ between wavevectors of the "plane-wave" modes and are thus denoted by g_θ. See Cross[63] for details on the mode normalization and the values of the coefficients.

It is possible to examine the stability of each of these patterns with respect to one another, for example by introducing perturbations into the above amplitude equations (with the appropriate numerical values of the coupling coefficients) and performing a linear stability analysis. The result is that rolls win and are expected to be the pattern selected near onset.

One measure of the nonlinear development of the rolls as Rayleigh number increases is the excess transport of heat from the bottom plate to the top due to convection. This is expressed in nondimensional form as $\mathrm{Nu}-1$, where the "Nusselt number" Nu is the ratio of total heat transport to the purely conductive transport. Suppose a parcel of fluid near the bottom has a temperature $(T+\theta)$; the amount of heat in the parcel (relative to some nearby reference temperature), is $C(T+\theta)$, where C is the heat capacity of the parcel. Neglecting losses due to thermal diffusion, we would expect this heat to be transported across the gap d at a rate $wC(T+\theta)/d$. The net transport would be given by an average over the cell, $(C/d)\langle w(T+\theta)\rangle$, where the brackets indicate an averaged quantity. Since T is uniform across any layer in the cell while w varies sinusoidally, $\langle wT\rangle = 0$. Thus we arrive at the conclusion that the net excess heat transport due to convective motions is proportional to the average $\langle w\theta\rangle$ of the product of the z-velocity and temperature deviations from the conducting state. See Busse[32] for a detailed argument. The quantities w and T are "in phase," so the average of the product gives a positive number. In fact, both w and θ are characterized by an amplitude $|A|$, which is proportional to $\varepsilon^{1/2}$. Thus we arrive at the important scaling relation:

$$\mathrm{Nu}-1 \sim |A|^2 \sim \varepsilon.$$

This linear dependence of convective transport on the distance ε above threshold has been well verified by experiment (see Figure 38).[11,12]

Returning to the pattern selection problem, there is a loophole that can produce hexagons instead. The fully nonlinear Boussinesq equations with symmetric boundary conditions (e.g., rigid, rigid) have an interesting symmetry property that is not necessarily true in general. Suppose $u(.,z), v(.,z), w(.,z), p(.,z), \theta(.,)$ is a solution; here "$.$," represents the other variables x, y, and t. Then another solution consists of the fields reflected through the midplane $z=0$ such that the temperature field changes sign as well: $u(.,-z)$, $v(.,-z)$, $-w(.,-z)$, $p(.,-z)$, and $-\theta(.,-z)$. Note especially the minus signs in front of w and θ. This has no effect on rolls and squares other than translation in the xy-plane, but it changes a hexagonal cell with warm fluid rising in the center and cool fluid falling at the edges into a hexagonal cell with warm fluid rising at the edges and cool fluid falling in the center (Figure 39). The consequence of this symmetry and the form of the eigenmodes of the linearized equations is that the amplitude equations must look the same if A_i is replaced with $-A_i$ $(i=1,2,3)$.

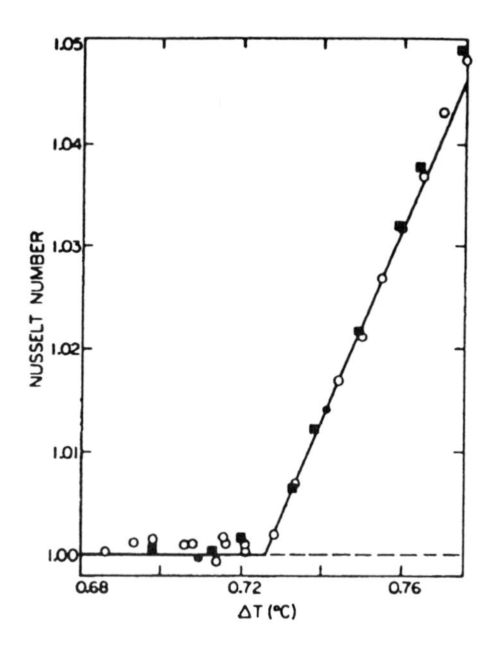

FIGURE 38 Nusselt number Nu (dimensionless heat transport) versus temperature near convection threshold. The fluid was a 5.1mm layer of water at 25°C. Note the sharp onset evident at DTc = 0.726°C and the evident linear relation $\nu - 1 = \alpha(\Delta T - \Delta T_c)$.[241]

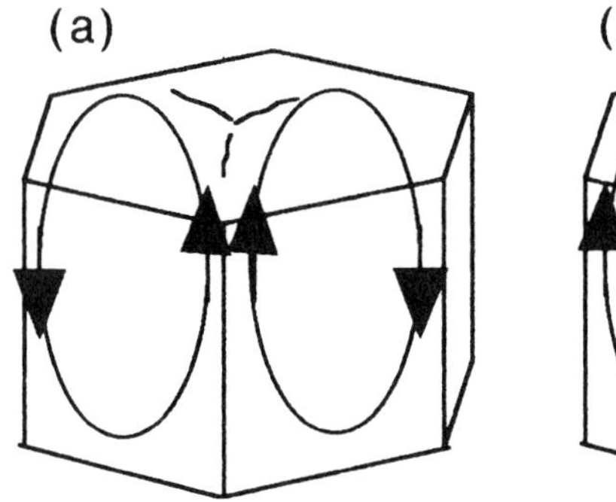

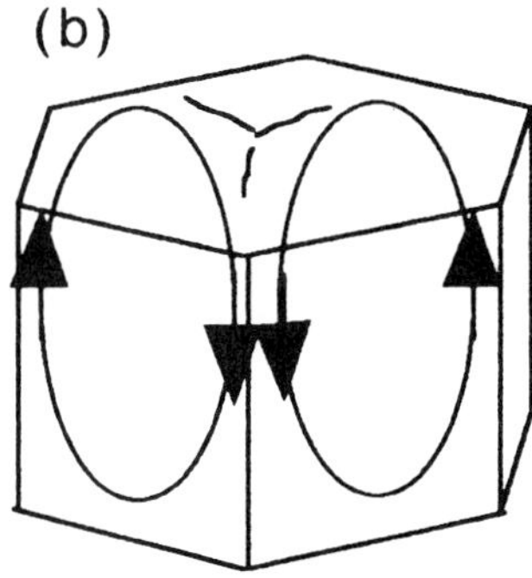

FIGURE 39 Two types of hexagonal convection cells: (a) fluid flows up the center and down the sides; (b) fluid flows up the sides and down the center.

Suppose this symmetry is broken, through nonsymmetric boundary conditions (solid wall at bottom and free surface at top), or by "non-Boussinesq" effects such as significant temperature dependence of the fluid viscosity. Then, it turns out that another term must be added to the amplitude equations describing the combination of modes that can form hexagons:

$$\tau_0 \frac{dA_1}{dt} = \varepsilon A_1 + \alpha A_2^* A_3^* - [g_0|A_1|^2 + g_{\pi/3}(|A_2|^2 + |A_3|^2)]A_1 \,, \tag{67a}$$

$$\tau_0 \frac{dA_2}{dt} = \varepsilon A_2 + \alpha A_1^* A_3^* - [g_0|A_2|^2 + g_{\pi/3}(|A_1|^2 + |A_3|^2)]A_2 \,, \tag{67b}$$

$$\tau_0 \frac{dA3}{dt} = \varepsilon A_3 + \alpha A_1^* A_2^* - [g_0|A_3|^2 + g_{\pi/3}(|A_1|^2 + |A_2|^2)]A_3 \,. \tag{67c}$$

The parameter α measures the strength, say, of the non-Boussinesq effects. For small but nonzero α, the nature of the bifurcation is radically changed and infinitesimal amplitude rolls and hexagons are both initially unstable. However, the hexagon state can be reached at finite amplitude and is the first to appear. As Rayleigh number is increased, though, the hexagons become unstable and rolls appear instead.

Returning to the Boussinesq case, we have established that rolls are the preferred state near transition; next, we will want to know about the stability of the roll state itself as Rayleigh number is increased and as the wavenumber k is varied between the marginal stability boundaries. Clever and Busse[48,49] (see also Busse and Clever[31]) computed the flow fields into the strongly nonlinear regime by expanding the fields in a complete basis of functions, truncating the expansion, and

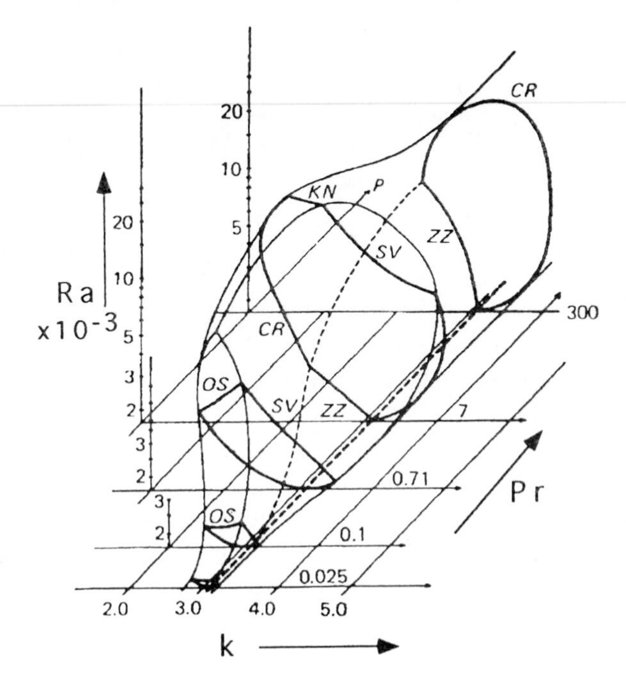

FIGURE 40 Stable parallel rolls occur within a volume of parameter space called the "Busse balloon": the paramters are Rayleigh number Ra, Prandtl number Pr, and wavenumber k. Numerical computions showd several types of instability once the sides of the ballon are crossed: oscillatory (OS), skewed varicose (SV), cross-roll), knot (KN), and zigzag (ZZ). [After Busse.[32]]

substituting back into Eq. (59) (this is an oversimplified description of the "Galerkin method," see Busse[33] for a discussion of the usefulness of this approach to a wide range of problems). Once the nonlinear states were computed, they were perturbed by various types of disturbances and the roll stability determined as a function of Rayleigh number, Prandtl number, and wavenumber. The region of stability for parallel rolls in this three-dimensional parameter space occupies a volume known as the Busse balloon (see Figure 40). Various boundaries of the balloon correspond to different modes of instability. The most dangerous mode strongly depends on Prandtl number. This is to be expected, since for low Prandtl number inertial effects dominate the nonlinearity, while for high Prandtl number the nonlinearity is dominated by thermal effects.

A good physical discussion of the various instabilities near threshold and in the low and high-Prandtl number limits far above threshold is given by Manneville.[167] Here we catalog some of the modes, as shown in Figure 41. Rolls with wavenumbers close to the marginal stability boundary are unstable to large length-scale modulations (Eckhaus instability, Figure 41(a)) that end up replacing the roll pattern with a parallel set of rolls with a stable wavelength. Alternatively, a cross-roll instability changes wavelength by replacing the unstable rolls with a perpendicular set of rolls (Figure 41(b)). A final mechanism for wavenumber adjustment is through local reorientation of the rolls in what is called the zig-zag instability (Figure 41(c)). For the rolls that fall within the wavenumber band stable to these perturbations, however, there will eventually be a secondary instability to a qualitatively different type of flow when the Rayleigh number is increased to a large enough value. When temperature dominates the nonlinear behavior (high Prandtl number), thermal boundary layers form near the top and bottom walls; these layers contain steep temperature gradients that can themselves become unstable. Small rolls develop perpendicular to the large rolls; the planform becomes rectangular while the flow remains steady in time (Figure 41(d)). This is called "bimodal convection." At low Prandtl numbers, on the other hand, waves appear and travel on the roll boundaries (Figure 41(e)), somewhat reminiscent of wavy vortex flow in the Couette-Taylor system. These traveling waves (Busse oscillations) have been the subject of some beautiful experiments on "waves-in-a-box"[62]—comparable to the study of spiral vortices in the Couette-Taylor problem and waves in binary fluid convection. Another mode of instability, which is steady in time, is the "skew-varicose" deformation of the rolls, where the rolls are deformed into an undulating pattern of skinny and fat sections (Figure 41(f)). These types of instability are well-documented experimentally, but various artifices were often necessary to prepare the desired roll state in the first place. For example, rolls of a certain wavelength and uniform orientation could be ensured by supplementing the heating from the bottom plate with a stripes of heat-generating light—a method known as thermal imprinting. As we will see below, maintenance of a uniform pattern of rolls is problematic, so in some cases the instabilities had to be explored within a time-window before more subtle spatio-temporal effects controlled the pattern. See Cross and Hohenberg[66] for further discussion.

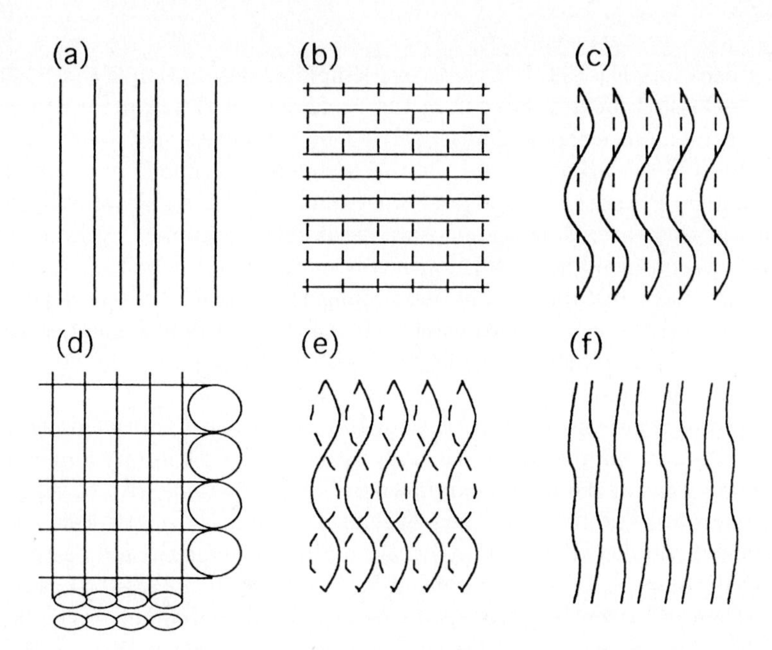

FIGURE 41 Instabilities of parallel rolls. (a) Eckhaus instability: a long-wavelength modulation eventually replaces the original rolls with a new set of rolls with a different wavelength. (b) Cross-roll instability: the original rolls (dashed) are unstable to rolls formed at 90 degrees with a different wavelength. (c) Zigzag instability: rolls deform so that their boundaries form a spatially sinusoidal pattern. (d) Bimodal instability: the convection becomes three-dimensional as fluid near the top and bottom boundaries also starts to overturn. (e) Busse oscillations: the flow becomes time dependent, with the dashed and solid curves showing snapshots at different times of the roll pattern; the deformations of the rolls may form waves traveling parallel to the roll axes. (f) Skew varicose: adjacent roll boundaries deform to give thicker and thinner portions of the rolls.

Indeed, for moderate to large-aspect ratio systems, there are some serious experimental wrinkles in the observations of convection patterns near threshold. The first of these is that the sidewalls joining the top and bottom plates can exert an overriding influence on the pattern chosen by the convecting rolls. Such sidewall forcing leads, for example, to concentric rolls in cylindrical containers.[143,144] It also leads to a rounding of the transition into an imperfect bifurcation, with rolls first appearing near the walls. However, by careful thermal bonding of the sidewalls to the endplates, horizontal temperature inhomogeneities induced by the sidewalls may largely be eliminated. There is still an effect due to the tendency of convection rolls to minimize dissipation by aligning perpendicular to the walls; rolls

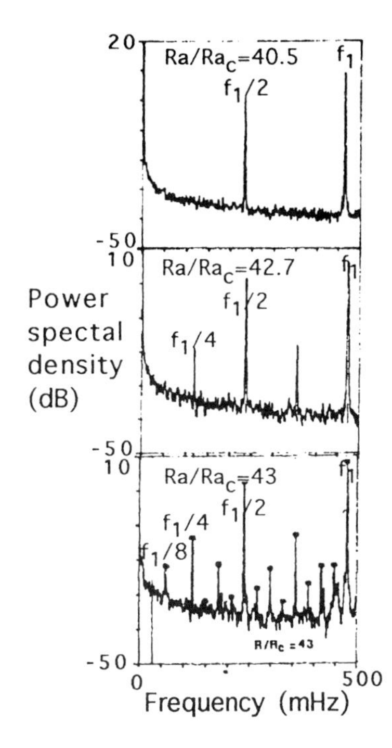

FIGURE 42 Period doubling in a small convection cell ($d \times 1.2d \times 2.4d$, where $d = 1.25$mm) using liquid helium at 3.5K (pressure 3 bar). [After Libchaber and Maurer.[162]]

in rectangular containers tend to align parallel to the short side of the container. However, in large-aspect-ratio systems, the patterns are no longer exclusively determined by the sidewall geometry. The resulting flows can have astonishing behavior: for Rayleigh numbers not far above threshold, irregular time-dependence can appear.[3] This is apparently due to the nucleation, drifting, and annihilation of dislocations in the roll patterns, particularly evident in cylindrical containers in which the perpendicular alignment of rolls to walls is frustrated. Under some conditions, this irregular time-dependence continues indefinitely; in other cases, the pattern may settle into a steady-state mix of domains of rolls oriented in different directions (Figure 37(b)).[4,114,119] The complex and extremely delicate determination of such patterns and associated spatio-temporal chaos is discussed by Newell,[184] Manneville,[167] Newell et al.,[185] and Cross and Hohenberg.[66]

One may go to the other extreme and do experiments with very small aspect ratios, of order 1. In such cases, we expect that the number of active modes becomes very limited and the possibility exists for observing the development of purely temporal chaos. Indeed, Libchaber and Maurer[162] observed a period-doubling route to chaos in just such a system—see Figure 42. Other classic routes to chaos can be observed, such as the Ruelle-Takens scenario of quasi-periodicity leading to chaos,[80] or the appearance of intermittency, that is, periodic oscillations interspersed with irregular "bursts" at random intervals.[20] See Manneville[167] for further discussion.

4.3 HIGHER INSTABILITIES, CHAOS, AND TURBULENCE

When the Rayleigh number is increased further (beyond secondary instability), the flow eventually becomes time-dependent. Depending on the Prandtl number, this might start out as periodic in time, but for further increases of Ra the flow becomes strongly irregular. Eventually, at large Rayleigh number, the motion is turbulent, characterized by strong three-dimensionality and a wide range of spatial scales. An early phase-diagram analogous to Figure 9 for the Couette-Taylor system was prepared by Krishnamurti.[148] This diagram (Figure 43) is undoubtedly an oversimplified picture, as we have seen above that a large number of complications can arise in the patterns formed even when the Rayleigh number is close to the critical value Ra_c.

At high Rayleigh numbers, temperature gradients become strongly concentrated in boundary layers near the walls. An analogous concentration of velocity gradients occurs for high Reynolds number flow in the Couette-Taylor system. In the convection system, thermal plumes (see Figure 44) may erupt from the boundary layer,[266] a manner of convection that is common in natural situations and used to advantage by soaring birds and glider pilots. Once again, we have reached the threshold of highly complex dynamics, so we move on.

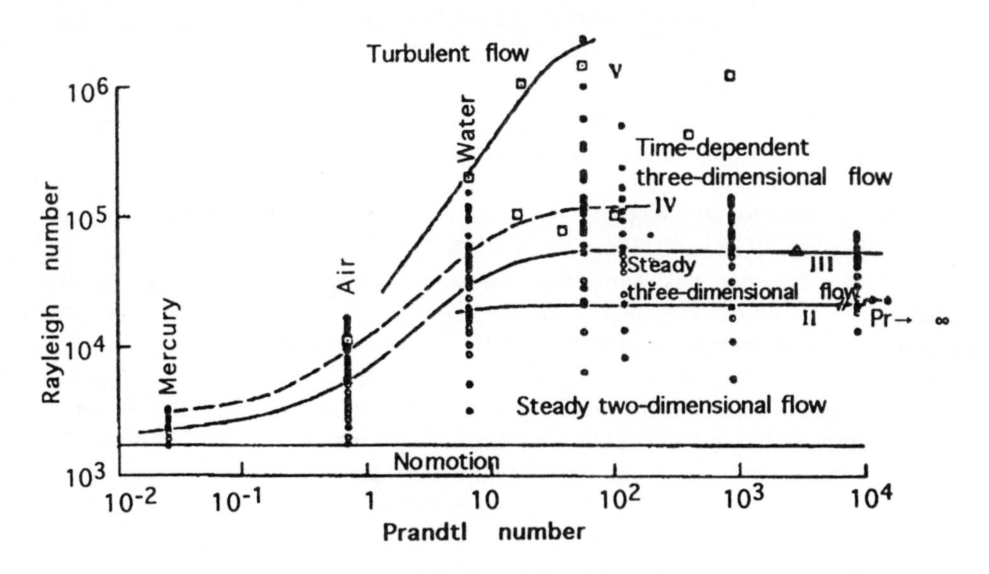

FIGURE 43 Phase diagram of flow states in convection experiments on a very broad scale. Note the Prandtl numbers associated with various fluids. [From Krishnamurti.[148]]

FIGURE 44 A kitchen experiment to demonstrate convection plumes at high Rayleigh number. Food coloring dye is carefully placed at the bottom of a Pyrex container of water. This is placed on a hot plate and covection plumes soon emerge.

4.4 VARIATIONS

There are many variations on the convection problem. Ironically, the free-surface convection patterns studied by Bénard turned to be examples of surface-driven convection, now known as the Bénard-Marangoni instability. This was not fully realized until fifty years later, when Block[23] and Pearson[202] established a theoretical description. The mechanism for driving the convection in this case is connected to the temperature dependence of the surface tension. A coupling between temperature and surface tension generates cellular patterns—apparently hexagonal at onset (Figure 45)—such that fluid is drawn across the surface from regions of low surface tension to regions of high surface tension, creating a circulation pattern in the fluid layer. This occurs for thin fluid layers and gravity plays no role; when the layer thickness increases, there can be a competition between surface tension effects and buoyancy-driven convection. It would be nice to remove gravity entirely, which is why this experimental system has become a prime candidate for experiments aboard the space shuttle.

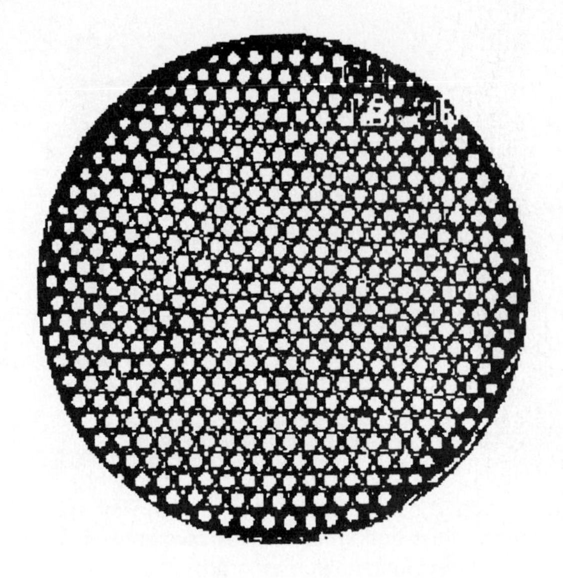

FIGURE 45 Hexagonal cell pattern in Benard-Marangoni convection. [Image courtesy of M. F. Schatz and H. L. Swinney.]

FIGURE 46 A localized region (upper left) of convection rolls in an annular container. The fluid is a binary mixture of ethanol and water. (Niemela et al.[187])

Returning to the case of buoyancy-driven convection between parallel plates, there are even many variations for this system. The effects of rotation and of magnetic fields (for conducting fluids) are discussed by Chandrasekhar[38] and continue to be investigated; see Busse[32] for an interesting discussion of time-dependence that arises as a consequence of rotation via the so-called Küppers-Lortz instability.[150]

Another important variation is convection in binary fluid mixtures. Concentration gradients can couple to thermal gradients and the instability mechanism can be significantly altered. As a result, it becomes possible for the conducting state to make a transition to time-dependent flow in the form of traveling waves or standing waves; the "exchange of stabilities" principle no longer holds. One difficulty with this system is the fact that the instability can be subcritical; on the other hand, the boundary conditions for treating the traveling waves via the coupled Ginzburg-Landau equations (Eq. 55(a) and (b)) have been well established. Many predictions[64,65] have clear correspondence in the experiments.[91,141,142] These include the phenomena of confined states—localized regions in which the wave amplitude is large—and blinking states mentioned in Section(3.9). Several experimenters have studied such waves in an annular geometry (see Figure 46); see Cross and Hohenberg[66] and references therein.

5. SHEAR FLOW INSTABILITY AND WAKES

5.1 GENERAL IDEAS ABOUT SHEAR FLOWS

The previous topics dealt with instabilities arising from the imposition of stresses that created body forces, either through buoyancy or centrifugal effects. Now we consider a different class of problems, in which the fundamental balance between inertia, applied pressure gradients, and viscous drag is upset. Quite a large variety of flows are described by a velocity pointing primarily in one direction (say the x-direction) and gradients primarily in a perpendicular direction (say the y-direction). We call such flows *shear flows*, with the idea that any rectangular region lying in the xy-plane placed in the flow will be sheared into an ever-lengthening parallelogram as it is carried along by the flow (see Figure 47). Because such flows are so common (see Figure 48), they have dominated the literature on hydrodynamic instabilities and transition. The subject is extremely challenging, having in its early development occupied the attention of such scientists as Lord Rayleigh, Lord Kelvin, H. von Helmholtz, G. I. Taylor, L. Prandtl, A Sommerfeld, and even the physicist better known for his work in quantum mechanics: W. Heisenberg.

Several experiments stand out in this subject as "classics." O. Reynolds[215] achieved a synthesis that was a milestone in fluid dynamics research through his realization that transition in pipe flow depended on the combination of variables that bears his name: the Reynolds number $\mathrm{Re} \equiv Ur/\nu$. ($U$ is the centerline velocity, r the radius, and ν the kinematic viscosity of the fluid.) In this century, the fundamental theoretical achievement of Tollmein[269] and Schlichting[232] in predicting instability in boundary layer flow through the growth of infinitesimal disturbances was not accepted for more than ten years until confirmation came from the beautiful

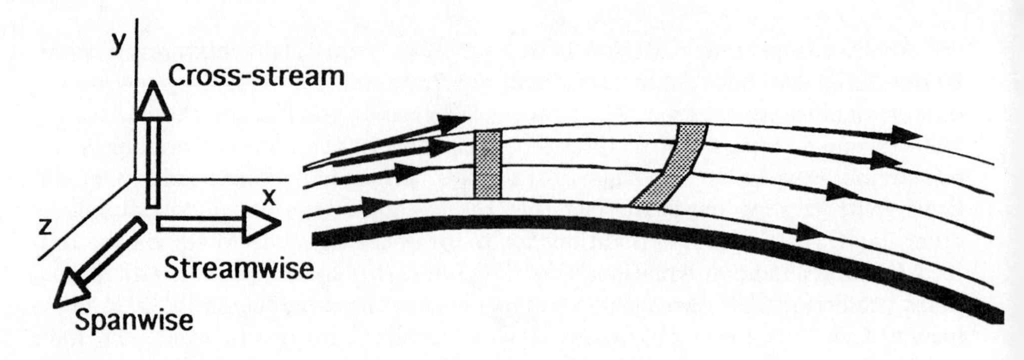

FIGURE 47 A general schematic of shear flow: notice how an imaginary parcel of fluid (shaded rectangle) is deformed as it progresses downstream.

experiments of Schubauer and Skramstad.[235] Even new theoretical developments refer to the data of this landmark set of experiments. Nowadays, carefully refined and instrumented experiments, advanced numerical computations, and analytical approaches such as singular perturbation theory play an important role in generating insight into the complex process of transition to turbulence in such flows. A good account of the problem from both a physical and historical point of view is found in Schlichting,[233] while recent reviews include those of Sherman,[238] Bayly, Orzag, and Herbert,[9] Morkovin,[179] Maslowe,[171] and last but not least, the extensive treatment in the book by Drazin and Reid.[79]

We will be concerned with how energy and momentum in a basic shear flow can be transferred to secondary motions, which in turn typically evolve downstream through a chain of instabilities into fully developed turbulence. It will also be useful to think of the vorticity of these flows, defined as $\omega \equiv \nabla \times \mathbf{u}$. The vorticity can be thought of as defining the local axis and rate of rotation of a tiny element of fluid at a given location in the flow. Since the basic flow has velocity in the x-direction and gradients in the y-direction, we can describe it by one function $U(y)$ and the find that the vorticity is simply $-\partial U/\partial y$ pointing along the z-direction (see Figure 47). We will see that instability leads to a systematic concentration and reorientation of vorticity. Understanding such processes might help us understand turbulence itself, whose disordered motion is sustained through a continual concentration (stretching) and re-orientation of vorticity.[264]

Why is the shear-flow transition process so complex? Part of the difficulty arises from the fact that such flows are open systems, with flow entering one end of a system and leaving the other. Instabilities in the flow are generally of the convective type: small disturbances at the inlet due to external vibrations or fluctuations in the incoming flow velocity (so-called free-stream disturbances) become amplified as they travel downstream. Unless a controlled external disturbance is imposed, instability does not lead to a dominant secondary mode of motion through the whole system, but instead produces a randomly phased ensemble of the fastest growing "normal modes" of the system passing by any given observation point. As we shall see, the boundaries can play a subtle role in inducing transition, so imperfections (roughness) in the boundaries can lead to further havoc. Pressure disturbances can propagate through the flow, even from downstream, providing a further coupling to noisy surroundings and a possible means of feedback into a growing disturbance. Considerable research continues to be done to explore the "receptivity" of transitional flows to controlled disturbances of these various types.

Linear stability analysis of such flows is itself quite tricky, because of the singular behavior of the linearized equations at points within the flow: at Reynolds numbers where transition might occur, the viscous terms are small for much of the flow region but act as "singular perturbations" in the sense that there are layers within the flow where viscosity, no matter how small, has a dominant effect. Even when, by an analytic or numerical tour de force, the normal modes and eigenvalues are found, it is now evident that the transitions are often subcritical: finite disturbances can cause instability well before the linear theory predicts transition. In fact, some flows (for example, pipe flow) are always stable according to linear theory, but observation shows that turbulence arises even for fairly moderate Reynolds numbers (around 2000 for pipe flow) in a "typical" situation. The early onset of turbulence, even below the subcritical regime determined by weakly nonlinear theories, is called the phenomenon of "bypass" transition.

Even when experimental ingenuity has overcome these obstacles (see, for example, Nishioka et al.,[188] by reducing background noise to levels less than 0.05% of typical values of the flow variables and by artificially creating a disturbance that dominates the residual background noise, there is a fundamental difficulty with transition in shear flows. The induced secondary motions do not reach saturation as they travel downstream, but instead become themselves unstable to disturbances—leading to a cascade of instabilities and finally to fully developed turbulence as the flow moves farther and farther downstream. Unlike the Couette-Taylor experiment, one cannot arrest the transition process for closer examination of the intermediate flows. This may be understandable for flows that are not strictly parallel, such as boundary layers or wakes, where the basic flow itself is changing and one can have the effective Reynolds number grow in the downstream direction. However, even for flow between parallel walls (plane Poiseuille flow), this spatial evolution occurs.

Nonetheless, it has been possible to sort out different stages in the transition to turbulence and make comparisons between careful experiments, numerical simulations that are now capable of handling wide ranges of spatial scales, and analytic treatments that glue the whole picture together. We will see, for example, that analysis leads to a fundamental distinction between two types of instability in shear flows. One type, usually referred to as an inviscid instability, is dominant in such flows as mixing layers, jets, and wakes (see Figure 48) and is associated with the existence of a maximum in vorticity in the flow profile; at such a maximum, the basic flow profile has an inflection point: $\partial^2 U/\partial y^2 = 0$. This sort of instability occurs even when viscosity is ignored, and corrections for the inclusion of viscosity can often be treated in a simple way. The other type of instability, surprisingly, depends on viscosity and is associated with flows that are bounded by a solid wall, such as channel flow or boundary layer flow. Here, viscosity is a sly broker in the exchange of energy between the basic flow and a disturbance, and is necessary for the latter to grow. Remarkably, the agent that delayed centrifugal and buoyant instabilities by dissipating the imbalance of stresses, is here found to be a double-agent, dissipating energy on one hand and aiding its exchange on the other.

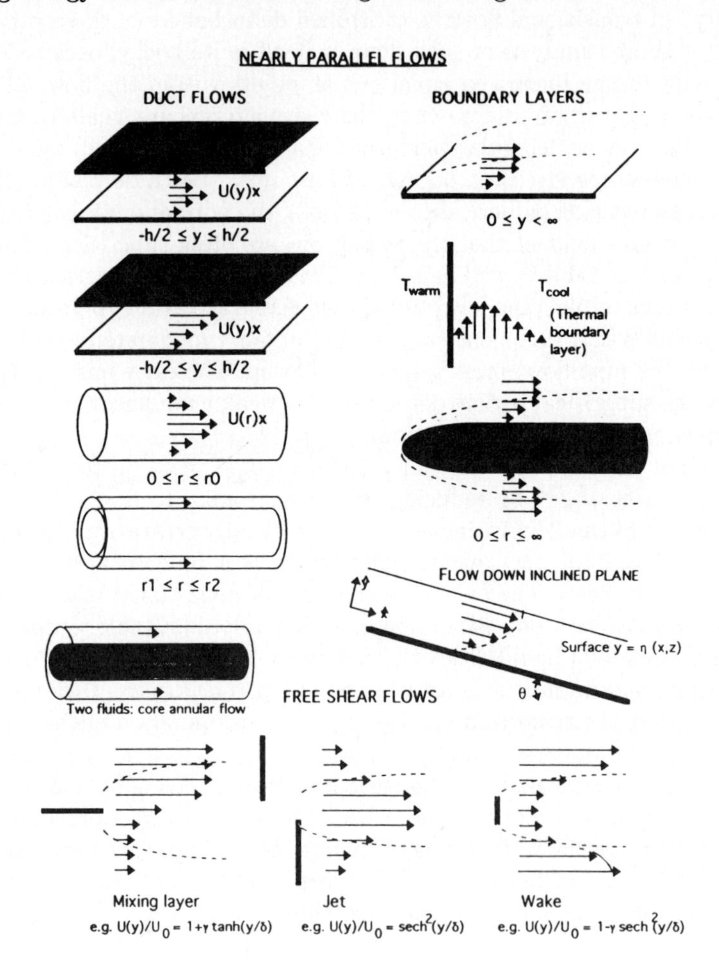

FIGURE 48 A visual catalog of shear flows, both in planar and cylindrical geometries.

5.2 A LINEAR VIEW: THE ORR-SOMMERFELD EQUATION AND TS WAVES

To proceed, we must ourselves dissipate effort in some mathematics. We describe our base flow, under the approximation that it is nearly parallel, by the fields:

$$\mathbf{U} \approx U(y)\hat{\mathbf{x}}, P \approx P(x). \tag{68}$$

We are presently restricting attention to two-dimensional flows and ignoring downstream development (x-dependence) of the basic flow. Indeed, if viscosity is also ignored and the geometry does not vary with x, it turns out that any profile $U(y)\hat{\mathbf{x}}$ and a constant pressure P_0 satisfy the resulting flow equations. In more general situations and with viscosity included, we will assume that the streamwise variation and curvature effects may first be ignored: the base flow is considered to be "locally" parallel. Corrections to this parallel-flow approximation can be made later in the analysis. Of course, for some flows (flows through uniform ducts), the base flow is exactly parallel.

In the following, lengths are scaled by some characteristic length h, for example, the half-width of the channel for flow between parallel plates; velocity is scaled by a characteristic velocity U_0, such as the maximum or centerline velocity $U_{\max}$; the time scale is h/U_0 and the pressure scale is ρU_0^2. Linear stability analysis proceeds by considering the perturbed velocity and pressure fields: $U(y)\hat{\mathbf{x}} + \mathbf{u}(x, y, z, t), P(x) + p(x, y, z, t)$. For infinitesimal disturbances $\mathbf{u} = (u, v, w)$ and p, the linearized equations are:

$$\frac{\partial \mathbf{u}}{\partial t} + \frac{U(y)\partial \mathbf{u}}{\partial x} + v\frac{dU}{dy}\hat{\mathbf{x}} = -\nabla p + (1/\mathrm{Re})\nabla^2 \mathbf{u}; \tag{69a}$$

$$\nabla \bullet \mathbf{u} = 0. \tag{69b}$$

$\mathbf{u}(x, y, z)$ must vanish at the boundaries $y = y_1$ and y_2 (one or both of which may be at infinity). Homogeneity of the equations in x, z, and t suggests solutions of the form:

$$(\mathbf{u}, p) = [\hat{\mathbf{u}}(y), \hat{p}(y)]e^{i(\alpha x + \beta z - \alpha c t)}, \tag{70}$$

with four unknown functions of y: $\hat{\mathbf{u}} \equiv \hat{u}, \hat{v}, \hat{w}$ and $\hat{p}$. Note that $-i\alpha c = \alpha c_i + i\alpha c_r$ is a new name for the eigenvalue $s = \sigma - i\omega$ used in previous problems (this new notation is customary in shear flow literature). Squire's transformation,

$$\tilde{\alpha} \equiv (\alpha^2 + \beta^2)^{1/2}, \tilde{c} \equiv c, \tilde{\alpha}\tilde{u} \equiv \alpha\hat{\mathbf{u}} + \beta\hat{w}, \tilde{v} \equiv \hat{v}, \frac{\tilde{p}}{\tilde{\alpha}} \equiv \frac{\hat{p}}{\alpha}, \tilde{\alpha}\tilde{\mathrm{Re}} \equiv \alpha \mathrm{Re}, \tag{71}$$

permits the system of equations to be reduced to an equivalent two-dimensional problem solving for $\tilde{\mathbf{u}} \equiv (\tilde{u}, \tilde{v}, 0)$ and $\tilde{p}$. For every three-dimensional disturbance satisfying the eigenvalue relation $f(\mathrm{Re}, \alpha, \beta, c) = 0$, there is a two-dimensional disturbance satisfying $f(\alpha\mathrm{Re}/\tilde{\alpha}, \tilde{\alpha}, 0, c) = 0$. If the three-dimensional disturbance becomes marginally stable ($c_i = 0$) for α and Re, then the corresponding two-dimensional disturbance is marginally unstable for $\tilde{\alpha}$ and $\tilde{\mathrm{Re}} = \alpha\mathrm{Re}/\tilde{\alpha}$. Since $\alpha < \tilde{\alpha}$, we have $\tilde{\text{Re}} < \text{Re}$, that is, the two-dimensional disturbance becomes unstable at lower Reynolds number. This important result, that the earliest instability occurs for two-dimensional modes, is known as Squire's Theorem.

Having reduced the problem back to two dimensions (and dropping the $\sim$ notation), we are able to write u and v in terms of a stream function ψ.

$$u = \frac{\partial \psi}{\partial y} \quad \text{and} \quad v = -\frac{\partial \psi}{\partial x}.$$

This achieves an important economy, because the assumption that such a function can be found allows us to automatically satisfy Eq. (69b), the continuity (conservation of mass) equation:

$$\nabla \bullet \mathbf{u} = \frac{\partial u}{\partial x} + \frac{\partial v}{\partial y} = \frac{\partial^2 \psi}{\partial x \partial y} - \frac{\partial^2 \psi}{\partial y \partial x} = 0. \tag{72}$$

Again, we expect from the invariance of the basic flow in both time and the x-direction, that we can write:

$$\psi = \phi(y) e^{i\alpha(x-ct)}. \tag{73}$$

The use of the symbol ϕ for the complex amplitude rather than something like $\hat{\psi}$, has become a standard notation in the literature.

Defining $D \equiv d/dy$ and some manipulation yields the Orr-Sommerfeld equation:

$$(U-c)(D^2-\alpha^2)\phi - U''\phi + \frac{i}{\alpha \text{Re}}(D^2-\alpha^2)^2\phi = 0. \tag{74}$$

The boundary conditions are:

$$\text{channel: } \alpha\phi = D\phi = 0 \text{ at } y = \pm 1; \tag{75a}$$

$$\text{boundary layer: } \alpha\phi = D\phi = 0 \text{ at } y = 0, \infty; \tag{75b}$$

$$\text{mixing layer, jets, wakes: } \alpha\phi = D\phi = 0 \text{ at } y = \pm\infty. \tag{75c}$$

The pressure $p(x, y, t)$ satisfies Poisson's equation:

$$\nabla^2 p = -2U' \frac{\partial v}{\partial x}. \tag{76}$$

Letting $p(x, y, t) = \hat{p}(y) e^{i(\alpha x - \alpha ct)}$ and using $v = -(\partial/\partial x)[\phi(y) e^{i\alpha(x-ct)}]$,

$$(D^2 - \alpha^2)\hat{p} = -2\alpha^2 U' \phi. \tag{77}$$

The Orr-Sommerfeld equation (Eq. 74), though linear, has been "notoriously difficult to solve."[198] Before discussing this further, we consider the inviscid limit $\mathrm{Re} \to \infty$. The result is the Rayleigh equation:

$$(U - c)(D^2 - \alpha^2)\phi - U''\phi = 0. \tag{78}$$

As was noted above, viscosity can act as a singular perturbation. By taking the limit $\mathrm{Re} \to \infty$, the highest order derivatives have been dropped. Thus there must be fewer boundary conditions for the Rayleigh equation: we are left with $\phi = 0$ at the boundaries (corresponding to vanishing normal velocity component). More subtle consequences of ignoring the viscous term will be discussed below. However, the inviscid equation itself yields important insights. The most important of these is Rayleigh's stability criterion: a necessary condition for a flow to be inviscidly unstable ($c_i > 0$) is that there be an inflection point in the base flow profile: $U''(y)$ must change sign for some y within the flow. This is equivalent to saying that the flow must contain an extremum in the vorticity, since for our parallel flows, $\omega_z = U'(y)$. Further necessary conditions, Fjørtoft criteria, are:

1. $U''(y_I) = 0$ at the inflection point y_I must correspond to a local *maximum* in the magnitude of the vorticity $|U'(y_I)|$ rather than a minimum;
2. $U''(y)[U(y) - U(y_I)] < 0$ somewhere in the flow.

An evaluation of various flow profiles in terms of these conditions is shown in Figure 49. It should be stressed that these are *necessary* conditions: if the conditions are not met, the profile $U(y)$ is inviscidly stable; if, however, the conditions *are* met, this is no guarantee that the profile is unstable.

Some actual flow profiles (for example: mixing layers, wakes, jets, and boundary layers with adverse pressure gradients) satisfy Rayleigh's and Fjørtoft's criteria as candidates for being unstable in the inviscid limit. It is often the case that such flows do in fact become unstable over wavenumber ranges in accord with results from Rayleigh's equation and that the actual viscous eigenmodes with positive growth rate may be determined as a regular perturbation of the inviscid eigenmodes with positive growth rate (expanding in powers of $1/\mathrm{Re}$). In such flows, viscosity acts mainly to delay instability or to restrict the range of unstable wavenumbers.

Other flows, particularly duct flows and boundary layers, are predicted to be stable in the inviscid limit. Observation shows, however, that instability does occur and for such flows, viscosity can play a subtle role in extracting energy from the mean flow to feed a growing eigenmode. It seems unfashionable these days to be too concerned with a discussion of how viscosity actually mediates transition. In his excellent chapter on instabilities, Sherman[238] discusses what he calls "word pictures"

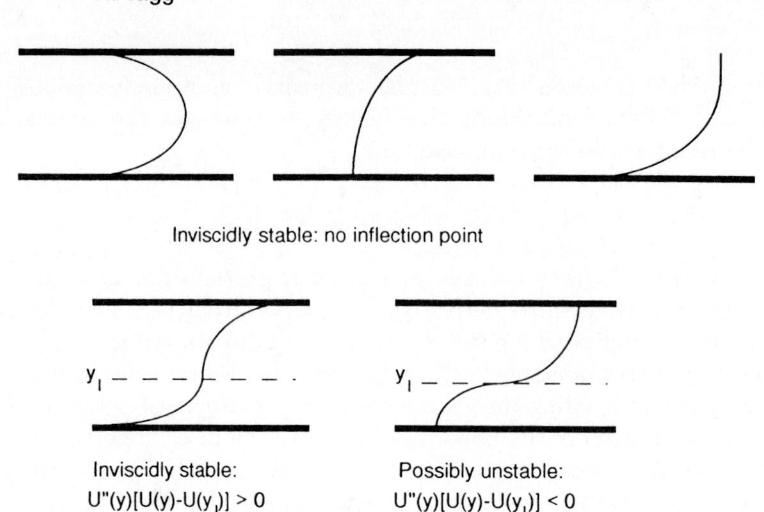

FIGURE 49 Inviscid stability or instability of two-dimensional shear flow profiles, according to Rayleigh's and Fjørtoft's criteria.

of the role of viscosity, noting that some aspects of a plausible mechanism don't quite match the details of a numerical investigation. There are several approaches to a heuristic discussion of the onset of instability. One, given by Lighthill[163] and discussed by Sherman examines a train of positive and negative vorticity disturbances and the flow they induce in surrounding fluid. Near the wall, there will be a response of opposite vorticity in order to maintain the no-slip boundary condition. This vorticity can diffuse, due to viscosity, back into the stream with just the right timing to augment the next vortex "blob" in the train. In other words, the viscous diffusion of vorticity from the walls can provide positive feedback to a layer of concentrated alternating vorticity disturbances. Another approach, discussed extensively by Lin[164] (see also an early paper by Taylor[262]), is to concentrate on the energy of the disturbance and to show how so-called Reynolds stress feeds energy into the disturbance. The Reynolds stress is the average of the product of streamwise and cross-stream velocities, $\langle uv \rangle$, and describe the flux of momentum across the shear. Viscosity can produce a phase shift in the net momentum transport between the wall and the stream so as, again, to augment a traveling periodic disturbance. It would be interesting to explore the possibility that related mechanisms coupling diffusive and advective (or induced) transport in other nonequilibrium system produce oscillatory instabilities.

The modes brought about by the viscous mechanism are usually referred to as Tollmien-Schlichting waves, after the scientists who computed approximate solutions for the case of the Blasius boundary layer (the boundary layer over a flat plate placed parallel to a free stream with zero streamwise pressure gradient). The accurate solution of the Orr-Sommerfeld equation in such cases has been an extremely delicate task (both analytically and numerically) that has gone on for more than 70 years. An instructive account of analytical approaches may be found in the book by Schlichting,[233] with a considerably extended discussion found in Drazin and Reid.[79] The "shooting method" for numerically solving such two-point boundary value problems, which can be applied in a relatively straightforward way to the Couette-Taylor problem,[216] must here be carefully adapted to ensure that the "trial functions" the method generates remain sufficiently independent as they are integrated across the domain (see, for example, Betchov and Criminale,[22] Appendix III). Alternatively, the whole problem can be re-cast by expanding the disturbance field in a suitable basis of functions: the application of Chebyshev polynomials to accurate solution of the Orr-Sommerfeld equation is discussed by Orszag.[194] Such approaches have finally led, in the last twenty years, to precise results for critical Reynolds numbers, wavenumbers, and frequencies for various flows; some results are summarized in Table 3.

As mentioned earlier, a major part of the difficulty with the Orr-Sommerfeld equation is the existence of "critical layers" in the flow. There is at least one point y where the phase velocity cr of a neutral mode (one that is not growing or decaying) matches the velocity $U(y)$. The first term in the Orr-Sommerfeld equation will therefore vanish; near this value of y, e viscous term containing the factor 1/Re will always be important, no matter how high the Reynolds number. In addition, if there are solid boundaries, there will be "boundary layers" where viscosity also remains important. Sometimes, these layers overlap. For a discussion of the importance of critical layers in the analysis of shear layer instabilities, see Maslowe.[170]

TABLE 3 Critical Reynolds numbers and wavenumbers for linear instability of some parallel shear flows. The characteristic length scales are: "displacement thickness" δ^* for a boundary layer; separation $2h$ between plates in plane Poiseuille or plane Couette flow; radius r for pipe flow. The velocity U is the free-stream velocity for the boundary layer; in other cases, it is the center-line velocity. The fluid has kinematic viscosity ν.

Type	Re	Re_c	k_c
Blasius boundary layer	$U\delta^*/\nu$	520	0.30
Plane Poiseuille flow	Uh/ν	5772	1.02
Plane Couette flow	Uh/ν	∞	—
Pipe flow	Ur/ν	∞	—

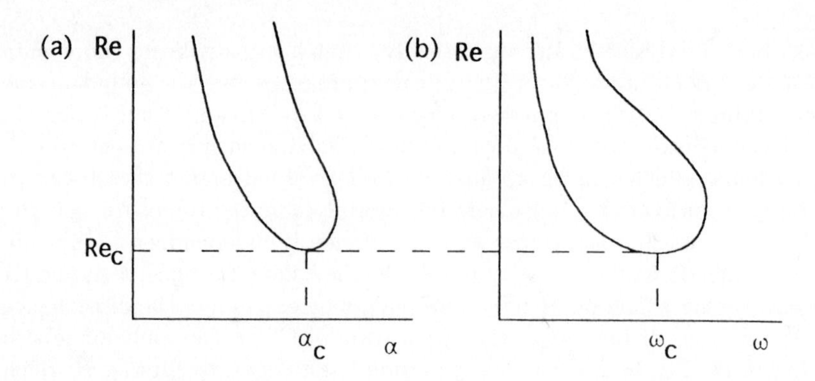

FIGURE 50 Marginal stability curves (schematic) for (a) temporal growth; (b) spatial growth.

It is now necessary to introduce a critical distinction between two types of stability analysis. Implicit in the above discussion was the idea that we examined the growth in *time* of a disturbance of given real wavenumber α. For a given α we compute the complex eigenvalue $s = -i\alpha c = \sigma - i\omega$ and, by looking for the onset of instability where the growth rate $\sigma = \alpha c_i$ passes through 0, we establish a marginal stability curve (see Figure 50(a)) in the Reynolds number—wavevector (Re $-$ α) plane, and also find the phase velocity $c_r = \omega/\alpha$ as a function of Re and α. The critical values, obtained for minimum Reynolds number, are Re_c, α_c, and ω_c.

Open flows, however, are typically characterized by disturbances generated near the inlet and propagating downstream. As we have commented earlier, instabilities under such conditions are frequently of the convective type: exponentially growing wavepackets originating in localized disturbances are quickly carried away from their origin, with the consequence that random phase in the upstream disturbances will lead to incoherence in the growing modes as they propagate downstream. Moreover, the response to the disturbances might be washed out of the system before it is detected, even though the system might be above the threshold for instability. Thus an experimental strategy—used for example in the pioneering work of Schubauer and Skramstad[235] on boundary layer flows—is to cause time-periodic disturbances at a fixed upstream location and observe the *spatial* growth of the response. In the Orr-Somerfeld equation, we now fix c to be real, $c = c_r$; equivalently, s is pure imaginary: $s = i\omega$. We now permit the wavenumber α to be complex: $\alpha = \alpha_r + i\alpha_i$. Rearranging the form of the disturbance streamfunction (Eq. (73)) clarifies what happens:

$$\psi = \phi(y)e^{i\alpha(x-ct)} = \phi(y)e^{i(\alpha_r x - c_r t)}e^{-\alpha_i x}. \tag{79}$$

If $\alpha_i < 0$, the mode grows spatially with increasing x. Marginal stability is now established when $\alpha_i = 0$, which generates a curve (see Figure 50(b)) in the Reynolds number-frequency (Re$-\omega$) plane. We now also look for the real part of the wavenumber as a function of Re and ω; the critical values, obtained for minimum Reynolds number, are Re_c, α_c, and ω_c (where it is understood that α_c is a real number).

The same critical values are obtained in both types of analysis, but how do we relate the growth rates σ and $-\alpha_i$ when we are above threshold? A first approximation to this is given by the Gaster transformation[93]:

$$\alpha_i(\sigma \equiv 0) = \frac{\sigma(\alpha_i \equiv 0)}{-\partial\omega/\partial\alpha_r}. \tag{80}$$

An argument for this result, based on expanding s as an analytic function of α, may be found in Appendix B.

A variation on the approach of periodic forcing in time to launch spatially developing flows is to cause an "impulses" in both space and time and to observe the resulting wavepacket. Indeed, this is the point of view of Figure 16 on the distinction between convective and absolute instabilities. A vivid experimental demonstration of the impulse response of a boundary layer was given by Gaster and Grant[96] (see also Gaster[95]); a simulation of the evolving wavepacket is shown in Figure 51.

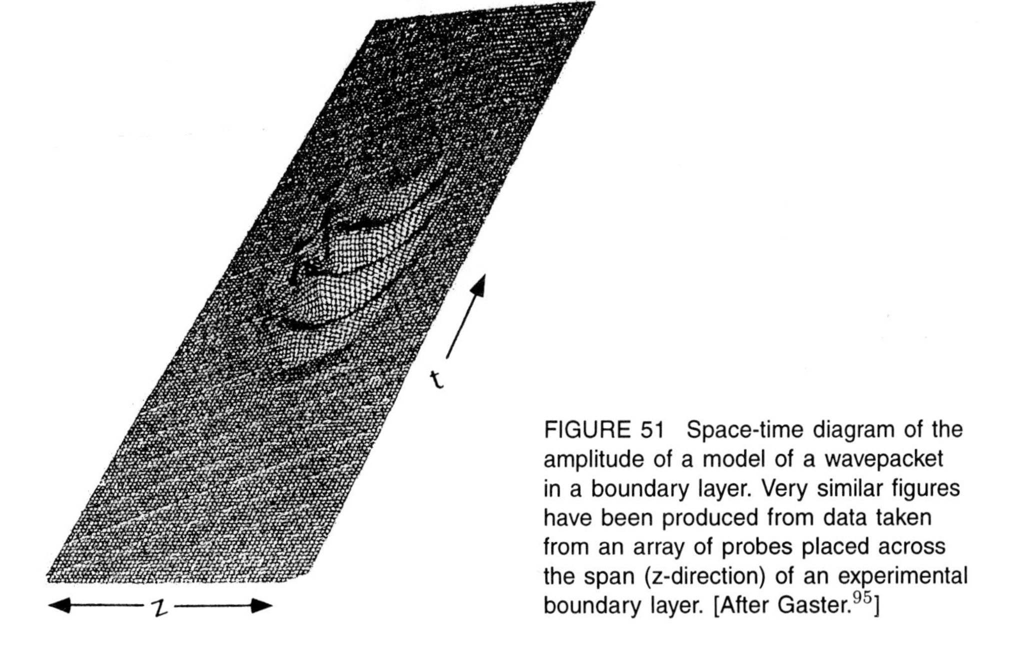

FIGURE 51 Space-time diagram of the amplitude of a model of a wavepacket in a boundary layer. Very similar figures have been produced from data taken from an array of probes placed across the span (z-direction) of an experimental boundary layer. [After Gaster.[95]]

Lately, there has been a surge of interest in a work aimed at identifying disturbances that can undergo a transient algebraic growth in time[270]; it may be that exponential damping eventually wins, but the intervening growth in energy can be several orders of magnitude. This growing transient could provide the altered base flow from which secondary instabilities and bypass transition to turbulence could occur. In other words, linear stability analysis may not be the best way to describe transition in many flows! Surprisingly, it is found that an initial disturbance formed from particular superpositions of the eigenmodes evolving according to the *linearized* equations of motion can extract energy from the mean flow and grow for a finite time, even though the component eigenmodes are all exponentially damped. For this to occur, the linear operators analogous to L described in Eq. (29) and Appendix A must have particular characteristics[111]: the eigenmodes of such operators are not orthogonal and they can interfere with one another in such a way as to bring about energy growth.

In this approach, it has been possible[86] to find "optimal" superpositions that achieve a maximum growth in a given amount of time and to relate these disturbances both to physical mechanisms for extracting energy from the mean flow and to the experimentally observed "streaks" that appear in the downstream development of transitional flows. Indeed, one of the appealing features of this analysis is its correspondence with physical mechanisms for transition, such as the "lift up" mechanism proposed years earlier by Landahl[153]: disturbances in the form of streamwise vortices exchange fluid between slow moving and fast moving layers in the shear flow, leading to stretching and intensifying of the vortices and producing algebraic growth in energy of the disturbance.

A further observation is that one must be careful in the interpretation of Squire's theorem. The theorem shows that, amongst individual eigenmodes, a two-dimensional mode is the "most dangerous"; however, the recent developments show that among *combinations* of the eigenmodes, it is three-dimensional disturbances that grow the fastest. The question then is whether and which such optimal disturbances are selected from the "soup" of noise and fluctuations present in any given experiment. See, for example, the discussion by Farrell and Ioannou.[87,88]

We will now proceed with a brief discussion of various types of parallel flows, isolating useful new points of view generated in each case and considering—in a very limited way—the complicated nonlinear processes that lead to turbulence. Needless to say, this is an intensely active subject and one should not become too complacent that the established wisdom will remain intact.

5.3 MIXING LAYERS AND JETS

As a prototype of shear instability arising from inviscid mechanisms in flows with inflectional profiles, we consider the Kelvin-Helmholtz instability. A region of inviscid fluid moves with uniform velocity $U_1\hat{\mathbf{x}}$ beneath a region of the same fluid moving with uniform velocity $U_2\hat{\mathbf{x}}$ (see Figure 52(a)). Indeed, the fluids could be moving with equal but *opposite* velocities. This then is an idealized, infinitesimally thin shear layer, with vorticity concentrated in the plane separating the two layers. In this limiting case, the growth of disturbances $y = \zeta_0 e^{i(k_x x + k_z z) + st}$ may be treated in the same way as surface waves on flat fluid interfaces, except that there is no density difference and no surface tension. We will discuss surface waves more thoroughly in Section 6.1 below, but for now concentrate on the results adapted to the shear layer. Note that for the more realistic cases of a shear profile of nonzero thickness, one resorts to the Orr-Sommerfeld equation for the stability analysis.

Linear analysis of the ideal infinitesimally thin shear layer gives the dispersion relation:

$$s = \pm k_x \frac{U_2 - U_1}{2} - i k_x \frac{U_1 + U_2}{2}. \tag{81}$$

The result is a traveling wave with x-component of the phase velocity equal to the mean velocity of the fluid layers; for *any* velocity difference, one of the two eigenvalues gives an unstable mode: regardless of the disturbance wavenumber, the wave grows, and the shear layer deforms (see Figure 52(b)). If viscosity is present, the infinitely sharp layer is diffused into a continuous velocity change over some scale d. Sometimes this is modeled by the profile $U(y) = U_1 + (U_2 - U_1)\tanh(y/d)$.

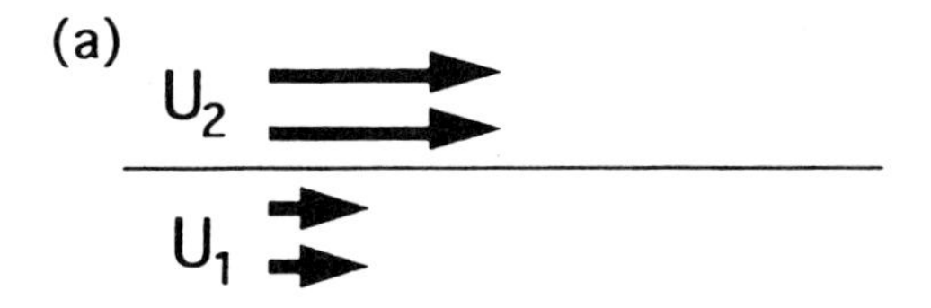

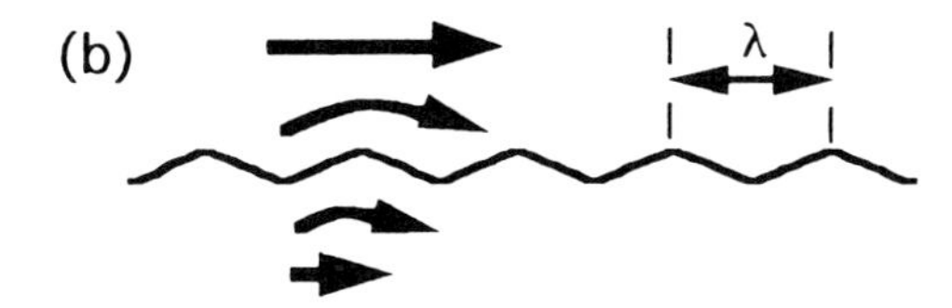

FIGURE 52 Mixing layer: (a) undeformed; (b) sinusoidal perturbation of wavelength λ; (c) nonlinear deformation and roll-up.

The critical Reynolds number remains at zero, but instabilities are prevented for short wavelength disturbances whose wavelength is of order the thickness of the layer d or less.

One approach to a physical understanding of the origins of this instability concentrates on the interaction between the velocity field and pressure. The so-called Bernoulli effect states in regions where the fluid accelerates to higher velocities, the pressure drops, and vise versa. This is essentially a statement of conservation of energy, reflecting that work was done by the streamwise decrease in pressure in order to increase the kinetic energy of the fluid. As the interface between the two moving fluids deforms into a wavy pattern, fluid on one side must accelerate around the crests of the wave, decreasing pressure on that side. The opposite situation holds for the other fluid layer, so that a pressure imbalance is created that tend to make the boundary deform even further. Eventually, the picture becomes more complicated: the interface rolls up into the beautiful patterns seen illustrated at the beginning of this chapter in Figure 52(c).

The deformation and ensuing roll-up of the layer may be explored through a discrete vortex model.[238] Consider the shear layer, which is a vortex sheet, as a linear array of vortices (see Figure 53) with circulation $\Gamma = (U_1 - U_2)$ and undistorted positions $(x_0, y_{n0}) = (na, 0)$, where $-\infty < n < \infty$ and a is the uniform spacing of the vortices along the x-axis. The vortices interact through the Biot-Savart law, a hydrodynamic analogue of the induction of magnetic fields by lines of current (here, instead, we think of velocity fields induced by vortex lines):

$$\begin{pmatrix} \frac{dx_m}{dt} \\ \frac{dy_m}{dt} \end{pmatrix} = \frac{\Gamma}{2\pi} \sum_{\substack{n=-\infty \\ n \neq m}}^{\infty} \frac{1}{(x_m - x_n)^2 + (y_m - y_n)^2} \begin{pmatrix} -(y_m - y_n) \\ (xm - xn) \end{pmatrix}. \tag{82}$$

There is no motion of the vortices as long as they remain exactly on the x-axis and uniformly spaced. If their initial positions are perturbed by $(\Delta x_{n0}, \Delta y_{n0}) = (0, \varepsilon \cos kx)$, then it is found that the array curls up into an intricate patterns of billows very much like that seen in real shear layers—see Sherman[238] (pp. 441–444 and references therein) for results, pitfalls, and important practical details of such computations.

Returning to the real problem of evolution of vortex structures in mixing layers, there has been considerable interest in the nonlinear development of the layer. A detailed review may be found in a 1986 paper by Ho and Huerre[117] (see also Michalke[176]). It is, of course, necessary to take into account the typically convective nature of the instability, that makes the downstream development strongly susceptible to upstream perturbations; see Huere and Monkewitz.[120] Sometimes, however, it is possible to obtain feedback from pressure disturbances propagated upstream from an obstacle in the flow[217]; this can lead to sustained oscillation that is called a "global instability."

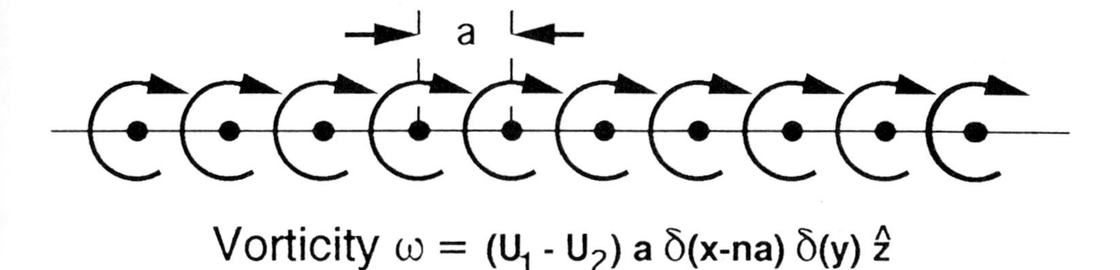

FIGURE 53 Point vortex array as a model of a mixing layer.

As a mixing layer develops, a significant subharmonic soon appears in observations of the time-dependence of motion past a fixed measurement point. Related to this is the observation that vortices begin to interact, typically in pairs, and ultimately the vortex pairs merge—a process that is significant of the spreading of the mixing layer into the surrounding flows. As the mixing layer develops, the flow also acquires a three-dimensional character through the production of vortices pointing in the streamwise direction.

Finally, a surprising fact, shown so vividly in Figure 4 is that large-scale structures reminiscent of the mixing-layer vortices, persist even in the presence of turbulence.[29] Even when they are not so visible, the presence of so-called "coherent structures" in turbulence may be inferred from careful measurement strategies that "educe" large scale features from the background of turbulent fluctuations. Such coherent motions can play a very important role in the transport of material and momentum across the layer.

The phenomena associated with mixing layers are important in processes that bring two fluids together to create a mixture, including combustion processes. One may also see manifestations of the Kelvin-Helmholtz instability in the atmosphere, where a shear layer contain clouds will show the characteristically rolled-up billows like Figure 52(c) (see, for example, the photo on page 21 of Drazin and Reid's book[79]). Toy mixing-layer experiments consisting of two immiscible fluids that flow past one another when the container is tilted may sometimes be found in novelty shops—again, an indication of how easy it is to find fluid instabilities.

5.4 WALL-BOUNDED SHEAR LAYERS: BOUNDARY LAYERS, CHANNELS, AND PIPES

As discussed above, viscosity actually helps produce instability in wall-bounded flow profiles that do not contain inflection points. An exception is the case of boundary layers where the flow must locally work against an adverse pressure gradient: such profiles can have inflection points, reducing the instability threshold considerably. Omitting such cases, we examine some of the ideas that arise in the study of wall-bounded flows.

A boundary layer is formed when a stream of uniform velocity U_0 impinges on a flat plate (see Schlichting[233]). Viscosity prevents the fluid from moving next to the plate, so there is an thin layer across which the fluid velocity must change from zero to the free-stream velocity. As the flow proceeds to a distance x downstream from the leading edge of the plate, the thickness Δ of this layer grows as $(x)^{1/2}$. The Reynolds number $U_0\Delta/\nu$ therefore grows in the downstream direction, so that the flow essentially performs its own scan of Reynolds number. Interestingly, for a fixed perturbing frequency ω, the marginal stability profile shown in Figure 54 is such that the flow becomes unstable at some wavenumber on the "lower branch" of the marginal curve, but restabilizes at higher Reynolds number as the upper branch

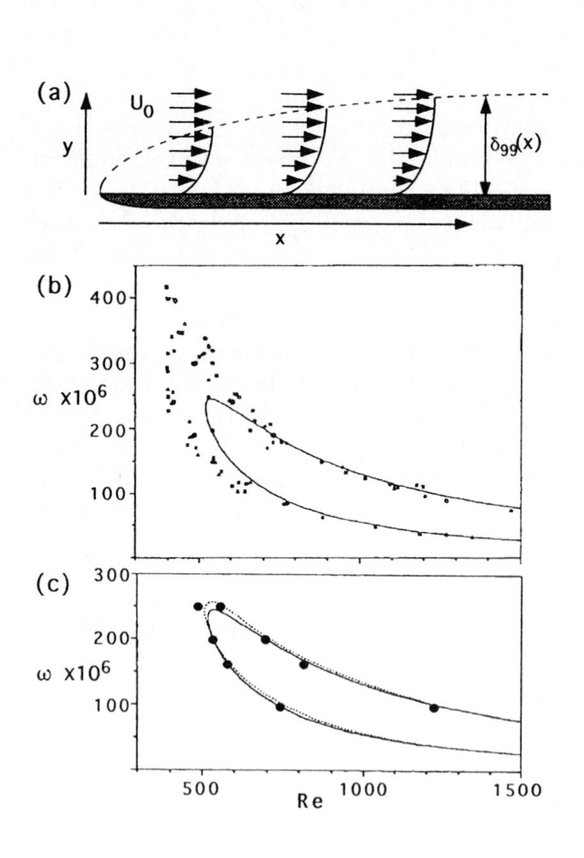

FIGURE 54 Boundary layer instability. (a) Boundary layer profile. (b) Marginal stability boundary: the axes are switched compared to Figure 50, as is the convention in shear flow literature. Within the "tongue," the boundary layer is unstable to waves excited by a thin ribbon vibrating at frequency ω. Data from several sources[127,136,221,235] is compared to linear theory (solid curve). The discrepancy at low Reynolds number is apparently due to a streamwise pressure profile that is different from the uniform pressure of an ideal boundary layer.[136] (c) Much better agreement is obtained when a downstream flap (not shown in (a)) is used to ensure that the pressure profile is uniform except very close to the leading edge. The solid curve is for linear theory with the parallel flow approximation, while the dotted curve includes corrections for non-parallel flow.[95] [After Klingmann et al.[136]]

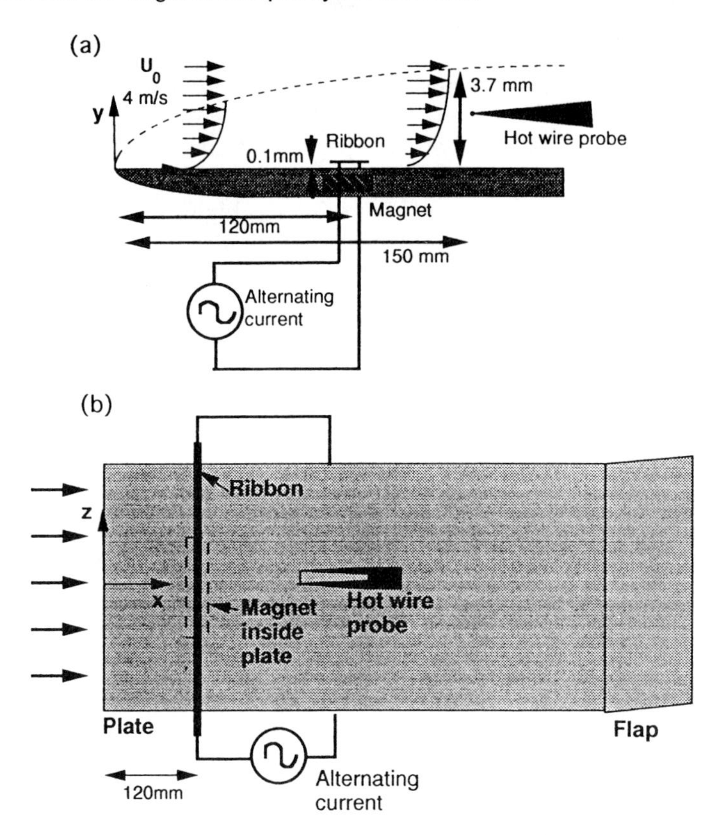

FIGURE 55 Schematic of the boundary layer experiments of Klingmann et al.[136] (a) Cross-section: the ribbon is made from phosphor bronze, is 0.05 mm thick, and is attracted to or repelled from the magnet embedded in the plate as current is run through the ribbon. The hot wire probe detects resulting fluctuations in the velocity in the boundary layer: the sensing wire, supported by two needles, is 5 μm in diameter and 1 mm long. Note that the horizontal and vertical scales are much different. (b) Plan view (with a different horizontal scale). The flap on the trailing edge was used to control the pressure profile along the plate.

is crossed. The shape of the marginal curve has this "doubling-back" characteristic because, for high-Reynolds numbers, the flow approaches the inviscid limit, where the profile is stable because it contains no inflection points! In practice, once two-dimensional TS waves are launched upon crossing the lower branch at some point x, there follows an evolution downstream into three-dimensional motion and turbulence.

The classic experiments of Schubauer and Skramstad firmly established the existence of TS-waves in boundary layers. They placed a metallic ribbon across the span of the incoming flow (Figure 55); the ribbon could be vibrated at precise frequencies. The downstream response could be measured, as a function of frequency, with hot wire probes; these respond very sensitively to the cooling action of fluid passing the very thin wire of each probe and give a signal that measures the strength of the fluid velocity. The comparison of Schubauer and Skramstad's data with the marginal stability curve (Figure 54) was highly encouraging, although exact quantitative agreement was not obtained. This disagreement has sometimes been attributed to the "parallel flow" assumption that ignores the slow streamwise change in thickness of the boundary layer profile. Corrections for this assumption (for example, Saric and Nayfeh[228]) seemed to bring better agreement, but the situation may have more to do with control of the conditions (such as the pressure uniformity) at the leading edge.[136]

Leaving these details aside, we next consider the important question of the characteristics of the three-dimensional motion that eventually appears. First, Figure 56 shows a profile of the intensity of fluctuations in the streamwise coordinate as a function of distance y away from the wall. These fluctuations—still a two-dimensional pattern—grow slowly downstream, because the spatial growth rate

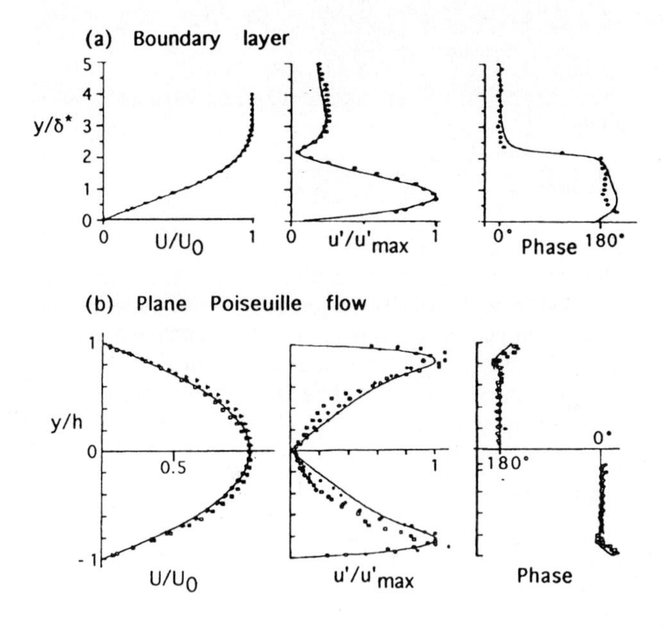

FIGURE 56 Profiles of mean flow and fluctuations in two types of wall-bounded shear flow. (a) Boundary layer flow. From left to right: mean flow U as a fraction of free-stream velocity U_0 (solid curve is the Blasius profile); root-mean-square fluctuation u' of the streamwise velocity component compared to the maximum fluctuation $u'\max$, where $u'\max < 0.6\%$ of U_0; and phase of the fluctuation. (Solid curves are from linear stability theory for fluctuations and phase.) [After Klingmann et al.[136]].(b) Plane Poiseuille flow: mean flow; fluctuations; phase. [After Nishioka et al.[188]].

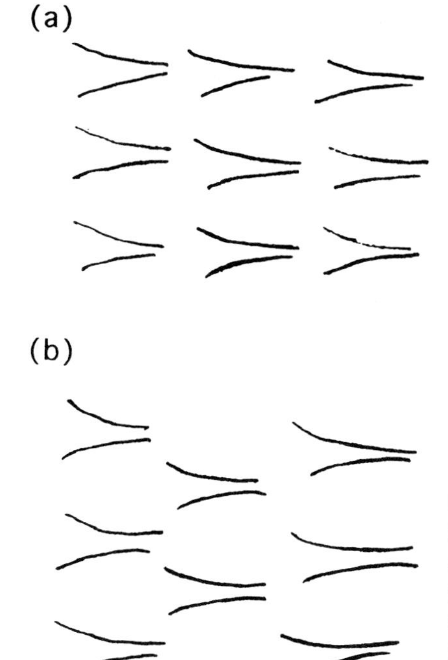

FIGURE 57 Schematic of two types of patterns formed as concentrations of vorticity turn streamwise in transitional wall-bounded shear flows. (a) K-type, where the L-structures have the same period (fundamental) as the TS waves. (b) H-type or C-type, where the L-structures have twice the period (subharmonic) as the TS waves.

αi is so small. Eventually, however, a non-uniformity appears in the spanwise (z) direction. In a remarkable experiment that produced data that, after thirty years, is still used by theorists, Klebanoff et al.[135] brought the otherwise erratic behavior of the three-dimensional disturbances under control by placing small strips of scotch-tape at intervals under the vibrating ribbon driving the instability. These strips introduced a very slight variation in the flow speed across the span, with the result that well-defined spanwise variations appeared on the TS-waves further downstream. Flow visualization reveals an aligned array of Λ-shaped features, called the K-mode (see Figure 57(a)), that point in the downstream direction; the features have been associated with the emergence of pairs of streamwise vortices. This pattern decays, within a few wavelengths, into regions of fine-scale turbulence. By measuring the fluctuations in the streamwise velocity as a function of coordinate z across the span, Klebanoff et al. obtained the data shown in Figure 58. There are peaks of large fluctuations interspersed with valleys of much more moderate fluctuations: this has been called "peak-valley splitting" in the strength of the TS-waves. Further downstream, probes placed at spanwise positions corresponding to peaks detect sudden, high-frequency fluctuations in the velocity in the form of large spikes seen on the measurement trace (see Figure 59). This is the signature of the cascade to fine-scale turbulence and is thought to arise from a fast instability of inflection points that have now developed locally within the flow.

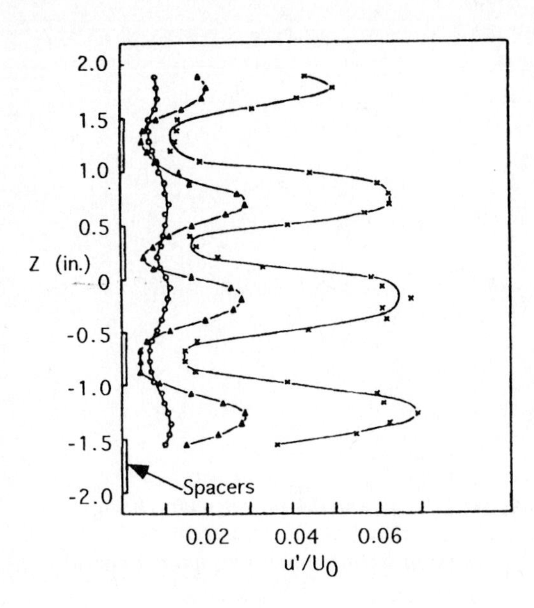

FIGURE 58 Spanwise variation in the strength of fluctuations in boundary layer flow. The fluctuations are excited by a vibrating ribbon beneath which thin cellophane spacers have been placed at one inch intervals. Three-dimensionality is thus induced in a controlled way into the initially two-dimensional TS-waves. The intensity of the fluctuation forms "peaks and valleys" that grow with distance downstream from the ribbon: measurements were made at 3 (E), 6(C), and 7.5(I) inches from the ribbon. The Reynolds number at the ribbon (based on boundary layer displacement thickness) was 1635 and the dimensionless excitation frequency was 57.6 x 10^{-6}. [After Klebanoff et al.[135]]

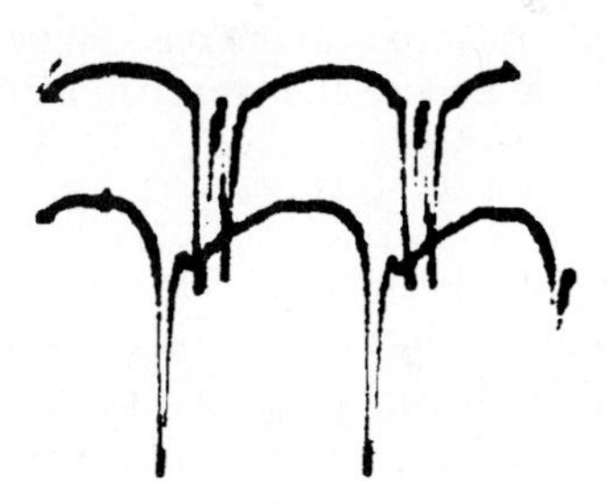

Signal from two slightly-separated probes.

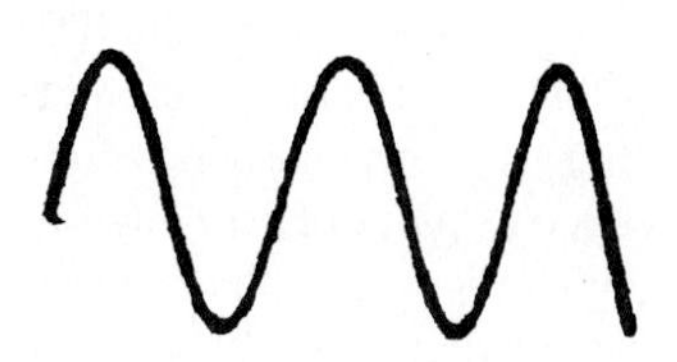

Signal to vibrating ribbon.

FIGURE 59 Spikes in hot-wire probe velocity signals (top two traces) signify fast excursions in velocity associated with the formation and development of streamwise or hairpin vortices in boundary layer flow. Upstream, the velocity measurements appear nearly as sinusoidal as the signal (bottom trace) to the vibrating ribbon. [Klebanoff et al.[135]]

Many years later, a completely different scenario was found[129,229]: instead of an array of Λ structures aligned with one another at each streamwise wavelength, a staggered array (see Figure 57(b)) was found. Velocity measurements at a single point in the flow showed a spectral component at half the frequency of the original TS-fluctuations. Thus it appears that this three-dimensional pattern grows as a subharmonic (spatially and temporally) of the TS waves. This mode is associated with various names (depending on the theoretical description), including "C-mode," "H-mode," and "N-mode."

The theory of these three-dimensional modes rests on two fundamentally different approaches. One is the idea of wave interactions, beginning with a model of Benney and Lin[18] that described the interaction of the TS wave of wavenumber $\alpha\hat{\mathbf{x}}$ with oblique waves with wavenumber $\alpha\hat{\mathbf{x}} \pm \beta\hat{\mathbf{x}}$ to produce the observed spanwise variations. However, the frequencies cannot be combined to keep the waves synchronized. However, another way of describing wave interaction that seems to work for the subharmonic mode was discussed by Craik,[59] involving oblique waves with half the axial wavenumber: $\alpha\hat{\mathbf{x}} \pm \beta\hat{\mathbf{z}}$. By choosing an appropriate value for the spanwise wavenumber β, it is possible to arrange that the sum of the frequencies of the oblique waves matches the frequency of the TS wave. Note that the sum of the oblique wavenumbers also matches the TS wavenumber. This is called three-wave resonance, a topic we will visit again in the context of surface waves.

A much different approach by Herbert has been described in his recent review[113]: the onset of three-dimensionality is treated as a secondary instability of the TS waves. In doing the linear stability theory of such a transition, one must replace the original (near) translational invariance of the original boundary-layer flow with a base state (boundary layer profile modified by TS waves) that is spatially periodic, with wavelength $\lambda = 2\pi/\alpha$. This leads to linear differential equations with periodic coefficients, whose solution is based on the ideas of Floquet theory (see, for example, Stoker,[243] p. 193ff). Instead of looking for eigenmodes whose x-dependence goes simply as $\exp(i\gamma x)$, one instead finds modes with x-dependence $\exp(i\gamma x)F(x)$, where the function $F(x)$ is periodic with wavelength $\lambda : F(x + \lambda) = F(x)$. The parameter γ, which could be complex if we consider spatially growing modes, has an interesting restriction:

$$\frac{-\alpha}{2} < \gamma_r \leq \frac{\alpha}{2}. \tag{83}$$

Adding any integer multiple of α to γ_r simply produces a periodic factor that can be absorbed into the function $F(x)$. Interestingly, this same theoretical machinery is fundamental to the quantum mechanical treatment of electron wavefunctions in crystalline lattice, and the associated eigenmodes are called "Bloch waves." The inequality (Eq. (83)) is said to restrict γ to the first Brillouin zone, to use solid-state physics language. In fact, this approach was implicit in the understanding of secondary instability of Taylor vortices and convection rolls, although we glossed over this aspect in our earlier discussion.

Herbert concentrates on three cases: $\gamma = 0$, $\gamma = \alpha/2$, and $\gamma = \tilde{\gamma}$, where $\tilde{\gamma}$ takes on any other value in the first Brillouin zone. We'll omit discussion of this last case, referring the reader to the article by Herbert.[113] The case $\gamma = 0$ leads to a mode with the same period as the underlying TS state and hence might be associated with the K-mode. The case $\gamma = \alpha/2$ corresponds to a spatial subharmonic and might then correspond to the second mode of breakdown. These two types of three-dimensional disturbance are then driven by a process called "parametric resonances" (we will revisit this again below in discussing Faraday waves), with the subharmonic resonance more likely to occur for smaller amplitude disturbances. Many aspects of this behavior, including the dependence on disturbance amplitude, appear to agree with experiment and are recovered in the spatial numerical simulations of Saiki et al.[226] for the related problem of plane channel flow. Because of the natural connection between the idea of secondary instability of TS waves to the secondary instabilities in the Couette-Taylor and Rayleigh-Bénard problems, I find this approach very appealing.

Overall, the three-dimensional secondary instabilities of TS-waves remain a tricky and active subject. A word of caution has been voiced by Kachanov[128] in an article that is full of physical insights on the process of breakdown of TS waves to turbulence. The two essential modes of breakdown may occur as a result of several mechanisms, with perhaps several of the above-mentioned ideas playing a role. In addition to Kachanov's review, the reader may consult the articles by Bayly et al.[9] and Herbert[113] as well as the book by Craik.[60]

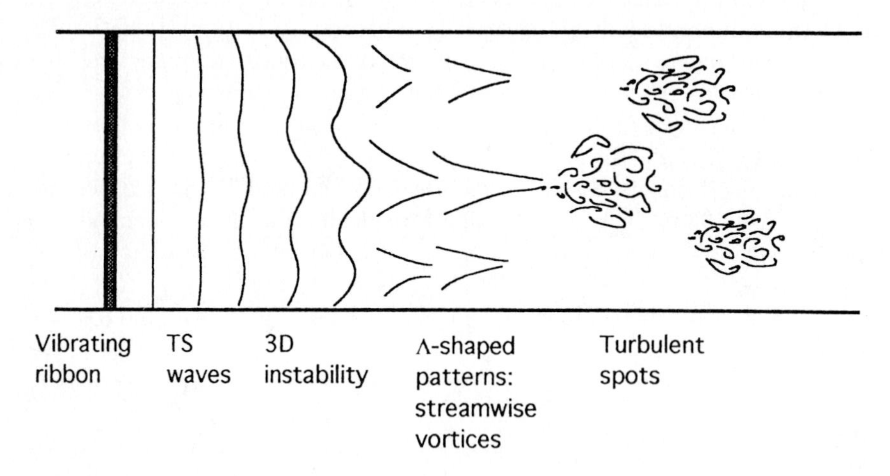

FIGURE 60 A schematic of the evolution of wall-bounded shear flows towards turbulence.

FIGURE 61 A turbulent or "Emmons" spot in a flat plate boundary layer.[36]

A summary sketch of the transition process in boundary layers and other wall-bounded shear flows is shown in Figure 60. Once turbulence begins in earnest, there can be the striking phenomena of turbulent "spots": confined regions of fine-scale turbulence separated by well-defined boundaries from quasi-laminar flow (see Figure 61). The laminar region is not, however, without structure. As mentioned earlier, it would be interesting to try to find a connection between the formation of these spots and the turbulent bursts that occur between counter-rotating cylinders in the Couette-Taylor system.

As another important case of wall-bounded shear flows, we must also look at the case of flow within ducts, such as pipes and channels. Reynolds landmark experiments on the existence of a transition to turbulence were conducted in a pipe flow apparatus shown in Figure 62(a). A schematic of a modern water-tunnel apparatus is shown, for comparison, in Figure 62(b). This is in fact an apparatus for studying flows between parallel plates. This prototype has key advantages for studying duct flows: the geometry is simple and the basic flow is rigorously parallel

(a)

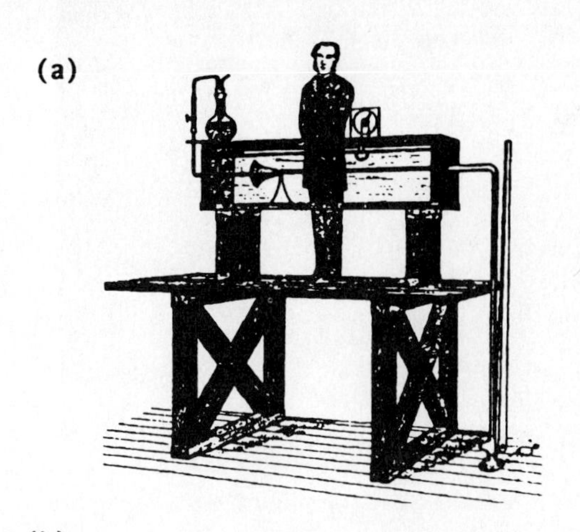

(b)

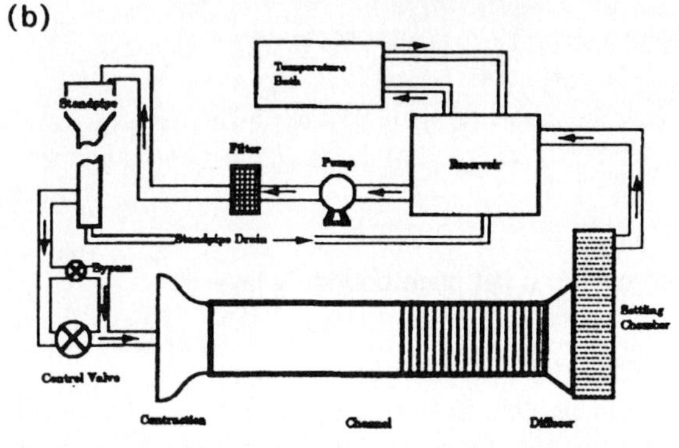

FIGURE 62 Pipe and channel flow experiments "then and now." (a) Reynolds' apparatus for studying the onset of turbulence in pipe flow.[215] (b) Schematic of a modern water tunnel with rectangular cross-section.[230]

and can be written down in closed form, assuming the channel has infinite span. The so-called plane Poiseuille flow profile is given by:

$$u = U_0[1 - (y/d)^2], \quad v = 0, \quad w = 0; \tag{84a}$$

$$p = p_0 - \frac{2\rho\nu U_0}{d^2}. \tag{84b}$$

Here the separation between the plates is $2d$, so $-d \leq y \leq d$; U_0 is called the centerline velocity. The Reynolds number is defined as $\mathrm{Re} = U_0 d/\nu$.

Linear stability computations based on the Orr-Sommerfeld equation yields a critical Reynolds number of $\mathrm{Re}_c = 5772$. Actually, accurate computation of this value is a delicate matter and was not accomplished until the late 1960s and 1970s (see Orszag[194]). Even more delicate was the experimental verification, which first came with very precise experiments of Nishioka et al.[188] In these experiments, the residual disturbance level in the flow entering the channel was reduced to below 0.1%. In such a quiet channel, Nishioka et al. could observe plane Poiseuille flow all the way up to Re ~ 8000, well beyond the predicted instability. We now understand that this is because the instability is of the convective type, so that growing disturbances could not be observed before they were washed out of the channel. By introducing disturbances by the technique of putting a vibrating ribbon across the upstream flow, Nishioka et al. were able to determine the marginal curve shown in Figure 63, which agrees well with theory.

However, it is well known that instability occurs all-too-readily in channels at Reynolds numbers as low as 1000. We might first try to explain this as the consequence of a subcritical instability. Nonlinear computations by Herbert[112] and others have demonstrated that the transition is in fact subcritical, but a plot of saturated secondary TS mode amplitude versus Reynolds number Re and wavenumber α reaches a minimum value of Re $= 2935$ at wavenumber $\alpha = 1.32$. This plot, which looks like a nosecone, is shown in Figure 64. The tip of the nosecone is well above the Reynolds number of 1000 where a transition may already be seen.

There are several possibilities for explaining the lower value of transition that can be obtained. One is the idea of transient algebraic growth of disturbances that was discussed above in our introduction to shear flow instabilities. Such an event, perhaps triggered initially by moderate background disturbances, can so alter the flow that further instabilities occur and turbulence results. Another idea is based on

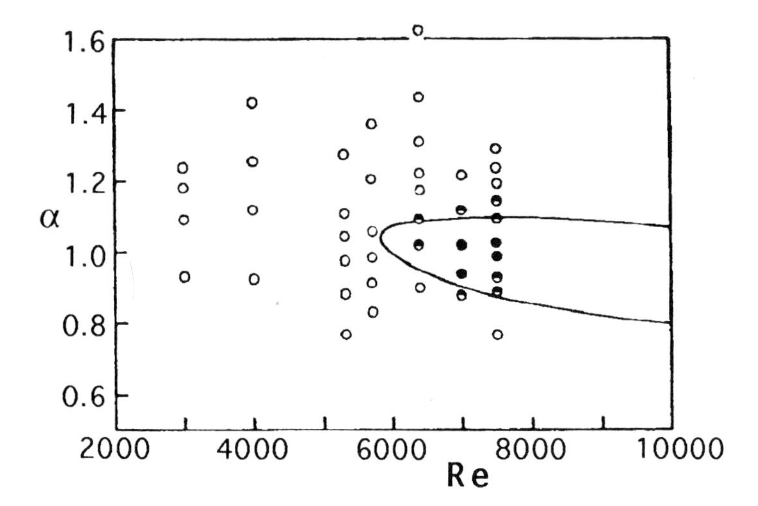

FIGURE 63 Marginal stability curve for plane Poiseuille flow. Circles are experimental points from Nishioka et al.[188]: open circles denote decyaing disturbances, filled circles denote amplifying diturbances, and half-filled circles are nearly neutral with respect to growth and decay. The theoretical marginal curve is due to Ito.[123] [After Nishioka et al.[188]]

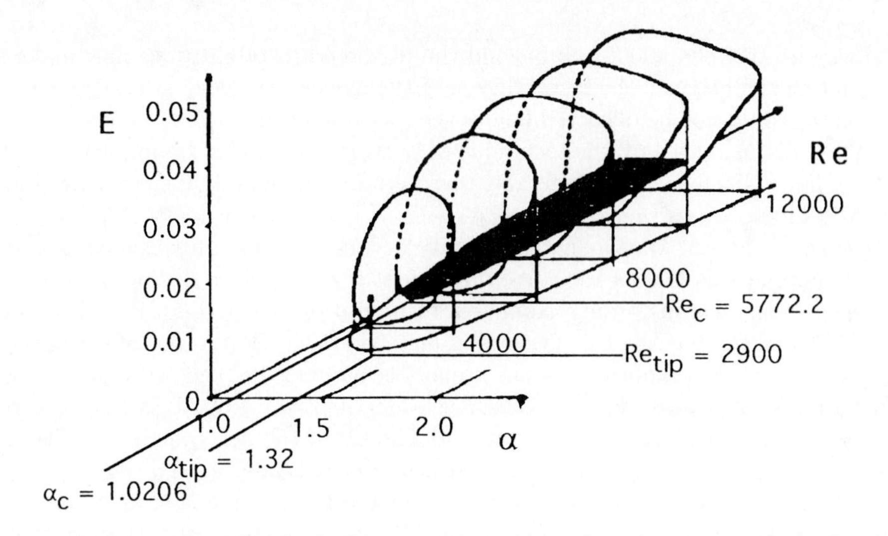

FIGURE 64 "Nose cone" locating nonlinear two-dimensional secondary flows in plane Poiseuille flow by their energy E as a function of Reynolds number Re and wavenumber a. The plane $E = 0$ corresponds to the undisturbed plane Poiseuille flow. Within the shaded region beginning at Reynolds number 5772.2, this flow is unstable to infinitesimal disturbances; this corresponds to the region bounded by the neutral stability curve of Figure 63. However, at lower values of the Reynolds number (down to 2900), finite disturbances can still lead to a secondary flow state. The secondary flows on the underside of the nose cone are unstable, while the flows on the topside are stable. [After Bayly et al.,[9] based on Herbert.[112] Values of the critical Reynolds number and wavenumber ac are from Orszag.[194]]

a numerical result[195]: even below Re_c, TS waves decay very slowly and down to a Reynolds numbers of about 1000, a decaying TS wave of large enough amplitude can become unstable to three-dimensional disturbances that in turn lead to turbulence. However, the required amplitude of the precursor TS wave may be unrealistically large compared to disturbances in typical flow systems. It is likely that further experimentation will be needed to clarify these matters.

Returning to the case of well-controlled experiments that launch TS-waves, the ensuing development of the flow has many similarities to boundary layer flows, including the appearance of Λ-patterns with associated streamwise vorticity and the existence of turbulent spots. Thus it seems that these phenomena are generic to wall-bounded flows.

So far, we would conclude that duct flows have linear instability at high Reynolds numbers but are characterized by real transitions occur much sooner, at Reynolds numbers of order a 1000. In fact, linear stability gives $\mathrm{Re}_c = \infty$ for flow in a circular pipe and flow in the so-called plane Couette problem (where one wall slides at uniform velocity and the other is stationary). Nonetheless, transition occurs quite readily, although it has been possible to extend smooth pipe flow up to Re $\sim 50,000$.

Sometimes one desires to go to the other extreme. For purposes of mixing and heat transport, it would be nice to excite a transition in a channel at fairly low Reynolds number in order to obtain a stable secondary flow that exists over a useful range of Reynolds numbers. Such a secondary flow could efficiently advect chemicals or heat from one side of the channel to the other without requiring high Reynolds numbers and consequent high drag due to turbulence. Patera and coworkers (see, for example, Karniadakis et al.[130]) had the intriguing idea of achieving this goal by building channels that contained streamwise spatially periodic disturbances ("eddy promoters"), such as grooves or an array of cylinders; they explored these geometries numerically and experimentally and found that the transition could indeed be drastically reduced and a saturated secondary flow could occur. Experiments by Schatz et al.[231] with the geometry shown in Figure 65 were in good quantitative agreement with the numerical computations; these experiments further established that the secondary flow occurred via a supercritical Hopf bifurcation. The instability was still of the convective type, suggesting a natural extension of the fundamental ideas of convective and absolute instabilities to spatially periodic domains. Because they produce a stable, saturated secondary flow, this system and the cylinder wake flows to be discussed next are the two principal cases in open flows that most resemble the behavior in closed flows like the Couette-Taylor and Rayleigh-Bénard problems.

5.5 WAKES

Wakes can be dangerous: they can upset small aircraft that follow larger aircraft, they can interact with the object producing them so as to cause large and destructive mechanical vibrations, and wakes behind boulders in fast-moving rivers contain recirculating eddies that can trap unfortunate rafting enthusiasts. Trying to understand wakes can also be dangerous: controversies have arisen over the Reynolds number dependence of wake vortex shedding frequencies and over the interpretation of time-series measurements as the onset of chaos in wake flows. Thus we will deal with this important topic rather cautiously. Fortunately, wakes have also been the topic of some beautiful experiments, computations, and theories; one review is by Oertel.[193] As we saw in Figure 3, wake flows are among the visually most striking phenomena in fluid dynamics.

Three conventional objects for generating wakes in laboratory studies are cylinders, spheres, and flat plates.[6] Cylinder wakes have by far been the most studied

[6] Note that we are not talking about free-surface wakes, that is, the v-shaped waves that propagate from the stern of a motorboat and cause consternation amongst genteel folk like me who prefer sailing and rowing. Instead, we are looking at the internal disturbances created when fluid must pass around a body.

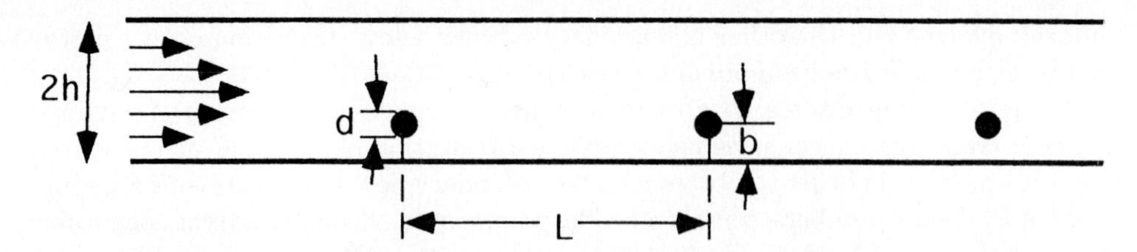

FIGURE 65 Eddy promoter geometry: a periodic array of cylinders of diameter d is spaced at intervals L and a distance b above one wall of a plane channel of width $2h$. The presence of these cylinders induces transition to a time dependent flow at a Reynolds number much lower than is the case for an ordinary channel. Moreover, the transition is now supercritical. In the experiments and simulations reported in Schatz et al.,[231] $d/h = 0.4$, $b/h = 0.5$, and $L/h = 6.66$; this gave a transition Reynolds number close to 130, compared to the value 5772 for a plane channel.

and this is the geometry we will focus upon; they are prototypes for so-called "bluff-body" wakes and start out (at low enough Reynolds number) as essentially two-dimensional flows, like the shear flows discussed above. Note that we can think of the cylinder moving through a stationary fluid or of the fluid moving past a stationary cylinder. Experimentally, the first point of view is accomplished in a towing tank, while the second may be established in a wind- or water-tunnel. Interpretation of streamlines of flow is much easier assuming a stationary cylinder, as we will do in the following, but it is useful to remember that our findings are also relevant to the important cases of objects (like cars and airplanes) moving through fluids at rest.

We assume that far from the cylinder, fluid flows uniformly in the x-direction with velocity $U_0\hat{\mathbf{x}}$; the same coordinate system is used as in Figure 47. The cylinder has diameter d and is assumed, at first, to have infinite span. Then we define the Reynolds number by $\text{Re} \equiv U_0 d/\nu$, where ν is the kinematic viscosity of the fluid. Figure 66 gives a schematic illustration of the development of the cylinder wakes as the Reynolds number is increased.

When $\text{Re} = 0$, the system has reflection symmetry "fore" and "aft" of the cylinder. Is this symmetry immediately broken when the fluid flow is turned on? Flow visualization photos for very low Reynolds number ($Re \ll 1$)—the "Stokes" flow regime where viscosity dominates the flow—look quite symmetric. By looking at the streamlines, you cannot tell which direction the fluid is moving. Actually, the Stokes equation, obtained by neglecting the inertial term $(\mathbf{u} \bullet \nabla)\mathbf{u}$ in Eq. (7), leads to a solution that is not uniformly valid for all radii from the surface of the cylinder to infinity. The reason for this, physically, is that the cylinder perturbs the flow at quite large distances, so that at sufficiently large radius, the inertial terms can no longer be neglected. For detailed discussions of the theory of the very low Reynolds number regime, see Illingworth.[122]

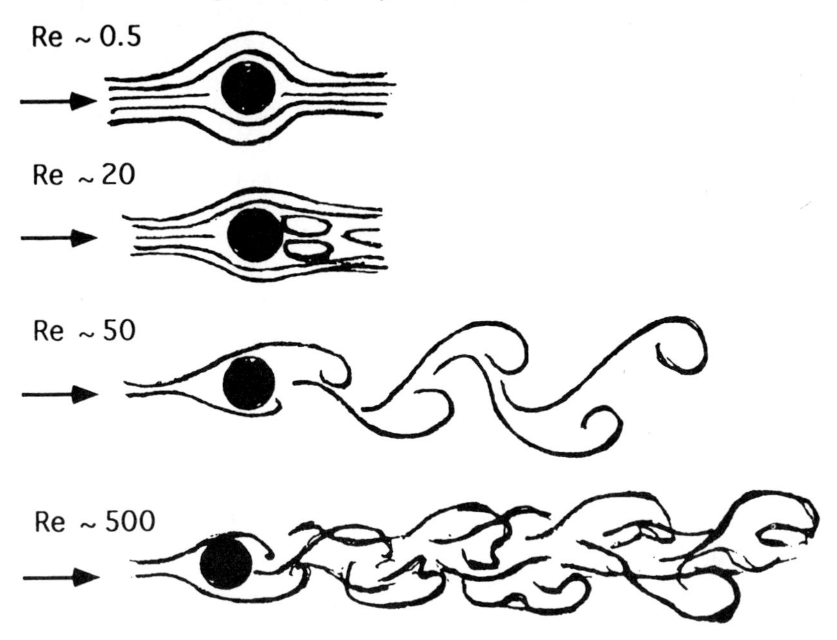

FIGURE 66 Schematic of the evolution of the wake of a circular cylinder for increasing Reynolds number. At very low Reynolds number, say Re $\approx$ 0.5, the flow streamlines appear symmetric fore and aft of the cylinder. At Reynolds number 20, there are clear recirculating eddies behind the cylinder, but the flow remains steady in time. At Reynolds number 50, vortices are now periodically shed from the region near the cylinder, producing what is sometimes known as a "von Karman vortex street" in the wake of the cylinder. At Reynolds number 500, the wake is turbulent, although some larger structures may still be evident.

At Re $\sim$ 4 recirculation eddies are evident. Experiments by Coustanceau and Bouard[57,58] and computations by Fornberg[92] show that the length of the recirculating eddies grows linearly with Reynolds number. Extrapolation back to zero length can then be used to estimate the Reynolds number where separation of flow from the cylinder surface might first occur, forming the eddies. However, the existence of such a transitional value of Re remains somewhat in question.

What happens next is also a little ambiguous in the literature. Plaschko et al.[205] cite an early work by Homann[118] in which oscillation of the eddies was observed to begin at $Re \sim 22$. There was, however, no shedding of vortices into the downstream flow. Also above this Reynolds number, Nishioka and Sato[189] observed that disturbances imposed by small cross-stream vibrations of the cylinder would grow spatially for several diameters downstream before finally decaying. By now, however, it is well established that above Re $\sim$ 46.7, vortex shedding begins and a self-sustaining oscillatory pattern appears that can extend hundreds of diameters downstream.[7] The actual onset of shedding, is, however, sensitive to the aspect ratio (length/diameter) of the cylinder and experimental data must be systematically extrapolated to a value for infinite aspect ratio. Numerical computation of the transition is in fairly good agreement with experiment; see, for example, Jackson.[124]

Once shedding begins, an obvious quantity for experimental measurement is the frequency f with which vortices pass a probe placed downstream of the cylinder. A well-defined shedding frequency can be detected over a wide range of Reynolds numbers, far into the regime of turbulent flow[266]; indeed, measurements of this frequency have become the basis of so-called vortex flowmeters. The nondimensional form for the vortex frequency is called the Strouhal number and may be defined as

$$S \equiv f \frac{d}{U_o}. \tag{85}$$

Roshko[219,220] found an empirical relation between Strouhal number and Reynolds number,

$$S = 0.212 - \frac{4.5}{\mathrm{Re}}, \tag{86}$$

fitting data in the range $40 < \mathrm{Re} < 150$. The nature of this relation is clarified if we use viscous time scales, defining a "Roshko" number

$$Ro \equiv f \frac{d^2}{\nu} = \mathrm{Re} \cdot S, \tag{87}$$

so that Eq. (86) becomes

$$Ro = 0.212\mathrm{Re} - 4.5. \tag{88}$$

This linear relation is precisely the result one expects from a supercritical Hopf bifurcation described by the cubic order Landau equation (Eq. (34)).

Tritton[265] observed discontinuities in the plot of Strouhal number versus Reynolds number at $\mathrm{Re} \sim 90$ and he speculated that the mechanism of shedding changed at the point where these discontinuities were observed. Debate emerged when Gaster[94] did experiments that indicated that such discontinuities might be due to spanwise nonuniformity in the base flow. The situation seems to have been resolved recently by Williamson,[279] Eisenlohr and Eckelmann,[85] and Hammache and Gharib[108]: jumps to lower frequency could be accounted for by vortices emerging obliquely rather than parallel to the cylinders (see Figure 67(b)). If the angle away from parallel shedding is θ, then Williamson found that all of the

[7] Caution is needed, however, in interpreting flow visualization patterns in wake flows. Far enough downstream, the dye streaks no longer represent actual vortices but are instead "fossil records" of vortex motion nearer the cylinder.

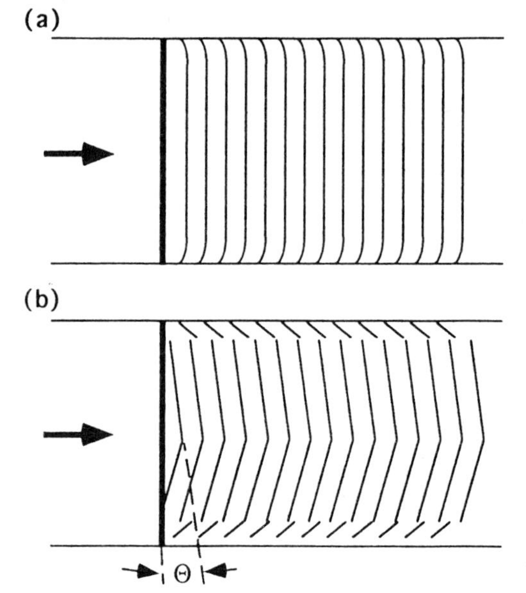

FIGURE 67 Vortex shedding patterns: the heavy line marks the cylinder causing shedding as flow comes past from the left. In experiments, the lines marking the vortices are visualized by concentration in the vortices of dye released from the cylinder wall. (a) Parallel shedding. (b) Oblique shedding forming a chevron pattern.

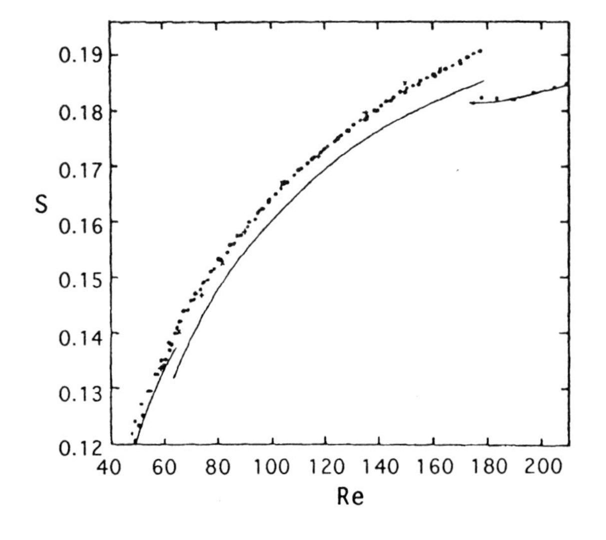

FIGURE 68 Strouhal number (dimesnionless vortex shedding frequency) versus Reynolds number: data collapse onto a single curve below Reynolds number 180, so long as the measured Strouhal number S_{meas} is renormalized by $S = S_{\text{meas}}/\cos\theta$, where θ is the shedding angle. Otherwise, the data falls onto separate curves (the solid curves) with a break near Reynolds number 63. [From Williamson.[279]]

data could be fit (see Figure 68) by a "universal" curve $f_{\|}(\mathrm{Re})$ by the conversion $f(\mathrm{Re}) = f_{\|}(\mathrm{Re}) \cos\theta$. These workers also found that by manipulating the boundary conditions at the ends of the cylinders, it was possible to obtain vortices parallel to the cylinders from the onset of vortex shedding at about Re = 47 up to Re = 180 (see Figure 67(a)).

The nature of the instability that produces vortex shedding has some interesting subtleties. We commented earlier that it was important, especially in open flows, to distinguish between convective and absolute instabilities. This distinction was strictly defined only for parallel flows with translational invariance in the x-direction, although evidently applicable to *nearly* parallel flows such as boundary layers. How, then, should the wake flow instability be classified? One idea[139] is to examine the time-averaged velocity profile across the wake at a given streamwise position x and determine its "local" stability properties as if the profile described a truly parallel flow. The local profile is plugged into the Orr-Sommerfeld equation and the stability characteristics examined. Then it is necessary to establish a criterion for when a region of local instability can make the whole flow unstable; when this occurs, the flow is said to be "globally" unstable. Various refinements of this idea have been developed (see Huerre and Monkewitz[121] for a review), including the exploration of models based on complex Ginzburg-Landau equation that allowed the parameters to vary slowly in space.[44] The idea that emerges is that a sufficiently large region of local *absolute* instability must develop in the near wake just behind the cylinder before self-sustained oscillations appear downstream in the form of vortex shedding. In a sense, the finite domain of absolute instability acts as an oscillator that drives the flow well into the region where local stability analysis predicts either convective instability or even stable base flow. Note that nowhere in this discussion has instability been connected to events happening at the surface of the cylinder itself; it seems that it is the shear in the wake profile, which includes inflection points, that is responsible for instability.

Strictly speaking, this point of view assumes that the streamwise development of the average profile is slow compared to the characteristic wavelength of the instability—an assumption that is not entirely valid for wake flows.[81] However, the qualitative picture seems correct and even justifies the description of global instability to vortex shedding as a supercritical Hopf bifurcation described by a Landau equation whose coefficients do *not* depend on streamwise position x. This has been confirmed by some beautiful experiments reported by Mathis et al.,[172] Provensal et al.,[208] and Strykowski and Sreenivasen.[246] These experiments are good examples of the delicate manner in which one establishes that a transition corresponds to a supercritical Hopf bifurcation; to illustrate the generality of the approach, similar techniques were employed for the eddy promoter flow[231] mentioned earlier. In Figure 69 we show the response of the wake flow to pulselike disturbances and steps in the Reynolds numbers; from such measurements the subcritical decay rate and supercritical growth rate can be determined, producing the plot of the imaginary part of the eigenvalue σ given in Figure 70. From this plot, the transition Reynolds number Re_c is determined by the intercept ($\sigma = 0$) with the Reynolds number axis. The experiments also established the characteristic square-root dependence of the saturated amplitude on $\mathrm{Re} - \mathrm{Re}_c$ (see Figure 71). As noted above, the initial linear relation between shedding frequency and Reynolds number is also consistent with the Landau equation, although the correction to the slope due to the cubic saturation term was found to be negligible. Eventually, of course, nonlinear effects must be important, as indicated by the curvature in Williamson's plot of Roshko number versus Reynolds number (see Figure 68).

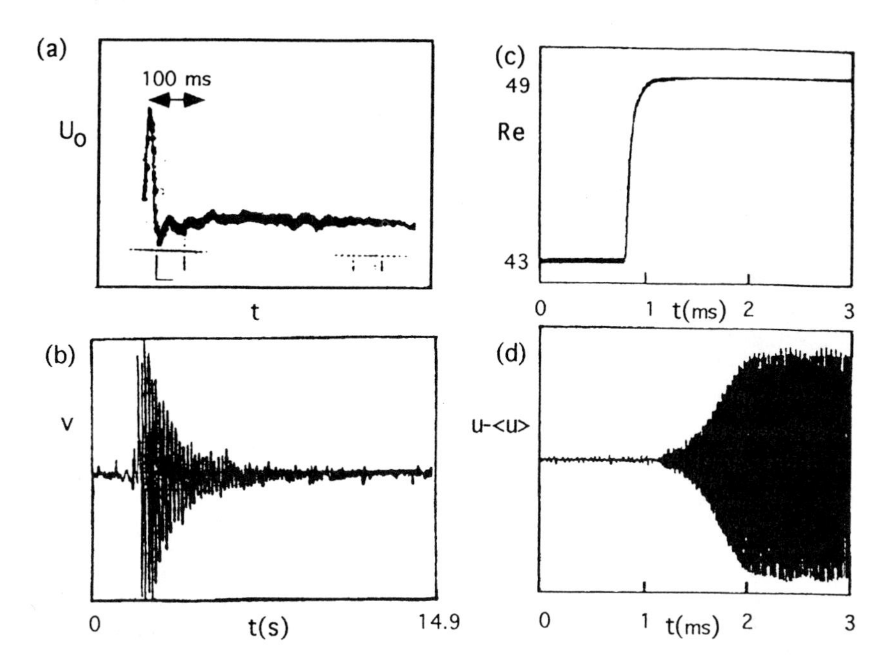

FIGURE 69 Response of flow behind a cylinder to impulses and steps in the upstream flow velocity. (a) A short impulse is applied at a subcritical Reynolds number $\mathrm{Re} = \mathrm{Re}_c - 1.4$. (b) The response in the transverse velocity component is measured using laser Doppler velocimetry five diameters downstream from the cylinder. Note that the response is an exponentially decaying envelope. [After Provansal et al.[208]] (c) In another approach, the Reynolds number is suddenly stepped past the critical value. (d) The response shown is from measurements of the fluctuation in the streamwise velocity component using a hot wire probe ten diameters from the cylinder. After an initial exponential growth in amplitude, the fluctuations eventually saturate as a steady oscillation. [From Strykowski and Sreenivasan.[246]]

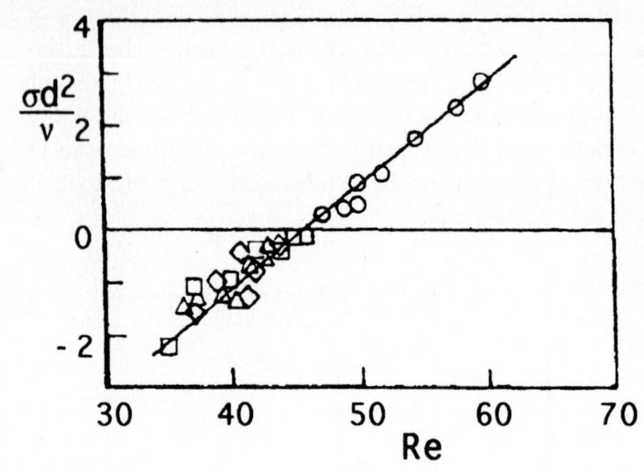

FIGURE 70 Decay and growth rates inferred from data such as that shown in Figure 69. The Reynolds number for zero growth rate is the critical Reynolds number, here measured to be 46 for a cylinder aspect ratio (length/diameter) of 60. [From Strykowski and Sreenivasan.[246]]

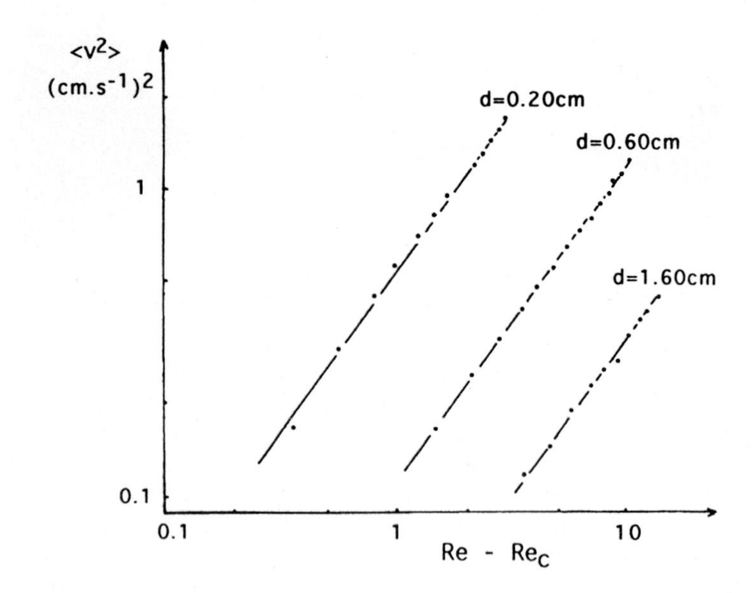

FIGURE 71 Mean square fluctuation in the transverse velocity versus Reynolds number for the wake of cylinders of different diameters. Each line in this log-log plot has a slope 1.00 ± 0.02. Thus the amplitude of the fluctuation is demonstrated to follow the Landau-type dependence of (Re - Re_c)$1/2$. [After Mathis et al.[172]]

The array of vortices aligned parallel or oblique to the cylinder becomes an ideal candidate for all sorts of questions on pattern formation. It is observed (see, for example, Williamson[280] and references therein) that in the far wake, parallel vortices become unstable to a two-dimensional pattern of much longer wavelength. It is tempting to speculate that this as an Eckhaus or "modulational" instability as the initial wavenumber is no longer stable for the evolving wake profile; however, I don't know if this is the case. The situation is complicated when the upstream wake contains oblique vortices, but there is still a shift to large length-scale features; Williamson has argued that this is not due to vortex merging of the sort that occurs in mixing layers.

Meanwhile, in the near wake, there has been a lot of discussion recently of the formation of "cells" or domains of vortices that have different shedding angles (see Figure 67(b)). One aim is to understand the effect of the side walls supporting finite length cylinders, whose influence can spread to the center of the wake. Most of the work is based on models that describe the vortex shedding as a system of oscillators coupled in the spanwise direction[190,5,200]; this coupled-oscillator approach was first proposed by Gaster[94] to describe the shedding from tapered cylinders. It appears that a connection of these models to the underlying Navier-Stokes equations remains an open challenge. A different model is described by Lefrançois and Ahlborn[159]: it is based on a "diffraction theory" whereby a given vorticity field induces a new vortex pattern—in a manner analogous to Huygen's principle for describing the advance of phase fronts in light waves. One important experimental tool to support such models is the ability to profile the velocity across the span with a scanning laser doppler system reported by Yang et al.[282] Overall, the subject of pattern formation in low Reynolds number wakes is undergoing rapid development and ought to be the subject of a critical review in the near future.

The next step in the development of wakes flows as Reynolds number is increased is somewhat obscured by the spatio-temporal effects discussed above. This is analogous to the evolution of buoyancy-driven convection, where the idealized evolution of two-dimensional convection rolls with increasing Rayleigh number must be reconciled with dynamics associated with the behavior of patterns in large but finite containers. In any case, it appears that near $Re_{3D} \sim 180$,[278] parallel vortex wakes undergo a secondary instability leading to spanwise variations of the velocities and creating an intrinsic three-dimensional flow pattern that is not driven by wall effects. A good "order parameter" is the intensity of fluctuations of the spanwise velocity component: this is zero below the secondary transition and appears, in numerical simulations, to saturate above the transition.[131] The question of a preferred spanwise wavelength has been examined computationally by Noack et al.,[191] giving a critical wavelength of 1.8d and a critical Reynolds number $Re_{3D} = 170$. Experimental flow visualization by the same authors is consistent with these results, showing pairs of streamwise vortices superimposed on the primary wake structure. Another interesting finding by Noack et al. was that the two-dimensional wake flows are neutrally stable to very large wavelength spanwise perturbations, which is identified with the tendency of wake flows to form the above-mentioned "cells" of oblique vortices.

The numerical work of Karniadakis and Triantafyllou[131] also indicates a jump in frequency of shedding at the onset of short wavelength three-dimensional patterns, followed by a period-doubling cascade to chaotic flow. Experimental verification of this scenario must still be done; period doubling was reported, however, for a forced, tapered cylinder by Rockwell et al.[218] The onset of chaotic dynamics in wake flows has long been a tricky subject. Sreenivasen[239] reported—for a pockets of Reynolds number values such as the range 54 to 102—the appearance of a second incommensurate frequency, then a third, and finally broad band spectra. This seemed to indicate the onset of chaos in the wake just behind the cylinder via the Ruelle-Takens-Newhouse scenario similar to that observed in the Couette-Taylor system. Above such a range, order re-emerged giving only one frequency (and its harmonics) in the flow. However, experiments by van Atta and Gharib[272] showed that nonlinear coupling of the wake to elastic vibrations of the cylinder could account for the additional frequencies; when their cylinder was prevented from vibrating, no additional frequencies were found. Sreenivasen[240] states that no cylinder vibration was detected in his experiments, but acknowledges that the wall-induced effects giving rise to domains of shedding patterns are the probable origin of quasi-periodicity. Indeed, the work of Williamson[279] and others on competing oblique-wave cells has clearly identified quasi-periodicity and broad-band spectra. This is always the danger in trying to interpret the dynamics of large aspect ratio systems: spatio-temporal dynamics over long length scales can masquerade as temporal chaos if examined with single-point probes.

At higher Reynolds numbers, cylinder wakes become increasingly turbulent, with strong mixing and irregular time-dependence. However, as noted above, coherence remains embedded in the flow: a sharp frequency associated with vortex shedding is still detected. Early work on the transition to turbulence is given by Bloor.[24] At high Reynolds numbers, one notable feature is a sharp drop in drag on the cylinder that occurs at $Re \sim 3 \times 10^5$; this is explained by Tritton[266] as the onset of turbulence in the boundary layer that has formed next to the cylinder. As a consequence, the mean flow separates much farther around the cylinder, producing a narrower wake and a smaller momentum deficit in the flow behind the cylinder.

Overall, the wake flow problem remains hazardous but enticing. It has been a source of rich new ideas such concerning the connection of local stability analysis to global instability and the application of coupled-oscillator models to understanding large-scale three-dimensional effects. It seems to be one example of an open-flow system for which an ordered sequence of transitions to turbulence may occur; in this way, wake flows resemble closed-flow systems such as the Couette-Taylor problem and the Rayleigh-Bénard convection problem. Delicate issues concerning pattern formation and wall effects must still be resolved. Computations and experiments are sensitive to the precise nature of boundary conditions. Finally, the detailed understanding of the transition to turbulence remains a subject quite open to further investigation.

6. MOTION OF INTERFACES

6.1 DESCRIBING WAVES: NONLINEARITY IN THE BOUNDARY CONDITIONS

There are some fluid motions that are readily accessible to the constructively lazy experimentalist, that is, one who is a keen observer but doesn't build apparatus. One such type of motion can be obtained by turning on a water faucet. Another can be observed by looking at waves on fluid surfaces. Both of these involve fluid interfaces and we will now examine the origin of complexity in the shape of such interfaces. Aside from the fact that such problems are readily accessible to observation, we will discover that they also introduce nonlinearity in a new way. Caution should be exercised, though: there will be an overwhelming temptation to build some experimental apparatus after all!

We must first make a brief mathematical detour. Nonlinearity entered the above problems through the dynamical term $(\mathbf{u} \bullet \nabla)\mathbf{u}$. One might ask if there are circumstances where, by a change of variables, one could arrive instead at a linear equation. We will show one way this is possible, but we will find that the entire problem is still nonlinear because of the boundary conditions! Consider the vorticity field ω, defined as $\omega \equiv \nabla \times \mathbf{u}$. By taking the curl of the Navier-Stokes equation, using the identity

$$\nabla \times (\omega \times \mathbf{u}) = \omega \nabla \bullet \mathbf{u} + (\mathbf{u} \bullet \nabla)\omega - \mathbf{u}\nabla \bullet \omega - (\omega \bullet \nabla)\mathbf{u}, \tag{89}$$

the fact that $\nabla \bullet \omega = \nabla \bullet (\nabla \times u) = 0$, and the continuity equation $\nabla \bullet \mathbf{u} = 0$ (whew!), we get

$$\frac{D\omega}{Dt} = (\omega \bullet \nabla)\mathbf{u} + \nu \nabla^2 \omega. \tag{90}$$

Let us consider the approximation that the fluid is inviscid: $\nu = 0$. Then the last term vanishes. Now suppose that we start with a fluid motion that has no vorticity: $\omega = 0$ everywhere. Then the above equation implies that $\frac{D\omega}{Dt} = 0$, so there is no way for the fluid to *acquire* vorticity. A solution for all times is $\omega = 0$, or equivalently, $\nabla \times \mathbf{u} = 0$. A well-known theorem in mathematical physics then tells us that we can write such a velocity field $\mathbf{u}$ as the gradient of some *potential* ϕ: $\mathbf{u} = \nabla\phi$. Finally, then, the continuity equation $\nabla \bullet \mathbf{u} = 0$ leads us to the following linear differential equation (Laplace's equation):

$$\nabla^2 \phi = 0. \tag{91}$$

Thus by assuming that the fluid is inviscid and that we start with a motion that contains no vorticity, we have gone from a nonlinear differential equation to a linear differential equation. We call this a potential flow problem, whose solution was developed to a high art in the nineteenth century and culminating, for example, in the masterful treatise by Lamb.[152] Such analysis was successful in explaining important problems having to do with the lift on airfoils, but was remarkably unsuccessful in explaining the accompanying drag. The resolution of this difficulty came from the realization that adjacent to solid surfaces where the no-slip condition must be applied, there are large gradients and the viscosity term is no longer a small perturbation—hence arose the subject of boundary layer theory (see Schlichting[233]). The moral of the story is that if we are to retain the simplicity of a problem described entirely in terms of potential flow, we must be careful to avoid situations where large velocities develop near solid surfaces. The motion of the interface between two fluids is just such an admissible problem. However, the "simplicity" obtained in Eq. (91) will be short-lived.

Once we are able to find ϕ we are able to determine $\mathbf{u}$ from the relation $\mathbf{u} = \nabla\phi$. Remember, however, that we also have pressure as an unknown. How do we find the pressure field? Substituting $\mathbf{u} = \nabla\phi$ into the momentum equation (without the viscous term, but including a body force due to gravity pointing in the $-z$ direction), we find:

$$\nabla\left[\frac{\partial\phi}{\partial t} + \frac{1}{2}(\nabla\phi)^2 + gz + \frac{p}{\rho}\right] = 0, \tag{92}$$

which means that the quantity inside the brackets must be independent of spatial position so that

$$\frac{\partial\phi}{\partial t} + \frac{1}{2}(\nabla\phi)^2 + gz + \frac{p}{\rho} = C(t), \tag{93}$$

where $C(t)$ is some function of time. This is a form of Bernoulli's equation.

While the potentials on either side of an interface might satisfy Laplace's equation (Eq. (91)), the problem is not completely posed until we have specified boundary conditions. Imagine that we have two layers of fluid with an undeformed interface at $z = 0$ (see Figure 72): fluid of density ρ_1 lies underneath fluid of density ρ_2. In the following, let the subscript $i = 1$ or 2 refer to the respective

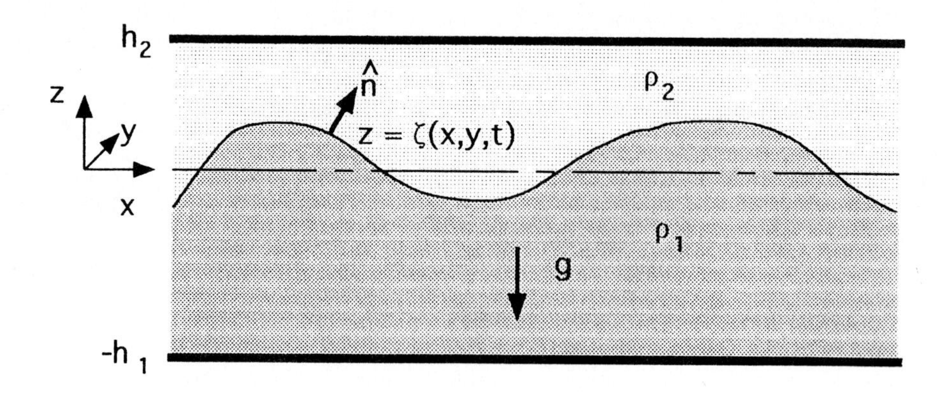

FIGURE 72 Deformation of the interface between a fluid of density ρ_1 lying beneath a fluid of density ρ_2.

fluids. As boundary conditions, we could specify that each fluid has a uniform velocity $\mathbf{u}_i$ far away from the interface, which is in fact the situation for the Kelvin-Helmholtz instability that was discussed at the beginning of the discussion of shear flow instabilities. Instead, we will assume the fluids are at rest far away from the interface, so $\phi_i \to 0$ as $z \to \pm\infty$. Alternatively, if the fluids eventually come into contact with walls at $z = \pm h_i$, we will require that the fluid not penetrate the walls: $w = \partial\phi/\partial z = 0$ at $z = \pm h_i$. Note that are assuming that the fluid layers have infinite extent in the x and y directions. The question remains: will an initially flat interface itself remain stable? We imagine deformations of the interface and, consequently, we must write appropriate conditions on the fluid adjacent to the interface. Here is where the nonlinearity enters and it is not a pretty sight.[8]

First, there is a purely kinematic condition: we don't allow the interface to open up into a gap between the fluids. This means that motion of the interface must match the motion of the fluid on either side. Suppose we describe the height of a planar interface by the equation $z = \zeta(x, y, t)$ (see Figure 72). Equivalently, we can describe a function $f(x, y, z, t) \equiv z - \zeta(x, y, t)$. This function must always have the value zero on the interface, which means that its convective derivative must vanish at the interface:

$$\left.\frac{Df}{Dt}\right|_{(x,y,z=\zeta(x,y,t),t)} = 0, \tag{94}$$

which implies

$$\left(\frac{\partial}{\partial t} + \mathbf{u}_i \bullet \nabla\right)(z - \zeta(x, y, t)) = 0. \tag{95}$$

This condition holds for both $i = 1$ and $i = 2$. Since $\mathbf{u} = \nabla\phi$, this implies:

$$\frac{\partial \phi_i}{\partial z} = \frac{\partial \zeta}{\partial t} + \frac{\partial \phi_i}{\partial x}\frac{\partial \zeta}{\partial x} + \frac{\partial \phi_i}{\partial y}\frac{\partial \zeta}{\partial y}, \tag{96}$$

where all terms must be evaluated at the interface $[x, y, z = \zeta(x, y)]$.

Second, there is a dynamical condition that arises if there is surface tension: the change in pressure across the fluid interface must equal twice the surface tension times the mean curvature of the interface. For example, the pressure jump across the surface of a spherical drop of radius r (and consequently of curvature $1/r$) is $2\sigma/r$. More generally, if a surface has unit normal $\mathbf{n}$, the pressure drop is $\sigma\nabla \bullet \mathbf{n}$.

$$p_2 - p_1 = -\gamma\nabla \bullet \hat{\mathbf{n}} = -\gamma\nabla \bullet \frac{(-\zeta_x, -\zeta_y, 1)}{[1 + (\zeta_x)^2 + (\zeta_y)^2]^{1/2}}. \tag{97}$$

[8] I made the mistake of treating this topic early in the actual lectures and lost half of the students by the next day.

Substituting for p from the Bernoulli equations, we obtain

$$\rho_2 \left[\frac{\partial \phi_2}{\partial t} + \frac{1}{2}(\nabla \phi_2)^2 + g\zeta \right]$$
$$- \rho_1 \left[\frac{\partial \phi_1}{\partial t} + \frac{1}{2}(\nabla \phi_1)^2 + g\zeta \right] = -\gamma \nabla \bullet \frac{(-\zeta_x, -\zeta_y, 1)}{[1 + (\zeta_x)^2 + (\zeta_y)^2]}^{1/2}). \tag{98}$$

Here, temporarily, we have used subscripts to denote partial derivatives. For example, $\zeta_x \equiv \partial \zeta / \partial x$. All of this is perfectly horrible, made worse by the fact that the derivatives must be evaluated at the interface $z = \zeta(x, y)$, which is one of the unknowns!

The problem actually becomes manageable, though algebraically tedious, if we linearize it by assuming small disturbances. As usual, this means we neglect nonlinear combinations of ϕ_1, ϕ_2, and ζ with each other and with each other's derivatives. It *also* means that we can evaluate all of the derivatives at the position $z = 0$, since corrections for nonzero ζ are also higher order. We also take advantage of the translational invariance of the problem in the x and y directions to achieve, once again, a separation of variables. Thus we try to see what happens, in the linearized equations, to a disturbance of the form

$$\Psi = \begin{pmatrix} \phi_1 \\ \phi_2 \\ \zeta \end{pmatrix} = \begin{pmatrix} \hat{\phi}_1(z) \\ \hat{\phi}_2(z) \\ \hat{\zeta}_0 \end{pmatrix} \exp(ik_x x + ik_y Y) \exp(st) + \text{c.c.} \tag{99}$$

(c.c. stands for "complex conjugate.") We find that this solution works provided the following dispersion relation is satisfied (here $k \equiv (k_x^2 + k_y^2)^{1/2}$):

$$s^2 = \frac{(\rho_2 - \rho_1)gk - \gamma k^3}{\rho_1 \coth kh_1 + \rho_2 \coth kh_2} \quad \ldots \text{for finite layers;} \tag{100}$$

$$s^2 = \frac{(\rho_2 - \rho_1)gk - \gamma k^3}{\rho_1 + \rho_2)} \quad \ldots \text{for semi-infinite layers.} \tag{101}$$

A whole bunch of nice consequences result from this equation. For example, we are reassured that a light fluid resting on top of a heavy fluid ($\rho_2 < \rho_1$) does not become unstable: Eq. (101) shows that $s^2 < 0$, so $s = i\omega$ is pure imaginary and the modes are neutrally stable traveling waves (they neither grow nor decay in the absence of viscosity):

$$\zeta(x, y, t) = \zeta_0 \exp[i(k_x x + k_y Y - \omega t)]\,. \tag{102}$$

Of course, we could have spared ourselves all the math and simply looked out the window at ripples on the fountain pond!

For the case $\rho_2 = 0$, $\rho_1 \equiv \rho$, and semi-infinite layers, we recover the linear dispersion relation for surface or "water" waves:

$$\omega^2 = gk - (\gamma/\rho)k^3 \qquad \text{capillary-gravity waves.} \tag{103}$$

The two limiting cases are:

$$\omega \approx (gk)^{1/2} \qquad \text{long wavelength: gravity waves} \tag{104}$$

$$\omega \approx \left(\frac{\gamma k^3}{\rho}\right)^{1/2} \qquad \text{short wavelength: capillary waves (ripples);} \tag{105}$$

Both terms in the dispersion relation have equal size when $\lambda \equiv 2\pi/k = 2\pi(\gamma/\rho g)^{1/2} \approx$ 1.7 cm (for pure water). The combination $(\gamma/\rho g)^{1/2}$ is called the capillary length. A full treatment of this problem, including damping due to viscosity, is found in Chandrasekhar.[38] When the densities of both fluids are significant, the modes are called internal waves. Note that for both types of waves, capillary and gravity waves, there is strong "dispersion": the phase speed of a wave, $c = \omega/k$, depends on the wavenumber.

6.2 RAYLEIGH-TAYLOR AND RICHTMYER-MESHKOV INSTABILITIES

From above, we can also conclude that we are in trouble if a heavy fluid lies on top of a light fluid. Note that this is a perfectly viable equilibrium state: hydrostatic pressure in the underlying fluid can hold up the heavy fluid on top. However, woe to the person standing underneath a thick layer of ceiling paint if the paint acquires the slightest ripple! The prediction is that for small enough k, the right-hand side of Eq. (101) is positive and one of the roots for s is then a real positive number, leading to exponential growth of the disturbance. In other words, instability occurs when the wavelength is larger than a cutoff $\lambda_{\text{watch-out!}} = 2\pi[\gamma/(\rho_2 - \rho_1)]^{1/2}$. Recall from above that for water and air, this has the value 1.7 cm. This explains a party trick: water may be held in an inverted glass even though the bottom of the glass is only covered with a gauze or screen, as long as the mesh has a scale finer than $\lambda_{\text{watch-out!}}$.

The application of this idea can be extended in some clever ways. For example, G. I. Taylor pointed out that accelerating a pair of fluid layers downwards at a rate a is equivalent to changing the gravitational parameter g to $g-a$ (you feel lighter in an elevator as it starts to accelerate downwards). In fact, if a is greater than g, we can destabilize a light fluid over a heavy fluid! This might seem to be a novelty, except that it has an important application in the attempt to induce fusion by compressing fuel pellets by laser-induced ablation of the outer surface of the pellets. Successful

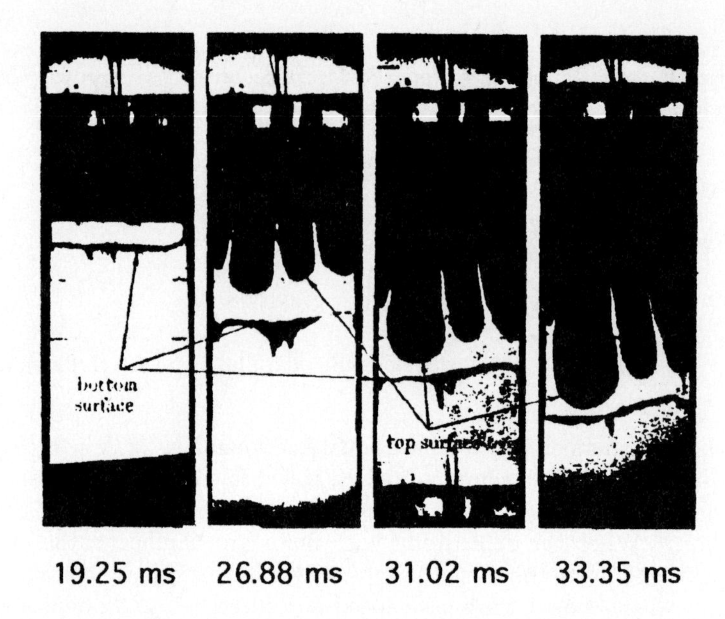

FIGURE 73 Late stages of the Rayleigh-Taylor instability in a layer of water accelerated rapidly (20.7g): the trailing air-water interface, given an initial sinusoidal deformation, has deformed into bubbles of air penetrating the water; the bubbles are separated by narrow sheets or "spikes" of water.[161]

compression requires extraordinary symmetry, which unfortunately can be broken by the aforementioned instability of a light fluid accelerating into a heavy fluid. This is known as the Rayleigh-Taylor instability.

Once the instability occurs, the growth of the interface can become very complex indeed, generating interpenetrating spikes and bubbles. A review of this topic may be found in Sharp,[237] while experiments to study the early stages of nonlinear growth are described in Fermigier et al.[90] A related instability (Richtmyer-Meshkov instability) occurs when a shock wave penetrates a fluid interface. In this case, it does not matter which fluid is more dense relative to the direction of the shock wave: instability occurs in either case. In Figure 74 we show a beautiful pattern of vortex plumes ejected into a surrounding lighter fluid from a curtain of dense gas that was penetrated by a shock wave. The curtain was prepared with an intentionally corrugated interface in order to generate such a regular array.[125]

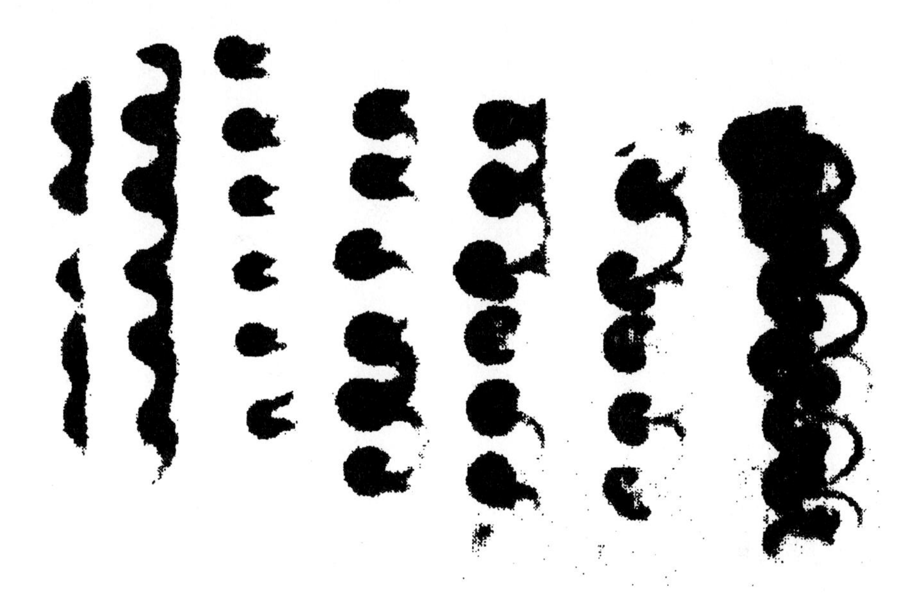

FIGURE 74 One form of development of the Richtmyer-Meshkov instability as a shock crosses the interfaces between an initially corrugated curtain of SF_6 surrounded by air. The profile develops into "mushrooms" that have caps oriented upstream from the shock direction. Each profile is $100_{\mu s}$ apart in time. [Jacobs et al.[125]]

6.3 RESONANT WAVE INTERACTION

The subject of nonlinear surface waves is an old, venerable, and difficult problem that continues to be very active, perhaps even enjoying a new growth of interest. It is a subject that is rich in unexpected phenomena, such as the existence of solitons. These are fully nonlinear, spatially localized disturbances that can propagate over great distances without decay and can even pass through one another without being substantial change—even though, again, the amplitudes are such that nonlinearity is definitely important. Unfortunately, I won't be able to discuss solitons further—suggesting references like Drazin and Johnson.[78]

One general aspect of surface waves that should be mentioned is the concept of resonant wave interact, which we have already encountered in the discussion of boundary layer instabilities. If you look at waves in nature, you may wonder at how it seems that waves beget waves beget waves. That is, under more controlled laboratory conditions, one can observe that an initial train of parallel waves soon develops additional sets of waves traveling in different directions.

Consider a small-amplitude surface wave as the base state:

$$\Psi_1 = A_1 \Psi_{10} e^{i(\mathbf{k}_1 \bullet \mathbf{r} - \omega_1 t)}. \tag{106}$$

We can consider the stability of this wave with respect to interaction with other modes. One important type of interaction is the resonant triad, arising from the nonlinear mixing of two waves

$$\Psi_2 = A_2 \Psi_{20} e^{i(\mathbf{k}_2 \bullet \mathbf{r} - \omega_2 t)} \quad \text{and} \quad \Psi_3 = A_3 \Psi_{30} e^{i(\mathbf{k}_3 \bullet \mathbf{r} - \omega_3 t)} \tag{107}$$

to form a wave varying like $e^{i[(\mathbf{k}_2+\mathbf{k}_3)\bullet\mathbf{r}-(\omega_2+\omega_3)t]}$. This will be seen by the original wave $(\mathbf{k}_1, \omega_1)$ as a forcing term that is resonant if:

$$\mathbf{k}_1 = \mathbf{k}_2 + \mathbf{k}_3 \tag{108a}$$

$$\omega_1(\mathbf{k}_1) = \omega_2(\mathbf{k}2) + \omega_3(\mathbf{k}_3). \tag{108b}$$

This is a stringent condition on the dispersion relations and may not be satisfied for any choice of $\mathbf{k}_1, \mathbf{k}_2, \mathbf{k}_3$. Indeed, for gravity waves $\omega = (gk)^{1/2}$ the resonant conditions cannot be satisfied at this order and a higher-order four-wave resonance must be sought instead.[204] For gravity-capillary waves, however, the resonant triad conditions can be met and the consequences have been explored in some beautiful experiments by Henderson and Hammack.[110] For reviews of wave interactions, with discussion of amplitude equations, see Phillips[204] and Craik.[60]

6.4 FARADAY CRISPATIONS

In 1831 Michael Faraday observed pronounced deformations of the surface of liquid in a dish that oscillated up and down. Furthermore, the frequency of the surface patterns appeared to be *half* that of the driving frequency. This was the first record of the phenomenon of parametric instability, which has since been found and exploited in a wide variety of systems (mechanical, electrical, optical, etc.). We have already discussed this topic in the context of boundary layer instabilities; we will now gain further insight into the beautiful topics stemming from Floquet theory of equations with periodic coefficients.

Benjamin and Ursell[17] found that the linearized analysis of Faraday's experiment with a vibrating layer of fluid leads to Mathieu's equation (Eq. (114) below). Let the undisturbed surface lie at $z = 0$ and assume the vertical position of the bottom of the dish vibrates in time as $z_0(t) = -h + \varepsilon \cos \Omega t$. As with G. I. Taylor's analysis of accelerating fluid layers, we can transform to a noninertial frame by replacing gravity g with an effective gravity $g + \varepsilon\Omega_2 \cos \Omega t$. The nomenclature "parametric instability" arises because the forcing of this system arises through one of the *parameters* (gravity). The surface is described in the noninertial frame by

$$z = \zeta(x, y, t) = \zeta_0(t) S_{lm}(x, y). \tag{109}$$

The indices l and m are integers and arise as follows. For rectangular containers of dimensions a and b

$$\begin{aligned} S_{lm}(x,y) &= \cos(k_x x)\cos(k_y y) \\ &= \cos\left(\frac{l\pi x}{a}\right)\cos\left(\frac{l\pi x}{b}\right), \end{aligned} \tag{110}$$

where the second part of the equation shows that the wavenumber $k^2 \equiv (k_x^2 + k_y^2)^{1/2}$ must be quantized according to $k^2 = k_{lm}^2 = (l\pi/a)^2 + (m\pi/b)^2$ in order to satisfy the boundary condition that at each wall the surface is flat in the direction normal to the wall (a kinematic consequence of the condition that the normal component of the velocity must vanish). For cylindrical containers of radius r_0

$$S_{lm}(r,\theta) = J_l(k_{lm}r)\cos(l\theta + \phi) \tag{111}$$

where J_l is a Bessel function and the same boundary condition requires that k_{lm} is the mth zero of $J_l'(k_{lm}r_0)$.

The amplitude $\zeta_0(t)$ satisfies the equation

$$\frac{d^2\zeta_0}{dt^2} + \beta\frac{d\zeta_0}{dt} + (\omega_{lm}^2 + \varepsilon\omega_2 k_{lm}\tanh k_{lm}h\cos\Omega t)\zeta_0 = 0, \tag{112}$$

where β is phenomenological damping coefficient and, from Eq. (100),

$$\omega_{lm}^2 = k_{lm}\left(g + \frac{\gamma k_{lm}^2}{\rho}\right)\tanh k_{lm}h. \tag{113}$$

Defining $\tilde{\zeta}_0 \equiv e^{\beta t}\zeta_0$ and $\tau \equiv \Omega t/2$ we have Mathieu's equation:

$$\frac{d^2\tilde{\zeta}_0}{d\tau^2} + (p - 2q\cos 2\tau)\tilde{\zeta}_0 = 0, \tag{114}$$

where

$$p \equiv 4\frac{\omega_{lm}^2 - \beta^2}{\Omega_2}; \tag{115a}$$

$$q \equiv 2\varepsilon k_{lm}\tanh k_{lm}h. \tag{115b}$$

Mathieu's equation has the remarkable property[174] that for a given level of forcing q there are "tongues" $b_n(q) < p < a_n(q)$ ($n = 0,1,2,3,\ldots$ with $b_0 = -\infty$) within which a solution $\tilde{\zeta}_0$ diverges exponentially with time—see Figure 75. Close to the level of zero forcing, the unstable values of p will be the sequence $1, 4, 9, \ldots, n^2, \ldots$.

When p lies in one of the resonance tongues, the corresponding solution has the Floquet form of an exponential times a function with the same period as the

$$\tilde{\zeta}_0(t) = e^{(\mu+i)\Omega t/2}\phi_n(t) \quad \text{for } n \text{ odd: subharmonic relative to } \cos\Omega t, \quad \text{(Floquet exponent } (\mu+i)\Omega/2) \tag{116a}$$

$$\tilde{\zeta}_0(t) = e^{\mu\Omega t/2}\phi_n(t) \quad \text{for } n \text{ even: harmonic; relative to } \cos\Omega t, \quad \text{(Floquet exponent } \mu\Omega/2) \tag{116b}$$

with $\phi_n(t + 2\pi/\Omega) = \phi_n(t)$ and $\mu(p,q) > 0$ (real). The first of these modes is subharmonic, because time t must advance by two periods $(2 * 2\pi/\Omega)$ to recover the same value for $\tilde{\zeta}_0(t)$.

Thus the actual surface deformation is given by

$$\begin{aligned} \zeta_0(t) &= e^{(\mu\Omega/2-\beta)t} e^{i\pi\Omega t/2}\phi_n(t) \\ \text{or} \quad \zeta_0(t) &= e^{(\mu\Omega/2-\beta)t}\phi_n(t) \end{aligned} \tag{117}$$

and is unstable provided the frequency ratio $4(\omega_{lm}^2 - \beta^2)/\Omega_2$ lies inside one of the unstable bands and

$$\mu(p,q) > \frac{2\beta}{\Omega}. \tag{118}$$

Contours of constant μ can be plotted in the p, q plane (see Figure 75); the inequality (Eq. (118)) will require a threshold value of q that depends on reduced frequency p. Recalling that q is proportional to ε, we see that this means that the forcing amplitude ε must reach a threshold that depends on drive frequency Ω. Note that in the limit of zero damping ($\beta = 0$) there are resonant frequencies $\omega_{lm} = n\Omega/2$ at which there is an unstable solution for vanishingly small forcing $\varepsilon = 0$. In particular, we see that for $n = 1$, the system responds at half the drive frequency, which is just the subharmonic response observed by Faraday. This is usually the easiest mode to excite because it turns out to have the lowest threshold amplitude for a given amount of damping. However, by harder driving, it is possible to excite other modes—which caused some consternation when other observers noticed that Faraday's shaking pail of water seemed to respond with the *same* frequency as the vibrations (not a subharmonic). One final note concerning the utility of such systems: in a real system with small damping, a small modulation signal adjusted to resonance can cause a large response (assuming nonlinearities saturate the unstable solution) and we have what is called a parametric amplifier.

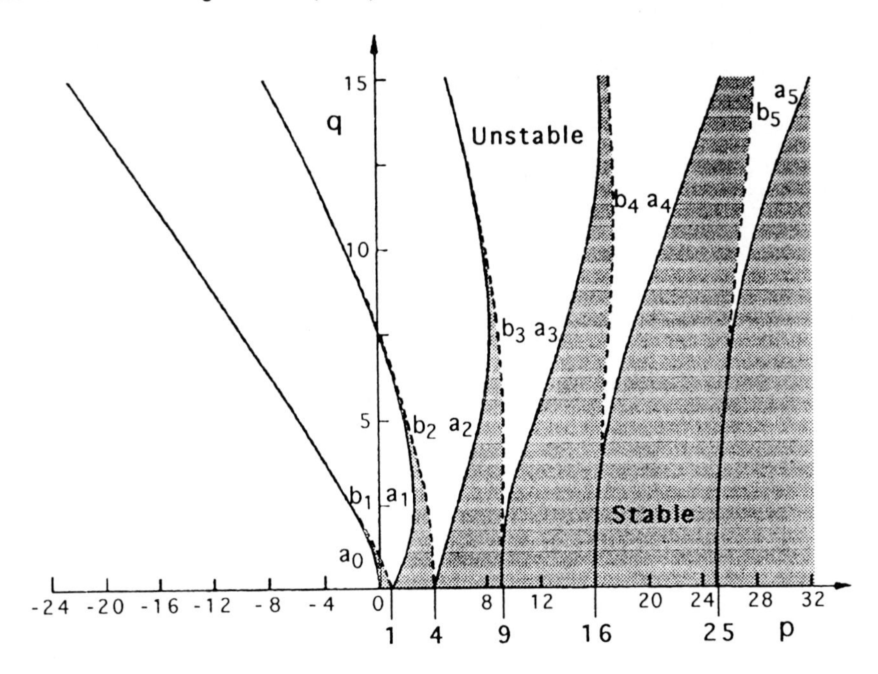

FIGURE 75 Tongues defining regions of stable (shaded) and unstable (white) solutions of Mathieu's equation.

If the Faraday experiment is done in relatively small containers, the mode frequencies will be distinct so that only one mode might become parametrically unstable for carefully selected amplitude ε and frequency Ω. However, as ε is increased, bands of instability for different modes overlap and there is competition. Since the full equations are nonlinear and since symmetry-caused degeneracy ensures that three or more modes interact when bands overlap, one has the ingredients for chaotic dynamics . This is exactly what has been observed by Ciliberto and Gollub.[47] In their experiment (see Figure 76), a cylinder of radius $r_0 = 6.35$ cm held a water layer of depth $h \approx 1$ cm. The (4,3) mode was excited for a frequency $\Omega/2\pi = 15.77$ Hz at a drive amplitude $\varepsilon \approx 37\mu m$ (quite small!). Nearby, at 16.21 Hz, a (7,2) mode could be excited with nearly the same drive amplitude. These results are shown in Figure 77. At a small range of intermediate frequencies and drive amplitudes $> 100\mu m$, the bands overlapped and chaotic dynamics was found. For such systems, the nonlinear perturbation analysis, though difficult, is tractable thanks to the clearly separated modes. Thus considerable progress has been made in a first-principles quantitative understanding of the chaotic dynamics of this fluid system. Recent reviews may be found in Miles and Henderson[177] and Gollub.[99]

6.5 BREAKUP OF LIQUID JET: CAPILLARY INSTABILITY

In the above examples, surface tension played a stabilizing role, as was the case when it caused a cutoff of short-wavelength modes in the Rayleigh-Taylor instability. In base states that are start out with curved surfaces, however, surface tension may be destabilizing. If we ignore damping, this may be understood by energy considerations: a base state will be unstable if a new surface of lower energy can be found. This is indeed the case, for example, in the breakup of a cylindrical liquid column into drops (see Figures 78 and 79).

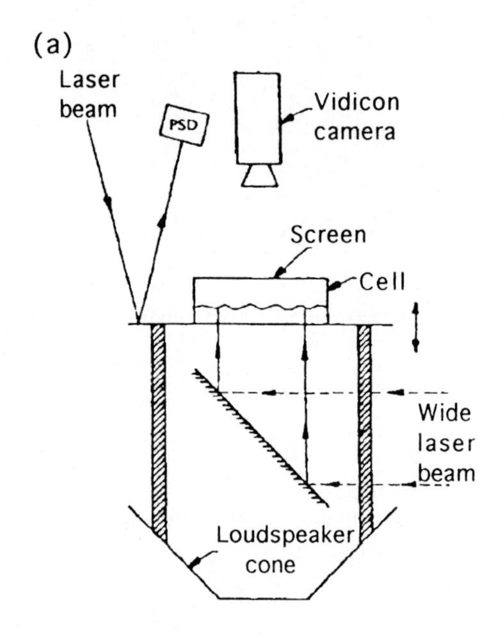

FIGURE 76 Parametrically forced surface waves. (a) Apparatus, showing how wave patterns are visualized by the shadows produced by projecting a wide collimated laser beam through the fluid layer. A position sensing detector (PSD) monitors deflections of a second laser beam in order to measure the amplitude of oscillation imposed on the cell by the supporting loudspeaker cone. (b) An example pattern: a (7,2) mode, corresponding to values for the integers (l,m) that define surface modes in Eq. (111). Note that the pattern appears to be seven-fold symmetric in the angular direction. [From Ciliberto and Gollub.[47]]

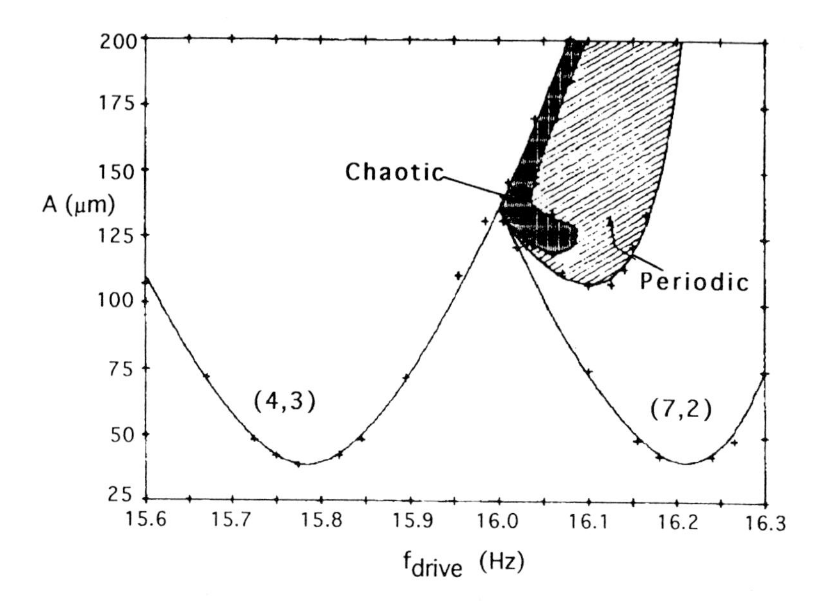

FIGURE 77 Phase diagram of surface states in an oscillating fluid layer. The control parameters are frequency f and amplitude A of the forcing. Above the curves, the surface is deformed into modes of vibration: the modes are indentified by a pair of integers l and m (see Eq. (111)). The shaded region shows where interaction of two different modes produces new dynmaics: either slow periodic variation or chaotic variation of the patterns. [From Ciliberto and Gollub.[47]]

The analysis of surface deformations is similar to that for surface waves, except we now have a cylindrical interface defined by $r = a + \zeta(z, \theta)$. For perturbations $\zeta(z) = \zeta_0 e^{i(kz+m\theta)+\mathrm{st}}$, a linear analysis gives:

$$s^2 = \frac{\sigma}{\rho a^3} \frac{kaI_n'(ka)}{I_n(ka)}(1 - k^2a^2 - m^2), \tag{119}$$

where $I_n(ka)$ is a modified Bessel function of order n. Only the term $(1 - k^2a^2 - m^2)$ determines the sign of s^2 and we see, then, that s is real and instability occurs if $m = 0$ (axisymmetric disturbances) and $k < 1/a$. In other words, the column is unstable for all wavelengths $\lambda > 2\pi a$, the circumference of the column. The fastest growing mode occurs for $ka = 0.6970$, or $\lambda = 1.435(2\pi a)$, with a growth rate $s = 0.3433(\sigma/\rho a^3)^{1/2} \approx (0.12 \text{ seconds})^{-1}$ for a water column with radius $a = 0.5$ cm and surface tension $\sigma = 73$ dynes/cm. In experiments, the column is produced as a jet exiting a nozzle with velocity U. Forcing the column with a speaker oscillating at frequency ω_0 radians/sec gives a well-defined wavenumber $k = U/\omega_0$. The growth rate may be inferred by fitting an exponential envelope

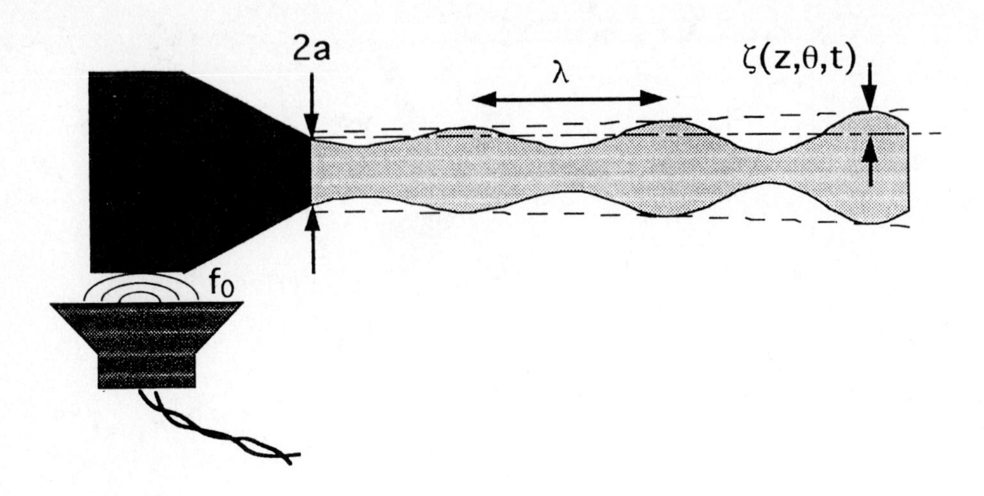

FIGURE 78 Breakup of a liquid column emerging from a nozzle. Note that the nozzle is excited by a loudspeaker at frequency f_0. Correspondingly, the excited wavelength $\lambda = U \times f_0$. The dashed curve shows an exponentially growing envelope to the surface deformations.

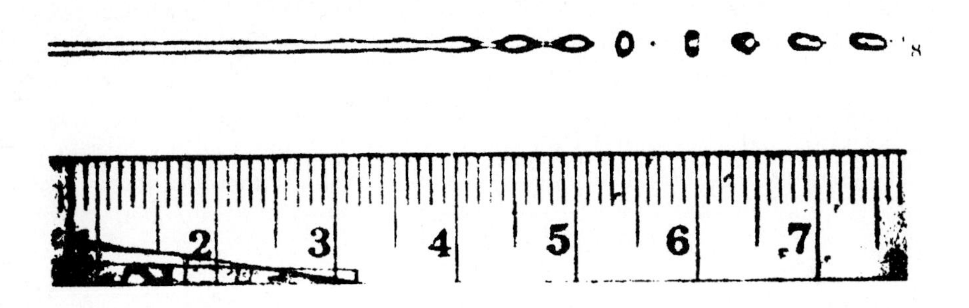

FIGURE 79 Breakup of a water column when the excitation generates a disturbance such that $2\pi/\lambda = 0.262$ (see Figure 80), where a is the column radius and λ is the wavelength of the disturbance. Note the small droplets that are formed between the larger drops when the connecting fluid pinches off. The larger drops are undergoing large amplitude oscillation. [From Donnelly and Glaberson.[76]]

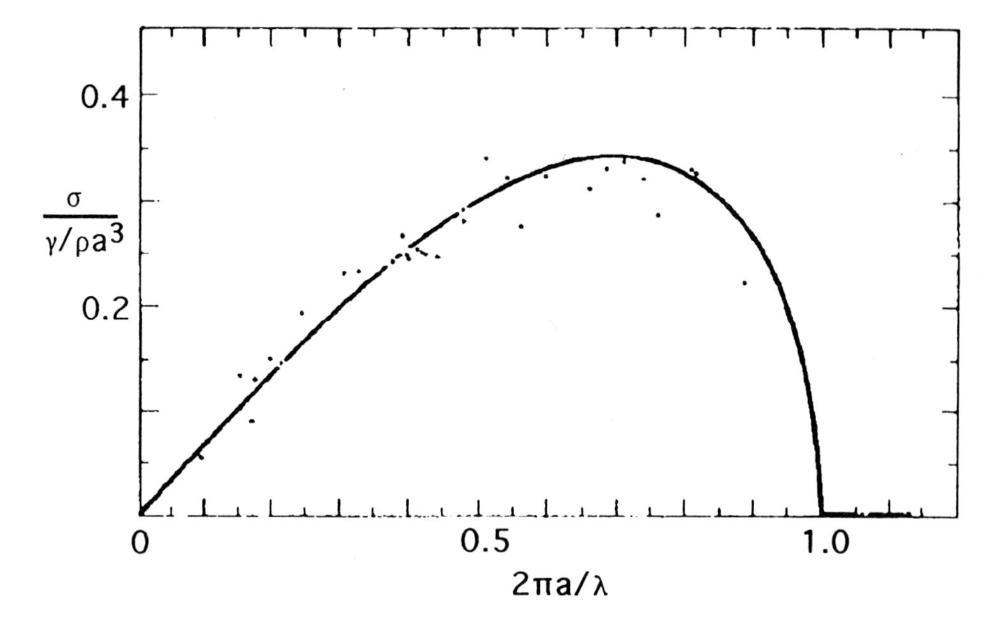

FIGURE 80 Growth rate as a function of wavenumber for a liquid column. The points are experimental data while the solid curve is the dispersion relation from Eq. (119). [From Donnelly and Glaberson.[76]]

to the jet profile; good agreement is found[76] comparing the initial growth with the inviscid theory—see Figure 80. Further discussion of this problem is found in Drazin and Reid[79] and Chandrasekhar.[38] For experimental investigation of the nonlinear regime, see Goedde and Yuen[98] and more recently in Chaudhary and Redekopp[40] and Chaudhary and Maxworthy.[41,42]

6.6 VISCOUS FINGERING

An important class of nonequilibrium phenomena concerns the shape of a propagating interface separating one medium from another. Examples of this situation include:

- motion of one fluid into another, where the two fluids have different viscosities (viscous fingering); this arises, for example, in secondary oil recovery when water is used to drive oil through the porous beds towards extraction wells;
- advance of a region undergoing the chemical reactions of combustion into an unignited phase (flame fronts); this is of obvious importance for the efficiency of combustion processes;

- expansion of a domain undergoing crystallization (directional solidification); the resulting structures may strongly influence the metallurgical properties of commercial alloys;
- the motion of an interface of a fluid that is wetting a solid substrate; this concerns technological problems of coating, draining, and drying.

The phenomenology of these problems is very rich, often producing intricate structures as a result of fingering instabilities. The moving interface deforms into fingers that penetrate the second medium; these fingers may themselves experience fingering instabilities, etc. One system that has received considerable attention is the displacement of fluids in a Hele-Shaw cell (Figure 81): a less viscous fluid (for example, air) pushes a more viscous fluid (for example, oil or glycerine) through a very narrow channel between two parallel plates. This prototypical system has many of the characteristics of fluid displacement in porous media, a topic of obvious importance with respect to secondary oil recovery. A cornucopia of fingering patterns generated in Hele-Shaw cells is assembled in Figure 82, where various kinds of fluids (Newtonian and non-Newtonian, immiscible and miscible, etc.) have been used. A wide-ranging discussion of such phenomena is found in the book by Vicsek,[274] which even gives some hints on making your own Hele-Shaw apparatus. I have seen such experiments conducted on overhead projectors, using a hand-operated syringe to pump in the air.

When confronted with such phenomena, the clever scientist tries to abstract the situation into the simplest configuration that retains the essentials of the dynamics to be studied. In this case, this abstraction is called the Saffman-Taylor problem, which concerns the motion of a less viscous fluid into a rectangular channel containing a more viscous fluid. One seeks to learn if and how a straight-line interface between the fluids deforms. Is there a most dangerous "wavelength" of deformation of the surface? What happens at long times when the deformation may no longer be considered infinitesimal? The reader who has made it this far will know to expect that even in this seemingly ultra-simple configuration, complications arise. Indeed, the shape and size of the finger that is ultimately formed is governed by a selection process that has only recently been understood.

We will set up the problem, describe some of the principal features, and then refer the reader to the literature for the intricate details involved in coming to an adequate solution. The gap b between the top and bottom walls of the cell is very small, so the flow is two-dimensional and is dominated by viscosity. The velocity—defined here to be an average across the narrow gap—is proportional to the pressure gradient, which is called Darcy's law:

$$\mathbf{u}_i = -\frac{b2}{12\mu_i}\nabla p_i, \qquad (120)$$

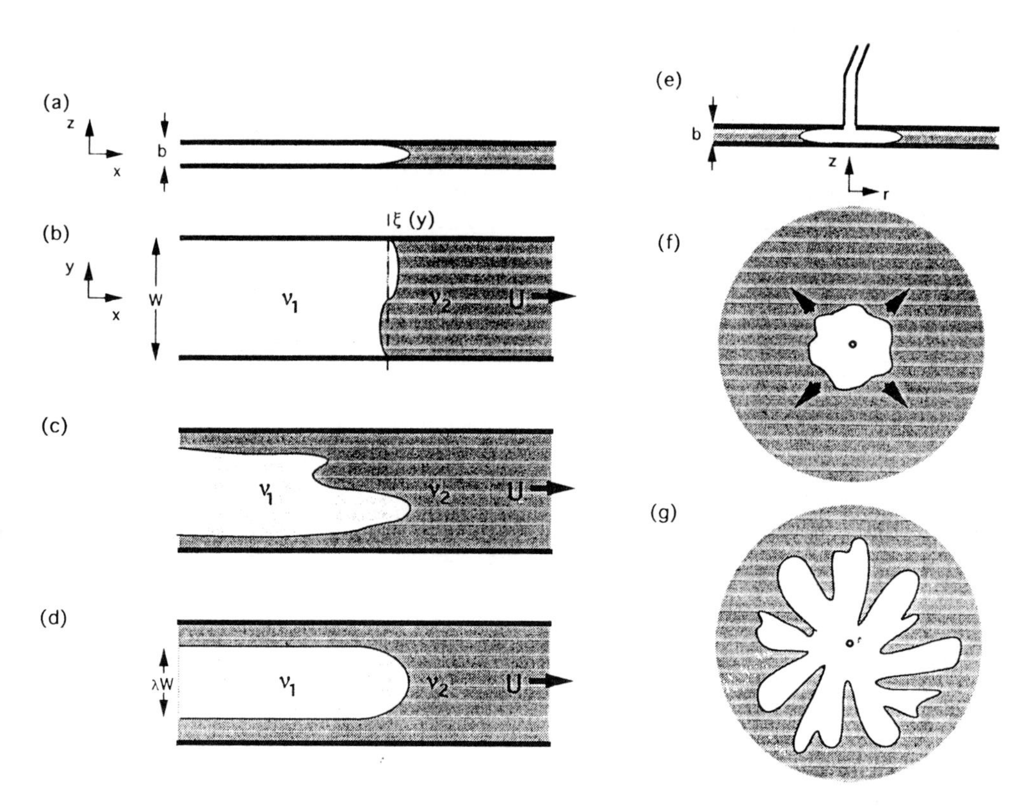

FIGURE 81 Motion of the interface between two fluids in a Hele-Shaw cell: a less viscous fluid (viscosity n_1, shown with lighter shading) displaces a more viscous fluid (viscosity n_2, shown with darker shading). (a)-(d) Rectangular geometry. (a) A side view and (b) a top view at an early stage of deformation of the interface. (c) Intermediate stage and (d) late stage of deformation: ultimately a single finger results whose width is a fraction l of the width of the cell. (e)-(g) Circular geometry. (e) A side view, (f) a top view at an early stage of deformation. (g) A later stage of deformation showing many fingers with different scales.

where $i = 1$ or 2 distinguishes the two fluids and μ_i is the dynamic viscosity. Gravity is neglected on the assumption that the cell is horizontal. Incompressibility, expressed as $\nabla \bullet \mathbf{u}_i = 0$, combined with Darcy's law then implies that the pressure will satisfy Laplace's equation in each fluid:

$$\nabla^2 p_i = 0. \tag{121}$$

Once again, we have gone form nonlinear Navier-Stokes equations to a circumstance where the essential field (this time, the pressure) is described by a linear equation. As with surface waves, this hopeful simplicity will be short-lived.

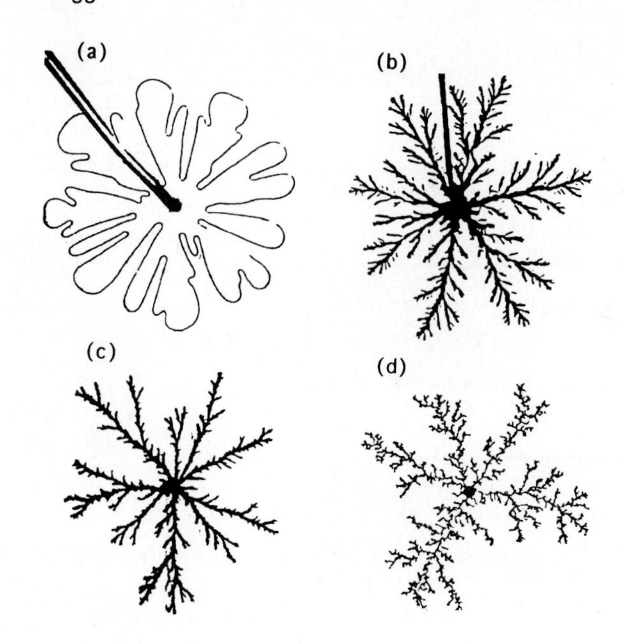

FIGURE 82 A cornucopia of viscous fingering patterns. Here the variation is brought about by choice of fluids, but it is important to keep in mind that the type of pattern may also depend on rate of injection. (a) Two Newtonian, immiscible fluids: air injected into glycerine.[201] (b) Two Newtonian, miscible fluids: water injected into glycerine.[43] (c) One of the fluids is non-Newtonian: water injected into polymer solutions.[67] (d) Displacement occurs in a random porous medium: air injected into oil in a system filled with random pores.[160]

The base flow is simply the case where both fluids and the interface moves with velocity $\mathbf{U} = U_0\hat{\mathbf{x}}$. Observing from a frame of reference moving with the undisturbed interface, we want to know what happens if there is a deformation

$$x = \xi(y) = \xi_0 \cos(ky)e^{\sigma t}. \tag{122}$$

We omit discussion of the kinematic and pressure jump conditions at the interface (which has interfacial tension γ), noting that in simplest form[9] they are similar to those given in the discussion of surface waves. Linearized analysis gives the result:

$$\sigma = \left[(\mu_2 - \mu_1)U - \left(\frac{\gamma b^2}{12}\right)k^2\right]\frac{k}{(\mu_1 + \mu_2)}. \tag{123}$$

[9] As discussed in Bensimon et al.,[19] there are more appropriate pressure drop conditions that include the effect of the velocity of the interface as it slips along the bounding plates.

This implies that $\sigma > 0$ and instability occurs for wavenumbers

$$k^2 < \frac{12(\mu_2 - \mu_1)}{\gamma b^2} U. \tag{124}$$

If the flow is confined by side walls a distance W apart (where free-slip boundary conditions are usually assumed as a good approximation), the wavenumbers k are restricted to values $k = 2\pi n/W$. Taking $n = 1$ for the smallest wavenumber and defining the dimensionless parameter

$$d_0 \equiv \frac{b^2}{\Omega_2} \frac{\gamma}{(\mu_2 - \mu_1)U}, \tag{125}$$

we see that instability occurs if the velocity is large enough that $d_0 < (3/\pi^2)$.

So far, so good. However, the nonlinear problem turns out to be one of very great subtlety. This is due to the fact that surface tension is a singular perturbation,

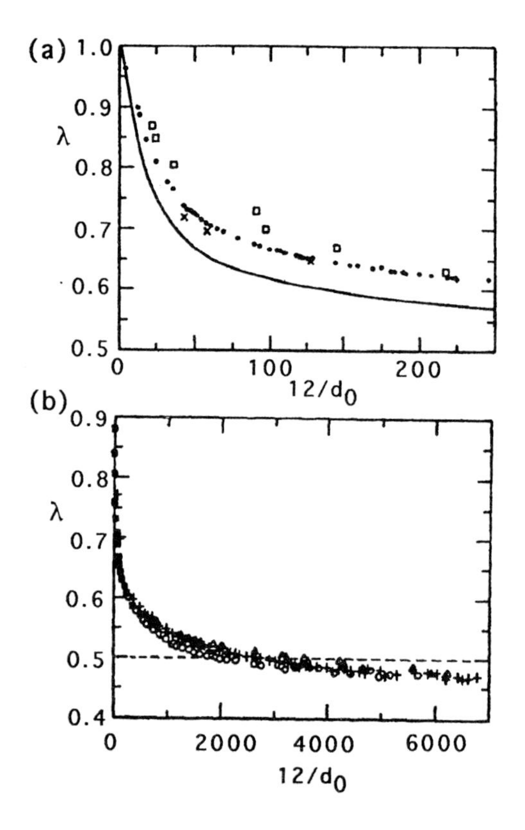

FIGURE 83 Experimental measurements of the dimensionless width λ of a viscous finger in a rectangular channel as a function of the parameter $12/d_0$, where d_0 (defined in Eq. (125)) compares surface tension and viscous forces. (a) The lower range of $12/d_0$, i.e., lower fluid speeds: squares are data from Saffman and Taylor (1958), the solid curve is from numerical computations of McLean and Saffman,[175] and the points and crosses are data from Tabeling et al. (1987). The last authors explain some of the discrepancy in results by the existence of a thin film of the more viscous fluid (oil) coating the walls after the bubble (air) has penetrated. (b) A higher range of $12/d_0$, i.e., higher flow velocities: data from Tabeling et al. (1987) for different aspect ratios w/b. Note that the width λ does not approach an asymptotic value of 1/2, but instead continues to slowly decrease.

so that solutions for small surface tension do not asymptotically approach those for zero surface tension. A one-parameter family of steady-state solutions was found for zero surface-tension by Saffman and Taylor[225]:

$$\frac{\xi(y)}{W} = \frac{1-\lambda}{\pi} \ln \cos \frac{\pi y}{\lambda W}, \tag{126}$$

where the parameter λ is the width of the finger relative to the width of the cell (As $y \to \pm\lambda W/2$, $\xi(y) \to -\infty$). However, there is no indication which value of λ will be chosen by an experiment. Furthermore, Eq. (124) may be interpreted to say that as surface tension vanishes, even very short wavelength perturbations destabilize a front. Relative to such short wavelength perturbations, the Saffman-Taylor front looks locally flat and ought to be subject to the same instability. Experiments show, however, that viscous fingers are stable and that a width parameter λ is uniquely determined by the velocity. Figure 83 shows data of Tabeling et al.[253] on the selected finger width. Considerable progress has been made recently in understanding the intricacies of this problem, whose discussion I will escape by recommending the reviews by Saffman[223,224] and Bensimon et al.[19]

7. EPILOGUES

7.1 FLUIDS ARE DUMB

As the time to give these lectures drew close and then passed, I developed a considerable "angst" over trying to make a connection between what I understood to be complexity in the context of fluid flows and the fascinating area of adaptive complexity as it is being applied to models of biological and economic systems. This angst is not without foundation: an essay by Anderson and Stein[7] takes issue with the speculations of Prigogine and others that suggest that complexity and the idea of "dissipative structures" in nonequilibrium systems like fluid flows is relevant to the understanding of life. In the essay, the key issue seemed to be that living nonequilibrium systems are quite stable, whereas the structures arising through symmetry-breaking transitions in inanimate nonequilibrium systems are ultimately chaotic and unstable unless dominated by confinement to relatively small geometries. While this last observation seems to be true—indeed, we see that for shear flows one cannot seem to overcome the ultimate evolution to turbulence once transition begins—I think that, even in the limit of infinitely large systems, one can find circumstances where there is a separation of scales: microscopic scale $\ll$ pattern wavelength (e.g., size of a convection cell) $\ll$ coherence length $\ll$ size of system (with corresponding hierarchy of time scales), so that in this sense the idea of "dissipative structures" still seems to hold water (pun intended). Consequently, one could argue that the difference between the stability of structure of a life form (like the author) and a convection cell is not so categorical.

Nonetheless, I remain bewildered about how to make a connection and conclude that one further difficulty could be the following profound insight: fluids are *dumb*. Fluid motions, though they might generate some sort of structure, do not seem to learn as they go, sustaining themselves by adapting to and co-opting the surrounding environment. Even this observation may not be so categorical: simulations (Phil Marcus) and experiment (Harry Swinney and collaborators) seek to explain the extraordinary long "life" of the red spot of Jupiter by the spontaneous

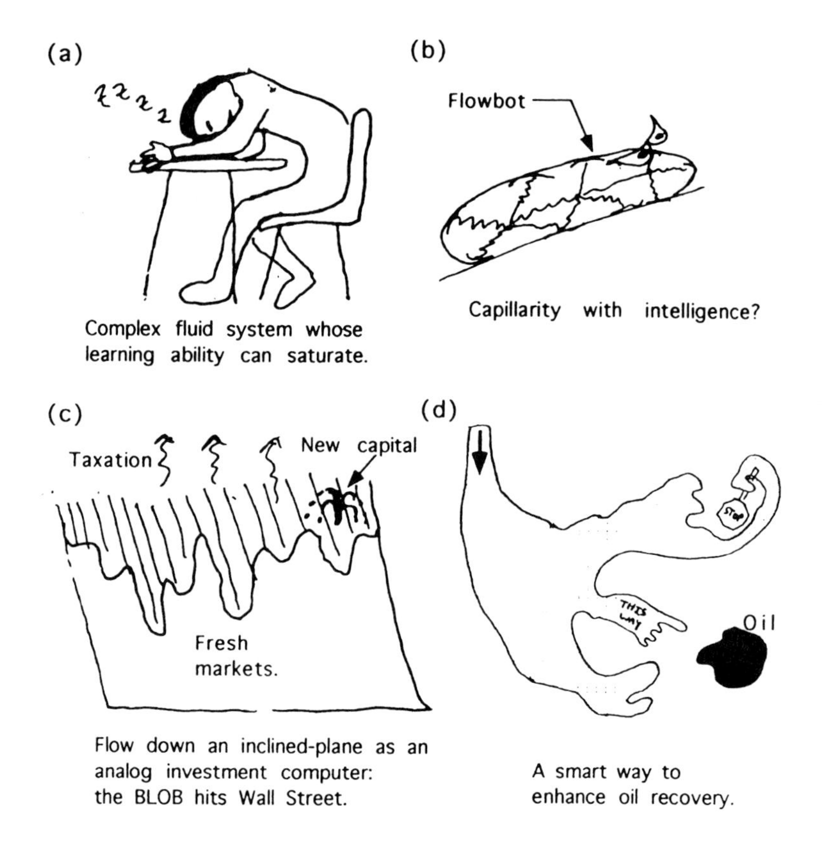

FIGURE 84 If only fluids were smart...

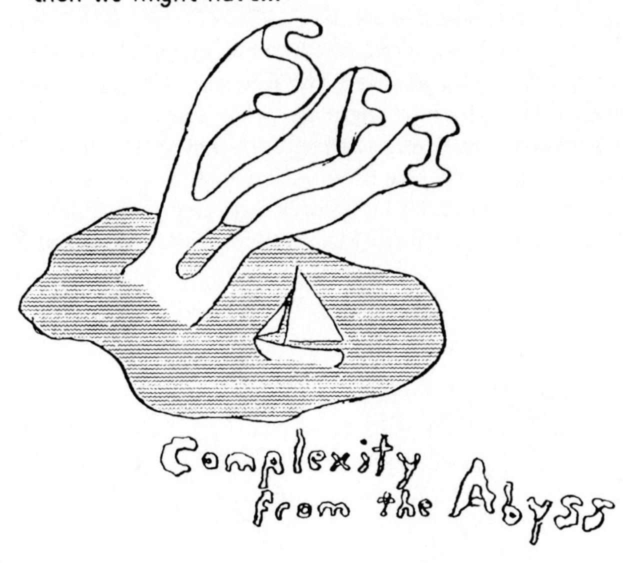

FIGURE 85 Inspired by the Hollywood movie *The Abyss.*

organization of a giant vortex out of an initial smear of vorticity in a sheared, rotating flow. Still, it does not appear that any fluid flow spontaneously holds itself at the brink between order and chaos, thereby differentiating and evolving into a complex structure that is sustained amidst random external perturbations.

Having thus vastly overreached the explanatory capabilities of my topic, I am left with only one resort: to take a humorous look at it. Thus, I came up with the following selection of cartoons as the final treat for students who made it through these lectures.

7.2 THE ENCHANTMENT OF VORTICES

This chapter began with the idea of enchantment, so that is how they will end. This is appropriate for their setting in New Mexico, a state which calls itself the "Land of Enchantment." Vortices—concentrated swirls of motion—hold a fascination for humans perhaps second only to fire. A person can gaze at these motions and conjure up all sorts of notions of sprites, demons, etc.

I am a new player in this game and my thinking about it is not that sophisticated. However, I hate to see the numericists having most of the fun, so I am devising experiments to study the effects and dynamics of concentrated vortices in turbulent flow. Thinking that such vortex filaments control turbulent motion may be as mistaken as thinking that the tail wags the dog. However, an understanding of how such vortices continually intensify and dissipate may begin to approach the ideas about complex processes that are studied in other fields at the Santa Fe Institute.

Vortices are, of course, intense fluid motions of great practical importance: unwanted in the case of tornadoes or aircraft trailing vortices, or desired in the case of hydrocyclone dust separators. One aim is to learn how to tickle these vortices, either into nonexistence or into more useful motions. The other aim is to understand vortices as organized, complex structures that seem to have unusual persistence in the midst of random perturbations.

ACKNOWLEDGMENTS AND DEDICATION

I wish to thank Dan Stein and the Santa Fe Institute staff for their interest and invitation to present this material; their patience has been stretched well into the nonlinear regime waiting for a final manuscript. I am especially grateful to the participants of the 1993 Complex Systems Summer School for conversations and comments; also to David Campbell for discussions and introductions to the fascinating people of the Complexity Institute. In years past, I have learned more than can be measured from Harry Swinney and his students and collaborators. I have also enjoyed collaboration with the research group of Russell Donnelly. Currently, my experimental work is supported by a grant from the Research Corporation, whose assistance in starting a new laboratory is gratefully acknowledged.

I started this article with a tongue-in-cheek description of the dangers associated with the study of fluid instabilities. More seriously now, I call attention to those cases where unexpected, sudden, intense organized motions of fluid can be truly catastrophic. We know, of course, of the annual devastation produced by tornadoes. There are also several airplane crashes—one only a couple of weeks ago in South Carolina—that may be attributed to sudden wind shear. And now, as I finish this manuscript, I have heard the news of fourteen firefighters killed tragically in a firestorm on a mountain ridge near Glenwood Springs, Colorado. I hope this chapter encourage thought about how to use the knowledge we gain about complexity and self-organization in fluid flows to give such people the edge they need—even a few seconds or minutes—to escape disaster and stay alive.

APPENDIX A: OPERATOR FORM OF NAVIER-STOKES EQUATIONS

The Navier-Stokes and continuity equations may be written in operator form:

$$\mathcal{L}\Psi + \mathcal{N}(\Psi, \Psi) = \mathcal{M}\frac{\partial \Psi}{\partial t}, \tag{A1}$$

where

$$\Psi \equiv \begin{pmatrix} u_x \\ u_y \\ u_z \\ p \end{pmatrix}, \tag{A2}$$

$$L \equiv \begin{pmatrix} \frac{1}{\text{Re}}\nabla^2 - \mathbf{U} \bullet \nabla - \frac{\partial U_x}{\partial x} & -\frac{\partial U_x}{\partial y} & -\frac{\partial U_x}{\partial z} & -\frac{\partial}{\partial x} \\ -\frac{\partial U_y}{\partial x} & \frac{1}{\text{Re}}\nabla^2 - \mathbf{U} \bullet \nabla - \frac{\partial U_y}{\partial y} & -\frac{\partial U_y}{\partial z} & -\frac{\partial}{\partial y} \\ \frac{\partial U_z}{\partial x} & \frac{\partial U_z}{\partial y} & \frac{1}{\text{Re}}\nabla^2 - \mathbf{U} \bullet \nabla - \frac{\partial U_z}{\partial z} & -\frac{\partial}{\partial z} \\ \frac{\partial}{\partial x} & \frac{\partial}{\partial y} & \frac{\partial}{\partial z} & 0 \end{pmatrix}, \tag{A3}$$

$$M \equiv \begin{pmatrix} 1 & 0 & 0 & 0 \\ 0 & 1 & 0 & 0 \\ 0 & 0 & 1 & 0 \\ 0 & 0 & 0 & 0 \end{pmatrix}, \tag{A4}$$

and the nonlinear term

$$N(\Psi^{(1)}, \Psi^{(2)}) = -\begin{pmatrix} \mathbf{u}^{(1)} \bullet \nabla u_x^{(2)} \\ \mathbf{u}^{(1)} \bullet \nabla u_y^{(2)} \\ \mathbf{u}^{(1)} \bullet \nabla u_z^{(2)} \\ 0 \end{pmatrix}. \tag{A5}$$

Note further that N is a bilinear operator:

$$\begin{aligned} N(\Psi^{(1)} + \Psi^{(2)}, \Psi^{(3)} + \Psi^{(4)}) = N(\Psi^{(1)}, \Psi^{(3)}) + N(\Psi^{(1)}, \Psi^{(4)}) + N(\Psi^{(2)}, \Psi^{(3)}) \\ + N(\Psi^{(2)}, \Psi^{(4)}); \end{aligned} \tag{A6}$$

$$N(a\Psi^{(1)}, b\Psi^{(2)}) = abN(\Psi^{(1)}, \Psi^{(2)}). \tag{A7}$$

As we have seen, further nonlinearity may enter through the boundary conditions.

APPENDIX B: EIGENVALUE EXPANSIONS AND GASTER'S TRANSFORMATION

Let us write down an approximate expression for the relation between eigenvalue s and wavenumber α assuming *both* are complex, by expanding around the marginal stability condition:

$$s = \frac{\partial s}{\partial \varepsilon}\varepsilon + \frac{\partial s}{\partial \alpha}(\alpha - \alpha_c) + \frac{\partial s^2}{\partial^2 \alpha}(\alpha - \alpha_c)^2 + \dots . \tag{B1}$$

Here the real number $\varepsilon \equiv (\mathrm{Re} - \mathrm{Re}_c)/\mathrm{Re}_c$, as before, and we expect that the range of wavenumbers within the marginal stability curve scales according to $|\alpha - \alpha_c| \sim \varepsilon^{1/2}$. We are treating treat s as an analytic function of α in the neighborhood of $\varepsilon = 0$ and $\alpha = \alpha_c$; we can take derivatives of $s = \sigma - i\omega$ along any direction in the complex-α plane, including along the real-α axis. Thus we can write:

$$\begin{aligned} \sigma - i\omega = &- i\omega_c + \left(\frac{\partial \sigma}{\partial \varepsilon)} - i\frac{\partial \omega}{\partial \varepsilon}\right)\varepsilon - i\frac{\partial \omega}{\partial \alpha_r}(\alpha_r + i\alpha_i - \alpha_c) \\ &+ \frac{1}{2}\left(\frac{\partial^2 \sigma}{\partial \alpha_{r^2}} - i\frac{\partial^2 \omega}{\partial \alpha_{r^2}}\right)(\alpha_r + i\alpha_i - \alpha_c)^2 + \dots \end{aligned} \quad . \tag{B2}$$

We have used the fact that $\partial \sigma / \partial \alpha_r$ vanishes at the critical point, because this is where the growth rate σ reaches a maximum with respect to variations in the real wavenumber α_r. Sorting out the real part of Eq. (B2), we find:

$$\sigma = \frac{\partial \sigma}{\partial \varepsilon}\varepsilon + \frac{\partial \omega}{\partial \alpha_r}\alpha_i + \frac{1}{2}\frac{\partial^2 \sigma}{\partial \alpha_{r^2}}[(\alpha_r - \alpha_c)^2 - \alpha_i^2] - \frac{\partial^2 \omega}{\partial \alpha_{r^2}}(\alpha_r - \alpha_c)\alpha_i + \dots \tag{B3}$$

Temporal instability assumes α_i is identically zero; letting $k_r = k_c$, we have to leading order that:

$$\sigma(\alpha_i \equiv 0) \approx \frac{\partial \sigma}{\partial \varepsilon}\varepsilon. \tag{B4}$$

Spatial instability assumes σ is identically zero, so (with $k_r = k_c$):

$$0 = \frac{\partial \sigma}{\partial \varepsilon}\varepsilon + \frac{\partial \omega}{\partial \alpha_r}\alpha_i - \frac{1}{2}\frac{\partial^2 \sigma}{\partial \alpha_{r^2}}\alpha_i^2 + \dots . \tag{B5}$$

To lowest order,

$$\alpha_i(\sigma \equiv 0) \approx -\frac{(\partial \sigma / \partial \varepsilon)\varepsilon}{\partial \omega / \partial \alpha_r}. \tag{B6}$$

Thus, comparing Eq. (B4) and Eq. (B6), we see that to leading order,

$$\alpha_i(\sigma \equiv 0) = \frac{\sigma(\alpha_i \equiv 0)}{-\partial \omega / \partial \alpha_r}. \tag{B7}$$

See Gaster[93] for a more general discussion.

REFERENCES

1. Acheson, D. J. *Elementary Fluid Dynamics.* Oxford: Oxford University Press, 1990.
2. Ahlers, G. "Experiments on Bifurcation and One-Dimensional Patterns in Nonlinear Systems Far from Equilibrium." In *Lectures in the Sciences of Complexity*, edited by D. L. Stein. Santa Fe Institute Studies in the Sciences of Complexity, Lectures Volume I, 175–224. Reading, MA: Addison-Wesley, 1989.
3. Ahlers, G., and R. P. Behringer. "Evolution of Turbulence from the Rayleigh-Bénard Instability." *Phys. Rev. Lett.* **40** (1978): 712–716.
4. Ahlers, G., D. S. Cannell, and V. Steinberg. "Time Dependence of Flow Patterns near the Convective Threshold in a Cylindrical Container." *Phys. Rev. Lett.* **54** (1985): 1373–1376.
5. Albarede, P., and P. A. Monkewitz. "A Model for the Formation of Oblique Shedding and 'Chevron' Patterns in Cylinder Wakes." *Phys. Fluids A* **4** (1992): 744.
6. Andereck, C. D., S. S. Lui, and H. L. Swinney. "Flow Regimes in a Circular Couette System with Independently Rotating Cylinders." *J. Fluid Mech.* **164** (1986: 155–183.
7. Anderson, P. W., and D. L. Stein. "Broken Symmetry, Emergent Properties, Dissipative Structures, Life: Are They Related?" In *Basic Notions of Condensed Matter Physics*, edited by P. W. Anderson, 262–286. Menlo Park, CA: Benjamin-Cummings, 1984.
8. Babcock, K. L., G. Ahlers, and D. S. Cannell. "Noise Sustained Structure in Taylor-Couette Flow with Through-Flow." *Phys. Rev. Lett.* **67** (1991): 3388–3391.
9. Bayly, B. J., S. A. Orszag, and T. Herbert. "Instability Mechanisms in Shear-Flow Transition." *Ann. Rev. Fluid Mech.* **20** (1988): 359–391.
10. Behringer, R. P. "Rayleigh-Bénard Convection and Turbulence in Liquid Helium." *Rev. Mod. Phys.* **57** (1985): 657–688.
11. Behringer, R. P., and G. Ahlers. "Heat Transport and Critical Slowing Down near the Rayleigh-Bénard Instability in Cylindrical Containers." *Phys. Lett.* **62A** (1977): 329–331.
12. Behringer, R. P., and G. Ahlers. "Heat Transport and Temporal Evolution of Fluid Flow near the Rayleigh-Bénard Instability in Cylindrical Containers." *J. Fluid Mech.* **125** (1982): 219–258.
13. Bénard, H. "Les tourbillons cellulaires dans une nappe liquide transportant de la chaleur par convection en régime permanent." *Ann. Chim. Phys.* **23** (1901): 62–144.
14. Benjamin, T. B. "Bifurcation Phenomena in Steady Flows of a Viscous Fluid. II. Experiments." *Proc. Roy. Soc. London*, Ser. A **359** (1978): 27–43.

15. Benjamin, T. B. "Bifurcation Phenomena in Steady Flows of a Viscous Fluid. I. Theory." *Proc. Roy Soc. London*, Ser. A **359** (1978): 1–26.
16. Benjamin, T. B., and T. Mullin. "Notes on the Multiplicity of Flows in the Taylor Experiment." *J. Fluid Mech.* **121** (1982): 219–230.
17. Benjamin, T. B., and F. Ursell. "The Stability of the Plane Free Surface of a Liquid in Vertical Periodic Motion." *Proc. Roy. Soc. London*, Ser. A **225** (1954): 505–515.
18. Benney, D. J., and C. C. Lin. "On the Secondary Motion Induced by Oscillations in a Shear Flow." *Phys. Fluids* **3** (1960): 656–657.
19. Bensimon, D., L. P. Kadanoff, S. Liang, B. Shraiman, and C. Tang. "Viscous Flows in Two Dimensions." *Rev. Mod. Phys.* **58** (1986): 977–1000.
20. Bergé, P., M. Dubois, P. Manneville, and Y. Pomeau. "Intermittency in Rayleigh-Bénard Convection." *J. Phys.(Paris) Lett.* **40** (1979): L-505.
21. Bers, A. "Space-Time Evolution of Plasma Instabilities—Absolute and Convective." In *Handbook of Plasma Physics*, Vol. 1: Basic Plasma Physics I, edited by A. A. Galeev and R. N. Sudan, 451–517. North-Holland, 1983.
22. Betchov, R., and W. O. Crimnale. *Stability of Parallel Flows.* New York: Academic, 1967.
23. Block, M. J. "Surface Tension as the Cause of Bénard Cells and Surface Deformation in a Liquid Film." *Nature* **178** (1956): 650–651.
24. Bloor, M. S. "The Transition to Turbulence in the Wake of a Circular Cylinder." *J. Fluid Mech.* **19** (1964): 290–304.
25. Brand, H. R. "Phase Dynamics—A Review and Perspective." In *Propagation in Systems far from Equilibrium*, edited by J. Wesfreid, H. Brand, P. Manneville, G. Albinet, and N. Boccara, 206–224. Berlin: Springer-Verlag, 1988.
26. Brandstater, A., J. Swift, H. L. Swinney, A. Wolf, J. Doyne Farmer, E. Jen, and P. J. Crutchfield. "Low-Dimensional Chaos in a Hydrodynamic System." *Phys. Rev. Lett.* **51** (1983): 1442–1445.
27. Brandstater, A., and H. L. Swinney. "Strange Attractors in Weakly Turbulent Couette-Taylor Flow." *Phys. Rev. A* **35** (1987): 2207–2220.
28. Brown, R. A. "The Shape and Stability of Three-Dimensional Interfaces." Ph.D. Thesis, University of Minnesota, 1979.
29. Brown, G. L., and A. Roshko. "On Density Effects and Large Structure in Turbulent Mixing Layers." *J. Fluid Mech.* **64** (1974): 775–816.
30. Busse, F. H. "Non-linear Properties of Thermal Convection." *Rep. Prog. Phys.* **41** (1978): 1929–1967.
31. Busse, F. H., and R. M. Clever. "Instabilities of Convection Rolls in a Fluid of Moderate Prandtl Number." *J. Fluid Mech.* **91** (1979): 319–335.
32. Busse, F. H. "Transition to Turbulence in Rayleigh-Bénard Convection." In *Hydrodynamic Instabilities and the Transition to Turbulence*, 2nd ed., edited by H. L. Swinney and J. P. Gollub, 97–137. Berlin: Springer-Verlag,1985.
33. Busse, F. H. "Numerical Analysis of Secondary and Tertiary States of Fluid Flow and Their Stability Properties." *Appl. Sci. Res.* **48** (1991): 341–351.

34. Bust, G. S., B. C. Dornblaser, and E. L. Koschmieder. "Amplitudes and Wavelengths of Wavy Taylor Vortices." *Phys. Fluids* **28** (1985): 1243–1247.
35. Buzug, Th., K. Pawelzik, J. von Stamm, and G. Pfister. "Mutual Information and Global Strange Attractors in Taylor-Couette Flow." *Physica D* **72** (1994): 343–350.
36. Cantwell, B., D. Coles, and P. Dimotakis. "Structure and Entrainment in the Plane of Symmetry of a Turbulent Spot." *J. Fluid Mech.* **87** (1978): 641–672.
37. Carr, J. *Applications of Centre Manifold Theory.* New York: Springer-Verlag, 1981.
38. Chandrasekhar, S. *Hydrodynamic and Hydromagnetic Stability.* Oxford: Oxford University Press, 1961; Dover (reprint), 1981.
39. Chandrasekhar, S. *Ellipsoidal Figures of Equilibrium.* New Haven: Yale University Press, 1969.
40. Chaudhary, K. C., and L. G. Redekopp. "The Nonlinear Capillary Instability of a Liquid Jet. Part 1. Theory." *J. Fluid Mech.* **96** (1980): 257–274.
41. Chaudhary, K. C., and T. Maxworthy. "The Nonlinear Capillary Instability of a Liquid Jet. Part 2. Experiments on Jet Behaviour Before Droplet Formation." *J. Fluid Mech.* **96** (1980): 275–286.
42. Chaudhary, K. C., and T. Maxworthy. "The Nonlinear Capillary Instability of a Liquid Jet. Part 3. Experiments on Satellite Drop Formation and Control." *J. Fluid Mech.* **96** (1980): 287–297.
43. Chen, J.-D. "Growth of Radial Viscous Fingers in a Hele-Shaw Cell." *J. Fluid Mech.* **201** (1989): 223–242.
44. Chomaz, J. M., P. Huerre, and L. G. Redekopp. "Bifurcations to Local and Global Modes in Spatially Developing Flows." *Phys. Rev. Lett.* **60** (1988): 25–28.
45. Chossat, P., Y. Demay, and G. Iooss. "Interaction de modes azimutaux dans le probème de Couette-Taylor." *Arch. Ration. Mech. Anal.* **99** (1987): 213–248.
46. Chossat, P., and G. Iooss. *The Couette-Taylor Problem.* New York: Springer-Verlag, 1994.
47. Ciliberto, S., and J. P. Gollub. "Chaotic Mode Competition in Parametrically Excited Surface Waves." *J. Fluid Mech.* **158** (1985): 381–398.
48. Clever, R. M., and F. H. Busse. "Transition to Time-Dependent Convection." *J. Fluid Mech.* **65** (1974): 625–645.
49. Clever, R. M., and F. H. Busse. "Large Wavelength Convection Rolls in Low Prandtl Number Fluids." *J. Appl. Math. Phys. (ZAMP)* **29** (1978): 711–714.
50. Cliffe, K. A., and T. Mullin. "A Numerical and Experimental Study of Anaomalous Modes in the Taylor Experiment." *J. Fluid Mech.* **153** (1985): 243–258.
51. Cliffe, K. A. "Numerical Calculations of the Primary-Flow Exchange Process in the Taylor Problem." *J. Fluid Mech.* **197** (1988): 57–79.
52. Coles, D. "Transition in Circular Couette Flow." *J. Fluid Mech.* **21** (1965): 385–425.
53. Coles, D. "A Note on Taylor Instability in Circular Couette Flow." *J. Appl. Mech.* **89** (1967): 529–534.

54. Coughlin, K. T., and P. S. Marcus. "Modulated Waves in Taylor-Couette Flow. Part 1. Analysis." *J. Fluid Mech.* **234** (1992): 1–18.
55. Coughlin, K. T., and P. S. Marcus. "Modulated Waves in Taylor-Couette Flow. Part 2. Numerical Simulation." *J. Fluid Mech.* **234** (1992): 19–46.
56. Coughlin, K. T., P. S. Marcus, R. P. Tagg, and H. L. Swinney. "Distinct Quasiperiodic Modes with Like Symmetry in a Rotating Fluid." *Phys. Rev. Lett.* **66** (1991): 1161–1164.
57. Coutanceau, M., and R. Bouard. "Experimental Determination of the Main Features of the Viscous Flow in the Wake of a Circular Cylinder in Uniform Translation. Part 1. Steady Flow." *J. Fluid Mech.* **79** (1977): 231–256.
58. Coutanceau, M., and R. Bouard. "Experimental Determination of the Main Features of the Viscous Flow in the Wake of a Circular Cylinder in Uniform Translation. Part 2. Unsteady Flow." *J. Fluid Mech.* **79** (1977): 257–272.
59. Craik, A. D. D. "Nonlinear Resonant Instability in Boundary Layers." *J. Fluid Mech.* **50** (1971): 393–413.
60. Craik, A. D. D. *Wave Interactions and Fluid Flows.* Cambridge: Cambridge University Press, 1985.
61. Crawford, J. D., and E. Knobloch. "Symmetry and Symmetry-Breaking Bifurcations in Fluid Dynamics." *Ann. Rev. Fluid Mech.* **23** (1991): 341–387.
62. Croquette, V., and H. Williams. "Nonlinear Competition Between Waves on Convective Rolls." *Phys. Rev. A* **39** (1989): 2765.
63. Cross, M. C. "Derivation of the Amplitude Equation at the Rayleigh-Bénard Instability." *Phys. Fluids* **23** (1980): 1727.
64. Cross, M. C. "Traveling and Standing Waves in Binary-Fluid Convection." *Phys. Rev. Lett.* **57** (1986): 2935.
65. Cross, M. C. "Structure of Nonlinear Traveling-Wave States in Finite Geometries." *Phys. Rev. A* **38** (1988): 3593.
66. Cross, M. C., and P. C. Hohenberg. "Pattern Formation Outside of Equilibrium." *Rev. Mod. Phys.* **65** (1993): 851–1112.
67. Daccord, G., J. Nittmann, and H. E. Stanley. "Radial Viscous Fingers and Diffusion-Limited Aggregation: Fractal Dimension and Growth Sites." *Phys. Rev. Lett.* **56** (1986): 336–339.
68. Davey, A. "The Growth of Taylor Vortices in Flow Between Rotating Cylinders." *J. Fluid Mech.* **14** (1962): 336–368.
69. Deissler, R. J. "External Noise and the Origin and Dynamics of Structure in Convectively Unstable Systems." *J. Stat. Phys.* **54** (1989): 1459.
70. Demay, Y., and G. Iooss. 1984 [need reference].
71. Demay, Y., G. Iooss, and P. Laure. "Wave Patterns in the Small Gap Couette-Taylor Problem." *Euro. J. Mech. B/Fluids* **11** (1992): 621.
72. DiPrima, R. C., and H. L. Swinney. "Instabilities and Transition in Flow Between Concentric Rotating Cylinders." In *Hydrodynamic Instabilities and the Transition to Turbulence*, 2nd ed., edited by H. L. Swinney and J. P. Gollub, 139–180. New York: Springer-Verlag, 1985.

73. Dominguez-Lerma, M. A., G. Ahlers, and D. S. Cannell. "Effects of 'Kalliroscope' Flow Visualization Particles on Rotating Couette-Taylor Flow." *Phys. Fluids* **28** (1985): 1204–1206.
74. Donnelly, R. J. "Experiments on the Stability of Viscous Flow Between Rotating Cylinders I. Torque Measurements." *Proc. Roy. Soc. London*, Ser. A **46** (1958): 312–325.
75. Donnelly, R. J., and N. J. Simon. "An Empirical Torque Relation for Supercritical Flow Between Rotating Cylinders (With an Appendix by G. K. Batchelor)." *J. Fluid Mech.* **7** (1960): 401–418.
76. Donnelly, R. J., and W. Glaberson. "Experiments on the Capillary Instability of a Liquid Jet." *Proc. Roy. Soc. London*, Ser. A **290** (1966): 547–556.
77. Donnelly, R. J., K. Park, R. Shaw, and R. W. Walden. "Early Nonperiodic Transitions in Couette Flow." *Phys. Rev. Lett.* **44** (1980): 987–989.
78. Drazin, P. G., and R. S. Johnson. *Solitons: An Introduction.* Cambridge: Cambridge University Press, 1989.
79. Drazin, P. G., and W. H. Reid. *Hydrodynamic Stability.* Cambridge: Cambridge University Press, 1981.
80. Dubois, M., and P. Bergé. "Instabilités de couche limite dans un fluide en convection. Evolution vers la turbulence." *J. Physique* **42** (1981): 167–174.
81. Dusek, J., P. Fraunie, and P. Le Gal. "Local Analysis of the Onset of Instability in Shear Flows." *Phys. Fluids* **6** (1994): 172.
82. Eckhaus, W. *Studies in Nonlinear Stability Theory.* Berlin: Springer-Verlag, 1965.
83. Edwards, W. S. "New Stability Analyses for the Couette-Taylor Problem." Ph.D. Thesis, University of Texas at Austin, 1990.
84. Edwards, W. S., R. P. Tagg, B. C. Dornblaser, H. L. Swinney, and L. S. Tuckerman. "Periodic Traveling Waves with Nonperiodic Pressure." *Eur. J. Mech. B/Fluids* **10** (suppl.) (1991): 205–210.
85. Eisenlohr, H., and H. Eckelmann. "Vortex Splitting and Its Consequences in the Vortex Street Wake of Cylinders at Low Reynolds Number." *Phys. Fluids A* **1** (1989): 189–192.
86. Farrell, B. F. "Optimal Excitation of Perturbations in Viscous Shear Flow." *Phys. Fluids* **31** (1988): 2093.
87. Farrell, B. F., and P. J. Ioannou. "Optimal Excitation of Three-Dimensional Perturbations in Viscous Constant Shear Flow." *Phys. Fluids A* **5** (1993): 1390–1400.
88. Farrell, B. F., and P. J. Ioannou. "Perturbation Growth in Shear Flow Exhibits Universality." *Phys. Fluids A* **5** (1993): 2298–2300.
89. Fenstermacher, P. R., H. L. Swinney, and J. P. Gollub. "Dynamical Instabilities and the Transition to Chaotic Taylor Vortex Flow." *J. Fluid Mech.* **94** (1979): 103–28.
90. Fermigier, M., L. Limat, J. E. Wesfried, P. Boudinet, and C. Quilliet. "Two-Dimensional Patterns in Rayleigh-Taylor Instability of a Thin Layer." *J. Fluid Mech.* **236** (1992): 349–383.

91. Fineberg, J., E. Moses, and V. Steinberg. "Spatially and Temporally Modulated Traveling-Wave Pattern in Convecting Binary Mixtures." *Phys. Rev. Lett.* **61** (1988): 838.
92. Fornberg, B. "A Numerical Study of Steady Viscous Flow Past a Circular Cylinder." *J. Fluid Mech.* **98** (1980): 819–855.
93. Gaster, M. "A Note on the Relation Between Temporally-Increasing and Spatially-Increasing Disturbances in Hydrodynamic Stability." *J. Fluid Mech.* **14** (1962): 222–224.
94. Gaster, M. "Vortex Shedding from Slender Cones at Low Reynolds Numbers." *J. Fluid Mech.* **38** (1969): 565–576.
95. Gaster, M. "A Theoretical Model of a Wave Packet in the Boundary Layer on a Flat Plate." *Proc. Roy. Soc. Lond.*, Ser. A **347** (1975): 271–289.
96. Gaster, M., and I. Grant. "An Experimental Investigation of the Formation and Development of a Wave Packet in a Laminar Boundary Layer." *Proc. Roy. Soc. Lond.*, Ser. A **347** (1975): 253–269.
97. Ginzburg, V. L., and L. D. Landau. "On the Theory of Superconductivity." *Zh. Eksp. Teor. Fiz.* **20** (1950): 1064.
98. Goedde, E. F., and M. C. Yuen. "Experiments on Liquid Jet Instability." *J. Fluid Mech.* **40** (1970): 495–511.
99. Gollub, J. P. "Nonlinear Waves: Dynamics and Transport." *Physica D* **51** (1991): 501–511.
100. Gollub, J. P., and H. L. Swinney. "Onset of Turbulence in a Rotating Fluid." *Phys. Rev. Lett.* **35** (1975): 927–930.
101. Gollub, J. P., A. R. McCarriar, and J. E. Steinman. "Convective Pattern Evolution and Secondary Instabilities." *J. Fluid Mech.* 125, (1982): 259–281.
102. Golubitsky, M., and I. Stewart. "Symmetry and Stability in Taylor-Couette Flow." *SIAM J. Math. Anal.* **17** (1986): 249–88.
103. Golubitsky, M., and W. F. Langford. "Pattern Formation and Bistability in Flow Between Counterrotating Cylinders." *Physica D* **32** (1988): 362–92.
104. Gorkov, L. P. "Stationary Convection in a Plane Liquid Layer near the Critical Heat Transfer Point." *Zh. Eksp. Teor. Fiz.* [English transl.: Sov. Phys.—JETP] **6** (1957): 311–315.
105. Gorman, M., and H. L. Swinney. "Visual Observation of the Second Characteristic Mode in a Quasiperiodic Flow." *Phys. Rev. Lett.* **43** (1979): 1871–1875.
106. Gorman, M., and H. L. Swinney. "Spatial and Temporal Characteristics of Modulated Waves in the Circular Couette System." *J. Fluid Mech.* **117** (1982): 123–142.
107. Gorman, M., H. L. Swinney, and D. A. Rand. "Doubly Periodic Circular Couette Flow: Experiments Compared with Predictions from Dynamics and Symmetry." *Phys. Rev. Lett.* **46** (1981): 992–995.
108. Hammache, M., and M. Gharib. "An Experimental Study of the Parallel and Oblique Vortex Shedding from Circular Cylinders." *J. Fluid Mech.* **232** (1991): 567–590.

109. Heinrichs, R. M., D. S. Cannell, G. Ahlers, and M. Jefferson. "Experimental Test of the Perturbation Expansion for the Taylor Instability at Various Wavenumbers." *Phys. Fluids* **31** (1988): 250–255.
110. Henderson, D. M., and J. L. Hammack. "Experiments on Ripple Instabilities. Part 1. Resonant Triads." *J. Fluid Mech.* **184** (1987): 15–41.
111. Henningson, D. S., and S. C. Reddy. "On the Role of Linear Mechanisms in Transition to Turbulence." *Phys. Fluids* **6** (1994): 1396–1398.
112. Herbert, T. "Periodic Secondary Motions in a Plane Channel." In *Proc. Intl. Conf. Njmer. Methods Fluid Dyn.*, edited by A. I. van de Vooren and P. J. Zandbergen, 235–240. Berlin: Springer-Verlag, 1976.
113. Herbert, T. "Secondary Instability of Boundary Layers." *Ann. Rev. Fluid Mech.* **20** (1988): 487–526.
114. Heutmaker, M. S., P. N. Fraenkel, and J. P. Gollub. "Convection Patterns: Time Evolution of the Wave-Vector Field." *Phys. Rev. Lett.* **54** (1985): 1369–1372.
115. Heutmaker, M. S., and J. P. Gollub. "Wave-Vector Field of Convective Flow Patterns." *Phys. Rev. A* **35** (1987): 242–260.
116. Hirst, D. "The Aspect Ratio Dependence of the Attractor Dimension in Taylor-Couette Flow." Ph.D. Thesis, University of Texas, Austin, 1987.
117. Ho, C.-M., and P. Huerre. "Perturbed Free Shear Layers." *Ann. Rev. Fluid Mech.* **16** (1984): 365–424.
118. Homann, F. "Einfluss gorsser Zähigkeit bei der Strömung um Zylinder." *Forsch Gebiet. Ing.* **7** (1936): 1.
119. Huerre, P. "Spatio-Temporal Instabilities in Closed and Open Flows." In *Instabilities and Nonequilibrium Structures*, edited by E. Tirapegui and D. Villarroel, 141–177. D. Reidel, 1987.
120. Huerre, P., and P. A. Monkewitz. "Absolute and Convective Instabilities in Free Shear Layers." *J. Fluid Mech.* **159** (1985): 151–168.
121. Huerre, P., and P. A. Monkewitz. "Local and Global Instabilities in Spatially Developing Flows." *Ann. Rev. Fluid Mech.* **22** (1990): 473–537.
122. Illingworth, C. R. "Flow at Small Reynolds Number." In *Laminar Boundary Layers*, edited by L. Rosenhead, 163–197. Oxford: Oxford University Press, 1963.
123. Ito, N. *Trans. Japan Soc. Areo. Space Sci.* **17** (1974): 65.
124. Jackson, C. P. "A Finite Element Study of the Onset of Vortex Shedding in Flow Past Variously Shaped Bodies." *J. Fluid Mech.* **182** (1987): 23–45.
125. Jacobs, J. W., D. K. Klein, D. G. Jenkins, and R. F. Benjamin. "Instability Growth Patterns of a Shock-Accelerated Thin Fluid Layer." *Phys. Rev. Lett.* **70** (1993): 583–586.
126. Joseph, D. D. *Stability of Fluid Motions, I and II.* New York: Springer-Verlag, 1976.
127. Kachanov, Yu. S., V. V. Kozlob, and V. Ya. Levchenko. "Experimental Investigation of the Influence of Cooling on t he Stability of Laminar Boundary Layer." *Rep. Siberian Div. Acad. Sci. USSR* **8** (1974).

128. Kachanov, Y. S. "Physical Mechanisms of Laminar-Boundary-Layer Transition." *Ann. Rev. Fluid Mech.* **26** (1994): 411–482.
129. Kachanov, Yu. S., and V. Ya. Levchenko. "The Resonant Interaction of Disturbances at Laminar-Turbulent Transition in a Boundary Layer." *J. Fluid Mech.* **138** (1984): 209–247.
130. Karniadakis, G. Em., B. B. Mikic, and A. T. Patera. "Minimum-Dissipation Transport Enhancement by Flow Destabilization: Reynolds Analogy Revisited." *J. Fluid Mech.* **192** (1988): 365–391.
131. Karniadakis, G. E. and G. S. Triantafyllou. "Three-Dimensional Dynamics and Transition to Turbulence in the Wake of Bluff Objects." *J. Fluid Mech.* **238** (1992): 1–30.
132. King, G. P., and H. L. Swinney. "Limits of Stability and Irregular Flow Patterns in Wavy Vortex Flow." *Phys. Rev. A* **27** (1983): 1240–1243.
133. King, G. P., Y. Li, W. Lee, H. L. Swinney, and S. Marcus. "Wave Speeds in Wavy Taylor-Vortex Flow." *J. Fluid Mech.* **141** (1984): 365–390.
134. Kirchgassner, K., and P. Sorger. "Branching Analysis for the Taylor Problem." *Qtr. J. Mech. Appl. Math.* **22** (1969): 183–209.
135. Klebanoff, P. S., K. D. Tidstrom, and L. M. Sargent. "The Three-Dimensional Nature of Boundary-Layer Instability." *J. Fluid Mech.* **12** (1962): 1–34.
136. Klingmann, B. G. B., A. V. Boiko, K. J. A. Westin, V. V. Kozlov, and P. H. Alfredsson. "Experiments on the Stability of Tollmein-Schlichting Waves." *Eur. J. Mech. B/Fluids* **12**(1993): 493–514.
137. Knobloch, E. "System Symmetry Breaking and Shil'nikov Dynamics." In *Pattern Formation: Symmetry Methods and Applications.* Fields Institute Communications. American Mathematical Society, 1994.
138. Knobloch, E., and J. De Luca. "Amplitude Equations for Travelling Wave Convection." *Nonlinearity* **3** (1990): 975.
139. Koch, W. "Local Instability Characteristics and Frequency Determination of Self-Excited Wake Flows." *J. Sound Vib.* **99** (1985): 53–83.
140. Kogelman, S. and R. C. DiPrima. "Stability of Spatially Periodic Supercritical Flows in Hydrodynamics." *Phys. Fluids* **13** (1970): 1–11.
141. Kolodner, P., and C. M. Surko. "Weakly Nonlinear Traveling-Wave Convection." *Phys. Rev. Lett.* **61** (1988): 842.
142. Kolodner, P., C. M. Surko, and H. Williams. "Dynamics of Traveling Waves near the Onset of Convection in Binary Fluid Mixtures." *Physica D* **37** (1989): 319.
143. Koschmieder, E. L. "Bénard Convection." *Adv. Chem. Phys.* **26** (1974): 177–212.
144. Koschmieder, E. L. "Stability of Supercritical Bénard Convection and Taylor Vortex Flow." *Adv. Chem. Phys.* **32** (1975): 109–133.
145. Koschmieder, L. *Bénard Cells and Taylor Vortices.* Cambridge: Cambridge University Press, 1993.
146. Kose, K. "Spatial Mapping of Velocity Power Spectra in Taylor-Couette Flow Using Ultrafast NMR Imaging." *Phys. Rev. Lett.* **72** (1994): 1467.

147. Kramer, L., and W. Zimmermann. "On The Eckhaus Instability for Spatially Periodic Patterns." *Physica D* **16** (1985): 221–232.
148. Krishnamurti, R. "Some Further Studies on the Transition to Turbulent Convection." *J. Fluid Mech.* **60** (1973): 285–303.
149. Kundu, P. K. *Fluid Mechanics.* San Diego: Academic Press, 1990.
150. Küppers, G., and D. Lortz. "Transition from Laminar Convection to Thermal Turbulence in a Rotating Fluid Layer." *J. Fluid Mech.* **35** (1969): 609–620.
151. L'vov, V. S., A. A. Predtechenskii, and A. I. Chernykh. "Bifurcation and chaos in the System of Taylor Vortices—Laboratory and Numerical Experiment." In *Nonlinear Dynamics and Turbulence*, edited by G. I. Barenblatt, G. Iooss, and D. D. Joseph, 238–280. Boston: Pitman, 1983.
152. Lamb, S. H. *Hydrodynamics*, 6th ed. Cambridge: Cambridge University Press, 1932.
153. Landahl, M. T. "Wave Breakdown and Turbulence." *SIAM J. Appl. Math* **28** (1975): 735.
154. Landau, L. D. "On The Problem of Turbulence." *C. R. (Dokl.) Acad. Sc. URSS* **44** (1944): 311–314.
155. Landau, L. D., and E. M. Lifshitz. *Fluid Mechanics*, 1st ed. Oxford: Pergamon, 1959.
156. Langford, W. F., R. Tagg, E. Kostelich, H. L. Swinney, and M. Golubitsky. "Primary Instabilities and Bicriticality in Flow Between Counter-Rotating Cylinders." *Phys. Fluids* **31** (1988): 776–785.
157. Lathrop, D. P., J. Fineberg, and H. L. Swinney. "Turbulent Flow Between Concentric Rotating Cylinders at Large Reynolds Number." *Phys. Rev. Lett.* **68** (1992): 1515–1518.
158. Lathrop, D. P., J. Fineberg, and H. L. Swinney. "Transition to Shear-Driven Turbulence in Couette-Taylor Flow." *Phys. Rev. A* **46** (1992b): 6390.
159. Lefrançois, M., and B. Ahlborn. "Phase Front Analysis of Vortex Streets." *Phys. Fluids* **6** (1994): 2021.
160. Lenormand, R. "Pattern Growth and Fluid Dispements Through Porous Media." *Physica A* **140** (1986): 114–123.
161. Lewis, D. J. "The Instability of Liquid Surfaces When Accelerated in a Direction Perpendicular to Their Planes. II." *Proc. R. Soc. London* **Ser. A 202** (1950): 81–96.
162. Libchaber, A., and J. Maurer. "Une expérience de Rayleigh-Bénard en géométrie réduite; multiplication, accrochage et démultiplication de fréquencies *J. Physique Colloq.* **41-C3** (1980): 51–56.
163. Lighthill, M. J. "Introduction. Real and Ideal Fluids. Section II.3. Instability and Turbulence." In *Laminar Boundary Layers*, edited by L. Rosenhead. Oxford: Oxford University Press, 1963.
164. Lin, C. C. "Some Physical Aspects of the Stability of Parallel Flows." *Proc. Nat. Acad. Sci. (USA)* **40** (1954): 741–747.
165. Lin, C. C. *The Theory of Hydrodynamic Stability.* Cambridge: Cambridge University Press, 1955. Reprinted with corrections in 1966.

166. Malkus, W. V. R., and G. Veronis. "Finite Amplitude Cellular Convection." *J. Fluid Mech.* **4** (1958): 225–260.
167. Manneville, P. *Dissipative Structures and Weak Turbulence.* San Diego: Academic Press, 1990.
168. Marcus, P. S. "Simulation of Taylor-Couette Flow. I. Numerical Methods and Comparison with Experiment." *J. Fluid Mech.* **146** (1984): 45–64.
169. Marcus, P. S. "Simulation of Taylor-Couette Flow. II. Numerical Results for Wavy-Vortex Flow with One Travelling Wave." *J. Fluid Mech.* **146** (1984): 65–113.
170. Maslowe, S. A. "Critical Layers in Shear Flows." *Ann. Rev. Fluid Mech.* **18** (1986): 405–432.
171. Maslowe, S. A. "Shear Flow Instabilities and Transition." In *Hydrodynamic Instabilities and Transition to Turbulence*, 2nd ed., edited by H. L. Swinney and J. P. Gollub, 181–228, 295–297. New York: Springer-Verlag, 1985.
172. Mathis, C., M. Provansal, and L. Boyer. "The Bénard-Von Kármán Instability: An Experimental Study near Threshold." *J. Phys. Lett.* **45** (1984): 483–491.
173. Matisse, P., and M. Gorman. "Neutrally Buoyant Anisotropic Particles for Flow Visualization." *Phys. Fluids* **27** (1984): 759–760.
174. McLachlan, N. W. *Theory and Application of Mathieu Functions.* Oxford: Oxford University Press, 1947.
175. McLean, J. W., and P. G. Saffman. "The Effect of Surface Tension on the Shape of Fingers ina Hele-Shaw Cell." *J. Fluid. Mech.* **102** (1981): 455.
176. Michalke, A. "The Instability of Free Shear Layers." *Prog. Aerospace Sci.* **12** (1972): 213–239.
177. Miles, J. W., and D. Henderson. "Parametrically Forced Surface Waves." *Ann. Rev. Fluid Mech.* **22** (1990): 143–65.
178. Monkewitz, P. A. *Eur. J. Mech. B* **9** (1990): 395.
179. Morkovin, M. V. "Guide to Experiments on Instability and Laminar-Turbulent Transition in Shear Flows." In *Notes AIAA Prof. Study Ser: Instabilities and Transition to Turbulence*, 1985.
180. Mullin, T. "Mutations of Steady Cellular Flows in the Taylor Experiment." *J. Fluid Mech.* **121** (1982): 207–218.
181. Mullin, T. "Finite-Dimensional Dynamics in Taylor-Couette Flow." *IMA J. Appl. Math.* **46** (1991): 109–119.
182. Mullin, T. "Finite-Dimensional Dynamics and Chaos in Fluid Flows." In *Introduction to Plasma Physics*, edited by R. O. Dendy. Cambridge: Cambridge University Press, 1993.
183. Nakayama, Y., and the Japan Society for Mechanical Engineers, eds. *Visualized Flow.* Oxford: Pergammon, 1988.
184. Newell, A. C. "The Dynamics of Patterns." In *Lectures in the Sciences of Complexity*, edited by D. L. Stein. Santa Fe Institute Studies in the Sciences of Complexity, Lectures Volume I, 107–174. Redwood City, CA: Addison-Wesley, 1989.

185. Newell, A. C., T. Passot, and J. Lega. "Order Parameter Equations for Patterns." *Ann. Rev. Fluid Mech.* **25** (1993): 399–453.
186. Newell, A. C., and J. A. Whitehead. "Finite Bandwidth, Finite Amplitude Convection." *J. Fluid Mech.* **38** (1969): 279.
187. Niemela, J. J., G. Ahlers, and D. S. Cannell. "Localized Traveling Wave States in Binary Fluid Convection." *Phys. Rev. Lett.* **64** (1990): 1365.
188. Nishioka, M., S. Iida, and Y. Ichikawa. "An Experimental Investigation of the Stability of Plane Poiseuille Flow." *J. Fluid Mech.* **72** (1975): 731–751.
189. Nishioka, M., and H. Sato. "Mechanism of Determination of the Shedding Frequency of Vortices Behind a Cylinder at Low Reynolds Numbers." *J. Fluid Mech.* **89** (1978): 49–60.
190. Noack, B. R., F. Ohle, and H. Eckelmann. "On Cell Formation in Vortex Streets." *J. Fluid Mech.* **227** (1991): 293–308.
191. Noack, B. R., M. König, and H. Eckelmann. "Three-Dimensional Stability Analysis of the Periodic Flow Around a Circular Cylinder." *Phys. Fluids A* **5** (1993): 1279–1281.
192. Normand, C., Y. Pomeau, and M. G. Verlade. "Convective Instability : A Physicist's Approach." *Rev. Mod. Phys.* **49** (1977): 581–624.
193. Oertel, H., Jr. "Wakes Behind Blunt Bodies." *Ann. Rev. Fluid Mech.* **22** (1990): 539–564.
194. Orszag, S. A. "Accurate Solution of the Orr-Sommerfeld Stability Equation." *J. Fluid Mech.* **50** (1971): 689–703.
195. Orszag, S. A., and A. T. Patera. "Subcritical Transition to Turbulence in Plane Channel Flows." *Phys. Rev. Lett.* **45** (1980): 989–993.
196. Packard, N. H., J. P. Crutchfield, J. D. Farmer, and R. S. Shaw. "Geometry from Time Series." *Phys. Rev. Lett.* **45** (1980): 712.
197. Palm, E. "Nonlinear Thermal Convection." *Ann. Rev. Fluid Mech.* **7** (1975): 39–61.
198. Panton, R. L. *Incompressible Flow.* New York: Wiley, 1984.
199. Park, K., and R. J. Donnelly. "Study of the Transition of Taylor Vortex Flow." *Phys. Rev. A* **24** (1981): 2277–2279.
200. Park, D. S., and L. G. Redekopp. "A Model for Pattern Selection in Wake Flows." *Phys. Fluids* **4** (1992): 1697–1706.
201. Paterson, L. "Radial Fingering in a Hele Shaw Cell." *J. Fluid Mech.* **113** (1981): 513–529.
202. Pearson, J. R. A. "On Convection Cells Induced by Surface Tension." *J. Fluid Mech.* **4** (1958): 489–500.
203. Pfister, G., H. Schmidt, K. A. Cliffe, and T. Mullin. "Bifurcation Phenomena in Taylor-Couette Flow in a Very Short Annulus." *J. Fluid Mech.* **191** (1988): 1–18.
204. Phillips, O. M. "Nonlinear Dispersive Waves." *Ann. Rev. Fluid Mech.* **6** (1974): 93–110.
205. Plaschko, P., E. Berger, and R. Peralta-Fabi. "Periodic Flow in the Near Wake of Straight Circular Cylinders." *Phys. Fluids A* **5** (1993): 1718–1724.

206. Plateau, J. "Experimental and Theoretical Researches on the Figures of Equilibrium of a Liquid Mass Withdrawn Form the Actions of Gravity." In *The Annual Report of the Board of Regents of the Smithsonian Institution* Washington, DC: Government Printing Office, 1863.
207. Pomeau, Y., and P. Manneville. "Stability and Fluctuations of a Spatially Periodic Convective Flow." *J. Phys. (Paris) Lett.* **40** (1979): 610.
208. Provansal, M., C. Mathis, and L. Boyer. "Bénard-von Kármán Instability: Transient and Forced Regimes." *J. Fluid Mech.* **182** (1987): 1–22.
209. Raffai, R., and P. Laure. "The Influence of an Axial Mean Flow on the Couette-Taylor Problem." *European J. Mechanics. B/Fluids* **12** (1993): 277.
210. Rand, D. "Dynamics and Symmetry. Predictions for Modulated Waves in Rotating Fluids." *Arch. Ration. Mech. Anal.* **79** (1982): 1–37.
211. Rand, R. H., and D. Armbruster. *Perturbation Methods, Bifurcation Theory and Computer Algebra.* New York: Springer-Verlag, 1987.
212. Rayleigh, L. "On the Stability, or Instability, of Certain Fluid Motions." *Proc. London Math. Soc.* **11** (1880): 57–70.
213. Rayleigh, L. "On the Dynamics of Revolving Fluids." *Proc. R. Soc. London, Ser. A* **93** (1916): 148–154.
214. Rayleigh, L. "On Convection Currents in a Horizontal Layer of Fluid when the Higher Temperature is on the Under Side." *Phil. Mag.* **32** (1916): 529–546.
215. Reynolds, O. "An Experimental Investigation of the Circumstances Which Determine Whether the Motion of Water Shall be Sinuous, and of the Law of Resistance in Parallel Channels." *Phil. Trans. Roy. Soc.* **174** (1883): 935–982.
216. Roberts, P. H. "The Solution to the Characteristic Value Problems (Appendix to R. J. Donnelly and K. W. Schwarz)." *Proc. R. Soc. London, Ser. A* **283** (1965): 550–556.
217. Rockwell, D. "Oscillations of Impinging Shear Layers." *AIAA J.* **21** (1983): 645–664.
218. Rockwell, D., F. Nuzzi, and C. Magnes. "Period Doubling in the Wake of a Three-Dimensional Cylinder." *Phys. Fluids A* **3** (1991): 1477–1478.
219. Roshko, A. "On the Development of Turbulent Wakes from Vortex Streets." Tech. Note 2913, NACA, 1953.
220. Roshko, A. "On the Development of Turbulent Wakes from Vortex Streets." Report No. 1191, NACA, 1954.
221. Ross, J. A., F. H. Barnes, J. G. Burns, and M. A. S. Ross. "The Flat Plate Boundary Layer. Part 3. Comparison of Theory with Experiment." *J. Fluid Mech.* **43** (1970): 819–832.
222. Ruelle, D., and F. Takens. "On the Nature of Turbulence." *Commun. Math. Phys.* **20** (1971): 167–192.
223. Saffman, P. G. "Viscous Fingering in Hele-Shaw Cells." *J. Fluid Mech.* **173** (1986): 73–94.
224. Saffman, P. G. "Selection Mechanism and Stability of Fingers and Bubbles in Hele-Shaw Cells." *IMA J. Appl. Math.* **46** (1991): 137–145.

225. Saffman, P. G., and G. I. Taylor. "The Penetration of a Fluid into a Porous Medium or Hele-Shaw Cell Containing a More Viscous Liquid." *Proc. R. Soc. London, Ser. A* **245** (1958): 312–329.
226. Saiki, E. M., S. Biringen, and G. Danabasoglu. "Spatial Simulation of Secondary Instability in Plane Channel Flow: Comparison of K- and H-Type Disturbances." *J. Fluid Mech.* **253** (1993): 485.
227. Saltzman, B., ed. *Theory of Thermal Convection.* New York: Dover, 1962.
228. Saric, W. S., and A. H. Nayfeh. "Nonparallel Stability of Boundary-Layer Flows." *Phys. Fluids* **18** (1975): 945–950.
229. Saric, W. S., and A. S. W. Thomas. "Experiments on the Subharmonic Route to Turbulence in Boundary Layers." In *Turbulence and Chaotic Phenomena in Fluids*, edited by T. Tatsumi, 117–122. New York: Elsevier/North-Holland, 1984.
230. Schatz, M. F. "Transition in Plane Channel Flow with Spatially Periodic Perturbations." Ph.D. Thesis, University of Texas at Austin, 1991.
231. Schatz, M. F., R. P. Tagg, H. L. Swinney, P. F. Fischer, and A. T. Patera. "Supercritical Transition in Plane Channel Flow with Spatially Periodic Perturbations." *Phys. Rev. Lett.* **66** (1991): 1579–1582.
232. Schlichting, H. "Zur Entstehung der Turbulenz bei der Plattenströmung." *Nachr. Ges. Wiss. Göttingen, Math. Phys. Klasse*, 182–208, 1933.
233. Schlichting, H. *Boundary Layer Theory*, 6th ed., Ch. 16 and 17. New York: McGraw-Hill, 1968.
234. Schlüter, A., D. Lortz, and F. Busse. "On the Stability of Steady Finite Amplitude Convection." *J. Fluid Mech.* **23** (1965): 129–144.
235. Schubauer, G. B., and H. K. Skramstad. "Laminar Boundary-Layer Oscillations and Transition on a Flat Plate." *J. Res. Natl. Bur. Stand.* **38** (1947): 251–292.
236. Segel, L. A. "Distant Sidewalls Cause Slow Slow Amplitude Modulation of Cellular Convection." *J. Fluid Mech.* **38** (1969): 203–224.
237. Sharp, D. H. "An Overview of Rayleigh-Taylor Instability." *Physica D* **12** (1984): 3–18.
238. Sherman, F. S. *Viscous Flow*, Ch. 13. New York: McGraw-Hill, 1990.
239. Sreenivasan, K. R. "Transitional and Turbulent Wakes and Chaotic Dynamical Systems." In *Nonlinear Dynamics of Transcritical Flows*, Proc. of a DFVLR International Colloquium, edited by H. L. Jordan, H. Oertel, and K. Robert, 123-154. Berlin: Springer-Verlag, 1985.
240. Sreenivasen, K. R. "The Utility of Dynamical Systems Approaches: Comment 3." In *Whither Turbulence? Turbulence at the Crossroads*, edited by J. L. Lumley, 269–291. Berlin: Springer, 1990.
241. Steinberg, V., G. Ahlers, and D. S. Cannell. "Pattern Formation and Wave-Number Selection by Rayleigh-Bénard Convection in a Cylindrical Container." *Physica Scripta* **32** (1985): 534–547.
242. Stern, C., P. Chossat, and F. Hussain. "Azimuthal Mode Interaction in Counter-Rotating Taylor-Couette Flow." *Eur. J. Mech. B/Fluids* **9** (1990): 93.

243. Stoker, J. J. *Nonlinear Vibrations.* New York: Interscience, 1950.
244. Stork, K., and U. Müller. "Convection in boxes: experiments." *J. Fluid Mech.* **54** (1972): 599–611
245. Straughan, B. *The Energy Method, Stability, and Nonlinear Convection.* New York: Springer-Verlag, 1992.
246. Strykowski, P. J., and K. R. Sreenivasan. "On the Formation and Suppression of Vortex 'Shedding' at Low Reynolds Numbers." *J. Fluid Mech.* **218** (1990): 71–107.
247. Stuart, J. T. "On the Non-linear Mechanics of Hydrodynamic Stability." *J. Fluid Mech.* **4** (1958): 1–21.
248. Stuart, J. T. "On the Nonlinear Mechanics of Wave Disturbances in Stable and Unstable Parallel Flows Part 1. The Basic Behaviour in Plane Poiseuille Flow." *J. Fluid Mech.* **9** (1960): 353–370.
249. Stuart, J. T. "Hydrodynamic Stability." In *Laminar Boundary Layers*, edited by L. Rosenhead, 492–579. Oxford: Oxford University Press, 1963.
250. Stuart, J. T., and R. C. Diprima. "On the Mathematics of Taylor-Vortex Flows in Cylinders of Finite Length." *Proc. R. Soc. London*, Ser. A **372** (1980): 357–365.
251. Swinney, H. L., and J. P. Gollub, eds. *Hydrodynamics Instabilities and the Transition to Turbulence*, 2nd ed. New York: Springer-Verlag, 1985.
252. Tabeling, P., and C. Trakas. "Spiral Vortices in a Taylor Instability Subjected to an External Magnetic Field." *J. Phys. (Paris) Lett.* **45** (1984): L159–L167.
253. Tabeling, P., G. Zocchi, and A. Libchaber. "An Experimental Study of the Saffman-Taylor Instability." *J. Fluid Mech.* **177** (1987): 67–82.
254. Tagg, R. Unpublished work on the "alternating spirals" state. 1988.
255. Tagg, R. "A Guide to Literature Related to the Taylor-Couette Problem." In *Ordered and Turbulent Patterns in Taylor-Couette Flow*, edited by C. D. Andereck and F. Hayot, 303–354. New York: Plenum, 1992.
256. Tagg, R. "The Couette-Taylor Problem." In *Nonlinear Science Today*, edited by R. Berhinger and P. Holmes, 1994.
257. Tagg, R., L. Cammack, A. Croonquist, and T. G. Wang. "Rotating Liquid Drops: Plateau's Experiment Revisited." Report No. 900-954, Jet Propulsion Laboratory, Pasadena, CA, 1978.
258. Tagg, R., W. S. Edwards, and H. L. Swinney. "Convective Versus Absolute Instability in Flow Between Counterrotating Cylinders." *Phys. Rev. A* **42** (1990): 831–837.
259. Tagg, R., W. S. Edwards, H. L. Swinney, and P. S. Marcus. "Nonlinear Standing Waves in Couette-Taylor Flow." *Phys. Rev. A* **39** (1989): 3734–3737.
260. Tagg, R., D. Hirst, and H. L. Swinney. Videotape of experiments on Taylor-vortex/spiral-vortex mode interaction. 1988.
261. Takens, F. "Detecting Strange Attractors in Turbulence." In *Dynamical Systems and Turbulence*, edited by D. A. Rand and L. S. Young. Lecture Notes in Mathematics, Vol. 898, 366. New York: Springer-Verlag, 1981.

262. Taylor, G. I. "Eddy Motion in the Atmosphere." *Phil. Trans. Roy. Soc. A* **215** (1915): 1–26.
263. Taylor, G. I. "Stability of a Viscous Liquid Contained Between Two Rotating Cylinders." *Phil. Trans. Roy. Soc. London, Ser. A* **223** (1923): 289–343.
264. Tennekes, H., and J. L. Lumley. *A First Course in Turbulence.* Cambridge, MA: MIT Press, 1972.
265. Tritton, D. J. "Experiments on the Flow Past a Circular Cylinder at Low Reynolds Number." *J. Fluid Mech.* **6** (1959): 547–567.
266. Tritton, D. J. *Physical Fluid Dynamics*, 2nd ed. Oxford: Oxford University Press, 1988.
267. Tsameret, A., and V. Steinberg. "Noise-Modulated Propagating Pattern in a Convectively Unstable System." *Phys. Rev. Lett.* **67** (1991): 3392–3395.
268. Tsiveriotis, K., and R. A. Brown. "Bifurcation Structure and the Eckhaus Instability." *Phys. Rev. Lett.* **63** (1989): 2048–2051.
269. Tollmein, W. "Über die Entstehung der Turbulenz." Nachr. Ges. Wiss. Göttingen, Math. Phys. Klasse, 21–24. 1929. Also Tech. Mem. 609, NACA, 1931.
270. Trefethen, L. N., A. E. Trefethen, S. C. Reddy, and T. A. Driscoll. "Hydrodynamic Stability Without Eigenvalues." *Science* **261** (1993): 578–584.
271. Tuckerman, L., and D. Barkley. "Bifurcation Analysis for the Eckhaus Instability." *Physica D* **46** (1990): 57–86.
272. van Atta, C. W., and M. Gharib. "Ordered and Chaotic Vortex Streets Behind Circular Cylinders at Low Reynolds Numbers." *J. Fluid Mech.* **174** (1987): 113–133.
273. Van Dyke, M. *An Album of Fluid Motion.* Stanford: Parabolic Press, 1982.
274. Vicsek, T. *Fractal Growth Phenomena*, 2nd ed. Singapore: World Scientific, 1992.
275. Walden, R. W., and R. J. Donnelly. "Reemergent Order of Chaotic Circular Couette Flow." *Phys. Rev. Lett.* **42** (1979): 301–304.
276. Wang, T. G., E. H. Trinh, A. P. Croonquist, and D. D. Elleman. "Shapes of Rotating Free Drops: Spacelab Experimental Results." *Phys. Rev. Lett.* **56** (1986): 452–455.
277. Watson, J. "On the Nonlinear Mechanics of Wave Disturbances in Stable and Unstable Parallel Flows Part 2. The Development of a Solution for Plane Poiseuille Flow and for Plane Couette Flow." *J. Fluid Mech.* **9** (1960): 371–389.
278. Williamson, C. H. K. "The Existence of Two Stages in the Transition to Three-Dimensionality of a Cylinder Wake." *Phys. Fluids* **31** (1988): 3165–3168.
279. Williamson, C. H. K. "Oblique and Parallel Modes of Vortex Shedding in the Wake of a Circular Cylinder at Low Reynolds Numbers." *J. Fluid Mech.* **206** (1989): 579–627.
280. Williamson, C. H. K., and A. Prasad. "Wave Interactions in the Far Wake of a Body." *Phys. Fluids A* **5** (1993): 1854.
281. Wu, M., and D. Andereck. "Phase Dynamics in the Taylor-Couette System." *Phys. Fluids A* **4** (1992): 2432.

282. Yang, P., H. Mans;y, and D. R. Williams. "Oblique and Parallel Wave Interaction in the Near Wake of a Circular Cylinder." *Phys. Fluids A* **5** (1993): 1657–1660.
283. Zaleski, S., P. Tabeling, and P. Lallemand. "Flow Structures and Wave-Number Selection in Spiraling Vortex Flows." *Phys. Rev. A* **32** (1985): 655–658.
284. Zhang, L., and H. L. Swinney. "Nonpropagating Oscillatory Modes in Couette-Taylor Flow." *Phys. Rev. A* **31** (1985): 1006–1009.

Kurt Thearling
Thinking Machines Corporation, 245 First Street, Cambridge, MA 02142;
e-mail: kurt@think.com

Massively Parallel Architectures and Algorithms for Time Series Analysis

With the recent development of massively parallel computing, extremely large amounts of processing power and memory are available for the analysis of complex data sets. At the same time, the complexity and size of these data sets has been increasing. Both of these trends are expected to continue for the foreseeable future. This paper will provide a general overview of massively parallel architectures and algorithms for the analysis of time series data. Two distinct approaches to this problem, computational and memory-based, will be described.

1. INTRODUCTION

The last decade has seen a revolution in large-scale computation. The massively parallel processing (MPP) paradigm, originally seen as an outsider in the supercomputer race, is widely recognized as the technology of the future. In this paper we will discuss a number of approaches to time series data analysis using massively parallel computers. We will first review some of the current levels and trends in MPP technology and then discuss how this technology can be leveraged for data analysis. Two disparate approaches to data analysis will be investigated, memory-based and computational algorithms. We will conclude with some predictions for the (near) future of parallel computing.

2. CURRENT MPP TECHNOLOGY

In a massively parallel processing system, current levels of technology allow for

- thousands of processors per system,
- hundreds of megabytes of RAM per processor,
- gigabytes of disk storage per processor,
- tens of megabytes/sec interprocessor communication bandwidth (per processor), and
- hundreds of MIPS/MFLOPS per processor.

At this time (1993), the largest installed MPP system in the world is a 1024 processor Connection Machine CM-5 (the system can scale to 16,384 processors). Each processor has a peak performance of 128 MFLOPS and has at least 32MB of memory. This equates to a peak performance for the machine of 128 GFLOPS and a total memory of 32 gigabytes. The total interprocessor communication bandwidth is 5 GB/sec.

One key to the advance of MPP technology is its scalability. If an application needs more MIPS or megabytes, additional processors can be added to help solve the problem. If the system is designed intelligently, the overall performance of the system (global communication bandwidth, MIPS, MFLOPS, etc.) will scale up linearly with the system size. It should be noted, though, that the degree to which performance can be extracted from an MPP system is very algorithm dependent.

Undoubtedly the level of computing power available in a large MPP system will increase dramatically over time. Processor speeds and memory sizes are doubling approximately every eighteen months and the increase can be quickly adopted by MPP manufacturers. This means that the age of a Teraflop/Terabyte computer is not far off. Extremely large amounts of data will be able to be analyzed using this amount of processing power. This has changed the way that data analysis is carried out. The fact that large amounts of data are available has created the situation in which pure number crunching has, to some degree, given way to "memory-based" algorithms. This will be further discussed later in this paper.

One major development that is making MPP more usable is in the area of programming languages and tools. New programming languages such as Fortran 90 and C* have made the task of actually programming these machines much easier (actually they are extensions to existing programming languages, with the parallel architecture taken into consideration). In addition, tools are now available to help programmers design and debug their software. This is probably one of the most important advances, since traditionally the most difficult aspect of extracting performance from an MPP machine has been debugging the software.

There are two major types of parallel computers. The Single-Instruction-Multiple-Data (SIMD) architecture is characterized by the fact that each processor executes the same instruction simultaneously. Examples of this type of architecture are the Connection Machine CM-2 and the Maspar MP-1. In Multiple-Instruction-Multiple-Data (MIMD) computers, each processor operates autonomously and executes instructions independently of the other processors. Examples of this type of architecture are the Connection Machine CM-5, Intel Paragon, and Fujitsu VPP500.

A programming form related to SIMD is the *data parallel* programming model. In the data parallel programming model, all data is operated on in parallel using the same process. But unlike SIMD, data parallel programming allows different instructions to be issued by different processors simultaneously. Data parallel code can be run on both SIMD and MIMD machines, with improved performance possible on MIMD machines. The major reason to create a data parallel algorithm is the fact that they are simpler to understand and program. Experience over the past decade has shown that many tasks are amenable to the data parallel programming model.

Besides the instruction execution style (SIMD/MIMD), parallel computers are defined by the type of memory system that they use. Shared memory machines are exactly what their name implies; there is a single memory system that is shared by all of the processors. The Cray X-MP and Y-MP supercomputers are examples of shared memory computers. Distributed memory architectures spread the memory out over the individual processors in the system. If one processor requires a piece of data located in another processor's memory, that data must be transmitted over an interconnection network that links processors together. The CM-5 and Intel Paragon are both distributed memory machines.

For distributed memory machines, the interconnection network between processors is a key component. The network is the medium over which memory from one processor is transmitted to another processor. When designing such a network, the goal is to maximize interprocessor communication bandwidth.

3. MEMORY VS. COMPUTATION

There are two extremes in the use of massively parallel computing power: memory-based algorithms and computation-based algorithms. Memory-based algorithms make use of the large storage capacity (both RAM and disk) of MPP systems but may use few of the FLOPS. Some of the more common processing tasks that correspond to this extreme are memory-based reasoning, relational database operations, and text retrieval. Computation-based algorithms, on the other hand, make use of Gigaflop performance of parallel CPUs but may use little of the memory.

Examples of this type of processing are numerical optimization, genetic algorithms, backpropagation (as well as some other neural network learning algorithms), and statistical model building. In the next two sections, we will discuss problem-solving techniques spanning the range of these extremes.

4. MEMORY-BASED REASONING

Memory-based reasoning (MBR)[17] is a form of the *K* Nearest Neighbor (KNN) classification technique.[4] It differs from traditional KNN algorithms in a qualitative, not quantitative, way. Most successful previous applications of KNN made use of small (less than 10 megabytes) databases which were hand-tuned to maximize accuracy. These applications were limited in the amount of data that could be used by their ability to quickly search for neighbors within the database. By applying the KNN approach to much larger databases (hundreds/thousands of megabytes), massively parallel computing transforms KNN into MBR. MBR applications rely on the ability to leverage the information contained in extremely large databases. Typical applications often involve hundreds of megabytes of data. In addition, data is often multidimensional, involving different types of information. There is little or no manual processing of the data (e.g., the removal of incorrect training examples) before it is used.

A parallel nearest-neighbor search is used to look at all training examples simultaneously. Distance metrics can be hand-tuned to improve performance based on application specifics, but simple measures can often produce very accurate results. Initial results can be quickly achieved since there are no models to create. Also, confidence levels can be generated using relative distances to matching and non-matching neighbors.

Some previously successful applications of MBR include:

- Protein Structure Prediction[22]
- Optical Character Recognition[16]
- Cardiac Patient Viability Prediction[20]
- Census Data Classification[3]

Although these examples are not strictly time series problem, they do illustrate the potential for the analysis and prediction of very large amounts of data. One aspect of these problems that separates them from most time series analysis problems is the amount of data to be analyzed. The largest of the datasets in the SFI time series competition was approximately 600 kilobytes. Contrast this with data analysis problems which have previously been performed on the Connection Machine that involved hundreds of megabytes of data. Some applications are currently working with single databases on the order of tens of gigabytes, with the expectation that they will grow by a factor of ten in as little as two years. Even though the problems listed above may seem very different from time series forecasting, they actually involve similar techniques.

An example of an MBR approach from the area of time series analysis is the work of Farmer and Sidorowich.[5,6] In their work, Farmer and Sidorowich attempted to predict the behavior of a time series generated by a chaotic system. Their training set consisted of a time series of up to 10,000 sampled points along the attractor. The time series was then transformed into a reconstructed state space using a delay space embedding.[12,18] In the delay space embedding, each point in the state space is a vector X composed of time series values corresponding to a sequence of d delay lags: $x_1(t) = x(t), x_2(t) = x(t-\tau), ..., x_d(t) = x(t-(d-1)\tau)$. For a D-dimensional attractor, d must be at least as large as D.

To forecast a time series value, they first transformed the value into the state space representation. The nearest k ($> d$) neighbors in the state space representation were then located. A local linear map was created for the k neighbors and applied to the value to be forecast. The result of the mapping was the predicted value. Although higher dimensional maps (quadratic, etc.) could be used, Farmer and Sidorowich did not find significant improvements over the linear map. Using this approach, they were able to forecast time series values for a number of systems (Mackey-Glass differential equation, Rayleigh-Benard convection, and Taylor-Couette flow) much more accurately than standard forecasting techniques (global linear autoregressive).

Another piece of related research is Atkeson's MBR approach to the approximation of continuous functions.[1] In his work, Atkeson applied a locally weighted regression technique to the set of nearest neighbors to accurately predict the output of a continuous function.

As stated earlier, there have been several very large MBR applications implemented on a massively parallel computer. The following examples were both performed on a Connection Machine CM-2. Of particular importance is the large amount of data that was used to train the algorithms. It would have been very difficult (if not impossible) to make full use of such large data sets using a traditional computer architecture.

The first memory-based MPP algorithm involves optical character recognition.[16] Optical character recognition is the problem of taking a bit-mapped array of pixels and correctly classifying it into an alphanumeric character category. For pixel arrays of size 128×128, the problem has 16,384 dimensions. Smith and his colleagues[16] used 300,000 training examples to provide a knowledge base for an MBR system. This corresponds to 614 megabytes of data. Using very simple Hamming distance metrics, classifying an unknown character could be performed with an average accuracy of 99%. The technique also allowed for the introduction of concept of confidence by allowing the system to refuse to classify unknown characters whose nearest neighbors fell below a threshold distance. When the confidence measure was introduced, the system achieved 99.9% accuracy at 93% coverage (i.e., the system was not able to confidently classify 7% of the data).

Another application of MBR to large databases involved the classification of Census Bureau data.[3] In this case the problem involved the classification of free-text responses to questions about a respondent's occupation. These responses needed to be classified by occupation category (504 types) and industry category (232 types). Approximately 130,000 training examples were used, corresponding to 15 megabytes of data. When compared to an existing expert system used by the census bureau, the MBR approach achieved a 57% improvement for the occupation classification. The MBR system also achieved a 10% improvement for the industry classification over the expert system.

4.1 PARALLEL SEARCH FOR NEAREST NEIGHBORS

As we have stated, it is necessary in MBR systems to locate the k nearest neighbors for a point in the state space. A number of distance metrics can be used, including Hamming, Euclidian, and a host of others. For serial computers, a K-D Tree representation[11] can effectively reduce search complexity for the nearest neighbors when there is structure in data. But when there is little or unknown structure in data, searching all data elements in parallel may be the most effective solution.[25]

In experiments on financial data (daily S&P 500 closing prices), we have compared several K-D tree algorithms (which are difficult to parallelize) with simple parallel search examining portions of all data points. The experimental data contained 6,067 points (over 20 years worth of data) embedded into a five-dimensional delay space. The distance metric was simply Euclidean distance:

$$D = \sqrt{\sum_{i=1}^{n} (x'_i - x_i)^2}$$

where n is number of state space dimensions, x' is the point whose neighbors are being searched for, x is the current candidate neighbor, and x_i is the position of point x in dimension i. The total number of operations required to evaluate the distance between two points in the state space is n squaring operations, n subtraction operations, and $n - 1$ addition operations. The square root operation is unnecessary since the ordering of the distances is the same as the ordering of the distances squared.

When the K-D tree algorithm attempted to locate the five nearest neighbors for a test point (which was another point from the time series that was removed from the training set), an average of 99.6% of the training set data needed to be examined. In a refinement of the K-D tree search algorithm, the search began in the leaf cell that the test data point mapped to. It was hoped that this would help by initializing the set of nearest neighbors to a good set of candidates and thereby allow the traversal of the tree be pruned subsequently in the search. This technique did improve the performance of the search, but the improvement was not signification (an average of 99.5% of the training data was examined).

Finally, the K-D tree search was replaced by a much simpler technique. In this approach, every piece of data was examined to see if it was one of the nearest neighbors. But instead of computing the entire distance from the test point, the distance was computed incrementally. The (square of the) distance corresponding to each dimension was added until all of the dimensions were included. If at any time the partial distance was greater than (the square of) the furthest of the current set of k nearest neighbors, that data point was discarded. So, although each of the data points in the state space was evaluated, only a fraction of the entire evaluation (n squaring operations, n subtraction operations, and $n - 1$ addition operations) was performed. In experiments on the same S&P 500 training data, this technique performed only 45.6% of the possible operations (squaring, subtraction, and addition) necessary to locate neighbors from the training data.

This incremental search technique can be efficiently implemented in parallel using a local nearest neighbor heap for each processor and updating the entries incrementally. After local neighbors are located, a global sort is used to find global neighbors. In addition to the overall efficiency of this approach, the use of multiple processors can result in significant increases in the search performance.

5. COMPUTATIONAL APPROACHES

5.1 GENETIC ALGORITHMS

Genetic algorithms are an attempt to model the search for a problem solution as an evolutionary activity.[7,8] Candidate solutions to a problem are encoded in a bit string (e.g., phase space coordinates for a delay embedding). A population of these candidates (genomes) are individually evaluated to determine how well they solve the problem. Survival to the next generation is based on a candidate's "fitness." Once survivors are found, mates are chosen for each genome. Genetic operations of crossover and mutation are then used to breed new solutions for next generation, given solutions from the current generation. Crossover is performed by choosing two points along the genome pairs and swapping the values in between. Mutation (at a user-specified probability) is then considered for each bit in the resulting genomes. The fitness of a new generation of genomes is then evaluated and the entire process is repeated.

Packard[13] has previously used genetic algorithms to build predictive models for complex data sets (including, but not limited to, time series data). The goal of the evolutionary process is to locate predictable points within a state space of complex multidimensional data. A genome specifies points within the state space and the fitness measures the distribution of nearest-neighbor data points. A distribution of neighbors corresponding to a tight cluster receives a high fitness while a distribution that is spread out receives a low fitness. Speciation techniques are used to keep the population of genomes from converging to a global optimum.

Each generation produces a set of data points/distributions along with their associated fitnesses. This set of distributions with high fitness values correspond to locally predictable sections of the phase space. By continuing the evolutionary process, a set of high fitness distributions are produced. This set is then be used to generate predictive rules to model the system under investigation. Sections of the phase space that are not represented in the set of predictive rules are simply unpredictable (or they are sufficiently less predicable than the rules that evolved) and thus are not include in the model.

5.1.1 IMPLEMENTATION OF A PARALLEL GA ALGORITHM. In a parallel implementation, each processor is assigned a genome to evaluate. If there are more genomes than processors, the genomes are evenly distributed over the set of processors. Initially the genomes are set to random bit patterns. Each processor determines the fitness of its genomes by evaluating each genome's ability to solve the problem at hand. If there is more than one genome per processor, each processor evaluates the genomes serially. The performance of fitness evaluation is very dependent upon the application but linear speedup is possible for parallel evaluation (this assumes that evaluation cost is constant regardless of genome specification).

Interprocessor communication is then used to rank the genomes based on fitness and then select them for survival into the next generation. After global selection, survivors are then replicated in proportion to (a function of) their fitness values. Next, each survivor chooses a mate at random. Once mating is complete, new genomes are created using (two-point) crossover and mutation. The cost (in time) of each generation dominated by evaluation; evaluation is usually much more expensive than breeding a new generation.

5.1.2 PERFORMANCE OF A PARALLEL GA ALGORITHM. A genetic algorithm system has been implemented on the Connection Machine CM-5.[19] The user provides the representation of a candidate solution as a bit string (genome) and generates an evaluation function for the genome. This system then provides the basic evolutionary operators (selection, mating, crossover, and mutation) which are interfaced to the user's application dependent evaluation function. The current performance for a CM-5 is approximately 5 μsec/bit (per processor) for breeding a new generation of genomes. For example, assume a population of 1024 genomes of length 4096 bits on a 128-processor system. This corresponds to approximately 0.16 seconds to breed a new generation after the fitness evaluation has been done. In most cases the user's specified fitness evaluation will take considerably longer than the time taken to breed a new set of genomes.

5.2 NEURAL NETWORKS

A variety of neural network approaches have been applied to the forecasting of nonlinear time series. The work of Lapedes and Farber[9] as well as several of the papers included in Casdagli and Eubank[2] have demonstrated the ability of neural network models to successfully predict time series data. The work of Zhang and Hutchinson,[24] describes the application of MPP neural network algorithms to time series analysis. In this section we will review some of the work done in mapping neural network algorithms to massively parallel computers.[1] For a more thorough description, see Singer[15] and Nordström and Svensson[10] and the references contained therein.

Neural network architectures are characterized by a collection of processing nodes connected by a series of weighted links. The relationship between the individual node's inputs and outputs is typically a nonlinear function (such as a sigmoid). By carefully choosing the weights for the links, a neural network can carry out complex mappings from global inputs to global outputs. The complicated issue in carrying this process out is in computing the interconnection weights. Algorithms such as backpropagation[14] are often used to perform this task.

5.2.1 A NAIVE IMPLEMENTATION. To map a neural network architecture to a massively parallel processor, the first approach that comes to mind is simply map each node in the network to a processor. The connections between nodes map to messages between processors over the interconnection network. When the network is learning its connection weights, the inputs are passed forward through the network to the outputs. The observed outputs are then compared with the expected output and the differences are propagated backward from the outputs to the inputs. During the backpropagation phase, the weights are adjusted to minimize the error at the outputs.

Once the connection weights have been learned, the network is then run only in the forward direction. The performance of this process is measured in CPS (connections per second). Estimates for the performance of the forward direction of this naive implementation are 13 million CPS (for a 64K processor CM-2[15]). This compares to a speed of on the order of 250 thousand CPS for a typical workstation. The performance of the learning algorithm will be 25 to 50% of the forward direction performance due to the fact that both a forward and backward pass are required during learning.

[1]Code for implementing neural networks on the Connection Machine CM-2 and CM-5 can be obtained from Thinking Machines Corporation.

5.2.2 AN IMPROVED IMPLEMENTATION. A more sophisticated approach to mapping a neural network architecture involves taking into consideration specific aspects of the MPP architecture. Zhang et al.[21] used knowledge of the communication and computation details for the Connection Machine CM-2 when they designed their neural network algorithm. Instead of mapping a single network node to a processor, Zhang and his colleagues carefully mapped multiple network nodes to the same processor. This allowed for reduced communication cost and efficient computation of the network node outputs. In addition, they also performed learning in parallel by replicating entire copies of the neural network over the MPP processors. The performance that they achieved (on a 64K processor CM-2) was approximately 80 million CPS and 40 million WUPS (weight updates per second) during learning.

5.2.3 THE FASTEST MPP NEURAL NETWORK YET DEVELOPED. Finally, the fastest performance achieved to date for the implementation of a neural network on an MPP is the work of Farber, described in Singer.[15] In that implementation, each processor represented a training example for the neural network. The network was broadcast to each of the processors and training was done in parallel. Each of the individual backpropagation results were combined globally after each training phase. Since there is little communication involved, this algorithm can achieve very high performance: 325 million WUPS and 1.3 billion CPS (on a 64K processor CM-2). The only real disadvantage of this algorithm is that it needs an extremely large number of training examples to be efficient (at least as many examples as processors).

6. MEMORY PLUS COMPUTATION

The integration of computational and memory-based problem-solving techniques can sometimes be more successful than either of the techniques used individually. An example of an application of this type of approach is in the area of protein secondary structure prediction.[23] In this work, a combination of neural network, MBR, and statistical approaches was used. The inputs to one neural network are the output of an MBR system and the output of another neural network. The accuracy of this hybrid system was better than that of any individual system and the overall accuracy was better than any previously published algorithm.

7. THE (NEAR) FUTURE FOR PARALLEL PROCESSING

There are two major trends in computational data analysis. First, available computational processing power is increasing dramatically over time. Supercomputers will ride the wave of faster chips and denser silicon to make TeraFLOP computing a reality around the year 1995. Also following this trend will be memory capacity, which will hit the TeraByte level at about the same time. To simplify the task of programming these machines, the software tools and programming languages will become much more important than the hardware issues.

Second, the task that data analysis systems will face will center on dealing with the massive amounts of data, orders of magnitude larger than currently exists. Terabyte-sized databases will soon become a reality. This will require that the bulk of the processing power in use will be spent analyzing data rather than generating it. Data visualization (the graphical display of complex data in order to reveal qualitative features) will be very important. "Database mining" is a recently coined term often used to describe techniques which allow users to sift through terabytes trying to locate useful bytes of data. If carefully applied, the increase in computational power should help us offset the complexity of the data that will need to be analyzed.

8. CONCLUSIONS

Massively parallel processing power and memory capacity is rapidly approaching the TeraFLOP/TeraByte level. This will allow users to explore two extremes in problem solution space: memory-based and computational algorithms.

Memory-based algorithms (such as MBR), which leverage the knowledge contained in very large databases, are extremely simple to implement and provide useful results quickly. Although MBR techniques are relatively new, they hold great hope for the future. By using massively parallel processing, these large databases can generate efficient solutions to difficult problems.

Computation-dominated techniques such as genetic algorithms and neural networks are also amenable to the massively parallel programming paradigm. Extremely high performance has been achieved for parallel implementations of genetic and neural network algorithms. This performance can be used to efficiently carry out analysis on large, complex databases.

ACKNOWLEDGMENTS

The work presented in this paper was supported by Thinking Machines Corporation and the Santa Fe Institute. The author would like to thank David Waltz, Stephen Smith, Xiru Zhang, and Jim Hutchinson for conversations, ideas, and support.

Connection Machine is a registered trademark of Thinking Machines Corporation. CM-2, CM-5, and CM are trademarks of Thinking Machines Corporation. C* is a registered trademark of Thinking Machines Corporation.

REFERENCES

1. Atkeson, C. G. "Memory-Based Approaches to Approximating Continuous Functions." In *Nonlinear Modeling and Forecasting*, edited by M. Casdagli and S. Eubank, 503–521. Santa Fe Institute Studies in the Sciences of Complexity, Proc. Vol. XII. Reading, MA: Addison-Wesley, 1992.
2. Casdagli, M., and S. Eubank, eds. *Nonlinear Modeling and Forecasting.* Santa Fe Institute Studies in the Sciences of Complexity, Proc. Vol. XII. Reading, MA: Addison-Wesley, 1992.
3. Creecy, R. H., B. M. Masand, S. J. Smith, and D. L. Waltz. "Trading MIPS and Memory for Knowledge Engineering: Automatic Classification of Census Returns on a Massively Parallel Supercomputer." *Comm. ACM* August (1992).
4. Dasarathy, B. V. *Nearest Neighbor (NN) Norms: NN Pattern Classification Techniques.* Los Alamitos, CA: IEEE Computer Society Press, 1991.
5. Farmer, J. D., and J. J. Sidorowich. "Predicting Chaotic Time Series." *Phys. Rev. Lett.* **59(8)** (1987): 845.
6. Farmer, J. D., and J. J. Sidorowich. "Exploiting Chaos to Predict the Future and Reduce Noise." In *Evolution, Learning, and Cognition*, edited by Y. C. Lee. Singapore: World Scientific, 1988.
7. Goldberg, D. E. *Genetic Algorithms in Search, Optimization, and Machine Learning.* Reading, MA: Addison-Wesley, 1989.
8. Holland, J. *Adaptation in Natural and Artificial Systems.* Cambridge, MA: MIT Press, 1992.
9. Lapedes, A., and R. Farber. "Nonlinear Signal Processing Using Neural Networks: Prediction and System Modeling." Technical Report LA-UR-87-2662, Los Alamos National Laboratory, July, 1987.
10. Nordström, T., and B. Svensson. "Using and Designing Massively Parallel Computers for Artificial Neural Networks." *J. Parallel & Distrib. Comp.* **14** (1992): 26.

11. Omohundro, S. "Efficient Algorithms with Neural Network Behavior." *Complex Systems* **1** (1987): 273.
12. Packard, N. H., J.-P. Crutchfield, J. D. Farmer, and R. S. Shaw. "Geometry from a Time Series." *Phys. Rev. Lett.* **45** (1980): 9.
13. Packard, N. H. "A Genetic Learning Algorithm for the Analysis of Complex Data." *Complex Systems* **4** (1990): 543.
14. Rummelhart, D. E., G. E. Hinton, and R. J. Williams. "Learning Internal Representations by Error Propagation." In *Parallel Distributed Processing*, edited by J. McClelland and D. E. Rummelhart. Cambridge, MA: MIT Press, 1986.
15. Singer, A. "Implementations of Artificial Neural Networks on the Connection Machine." *Parallel Computing* **14** (1990): 305.
16. Smith, S., M. Bourgorn, K. Sims, and H. Voorhees. "Handwritten Character Classification Using Nearest Neighbor in Large Databases." *IEEE Transactions in Pattern Analysis and Machine Intelligence* **16** (1994): 965.
17. Stanfill, C., and D. L. Waltz. "Toward Memory-Based Reasoning." *Comm. ACM* **29** (1986): 1213.
18. Takens, F. "Detecting Strange Attractors in Fluid Turbulence." In *Dynamical Systems and Turbulence*, edited by D. Rand and L.-S. Young. Berlin: Springer-Verlag, 1981.
19. Thearling, K. "An Evolutionary Programming Package for the Connection Machine CM-5." Unpublished manuscript, 1992.
20. Waltz, D. L. "Applications of the Connection Machine." *IEEE Computer* **20(1)** (1987): 85.
21. Zhang, X., M. McKenna, J. P. Mesirov, and D. L. Waltz. "An Efficient Implementation of the Backpropagation Algorithm on the Connection Machine CM-2." In *Neural Information Processing Systems 2*, 801–809. Denver, CO, 1989.
22. Zhang, X., D. L. Waltz, and J. P. Mesirov. "Protein Structure Prediction by Memory-Based Reasoning," *Proceedings of the Case-Based Reasoning Workshop*, 1–5. Pensacola Beach, FL, May 1989.
23. Zhang, X., J. P. Mesirov, and D. L. Waltz. "A Hybrid System for Protein Secondary Structure Prediction." *J. Mol. Biol.* **225** (1992): 1049.
24. Zhang, X., and J. Hutchinson. "Practical Issues in Nonlinear Time-Series Prediction." In *Time Series Prediction: Forecasting the Future and Understanding the Past*, edited by A. Weigend and N. Gershenfeld, 219–242. Santa Fe Institute Studies in the Sciences of Complexity, Proc. Vol. XV. Reading, MA: Addison-Wesley, 1993.
25. Zhang, X., and K. Thearling. "Nonlinear Time-Series Prediction by Systematic Data Exploration on a Massively Parallel Computer." Santa Fe Institute working paper 94-078-045. Santa Fe, New Mexico, 1994.

Student Contributions

Eric J. Anderson
Department of Computer Science, Box 90129, Duke University, Durham, NC 27708-0129;
email: eric@cs.duke.edu

Numerical Investigations of the Krugman Population Model

In his 1992 paper,[6] Prof. Krugman develops a nonlinear model of economic forces affecting population concentration. He offers results of some preliminary numerical simulations that suggest that the model exhibits "emergent structure," or complex behavior, in several respects. Here, we present further results of simulations based on the Krugman model, amplified in several directions. We show that as the number of potential city-sites increases, the complexity of the system (as measured by the number of distinct steady states) increases dramatically. We also investigate the sensitivity of the steady states to parameters at higher numbers of city-sites. We examine the time-dependent behavior of the model, to demonstrate that the model often describes surprisingly varied nonequilibrium, transient behavior on the way to a steady state. We also illustrate the behavior of the model under a geometrically uniform lattice distribution. Finally, we offer some suggestions for future research.

This article arises from work done at the 1993 Complex Systems Summer School.

1. INTRODUCTION—SUMMARY OF THE KRUGMAN MODEL

In his 1992 paper,[6] Prof. Krugman proposes and analyzes a model of several of the economic forces affecting concentration of the worker population in an economy. The economic considerations behind the Krugman model are set forth in detail in his paper, and are summarized elsewhere in this volume. For our purposes, we focus on the definition of the model parameters, the equilibrium wage condition, and the equation of motion for the economy. The parameters for the model are

1. μ, the overall proportion of manufacturing workers,
2. τ, a transportation cost factor,
3. σ, a constant elasticity of substitution for manufactured goods, and
4. ρ, a time-scaling parameter.

The model posits a population at each of n locations, referred to as city-sites. The geometric aspects of the model are taken into account through the (assumed exogenous) factors

1. ϕ_j, the proportion of agricultural workers in location j, and
2. D_{jk}, the distance from location j to location k.

For the purposes of this model, the agricultural workers are assumed to be immobile; in all the simulations that follow, the distribution is assumed uniform. The distances D_{jk} reflect the geometry of the city-sites in the model. Except in one section below, we follow Prof. Krugman's approach of analyzing city-sites evenly spaced around a circle (or regular polygon), with transportation distances based on perimeter distance. In Section 4, Different Geometries, we look at a lattice geometry, where cities are distributed uniformly about a rectangular lattice. In either case, the transportation cost is taken into account through an (admittedly oversimplified) "iceberg" form, assuming that a proportion of goods exponentially related to distance simply fails to arrive ("melts away en route"). The specific proportion, and hence the effective transportation cost, is controlled by the model parameter listed above.

The model predicts the behavior of the following factors:

1. λ_j, the proportion of manufacturing workers in location j,
2. Y_j, the total income of location j,
3. w_j, the nominal wage in location j, and
4. ω_j, the real wage in location j.

The equations for equilibrium wage are given as

$$Y_j = (1-\mu)\phi_j + \mu\lambda_j w_j\,, \tag{1}$$

$$T_j = \left[\sum_{k=1}^{n} \lambda_k \left(w_k e^{\tau D_{jk}}\right)^{1-\sigma}\right]^{1/1-\sigma}, \tag{2}$$

$$w_j = \left[\sum_{k=1}^{n} Y_k \left(\frac{e^{\tau D_{jk}}}{T_k}\right)^{1-\sigma}\right]^{1/\sigma}, \tag{3}$$

$$\omega_j = w_j T_j^{-\mu}. \tag{4}$$

The first equation, for income Y_j, reflects simply a normalization. The second equation, for the true price index T_j, incorporates a constant elasticity of substitution and the decaying-exponential transportation cost factor described above. The third equation, derived in the appendix to the Krugman paper, reflects the short-run local market clearing condition that equalizes wages in a particular location. These first three equations provide the equilibrium wage condition. The final equation, for the real wage ω_j, applies the price index, in terms of a market basket that reflects both agricultural and manufacturing goods.

The equilibrium wage equations form a short-run equilibrium requirement in this model. The equation of motion for the economy then assumes a movement of workers in response to wage differentials, and posits that these occur on a slower time scale:

$$\frac{d\lambda_j}{dt} = \rho\lambda(\omega_j - \overline{\omega}), \tag{5}$$

where $\overline{\omega}$ is the (population-weighted) average real wage in the economy. In words, the rate of change of manufacturing population at location j is in proportion to the difference between the prevailing (equilibrium) wage at location j and the average real wage. The parameter ρ imposes a time scale on the equation of motion but does not otherwise affect the evolution of the economy.

In these simulations, we have generally used the following values:

1. $\mu = 0.2$,
2. $\tau = 0.2$,
3. $\sigma = 4.0$,
4. $\rho = 1.0$, and
5. ϕ_j distributed uniformly.

As initial conditions, we choose λ_j randomly and uniformly over $[0, 1]$, then scale the j to sum to unity.

The short-run wage equilibrium conditions in items 1–3 form a set of nonlinear equations that must be set to zero at each time step. This and other numerical aspects of the simulation are discussed below in Section 6, Numerical Methods.

2. EQUILIBRIUM STATES

In his original paper, Prof. Krugman investigated a two-city model (which can be solved analytically), and a twelve-city model. In the two-city model, he investigated the transitions between a two-city balanced population distribution and a single-city concentration. In the twelve-city model, he found through simulation that with most choices of parameters and initial conditions, the model evolved toward agglomeration into one large city. For a specific choice of parameters (achieved "after some experimenting"), he found evolution toward two- and occasionally three-city equilibria.

We have found for larger numbers of city-sites that the incidence of many-city equilibria is much more common. As in the twelve-city simulations Prof. Krugman examined, typically in our simulations the initial populations fluctuate, then settle down into a few populated cities and many depopulated city-sites. (See Section 4 for more discussion of the transient behavior; see also Section 6 on technical issues for a discussion of "small" cities.) In Figure 1, we examine the number of cities in the steady state as we change the number of potential city-sites in the simulation. Specifically, we count the number of populated cities in the steady state for each of one hundred simulations; we then plot the mean number of cities against the number of available city-sites in the simulation. The number of populated cities in the equilibrium state rises sharply as the number of city-sites increases.

In Figure 2, we expand the analysis in Figure 1 to look at the distribution of the number of cities in the steady state. It appears that in most cases the number of cities is sharply peaked about a specific value. For example, in the simulation based on 20 city-sites, 85 of the 100 simulations end in states with specifically four populated cities. Although the number of populated cities is apparently predictable within a quite narrow range, the variety of steady states still increases combinatorically. That is, the increase in the number of cities in the typical steady state in itself signals a dramatic increase in complexity. In Figure 3, we illustrate one aspect of this complexity. and examine which specific cities are populated in the steady states. For five particular simulation sizes, we plot the locations of each of the populated cities in the steady state. (We normalize the first such city to location number 1 in each case, so that the labels reflect offsets from a populated city.) We see that for a small number of city-sites, the two-city equilibria are generally located at antipodal points. For more city-sites, however, the distribution over potential city-sites is much more varied. At 16 city-sites, a three-cornered equilibrium is heavily favored (according to Figure 2), and the distribution of the locations of these cities has two well-defined peaks in Figure 3. But at 20 city-sites, where most steady states have four populated cities, the distribution of those cities is more diffuse; and at 24 city-sites, the distribution of cities is nearly uniform.

In Figure 2, we also strongly imply that the likelihood of complete agglomeration is sharply reduced as the size of the simulation increases. The likelihood of a single city agglomeration as a steady state seems to be negligible at 16 city-sites, and seems to vanish completely at 24.

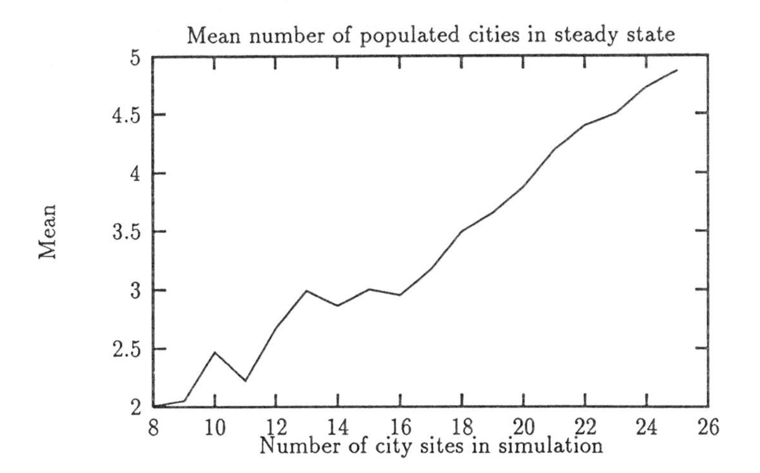

FIGURE 1 Mean number of populated cities in steady state.

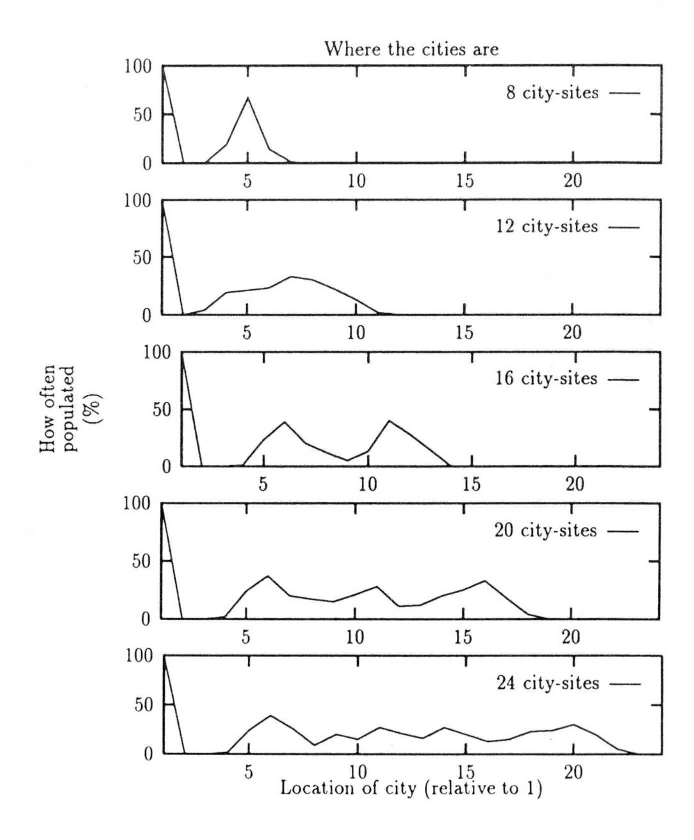

FIGURE 2 Number of cities in steady state. Complexity increases as number of sites increases.

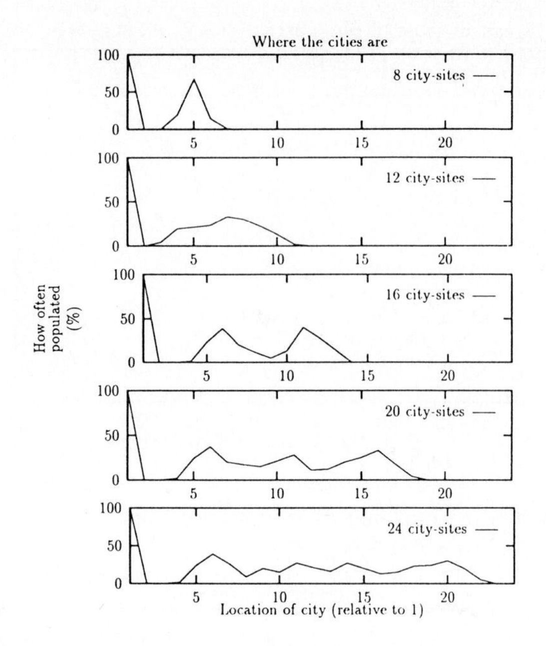

FIGURE 3 Location of city (relative to 1). Where the cities are.

3. SENSITIVITY TO MODEL PARAMETERS.

Our simulations suggest that larger model sizes continue to reflect considerable sensitivity of the results to the tuning of the model parameters. Specifically, we investigated the relationship between the transportation cost and the complexity of the steady states, as measured by the number of populated cities. Qualitatively, Prof. Krugman correctly noted[6] that high transport costs favor dispersion of the population. In Figure 4, we plot the mean number of cities in the steady state versus the transportation cost parameter τ (times 100). We see that the mean number of cities increases sharply as we increase τ over that range. This behavior holds generally over a wide range of simulation sizes. In Figure 5, we illustrate the specific distribution of steady states, indicating that not only does the mean number of cities increase, but the distribution also spreads out noticeably at higher values of T. In Figure 6, we further corroborate this intuition by showing a trend toward a wide and flat distribution of the location of populated cities as transportation costs increase.

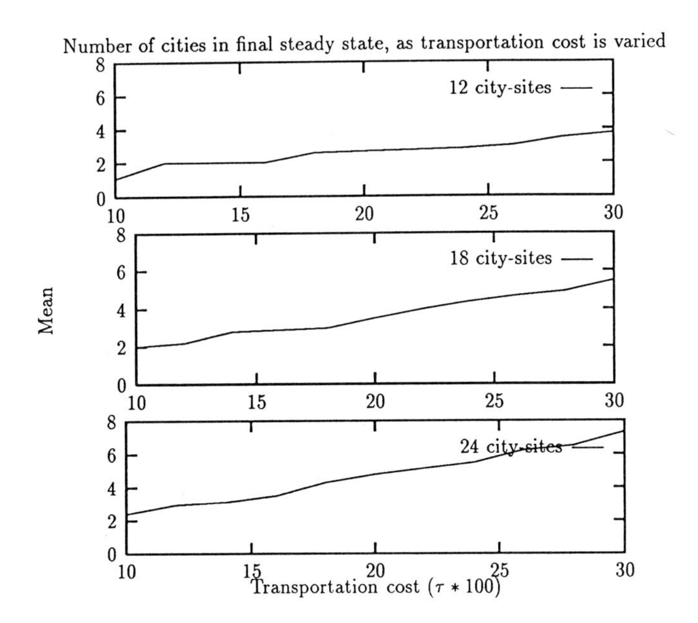

FIGURE 4 Number of cities in final steady state, as transportation cost is varied.

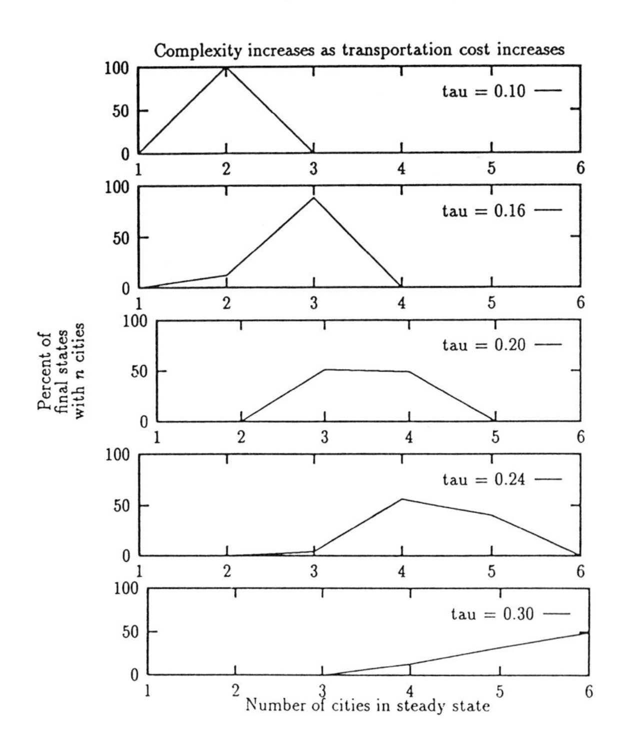

FIGURE 5 Complexity increases as transportation cost increases.

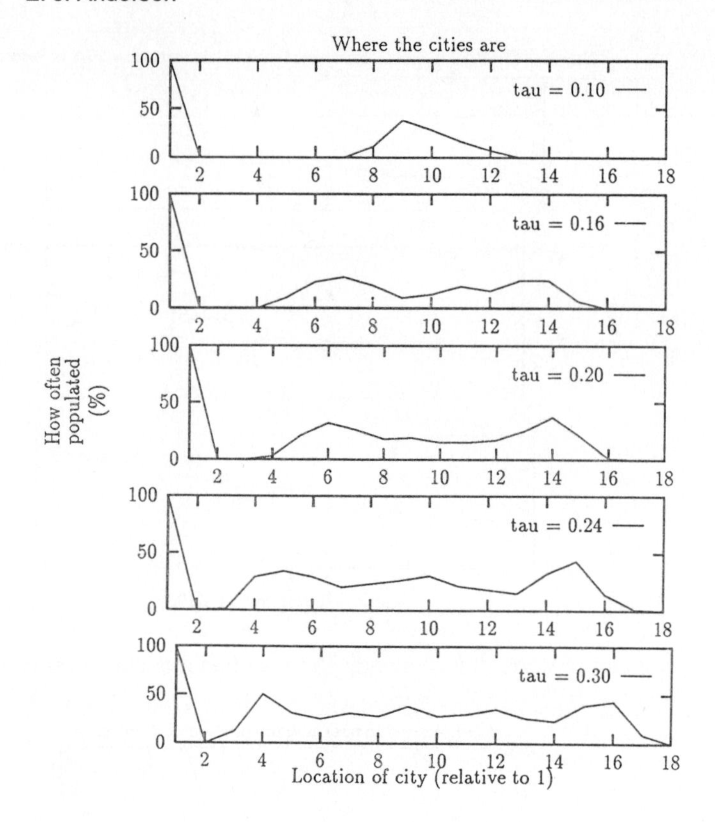

FIGURE 6 Where the cities are.

This analysis of the sensitivity of the types of steady states to a change in external parameters lends support to Prof. Krugman's discussion of the potential historical implications of the model. In that discussion, he notes several historical instances of populations of individual cities or regions that suddenly enter into periods of abrupt, seemingly irreversible decline. In the model, as the fundamental parameters are adjusted there are significant shifts in the types and variety of steady-state solutions. In particular, the typical number of populated cities can shift by one or two as a result of relatively modest changes in the external parameters. This behavior, observed in the model, would predict the rapid growth or decline of individual cities upon changes in the external parameters.

4. DIFFERENT GEOMETRIES

In the interests of simplicity, the original Krugman paper[6] investigated only one geometry, consisting of 12 cities evenly spaced around a circle, with transportation restricted to the circumference. We have also investigated as an alternative geometry a square of 25 cities, arranged in a square lattice with "Manhattan" distances between cities (a distance of one for each block north, south, east, or west that must be travelled). In this model, cities at the edges of the graph are not connected to their counterparts on the other side of the country; that is, the geometry is not toroidal. This geometry, therefore, has asymmetry between well-connected heartland cities on the one hand and more isolated rim cities on the other.

The conclusions are somewhat intriguing. Not unexpectedly, the model tends to concentrate (manufacturing) population in the well-connected interior region. In Table 1, we list the percentage of steady-state cities that were located in the very center of the region, immediately adjacent to the center, or diagonally adjacent to the center (at distance two from the center). As we show in Table 1, the most popular city in the final steady state is the center city, which is relatively closer to all other cities than cities on the rim. Indeed, no city on the rim survived in the steady states of these simulations.

TABLE 1

Tau	Center	Directly Adjacent	Diagonally Adjacent
0.1	94	6	
0.2	90	10	
0.3	21	38	41

TABLE 2

tau/cities	1	2	3	4
0.1	100			
0.2	90	10		
0.3	41	22	33	4

The inhomogeneous transportation cost also seems to inhibit the development of a widely scattered population structure. The tendency toward single-city agglomeration is much more pronounced than in the corresponding simulation with 25 cities distributed around a circle. In Table 2, we list the percentage of simulations that end with one, two, three, or four cities in the steady state. A far greater percentage of these states have only one heavily populated city.

5. TIME DYNAMICS

The original Krugman paper does not explicitly discuss the *time-dependent evolution* of cities toward a steady state. In one sense this is not surprising, as in the model there is no natural time scale in the equation of motion for the economy. The parameter ρ governs the speed with which workers respond to wage differentials; this parameter is not meaningful (and indeed is essentially undefined) when investigating only the characteristics of steady states.

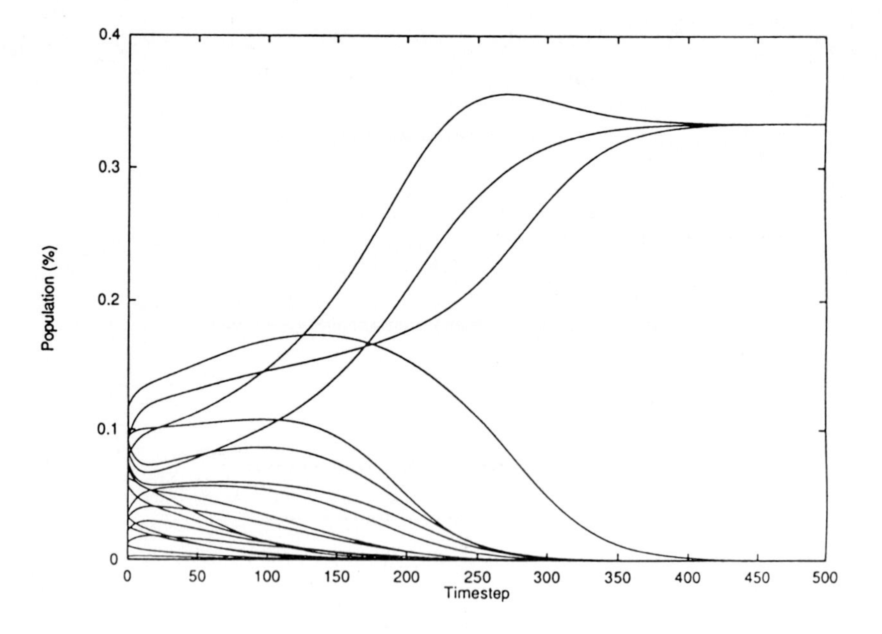

FIGURE 7 Time evolution of city populations.

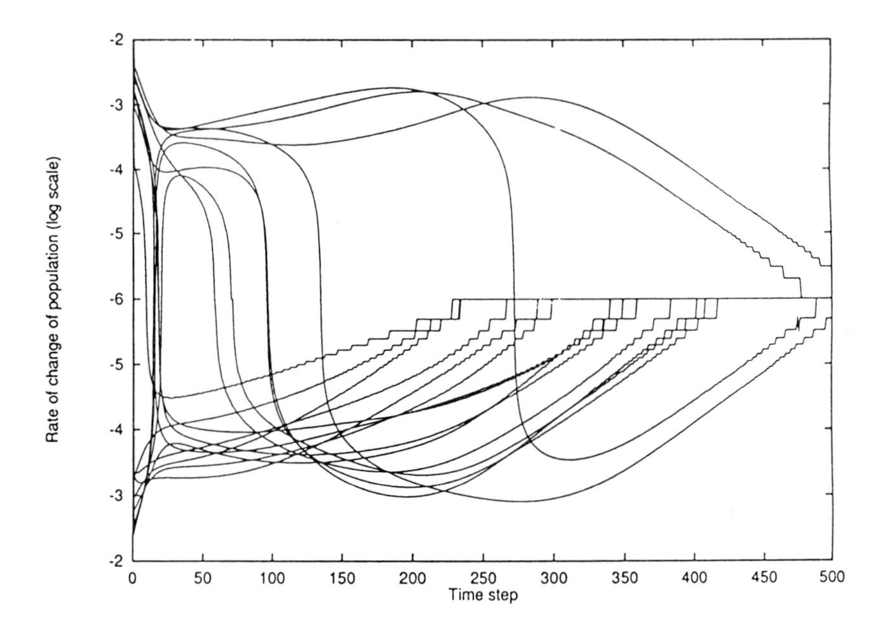

FIGURE 8 Rates of change of population.

In view of the relative inflexibility of worker relocation, it is likely to be important to study the time-dependent behavior of the equation of motion in addition to the final steady states. Figure 7 illustrates the time evolution of the populations of each city in an eighteen-city model. While the graph is somewhat confusing, we see some cities decaying quickly; other cities maintaining their population for a period of time, then dying off; still other cities increasing their population (and indeed one city becoming the most populous for a period of time) before dying off; and the three cities that form the equilibrium solution increasing more or less steadily from their initial values to the final one-third of the total. This graph suggests that the early behavior of populations is not necessarily indicative of long-term trends.

The analysis of the rates of change of the populations is if anything even more intriguing. In Figure 8, we plot (on a log scale) the rate of change $d\lambda/dt$ for each of the eighteen cities. Values below the line $d\lambda/dt = -6$ (equating to zero) correspond to negative values of $d\lambda/dt$, or declining population; these rates were converted in absolute value before taking the logarithm. Many cities quickly took on decaying values for $d\lambda/dt$, and these cities continued to decay steadily, though at an exponentially slowing pace. Other cities maintained a steady growth, then suddenly switched sign and decayed steadily thereafter. The contrast between the steady, slow decay of some cities and the rapid switch of others between periods of steady growth and periods of rapid decay is intriguing and not easily explained.

Assiging a suitable time scale for this model is not easy. In the model, the parameter λ can denote either populations or varieties of manufactured goods. If the parameter λ denotes a population level, one could presumably estimate from census figures the period of time over which a given rate of change of λ should correspond. A typical size of $d\lambda/dt$ for the eighteen-city model (with $\rho = 1$) is approximately 0.001 (Figure 8). Under that measure, a timestep of one would correspond to migration of one-tenth of one percent of the local population during that time period. To draw analogies to the American economy, one could relate cenus data on the pace of relocation of workers to the value of the parameter $d\lambda/dt$. If for a typical large or medium-sized city, one percent of workers relocate each year, it would be reasonable to identify a timestep in these simulations with a period of one calendar month; the evolution over five hundred timesteps then corresponds to approximately two generations. For a simillar study based on census data, see Dendrinos and Sonis.[5]

6. NUMERICAL METHODS

The numerical solution of the model for a steady state requires a time evolution of the differential equation. At each step, the "functions" $\omega_i - \overline{\omega}$, the equilibrium wage and equilibrium average wage, must be evaluated. Under the economic model, these wages are assumed to equilibrate on a time scale much faster than the time steps in the equation of motion. Therefore, we must solve at each time step the system of nonlinear equations corresponding to the wage equilibrium condition. This condition can be expressed as $F_j(w) = 0$, where

$$F_j = \left[\sum_{k=1}^{n} Y_k \left(\frac{e^{\tau D_{jk}}}{T_k} \right)^{1-\sigma} \right]^{1/\sigma} - w_j .$$

The system of n equations, where n is the number of city-sites in the model, is readily solved using Newton's method.[4] Here, the Jacobian of F is dense, in that each F_j depends on all the wages w_i, and moderately difficult to calculate, requiring (for fractional a, at least) several evaluations of an exponential function. (In one profiling, approximately seventy percent of the model computation time was spent in evaluating exponents.) For large numbers of cities, too, the cost of factoring the Jacobian can become noticeable. On the other hand, the local radius of convergence of Newton's method is evidently large enough in this problem to allow us to use the wage from one time step as an appropriate starting point for wages at the next; that is, the equilibrium wage does not seem to change dramatically from one time step to the next. As a result, Newton's method converges quickly without need to resort to line searches or other "global" Newton methods. For that matter, the Jacobian itself does not change dramatically from one time step to the next; we found

TABLE 3

Update frequency	Broyden	Newton	Damping factor	Iteration
1	163.22	464.8	1	590.21
4	161.26	235.11	0.25	2284.28
8	158.75	195.86	0.125	4887.44
10	158.9	188.29	0.1	5665.28
20	157.16	173.18		
40	156.69	165.31		
100	156.7	160.46		
500	156.25	157.71		

experimentally that it was not necessary to recalculate the Jacobian more than once every 40 Newton steps or so. In Table 3, we compare the computation times of three methods for solving the nonlinear equations: Newton's method, the quasi-Newton method due to Broyden (with factored updates), and the straightforward method of "damped iteration," applying the wage equilibrium equations repeatedly until convergence. We measured the (user) runtime of ten simulations of two thousand time steps each, on a DEC Alpha AXP 3000/500. (Times are in seconds.)

For the Newton-type methods, we examined the effect of bypassing the recomputation (Newton) or update (Broyden) of the Jacobian for the specified number of Newton steps. Typically, five to six Newton (six to eight quasi-Newton) steps were taken at each outer time step. We found, not surprisingly, that if the Jacobian was recalculated or updated at every step, Broyden's method was considerably more efficient than Newton's method. However, the two methods were less easily distinguished as the frequency of recalculation or update was decreased. Both methods were noticeably superior to simple iteration, but only by a factor of four or so. The iteration method was apparently not improved by a choice of damping factor.

The Jacobian is comfortably nonsingular at the equilibrium wage, and appears to converge to a matrix with most of its eigenvalues of modulus one. The differential equation of motion can be readily evolved to a steady state using any of several different methods. We chose to implement a fixed time-step Runge-Kutta method. While an adaptive time-step method would more quickly determine steady-state solutions, the analysis of time-dependent, nonequilibrium behavior suggested a fixed time step. Approximately a thousand timesteps sufficed to reduce most steady state cities to zero population; after five hundred timesteps, in many simulations some cities had populations below 2% but not entirely negligible.

Interestingly, a forward Euler method with large enough timestep (ρ approximately 10) can in some cases introduce enough inaccuracy to cause the system to evolve to a different steady state. This is one indication (albeit indirect) of the effect of noise on this system.

In summary, we use a second-order Runge-Kutta time stepping algorithm, at each time step solving for the equilibrium wage using a multidimensional quasi-Newton method with (occasional) factored update.

As a side note, if inter-city distances to nearest neighbors are specified, the remainder of the geometry can readily be completed using an all-pairs shortest path algorithm. This additive approach to inter-city distances is, we believe, more general than one that computes an inter-city distance matrix based on two-dimensional Euclidean distances.

7. OTHER TECHNICAL ISSUES

This simulation's focus on random "Monte Carlo" initial conditions, while convenient from the standpoint of computation, might well be criticized as unrealistic. In determining whether agglomeration is "likely" or more generally whether a particular distribution is "substantially" stable (i.e., stable in a significant neighborhood), one needs to apply a notion of measure of likelihood to various distributions of city populations. A (scaled) uniform distribution of the sort used here is not particularly appropriate; the rapid initial reconfiguration illustrated in Figures 7 and 8 suggest that these initial distributions are not typical of transient states generally. The impact of the model on a pre-existing set of populations would seem in practice to arise most often when examining the transition of one state to another upon the evolution of exogenous factors. One could argue in that case that the appropriate probability measure is over the class of possible states, including nonequilibrium states, that might obtain at the time of a change in exogenous factors. We argue above (see Figure 6) that for many values of parameters the variety of steady states is considerable. In that sense, a wide ranging probability measure might well be appropriate.

Another technical issue not discussed in detail above concerns the appropriate role of "small" cities. As noted in the discussion accompanying Figure 8, the decay rate for dying cities slows exponentially; and small cities take a long time to decay. Furthermore, some small cities can even have positive (but small) $d\lambda/dt$, possibly suggesting that on a very long time scale, putative steady states may conceivably ultimately reverse themselves. In these simulations, we have generally implemented a cutoff of 2% of the total population, below which the city is assumed ultimately to die out.

8. FUTURE WORK

This work can readily be expanded in several ways. Time-dependent population adjustment upon a sharp change in exogenous parameters is one area noted by Prof. Krugman but not specifically discussed here. The sensitivity of the model to the other model parameters, σ, μ, and the distributions, could be examined in detail.

Three model enhancements in particular potentially offer significant additional insight into the underlying mathematical behavior: stability analysis, expansion to a continuum model, and the effect of noise.

8.1 STABILITY ANALYSIS

A straightforward question (and one of pressing practical importance) that can be asked of this model is the presence of thresholds above and below which the behavior of the model can be predicted. Specifically, is there a concentrated population level above which an agglomeration is inevitable? and is there a sparse population level below which a city will inevitably wither? Because of the complexity of the model, these questions are probably difficult to approach analytically. Numerical investigation, on the other hand, would be complicated by the likely possibility that the system would take much longer to evolve to a steady state near the boundaries of stability regions. As a start, one could investigate the sign and magnitude of $d\lambda/dt$ for a single large city for various distributions of (small) populations of the remaining cities.

8.2 EXPANSION TO A CONTINUUM MODEL

As the number of city-sites increases, the Krugman model begins to resemble the discretization of a continuum model. An appropriate continuum model that incorporated spatial inhomogeneities might offer comparison with other continuum complex systems, particularly those illustrating "local defects."[2,3] Intuitively, the continuum population model would seem to have an appealing nonlocal interaction across many length scales, stemming from the underlying assumption in the model that the prospect of worker relocation is largely independent of the distance traveled.

8.3 NOISE

A major focus of the economic group's effort at the Complex Systems Summer School was the impact of noise on equilibrium solutions to nonlinear equations of motion. The effects of noise on this model could be easily measured, but perhaps less easily interpreted. At a basic level, the introduction of noise requires a shift in emphasis from specific steady states to long-term statistical averages. In addition to that conceptual shift, four basic changes to the steady-state analysis might be expected as a result of the influence of noise on this model. First, the actual population values for long-term steady state populations may shift (e.g., the "blowtorch theorem" described in Landauer[7]); these effects are typically proportional to the amount of noise. Second, the character of an equilibrium can shift from (meta)stable to unstable or vice versa, especially if the equilibrium is "shallow" to begin with. Third, noise can blur the boundaries of the basins of attraction, so that initial configurations that would converge to one steady state under a deterministic evolution could wind up in another as a result of stochastic evolution. Finally, where a model has a strongly attracting steady state (such as the Verhulst population model at population zero), wide enough swings due to noise can trigger absorption into the attracting steady state. Whether this particular effect would obtain in the Krugman model depends on the sizes of the basins of attraction of states with population zero and with population one.

For example, the effect of noise on a completely symmetric initial configuration could be investigated. For larger numbers of cities, this configuration is apparently typically unstable. Thus it would not be surprising to find that the evolution of an initially symmetric equilibrium under a stochastic equation of motion eventually ends up in an asymmetric, inhomogeneous state.

Another intriguing question raised by noise in the context of this model is the connection between stochastic evolution of the (continuous) differential equation and the nondeterministic evolution of a discrete model based on agent interactions. Does the behavior exhibited by a discrete model[1] also occur in a continuous differential equation with added noise? Can the effect of noise be duplicated merely by the discretization of the equation into individual actors, whether or not acting stochastically?

Intuitively, economic noise, whether in the form of limited information, variation in individual utility functions, or external environmental factors, seems an inextricable part of any comprehensive economic model.

9. CONCLUSIONS

We have demonstrated that the Krugman model has three of the most significant features commonly associated with a complex system: behavior that varies under the influence of exogenous parameters, a large number of stable equilibrium states, and rich nonequilibrium or transient dynamics. As Prof. Krugman commented, this model provides "an example of emergent structure in which the assumptions are not too close to the conclusions." These features become more pronounced as the model expands in size. For this reason, we think that the model forms an especially good example of a complex system.

ACKNOWLEDGMENTS

This work is the direct outcome of research conducted at the 1993 Complex Systems Summer School, by a group of participants including (in addition to the listed author) Philip Auerswald, Ann Bell, Jose Lobo, David McCormack, and Brett McDonnell. The work of Daniel Stein in organizing the Summer School itself, and his encouragement of this project, was also essential.

REFERENCES

1. Auerswald, P. E., and J. T. T. Kim. "Self-Organization in a Dynamic Spatial Model." This volume.
2. Coppersmith, S. N., and A. J. Millis. "Diverging Strains in the Phase-Deformation Model of Sliding Charge-Density Waves." *Phys. Rev. B* **44(15)** (1991): 7799–7807. See also Coppersmith, S. N. this volume.
3. Coppersmith, S. N. "Complex
4. Dennis, J. E., Jr., and R. B. Schnabel. *Numerical Methods for Unconstrained Optimization and Nonlinear Equations*, 168–190. Englewood Cliffs, NJ: Prentice Hall, 1983.
5. Dendrinos, D. S., and M. Sonis. *Chaos and Socio-Spatial Dynamics*, 120–126. New York: Springer-Verlag, 1990.
6. Krugman, P. "A Dynamic Spatial Model." NBER Working Paoer Series No. 4219, 1992.
7. Landauer, R. "Statistical Physics of Machinery: Forgotten Middle-Ground." *Physica A.* **194(1-4)** (1993): 551–562.

Philip E. Auerswald† and Jan Tai Tsung Kim‡
Department of Economics, University of Washington
†phil@santafe.edu
‡kim@vax.mpiz-koeln.mpg.d400.de

Transitional Dynamics in a Model of Economic Geography

1. INTRODUCTION

Of the many types of assumptions on which economic models are based, among the most powerful are those that concern the relative time and space scales at which such fundamental processes as learning, price adjustment, and relocation take place. A half-century of debate between Keynesians and Monetarists has demonstrated that economic models based on slightly differing assumptions regarding the rates at which these transitional processes occur can be invoked to argue for very different macroeconomic policies. Such distinguished economists as Young,[10] Schumpeter,[9] Hirshman,[4] Kaldor, and Arthur[1] have argued persuasively that transitional dynamics play an active and important role in the evolution of economic systems. However, until recently, economists lacked the tools to follow up these arguments with explicit models and careful study. If it is true that, as Robert Lucas[8] wrote, "progress in economic thinking means getting better abstract, analog economic models, not better verbal observations about the world," then progress in the study of transitional processes has been slow.

In this paper we argue that computationally intensive simulations of economic systems can be used to study transitional dynamics. We build upon previous work by Dixit and Stiglitz[2] and Krugman[6,7] to introduce a model of economic geography in which agglomeration of production, a zero-profit competitive equilibrium, and even business cycles arise out of the profit-maximizing choices made by firms and the utility-maximizing choices made by workers/consumers. We further describe results that imply, contrary to some to conclusions of much analytical work on business cycles, that "sticky" wages actually appear to stabilize, rather than destabilize, output and employment in a Krugman-type model of economic geography.

While our basic framework is that developed by Krugman, we abandon two time/space assumptions which are central to Krugman's results. These are: (1) economic profits are driven to zero by entry and exit of firms at a much more rapid rate than the real wage between regions is equalized by relocation of workers, and (2) the rate at which wages and prices adjust within regions to clear local goods and labor markets is much more rapid than the rate at which workers relocate. In place of these assumptions, we introduce a process of local wage tâtonnement by which wages and prices partially adjust in each period to clear markets, but market clearing conditions are not imposed. These changes clearly have the disadvantage of making the model analytically intractable. The advantage, however, is that we are able to represent explicitly the simultaneous effects of wage and price adjustment and of relocation on the emergence of spatial and temporal patterns.

2. THE MODEL

2.1 OUTLINE

The constructed economy consists of an array of regions containing firms of two types: farms and factories. Both types of firms produce output as a function of a single input, labor. The production function for farms is bimodal:

$$F_a L = \begin{cases} 0 & \text{if } L = 0; \\ Q_a & \text{if } L \geq 0. \end{cases} \tag{1}$$

In other words, farms will produce output Q_a during every time period as long as they have at least one worker. The marginal product of additional workers is zero. The production function for factories is

$$Q_m = \frac{L - \alpha}{\beta}, \tag{2}$$

$\alpha =$ units of labor devoted to overhead per period,

$\beta =$ units of labor per unit of output per period.

The marginal product of workers is thus $1/\beta$.

Both types of firms employ worker/consumers with identical Dixit/Stiglitz utility functions of the form

$$U(C_a, C_m) = C_m^{\mu} C_a^{1-\mu} \tag{3}$$

where

$$C_m = \left[\sum_{i=1}^{N} c_i^{(\sigma-1)/\sigma}\right]^{\sigma/(\sigma-1)}, \tag{4}$$

σ = elasticity of substitution (constant).

In every period, workers spend a share μ of their income on manufactured goods and the rest on agricultural goods.

Based on the assumption that the local labor market is competitive and workers are undifferentiated, wage levels are defined for regions rather than for firms or individuals. In other words, at every time period, all firms in a given region offer the same wage. However, *we do not assume that wages adjust rapidly enough to clear the labor market in every time step.* Rather, wages adjust from period to period. Wages increase when there is excess demand in the labor market and decrease when there is excess supply. Workers move deterministically to the neighboring region that offers the highest average utility. The probability that a given firm will survive from one period to the next is a function of its profits.

2.2 INITIALIZATION

At time $t = 0$ the system is initialized in the following manner. Regions are allocated an identical number of workers and farms. An initial set of firms is located, one firm at a time and at random, on the landscape. Employee numbers are then adjusted to ensure that the total number of employees in a region does not exceed the number of workers in that region. Any single-period shortage in the labor market is resolved by rationing.

2.3 SINGLE-PERIOD PROFIT MAXIMIZATION BY FIRMS

The qualitative features of the evolution of the economy are determined by the profit-maximizing choices of firms and the utility-maximizing choices of consumers. Given the wage at the start of the period, firm i located in region j chooses prices in each period to maximize

$$\Pi = p_{i,j}Q_{m,i} - w_j L = p_{i,j}Q_{m,i} - w_j(\alpha + \beta Q_{m,i}), \tag{5}$$

$p_{i,j}$ = the price received by firm i in its home region j for a unit of its output;

w_j = the wage in region j.

Solution of this problem leads to the following relation for price as a function of the local wage:

$$p_{i,j}^{*} = \left(\frac{\sigma}{\sigma - 1}\right) w_j \beta. \tag{6}$$

Firms do not look ahead and they do not remember the past. They simply follow the single-period rule for profit maximization.

2.4 SINGLE-PERIOD UTILITY MAXIMIZATION BY CONSUMERS

Given local prices of goods from all regions, a consumer at region k chooses the bundle of goods which maximizes his/her utility function given in Eq. (3) above, subject to the budget constraint given by

$$\sum_{i=1}^{N} p_{i,k} q_i \leq w_k, \tag{7}$$

$$\begin{aligned} p_{i,k} &= \text{price of firm } i\text{'s output at region } k \\ &\quad \text{(inclusive of transportation costs per region);} \\ q_i &= \text{amount purchased from firm } i; \\ N &= \text{number of firms (same as number of products).} \end{aligned}$$

Solution of the consumer's single-period utility maximization problem leads to the following expression for quantities as a function of local prices:

$$q_i^* = \left(\frac{p_{i,k}^{-\sigma}}{\sum_i p_{i,k}^{1-\sigma}} \right) \mu w_k.$$

Demands by all consumers from all regions for the output of a given firm are summed to yield total demand from that firm for that period:

$$Q_i(P) = \sum_{k=1}^{R} \sum_{p=1}^{P} q_{ik,p}, \tag{9}$$

$$\begin{aligned} R &= \text{total number of regions;} \\ P &= \text{total population of region } k. \end{aligned}$$

2.5 LABOR RATIONING AND WAGE TÂTONNEMENT

Firms will hire only the workers needed to produce the output demanded. Consequently, Eq. (9) determines labor demand. However, given the "stickiness" of the wage in this model, there is no reason for the labor demand of firms to equal the labor supply. If there is a shortage of labor, the number of employees of each firm is linearly adjusted so that the total number of employees is less than, or equal to, the number of workers. If there is a surplus of labor, all demands for labor are satisfied.

Wages adjust to eliminate excess demand or supply in the labor market. The rate of adjustment is governed by the user-defined parameters: the percentage by which a regional wage is increased per period in a region when there is no unemployment ($\delta+$) and the fractional decrease in the wage per period for each percentage point of unemployment ($\delta-$). At the beginning of each time step, new wages at region k are set as

$$w_{k,t+1} = w_{k,t}(1 + \delta^+ - [\delta^- \cdot \varepsilon_{k,t}]), \tag{10}$$

$$\varepsilon = \text{the percentage of unemployment at region } k, \text{ time } t.$$

2.6 WORKER TRANSITION

Workers base their transition decision on *local* information alone. They do not form expectations, but rather move deterministically at period t to the *neighboring* region that offered the highest utility to its residents at period $t-1$. This is the only type of transition that takes place. There is no relocation by firms or transition by workers to any region for any other reason. Since all individuals share the same utility function and all firms within a region offer the same wage, the desirability of a move to region k from region j may be calculated in terms of the following ratio:

$$\rho_{j,k} = \frac{U^*(c_a, c_m; w_k, p_k)}{U^*(c_a, c_m; w_j, p_j)}, \tag{11}$$

$$U^*(c_a, c_m; w_j, p_i) = \text{the maximum utility obtainable at region } i \text{ given } w_i \text{ and } p_i.$$

Even when a neighboring region does offer better prospects for workers, all workers do not move at once. The number of workers who move is given by the expression

$$N_j \left(\frac{\rho_{j,k} - 1}{\rho_{j,k}} \right). \tag{12}$$

For example, if the next best region has twice the aggregate utility level of the current region, half the workers will move; if the best neighboring region has thrice the aggregate utility of the worker's current location, then two-thirds of the workers will move.

2.7 FIRM SURVIVAL AND LABOR DEMAND TRANSITION

At the end of each time step, firms calculate profits (Eq. 5). All firms that earn positive profits survive into the next period. Firms that earned negative profits have a probability of survival $p(\Pi)$ given by

$$p(\Pi) = e^{\Pi \cdot v}, \tag{13}$$

$$v = \text{a user-defined coefficient greater than } 0.$$

A firm that does not survive is removed from the simulated economy. Firms that do survive then calculate their labor demand for the next period. Since firms are uncertain regarding the demand for their output in the next period, they plan to hire the number of workers needed to produce the arithmetic mean of current production and current demand.

3. RESULTS

In this section we present some persistent features of the behavior of the model. Our results take the form of outcomes from four representative runs (drawn from an set of more than one hundred).

All the results presented are from runs of 3,000 time steps. The time scale of a given run is determined by the rate of wage tâtonnement. While there is a good deal of debate concerning the speed with which firms and workers can negotiate some adjustment in wages in response to disequilibrium in the labor market, let us say that on average this process takes between a day and a month. Then the runs of 3,000 time steps presented below can be interpreted as having a natural time scale of anywhere from 10 years to 300 years.

In order to simplify the presentation, the runs are described as variants on two basic scenarios: a low-productivity economy (high transport costs, low differentiation between products, and low marginal product of labor) and a high-productivity economy (lower transport costs, high differentiation between products, and high marginal product of labor). A complete listing of parameter values for these two scenarios is given in the appendix.

3.1 AGGLOMERATION

For a wide range of parameter values in both scenarios I and II in the Appendix, large-scale agglomeration occurs. An example of the agglomeration of firms for the case of a high-productivity economy is given in Figures 1–4. At time step 0 (see Figure 1, representing the first iteration of the model), firms are fairly equally spread among the 121 regions. Within only a few hundred time steps, worker transition has stopped altogether. By the thousandth time step (see Figure 2), more than 95% of the firms are located in less than 5% of the territory. While the distribution of firms shifts over time (reflecting the stochastic nature of the process by which firms are introduced into the economy), the degree of firm concentration is not diminished.

The key parameter affecting the magnitude of agglomeration is transportation cost. Lowering transportation costs by a factor of 1,000 sharply reduces the concentration of workers. The extent of agglomeration for a high-productivity economy's low transport costs is illustrated by Figures 5–7.

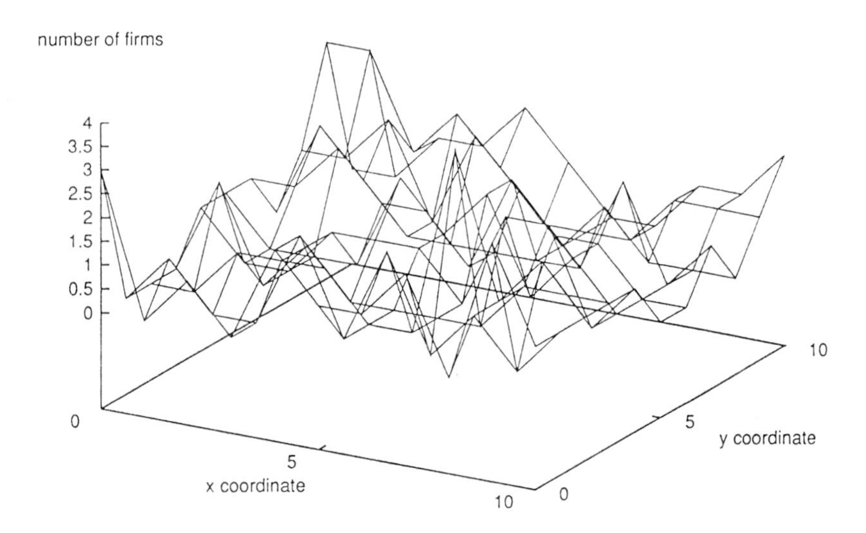

FIGURE 1 High-productivity economy, time step 0.

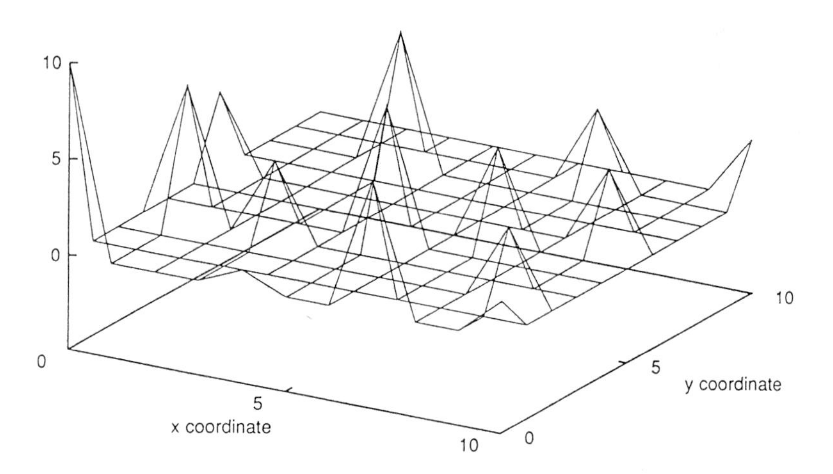

FIGURE 2 High-productivity economy, time step 1,000.

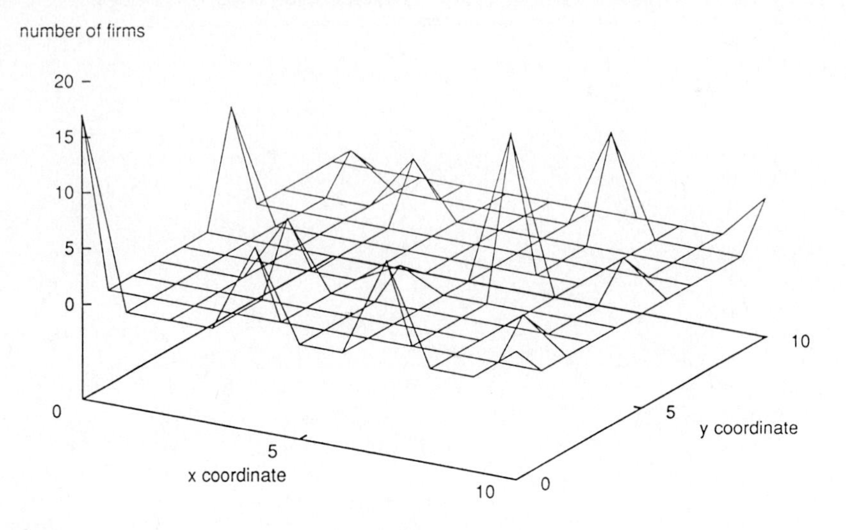

FIGURE 3 High-productivity economy, time step 2,000.

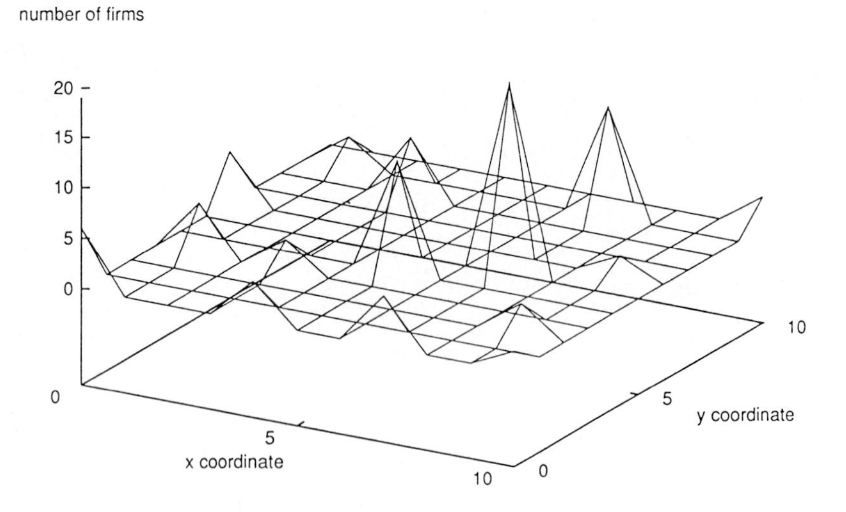

FIGURE 4 High-productivity economy, time step 3,000.

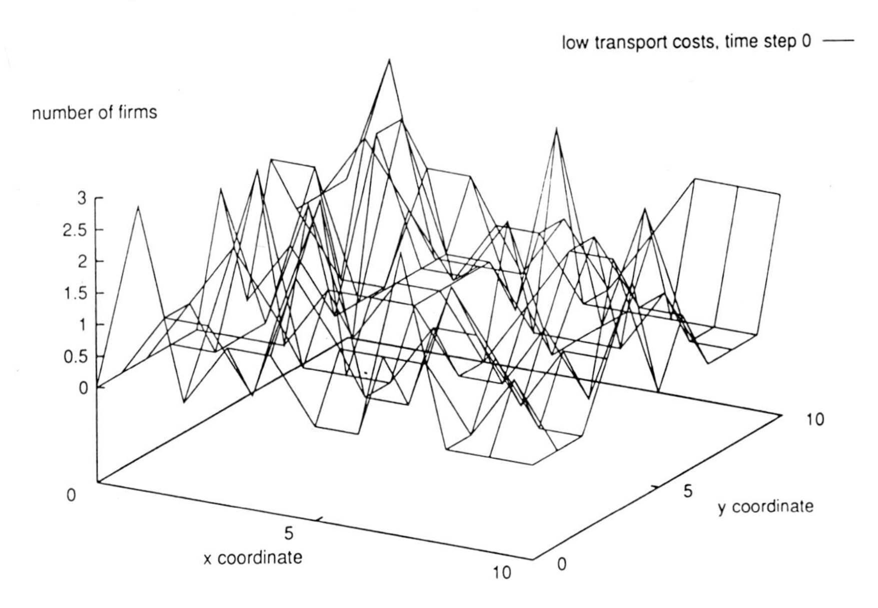

FIGURE 5 Low transport costs, time step 0.

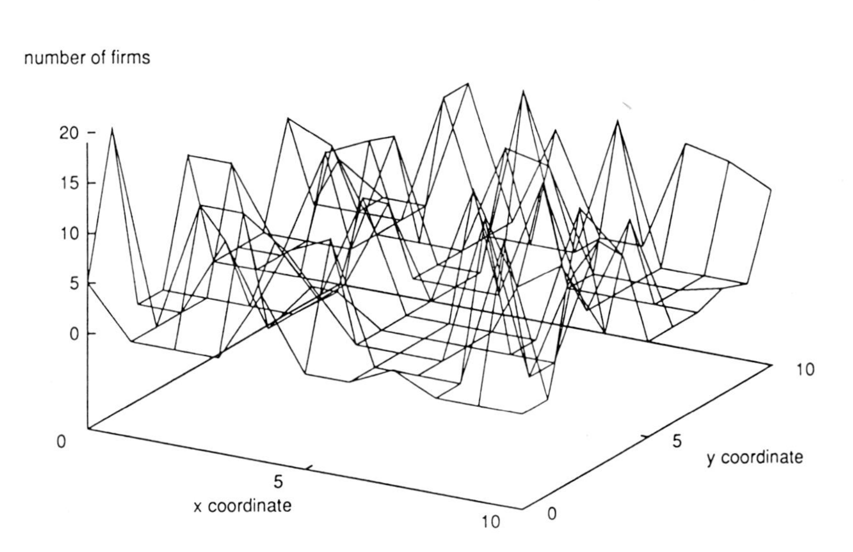

FIGURE 6 Low transport costs, time step 1,000.

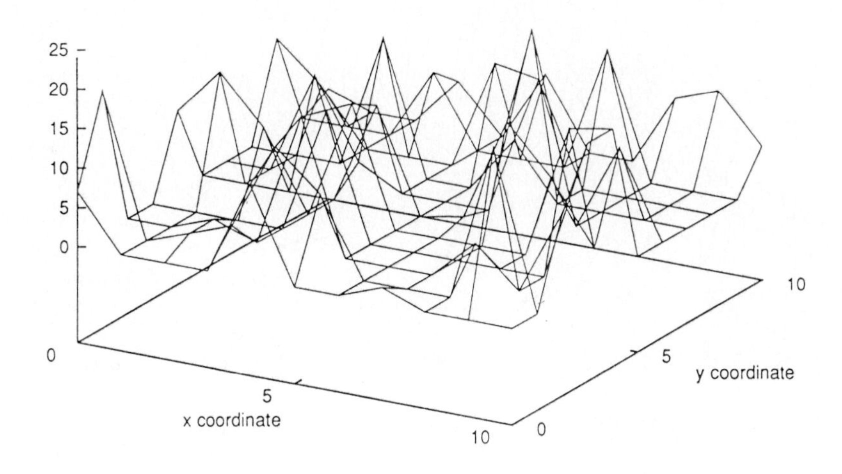

FIGURE 7 Low transport costs, time step 3,000.

3.2 THRESHOLD EFFECTS

The transition from an agricultural to a manufacturing economy is highly nonlinear. In Figure 8 we show an endogenously driven "industrial revolution" occurring in a low-productivity economy. This phenomenon appears to be attributable the combination of increasing returns to scale at the firm level, and increasing returns to agglomeration at the regional level.[1]

3.3 FLUCTUATIONS IN OUTPUT AND THE BENEFITS OF STICKY WAGES

When wages change rapidly to clear labor markets, worker transition consistently ceases after roughly 100 time steps. Nonetheless, *long after the population distribution has stabilized, the economy experiences persistent large-scale fluctuations in output (see Figure 9) and employment.* The source of these fluctuations is the sensitivity of the wage rate to local unemployment.

[1] A somewhat similar dynamic is found in Kelly[5] and Durlauf.[3]

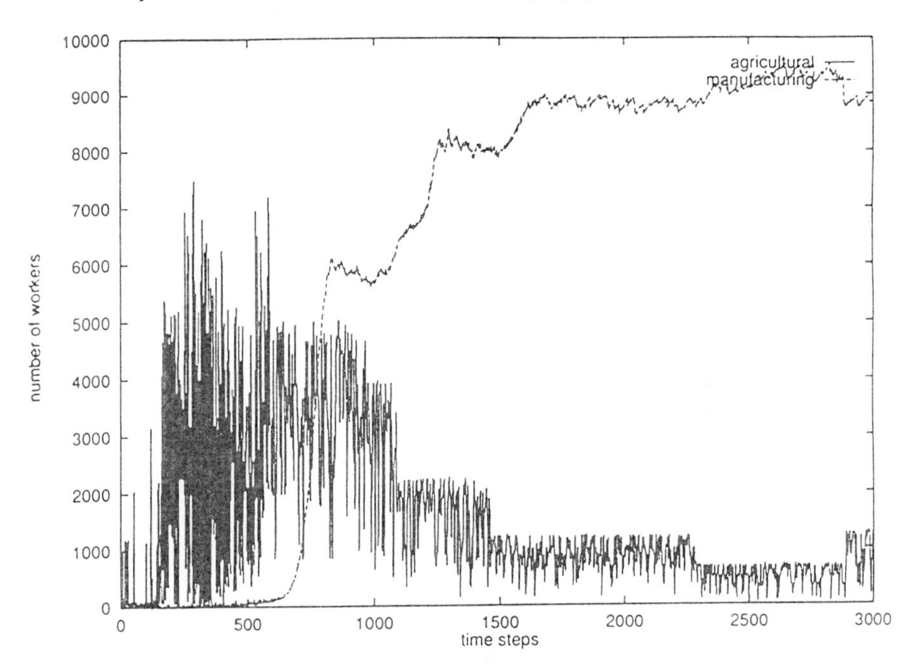

FIGURE 8 Endogenously driven industrial revolution.

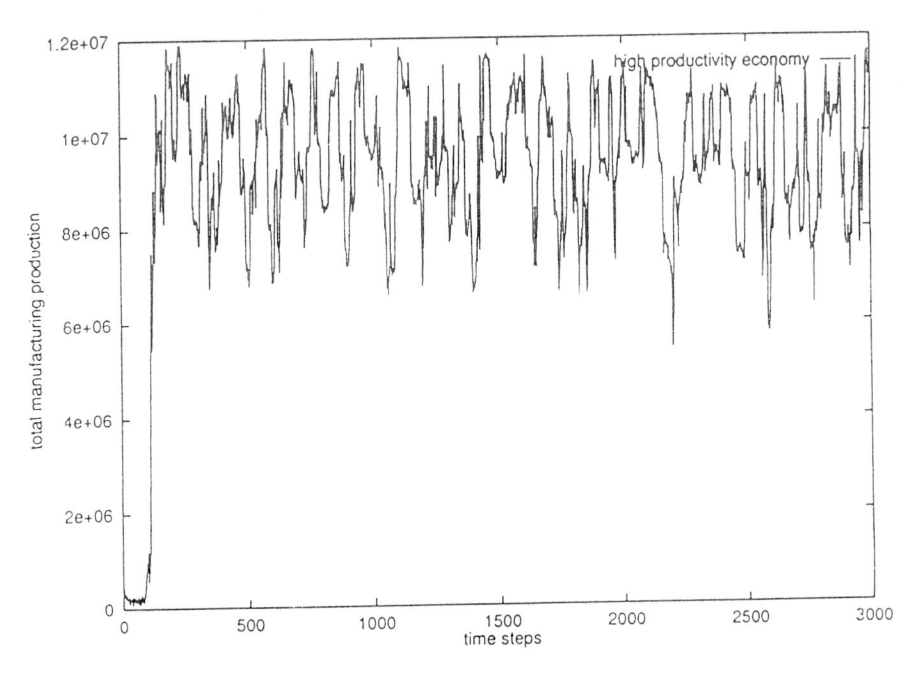

FIGURE 9 Large-scale fluctuations in output and employment.

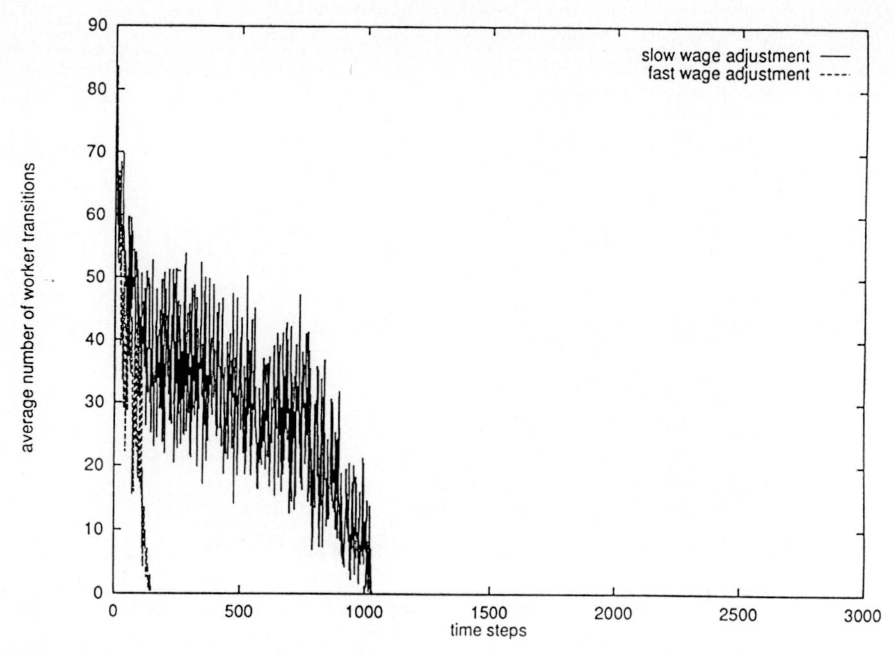

FIGURE 10 Extended duration of worker transition in a high-productivity economy with wage adjustment slowed by a factor of 10.

In order to examine the effect on the economy of changes in the time scale of wage adjustment, we simulated the high-productivity economy with the rate of wage adjustment slowed by a factor of ten ($\delta^+ = 0.0004$ and $\delta^- = 0.00002$). Doing so not only extended the duration of worker transition by a factor of roughly ten (Figure 10), but also *smoothed the fluctuations in employment and output, and lowered the equilibrium rate of unemployment.* The stabilizing effect of a "sticky wage" is represented in Figure 11 (the average rate of unemployment in both fast and slow adjustment cases) and Figure 12 (the average wage in both cases). Even more surprising is the fact that the unemployment rate in the slow adjustment, "sticky wage," case oscillates near a value of 6%, well below the analytically derived prediction of 20%.[2]

[2]The expected steady state level of unemployment is found by solving Eq. (10) with $\delta^+ = 0.0004$ and $\delta^- = 0.00002$.

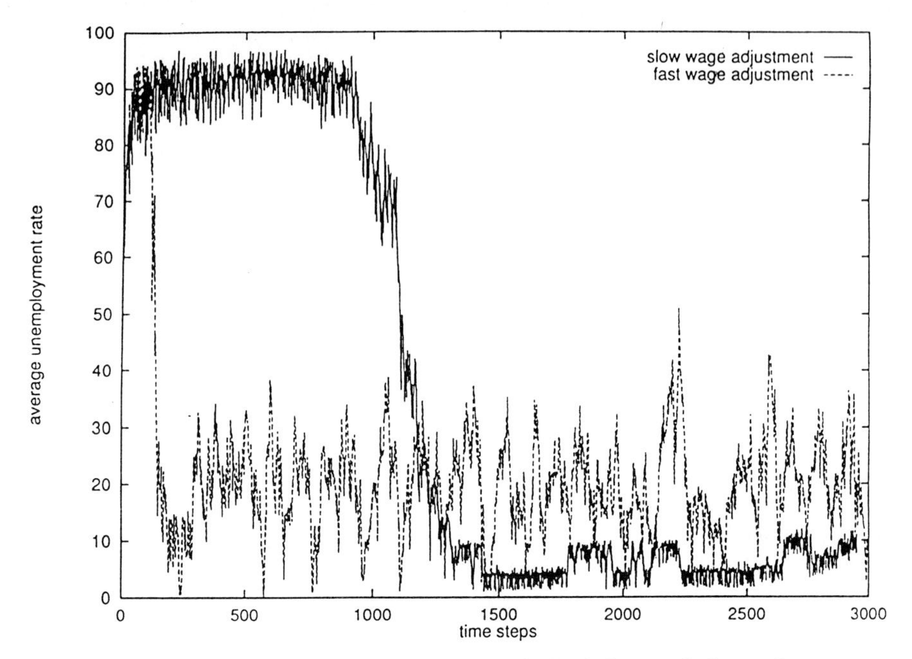

FIGURE 11 The average rate of unemployment in both fast and slow adjustment cases.

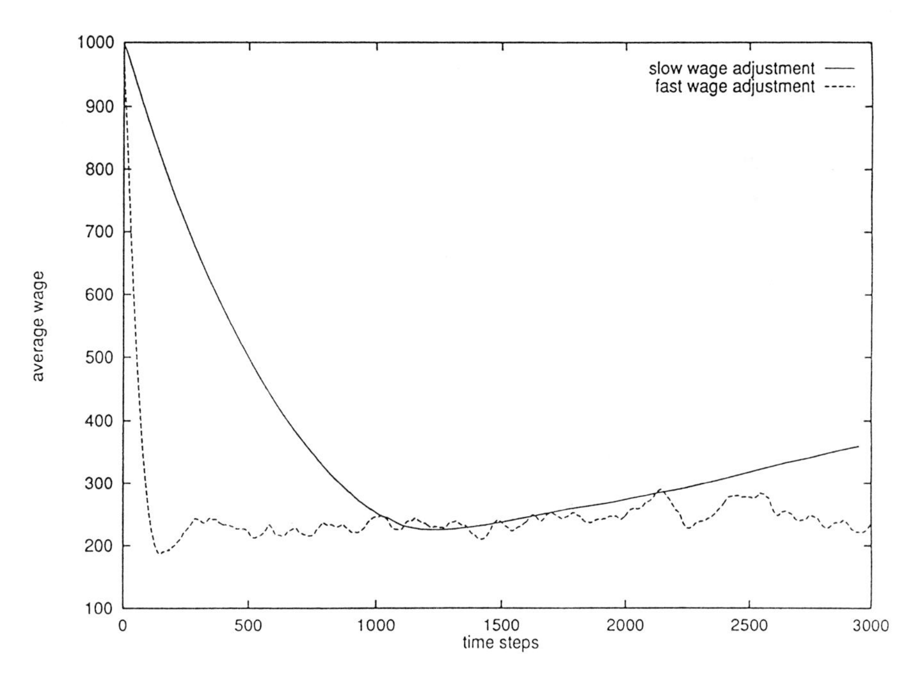

FIGURE 12 The average wage in both fast and slow adjustment cases.

4. CONCLUSION

In a series of papers published over the past ten years, Paul Krugman, W. Brian Arthur, and others have begun the process of establishing the microeconomic foundations on which the long-neglected field of economic geography may be reconstructed. Among the main themes of this work has been the description of situations under which economic/geographical structures at a large scale may appear out of interactions at a small scale. In this paper we have attempted to take this line of inquiry a step further by describing the richness of the transitional dynamics and emergent structures which can be generated by even a very simple system of interdependent economic agents. The work presented in this paper supports the contention that the time scale of transitional process can have important implications for macroeconomic outcomes. Our results further suggest that simulations based on maximizing behavior by such agents can provide economists with a much needed tool for use not only in the study of economic geography, but also in the study of transitional dynamics, the development of hypotheses concerning rates of economic adjustment in various environments, and the explicit modeling of self-organization in economic systems. While more work is needed to establish the causal link between variations in underlying parameters and observed behaviors of the model described above, we believe that the preliminary results we have presented argue for further work in this direction.

ACKNOWLEDGMENTS

This paper is the produce of work initiated at the 1993 Santa Fe Institute Complex Systems Summer School. We are grateful to Eric Anderson, Ann Bell, James Kittock, José Lobo, Brett MacDonald, David McCormack, Arjendu Pattanayak, and Richard Startz for helpful conversations. We thank the Santa Fe Institute for sponsoring the summer school.

APPENDIX: PARAMETER VALUES

SCENARIO I: A "PRE-INDUSTRIAL" ECONOMY

Productivity is low ($\alpha = 3$ and $\beta = 1$). Manufactured products are relatively undifferentiated ($\sigma = 1.5$). Transport costs are high ($\tau/\overline{p} \approx 100$, where τ is the per unit cost of transport from one region to the next and $\overline{p}$ is the value around which the average price of a manufactured good oscillates after worker transition has ceased). The wage responds rapidly to disequilibrium in the labor market ($\delta^+ = 0.004$ and $\delta^- = 0.00002$ in Eq. 10). The share of expenditure on manufactured goods is 40%. The unemployed receive a small, fixed income which may be thought of as the return from subsistence farming. The value of the firm survival coefficient, ν, is 0.001.

SCENARIO II: AN "INDUSTRIALIZED" ECONOMY

Productivity is high ($\alpha = 3$ and $\beta = 1$. Transport costs are lower in real terms than in scenario I ($\tau/\overline{p} \approx 10$). Products are highly differentiated ($\sigma = 5$). The share of expenditure on manufactured goods is 60%. Other parameters are unchanged from the previous scenario.

REFERENCES

1. Arthur, W. B. "'Sillicon Valley' Locational Clusters: When Do Increasing Returns Imply Monopoly?" *Math. Soc. Sci.* **19** (1990): 235–251.
2. Dixit, A., and J. Stiglitz, J. "Monopolistic Competition and Optimum Product Diversity." *Am. Econ. Rev.* **67(3)** (1977): 297–308.
3. Durlauf, S. "Nonergodic Economic Growth." *Rev. Econ. Stud.* **60** (1993): 349–366.
4. Hirshman, A. O. *The Strategy of Economic Development.* New Haven: Yale University Press, 1958.
5. Kelly, M. "Division of Labor and the Extent of the Market: On the Dynamics of Industrialization." Working Paper 93-02, Center for Analytical Economics, Cornell University, 1993.
6. Krugman, P. "Increasing Returns and Economic Geography." *J. Pol. Econ.* **99(3)** (1990): 483–499.
7. Krugman, P. "A Dynamic Spatial Model." Working Paper No. 4219, NBER, 1992.
8. Lucas, R. "Methods and Problems in Business Cycle Theory." *J. Money, Credit, & Banking* **12(4)** (1980): 696–713.
9. Schumpeter, J. A. *Capitalism, Socialism and Democracy.* New York: Harper and Row, 1948.
10. Young, A. "Increasing Returns and Economic Progress." *Econ. J.* (December 1928): 527–542.

Ann M. Bell
Department of Economics, University of Wisconsin, 1180 Observatory Drive, Madison, WI, 53706, e-mail: abell@macc.wisc.edu

Dynamically Interdependent Preferences in a Lattice Economy

1. INTRODUCTION

Economic models typically consider a very small number of agents that operate within narrowly defined goals and in rigid institutional structures. Actual economies consist of millions of interacting individuals and institutions. Economic models typically assume that all agents have fixed preferences. Actual individuals have been known to change their preferences over time. Recent advances in the mathematical modeling and computer simulation of complex systems greatly expand the kinds of economic behavior that can be analyzed and open up new possibilities for reconciling theory and reality.

This chapter presents a model of a bandwagon effect in which individuals' preferences for consumer goods evolve in response to the consumption decisions of their neighbors. The bandwagon effect is essentially a positive feedback mechanism where agents' preferences for a good increase whenever the agents in their local neighborhoood consume more of it relative to other goods. The effect of interdependent preferences on consumer demand was addressed by Pollak.[5] The model here considers both the demand and supply aspects in a competitive general equilibrium framework. Different regimes of production govern the economy-wide availability of goods. Because an agent's consumption depends on both his or her preferences and the price of the good, agents interact both globally and locally.

The model incorporates existing economic theory into a dynamical system, thereby allowing the parameters that control agents' preferences for goods to vary endogenously over time. The equations describing the standard single-period competitive general equilibrium[1,6] are embedded in a lattice structure that governs the dynamic interdependence of preferences. Mathematically, the model is an example of a coupled map lattice; i.e., it is a dynamical system with discrete time and space components but continuous state variables.[3,4] Despite the continuous state space and the global interactions, a useful analogy can be drawn between this model and cellular automata governed by voting rules.[2]

The discussion focuses on the "macroeconomic," or collective, behavior of the model as exhibited in computer simulations, in particular, on the evolution of the mean value of preferences over the lattice. Prices play a crucial role in mediating the interaction between the available production technology and consumers preferences. When the amount of the two consumer goods available is fixed, as in a simple exchange economy, the relative proportion of one good to the other has the greatest influence on the steady-state value of average preferences. In contrast, when the amounts of the two goods produced can be varied using the same constant returns to scale technology for both goods, the initial distribution of consumer preferences determines the final average value of preferences. Decreasing returns to scale technology represents an intermediate case. The production of goods varies in response to changes in average preferences but the cost of producing additional goods depends positively on the level of output. The initial distribution of preferences influences the final steady state but average preferences evolve so as to minimize the cost of production, or equivalently, to maximize the total quantity of goods produced in the economy.

2. STRUCTURE OF THE MODEL

Agents are located on a rectangular lattice with periodic boundary conditions, i.e., on a torus. In the first time period, agents begin with randomly chosen preferences for consumer goods and an endowment of either goods or labor. They participate in a centralized market where they trade their endowments for consumption goods at the set of prices which clears the market, that is, which sets supply equal to demand for all goods. The agents then look around their own neighborhood on the lattice and see what their neighbors have bought. If consumption of a good exceeds 50 percent of the total in an agent's neighborhood, then his or her preferences for the good increase. The psychology of this is simple: the more agents see of a particular good the more they like it, even if their consumption this period already exceeds the neighborhood average. The next period, agents begin with the same endowment as before but with new preferences, and the whole procedure repeats.

In Table 1, we summarize the competitive general equilibrium model. (Equations (t1)–(t40) appear in Table 1.) There are two consumption goods, x_1 and x_2, which are measured in comparable units; i.e., x_1 is red t-shirts, x_2 is blue t-shirts. In keeping with neo-classical economic theory, each agent is represented by a utility function which assigns a value to every possible consumption bundle. (See Eqs. (t1), (t8), and (t22).) The agent maximizes this function subject to a budget constraint (see Eqs. (t2), (t9), and (t23)), which requires that the value of goods purchased must equal the value of the agent's endowment taking market prices as given. The solution to this maximization problem is a set of demand functions that express each individual's demand for a good as a function of the prices of all of the goods in the economy (see Eqs. (t3), (t4), (t10), (t11), (t24), and (t25)).

In an exchange economy, the total amount of goods available is determined exogenously by the agents' endowments of consumer goods. In a production economy, agents are endowed with labor and with shares of ownership of the firms. The firms purchase labor and produce goods for consumption in order to maximize profits (see Eqs. (t12), (t15), (t26), and (t31) in Table 1) subject to the available production technology (see Eqs. (t13), (t16), (t27), and (t32)). Consequently, the total and relative amounts of the goods produced are determined endogenously through the interaction of technology and consumer demand. Two different production technologies are considered here: constant returns to scale (see Eqs. (t13) and (t16)) where the amount of labor needed to produce additional goods is independent of the level of output and decreasing returns to scale (see Eqs. (t27) and (t32)) where the the amount of labor needed to produce additional goods increases with the level of output. In a constant-returns-to-scale economy, the price (see Eq. (t21)) is fixed by the technological parameters, and firms earn no profits in equilibrium. In a decreasing-returns-to-scale economy, the price (see Eq. (t39)) varies with the level of demand and firms earn profits in equilibrium (see Eqs. (t30) and (t35)).

Equilibrium prices (see Eqs. (t7), (t21), and (t36–38)) coordinate agent's behavior economy wide by setting the sum of all the individual demand functions equal to the total quantity of goods available (see Eqs.(t5–6), (t18–20), and (t36–38)). Restrictions on the types of utility and production functions used ensure the existence of market clearing equilibrium prices. In this case, the parameter a which reflects the agents' preferences for the two goods must lie in the closed interval [0,1] to ensure the existence of a market equilibrium.

Each agent's neighborhood consists of the agent and his or her eight nearest neighbors. After the equilibrium level of production and individual consumption bundles are determined, agents adjust their preferences upwards or downwards proportional to the deviation of neighborhood consumption of x_1 from 50 percent.

TABLE 1 Competitive General Equilibrium

Exchange Economy	
$\max u(x_1, x_2) = x_1^{a_t} x_2^{1-a_t}\ 0 \le at \le 1$	(t1)
$s.t.\ p_1x_1 + p_2x_2 = p_1w_1 + p_2w_2$	(t2)
$x_1(p_1, p_2) = a_t(w_1 + (p_2/p_1)w_2)$	(t3)
$x_2(p_1, p_2) = (1 - a_t)(p_2/p_1)(w_1 + (p_2/p_1)w_2)$	(t4)
equilibrium conditions	
$\Sigma_{\text{lattice}} x_1(p_1, p_2) = \Sigma_{\text{lattice}} w_1$	(t5)
$\Sigma_{\text{lattice}} x_2(p_1, p_2) = \Sigma_{\text{lattice}} w_2$	(t6)
$p_1^*/p_2^* = (w_2 \Sigma_{\text{lattice}} a_t)/w_1 \Sigma_{\text{lattice}}(1 - a_t)$	(t7)
Constant Returns to Scale	
$\max u(x_1, x_2) = x_1^{a_t} x_2^{1-a_t}\ 0 \le a_t \le 1$	(t8)
$s.t.\ p_1x_1 + p_2x_2 = wL$	(t9)
$x_1(p_1, p_2, w) = a_t(w/p_1)$	(t10)
$x^2(p_1, p_2, w) = (1 - a_t)(w/p_2)$	(t11)
firm1	
$\max \pi_1 = p_1x_1 - wL_1$	(t12)
$s.t.\ x_1 = b_1L_1$	(t13)
$w = p_1b_1$	(t14)
firm2	
$\max \pi_2 = p_2x_2 - wL_2$	(t15)
$s.t.\ x_2 = b_2L_2$	(t16)
$w = p_2b_2$	(t17)
equilibrium conditions	
$\Sigma_{\text{lattice}} x_1(p_1, p_2, w) = b_1L_1$	(t18)
$\Sigma_{\text{lattice}} x_2(p_1, p_2) = b_2L_2$	(t19)
$L_1 + L_2 = nL$	(t20)
$p_1^*/p_2^* = b_2/b_1$	(t21)

$$a_{t+1} = a_t + r\left(\frac{\Sigma_{nbhd}\ x_1(p_1^*, p_2^*, w^*)}{\Sigma_{nbhd} x_1(p_1^*, p_2^*, w^*) + \Sigma_{nbhd}\ x_2(p_1^*, p_2^*, w^*)} - 0.5\right) \tag{1}$$

$$\text{if } a_{t+1} < 0\,, \text{then } a_{t+1} = 0; \qquad \text{if } a_{t+1} > 1\,, \text{then } a_{t+1} = 1. \tag{2}$$

By substituting the equilibrium prices and wage into individual demand functions and the individual demand functions into the equation that updates preferences (Eq. (1)) the entire transition rule can be expressed as a function of the neighborhood average of preferences and the price:

TABLE 1 (continued) Competitive General Equilibrium

Decreasing Returns to Scale	
agent	
$\max u(x_1, x_2) = x_1^{a_t} x_2^{1-a_t} \quad 0 \leq at \leq 1$	(t22)
$s.t.\, p_1x_1 + p_2x_2 = wL + (\pi_1 + \pi_2/n)$	(t23)
$x_1(p_1, P_2, w) = (a_t/p_1)(w + (\pi_1 + \pi_2/n))$	(t24)
$x_2(p_1, p_2, w) = (1 - a_t/p_2)(w + (\pi_1 + \pi_2/n))$	(t25)
firm1	
$\max \pi_1 = p_1x_1 - wL_1$	(t26)
$s.t.\, x_1 = \sqrt{L_1}$	(t27)
$x_1(p_1, p_2, w) = p_1/2w$	(t28)
$L_1(p_1, p_2, w) = p_1^2/4w^2$	(t29)
$\pi_1(p_1, p_2, w) = p_1^2/4w$	(t30)
firm2	
$\max \pi_2 = p_2x_2 - wL_2$	(t31)
$s.t.\, x_2 = \sqrt{L_2}$	(t32)
$x_2(p_1, p_2, w) = p_2/2w$	(t33)
$L_2(p_1, p_2, w) = P_2^2/4w^2$	(t34)
$\pi_2(p_1, p_2w) = p_2^2/4w$	(t35)
equilibrium conditions	
$\Sigma_{\text{lattice}} x_1(p_1, p_2, w) = \sqrt{L_1}$	(t36)
$\Sigma_{\text{lattice}} x_2(p_1, p_2) = \sqrt{L_2}$	(t37)
$L_1 + L_2 = nL$	(t38)
$\frac{p_1^*}{p_2^*} = \sqrt{\Sigma_{\text{lattice}} a_t / \Sigma_{\text{lattice}} (1 - a_t)}$	(t39)
$w^* = 1/\sqrt{2(1 - a_t)(1 + nL)}$	(t40)

$$a_{t+1} = a_t + r\left(\frac{\Sigma_{nbhd} a_t}{\Sigma_{nbhd} a_t + (p_1/p_2)\Sigma_{nbhd}(1 - a_t)} - 0.5\right). \tag{3}$$

The price in turn depends on endowments, technological parameters, and the average value of preferences over the lattice (see Eqs. (t7), (t21), and (t39–40) in Table 1). Note that when there is constant-returns-to-scale production the price is constant and there are no global interactions; hence, the update rule (Eq. (2)) Eq. (3) resembles a continuous state version of a voting rule cellular automata.

3. SIMULATION RESULTS

At the level of individual agents, the three different types of economies exhibit the same generic behavior. Starting from the random inital state, a scattering of individual agent's preferences hit either zero or one within a few iterations. Groupings of zeros and ones typically emerge within ten iterations and expand over time to occupy the entire lattice. In every simulation performed, the lattice converged to an equilibrium consisting exclusively of zeros and ones. No limit cycles, random behaviors, or multiperiod orbits were observed. In Figure 1, we show the evolution of the preferences of 15 randomly chosen agents in a sample simulation of the exchange economy. Analytical results regarding the existence, stability, and other characteristics of equilibria are discussed by Bell.[2]

The average of preferences over the whole lattice is shown in Figure 1 by the crosses marked "representative agent." Many macroeconomic models utilize a single aggregated "representative agent" in place of multiple agents. The lattice-wide average of preferences is the only endogenous variable affecting prices and the relative levels of production of the two consumer goods. When an equilibrium consists exclusively of zeros and ones the avergage value of preferences equals the proportion of agents consuming x_1 exclusively. The evolution of average preferences for different regimes of production is discussed below.

3.1 CONSTANT-RETURNS-TO-SCALE PRODUCTION

Figure 2 shows the evolution of average preferences for ten simulations of the constant-returns-to-scale production economy with $b_1 = b_2$ starting from an expected mean of preferences of 0.5. Average preferences stay relatively constant as the lattice evolves. However, when the initial state of preferences is biased away

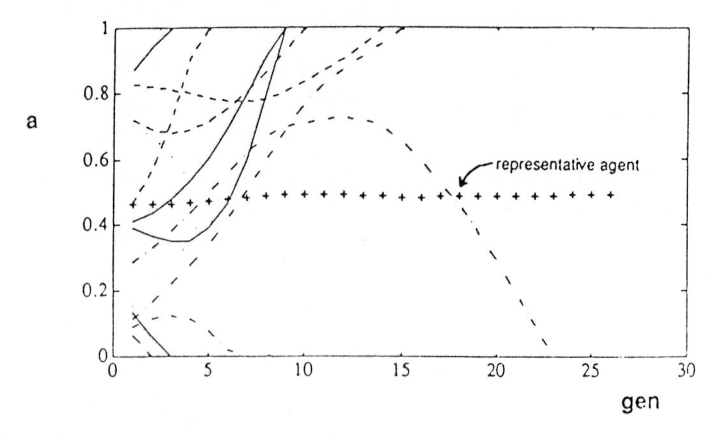

FIGURE 1 Paths of preference for individual agents.

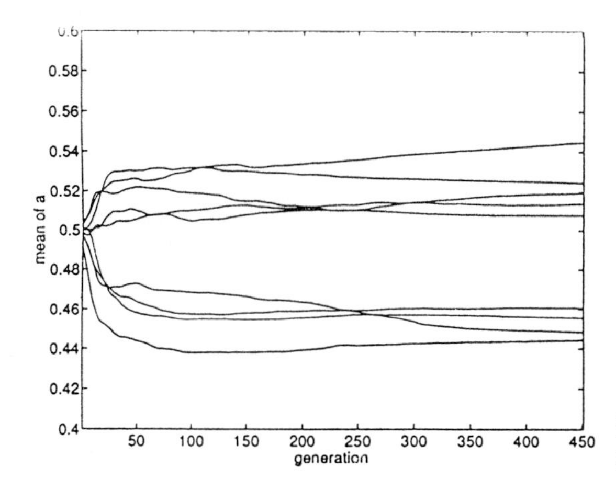

FIGURE 2 Constant-returns-to-scale production, initial mean 0.5.

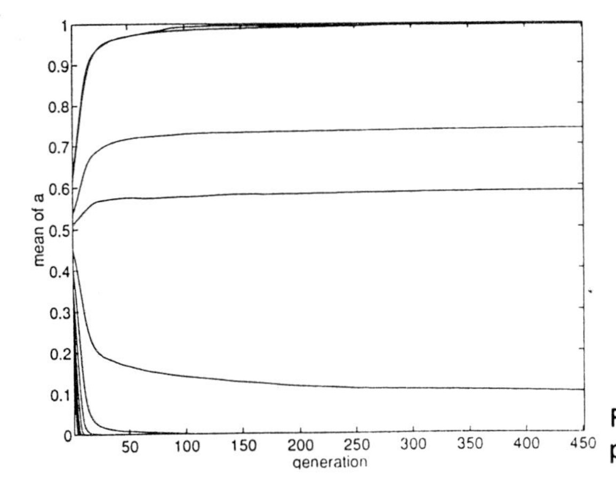

FIGURE 3 Constant-returns-to-scale production, variable intial means.

from 0.5, differences in the initial mean of preferences are accentuated over time in the constant-returns-to-scale economy. In Figure 3 we present simulation results for the constant-returns-to-scale production economy with $b_1 = b_2$ with initial states of preferences averaging from 0.15 to 0.63. Deviations of the initial state away from a mean of 0.5 are often compounded until the entire lattice consists entirely of ones or entirely of zeros. To understand this positive feedback mechanism, note that in the constant-returns-to-scale economy there are no global interactions in the update rule: the price does not depend on average preferences. The relative supply of the two goods varies as average preferences vary but the price remains constant. Because it is equally easy to produce both goods, $b_1 = b_2$, production places no constraints on the evolution of preferences. The update rule simply moves preferences in direction of the neighborhood average.

When b_1 is greater than b_2, that is, when it is relatively more efficient to produce x_1 than x_2, the update rule (Eq. (2)) is weighted in favor of x_1. When b_1 is greater than b_2, preferences for x_1 increase more quickly (or decrease more slowly) and the threshold of neighborhood average preferences which divides increases in next period preferences from decreases is lower. Figure 4 shows simulation results starting from the same initial state of preferences with a mean of 0.5 but with b_1 ranging in value from 1/8 to 8 and $b_2 = 1$. The evolution of preferences strongly favors the good with the lower cost of production.

3.2 EXCHANGE ECONOMY

The exchange economy with equal endowments of the two goods, $w_1 = w_2$, also tends to preserve an initial average of preferences of 0.5, as we show in Figure 5. However, the average behavior of the lattice in the exchange economy contrasts sharply with that of the constant-returns-to-scale production economy when the initial means of preferences is allowed to vary over a wider range. In Figure 6 we present simulation results for the exchange economy with the same initial states of preferences averaging from 0.15 to 0.63 used in Figure 3. Despite the differences in initial states the means of preferences tend back towards 0.5. Recall that in the exchange economy the relative supply of the two goods is fixed by the endowments

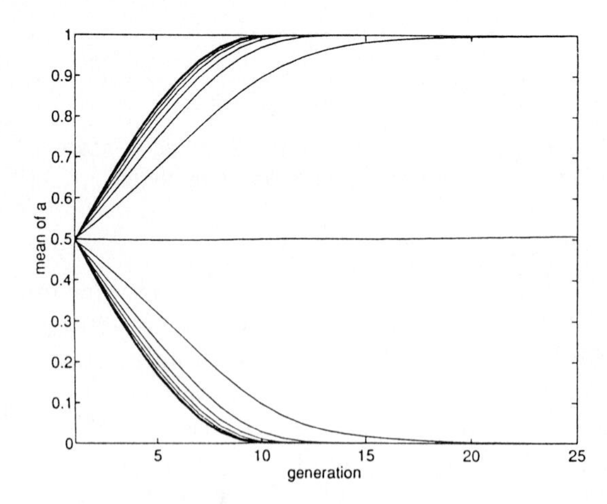

FIGURE 4 Constant-returns-to-scale production, variable prices, initial means 0.5.

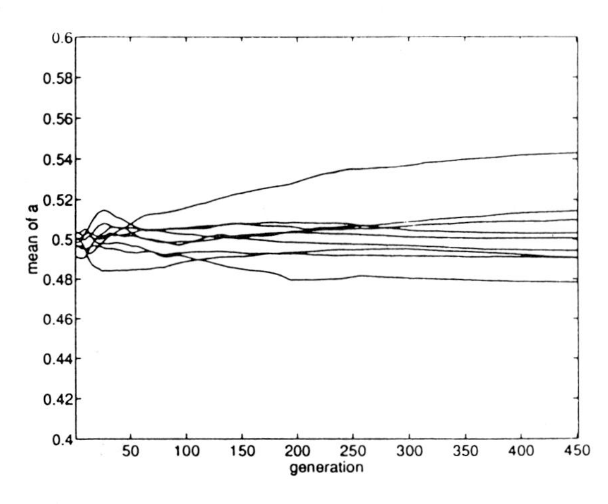

FIGURE 5 Exchange economy, initial mean 0.5.

and cannot vary in response to changes in average preferences. The price must vary in order to ensure that the market clears, even when on average the agents have much stronger preferences for one of the goods. The price operates as a negative feedback mechanism in the update rule and mediates the evolution of preferences.

Agents who prefer the relatively unpopular but low-priced good can consume more of it. Because the rule for updating preferences depends on the percent of total neighborhood consumption (Eq. (1)), these agents have a greater influence on local preferences than those who prefer the popular high-priced good. Next period,

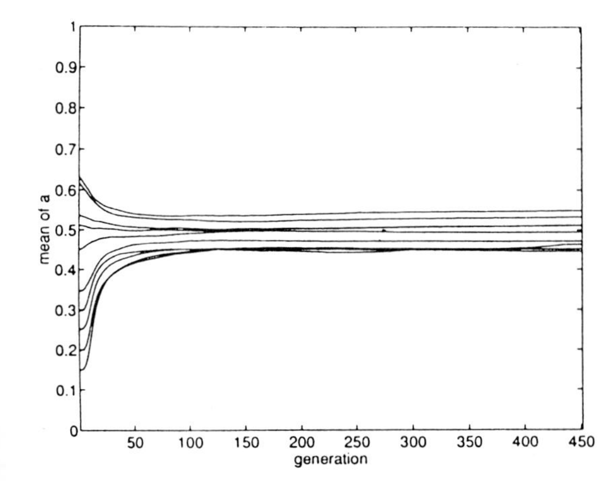

FIGURE 6 Exchange economy, variable initial means.

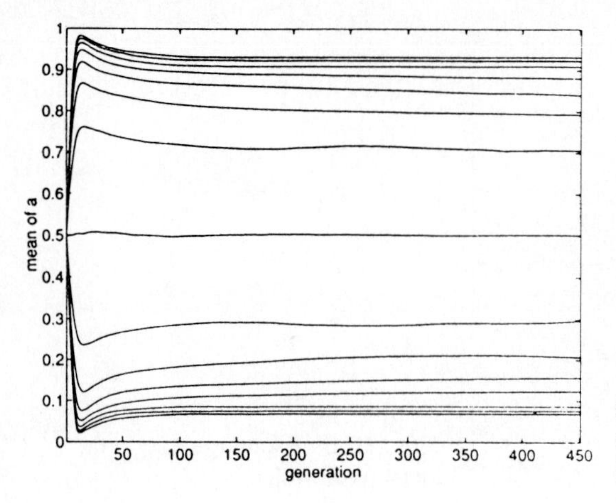

FIGURE 7 Exchange ecomony, variable endowments, initial mean 0.5.

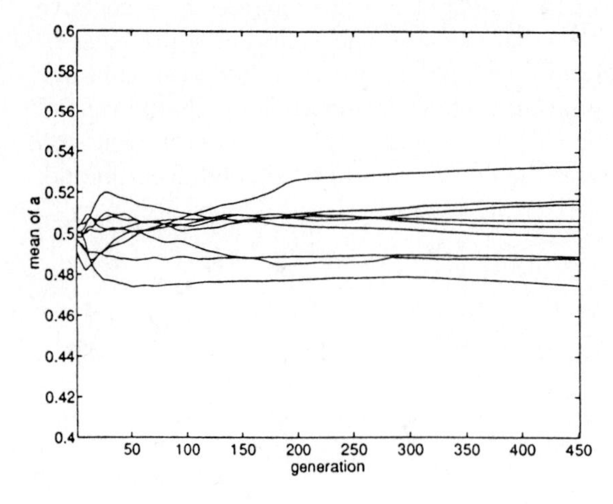

FIGURE 8 Decreasing-returns-to-scale production, initial mean 0.5.

their neighbors will demand more of the unpopular good. Over time the number of agents who prefer that good will increase, eventually pushing up the price and decreasing the influence of those agents on neighborhood preferences.

Average consumption mirrors the relative supply of the two goods: if agents are endowed with twice as much x_1 as x_2, then overall twice as much x_1 will be consumed. Combined with the price mechanism described above, this drives the average value of preferences towards the relative supply of $x, w_1/w_1 + w_2$. In the simulations in Figures 5 and 6 the endowment consists of equal amounts of the two goods. Correspondingly, the average value of preferences tends towards 0.5, with approximately half of the agents consuming x_1 exclusively and half x_2 exclusively, despite the range of initial starting conditions. In Figure 7 the relative endowments of x_1 to $x_2, w_1/w_2$, ranged from 1/8 to 8, and the average value of preferences evolved towards the predicted value of $w_1/w_1 + w_2$.

3.3 DECREASING-RETURNS-TO-SCALE PRODUCTION

In Figure 8 we show simulation results for the model with decreasing-returns-to-scale starting from the same initial configurations of preferences used in Figures 2 and 5. When the initial means of preferences is near 0.5 and the technology or production does not favor one good over another, the aggregate behavior of the model is the same for all three types of production: exchange economy, constant-returns-to-scale economy, and decreasing-returns-to-scale economy (Figure 9) represents an intermediate case beween the constant-returns-to-scale and the exchange economies. The supply of goods responds to changes in agents' preferences, but the cost of producing additional goods depends positively on the level of production. The price still affects the evolution of preferences as described for the exchange economy but not as strongly because some of the change in aggregate demand is offset by an increase in supply. Note that the price depends less strongly on average preferences because of the square root (Eq. (t39) in Table 1). In Table 2 we compare the initial and final means of preferences for the exchange and decreasing-returns-to-scale economies. Similar to the examples of the constant-returns-to-scale economy shown in Figure 4 the price weights the update rule in favor of the aggregate consumption

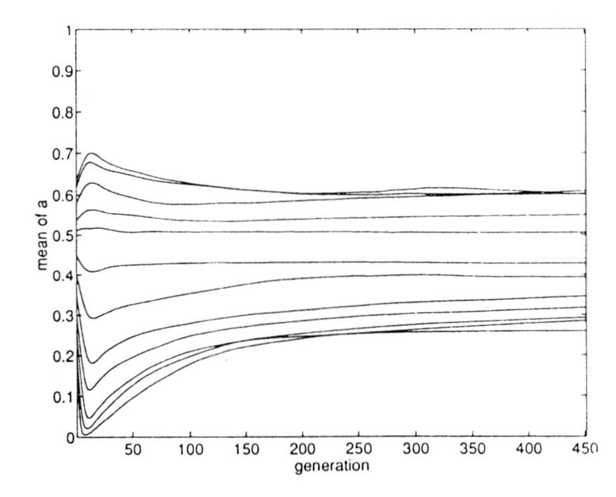

FIGURE 9 Decreasing-returns-to-scale production, variable intial means.

bundle that minimizes the cost of production, in this case equal amounts of the two goods. As demand for one good increases so does the marginal cost of production, making the good relatively more expensive and shifting the update rule back towards preferences for the other good.

In summary, computer simulations of these three models all tend toward a steady state composed exclusively of zeros and ones. The relative proportion of agents consuming only x_1, i.e., the average value of preferences, evolves towards the relative proportion of x_1 available in the case of the exchange economy and towards the value which corresponds to the cost minimizing aggregate consumption bundle in the case of constant- and decreasing-returns-to-scale economies.

3.4 CONSTANT-RETURNS-TO-SCALE PRODUCTION

In the constant-returns-to-scale economy, differences in the initial mean of preferences are accentuated as the lattice evolves. (See Figure 5). Deviations of the initial state away from a mean of 0.5 are often compounded until the entire lattice consists entirely of ones or entirely of zeros. To understand this positive feedback mechanism note that in the constant-returns-to-scale economy there are no global interactions in the update rule: the price does not depend on average preferences. The relative supply of the two goods varies as average preferences vary but the price remains constant. Because it is equally easy to produce both goods, $b_1 = b_2$, production places no constraints on the evolution of preferences. The update rule simply moves preferences in direction of the neighborhood average.

When b_1 is greater than b_2, that is, when it is relatively more efficient to produce x_1 than x_2, the update rule (see Eq. (2)) is weighted in favor of x_1. When b_1 is greater than b_2 preferences for x_1 increase more quickly (or decrease more slowly) and the threshold of neighborhood average preferences which divides increases in next-period preferences from decreases is lower. In Figure 4, we show simulation results starting from the same intitial state of preferences with a mean of 0.5 but with b_1 ranging in value from 1/8 to 8 and $b_2 = 1$. The evolution of preferences strongly favors the good with the lower cost of production.

3.5 EXCHANGE ECONOMY

The average behavior of the lattice in the exchange economy (see Figure 6) constrasts sharply with that of the constant-returns-to-scale production economy: despite the differences in initial states the mean of preferences tends back towards 0.5. Recall that in the exchange economy the relative supply of the two goods is fixed by the endowments and cannot vary in response to changes in average preferences. The price must vary in order to ensure that the market clears, even when on average the agents have much stronger preferences for one of the goods. The price operates as a negative feedback mechanism in the update rule and mediates the evolution of preferences.

Agents who prefer the relatively unpopular, and hence low-priced, good can consume more of it. Because the rule for updating preferences depends on the percent of total neighborhood consumption (see Eq. (1)), these agents have a greater influence on local preferences than those who prefer the popular, high-priced good. Next period, their neighbors will demand more of the unpopular good. Over time the number of agents who prefer that good will increase, eventually pushing up the price and decreasing the influence of those agents on neighborhood preferences. Average consumption mirrors the relative supply of the two goods: if agents are endowed with twice as much x_1 as x_2, then overall twice as much x_1 will be consumed. Combined with the price mechanism described above this drives the average value of preferences towards the relative supply x, $w_1/(w_1+w_2)$ In the simulations shown in Figures 3 and 6, the endowment consists of equal amounts of the two goods. Correspondingly, the average value of preferences tends towards 0.5, with approximately half of the agents consuming x_1 exclusively and half x_2 exclusively, despite the range of initial starting conditions. In Figure 9, we show that the relative endowments of x_1 to x_2, w_1/w_2 ranged from 1/8 to 8, and the average value of preferences evolved towards the predicted value $w_1/(w_1+w_2)$.

3.6 DECREASING-RETURNS-TO-SCALE PRODUCTION

The decreasing-returns-to-scale economy (see Figure 7) represents an intermediate case between the constant-returns-to-scale and the exchange economies. The supply of goods responds to changes in agents' preferences, but the cost of producing additional goods depends positively on the level of production. The price still affects the evolution of preferences as described for the exchange economy but not as strongly because some of the change in aggregate demand is offset by increases in supply. Note that the price depends less strongly on average preferences because of the square root (see Eq. (t39)). In Table 2, we compare the initial and final means of preferences for the exchange and decreasing-returns-to-scale economies. Similiar to the examples of the constant-returns-to-scale economy shown in Figure 8, the price weights the update rule in favor of the aggregate consumption bundle that minimizes the cost of production, in this case equal amounts of the two goods. As demand for one good increases so does the marginal cost of production, making the good relatively more expensive and shifting the update rule back towards preferences for the other good.

In summary, computer simulations of these three models all tend toward a steady state composed exclusively of zeros and ones. The relative proportion of agents consuming x_1, i.e., the average value of preferences, evolves towards the relative proportion of x_1 available in the case of the exchange economy and towards the cost minimizing aggregate consumption bundle in the the case of constant- and decreasing-returns-to-scale economies.

TABLE 2 Initial and Steady State Average Preferences

initial mean	steady state exchange economy	steady state decreasing returns
0.1501	0.4516	0.2853
0.1995	0.4519	0.2929
0.2526	0.4452	0.2596
0.2986	0.4626	0.3183
0.3480	0.4720	0.3460
0.4507	0.5115	0.3930
0.5116	0.4948	0.4287
0.5374	0.5118	0.5056
0.6174	0.5329	0.6023
0.6329	0.5491	0.6078

4. CONCLUSION

The observed macroeconomic properties of the model result from the interaction of the bandwagon effect and market-based price mechanism. The positive feedback of the bandwagon effect makes agents want to consume more of what everyone else is consuming, while the negative feedback inherent in the price mechanism gives a greater vote to those who consume a less popular good. The interaction of the two produces a global property that is, on the surface at least, unrelated to agent's individual maximizing and imitative behaviors: consumer's aggregate preferences evolve in such a way that the economy as a whole tends towards the point where the cost of production is minimized. This system-wide phenomena resonates with not only with the modern idea of emergent behavior, but also with Adam Smith's fabled "invisible hand" capable of producing an efficient global allocation of resources without any centralized decison-making process. The approach developed here may be useful in explaining more than consumer behavior. The development of new technoloies often involves both positive feedback mechanisms like increasing returns to scale and network externalities and negative feedback mechanisms like the decreasing returns encoded in the standard economic models of market economies. The clustering of like-minded consumers that occurs on a microecomic level and the cost minimizing trajectory that occurs on a macroeconomic level mirror the "Silicon Valley" phenomena often observed in technological development.

The results of this chapter also highlight one of the weaknesses of "representative agent" models that rely on the aggregation of many individuals into one. Although the characteristics of actual lattice configurations were not emphasized

Rajarshi Das
Colorado State University, Fort Collins, CO 80523

Evolution in Cellular Automata Rule Space

1. INTRODUCTION

Recent research in the study of complex systems has focused on the conditions under which a physical system can support the basic operations of information processing.[5,7,10,13] As models of physical systems, several of these studies use a class of formal models called cellular automata. It has been hypothesized that in order to perform nontrivial computation, the dynamical behavior of a cellular automaton (CA) should be "at the edge of chaos."[6,7] Another related aspect of this issue has concentrated on applying simulated evolution to a population of CA rules. When CA rules are evolved to perform nontrivial computation, it has been suggested that natural selection would preferentially select those CAs which "linear the transition to chaos." Although initial empirical results by Packard[10] tended to support both the hypotheses, later experiments by Mitchell, Hraber, and Crutchfield[9] lead them to question Packard's conclusions. However, both of these studies base their conclusions on the λ value of a CA rule. Unfortunately, the λ parameter is a measure that is only loosely correlated with the dynamical behavior of those CAs studied by Packard and by Mitchell et al.

This chapter re-examines the above questions by employing a parameter called the Z parameter which modulates CA behavior more closely.[14] Preliminary results presented here show that when CA rules are evolved to perform the computational task proposed by Packard, the rules does indeed cluster near Z values associated with the "edge of chaos" behavior. Nevertheless, caution is advised against accepting these results as evidence for the "edge of chaos" hypothesis and in this chapter I delineate further research that might give a better understanding of the whole issue.

The subsequent sections are organized as follows. A brief overview of the CA rule space is presented in Section 2, prior to reviewing the earlier works on evolving CA rules in Section 3. Section 4 introduces the Z parameter while Section 5 presents the results obtained from the new experiments which focus on the Z parameter.

2. CELLULAR AUTOMATA

CAs have been used to model a wide spectrum of phenomena in physics, chemistry, and biology.[12] Formally, a CA is a discrete spatially-extended dynamical system consisting of a finite number of cells in a D-dimensional fixed lattice. Each cell in a CA can exist in any one of Σ states. The input to each cell, at time t, is the state of its M neighbors. Hence, the set of all possible neighborhood configurations is Σ^M.

A transition function or rule table is also defined for a CA, which is a mapping from the set of neighborhood configurations to the cell states $\Delta : \Sigma^M \rightarrow \Sigma$. The state of each cell is synchronously updated at every time step using the transition function Δ. Since there are Σ choices for a cell state for each of the Σ^M neighborhood configurations, the total number of all possible transition functions of rules is $\Sigma^{(\Sigma^M)}$.

From a computational standpoint, the rule table of a CA can be thought of as a program, which can act on a large set of data in parallel and modify it according to some criterion. The initial conditions of a CA constitute the input data, while the output of the program is the final configuration of the CA. It has been proven that many forms of CAs can perform universal computation.[11]

Following,[9,10] in this chapter we only consider one-dimensional CA ($D = 1$) with two possible states per cell: 0 and 1 (i.e., $\Sigma = 2$). Each cell in a CA is influenced by its three nearest neighbors on each side. Since a cell is considered to be a member of its own neighborhood by convention, the neighborhood size $M = 7$. Thus each rule table has 2^7 elements, and the total number of all possible rule tables equals 2^{128}.

2.1 ANALYSIS OF CA RULE SPACE

The first extensively study of CA rule space was done by Wolfram[13] where he investigated the dynamical behavior of one-dimensional binary CA with $\Sigma = 2$ and $M = 3$. Wolfram qualitatively categorized the behavior of CA rules into four classes. Class I CA rules result in spatially homogeneous configurations, Class II rules create spatially inhomogeneous periodic structures, Class II rules produce chaotic behavior while Class IV rules display long transient behavior rich in propagating structures of all sizes. Wolfram also hypothesized that Class IV CA rules are capable of supporting computation. However, a significant disadvantage of Wolfram's classification scheme is that class membership is undecidable.[1]

Later, Langton introduced a parameter λ to categorize CA rule spaces.[7] An arbitrary sate s is chosen as the quiescent state and a transition function Δ is created such that there are n transitions to the state s. The remaining $\Sigma^M - n$ transitions are randomly and uniformly divided among the nonquiescent states. The parameter λ is equal to the fraction of nonquiescent output states in the rule table and equals $(\Sigma^M n)/\Sigma^M$. The rule table is fully homogeneous for $\lambda = 0$ and most heterogeneous for $\lambda = (1 - 1/\Sigma)$. Using statistical measures, Langton found that for CA with $\Sigma \geq 4$ and $M \geq 5$, the λ value of a rule correlates well with the behavior of the CA. Near a "critical value," λ_c, the rules engender complex spatio-temporal behavior with long transients as seen in Class IV rules. Langton called this the *phase transition region* and conjectured that such systems support the conditions necessary for performing computation. As λ is incremented towards $(1 - 1/\Sigma)$, the behavior obtained is similar to Class III rules, where only random patterns are produced. When λ is further increased from $(1 - 1/\Sigma)$ to 1.0, the four CA classes are re-encountered in succession, but this time in the reverse order.

For binary CA, where $\Sigma = 2$, λ is simply the density of 1s in the rule table. The behavior of binary CA is necessarily symmetric about $\lambda = 1/2$ because there are only two states and flipping 1s with 0s result in complimentary behavior. Thus there are two values of λ_c at which complex behavior might occur. However, in general Langton found λ to be a poor discriminator of behavior of CA rules with small values of Σ.

3. EVOLVING CELLULAR AUTOMATA RULES

Although Langton and others used statistical methods to correlate the behavior of CA with the λ parameter,[7,8] they did not attempt to relate the above measures to an independent measure of computation. To investigate the "edge of chaos" phenomenon, Packard suggested using a genetic algorithm (GA) to evolve CA rules that can perform a specified computational task.[10]

3.1 GENETIC ALGORITHM

Genetic algorithms are search strategies based on the principles of natural selection.[2] They have been successfully used in variety of optimization problems, and are best known for being robust search techniques that can search extremely large spaces and obtain globally competitive solutions quickly.

A GA maintains a population of possible solutions, each encoded in a string of bits (the genotype). Each individual in the population is associated with a fitness, i.e., a measure of the goodness of the solution (the phenotype). The individuals in the population undergo fitness based selection, where strings with more-fit solutions are allowed to reproduce preferentially into the next generation. In addition to the reproduction operator, other genetic operators like crossover and mutation are also used to create the next generation from the old population. The process can be repeated resulting in highly fit solutions in a few generations.

To evolve CA rules, we can consider the rule table as the genotype upon which the genetic operators can act. Since the rule table of a one-dimensional binary CA with the neighborhood size $M = 7$ consists of 2^7 bits, any binary string of length 128 bits can represent a transition function. The binary sequence ordering is imposed (as in Wolfram[13]) so that, the bit in the first position in the sting represents the state to which the neighborhood state 0000000 is mapped, the bit in the second position in the string is the state to which neighborhood state 0000001 is mapped, and so on. Since there are 2^{128} possible rules, the search space for the GA is indeed extremely large. Unlike many applications of GAs, evolving CA rules present an interesting genotype-phenotype distinction. Here, natural selection does not act on the genotype (the rule table) itself, but on the behavior (the phenotype) of the CA produced by the underlying rule table.

3.2 PREVIOUS RESULTS

Packard designed his experiments to test two hypotheses: (1) CA rules that have the capacity to support computation are likely to be found near λ_c; and (2) natural selection preferentially selects rules near λ_c when CA rules are evolved to perform complex computation. In his experiments, Packard chose a computational task which involved the following density classification problem: if there are more 1s than 0s in the initial configuration of the CA, then after some number of time steps, all cells in the CA attain a fixed value of 1; else all cells reach a final state of 0 (in which case there are more 0s than 1s in the initial pattern for an odd-sized lattice). Since the density of the configuration is a global parameter, and because each cell has only local information, this task is nontrivial. Thus to perform the density classification task, information transmission across space and time is necessary. By simulating natural evolution with a GA, Packard evolved CA rules for the density classification task.

To run the GA, the initial population consisted of randomly generated rules. However, the population was constrained such that a uniform distribution across λ

values between 0.0 and 1.0 was obtained. Each CA rule in the population was tested on a number of random initial patterns with different densities of 1s. The fitness of a CA rule was evaluated by determining how close the obtained final pattern was to the expected target pattern. As evolution progressed, Packard focused on the distribution of λ values in the population. He found that after 100 generations, the rules in the population clustered around either $\lambda \approx 0.25$ or $\lambda \approx 0.75$ regions. To determine the λ_c value of $\Sigma = 2, M = 7$ CA rules, Packard also measured their *difference spreading rate* as a function of λ. CA rules with low difference spreading rate quickly attain fixed or periodic configurations, and they typically exhibit Class I or Class II CA behavior. CA rules with high difference spreading rate, on the other hand, display Class III CA behavior. Class IV rules are supposed to be in the transition region and have intermediate difference spreading rates. Packard found that CA rules in this transition region have $\lambda \approx 0.25$ or $\lambda \approx 0.80$, giving him a rough estimate of the value of λ_c. Since the GA found rules with similar λ values, Packard presented these results as evidence for the hypotheses he sought to verify.

Mitchell et al. repeated Packard's experiments to find rather different results. Starting with a population of CA rules, uniformly distributed over the full range of λ values, they found that any individual run of the GA resulted in the rules that either clustered around $\lambda \approx 0.43$ or $\lambda \approx 0.57$. In no case did they find the rules to cluster around the λ_c regions as claimed by Packard. Moreover, since the computational task is symmetric with respect to 1s or 0s, they argued that the most successful rules for the task should have a λ close to 0.5.

The above conflicting results, along with Langton's observation regarding loose correlation between λ and CA behavior for small values of Σ, suggest using a different parameter that is better correlated with CA behavior. One such parameter has been suggested by Wuensche, who analyzed the "global dynamics" of CA and defined a Z parameter to modulate one-dimensional binary CA behavior.

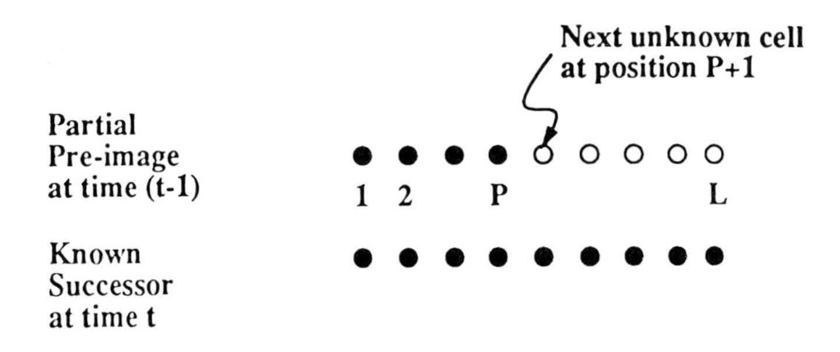

FIGURE 1 The reverse algorithm for determining preimages. Any cell whose state is known is represented by a "•," while a cell with an unknown state are shown as a "○."

4. THE Z PARAMETER

The global state or the configuration of a one-dimensional binary CA in a lattice of length L is defined by the state of its L cells. The global states of a CA can be arranged in a *state transition graph*, where each node in the graph represents a particular global state, and each directed edge represents the transition from one global state to another when a CA follows a given rule table. Since the system is deterministic, each global state has one and only one successor at the next time step. Thus each node in the state transition graph has an out-degree of one. However, a global state can have zero, one, or more than one predecessor or preimage. The in-degree (number of incoming arcs) of a node is referred to as the degree of preimaging of a global state. Global states without any preimage or incoming arcs are referred to as *Garden of Eden* states.

Given a rule table Δ, it is easy to calculate the successor state for any global state. However, calculating the preimage of a global state is more complex. Wuensche[14] described a reverse algorithm to determine the preimages of any global state of a CA in a lattice of length L with periodic boundary conditions and following a particular transition function. Given a known successor state and a partial preimage of length P as shown in Figure 1, the reverse algorithm uses the transition function to enumerate all possible partial preimages of length $(P+1)$. To bootstrap the algorithm, note that for $P = 1$, there are only two possible alternatives: 0 and 1. The algorithm is applied iteratively until the length of the partial preimage equals $(L+1)$. Since periodic boundary conditions are used, the state of cell at position 1 must equal the state of cell at position $(L+1)$. Any preimage of length $(L+1)$ that is produced by the reverse algorithm and satisfies this condition is a valid preimage of the known successor state.[1]

Wuensche also defined the Z parameter, which is the probability that the *next unknown cell* in a partial preimage has a unique value (i.e., having one solution: either 0 or 1) when using the reverse algorithm. The Z parameter can be directly computed from the rule table and it is related to the *maximum preimaging*: the greatest number of incoming arcs exhibited by a state transition graph representing the dynamics of the CA. Class I and Class II CA show high degree of convergence in their state transition graphs and are thus associated with a large value of maximum preimaging. For such CAs the Z value is small, since it is difficult to predict the value of the next unknown cell in a partial preimage. Class III CA, on the other hand, seemingly generate random nonperiodic patterns. Thus there is very little convergence in the state transition graph resulting in a low value of maximum preimaging. Thus in turn makes the prediction of the next unknown cell in the partial preimage relatively easy; i.e., the Z value is close to 1.0. Wuensche finds that for one-dimensional binary CA with $M = 3$ and 5, the behavior of a CA correlates

[1] Interested readers may refer to Wuensche[14] for details regarding the reverse algorithm for preimages and the calculations for the Z parameter.

well with the Z parameter. Rules with a Z value close to 0.25 usually belong to Class I CA, and those with a Z value near 0.5 exhibit Class II behavior. CA rules with a Z parameter near 0.75 are typically Class IV CA, while Class III CA have a Z value close to 1.0. Unlike the λ parameter, which is only concerned with the proportion of 1s and 0s in the rule table, the Z parameter takes into consideration the relative positions of 1s and 0s in the rule table; and is, therefore, expected to better correlate with CA behavior. The λ parameter is only an approximation of the Z parameter and it can be shown that $\lambda \geq Z/2$.

5. NEW EXPERIMENTS

In our experiments, we have attempted to evolve CA rules using a simpler and more widely used version of GA[3] than that used by Packard or Mitchell et al. The population size was fixed at 200, while the crossover and mutation probability was set to 0.6 and 0.001 respectively. The initial population approximated a uniform distribution across λ as shown in Figure 2. Each rule table in the population was tested on 30 to 50 randomly generated initial patterns with various densities of 1s. For each initial pattern, the transition function was applied for 512 time steps on a lattice of 209 cells, after which the resulting CA configuration was accepted as the final pattern. The fitness of a rule table was determined by summing over the

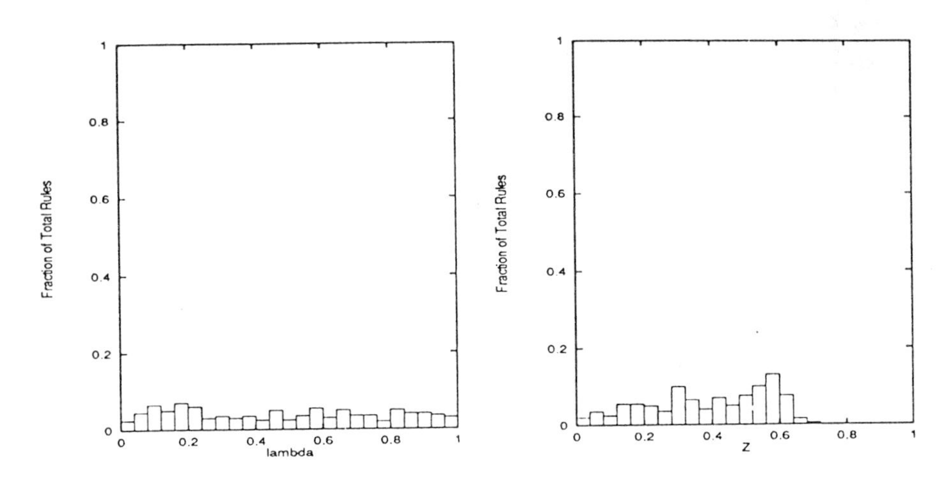

FIGURE 2 The histogram on the left depicts the typical distribution of λ in the initial population, while the histogram on the right presents the distribution of the Z parameter in the same initial population. Each histogram divides the range of λ and Z values from 0.0 to 1.0 into 25 equal sized bins.

Hamming match between the target pattern and the final pattern obtained from each of the initial patterns. We note that evaluating a CA rule table is inherently noisy, since we randomly sample the space of all possible initial configurations to test a given rule table.

Since we are interested in the Z parameter, in all of our experiments, we calculated the distribution of Z values of the CA rules in the population along with the distribution of λ values.

5.1 RESULTS

As the control case, we first evolved CA rules without any selection pressure at all. The results, obtained after 100 generations, are shown in Figure 3. As expected, the rules peaked sharply around $\lambda = 0.5$. When genetic operators act on a population free of any selection pressure, then the distribution of λ values in the population of binary strings approximates the binomial distribution curve which peaks at 0.5. We note that the corresponding Z distribution for the same population is clustered between 0.5 and 0.6.

We then evolved CA rules for the density classification task. After 100 generations, some runs of the GA, the λ value of the rules in the population are clustered near 0.46, while for other runs it is clustered around $\lambda = 0.58$ (Figures 4 and 5). These results are very similar to the empirical results reported by Mitchell et al.

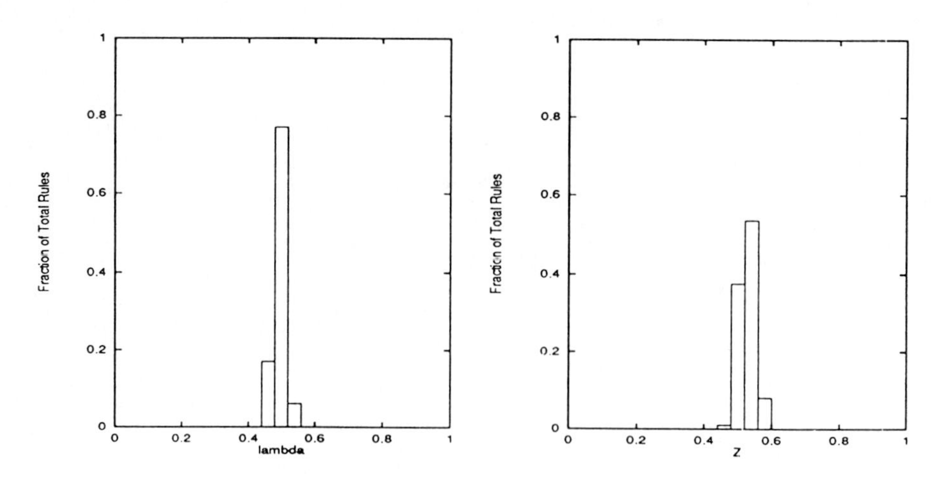

FIGURE 3 The results of evolving CA rules without any selection pressure are shown. The distribution of λ and Z values in the final generation are shown in the left and right histogram respectively.

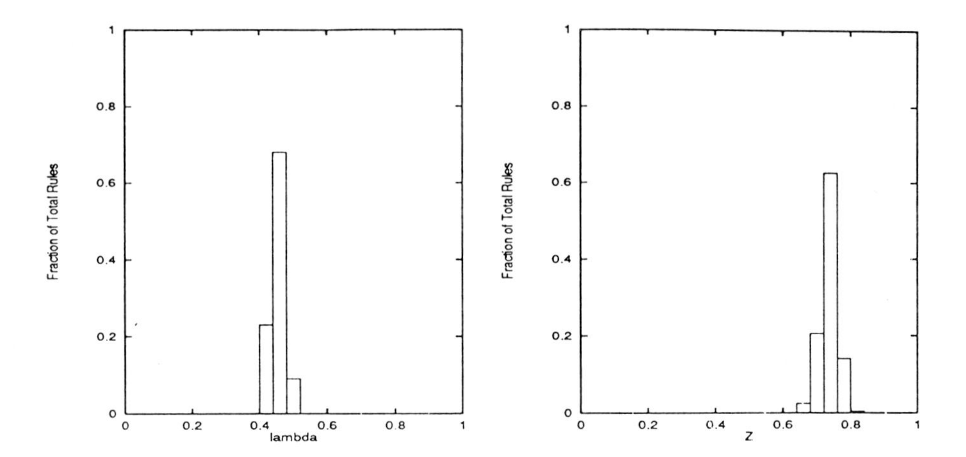

FIGURE 4 Typical results obtained after evolving CA rules that can perform the density classification task are shown. The histogram on the left presents the distribution of λ values in the final population, while the histogram on the right shows the distribution of the Z parameter in the same population. Note that the λ values peak near 0.46, and the Z values cluster near 0.74.

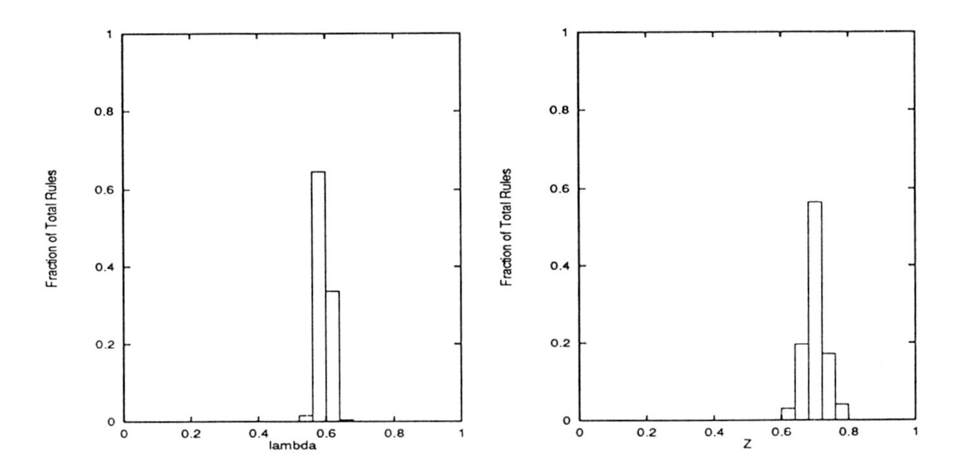

FIGURE 5 Another set of typical results after evolving CA rules that can perform the density classification task. The histogram on the left presents the distribution of λ values in the final population, while the histogram on the right shows the distribution of the Z parameter in the same population. Note that, in this case the λ values peak near 0.58 while the Z values cluster near 0.70.

For the same two population sets, we also plot the distribution of the Z parameter of the CA rules. We notice that in both cases the rules peaked near $Z = 0.7$ region; which is also the region where complex Class IV rules are usually found as claimed by Wuensche. Thus, these preliminary results would suggest that, for the density classification task, evolution does indeed select rules that "lie near the transition to chaos."

However, several questions remain unanswered. How sensitive are these results when the ordering of the bits in the rule table are changes? Why features in the rules near $Z = 0.7$ region permit the rules to attain high fitness in solving density classification task? Currently, we are conducting extensive experiments aimed to give us a better understanding of such questions.

6. CONCLUSIONS AND FUTURE WORK

In this chapter, we have presented preliminary results obtained from evolving CA rules for the density classification task. We show that the evolved rules cluster near Z values which have been associated with complex Class IV behavior. Although, the results presented in this chapter might seem to provide evidence for the "edge of chaos" hypothesis, a cautionary note needs to be added. Although the Z parameter correlates better with CA behavior than the λ parameter, more research is needed to fully understand its characteristics. We should note, however, that the Z parameter has its own deficiencies. As Wuensche notes,[14] there are a few exceptions when the Z parameter does not correlate with the CA behavior. Moreover, we have found that as the neighborhood size M is increased, the discriminating capacity of Z tends to decrease. Indeed, one might argue, it is impossible for a single scalar parameter like Z or λ to capture the full range of possible CA behavior. Instead of focusing only on the Z parameter, a broader approach that takes into account other features of the CA rules, like the Garden of Eden states density, might be more appropriate. Along similar lines, one might also use a hierarchical classification scheme which attempts to model GA behavior with n-step Markov measure.[4] In such a scheme, the λ parameter forms the lowest and simplest level of Markov approximation. Incrementing n results in increasingly faithful representations of CA behavior, along with an increased number of model parameters.

So far, all the conclusions in this chapter, and in Mitchell et al.[9] and Packard,[10] have been based on a single measure of computation: the one-dimensional binary CA's ability to perform the density classification task. It remains to be seen if similar results can be obtained not only for other computational tasks, but also for CA with states $\Sigma > 2$ and with dimensionality $D > 1$.

REFERENCES

1. Culik II, K., and S. Yu. "Undecidability of Cellular Automata Classification Schemes." *Complex Systems* **2** (1988): 177–190.
2. Goldberg, D. *Genetic Algorithms in Search, Optimization, and Machine Learning.* Reading, MA: Addison-Wesley, 1989.
3. Grefenstette, J. "GENESIS: A System for Using Genetic Search Procedures." In *Proceedings of the Conference on Intelligent Systems and Machines.* Rochester, MI, 1984.
4. Gutowitz, H. "A Hierarchical Classification of Cellular Automata." *Physica D* **45** (1990): 136–156.
5. Kauffman, S. "Requirements for Evolvability in Complex Systems: Orderly Dynamics and Frozen Components." *Physica D* **42** (1990): 135–152.
6. Kauffman, S. "Antichaos and Adaptation." *Sci. Am.* **August** (1991): 78–84.
7. Langton, C. "Computation at the Edge of Chaos: Phase Transitions and Emergent Computation." *Physica D* **42** (1990): 12–37.
8. Li, W., N. Packard, and C. Langton. "Transition Phenomenon in Cellular Automata Rule Space." *Physica D* **45** (1990): 77–94.
9. Mitchell, M., P. Hraber, and J. Crutchfield. "Revisiting the Edge of Chaos: Evolving Cellular Automata to Perform Computations." *Complex Systems* (1993).
10. Packard, N. "Adaptation Toward the Edge of Chaos." In *Dynamic Patterns in Complex Systems.* Singapore, World Scientific, 1988.
11. Smith, A. "Simple Computation-Universal Cellular Spaces." *J. Assoc. Comp. Machinery* **18** (1971): 339–353.
12. Toffoli, T., and N. Margolus. *Cellular Automata Machines.* Cambridge, MA: MIT Press, 1985.
13. Wolfram, S. "Universality and Complexity in Cellular Automata." *Physica D* **10** (1984): 1–35.
14. Wuensche, A., and M. Lesser, eds. *The Global Dynamics of Cellular Automata.* Santa Fe Institute Studies in the Sciences of Complexity, Ref. Vol. I. Reading, MA: Addison-Wesley, 1992.

Gyöngyi Gaál*
Brown University, Department of Neuroscience, Box 1953, Providence, RI 02912
*Present address: Division of Neurobiology, Department of Molecular and Cell Biology, 129 LSA, University of California, Berkeley, CA 84720

Prediction of External Stimuli from Neuronal Responses

INTRODUCTION

Earlier we compared four population coding algorithms to reconstruct stimuli of the external world in coordinate systems defined by preferred response vectors of neurons (such as receptive field profiles for sensory areas or preferred directions for motor areas[19,20]). One such algorithm was designed by Georgopoulous et al.[26,28] Another was derived from the tensor network theory.[41] A third was inspired by an iterative network using feedforward-feedback connections designed[8,9] for image compression and reconstruction. We tested also a fourth algorithm, a nonvectorial, nonlinear mapping to estimate movement direction from responses of neurons.[20] We showed that each algorithm, equally plausible in computer simulations, provide different predictions for the functional architecture of the coding network.

Here we show that in the temporal domain, the population coding algorithms are capable not only of stimulus reconstruction but also of short-term prediction.

Georgopoulous et al.[29] reported experimental results that are consistent with our predictions regarding the neural implementation of their population code to estimate the direction of arm movement from responses of directionally tuned motor cortical neurons. They found that the angle between preferred directions of pairs of neurons and the synaptic strength of their interconnections are negatively correlated. In contrast, the synaptic weight patterns found by Fetz[14] for neurons nearer to motor output might seem to be counterintuitive: neurons with similar response properties can also inhibit one another, while neurons with dissimilar responses can excite each other. We showed how such connection patterns might be interpreted in terms of sensorimotor transformations as the Jacobian matrix of transformations between different coordinate systems. We also demonstrated that cells with nonlinear responses (e.g., complex cells in visual cortex) might calculate invariants that are independent of coordinate systems.

PREFERRED RESPONSES OF NEURONS AND FIRST-ORDER DYNAMICS OF DYNAMICAL SYSTEMS

The dynamics of complex biological systems is governed by a nonlinear evolution equation of the type

$$\frac{d\mathbf{q}}{dt} = \mathbf{f}(\mathbf{q}) \tag{1}$$

where $\mathbf{q}$ is a dynamical variable vector (a point in the configuration space of the system).

Its components could be, e.g., firing rates of cells, transmembrane potential of individual neurons, concentration of neurotransmitters in extracellular space or blood or blood pressure. In Eq. (1) $\mathbf{f}()$ is a *nonlinear* vector-vector function and $d\mathbf{q}/dt$ is the rate of change of the dynamical variable vector (its time derivative).

We *do not have* the equations (and cannot even identify all the components of the dynamical variable).

There are two possible approaches to follow:

1. We can set up the equation by intuition to reproduce the experimentally observed phenomena (Neural Network approach).
2. Or "read it off" directly from the partially observed experimental time series (Dynamical Systems approach).

Walter Freeman (olfactory system; Freeman,[16] Skarda and Freeman[46]) and Friedrich, Fuchs, and Haken[18] (EEG in epilepsy) are following both approaches.

If the components of the dynamical variable vector are: external stimuli, firing rates of cells, final outcome (e.g., direction of arm movement, force), the equation expresses how external events influence firing rates, how cells influence each other's activity, and how the activity gives rise to a final outcome such as arm's movement.

If we knew the equation, we could investigate stability

$$\frac{d\mathbf{q}}{dt} = 0 \text{ at } \mathbf{q} = \mathbf{q}_0 \text{ stable points.}$$

The effect of small perturbations around the stable points could be expressed as:

$$q_i(\mathbf{q}_0 + \Delta\mathbf{q}) = q_i(\mathbf{q}_0) + \frac{\partial f_i}{\partial q_j}(\mathbf{q}_0)\Delta q_j + 0(\Delta q^2)\,. \tag{2}$$

(higher order terms)

The $\partial f_i/\partial q_j(\mathbf{q}_0)\Delta q_j$ terms, that is, the dot product of the vectors from the Jacobian matrix and the small changes in the dynamical variable vector, aare a generalized equation of the dot product of a cell's preferred direction vector and the actual movement direction vector which gives the firing rate of the cell as a cosine directional tuning function.[26,27,28]

This shows that the Jacobian matrix consists of receptive field weighting functions, preferred directions, and effective weights between pairs of neurons, depending on what components of the dynamical variable vector we focus on. In vector space they can be approximated by correlations (dot products).

This is what we are doing by calculating perievent time histograms, reverse correlations for stimulus-response in sensory areas, cross-correlations for neuronal firing rates, or spike triggered averaging between neuronal and muscle activity.

Even though the entire Jacobian matrix is accessible only in a computer simulation of biological processes, it is now also possible to experimentally determine multiple elements of the matrix. Eckhorn et al.[12] used a cross-correlation technique, the so-called Receptive Field cinematogram, to map several visual receptive fields at once. Georgopoulous et al.[29] determined preferred directions of several directionally tuned neurons in motor cortex using multielectrode recordings. Dinse et al.[10] investigated the temporal structure of multiple visual receptive fields. Nicolelis et al.[36] also determined the spatiotemporal structure of somatosensory responses of many-neuron ensembles in the thalamus. Wilson and McNaughton[48] (see also this volume) mapped multiple place fields in the hippocampus to explore the dynamics of the hippocampal ensemble code for space.

The Jacobian matrix consisting of the partial derivatives of the nonlinear function, that is the gradient of the nonlinear response function in external stimulus space, takes the role of linear receptive field weighting functions or preferred directions and the set of vectors constituting the Jacobian matrix can be regarded as a generalized intrinsic coordinate system of preferred responses. Such intrinsic neuronal coordinate systems are globally curvilinear in the external stimulus space (that is they depend on the location in stimulus space—e.g., the preferred directions of directionally tuned motor cortical neurons depend on arm locations, as in Caminiti et al.[1,2]). This approach makes it possible also to investigate the linear stability of steady states and periodic orbits from nonlinear dynamical systems approach. The determination of the eigenvectors and eigenvalucs of the Jacobian matrix of preferred responses will represent the time evolution of the neuronal dynamical system, stable and unstable manifolds can be revealed, and calculation of the Lyapunov exponents and entropies from the matrix of preferred responses would reveal the short-term predictability of the system.[22,37] Although seemingly different from the usual time delayed method used by the nonlinear dynamics community for state-space reconstruction (for a review see Casdagli et al.[4]) and time series analysis employing local principal value decomposition, related approaches have appeared also in neuroscience. Richmond et al.[43] creates a complete matrix consisting of neuronal responses given to a complete set of Walsh pattern stimuli (a special case of time delayed method, when the time delay equals a full stimulus period) and calculates principal components from such matrices in order to investigate temporal encoding of two-dimensional patterns by single units in primate inferior temporal cortex.

Higher order terms will show whether receptive field weighting functions, preferred directions or effective weights change in time ("learning rules").

PREDICTING THE FUTURE BY POPULATION CODING ALGORITHMS

Building on the ideas of Packard et al.[39] for reconstructing the attractor from a time series in a time-delayed embedding space, Farmer and Sidorowich[13] and Crutchfield and MacNamara[5] proposed local prediction methods that approximate the future trajectory of the current state with a simple function of the trajectories of the current state's nearest neighbors on the attractor. We compared our population coding algorithms in predicting five time steps in the future at once, using a weighted averaging of the future outcome vectors of nearest neighbors of the vector to be predicted. The coefficients for the weighted averaging in the prediction algorithm whose results are shown in Figures 1(c), 1(f), 1(i), and Figure 2(c) were derived from the population coding algorithm introduced by Georgopoulous et al.[27,28] and also analyzed by Daubechies[6] as well as Daubechies et al.[7] in the following way: The original population coding algorithm would use the dot product of the future outcome of the vector to be predicted and appropriate future outcome vectors of the corresponding nearest neighbors of the vector whose future is to be predicted. After summation, a renormalization would take place in order to obtain a length for the predicted vector which is comparable to the length of the vector obtained from the observed time series. (Note that two time steps but no iteration are needed for this algorithm.) As the future outcome (this is the vector to be determined) is not known, the coefficients were substituted by (1) the dot product of the vector used for the prediction and its appropriate nearest neighbors or (2) the dot product of the vector used for the prediction and the future outcome of its nearest neighbors.

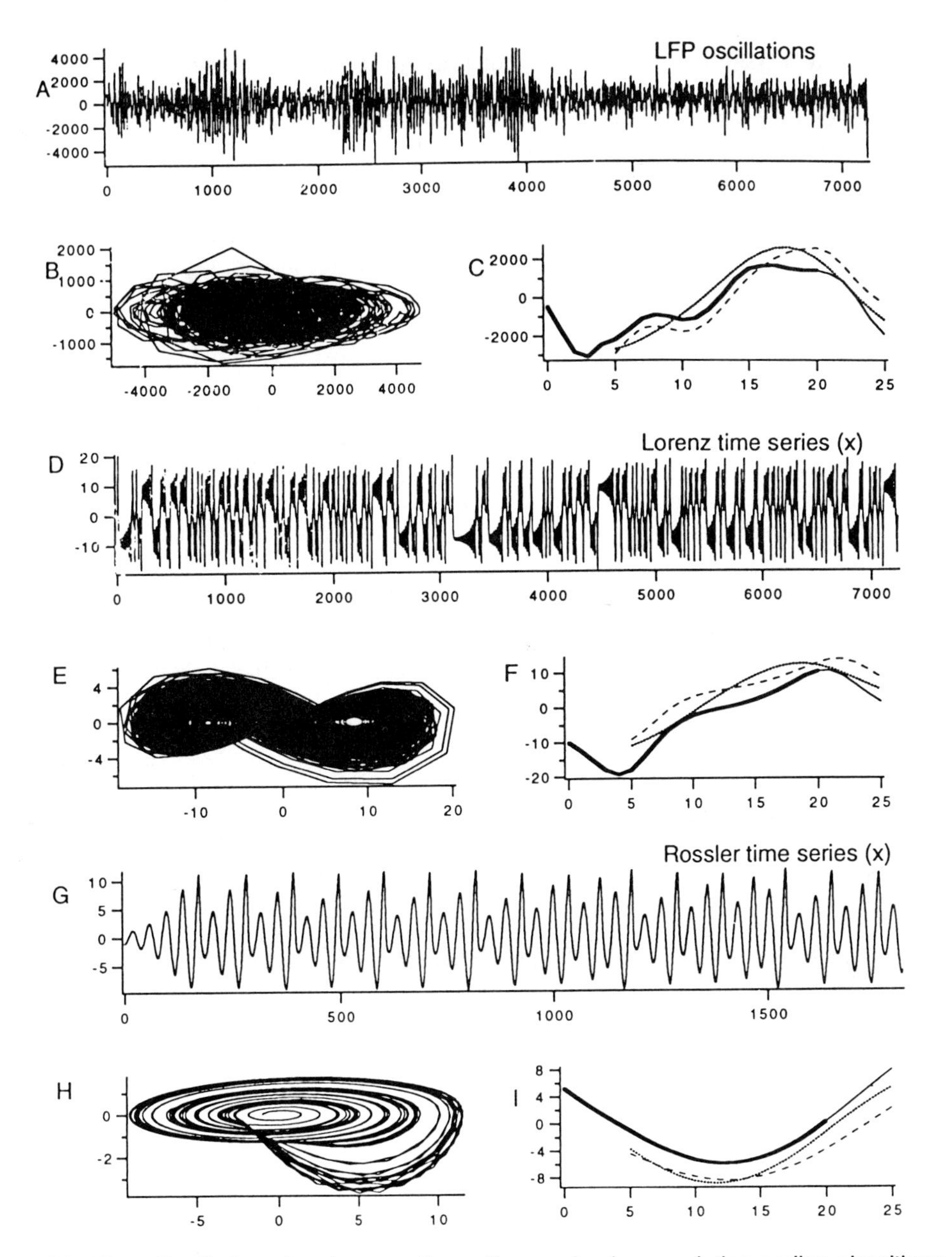

FIGURE 1 Prediction of various nonlinear time series by population coding algorithms. (a) Local field potential oscillations recorded in primate motor cortex during a visuomotor task (Gaál et al.,[21,22] see Figure 4 and 5, wire 20). (b) The time derivative of the local field potential is shown as a function (continued)

FIGURE 1 (continued) of the local field potential. (c) Prediction of a segment of the time series (starting from 1000) five steps into the future using 78 consecutive vectors whose dot product with the vector to be predicted is greater or equal than -1.0 (dotted line; the correlation coefficient -CC- between the predicted vector -PV- and the vector to be predicted -VTP- is 0.95); or the four nearest neighbors, whose correlation with the selected time segment is higher than 0.9 (CC between PV and VTP is 0.895. (d) The x time series of the Lorenz attractor. (e) The time derivative of the x function versus the x function. (f) The same prediction algorithm was applied as in Figure 1(c). Seventy-eight vectors were used to calculate the dotted line. CC between PV and VTP is 0.95. Four nearest neighbors (their CC with the selected time series whose future was to be predicted was higher than 0.85) were used to calculate the dashed line. CC between PV and VTP is 0.98. (g) The x time series of the Rössler attractor. (h) The time derivative of the x time series of the Rössler attractor vs. the x time series. (i) The same prediction algorithm was tested as in Figures 1(c) and 1(f). Seventy-eight neighbors were used to calculate the dotted line (CC between PV and VPT is 0.99). Four nearest neighbors were used to calculate the dashed line; their CC with the selected time series was higher than 0.96 (CC between PV and VTP is 0.98).

Although the algorithm gives only approximate results (see Figures 1 and 2) and it is inferior to other, more sophisticated prediction algorithms for noise-free data, its advantages are its simplicity, its speed, and its robustness against noise. All algorithms, independent of how accurate they can be for noise-free data, are bound to give erroneous predictions when noise has to be reckoned with in the nonlinear dynamical system. We illustrate the results of the prediction of nonlinear time series using the population coding algorithm introduced by Georgopoulous and colleagues here for several reasons: (1) the algorithm is plausible as a population code in the motor cortex to compute direction of arm movement and was also proposed to calculate stimuli in other brain areas, (2) it proved to be superior to other algorithms when stimulus estimation was carried out with noisy units in computer stimulations,[20] (3) it predicts nonlinear time series derived from multielectrode recordings in motor cortex of primates in a visuomotor task and filtered for local field potentials between 10 and 100 Hz, (4) it is able to predict nonlinear time series derived from integrating coupled differential equations such as the Lorenz,[33] or Rössler[44] time series, (5) it could also predict the responses of a nonlinear four-state spiking neuron[34] whose input was a chaotic time series (derived from the Lorenz system of equations), (6) it requires only two time steps to predict five time steps into the future, which makes it plausible to operate as a brain code, where it is important to identify fast algorithms and simple networks capable of short-term prediction. Psychophysical results indicate that such processes are operative in biological networks. For example, there are experimental implications that limb movements might be guided by predictive feedback control.[15,31,32]

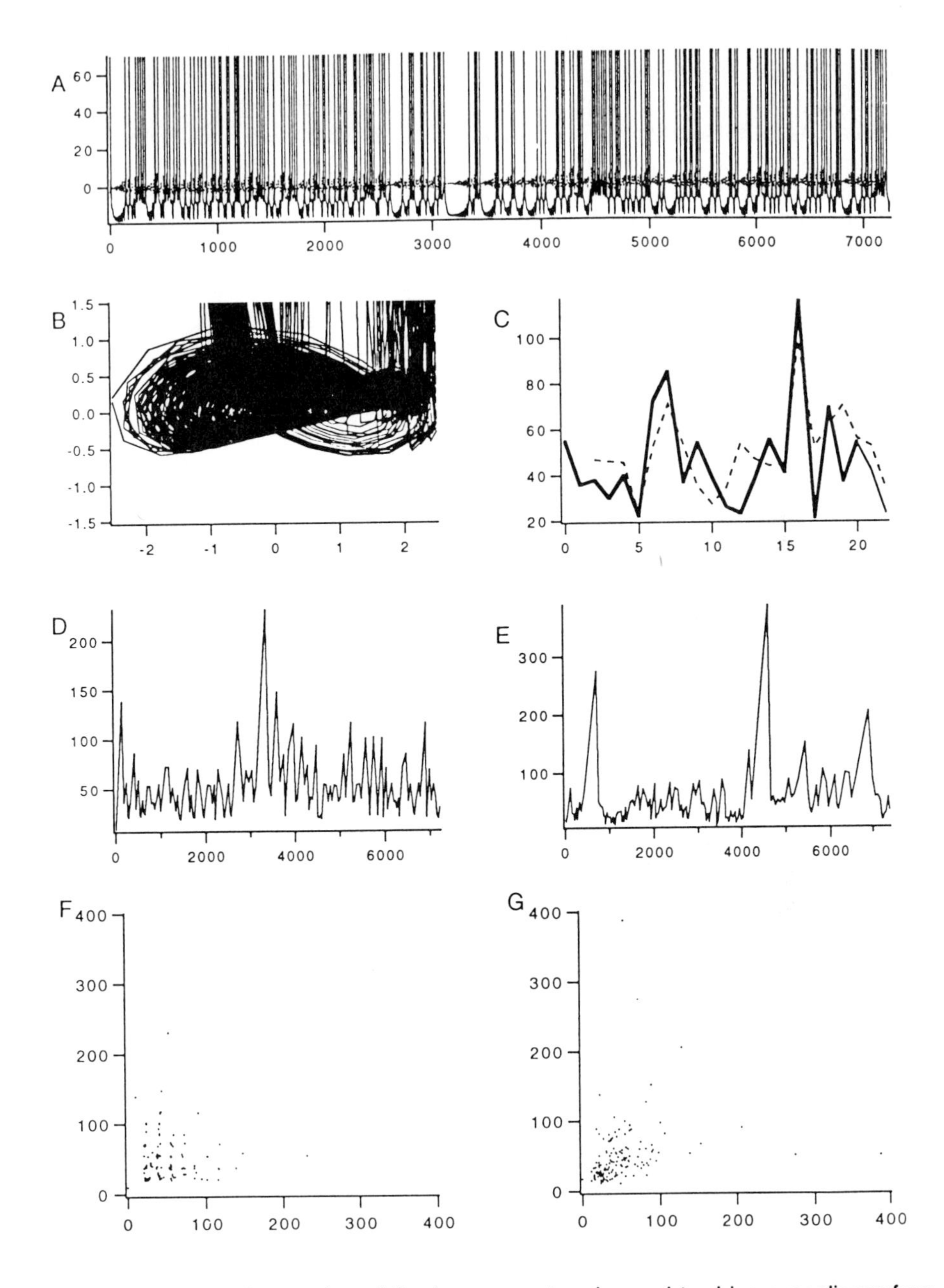

FIGURE 2 The x time series of the Lorenz system is used to drive a nonlinear four-point neuron.[34] (a) The intracellular membrane potential of the simulated neuron. Superimposed on the intracellular membrane (continued)

FIGURE 2 (continued) potential is the driving input itself (dotted line). (b) The image of the second time derivative of the intracellular membrane potential versus the first time derivative of the membrane potential at an expanded time scale still bears some resemblance to the original attractor (compare with Figure 1(e)). (c) Similar prediction method was used to predict the instantaneous frequency of the spike train two time steps into the future, as used for prediction in Figures 1(c), 1(f), 1(i). (d) The instantaneous frequency (the reciprocal of the interspike interval between consecutive spikes) of the model neuron driven by the Lorenz input. (e) The interspike interval of the real neuron, whose activity was recorded in a conscious primate during one trial period when the monkey was instructed to perform a visuomotor task, but skipped this particular trial.[23] (f) The interspike interval versus previous interspike interval of the model neuron, whose behavior could be predicted to some extent in Figure 2(c) (g) The interspike interval vs. previous interspike interval of the real neuron did not reveal any particular pattern using only 150 spikes (recorded in a 7.25-sec trial period). The behavior of the neuron could not be predicted from the previous pattern using such a low number of action potentials.

DISCUSSION: NONLINEAR DYNAMICAL SYSTEMS APPROACHES IN BRAIN RESEARCH IN THE FUTURE

The robust presence of local field potential and multiunit oscillations which are modifiable by stimuli, and the high degree of synchrony of oscillations across several cortical sites in different conditions[11,16,17,30,21,35,45] (see also Figure 1(a)) all imply that signals recorded in the brain using microelectrodes carry task-related information and that they are not simply random stochastic noise. We cannot assume that the dynamical system of cortical networks is linear. If it is nonlinear, it might turn out to be low- or high-dimensional deterministic chaotic system, and then it is possible to predict on a short-time scale. It is promising, that recent results or nonlinear dynamical analysis of human EEG signals indicate that low-dimensional deterministic chaotic processes might be operative in the human brain.[24] Gallez and Babloyantz state that the dynamical processes in the brain might be an intermediary between absolute determinism, which would have poor information processing capability and randomness, which would not be useful to control behavior. They argue that purely deterministic systems with a finite number of frequencies may only code for a limited amount of information, whereas the broad band spectra of chaotic dynamics have a richer information content. Freeman and his colleagues[46] also arrived at the conclusion that the high sensitivity of deterministic chaotic nonlinear systems to initial conditions might be exploited by neural networks of the brain in olfactory areas for fast and robust information processing.

If the system proves to be nonlinear, the question of how it can be so reliably predicted and controlled e.g., during a visuomotor task arises. Recent theoretical and experimental physics studies show promise in this respect, proving that a remarkable degree of control of nonlinear deterministic chaotic systems can be achieved by applying small perturbations[38] or self-controlling feedback.[42] These methods have already showed promise in physiology for controlling cardiac chaos.[25] An even more striking recent result related to driving of nonlinear deterministic chaotic systems is the perfect synchrony that can be achieved, reminiscent of synchrony between local field potential and multiunit oscillations on different wires in electrophysiological recordings, if their inputs are appropriate chaotic signals.[40] Pecora and Carroll[40] hypothesize that chaotic signals may be preferable drives to periodic signals in cases where increased robustness is advantageous. They also conjecture that the neural response to different stimuli need not be simple (periodic for example) for the same predictable state to be reached, provided that the driven neurons form stable responses to repeated stimuli. It may only be necessary that the same pattern emerge for similar stimuli in order to achieve useful coding in the nervous system, but the pattern might not have to be regular. It is tempting to speculate that the predictable task-related neural oscillations might reveal a neural implementation of the predictions of Pecora and Carroll[40] regarding stable chaotic driving and synchronization of physiological systems.

ACKNOWLEDGMENTS

This theoretical research was carried out in part at the University of Pennsylvania, supported by NIH/Fogarty 1 FO5 TWOO04265-01 BI-5 and IBRO/MacArthur training grants awarded to Gyöngyi Gaál. The author recorded and digitized the local field potentials shown in Figures 1(a), 1(b), 1(c) and extracellular unit activity shown in Figures 2(a) and 2(g) as a postdoctoral fellow at Brown University supported by NIH grant NS25074 awarded to John Donoghue.

REFERENCES

1. Caminiti, R., P. B. Johnson, and A. Urbano. "Making Arm Movements Within Different Parts of Space: Dynamic Aspects in the Primate Motor Cortex." *J. Neurosci.* **10** (1990): 2039–2058.
2. Caminiti, R., P. B. Johnson, C. Galli, S. Ferranina, and Y. Burnod. "Making Arm Movements within Different Parts of Space: The Premotor and Motor Cortical Representation of a Coordinate System for Reaching to Visual Targets." *J. Neurosci.* **11** (1991): 1182–1197.
3. Casdagli, M. "Nonlinear Prediction of Chaotic Time Series." *Physica D* **35** (1989): 335–356.

4. Casdagli, M., S. Eubank, J. D. Farmer, and J. Gibson. "State Space Reconstruction in the Presence of Noise." *Physica D* **51** (1991): 52–98.
5. Crutchfield, J. P., and B. S. McNamara. "Equation of Motion from a Data Series." *Complex Systems* **1** (1987): 417–452.
6. Daubechies, I. *Ten Lectures on Wavelets.* Philadelphia, PA: Society of Industrial and Applied Mathematics, 1992.
7. Daubechies, I., A. Grossman, and Y. Meyer. "Painless Nonorthogonal Expansions." *J. Math. Phys.* **27** (1986): 1271–1283.
8. Daugman, J. G. "Complete Discrete 2-D Gabor Transforms by Neural Networks for Image Analysis and Compression." *IEEE Trans. Acoust., Speech Signal Proc.* **36** (1988): 1169–1179.
9. Daugman, J. "Entropy Reduction and Decorrelation in Visual Coding by Oriented Receptive Fields." *IEEE Trans. Biomed. Eng.* **36** (1989): 107–114.
10. Dinse, H. R., K. Krüger, and J. Best. "A Temporal Structure of Cortical Information Processing." *Concepts in Neurosci.* **1** (1990): 199–238.
11. Eckhorn, R., R. Bauer, W. Jordan, M. Brosch, W. Kruse, M. Munk, and H. J. Reitboeck. "Coherent Oscillations: A Mechanism of Feature Linking in the Visual Cortex?" *Biol. Cyber.* **60** (1988): 121–130.
12. Eckhorn, R., F. Krause, and J. I. Nelson. "The RF-Cinematogram—A Cross-Correlation Technique for Mapping Several Visual Receptive Fields at Once." *Biol. Cyber.* **69** (1993): 37–55.
13. Farmer, J. D., and J. J. Sidorowitz. "Predicting Chaotic Time Series." *Phys. Rev. Lett.* **592** (1987): 845–848.
14. Fetz, E. E. "Are Movement Parameters Recognizably Coded in the Activity of Signal Neurons?" *Behav. & Brain Sci.* **15** (1992): 679–690.
15. Flament, D., and J. Hore. "Relations of Motor Cortex Neural Discharge to Kinematics of Passive and Active Elbow Movements in the Monkey." *J. Neurophys.* **60** (1988): 1268–1284.
16. Freeman, W. J. "Spatial Properties of an EEG Event in the Olfactory Bulb and Cortex." *Electroencephalogr. & Clin. Neurophysiol.* **44** (1978): 586–605.
17. Freeman, W. J., and B. W. van Dijk. "Spatial Patterns of Visual Cortical Fast EEG During Conditioned Reflex in a Rhesus Monkey." *Brain Res.* **422** (1987): 267–276.
18. Friedrich, R., A. Fuchs, and H. Haken. "Modelling of Spatio-Temporal EEG Patterns." In *Mathematical Approaches to Brain Functioning Diagnostics*, edited by I. Dvorak and A. V. Holden, 45–62. Great Britain: Biddles Ltd., 1991.
19. Gaál, Gy. "Population Coding by Simultaneous Activities of Neurons in Intrinsic Coordinate Systems Defined by Their Receptive Field Weighting Functions." *Neural Networks* **6(4)** (1993): 499–516.
20. Gaál, Gy. "Calculation of Movement Direction from Simultaneous Activities of Neurons in Intrinsic Coordinate Systems Defined by Their Preferred Directions." *J. Theor. Biol.* **162** (1993): 103–130.

21. Gaál, Gy., J. N. Sanes, and J. P. Donoghue. "Motor Cortex Oscillatory Neural Activity During Voluntary Movement in Macaca Fascicularis." *Soc. Neurosci. Abstr.* **18:355.14** (1992): 848.
22. Gaál, Gy., and J. P. Donoghue. "Predicting Synaptic Weight Patterns from Sensorimotor Coordinate Transformations Related to Population Coding of Movement Direction in Motor Cortex." 23rd Congress of the Society of Neuroscience, Washington, November 7-12, 1993.
23. Gaál, Gy., J. N. Sanes, and J. P. Donoghue." Oscillations in Neural Discharge and Local Field Potentials in Primate Motor Cortex During Voluntary Movement." *J. Neurophysiol.* submitted.
24. Gallez, D., and A. Babloyantz. "Predictability of Human EEG: A Dynamical Approach." *Biol. Cybern.* **64** (1991): 381–391.
25. Garfinkel, A., M. L. Spano, W. L. Ditto, and J. N. Weiss. "Controlling Cardiac Chaos." *Science* **257** (1992): 1230–1235.
26. Georgopoulos, A. P., J. F. Kalaska, R. Caminiti, and J. T. Massey. "On the Relations Between the Direction of Two-Dimensional Arm Movements and Cell Discharge in Primate Motor Cortox." *J. Neurosci.* **2** (1982): 1527–1537.
27. Georgopoulos, A. P., A. B. Schwartz, and R. E. Kettner. "Neuronal Population Coding of Movement Direction." *Science* **233** (1986): 1416–1419.
28. Georgopoulos, A. P., R. E. Kettner, and A. B. Schwartz. "Primate Motor Cortex and Free Arm Movements to Visual Targets in Three-Dimensional Space. II. Coding of The Direction of Movement by a Neuronal Population." *J. Neurosci.* **8** (1988): 2928–2937.
29. Georgopoulos, A. P., M. Taira, and A. Luskashin. "Cognitive Neurophysiology of the Motor Cortex." *Science* **260** (1993): 47–52.
30. Gray, C. M., and W. Singer. "Stimulus-Specific Neuronal Oscillations in Orientation Columns of Cat Visual Cortex." *PNAS* **86** (1989): 1698–1702.
31. Humphrey, D. R., and J. Tanji. "What Features of Voluntary Motor Control are Encoded in the Neuronal Discharge of Different Cortical Motor Areas? In *Motor Control: Concepts and Issues*, edited by D. R. Humphrey and H.-J. Freund. New York: John Wiley and Sons, 1991.
32. Kelso, J. A. S., S. L. Bressler, S. Buchanan, G. C. DeGuzman, M. Ding, A. Fuchs, and T. Holroyd. "A Phase Transition in Human Brain and Behavior." *Phys. Lett. A* **169** (1992): 134–144.
33. Lorenz, E. N. "Deterministic Nonperiodic Flow." *J. Atmos. Sci.* **20** (1963): 130–145.
34. MacGregor, R. J. *Neural and Brain Modeling.* London: Academic Press, 1987.
35. Murthy, V. N., and E. E. Fetz. "Coherent 25–35 Hz Oscillations in the Sensorimotor Cortex of Awake Behaving Monkeys." *PNAS* **89** (1992): 5670–5674.
36. Nicolelis, M. A. L., R. C. S. Lin, D. J. Woodward, and J. K. Chapin. "Induction of Immediate Spatiotemporal Changes in Thalamic Networks by Peripheral Block of Ascending Cutaneous Information." *Nature* **361** (1993): 533–536.

37. Ott, E. *Chaos in Dynamical Systems.* Cambridge, MA: Cambridge University Press, 1993.
38. Ott, E., C. Grebogi, and Y. A. Yorke. "Controlling Chaos." *Phys. Rev. Lett.* **64** (1990): 1196–1199.
39. Packard, N. H., J. P. Crutchfield, J. D. Farmer, and R. S. Shaw. "Geometry From A Time Series." *Phys. Rev. Lett.* **45** (1980): 712–716.
40. Pecora, L. M. and T. J. Carroll. "Driving Systems with Chaotic Signals." *Phys. Rev. A* **44** (1991): 2374–2383.
41. Pellionisz, A., and R. Llinas. "Brain Modeling by Tensor Network Theory and Computer Simulation. The Cerebellum: Distributed Processor for Predictive Coordination." *Neurosci.* **4** (1979): 323–348.
42. Pyragas, K. "Continuous Control of Chaos by Self-Controlling Feedback." *Phys. Lett. A* **170** (1992): 421–428.
43. Richmond, B. J., L. M. Optican, M. Podell, and H. Spitzer. "Temporal Encoding of Two-Dimensional Patterns by Single Units in Primate Inferior Temporal Cortex. I. Response Characteristics." *J. Neurophys.* **57** (1987): 132–146.
44. Rössler, O. E. "Chaotic Behavior in Simple Reaction Systems." *Z. Naturforsch. A* **31** (1976): 1168–1173.
45. Sanes, J. N., and J. P. Donoghue. *Oscillations in Local Field Potentials of the Primate Motor Cortex During Voluntary Movement*, Vol. 90, 4470–4474. Proceedings of the National Academy of Sciences of the United States of America, 1993.
46. Skarda, C. A., and W. J. Freeman. "How Brains Make Chaos in Order to Make Sense of the World." *Behav. Brain Sci.* **10** (1987): 161–180 and commentaries.
47. Sugihara, G., and R. M. May. "Nonlinear Forecasting as a Way of Distinguishing Chaos from Measurement Error in Time Series." *Nature* **344** (1990): 734–741.
48. Wilson, M. A., and B. L. McNaughton. "Dynamics of the Hippocampal Ensemble Code for Space." *Science* **261** (1993): 1055–1058.

Liane M. Gabora
Department of Biology, UCLA, Los Angeles, CA 90024-1606; e-mail: liane@cs.ucla.edu

Meme and Variations: A Computational Model of Cultural Evolution

1. INTRODUCTION

In order for a pattern of information to evolve, we need a way of generating variations of the pattern, and a way of *selectively replicating* it. In *Adaptation in Natural and Artificial Systems*,[11] Holland introduced a computational model of how these processes are carried out on patterns of genetic information in biological systems. In this chapter we introduce a computational model of how they are carried out on patterns of cultural information in a society of interacting individuals.

Ideas, like genomes, are patterns that evolve. However, their evolution is not subject to the same constraints, and employs different mechanisms. The generation of variation is less random in cultural evolution than in biological evolution; it reflects the accumulated knowledge of individuals, and the social structure in which they are embedded. In cultural evolution, selective replication is Lamarkian; an idea can be modified through experience after it has been phenotypically expressed. Genetic information is coded in a physical sequence of nucleotides, and contains the instructions for its own replication, whereas ideas are coded in patterns of neuron activation, and do not contain the instructions for their replication; they rely on human hosts to replicate them. Ideas that satisfy our needs or drives are preferentially learned and implemented. Thus the fitness landscape for the evolution of ideas is molded by our drives. By developing a minimal model that incorporates mechanisms which differentiate cultural evolution from biological evolution, we hope to broaden our understanding of how it is that something can evolve.

In *The Selfish Gene*,[8] Dawkins coined the word "meme" to refer to "ideas, catch-phrases, clothes-fashions, ways of making pots or building arches." He elaborates: "Just as genes propagate themselves in the gene pool by leaping from body to body via sperm or eggs, so memes propagate themselves in the meme pool by leaping from brain to brain." Our model is called *Meme and Variations* because it is based on the premise that new ideas are variations of old ones; they result from tweaking or blending existing ideas.[12] We will explore the impact of three phenomena that are unique to cultural evolution. The first is *knowledge-based operators*: brains detect regularity and build schemas that they use to adapt the mental equivalents of mutation and recombination to their meme substrates. The second is *imitation*: ideas spread when members of a society observe and copy one another. This is seen in both animal and human societies.[4,15,16] Imitation enables individuals to share complete or partial solutions to the problems they face. The third is *mental simulation*: individuals can imagine what would happen if a meme were implemented before resources are spent on it. This provides them with a rudimentary form of selection before the phenotypic expression of a meme.

We will adopt the terminology of biology. Each component of a meme is referred to as a locus, and alternative forms of a locus are referred to as alleles. The processes that generate variation—the cultural counterparts to mutation and recombination—are referred to as operators. Forward mutation is mutation away from the initial or wild-type allele, and backmutation is mutation from an alternative form back to a wild type. Changes in the relative frequencies of different alleles due to random sampling processes in a finite population are referred to as drift.[17] The set of all existing patterns is referred to as the population. Following Braitenberg,[6] an individual is referred to as a vehicle, and the set of all vehicles is referred to as the society.

2. THE MODEL

2.1 THE DOMAIN

Donald[9] has provided substantial evidence that the earliest form of culture was mimetic display. The memes in *Meme and Variations* are mental representations of how to implement various mating displays. A meme has six loci that correspond to six body parts: left forelimb, right forelimb, left hindlimb, right hindlimb, head, and tail. Each locus has a floating point activation between -0.5 and 0.5 which determines the amount of movement (angle of rotation from rest position) of the corresponding body part when the meme is implemented. A value of 0.0 corresponds to rest position; values above 0.0 correspond to upward movement, and values below 0.0 correspond to downward movement. Floating-point loci activations produce graded limb movement. However, for the purpose of mutation, loci are treated as if there are only three possible alleles at each locus: stationary, up, and down. Six loci with three possible alleles each gives a total of 729 possible memes.

2.2 THE NEURAL NETWORK

The neural network is an autoassociator; it learns the identity function between input and output patterns. It has six input/output units numbered 1 through 6, corresponding to the six body parts. It has six hidden units numbered 7 through 12, corresponding to the general concepts, "forelimbs," "hindlimbs," "left," "right," "movement," and "symmetry" (Figure 1).

Hidden units are linked with positive weights to input/output units that are positive instances of the concepts they represent, and linked with negative weights to input/output units that represent negative instances of the memes they represent (thus "left forelimb" is positively linked to "left" and negatively linked to "right").

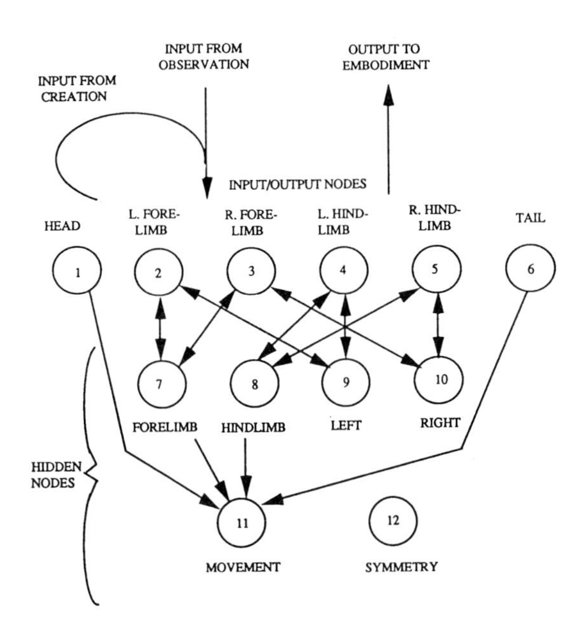

FIGURE 1 The neural network. Arrows represent connections with positive weights. For clarity, negative connections are not shown. Connections to the symmetry unit are also not shown.

Hidden units that represent opposite concepts have negative connections between them. The hidden units enable the network to encode the semantic structure of a meme, and their activations are used to bias the generation of variation (Section 2.5).

The neural network starts with small random weights between input/output nodes. Weights between hidden nodes, and weights between hidden nodes and input/output nodes, are fixed at +/−1.0. Memes are learned by training for 50 iterations using the generalized delta rule with a sigmoid activation function.[10] The relevant variables follow:

a_j = activation of unit j
t_j = the jth component of the target pattern (the external input)
w_{ij} = weight on line from unit i to unit j
$\beta = 0.15$
$\theta = 0.5$
$a_j = 1/(1 + e^{-[\beta \sum w_{ji} a_i + \theta]})$

For the movement node, we use the absolute value of a_i. The error signal, δ_j, is calculated as follows. For input/output units:

$$\delta_j = (t_j - a_j)a_j(1 - a_j)\,.$$

For hidden units:

$$\delta_i = a_j(1 - a_j) \sum \delta_j w_{ij}\,.$$

2.3 THE EMBODIMENT

The embodiment is a six-digit array that specifies the behavior of the six body parts. While the output of the neural network represents what the vehicle is thinking about, the embodiment represents what the vehicle is doing. A meme cannot be observed and imitated by other vehicles until has been copied from the neural network to the embodiment.

2.4 THE FITNESS FUNCTION

An optimal meme is one in which all body parts except the head are moving, and limb movement is symmetrical. (Thus if the left forelimb is moving up, the right forelimb is moving down, and vice versa.) This is implemented as follows:

F = fitness
$c = 2.5$
a_m = activation of movement hidden node
a_s = activation of symmetry hidden node

a_h = activation of head node

$$i = \begin{cases} 1 & \text{if } a_h = 0.0; \\ 0 & \text{otherwise} \end{cases}$$

$$F = \mu a_m + 2\mu a_s + \mu i$$

This fitness function corresponds to a relatively realistic display, but it also has some interesting properties. A vehicle that develops the general rule "movement improves fitness" risks over-generalization since head stability contributes as much to fitness as movement at every other limb. This creates a situation that is the cultural analog of *over-dominance* in genetics; the optimal value of this locus lies midway between the two extremes. We also have a situation analogous to bidirectional selection or *underdominance*; the optimal value of the tail locus lies at either of the two extremes. There is *epistasis*; the value of what one limb is doing depends on what its left-right counterpart is doing. Finally, since there is one optimal allele for the head, two optimal alleles for the tail, two optimal forelimb combinations, and two optimal hindlimb combinations, we have a total of eight different optimal memes. This enables us to perform a comparative analysis of diversity under different ratios of creation to imitation. Note that there are nonoptimal local maxima, corresponding to memes in which limbs are moving, but in the same direction.

2.5 THE MEMETIC ALGORITHM

The basic idea of the memetic algorithm is to translate knowledge acquired through the process of evaluating a new meme into educated guesses about what increases meme fitness. Each locus starts out with the allele for no movement, with an equal probability of mutating to each of the other two alleles (the alleles for upward and downward movement). A new meme is not learned unless it is fitter than the currently implemented meme, so we use the difference between these two memes to bias the direction of mutation.

Two rules of thumb are used. The first rule is: if the fitter meme codes for more movement, increase the probability of forward mutation and decrease the probability of backmutation. Do the opposite if the fitter meme codes for less movement. This rule of thumb is based on the assumption that movement can general (regardless of which particular body part is moving) can be beneficial or detrimental. This seems like a useful generalization since movement of *any* body part uses energy and increases the likelihood of being detected. It is implemented as follows:

a_{m1} = activation of movement unit for currently implemented meme
a_{m2} = activation of movement unit for new meme
$p(fmut)_i$ = probability of forward mutation at allele i (increased movement)
$p(bmut)_i$ = probability of backward mutation at allele i (decreased movement)
IF $(a_{m2} > a_{m1})$
    THEN $p(fmut)_i = \text{MAX}(1.0,\ p(fmut)_i + 0.1)$

ELSE IF $(a_{m2} < a_{m1})$
THEN $p(fmut)_i = \text{MIN}(0.0,\ p(fmut)_i - 0.1)$
$p(bmut)_i = 1 - p(fmut)_i$

The second rule of thumb biases the vehicle either toward or away from symmetrical limb movement. It has two parts. First, if in the fitter meme both members of one pair of limbs are moving either up or down, increase the probability that you will do the same with the other pair of limbs. Second, if in the fitter meme, one member of a pair of limbs is moving in one direction and its counterpart is moving in the opposite direction, increase the probability that you will do the same with the other pair of limbs. This generalization is also biologically useful, since many beneficial behaviors (walking, etc.) entail movement of limbs in opposite directions, while others (galloping, etc.) entail movement of limbs in the same direction. Space constraints do not permit a detailed explanation of the implementation of this rule; however, it is analogous to the implementation of the first rule.

In summary, each meme is associated with a measure of its effectiveness, and generalizations about what seems to work and what does not are translated into guidelines that specify the behavior of the memetic algorithm.

3. PROTOCOL FOR EVOLVING MEMES

Vehicles are in a two-dimensional wraparound 10×10 grid-cell world, one vehicle per cell. Each iteration, every vehicle has the opportunity to (1) acquire a new meme through creation or imitation, (2) update the mutation operator, and (3) implement the new meme.

Vehicles have an equal probability of creating and imitating. To create a new meme, the memetic algorithm is applied to the meme currently represented on the input/output layer of the neural network. For each locus, a vehicle decides whether mutation will take place. The probability of mutation is specified globally at the beginning of a run. If it decides to mutate, the direction of mutation is stochastically determined. If the new meme has a higher fitness than the currently implemented meme, the vehicle learns and implements the new meme.

To acquire a meme through imitation, a vehicle randomly chooses one of its eight neighbors at random and evaluates the fitness of the meme the neighbor is implementing. If its own meme is fitter than that of the neighbor, it chooses another neighbor, until it has either observed all eight neighbors or found one with a fitter meme. If no fitter meme is found, the vehicle does nothing. Otherwise, the neighbors' meme is copied to the input/output layer of the vehicle's neural network, and is learned.

Since in both creation and imitation a new meme is not acquired unless it is fitter than the currently implemented meme, the new meme provides information that is used by the memetic algorithm. For example, since we arbitrarily chose a fitness function in which movement is generally beneficial, if the new meme codes for more movement than the old meme, the probability of forward mutation will almost always increase.

In the no-mental-simulation condition, whether the new meme was acquired through creation or imitation, it must be implemented for at least one iteration before its fitness can be assessed. In this case mutation operators are updated the following iteration.

Finally, the new meme is copied from the neural network to the embodiment. The vehicle is now implementing the new meme.

4. RESULTS

The following experiments were conducted using a mutation rate of 0.17 per locus, a 1:1 creation to imitation ratio, and all cultural evolution strategies operative, unless otherwise indicated.

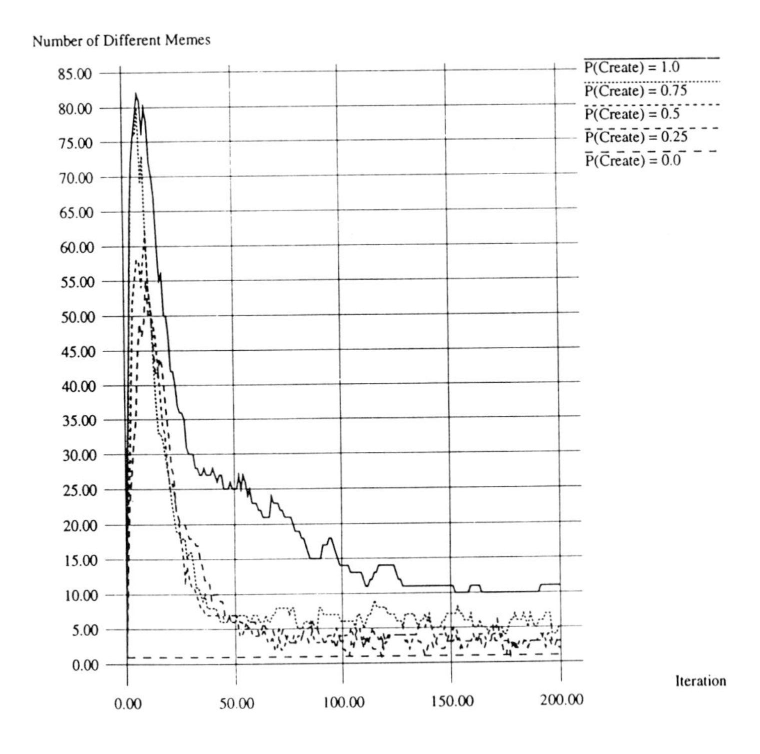

FIGURE 2 Effect of P(Create) to P(Imitate) ratio on meme diversity.

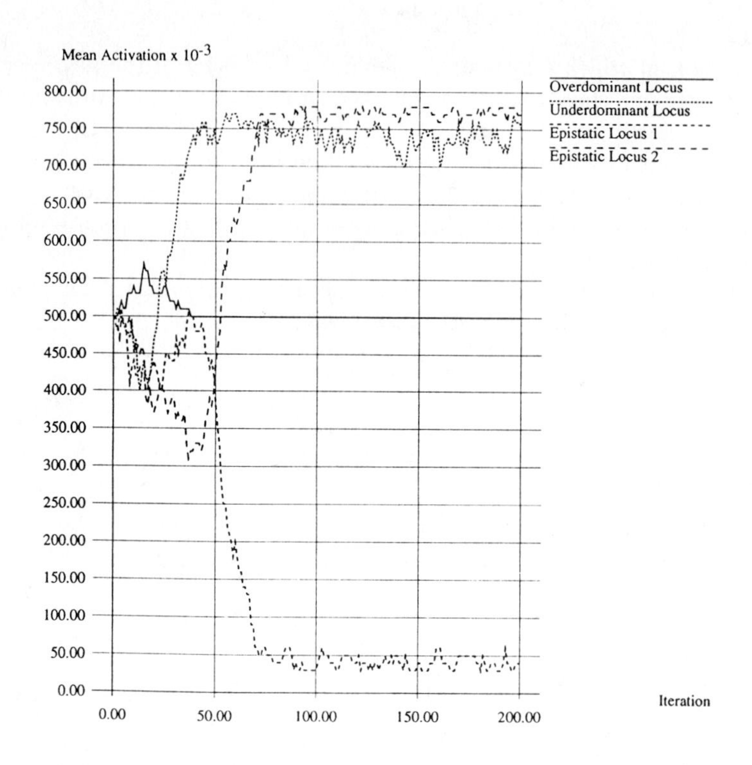

FIGURE 3 Activation over time for loci under various selection regimes. Epistatically linked alleles take longest to stabilize.

4.1 OUTLINE OF A RUN

Initially all vehicles are immobile. The immobility meme quickly mutates to a new meme that codes for movement of a body part. The new meme has a higher fitness and is preferentially implemented. Diversity increases as memes continue to mutate and spread through imitation. Diversity peaks when the first maximally-fit meme is found, and decreases as the society converges on maximally-fit memes (Figure 2). Stabilization takes longer for epistatically linked loci than over- or underdominant loci (Figure 3). The best performance is obtained with a high mutation rate: between 0.07 and 0.22 mutations per locus, or approximately one mutation per meme (Figure 4). This is probably because mental simulation ensures that poor memes are not implemented, and good memes are imitated by others, so they are unlikely to be lost through mutation from the society as a whole.

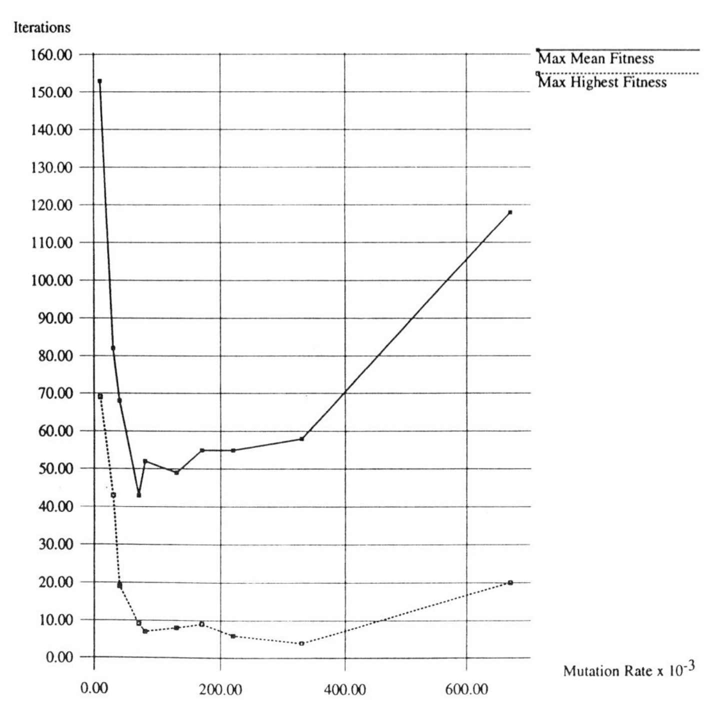

FIGURE 4 Effect of mutation rate.

Since we have eight optimal memes, there are many stable configurations for the distribution of memes. Drift is observed among equally-fit alleles (Figure 3), as predicted by Cavalli-Sforza and Feldman's mathematical model.[7]

4.2 EXPERIMENTS

The three cultural evolution strategies—mental simulation, imitation, and knowledge-based operators—were made inoperative one at a time to determine to their contribution to optimization speed and peak mean fitness. All three increase the rate of optimization, and mental simulation and imitation also increase peak mean fitness (Figures 5, 6, and 7).

The fitness of the fittest meme (Figure 8) increases as a function of the ratio of creation to imitation; however, the highest mean fitness is achieved when both creation and imitation are employed, in a ratio of approximately 2:1 (Figure 9). Since the vehicles with the fittest memes gain nothing by imitating others, there is a trade-off between average meme fitness and fitness of the fittest meme. Interestingly, meme diversity also varies with the ratio of creation to imitation (Figure 2), ranging

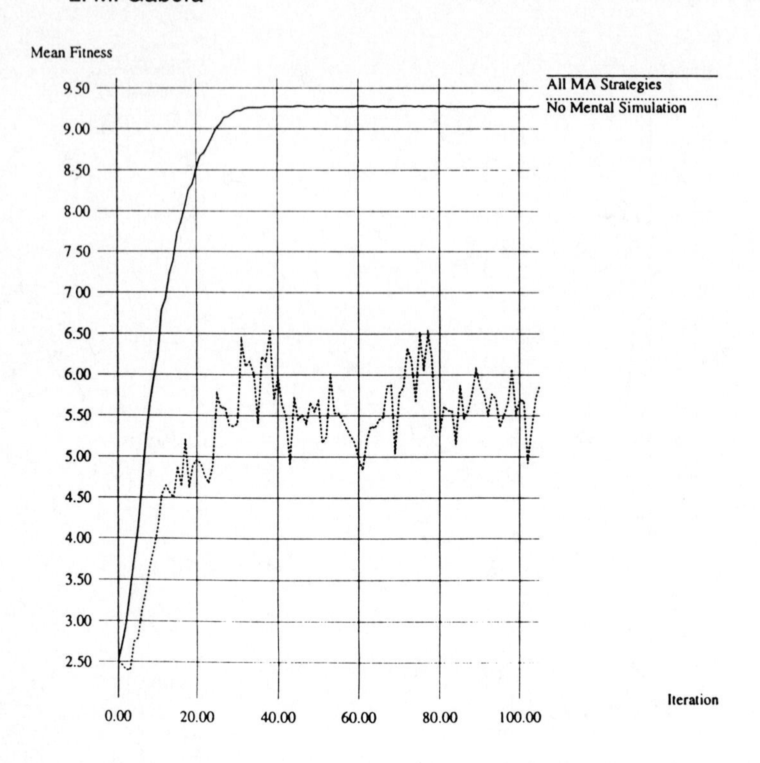

FIGURE 5 Meme fitness increases slowly and eratically without mental simulation.

from 1–2 memes when P(Create) = 0.25 to 10-11 memes when P(Create) = 1.0. When P(Create) = 0.75, the society converges on 7–8 memes; it finds all (or nearly all) of the fittest memes. A nice balance is struck between the diversifying effect of mutation and the converging effect of imitation. In future experiments in which the fitness landscape will fluctuate, maintaining diversity may prove to be more important than speed.

5. FUTURE PLANS

This program will soon run on a parallel machine, which will allow us to increase the size of the artificial society to several thousand individuals. We will examine the effect of erecting complete or semipermeable barriers between different societies, and the effect of migration.

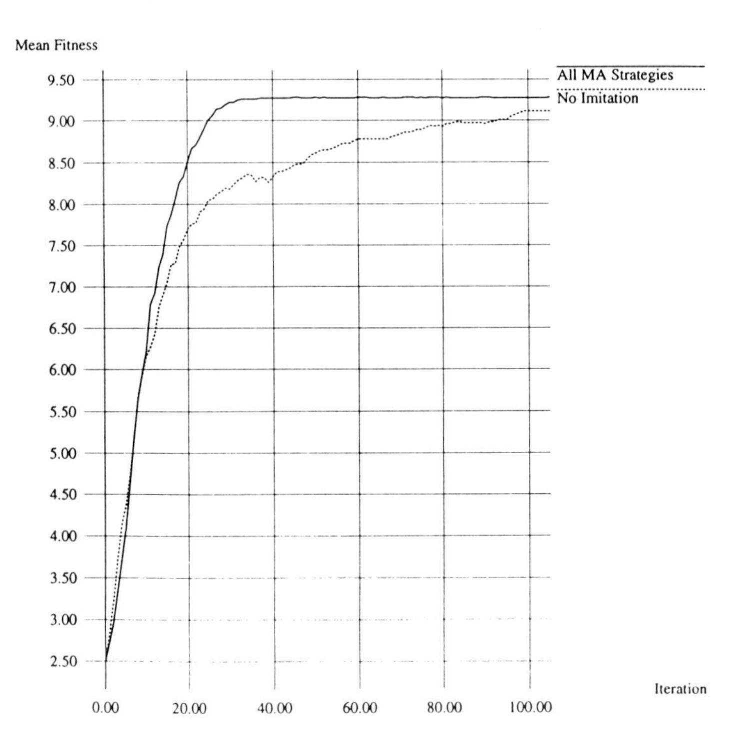

FIGURE 6 Imitation increases rate of optimization and peak meme fitness archieved.

A salient feature of human behavior is *self-tuning*: individuals modify their behavior according to how well it satisfies their needs or drives. Drives amount to conceptual niches that guide the evolution of culture. In future experiments vehicles will have two drives: to mate, and to acquire territory. The implementation of a meme will produce a response in neighboring vehicles that satisfies one drive or the other. When a drive is satisfied by implementing a display, its strength decreases. Fitness functions are not built into vehicles; they vary with the relative strengths of the drives. Memes are expected to specialize for one drive or the other; evolving along different trajectories toward two different basins of attraction.

Vehicles will be able to monitor their success with creation and imitation, and adjust their creation/imitation ratio accordingly. Individual differences will be introduced. Those that have a flawed memetic algorithm might specialize in imitation, while those that can not correctly translate the behavior of a neighbor into input to their neural network might specialize in creation. When vehicles can recognize one another and associate each other with the fitness of their memes, a hierarchical social structure could emerge in which some vehicles are ignored while others are imitated by many.

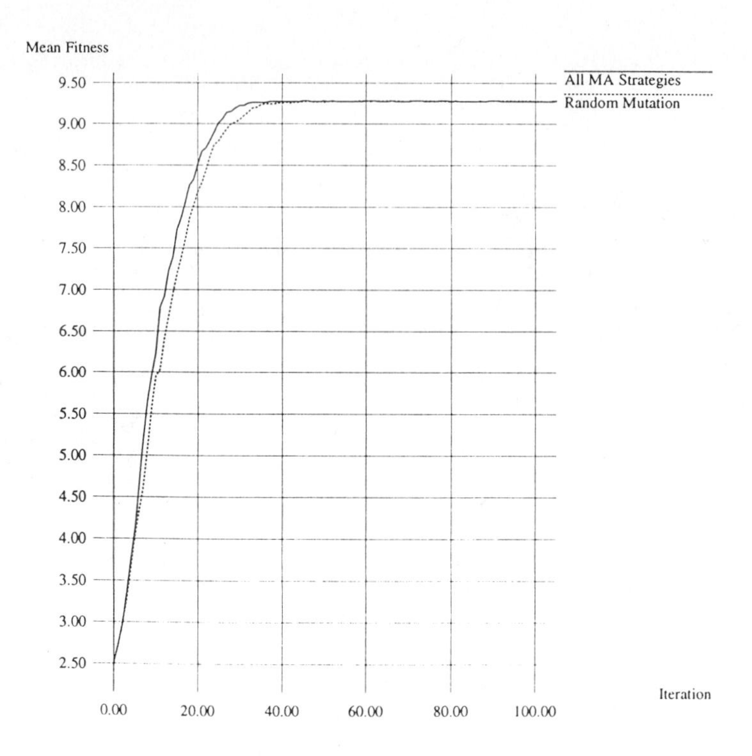

FIGURE 7 Knowledge-based mutation increases rate of optimization.

Another plan is to add recombination: vehicles will acquire new memes by combining a meme that is being imitated with a previously learned meme. This will allow them to specialize on different parts of a meme, and then share partial solutions. Vehicles will be able to monitor the relative effectiveness of mutation and recombination throughout a run, and adjust their frequencies accordingly.

6. COMPARISON WITH OTHER APPROACHES

This work differs from anthropological approaches to culture in that the goal is not to put together a detailed picture of how human culture evolved, but to abstract a general model of cultural evolution. Cavalli-Sforza and Feldman[7] have developed a mathematical theory of cultural evolution which has been extended by Lumsden and Wilson[14] and Boyd and Richerson.[5] The computational approach taken here allows us to model not only what happens, but the mechanisms that make it happen. It enables us to look for patterns that arise when these mechanisms are carried out in parallel in a society of interacting individuals, and to use relatively complex memes with more than one or two alleles.

Some interesting work has been done using a genetic algorithm to investigate the evolution of cooperation,[2] and the interaction between genetic evolution, learning, and culture.[3] Ackley found that Lamarkian evolution increases the efficiency of a genetic algorithm.[1] Hutchins and Hazelhurst used a computer model to explore the relationship between environment, internal representation of the environment, and cultural artifacts that mediate the learning of environmental regularity.[13] These studies model vertical (intergenerational) transmission. *Meme and Variations* differs from these approaches in that we model horizontal (intragenerational) transmission, and we look at the dynamics that emerges when a society of agents can each invent their own ideas and imitate others.

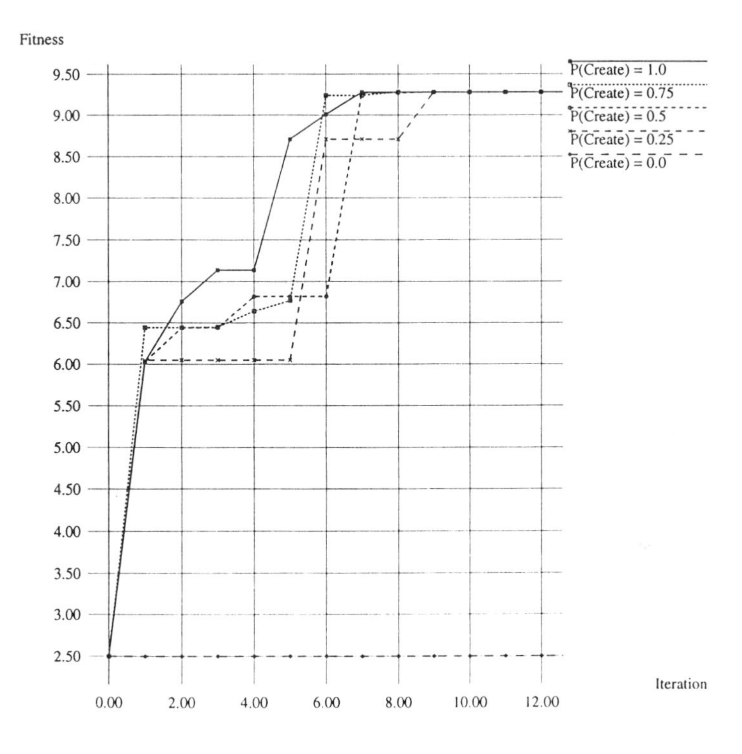

FIGURE 8 Effect of P(Create) to P(Imitate) ratio on fittest meme. The higher the probability of creation, the higher the fitness of the fittest meme.

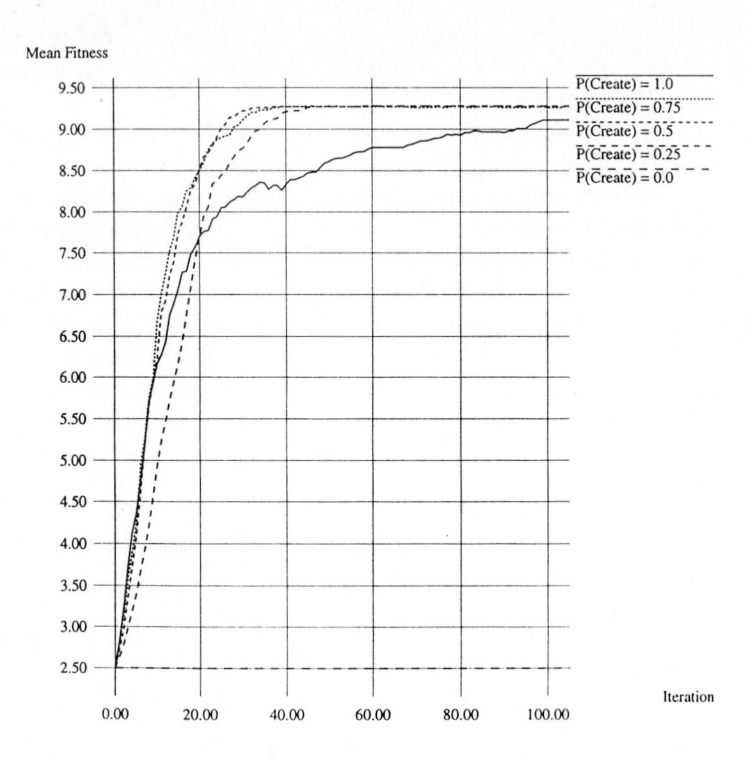

FIGURE 9 Effect of p(create) to p(imitate) ratio on mean fitness. For the society as a whole, the ideal ratio of creation to imitation is 2 or 3 : 1.

ACKNOWLEDGMENTS

I would like to thank David Chalmers and David Jefferson for useful comments on a draft of this manuscript, and Robert Boyd, Michael Dyer, and Charles Taylor for discussion. Thanks also to the Center for the Study of the Evolution and Origin of Life (CSEOL) at UCLA for inspiration and support.

REFERENCES

1. Ackley, D. "A Case for Distributed Lamarkian Evolution." Presentation at the Third Artificial Life Conference, Santa Fe, NM, 1992.
2. Axelrod, R. "Modeling the Evolution of Norms." Paper presented at the American Political Science Association Meeting, New Orleans, 1985.
3. Belew, R. K. "Evolution, Learning, and Culture: Computational Metaphors for Adaptive Algorithms." *Complex Systems* **4** (1990): 11–49.
4. Bonner, J. T. *The Evolution of Culture in Animals.* Princeton, NJ: Princeton University Press, 1980.
5. Boyd, R., and P. J. Richerson. *Culture and the Evolutionary Process.* Chicago, IL: The University of Chicago Press, 1985.
6. Braitenberg, V. *Vehicles.* Cambridge, MA: MIT Press, 1984.
7. Cavalli-Sforza, L. L., and M. W. Feldman. *Cultural Transmission and Evolution: A Quantitative Approach.* Princeton, NJ: Princeton University Press, 1981.
8. Dawkins, R. *The Selfish Gene.* Oxford: Oxford University Press, 1976.
9. Donald, M. *Origins of the Modern Mind.* Cambridge, MA: Harvard University Press, 1991.
10. Hinton, G. E. "Implementing Semantic Networks in Parallel Hardware." In *Parallel Models of Associative Memory.* Hillsdale, NJ: Erlbaum Press, 1981.
11. Holland, J. K. *Adaptation in Natural and Artificial Systems.* Ann Arbor, MI: University of Michigan Press, 1975.
12. Holland, J. H., K. J. Holyoak, R. E. Nisbett, and P. R. Thagard. *Induction.* Cambridge, MA: MIT Press, 1986.
13. Hutchins, E., and B. Hazelhurst. "Learning in the Cultural Process." In *Artificial Life II*, edited by C. G. Langton, C. Taylor, J. D. Farmer, and S. Rasmussen, 689–706. Santa Fe Instituute Studies in the Sciences of Complexity, Proc. Vol. X. Redwood City, CA: Addison-Wesley, 1992.
14. Lumsden, C., and E. O. Wilson. *Genes, Mind, and Culture.* Cambridge, MA: Harvard University Press, 1981.
15. Robert, M. "Observational Learning in Fish, Birds, and Mammals: A Classified Bibliography Spanning over 100 Years of Research." *Psych. Record* **40** (1990): 289–311.
16. Smith, W. J. *The Behavior of Communicating.* Cambridge, MA: Harvard University Press, 1977.
17. Wright, S. *Evolution and the Genetics of Populations.* Chicago, IL: University of Chicago Press, 1969.

Maureane Hoffman,* William Fortin, Arjendu Pattanayak,† and Dougald M. Monroe††**
*Department of Pathology, Duke University and Durham VA Medical Centers, Durham, NC 27705
**Florida Atlantic University, Boca Raton, FL
†University of Pattanayak, University of Texas, Austin, TX
††University of North Carolina, Chapel Hill, NC

Blood Coagulation is a Complex System

The understanding and control of blood clotting are of great importance in modern medicine. Both excessive bleeding and excessive clotting are well-recognized clinical problems. Stroke, myocardial infraction, venous thrombosis, and pulmonary embolism are common manifestations of abnormal coagulation. While many individual components of the coagulation system have been isolated and biochemically characterized, the way they work together *in vivo* is still imperfectly understood. Multiple protein coagulation factors present in blood plasma contribute to the formation of a blood clot. Based on studies of plasma clotting in test tubes, a cascade mechanism of coagulation was posed similar to that shown in Figure 1.[1,6] Each coagulation factor is named with a roman numeral. The activated form of each factor is indicated by appending an "a." There are three principal procoagulant complexes, IXa/VIIIa, VIIa/tissue factor, and Xa/VA. Each consists of a protease and its enzymatically inactive cofactor, and requires a phospholipid surface and calcium ions for optimal activity. At each step a protease cleaves the next protein in the cascade, which converts it into an active protease to continue the cascade. A procoagulant stimulus (such as tissue injury) is amplified at each step in this multistep pathway. In this scheme, separate "intrinsic" and "extrinsic" pathways are activated by different mechanisms, but converge into a final "common" pathway leading to

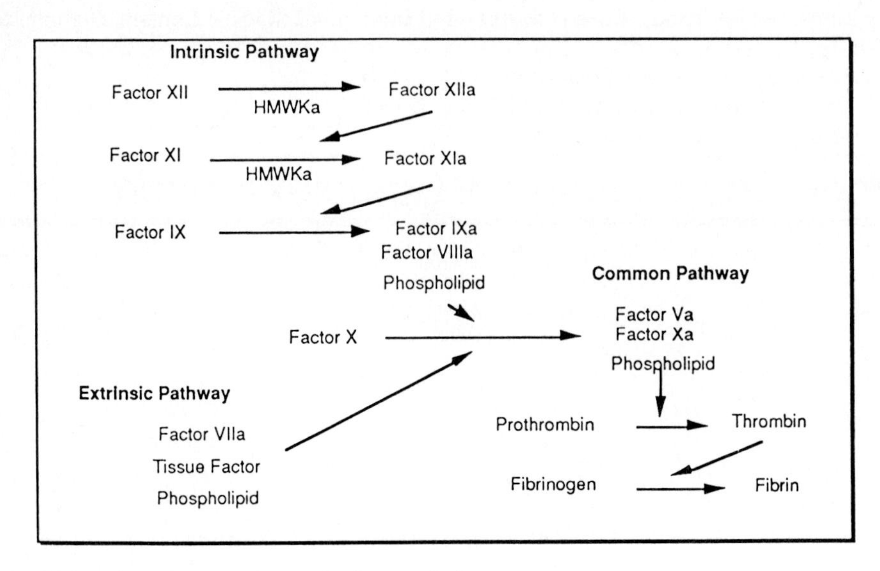

FIGURE 1 Cascade mechanism of coagulation.

thrombin generation. Thrombin cleavages fibrinogen to fibrin, which assembles into a fibrin clot. This scheme adequately explains plasma clotting *in vitro*, and was generally accepted by researchers and clinicians.

It has become increasingly clear that this model does not explain some features of coagulation *in vivo*. For example, it does not explain why patients lacking a factor in either the intrinsic (such as factor VIII or IX) or extrinsic pathway (such as factor VII) have a severe bleeding tendency. According to the scheme, a defect in one pathway should be compensated for by the other pathway. It is also now apparent that coagulation occurs on cell surfaces rather than in solution. Modifications to the cascade model viewed several of the reactions as requiring phospholipid as a cofactor. However, this view still fails to take into account the active role that living cells play in regulating the coagulation reactions. Cell membranes are more than inert phospholipid surfaces, they include specific receptors for coagulation factors. Cell membranes also change in response to internal and external signals, such as occurs when platelets are activated by thrombin. Therefore, there have recently been moves toward revising the coagulation scheme, and studying more physiologic models. This has led to the increased popularity of animal models of hemeostasis and thrombosis.

We have proposed a conceptual model that explains the bleeding tendency in most of the clinically recognized deficiencies. Our model incorporates those coagulation proteins whose deficiency states result in a bleeding tendency. Our model also recognizes the crucial role of cell surfaces in supporting and regulating coagulation. This model views coagulation as occurring in distinct stages (see Figure 2). Coagulation is initiated by exposure of tissue factor (TF) at a site of injury.

Tissue factor is a membrane protein that is not usually expressed on cells in contact with blood plasma. Once TF is exposed to blood, it binds factor VII to form an active proteolytic complex. The TF/VII complex immediately begins to activate coagulation factors IX and X. Activated factor X (factor Xa) then activates small amounts of thrombin before it is inhibited by plasma protease inhibitors. A specific inhibitor, tissue factor pathway inhibitor (TFPI), binds to factor Xa, then binds to the TF/VIIa complex and terminates the initiation step. The initiation step is equivalent to the "extrinsic" pathway of the classical model, with the exception that it only functions for a limited time, before being shut off. During the *amplification* step (Figure 3), thrombin acts as a feedback activator of the process: activating platelets, factors V, VII, and XI. Once activated, the platelets provide the surface on which activated factors assemble into functional coagulation complexes. If enough coagulation factors and platelets are activated before the thrombin is inhibited by plasma protease inhibitors, coagulation proceeds to the next stage.

Once a sufficient number of IXa/VIIIa and Xa/VA complexes have been assembled on the platelet surface, the *propagation* phase begins. In this stage, large-scale thrombin production takes place on the platelet surface. The thrombin formed by this process cleaves sufficient fibrinogen to result in fibrin clot formation.

Two features of this system provide regulatory control. First, coagulation only begins when a TF source contacts plasma and platelets. Second, there is an inhibitor of each step, and the procoagulant stimulus must be sufficient to overcome the resistance of the inhibitors for clotting to occur.

Note that this model proposes that the process of coagulation evolves in time, and that different cellular and protein components are active at different points during the process. We do not believe that there are abrupt transitions between stages, however. It is most likely that activity in one stage increases as the activity of the previous phase wanes. Thus, one phase would blend into the next in a

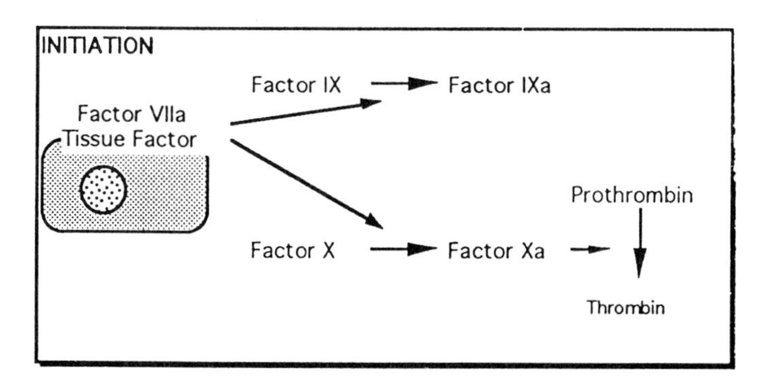

FIGURE 2 Initiation step.

smooth progression. There are multiple positive and negative regulatory interactions between components. These features imply that the coagulation process is inherently nonlinear and cannot be adequately modeled by steady-state systems. The high level of interaction between components indicates that the coagulation system is complex, and that studies of its individual components will not predict the behavior of the complete system.

We are taking a two-pronged approach to testing our conceptual model of coagulation. First, we constructed a cell-based experimental system that includes those elements of coagulation that are in the conceptual model. We used cultured human blood monocytes as a source of tissue factor, and freshly isolated human platelets. Protein components were purified from fresh frozen human plasma, and used at concentrations approximating those found in normal human plasma. The components of the experimental system are shown in Table 1. All of the coagulation factors included in the model are physiologically important, since their deficiencies lead to a bleeding tendency. Tissue factor pathway inhibitor (TFPI) and anti-thrombin (AT) are plasma protease inhibitors. TFPI inhibits factor Xa. Once TFPI binds Xa, it is also able to inhibit the TF/VIIa complex. This means that the TF/VIIa complex is inhibited only after it has had the chance to activate some factor IX and factor X. Anti-thrombin inhibits thrombin, factor Xa, and IXa. Patients who lack AT have fatal thrombosis in infancy. Thus, AT is an essential part of the coagulation system. No patients lacking TF or TFPI have been described. This suggests that absence of these proteins may be incompatible with life (or maybe we just haven't looked hard enough).

The activity of the model system was initiated by adding calcium. Aliquots of supernatant we then removed at timed intervals and the amounts of thrombin and factor Xa that had been formed were measured. We also measured the proportion

TABLE 1 Components of the Monocyte-Platelet Model System.

Cells	Proteins	Activator
monocytes, cultured to induce tissue factor expression	Prothrombin	Factor VIIa
	Factor V	Calcium
unactivated platelets	Factor VIII	
	Factor IX	
	Factor X	
	TFPI	
	Anti-Thrombin	

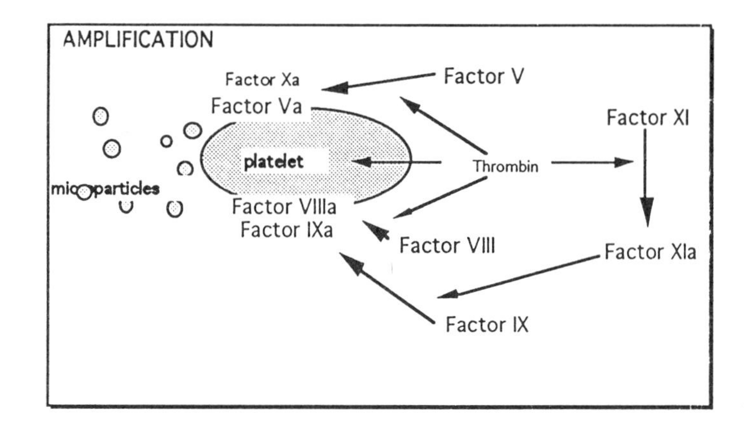

FIGURE 3
Amplification step.

of platelets that had been activated. We fount that TF-bearing monocytes promoted platelet activation by a mechanism that depended on thrombin generation, as would be predicted by the conceptual model. Also, monocytes and platelets were synergistic for thrombin generation. That is, more thrombin was generated when both monocytes and platelets were present in the system, then when the amounts produced by each alone were added together. This is also as we would predict if each component filled a distinct niche in the overall system, and both cellular components needed be present to achieve optimal activity. In addition to supporting our conceptual model, the experimental model system was found to mimic important aspects of normal homeostasis and hemophilia B (factor IX deficiency).[2] This suggests that the experimental and conceptual models are physiologically relevant, and may provide insight into mechanisms of coagulation *in vivo*.

Our second step will be to construct a mathematical/computer model of coagulation. A computer model will allow us to explore the qualitative effects of quantitative changes in procoagulant and inhibitor levels. Such studies would be time consuming and expensive if conducted in an experimental system. A computer model will also allow us to target our experimental work based on computational predictions.

Several mathematical/computer models have been published that deal with portions of the coagulation process.[3,4,5,7] In most cases, these reports have observed a nonlinear response to the initial procoagulant stimulus. We plan to use techniques similar to those used in some of the earlier models, but based on our conceptual model of coagulation. We are in the process of constructing a system of linked differential equations describing the generation and inhibition of crucial procoagulant components. These include the proteases Xa, IXa, and thrombin, as well as the cofactors Va and VIIIa. The rate of change in the concentration of each activated component at any given time is a function of its rate of

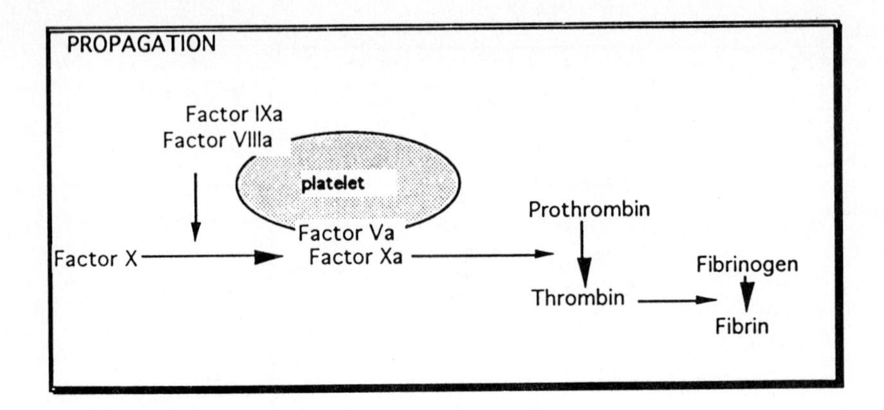

FIGURE 4 Propagation phase.

formation and its rate of inhibition or breakdown. The kinetics of factor Xa formation are the most complicated. Factor X is activated to Xa by the TF/VIIa complex during the initiation stage, and by the IXa/VIIIa complex during the propagation phase (Figure 4). Factor Xa is inhibited both by AT and by TFPI. The TFPI/Xa complex can inhibit the formation of more Xa by inhibiting the TF/VIIa complex. A system of equations describing the kinetics of the model system is shown below:

$$
\begin{aligned}
\frac{\partial[Xa]}{\partial t} =& k_1\left[\frac{TF}{VIIa}\right] + k_2[IXa]\left(\frac{[VIIIa]}{k_{a1}+[VIII]}\right)\pi - k_3[TFPI][Xa]\,, \\
& - k_4[AT][Xa]\,, \\
\frac{\partial[IXa]}{\partial t} =& k_5\left[\frac{Tf}{VIIa}\right] - k_6[AT][IXa]\,, \\
\frac{\partial[IIa]}{\partial t} =& k_7[Xa]\left(\frac{[Va]}{k_{a2}+[VA]}\right)\pi - k_8[AT][IIa]\,, \\
\frac{\partial[Va]}{\partial t} =& k_9[IIa]\left([V_p]+[V_0]\right) - k_{10}[V_a]\,, \\
\frac{\partial[VIIIa]}{\partial t} =& k_{11}[IIa][VIII] - k_{12}[VIII_a]\,, \\
\frac{\partial[TF/VIIa]}{\partial t} =& \left[\frac{TV}{VIIa}\right]_0 - k_{13}\left[\frac{TF}{VIIa}\right]\left([TFPI][Xa]k_{14}\right)\,.
\end{aligned}
$$

The concentrations of the reactants (in brackets) and the kinetic constants ($k_{1\ldots n}$) determine the rate of each reaction. Factors Va and VIIa are cofactors for factors Xa and IXa, respectively. They bind to the activated platelets surface with affinities $K_{a1\ldots an}$. The concentrations of the unactivated factors are assumed not to change over the period of the model. *In vivo* these factors are present in the plasma in great excess over the amounts consumed during the normal clotting process. We will also assume that the amount of TF is limiting, and that all of the TF available forms a complex with VII, which is rapidly autoactivated to VIIa. Tissue factor is not formed, but is assumed to be exposed at $t = 0$ to start the process. Therefore, the equation for TF/VIIa only includes the decrease in the amount of TF/VIIa present at $t = 0$ as it is inhibited by TFPI-Xa. Factor V is present in the plasma (V_0), and is also released from platelets when they are activated by thrombin (V_p). Activated platelets also provide the surface on which the assembly of the coagulation complexes occurs. The proportion of platelets activated at any given time (π : range 0 to 1.0) is a function of the proportion of platelets activated at $t = 0$, and the thrombin concentration: $\pi = \pi_0 + f[IIa]$. While other mediators can have effects on platelet function, we will assume for this model that platelet activation is strictly a function of thrombin concentration.

In order to analyze the behavior of the system using the tools of nonlinear dynamics, some of the nonlinear differential equations must be replaced by linear approximations. The difficult part of this project will be to decide which of the equations can be approximated without impairing the model, and to choose the appropriate substitutions. Fortunately, we have the means of experimentally validating our choices. We can directly compare the predictions of the computer model to experimental results in the cell culture model as we test different assumptions during the modeling process. Not only will the approach allow us to construct a valid model, but the process of checking the model will give us information on which steps are critical in hemeostasis.

We can also compare the predictions of our model(s) to the known clinical manifestations in patients with various factor deficiencies. For example, it is known that patients with levels of factor IX below 1% of normal have a severe bleeding tendency with a dramatically reduced rate of thrombin generation. If our cell culture and computer models do not predict that the reducing factor IX levels to $< 1\%$ will have a marked effect on thrombin generation, then we know that our model is flawed and must be revised. Thus, we have an ideal conceptual, experimental and clinical framework to ensure that our mathematical/computer model is grounded in reality, and that its predictions are likely to be of physiological relevance.

ACKNOWLEDGMENTS

The authors would like to thank Drs. Lee Segel and David Campbell for helpful discussions.

REFERENCES

1. Davie, E. W., and O. D. Ratnoff. "Waterfall Sequence for Intrinsic Blood Clotting." *Science* **145** (1964): 1310.
2. Hoffman, M., D. M. Monroe, and H. R. Roberts. "The Interaction of Monocytes and Platelets in a Cell-Based Model of Factor IX Deficiency." *Circulation* **86** (1992): 408.
3. Jesty, J., E. Beltrami, and G. Willems. "Mathematical Analysis of a Proteolytic Positive-Feedback Loop: Dependence of Lag Time and Enzyme Yields on the Initial Conditions and Kinetic Parameters." *Biochem.* **32** (1993): 6266.
4. Khanin, M. A., and V. V. Semenov. "A Mathematical Model of the Kinetics of Blood Coagulation." *J. Theor. Biol.* **136** (1989): 127.
5. Khanin, M. A., L. V. Leytin, and A. P. Popov. "A Mathematical Model of the Kinetic of Platelets and Plasma Hemostasis System Interaction." *Thromb. Res.* **64** (1991): 659.
6. Mac Farlane, R. G. "An Enzyme Cascade in the Blood Clotting Mechanism, and Its Function as a Biological Amplifier." *Nature* **202** (1964): 498.
7. Nesheim, M. E., R. P. Tracy, and K. G. Mann. "'Clotspeed,' a Mathematical Simulation of the Functional Properties of Prothrombinase." *J. Biol. Chem.* **259** (1984): 1447.

Jan T. Kim
Max-Planck-Institut für Züchtungsforschung, Carl-von-Linne-Weg 10, 50829 Köln, GERMANY

Using Distance Distributions to Measure Complexity of Populations

The idea that complexity increases in the course of evolution is widely spread among biologists and complexity researchers. A usual understanding of this notion is that recent species are more complex than ancient ones, in some general average sense. Another way of approaching this issue is to look at the evolution of taxonomic structures and their complexity. Distance distribution complexity is introduced as a method of approaching evolution of complexity in the latter way. It can be applied to any collection of entities for which a suitable distance measure is available; thus, distance distribution complexity can be measured in a wide variety of systems. Initial results from applying this measure to populations in genetic algorithms are presented. These demonstrate that distance distribution complexity meets key criteria for a complexity measure and conforms expectations derived from previous experience with the simulator used.

1. MOTIVATION

Most efforts to measure complexity in the course of evolution quantitatively focus on the complexity of individuals. Examples for such approaches are estimating the information content of their genomes with the c_0t method,[5] or measuring the complexity of strategies of ALife agents.[2] The complexity of populations and ecosystems is not taken into account with such approaches. However, an ecosphere with few species that have very high individual complexity is not necessarily more complex than an ecosphere populated with a large variety of species that have smaller individual complexities. It may even be argued that a diverse collection of interacting life forms is a key driving force for complexity increase.[7]

In previous work,[4] we have addressed the evolution of complexity at the level of populations by investigating the evolution of taxonomic structures in our ALife simulations. To do this, we developed the technique of monitoring distance distributions. These are the frequency distributions of values in the distance matrix of a population. The idea is that a higher complexity of a population is indicated by a more intricate taxonomic structure with more taxonomic categories. A taxonomic category is characterized by a degree of similarity that is shared among species in the same taxon of that category, whereas it is not shared with other species. Similarity (or rather dissimilarity) can be quantitatively expressed by distance measures. A taxonomic category therefore gives rise to a peak in the distance distribution, as described by Higgs and Derrida.[1] Populations with a rich taxonomic structure have multipeaked distance distributions, whereas populations with only one taxonomic category, such as randomly generated populations, have only one peak in the distance distribution. Such a distance distribution is shown in the left histogram in Figure 1, which shows the edit distance distribution of a population of 50 randomly generated strings of length 40 characters over an alphabet of size 256. Populations of genomes in a genetic algorithm are commonly found to have multipeaked distance distributions, indicating the presence of multiple taxonomic categories. In LindEvol, multipeaked structures in distance distributions turned out to be well correlated to other indicators of complexity, such as the occurrence of complex growth patterns. It was therefore desirable to transform distance distributions into a quantity that reflects the complexity of the taxonomic structures in the populations. Distance distribution complexity provides a method to do this.

2. INTRODUCTION OF DISTANCE DISTRIBUTION COMPLEXITY

Let $\{P = g_0, \ldots, g_{n-1}\}$ be a population of n individuals and $D(g_i, g_j)$ be a measure for the distance between g_i and g_j. Practically, the individuals will be given by a record of information about them, e.g., their genome or some set of taxonomically relevant data. The distance measure operates on these records and must yield integer or at least discrete values. Let $F(d)$ denote the number of times with which the value d appears in the lower left-hand triangle of the distance matrix $(D(g_i, g_j))$ of P. Since the number of elements in the lower lefthand triangle of the distance matrix is $n * (n-1)$, the relative frequency of a distance value d in the distance matrix is $f(d) = F(d)/(n*(n-1))$. Now, distance distribution complexity is defined to be the Shannon entropy[6] of the distribution f:

$$C_d := -\sum_d f(d)\log(f(d)).$$

For populations consisting of identical individuals only, all $D(g_i, g_j)$ are 0, so C_d equals 0. For populations that are completely random, the distance distribution has only one sharp, high peak at the expectation value for distances between unrelated individuals, this gives rise to a low C_d value. Large C_d values result from populations where distances are distributed over a wide range, i.e., that have a multipeaked distance distribution. Thus, distance distribution complexity poses a key property expected from a complexity measure[3]: It gives low values for both strongly ordered and very randomized populations, and peak values for populations with rich patterns of intercorrelation among individuals.

3. SOME RESULTS FROM ANALYZING LindEvol SIMULATIONS WITH DISTANCE DISTRIBUTION COMPLEXITY

Distance distribution complexity analysis was applied to evolving populations in the LindEovl simulation,[4] and on a GA simulator derived from LindEvol by replacing the fitness function. LindEvol simulates the coevolution of plant growth patterns encoded as simple L-systems in genomes. This is done by allowing the simulated plants to grow together in a two-dimensional lattice world for a vegetation period. Each genome is assigned a fitness value based on its performance in this simulated ecology. LindEvol fitness has a stochastic component resulting from a probabilistic simulation of light absorption and the random choice of the order of processing of plants in each time step. The population is then subjected to selection and mutation operations known from genetic algorithms. Selection is done by deleting the lowest ranking genome until the share of the population given by the selection rate is removed, then randomly selecting genomes from the surviving remainder for vegetative reproduction until the original population size is restored. Mutation is governed by the control parameters replacement rate, insertion rate and deletion rate, which control the probabilities of a character being replaced with a randomly chosen one, a randomly constructed gene being inserted between two characters, and a gene to be deleted respectively.

As the distance measure D for the computation of distance distribution complexity in LindEvol, the relative edit distance $e(g_i, g_j)$ was used. The edit distance $E(g_i, g_j)$ between two genomes g_i and g_j is the minimal number of editing operations (i.e., overwriting, inserting, or deleting a character) necessary to transform g_i into g_j. The definition of relative edit distance is $e(g_i, g_j) := E(g_i, g_j)/\max\{\text{length}(g_i), \text{length}(g_j)\}$. For the computation of distance distribuion complexity, the relative edit distance was discretized into intervals of 0.01 (i.e., 1%).

Plain edit distance is less suitable as a basis for distance distribution complexity analysis because it reflects the relatedness of sequences less faithfully. With the mutation operators controlled by per character mutation rates that are used in LindEvol, the edit distance between longer genomes builds up faster than the edit distance between shorter genomes as they diverge during evolution. Moreover, the expectancy value for the edit distance between two sequences is smaller if both are short than it is if at least one of the sequences is long. These observations demonstrate that plain edit distance does not only reflect the similarity between sequences, but also their length. Relative edit distance corrects for this effect.

As a first experiment, the parameter space of mutation rates and selection rates was scanned. This was done with sets of mutation rates of the form replacement rate $= 3^*m$ insertion rate $= m$, deletion rate $= m, m$ running from 0 to 0.2 in steps of 0.01. For each such set of mutation rates, runs were made with selection rates s from 0 to 0.9 in steps of 0.1. Each run was started with 50 randomly created genomes consisting of 40 characters that form 20 genes, and extended for 600 generations. The average C_d from the last 100 generation was determined, the first 500 generations were not considered to allow initial transients to settle down. The results are shown in Figure 2.

These results demonstrate that distance distribution complexity indeed has the property described in the previous section. For $m = 0.0$ and $s = 0.0$, the population remains as it was randomly initialized throughout the whole run. Consequently, C_d is low, but not zero, because not all distances equal the expectancy value. C_d is equal to 0 for all runs with $m = 0$ and $s = 0$. Without mutation, populations converge towards uniformity because species can only be lost, new ones cannot be generated.

It runs without selection and $m = 0$, the average C_d is larger than in the run with $s = 0$ and $m = 0$. This indicates that the usage of relative edit distance does not entirely eliminate the effects of varying genome lengths. The discretization method is at least in part responsible for this problem. With genomes of length 40, only relative edits distance values of the form $x/40$ can occur, this means that values other than 0%, 2%, 5% etc. are never found in the distance matrix. With genome length varied by insertions and deletions, such distance values can occur, so the number of nonzero frequencies in the distance distribution increases and higher C_d values result.

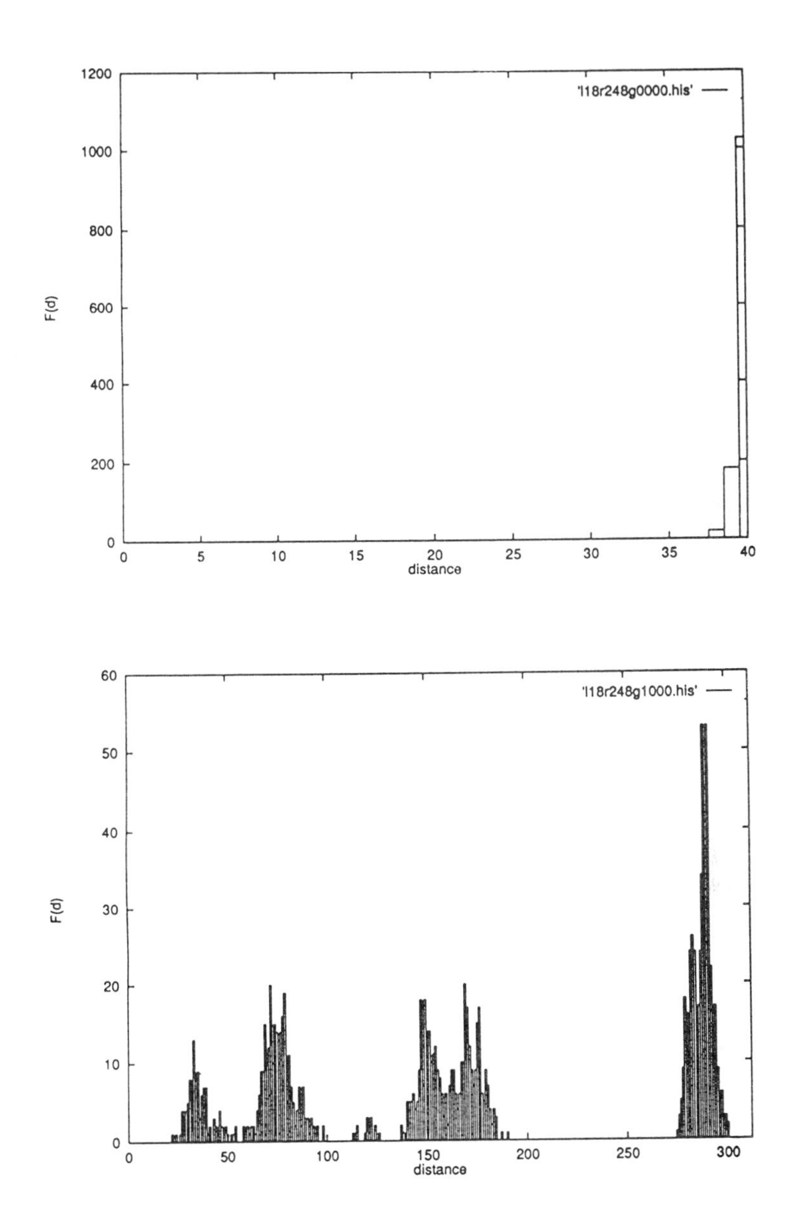

FIGURE 1 Sample distance distributions. The left histogram shows the distance distribution of a population of randomly generated genomes. The right histogram displays the distance distribution after 1000 generations of a LindEvol run.

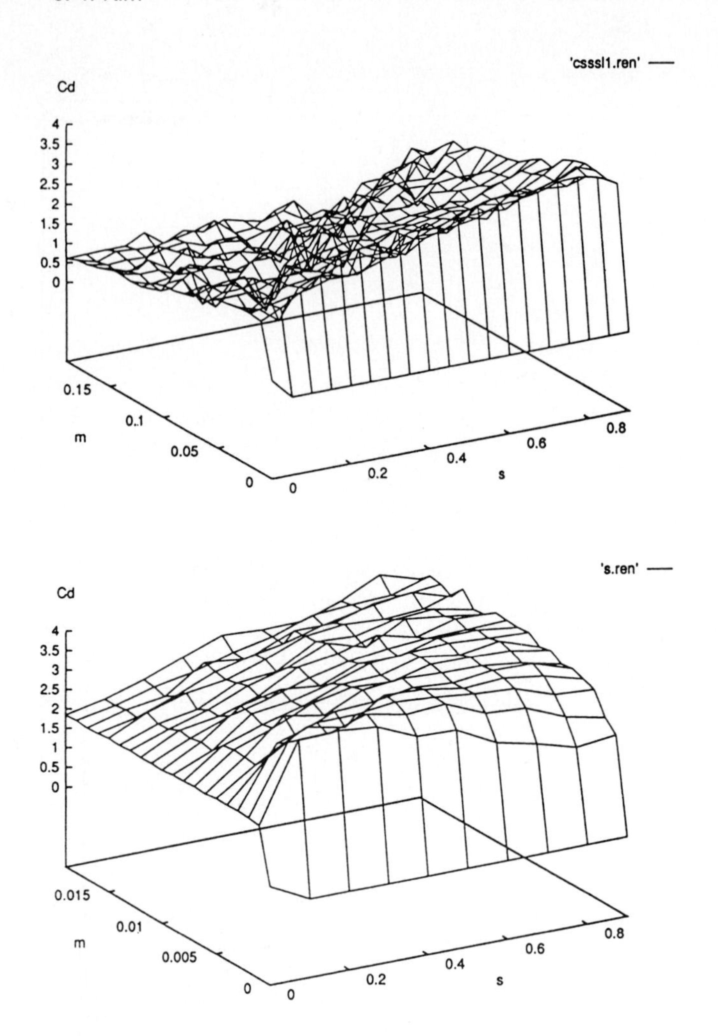

FIGURE 2 Exploration of the (m, s) plane of LindEvol. The left plot shows a coarse scan of the parameter space along the m axis. The right plot is a closeup of the low mutation rate part of the parameter space. Average C_d values are plotted on the vertical axis.

For runs with both mutation and selection, the result is that stronger selection leads to higher average C_d values. As mutation rates are turned on to $m = 0.01$, average C_d values go up to a maximum for all selection rates. Further increase of m caused average C_d values to decrease. With weak selection, average C_d approaches the level of runs with $s = 0$ quite quickly as m goes up. With strong selection, average C_d values decline much slower with increasing mutation.

Average C_d jumps up as mutation comes in because it is necessary for the formation of taxonomic structures. As mutation rates are turned up further, it becomes increasingly difficult for selection to preserve any clusters with elevated fitness levels from being lost to random mutation. Stronger selection can do this in the face of higher mutation rates.

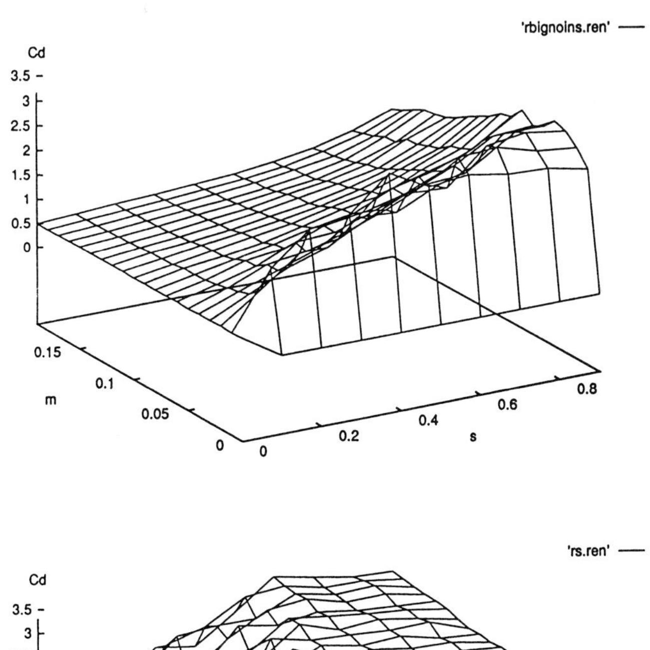

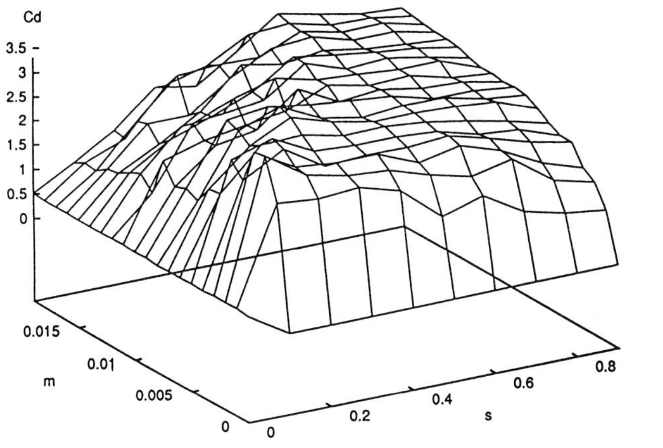

FIGURE 3 Average C_d plots of the (m, s) plane of a genetic algorithm that works the same way as LindEvol, but uses a static, uncorrelated fitness function instead of LindEvol fitness evaluation.

Because high selection rates drive populations towards uniformity strongly, one expects that the richness of taxonomic structures, and consequently average C_d, should decline as selection rates are pushed up to high levels. Also, for strong selection, the mutation rates giving rise to maximal average C_d should be larger than for weak selection. These effects are not visible in the left plot of Figure 2 because this experiment shows a course scan over a large part of the (m, s) plane of LindEvol control parameter space. The effects can be seen in the plot on the right side, which displays a scan of $m = 0$ to 0.02 and $s = 0$ to 0.9, i.e., a closeup of the low mutation rate region in Figure 2.

Average C_d values can be seen to reach a maximum at higher m levels when selection is stronger, as expected. The expected decline of average C_d at very high selection rates and low m values is not very prominent, and visible only for very low mutation rates. I assume that a decline on average C_d becomes clearly visible with high selection rates and low mutation rates only because the LindEvol actively selects for diversity instead of uniformity.

To test this assumption, some parameter scans were done in which the LindEvol method of fitness evaluation was replaced by a static, fully uncorrelated rugged fitness function, listed in the appendix. There were no length varying mutations (i.e., insertions and deletions) in this experiment. The results are shown in Figure 3. The decline in average C_d at low mutation rates and high selection rates is much more prominent than in the LindEvol data shown in Figure 2. Other than that, the surfaces in Figure 3 are much smoother than those in Figure 2. The main reason for this is that runs of LindEvol develop an individual history that can cause all quantities to differ radically between two runs performed with identical control parameter sets in extreme cases. With the static, fully uncorrelated fitness function such long-term history dependence is much less pronounced.

4. CONCLUSION AND OUTLOOK

Distance distribution complexity is a method to evaluate the taxonomic structuredness of a population quantitatively. The initial results discussed in this chapter indicate that distance distribution complexity has key properties that complexity measures are demanded to have, and that it can detect parts of control parameter space where complex structures emerge in LindEvol and other GA simulations.

The most problematic part in analyzing data with distance distribution complexity is the choice of the distance measure. The measure needs to yield values that are as closely correlated to the taxonomic similarity of two individuals as possible. Sources of errors, such as the dependence on genome length of edit distances between equally related genomes in LindEvol, need to be examined and eliminated if possible. The example of the LindEvol experiments presented in this chapter also demonstrate that unintended effects caused by discretization of distance values can distort the distance distribution complexity results.

Distance distribution complexity can be applied to a wide variety of ALife simulations, as well as to data from molecular biology. Investigations of both types are planned for the future. For cases where data are only available for one point in time, such as molecular data, comparing C_d values from randomly chosen subsets and from subsets assumed to form a cluster will be interesting: a method to use distance distribution complexity to validate clustering assumptions will possibly result from such an approach.

5. ACKNOWLEDGMENTS

I thank all those who have made Sun workstations available to the sixth Annual Complex Systems Summer School, where I started coding distance distribution complexity analysis into the LindEvol system and carried out some initial tests of the measure. Special thanks go to Dave Mathews, who managed the Sun cluster and installed the software I needed for my work, and to Mark Davis, who got me started with using Gnuplot and other software. I also thank all participants, lecturers, and organizers of the Summer School for the pleasant coevolution during the school.

APPENDIX: RANDOM FITNESS FUNCTION

The following C listing was used to generate the fully uncorrelated, static fitness function used for the control discussed in Section 3.

```
void fitness(void)
{
  unsigned long i;
  long j; long s;

  for (i = 0; i < psize; i ++)
  {

    s = 0;
    for (j = 0; j< genome[i].len; j++)
    {
      s += (unsigned long) genome[i], g[j];
    }
    for (j = 0; j< size of (long) *  20; j++)
    {
      if (s < 0)
      {
        s <<= 1;
          s ^= 0x1d872b41L;
        }
        else
        }
        s <<= 1'
        }
      }
      genome[i].fitness = ((unsigned long) s) % 100;
  }
}
```

REFERENCES

1. Higgs, P. G, and B. Derrida. "Genetic Distance and Species Formation in Evolution Populations." *J. Mol. Evol.* **35** (1992): 454–465.
2. Horn, J. "Measuring the Evolving Complexity of Stimulus-Respone Organisms." In *Towards a Practice of Autonomous Systems*, edited by F. J. Varela and P. Bourgine. Cambridge, MA: MIT Press, 1992.
3. Huberman, B. A., and T. Hogg. "Complexity and Adaptation." *Physica D* **22** (1986): 376–384.
4. Kim, J. T., and K. Stüber. "Patterns of Cluster Formation and Evolutionary Activity in Evolving L-Systems." Unpublished.
5. Kornberg, A. "Structure and Functions of DNA." In *DNA Replication*, Ch. 1. San Francisco, CA: W. H. Freeman, 1980.
6. Shannon, C. E., and W. Weaver. *The Mathematical Theory of Information.* University of Illinois Press, 1949.
7. Stanley, S. M. "An Ecological Theory for the Sudden Origin of Multicellular Life in the Late Precambrian." *Proc. Natl. Acad. Sci.* **70** (1973): 1486–1489.

James E. Kittock
Robotics Laboratory, Stanford University, Stanford, CA 94305; e-mail: jek@cs.stanford.edu

Emergent Conventions and the Structure of Multiagent Systems

This chapter examines the emergence of conventions through "co-learning" in a model multiagent system. Agents interact through a two-player game, receiving feedback according to the game's payoff matrix. The agent model specifies how agents use this feedback to choose a strategy from the possible strategies for the game. A global structure, represented as a graph, restricts which agents may interact with one another. Results are presented from experiments with two different games and a range of global structures. We find that for a given game, the choice of global structure has a profound effect on the evolution of the system. We give some preliminary analytical results and intuitive arguments to explain why the systems behave as they do and suggest directions of further study. Finally, we briefly discuss the relationship of these systems to work in computer science, economics, and other fields.

1993 Lectures in Complex Systems, Eds. L. Nadel and D. L. Stein,
SFI Studies in the Sciences of Complexity, Lect. Vol. VI, Addison-Wesley, 1995

1. INTRODUCTION

Conventions are common in human society, including such disparate things as standing in line and trading currency for goods. Driving an automobile is a commonplace task which requires many conventions—one can imagine the chaos that would result if each driver used a completely different set of strategies. This example is easy to extend into an artificial society, as autonomous mobile robots would also need to obey traffic laws. Indeed, it appears that conventions are generally necessary in multiagent systems: conventions reduce the potential for conflict and help ensure that agents can achieve their goals in an orderly, efficient manner.

Shoham and Tennenholtz[5] introduced the notion of **emergent conventions**. In contrast with conventions which might be designed into agents' behavior or legislated by a central authority, emergent conventions are the result of the behavioral decisions of *individual* agents based on feedback from *local* interactions. They later extended this idea into a more general framework, dubbed **co-learning**.[6] In the co-learning paradigm, agents acquire experience through interactions with the world, and use that experience to guide their future course of action. A distinguishing characteristic of co-learning is that each agent's environment consists (at least in part) of the other agents in the system. Thus, in order for agents to adapt to their environment, they must adapt to one another's behavior.

Here, we describe a modification of the co-learning framework as presented by Shoham and Tennenholtz[6] and examine its effects on the emergence of conventions in a model multiagent system.

SIMULATION MODEL

We assume that most tasks that an agent might undertake can only be performed in a limited number of ways; actions are thus chosen from a finite selection of **strategies**. A **convention** exists when most or all agents in a system are using one particular strategy for a given task. We will consider a simplified system in which the agents have only one abstract task to perform, and we will examine how a convention for this task can arise spontaneously through co-learning, using a very basic learning rule.

In our model system, each agent's environment is comprised solely of the other agents. That is, the only feedback that agents receive comes from their interactions with other agents. We model agent interactions using payoff matrices analogous to those used for two-person games in game theory. Each time two agents interact, they receive feedback as specified by the payoff matrix. It is this feedback the agents will use to select their action the next time they interact. Systems similar to this have been referred to as **iterated games**.[3,6,7] Each task has a corresponding two-person iterated game. In this chapter, we will consider two different games, which represent the distinct goals of "coordination" and "cooperation."

Our modification to the co-learning setting is the addition of an **interaction graph** which limits agent interactions. In the original study of emergent conventions, any pair of agents could interact[5]; we will restrict this by only allowing interactions between agents which are adjacent on the interaction graph. Our primary objective is to explore the effects of this global structure on the behavior of the system. In particular, we examine how the time to reach a convention scales with the number of agents in the system for different types of interaction graph.

The basic structure of our simulations is as follows. We select a game to use and specify an interaction graph. We create a number of agents, and each agent is given an initial strategy. The simulation is run for a finite number of "time steps," and during each time step, a pair of agents is chosen to interact. The agents receive feedback based on the strategies they used, and they incorporate this feedback into their memories. Using a learning algorithm we call the **strategy update rule**, each agent then selects the strategy it will use the next time it is chosen to interact. The system can either be run for a predetermined number of time-steps or be run until a convention has been reached.

OVERVIEW

In the following section, the structure of the simulation model is explained in more detail. In Section 3, we describe some results of experiments with these systems. Section 4 puts forth our preliminary analytic and intuitive understanding of these systems. In Section 5 we discuss possibilities for further research and the relationship of these experiments to work in a number of other fields.

2. SIMULATING AGENT SOCIETIES

In order to conduct experiments on artificial agent societies, we must choose a way to model them. Since the simulations detailed in this chapter are intended only to explore some basic issues, the model used is deliberately simple. We envision agents existing in an environment where they choose actions from a finite repertoire of behaviors. When an agent performs an action, it affects the environment, which in turn affects the agent. That is, an agent receives **feedback** as a result of its behavior. When investigating emergent conventions, we are primarily concerned with how the agents are affected by each others' behavior. Thus, in the present implementation of our system, all feedback that agents receive is due to their mutual interactions—the agents *are* each others' environment.

2.1 THE AGENT MODEL

In each simulation there is a fixed, finite number, N, of agents. Each agent has two defining characteristics, its **strategy** and its **memory**.

For a given task, s_k, the strategy of agent k, is chosen from a set $\Sigma = \{\sigma_1, \ldots, \sigma_S\}$ of S distinct abstract strategies. It is important to note that Σ does not represent any *particular* suite of possible actions; rather, it serves to model the *general* situation where multiple strategies are available to agents. We will only consider the two strategy case; similar systems with more than two strategy choices are discussed by Shoham and Tennenholtz.[5]

The memory of agent k, $\mathcal{M}_k$, is of maximum size μ, where μ is the **memory size**, a parameter of the agent model. An agent's memory is conveniently thought of as a set, each element of which is a **feedback event**. An event $\mathtt{m} \in \mathcal{M}_k$ is written as a triple, $\mathtt{m} = \langle \mathtt{t}(\mathtt{m}), \mathtt{s}(\mathtt{m}), \mathtt{f}(\mathtt{m}) \rangle$, where $\mathtt{f}(\mathtt{m})$ is the feedback the agent received when using strategy $\mathtt{s}(\mathtt{m})$ at time $\mathtt{t}(\mathtt{m})$. An agent uses the contents of its memory to select the strategy it will use the next time it interacts with another agent.

2.2 MODELLING INTERACTIONS

Once the structure of individual agents is specified, we must decide how the agents interact with their environment. We introduce the concept of an **interaction graph** to specify which agents can interact with one another, and we use the payoff matrix of a two-player game to determine agents' feedback.

2.2.1 INTERACTION GRAPH. In other, similar models, it was possible for any pair of agents to interact.[5,6] To explore the effects of incomplete mixing of agents, we specify an **interaction graph**, $\mathcal{I}$, which has N vertices, each representing one of the agents in the system. An edge connecting vertices i and j in $\mathcal{I}$ indicates that the pair of agents (i, j) may be chosen to interact by playing the specified game. Interacting pairs are chosen randomly and uniformly from all pairs allowed by $\mathcal{I}$. Note that when $\mathcal{I}$ is K_N, the complete N-vertex graph, the present model is equivalent to those which allowed complete mixing of the agents.

To facilitate investigation of the effects of the structure of $\mathcal{I}$, we define a class of graphs representing agents arranged on a circular lattice with a fixed interaction radius.

DEFINITION ($C_{N,r}$) $C_{N,r}$ is the graph on N vertices such that vertex i is adjacent to vertices $(i+j) \bmod N$ and $(i-j) \bmod N$ for $1 \leq j \leq r$. We call r the *interaction radius* of $C_{N,r}$.[1] $C_N \equiv C_{N,1}$ is the *cycle* on N vertices. Note that for $r \geq \lfloor \frac{N}{2} \rfloor$, $C_{N,r} = C_N = K_N$, the complete graph on N vertices.

See Figure 1 for an illustration of the definition. We note that while this is a somewhat arbitrary choice of structure (i.e., why not a two-dimensional grid or a tree structure?), it does yield interesting and illuminating results, while avoiding the added complexity we might expect from a more elaborate structure.

The interaction graph is a general way to model restrictions on interactions. Such restrictions may be due to any number of factors, including hierarchies, physical separations, communication links, security barriers, etc. Whatever its origin, the structure of $\mathcal{I}$ will be seen to have a substantial effect on the behavior of the systems we examine.

2.2.2 TWO-PLAYER GAMES. Once we have determined which pairs of agents are allowed to interact, we must specify what game the agents will play. We examine two games, the iterated cooperation game (ICG) and the iterated Prisoner's dilemma (IPD).

ICG is a "pure coordination" game,[3] with two possible strategies, labelled A and B. When agents with identical strategies meet, they get positive feedback, and when agents with different strategies meet, they get negative feedback. The payoff matrix is specified in Table 1. This game is intended to model situations where: (1) from the point of view of interaction with the world, the two available strategies are equivalent and there is no *a priori* way for agents to choose between them, and (2) the two strategies are mutually incompatible. A simple example of such a situation is driving on a divided road: one can either drive on the left or on the right, but it is suboptimal if some people do one and some people do the other. In this case, our goal is for the agents to reach any convention—either with strategy A or with strategy B.

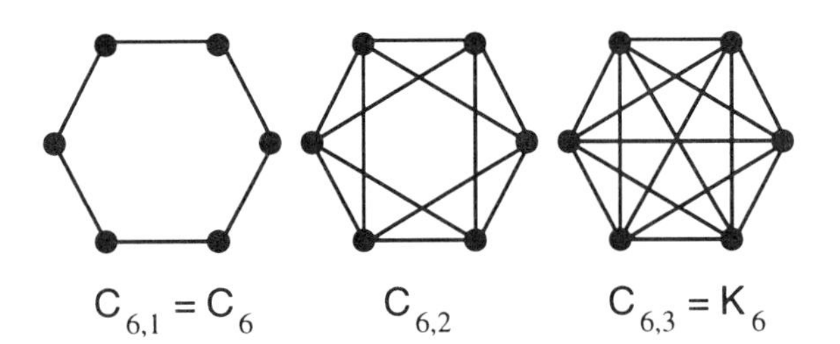

FIGURE 1 Relationship between C_N, $C_{N,r}$, and K_N.

[1]Not to be confused with the graph-theoretic radius, which is something else altogether.

TABLE 1 Payoff matrices for coordination game and Prisoner's dilemma.

	ICG			IPD	
	A	B		C	D
A	+1, +1	−1, −1	C	+2, +2	−6,+6
B	−1, −1	+1, +1	D	+6, −6	−5,−5

IPD has two available strategies, labelled C and D. The payoff matrix is detailed in Table 1. This game is designed to model situations where two agents benefit from cooperating (strategy C), but there is also the potential to get a large payoff by defecting (strategy D) if the other player cooperates. However, if both agents defect, they are both punished. Our goal here is for the agents to reach a convention with strategy C, which indicates the agents are all cooperating.

The Prisoner's dilemma has been examined extensively, in particular by Axelrod in his classic book.[1] The relationship between previous work with Prisoner's dilemma and the present experiments will be briefly discussed in Section 5.

2.2.3 STRATEGY SELECTION. Once the agent model and game are specified, our final step in defining the system is to determine how agents choose the strategy that they will use. In these experiments we use a version of the Highest Current Reward (HCR) strategy update rule.[6][2]

The **current reward** for a strategy is the total remembered feedback for using that strategy; i.e, for strategy σ the current reward is the sum of `f(m)` for all feedback events `m` in the agent's memory such that `s(m)` $= \sigma$. We can now define the HCR rule (in the two-strategy case) as: "If the other strategy has a current reward greater than that of the current strategy, change strategies." Note that HCR is performed after the feedback event from the interaction which has just occurred is added to the agent's memory. Once an agent's next strategy has been chosen, the agent's memory is updated discarding the oldest event.

Agents apply the HCR rule immediately after receiving feedback, and their strategies are considered to be updated instantaneously. Agents which were not chosen to interact at time t do nothing so their memories do not change.

[2] It should be noted that the present definition of memory is slightly different than that found in Shoham and Tennenholtz.[6] There, memory was assumed to record a fixed amount of "time" during which an agent might interact many, few, or no times. Here, memory refers explicitly to the number of previous interactions *in which an agent has participated* that it remembers.

3. EXPERIMENTAL RESULTS

Before we proceed with a look at some results from our simulations, a word is in order about how we compare the behavior of the various possible systems.

3.1 PERFORMANCE MEASURES

In the present situation, the most obvious performance criterion is "how well" the system reaches a convention. For an ICG system, the goal is to have all of the agents using the same strategy, but we do not care which particular strategy the agents are using. On the other hand, for an IPD system, we want all agents to be cooperating. Thus, we have different notions of **convergence** for the two systems. We define $\mathcal{C}_t$, the convergence of a system at time t, as follows. For ICG, the convergence is the fraction of agents using the majority strategy (either A or B); for IPD, the convergence is the fraction of agents using the cooperate strategy (C). Note that convergence ranges from 0.5 to 1 for ICG and from 0 to 1 for IPD. Given this definition of convergence, we can also define T_c, the **convergence time** for a simulation: the convergence time for a given level of convergence c is the earliest time at which $\mathcal{C}_t \geq c$. In this chapter, we will use "time" and "number of interactions" interchangeably. Thus, when we speak about "time t," we are referring to the point in the evolution of the system when t interactions have occurred.

We use two different measures of performance which arise from these definitions. The first measure is average time to a fixed convergence. In this case, we run simulations until a fixed convergence level is reached and note how long it took. When our concern is that the system reach a critical degree of convergence as rapidly as possible, this is a useful measure. However, some systems will never converge on practical timescales, and yet may have interesting behavior which will not be evident from timing data. The second measure we use is average convergence after a fixed time. We simply run simulations for a specified amount of time and note the convergence level at the end of the run. We find that this is often one of the most revealing measures. However, for systems where the convergence over time is not generally monotonic, this measure is effectively meaningless.

There are, of course, other possible measures of performance, such as probability of achieving a fixed convergence after a fixed time (used by Shoham and Tennenholtz[5,6]) and maximum convergence achieved in a fixed amount of time. We have chosen the measures which we found most revealing for the issues at hand.

3.2 EFFECTS OF INTERACTION GRAPH TOPOLOGY

Unless otherwise specified, the simulations were run with one hundred agents, each agent had an equal probability of having the two possible strategies as its initial strategy, and the data were averaged over one thousand trials with different random seeds. We used memory sizes of $\mu = 2$ and $\mu = 1$ for ICG and IPD, respectively (note that with $\mu = 1$, agents change strategies immediately upon receiving negative feedback); experimentation showed that these memory sizes were among the most efficient for the HCR rule.

We will first consider the extreme cases, where all agents are allowed to interact with one another ($\mathcal{I} = \mathrm{K}_N$) and where agents are only allowed to interact with their nearest neighbors on the one-dimensional lattice ($\mathcal{I} = \mathrm{C}_N$). One of our most interesting discoveries was the radically differing behavior of HCR with ICG and IPD as a function of the structure of $\mathcal{I}$. The experimental data are presented in Figure 2, which shows the time to achieve 90% convergence as a function of the number of agents for both games on K_N and C_N.

The performance of the HCR rule with ICG is reasonable for both cases of $\mathcal{I}$. The linear form of the data on the log-log plot indicates that ICG systems can be expected to converge in polynomial time on both K_N and C_N. For intermediate interaction radii, the performance of the ICG systems is somewhere between that for C_N and K_N.

For IPD, the story is different. Using the HCR rule and working with a system equivalent to our IPD system on K_N, Shoham and Tennenholtz write, "[HCR] is hopeless in the cooperation setting."[6] They had discovered what we see in Figure 2: convergence time for IPD on K_N appears to be at least exponential in the number of agents. HCR is redeemed somewhat by its performance with IPD on C_N, which appears to be polynomial, and is possibly linear. On $\mathrm{C}_{N,2}$ (not shown), the IPD system still manages to converge in reasonable time, but for all interaction radii greater than two, it once again becomes "hopeless."

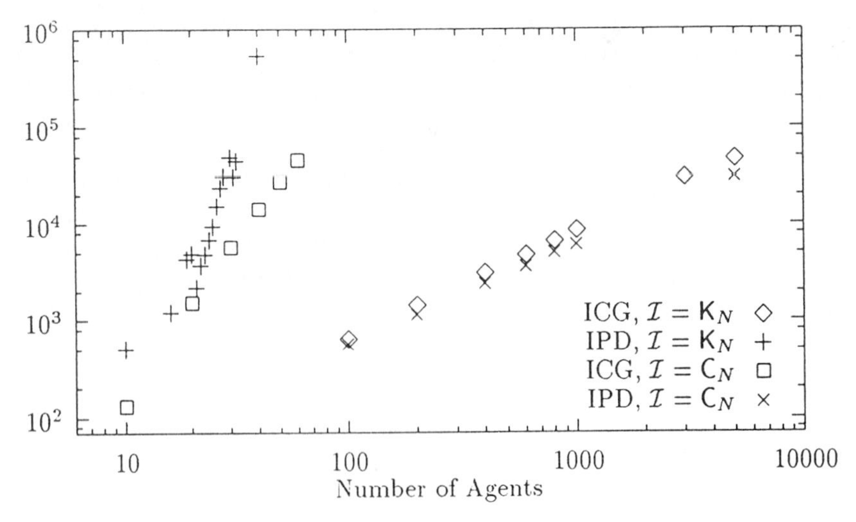

FIGURE 2 $T_{90\%}$ vs. N for ICG and IPD with different configurations of $\mathcal{I}$.

TABLE 2 Possible relationships between $T_{90\%}$ and the number of agents for different interaction graph structures.

	$\mathcal{I} = \mathrm{K}_N$	$\mathcal{I} = \mathrm{C}_N$
ICG	$T_{90\%} \propto N \log N$	$T_{90\%} \propto N^3$
IPD	$T_{90\%} \propto c^N$	$T_{90\%} \propto N$

In general, it appears that the particular choice of $\mathcal{I}$ has a drastic effect on the way system performance scales with system size.

Possible functional relations between expected values of $T_{90\%}$ and N are summarized in Table 2; they were derived from fitting curves to the simulation results and are merely descriptive at this stage.

4. ANALYSIS

In this section, we aim to give a flavor for some of the ways we can pursue an understanding of the behavior of both the coordination game and the Prisoner's dilemma.

4.1 ITERATED COORDINATION GAME

To begin our investigation of the relationship between ICG performance and the structure of $\mathcal{I}$, we look to Figure 3, which shows how performance (measured as convergence after a fixed time) varies with the interaction radius for agents on $\mathrm{C}_{100,r}$. Empirically, we find that performance increases with increasing interaction radius. Thus, we are lead to ask, what properties of $\mathcal{I}$ vary with the interaction radius? Two important ones are the **vertex degree** and the **graph diameter**.

The degree of a vertex is the number of edges containing that vertex. In the present case, all vertices of $\mathcal{I}$ have the same degree, so we can speak of the vertex degree, ν, of $\mathcal{I}$. For $\mathrm{C}_{N,r}$, $\nu = 2r$ (restricted to $2 \leq \nu \leq N-1$); as r increases, each agent can interact with more agents.

The diameter of a graph is the maximum shortest path between any two vertices, and it provides a lower limit on the time for information to propagate throughout the graph. As r increases, the diameter of $\mathrm{C}_{N,r}$ decreases, and we expect that the time for information to travel among the agents will decrease as well.

We speculate that either or both of these properties of $\mathcal{I}$ affect the observed performance. However, for $C_{N,r}$, the diameter is $\lceil \frac{N}{2r} \rceil = \lceil \frac{N}{\nu} \rceil$, so it is closely related to the vertex degree. To test the relative importance of graph diameter and vertex degree, it would be useful to construct a set of graphs for which one property (diameter or vertex degree) is constant, while the other property varies. Initially, we would also like to keep our graphs symmetric with respect to each vertex, to avoid introducing effects due to inhomogeneity.

It turns out to be straightforward to construct a symmetric graph of fixed vertex degree. As a test case, we define the class of graphs $\mathcal{D}_{N,\delta}$ such that an edge connects vertex i with vertices $i+1, i-1, i+\delta$, and $i-\delta$ (all mod N). For $2 \leq \delta < \lfloor \frac{N}{2} \rfloor$, the

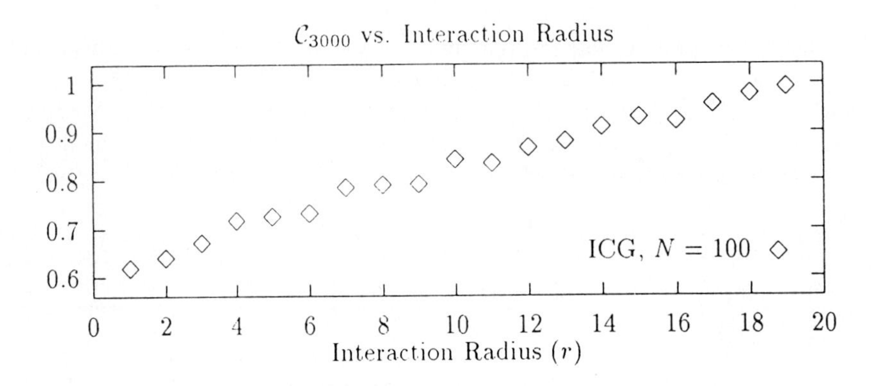

FIGURE 3 ICG: $\mathcal{C}_{3000}$ as a function of r, $\mathcal{I} = C_{100,r}$.

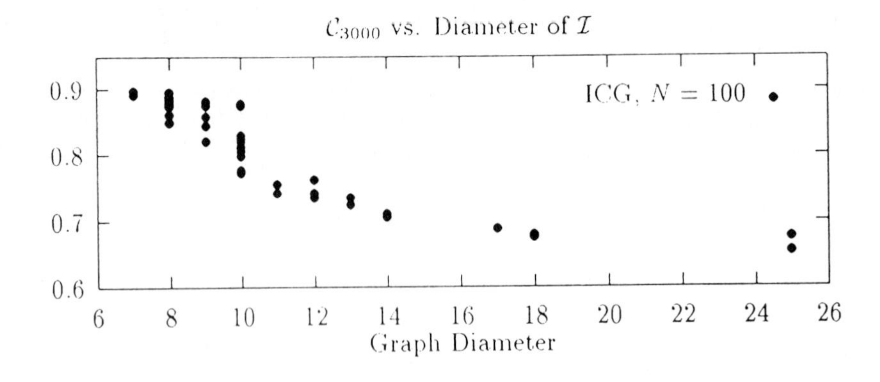

FIGURE 4 ICG: $\mathcal{C}_{3000}$ plotted against the diameter of $\mathcal{I}$, for $\mathcal{I} = \mathcal{D}_{100,\delta}$, $2 \leq \delta \leq 49$.

vertex degree is fixed at four, and a variety of diameters result (we can measure the diameter of each graph using, e.g. Dijkstra's algorithm[8]). Once we have measured the performance of ICG on each graph $\mathcal{D}_{N,\delta}$, we can plot performance against diameter, as seen in Figure 4.

We see that there is a correlation between the diameter of an interaction graph and the performance of an ICG system on that graph. We have hinted that this may be a function of the speed with which information can flow among the agents. However, more work is necessary to determine precisely how and why the diameter, vertex degree, and other graph properties of $\mathcal{I}$ affect the performance of an ICG system. It will also take further study to prove (or disprove) the relationships between expected convergence time for ICG and number of agents proposed in Table 2.

4.2 ITERATED PRISONER'S DILEMMA

It was seen in Section 3 that IPD behaves quite differently from ICG with respect to the structure of $\mathcal{I}$. For $r = 1$, IPD on $\mathrm{C}_{N,r}$ converges quite rapidly, but for large r, it does not converge on any reasonable time scale. We can get some intuition for why this is if we think in terms of "stable" cooperative agents. A cooperative agent is *stable* at a given time if it is guaranteed to interact with a cooperative agent. On K_N, we have an all-or-none situation: an agent can only be stable if all of the agents are cooperative. In contrast, on C_N an agent need only have its two neighbors be cooperative to be stable. As yet, we have not extended this notion to a formal analysis. However, for IPD on K_N, we can give an analytical argument for the relatively poor performance of HCR in our experiments (recall that $\mathcal{I} = \mathrm{K}_N$ shows a dramatic increase in convergence time as the number of agents is increased, as seen in Figure 2).

We begin by computing the expected change in the number of cooperative agents as a function of the convergence level. Since we are considering agents with memory size $\mu = 1$, whenever two agents using strategy D meet, they will both switch to using strategy C (because they will both get negative feedback, as seen in Table 1). When an agent using strategy C encounters an agent using strategy D, the agent with strategy C will switch to strategy D. When two agents with strategy C meet, nothing will happen. Now we can compute the expected change in the number of cooperative agents as a function of the probabilities of each of these meetings taking place:

$$\langle \Delta N_C \rangle = 2 \cdot p(\mathrm{DD}) - 1 \cdot [p(\mathrm{CD}) + p(\mathrm{DC})] .$$

These probabilities are functions of the number of agents using strategy C and hence the convergence level of the system. Thus, we can compute the expected change in the number of cooperative agents as a function of the convergence level; this function is plotted in Figure 5. Note that for $\mathcal{C} > 0.5$ we can actually expect the number of cooperative agents to *decrease*. This partially explains the reluctance of the IPD system to converge—as the system gets closer to total convergence it tends to move away, and towards 50% convergence. Note that before converging, the IPD system must pass through a state in which just two agents have strategy D. We can calculate that in this state, the probability that the system will move *away* from convergence is a factor of $O(N)$ greater than the probability that the system will move *towards* convergence.

Thus, we see that there is essentially a probabilistic barrier between the fully converged state and less converged states. We can approximate this situation by assuming that the system repeatedly attempts to cross the barrier all at once. The expected time for the system to achieve the fully convergent state is then inversely proportional to the probability of the system succeeding in an attempt: $\langle T_{conv} \rangle \propto p_{conv}^{-1}$. A straightforward analysis shows that $p_{conv} \leq O(c^{-N})$ for some $c > 1$, so $\langle T_{conv} \rangle \geq O(c^N)$, which correlates with what we saw experimentally.

On C_N, analysis is complicated by the fact that not all states with the same convergence level are isomorphic (for example, consider the case where agents have alternating strategies around the lattice versus the case where agents $1 \ldots \frac{N}{2}$ have one strategy and agents $\frac{N}{2} + 1 \ldots N$ have the other strategy). Thus, our analysis would require a different methodology than that used for K_N. Experimentally, we see that the order imposed by $\mathcal{I}$ allows the formation of stable groups of cooperating agents. These groups tend to grow and merge, until eventually all agents are cooperative. It is hoped that continued work with the idea of "stable" agents will lead to a more complete understanding of the relationship between performance and the topology of $\mathcal{I}$.

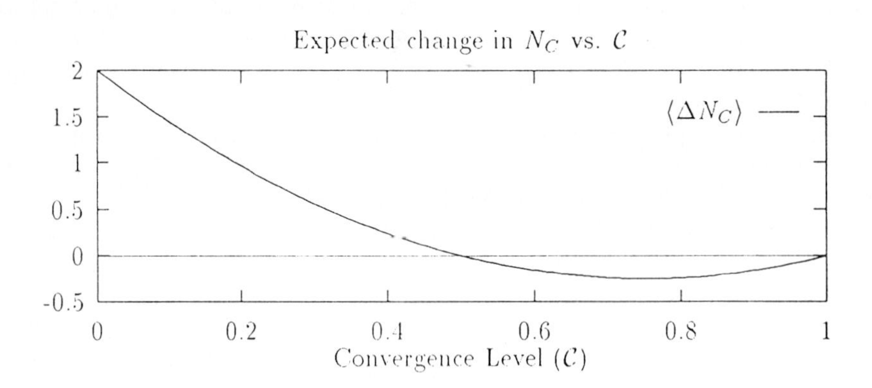

FIGURE 5 Expected change in number of cooperative agents as a function of the convergence level for IPD, $\mathcal{I} = K_N$.

As a final note, Shoham and Tennenholtz have proven a general lower bound of $N \log N$ on the convergence time of systems such as these,[6] which appears to contradict our assertion that $T_{90\%}$ appears to be proportional to N. However, in the general case $T_{100\%}$ need not be proportional to $T_{90\%}$, because the final stages of convergence may proceed much more slowly than the initial stages. Experimental data for IPD on C_N indicate that while $T_{90\%} \propto N$, $T_{100\%} \propto N \log N$.

The arguments presented in this section are an attempt to explain the results of the simulations; they are more *theory* than *theorem*. Deriving tight bounds on the performance for any of these systems is an open problem which will most likely require an appeal to both rigorous algorithmic analysis and dynamical systems theory.

5. DISCUSSION

We have seen that a wide variety of interesting and often surprising behaviors can result from a system which is quite simple in concept. Further analytic investigation is necessary to gain a clear theoretical understanding of the origins and ramifications of the complexity inherent in this multiagent framework. To move from the current system to practical applications will also require adding features to the model that reflect real-world situations, such as random and systemic noise and feedback due to other environmental factors. A number of possible applications to test the viability of this framework are presently under consideration, including distributed memory management and automated load balancing in multiprocessor systems.

The systems discussed here have ties to work in a number of other areas both within and outside of computer science. Co-learning has fundamental ties to the machine learning subfield of artificial intelligence. The Highest Current Reward strategy update rule provides another link to machine learning, as it is essentially a basic form of reinforcement learning. This leads to another possibly fruitful avenue of investigation: systems of substantially more sophisticated agents. Schaerf et al. have used co-learning with a more sophisticated learning rule to investigate load balancing without central control in a model multiagent system.[4] Our present framework also has ties to theoretical computer science, especially when we view either individual agents or the entire system as finite state machines.

Readers familiar with the Prisoner's Dilemma and its treatment in game theory and economics have probably noticed that our approach is markedly different. Our emphasis on feedback-based learning techniques violates some of the basic assumptions of economic cooperation theory.[1,6] In particular, we do not allow for any meta-reasoning by agents; that is, our agents do not have access to the payoff matrix and thus can only make decisions based on their experience. Furthermore, agents do not know the specific source of their feedback. They do not see the actions which other agents take and, indeed, have no means of distinguishing feedback due to interactions with other agents from feedback due to other environmental factors. In our framework, agents must learn solely based on the outcome of their actions. In some respects, this may limit our systems, but it also allows for a more general approach to learning in a dynamic environment. The current interest in economics with "bounded rationality" has led to some work which is closer in spirit to our model.[3]

The systems discussed in this chapter (and multiagent systems in general) are also related to various other dynamical systems. Ties to population genetics are suggested both by the resemblance of the spread of convention information through an agent society to the spread of genetic information through a population and by the possible similarity of the selection of behavioral conventions to the selection of phenotypic traits. There are also links to statistical mechanics, which are exploited more thoroughly in other models of multiagent systems which have been called "computational ecologies."[2] For a more thorough discussion of the relationship of the present framework to other complex dynamic systems, see Shoham and Tennenholtz.[6]

6. CONCLUSION

We have seen that the proper global structure is required if conventions are to arise successfully in our model multiagent system, and that this optimal structure depends upon the nature of the interactions in the agent society. Social structures which readily allow conventions of one sort to arise may be completely inadequate with regards to other conventions. Designing multiagent systems with the capacity to automatically and locally develop behavioral conventions has its own unique difficulties and challenges; emergent conventions are not simply a panacea for the problems of off-line and centralized social legislation methods. However, the study of emergent conventions is in its earliest stages and still has potential for improving the functionality of multiagent systems. Furthermore, the framework presented here invites creative design and investigative efforts which may ultimately borrow ideas from—and share ideas with—the broad range of subjects loosely grouped under the heading "complex systems."

ACKNOWLEDGMENTS

I would like to thank Yoav Shoham for introducing me to this interesting topic, and to Marko Balabanović and Tomás Uribe for reading and commenting on a draft of this chapter. This research was supported in part by grants from the Advanced Research Projects Agency and the Air Force Office of Scientific Research.

REFERENCES

1. Axelrod, R. *The Evolution of Cooperation.* New York: Basic Books, 1984.
2. Huberman, Bernardo, and Tad Hogg. "The Behavior of Computational Ecologies." In *The Ecology of Computation*, edited by Bernardo Huberman. Elsevier, 1988.
3. Kandori, M., G. Mailath, and R. Rob. "Learning, Mutation and Long Equilibria in Games." *Econometrica* **61** (1993): 29–56.
4. Schaerf, Andrea, Moshe Tennenholtz, and Yoav Shoham. "Adaptive Load Balancing: A Study in Co-Learning." Draft manuscript, 1993.
5. Shoham, Yoav, and Moshe Tennenholtz. "Emergent Conventions in Multi-Agent Systems: Initial Experimental Results and Observations." *KR-92*, 1992.
6. Shoham, Yoav, and Moshe Tennenholtz. "Co-Learning and The Evolution of Social Activity." Submitted for publication, 1993.
7. Sigmund, Karl. *Games of Life: Explorations in Ecology, Evolution, and Behaviour.* Oxford, 1993.
8. Skiena, Steven. *Implementing Discrete Mathematics.* Reading, MA: Addison-Wesley, 1990.

Jose Lobo
Graduate Program in Regional Science, Cornell University

Stochastic Fluctuations, Noise-Induced Transitions, and the Blowtorch Theorem: Does Noise Matter in Economics?

1. INTRODUCTION

Almost every natural macroscopic system is subjected to spontaneous, unpredictable environmental fluctuations (often termed noise). The existence of noise is, in fact, well known, and it serves as the starting point for the highly developed, theoretical and analytical framework of thermodynamics and statistical mechanics. Relatively novel, however, is the understanding—accrued over the past 20 years or so—that microscopic noise can generate and sustain macroscopic dynamics, acting, in effect, as a source of complexity. For a large class of physical nonlinear systems, external noise can often give rise to transitions where a spontaneous structuring of the system occurs. A well-known example is provided by Bernard convection cells.[24] This structuring is induced by external randomness and does not survive without the presence of external fluctuations.[14] It is also now understood that in systems with unstable equilibria, pattern formation can result from the amplification of random disturbances. In such systems, coherent spatio-temporal patterns emerge from and are sustained by noise.[15]

The importance of noise has also been highlighted by Landauer's memorably named "Blowtorch Theorem." The theorem states that the question of relative stability along a potential field—in which of various possible states a system is more likely to be—cannot, in general, be decided by only examining the neighborhoods of the competing states. The kinetics along the various pathways connecting the states must be taken into account, and these kinetics are crucially affected by stochastic fluctuations.[19,20] Noise has also come to be seen as playing an important role in biological phenomena, such as morphogenesis (the development of spatial order in living systems)[26] and various cellular processes.[2] Stochastic fluctuations—mutations—play a crucial role in the process of evolution: biological evolution can be understood as the selection and amplification of those random effects which increase a genotype's fitness.[18]

Are the above phenomena at all relevant for the understanding of economic systems? The surface similarities are immediately suggestive, as one can very generally describe an economic system in a manner highly evocative of physico-chemical and biological macroscopic systems. An economic system can be depicted as an ensemble of numerous individual but interconnected behaving units, each endowed with a limited behavioral repertoire and possessing incomplete information about the system's state and the behavior of the other units. Some economists are beginning to think of the workings of the market mechanism in terms familiar to those who study statistical mechanics: global behavior (determination of prices) results from the propagation of local behavior (supply of and demand for goods at spatially proximate areas).[5]

Empirical considerations also motivate the hypothesis that stochastic fluctuations can affect economic dynamics. There is by now ample evidence indicating that many important economic variables display oscillatory behavior which seems essentially random.[10,29] Furthermore, there is an aspect of economic behavior—smooth, gradual change punctuated by sharply discontinuous development—which is strongly reminiscent of phase transitions in macroscopic systems. For an instance of this, consider the process of job creation and destruction in industrialized economies. Undoubtedly individual firms, of all sizes, continuously adjust the composition of their payrolls in response to changes in their economic environment. At the aggregate level, however, significant changes in employment levels seem to occur as large, spatio-temporally concentrated, sectoral shifts. And significantly, once changes in sectoral employment take place, they are fairly permanent.[1,27]

Despite these enticing considerations, the role of noise in creating order, forming patterns, and inducing phase-transitions remains, by and large, unexplored territory within economics. Typically, the role of stochastic fluctuations in economic theorizing is limited to that of error terms in econometric models,[6,13] or as random shocks driving economic systems away from the theoretically expected equilibrium state.[1,17,22] The task of incorporating stochastic fluctuations and their effects into the explanatory repertoire of economic science constitutes a major research agenda, and will require much careful empirical and analytical work.

This discussion is much more limited in scope. The dynamical impact of noise in economics is here explored by considering what happens when noise is added to a deterministic law of motion in a particular, but broadly representative, economic model. The effect of noise in altering the dynamics of the model is studied by using a Fokker-Planck Equation (FPE), one of the most important and useful tools available to study noise in physico-chemical systems, but largely unknown in the economics community.

2. MATHEMATICAL PRELIMINARIES: STOCHASTIC DIFFERENTIAL EQUATIONS AND THE FOKKER-PLANCK EQUATION

To study the dynamical effects of noise one needs the mathematical machinery of stochastic differential equations (s.d.e.'s). In this section the mathematical tools employed in the rest of the discussion are very quickly reviewed and presented; for complete derivations, comprehensive explications and rigorous proofs the reader should consult the references provided. [The following presentation of stochastic differential equations follows closely that given by Horsthemke and Lefever.[15]]

Consider a system which can be modeled by an equation of the type:

$$\frac{dx}{dt} = f\lambda(x)\,, \tag{1}$$

where x is a variable characterizing the state of the system and λ denotes an external control parameter linked to the environment in which the system is embedded. [Although Eq. (1) is of dimensionality one, everything that is subsequently said can be easily generalized to higher dimensional systems.] In Eq. (1), an ordinary differential equation (o.d.e.), the control parameter λ is assumed to remain constant through time. Equation (1) can be rewritten as:

$$\frac{dx}{dt} = h(x) + \lambda g(x) \tag{2}$$

where the function describing the time evolution of X is desegregated into a component not affected by $\lambda, h(x)$, and a component directly affected by the control parameter, $g(x)$.

How is the behavior of a system modified by an environment subject to stochastic fluctuations? In some circumstances this question can be equivalently posed by inquiring what happens if the control parameter is not constant through time, but instead is affected by stochastic fluctuations? This can be modeled by replacing the deterministic control parameter λ in Eq. (2) with:

$$\lambda_t = \lambda + \sigma\xi_t \tag{3}$$

where ξ_t represents Gaussian White Noise and σ is the amplitude of the noise. Equation (2) then becomes:

$$\begin{aligned} \frac{dx_t}{dt} &= h(x_t) + \lambda g(x_t) + g(x_t)\sigma\xi_t \\ &= f(x_t) + g(x_t)_t\sigma\xi_t \end{aligned} \tag{4}$$

where $f(x_t)$ equals $[h(x_t) + \lambda g(x_t)]$. An equation such as Eq. (4) is known as an s.d.e.[11]

Turning the control parameter λ into a random variable has the effect of turning the state variable x into a random variable x_t. To therefore inquire into the behavior and time evolution of the system under study, one needs to calculate a time-differential for the probability density characterizing the stochastic variable x_t. With an expression for the time-differential of the probability density of x_t, one can then proceed to investigate the behavioral consequences of noise on the probability density. This is precisely what the FPE allows us to do. The full derivation of the FPE is rather technical and involved, and space constraints prevent its presentation here. Let us instead attempt to motivate the intuition behind the equation (for a derivation of the FPE see Reichl,[25] Doering,[9] or Mackey[23]; an exhaustive treatment is found in Risken[27]).

Recall that treating the state variable x as a random variable is a recognition that the system characterized by x can assume different values of x at different times. The greater or lesser likelihood that a specific value x_i of x_t will be instantiated is represented by the different probabilities assigned to each x_i. The following probability densities for a random variable x_t can defined:

$P(x_1, t) =$ the probability density that the stochastic variable x_t has value x_1 at time t_1; (5)

$P(x_1, t_1; x_2, t_2) =$ the joint probability density that $x_t$has value x_1 at time t_1 and x_2 at time t_2; (6)

$P(x_1, t_1 | x_2, t_2) =$ the conditional or transition probability that x_t has value x_2 at t_2 given that it had value x_1 at t_1. (7)

It can then be shown that $P(x_1, t_1; x_2, t_2)$ equals:

$$P(x_1, t_1)P(x_1, t_1 | x_2, t_2)\,. \tag{8}$$

From Eq. (8) one can in turn derive an expression relating the probability densities at different times:

$$P(x_2, t_2) = \int P(x_1, t_1) P(x_1, t_1 | x_2, t_2) dx_1 \,. \tag{9}$$

[An explanation and derivation of these equations can be found in any advanced or intermediate text on probability theory; for a reference see Billingsley[3] or DeGroot.[8]]

Whether or not the system is in state x_1 can be understood as resulting from two competing transition probabilities: the total probability for a transition out of state x_1 (into any of the other states which is possible for the system to be in) and the total probability that the system will change into state x_1 from any other of its possible states. Define $W(x_1, x_2)$ as the conditional probability density that the system changes from state x1 into state x_2 in the time interval $t_2 - t_1$. The partial time derivative of $P(x_2, t)$ is then given by the Master Equation:

$$\frac{\partial P(x_2 t}{\partial t} = \int [W(x_1, x_2) P(x_1, t) - W(x_2, x_1) P(x_2, t)] dx_1 . \tag{10}$$

The first term on the right-hand side of the Master Equation represents the transition into state x_2 from all other states, represented by x_1, while the second term on the right represents the transitions out of state x_2 into any other state x_1.

Although the meaning of the Master Equation can be considered intuitively clear, the equation is difficult to manipulate directly due to the presence of the integral. By making a series of plausible assumptions—such that x_t is a continuous variable whose changes take place in small jumps—and carrying out several far-from-obvious mathematical operations, a much more computationally tractable expression for the time differential of $P(x, t)$ is obtained:

$$\frac{\partial P(x_t)}{\partial t} = \frac{\partial}{\partial x} [-f(x_t) P(x_t) + \left(\frac{\sigma^2}{2}\right) \frac{\partial}{\partial x} g^2(x) P(x_t) \tag{11}$$

where $f(x_t), \sigma^2$, and $g^2(x)$ are the same as in Eqs. (3) and (4). Equation (11) is the Fokker-Planck Equation, an equation of motion for the probability distribution function of a state variable.

3. NOISE-INDUCED TRANSITION IN AN ECONOMIC MODEL

We now proceed to investigate the effects of noise on a specific economic model, proposed by Isard and Liossatos[16] to explain the process of economic agglomeration, that is, the geographical concentration of economic agents and activities. The model establishes the existence of a "social welfare function" of the form:

$$W(x, \alpha, \beta) = -\frac{x^4}{4} + \frac{\alpha x^2}{2} + \beta x + c \tag{12}$$

where the state variable, x, represents the population of an urban center. The rationale for the model can be given as follows. There are negative externalities associated with population growth: increased land rents, congestion, higher wages, pollution, etc. These "dispersion forces" are represented by the $-x^4/4$ term in Eq. (12). There are also positive externalities associated with the spatial concentration of economic activities: economies of scale, lower transportation costs, "knowledge-spillovers," etc. These "agglomeration forces," differing in their forcefulness and in their relation to city size, are represented by the $\alpha x^2/2$, βx and C terms. [For a review and discussion of the typically considered positive and negative externalities see Glaeser et al.[12]]

Taking the derivative with respect to x of Eq. (12) one obtains:

$$\frac{\partial w}{\partial x} = -x^3 + \alpha x + \beta\,. \tag{13}$$

The equilibrium solutions of Eq. (13) are the equilibrium states of:

$$\frac{dx}{dt} = x^3 + \alpha x + \beta \tag{14}$$

which is an ordinary differential equation (for purposes of our discussion the β term can be ignored). Notice that in Eq. (14) the control parameter associated with the positive externalities is a constant term. How realistic is this assumption? Let's consider one of the agglomeration forces, knowledge spillovers. By locating in close proximity to each other firms can learn from, copy or modify technological and organizational innovations, improvements, and inventions taking place in other firms. The process of innovation, itself related to the level and type of research carried out by firms, the type of personnel employed by the firm, and how the employees are organized to carry out their tasks, can hardly be considered a deterministic process. Coming up with a "good" idea is, after all, a good example of a stochastic phenomena. It would then seem more realistic to consider the control parameter α to be affected by stochastic fluctuations in the economic environment where the firms are located.

Replace α in Eq. (14) with:

$$\alpha_t = \alpha + \alpha\xi_t \tag{15}$$

where α is the time-average value of α_t. Equation (14) is then changed into:

$$\begin{aligned} dx_t &= \alpha x_t - (x_t)^3 dt + x_t \sigma \xi_t \\ &f(x_t) + g(x_t)\sigma\xi_t\,. \end{aligned} \tag{16}$$

The state variable x, representing urban population, has then become a stochastic variable, x_t.

The impact of noise on the behavior of x_t can also be studied by analyzing how the probability density of x_t changes with varying levels of noise. The FPE corresponding to Eq. (16) is:

$$\begin{aligned}\partial_t P(x,t) &= -\partial_x(\alpha x_t - (x_t)^3)P(x,t) + \left(\frac{\sigma^2}{2}\right)\partial_{xx}(x_t)^2 P(x,t) \\ &= \partial_x((x_t)^3 - \alpha x_t)P(x,t) + \left(\frac{\alpha^2}{2}\right)\partial_{xx}(x_t)^2 P(x,t)\,.\end{aligned} \tag{17}$$

As in standard stability analysis, one can calculate the form of the stationary probability density, P_s, by solving the above equation when $\partial_t P(x,t) = 0$. [The analysis in this section mirrors that found in Doering.[9]]

The calculation is as follows:

$$0 = ((x_t)^3 - \alpha x_t)P(x,t) + \left(\frac{\sigma^2}{2}\right)\partial_x(x_t)^2 P(x,t); \tag{18}$$

$$0 = \left(\frac{2}{\sigma^2}\right)\frac{x^3 - \alpha x}{x^2}(x^2 P_s) + \partial_x(x^2 P_s)\,. \tag{19}$$

Letting

$$\exp\left[\left(\frac{2}{\sigma^2}\right)\int_x \frac{(y^3 - \alpha y)}{y^2 dy}\right] = e^{\Phi(x)}\,, \tag{20}$$

then

$$0 = \Phi'(x)(x^2 P_s)e^{\Phi(x)} + e^{\Phi(x)}(x^2 P_s)' \tag{21}$$

where the prime sign indicates the derivative with respect to x. From Eq. (21) one gets:

$$0 = \frac{d}{dx}[e^{\Phi(x)}(x^2 P_s)\,. \tag{22}$$

An expression for P_s is then given by integrating Eq. (22):

$$P_s = Ne^{-\Phi(x)/x^2} \tag{23}$$

where N is just an integration constant. All that is left to do is to solve $\Phi(x)$ in order to get a more detailed function for P_s;

$$\Phi(x) = \left(\frac{2}{\sigma^2}\right)\int^x \left(\frac{y^3}{y^2}\right)dy - \left(\frac{2\alpha}{\sigma^2}\right)\int^x \left(\frac{1}{y}\right)dy\frac{x^2}{\sigma^2} - \left(\frac{2\alpha}{\sigma^2}\right)\ln x \tag{24}$$

So $e^{-\Phi(x)}$ then equals:

$$\exp\left[-\frac{x^2}{\alpha^2}\right]x^{2\alpha/\sigma 2} \tag{25}$$

and the stationary probability density therefore equals:

$$P_s = Nx^{2((\alpha/s2)-1)} \exp\left[-\frac{x^2}{\alpha^2}\right]. \tag{26}$$

Examining Eq. (26) it is easy to see that there are three different behavior regimes depending on whether the term $(\alpha/\sigma^2 - 1)$ is greater than, equal to, or less than, zero. This in turn depends on the magnitude of the noise amplitude, σ, relative to the control parameter α. The change in dynamical behavior precipitated by changes in the level of noise is an instance of a noise-induced transition. The behavioral implication of different levels of noise is clearly seen by examining the graphs of the stationary probability density for different values of σ^2 (Figures 1 - 6).

At small noise levels (Figures 1 and 2), the process fluctuates in a narrow region around its most probable value. When the magnitude of the noise is increased further, as in Figure 3, a qualitative change can be observed: not only is the expected value greater than that in the previous two figures, the probablity density is also flatter. There is another interesting and noticeable change in dynamics when the value of σ^2 approaches that of $\langle\alpha\rangle$. In Figure 4, with a noise amplitude of .9, there is a significantly higher probability for the state variable to have a value of zero. When $\sigma^2 = \langle\alpha\rangle = 1$, the zero value is the most probable value for the state variable (Figure 5). What happens if the noise is even stronger? When the fluctuations affecting the growth rate a are too strong, i.e., if $\sigma^2 \leq 2\langle\alpha\rangle$, the population (represented by the state variable x) eventually dies out, as shown on Figure 6.

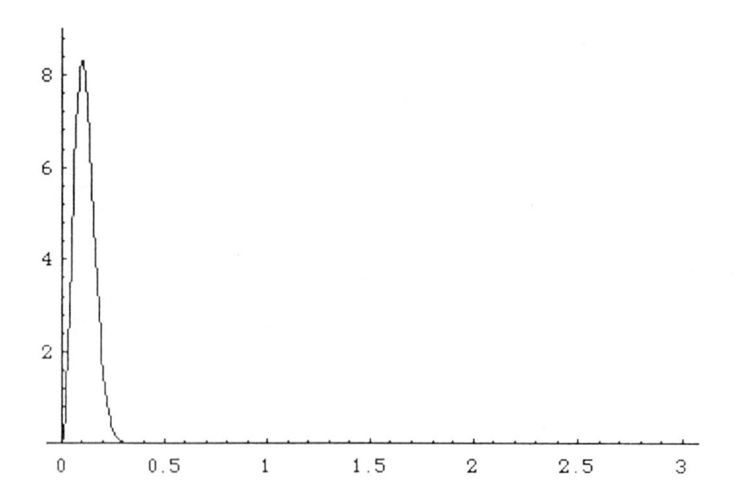

FIGURE 1 Stationary probbility density for $\sigma^2 = .01(\alpha = 1)$.

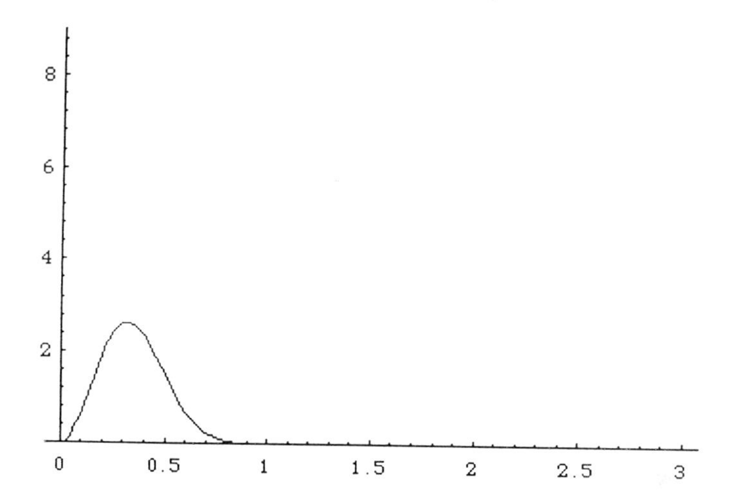

FIGURE 2 Stationary probbility density for $\sigma^2 = .1(\alpha = 1)$.

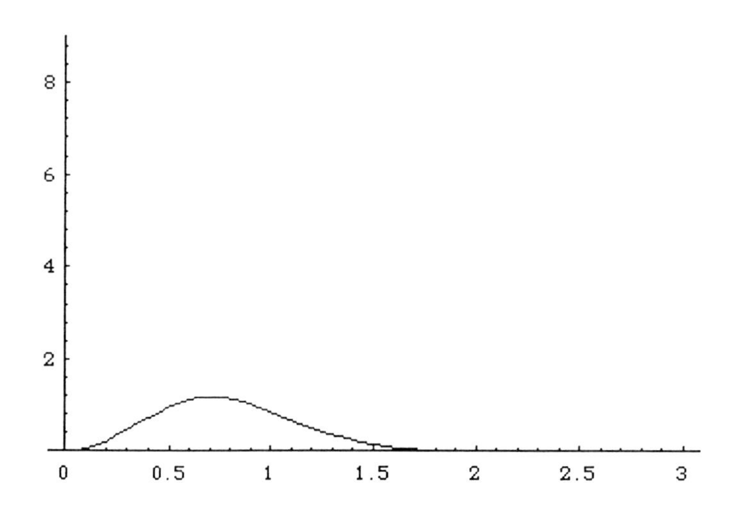

FIGURE 3 Stationary probbility density for $\sigma^2 = .5(\alpha = 1)$.

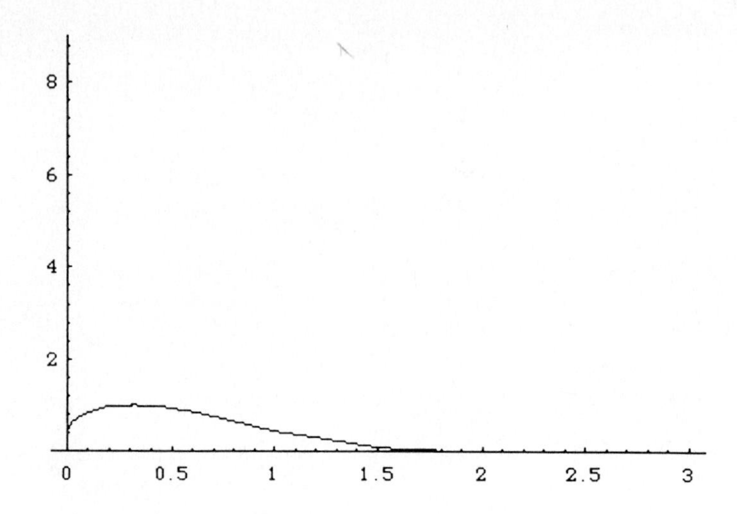

FIGURE 4 Stationary probbility density for $\sigma^2 = .5(\alpha = 1)$.

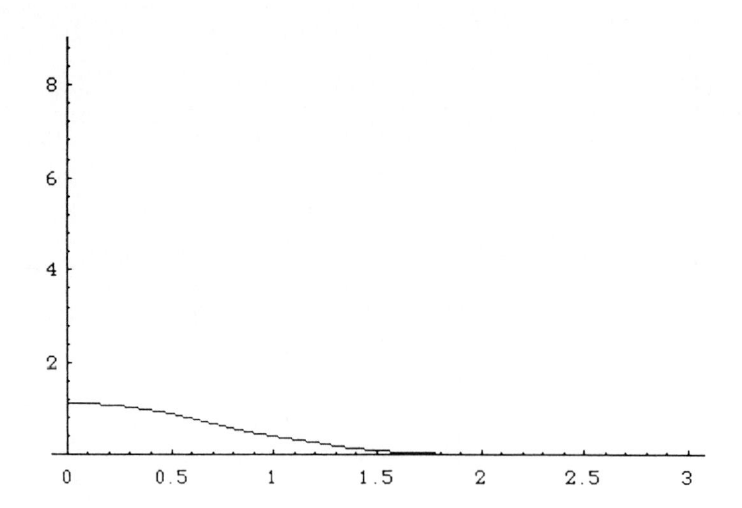

FIGURE 5 Stationary probbility density for $\sigma^2 = 1(\alpha = 1)$.

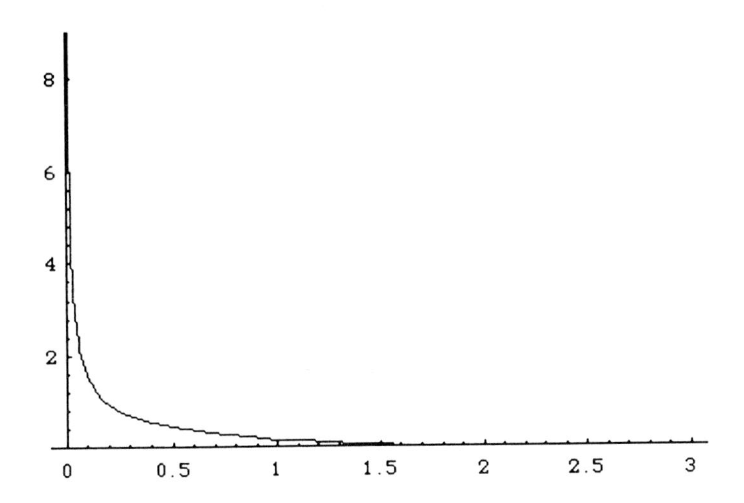

FIGURE 6 Stationary probbility density for $\sigma^2 = 1.5(\alpha = 1)$.

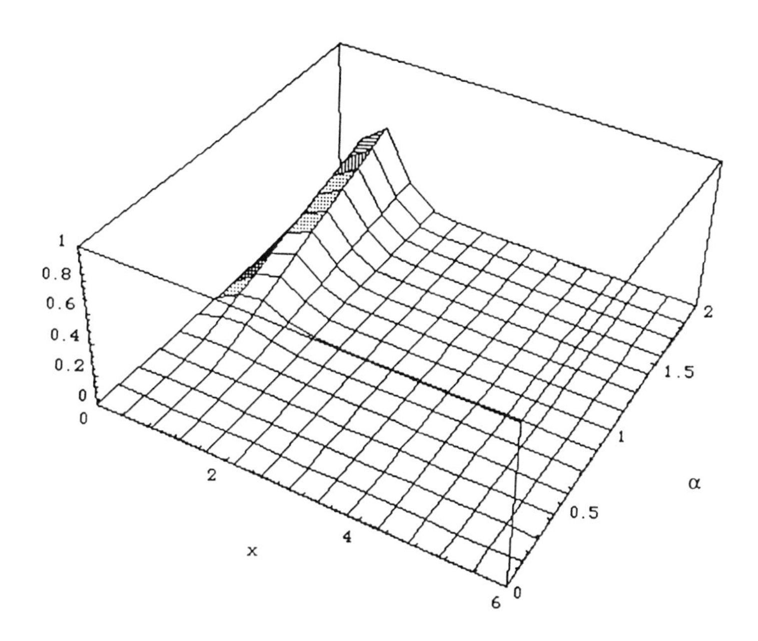

FIGURE 7 Stationary probability density for $\sigma^2 = 0.5$.

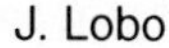

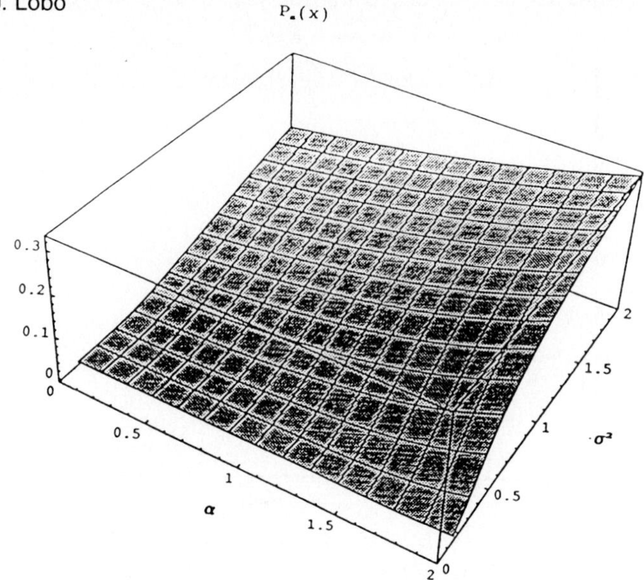

FIGURE 8 Stationary probability density for $\sigma^2 = 1$.

The effect of noise on the efficacy of the "agglomeration forces" is further examined in Figures 7 and 8. In both figures the stationary probability density is plotted against the control parameter and the state variable. Under "noisy" conditions the control (or growth) parameter must assume a high value in order to increase the probability that the state variable assumes a value greater than zero.

An economically plausible interpretation for the previous results can be readily provided: if the magnitude of "agglomeration forces" (represented by the control parameter α) fluctuate too wildly, economic agents find it very difficult to assess the benefits of locating near to other agents in a given geographic location. As a result, no spatial location is deemed to be economically more advantageous than any other, and the economic rationale for agglomeration thereby disappears. Under conditions of little noise, on the other hand, the possibilities for an economic system (in this case firms situated close to each other) to find optimally superior locations in configuration space are greatly reduced. The actual spatial pattern formed is then crucially dependent on the level of environmental noise.

4. WHAT IF NOISE DOES MATTER FOR ECONOMIC SYSTEMS?

In the previous section the possible dynamical effects of noise on economic behavior were explored by considering the consequences of introducing noise, in a meaningful and realistic manner, in a specific model. As shown, the presence of noise can significantly alter the behavior displayed by an economic system. One possible repercussion from this analysis is to cast doubt on the efficacy of econometric estimation and forecasting in the presence of noise. An economic model like the one presented in Eq. (12) is typically estimated using regression methods, with the estimated model then commonly used for forecasting purposes.[13] Yet if the noise-induced probability distribution function generating $\{x_t\}$ is like the ones depicted in Figures 4–6, econometric techniques will fail to produce reliable, unbiased estimates. The dynamical effects of noise, and their usual neglect in economic modeling, could be one of the reasons why economic forecasts tend to be inaccurate and fallacious.

The cautionary note about the consequences of ignoring the effects of noise when modelling economic phenomena can be more formally grounded by applying Landauer's "Blowtorch Theorem."[19,20] Landauer's message for the past 30 years has been that the occupation of different states of local stability along a potential field is not solely determined by the characteristics of the competing states, but also depends on the noise along the path connecting the competing states. By making it possible for the system to jump into a state of higher energy, noise allows a system to reside in a state that would otherwise have a lower probability of being occupied. This message can be of relevance to economic analysis, as economics is no stranger to potential fields (even if they don't go by that name).

An ordinary differential equation $dx/dt = f\lambda(x) = h(x) + \lambda g(x)$ can be given the alternative representation

$$\frac{dx}{dt} = -\partial_x V\lambda(x) \tag{27}$$

where $V\lambda(x)$, a potential, is equal to

$$-\int [h(x) + \lambda g(x)]dx\,. \tag{28}$$

In the case of the economic model discussed in Section 3, we have that $f\lambda(x) = \alpha x - x^3$, with $-x^3 = h(x)$ and $\alpha x = \lambda g(x)$. Then the potential is given by:

$$V\alpha(x) = -\int [\alpha x - x^3]dx \tag{29}$$

and $dx/dt = -\partial_x V_(x)$. It is easy to see how the maxima and minima of the function in Eq. (29) would be affected by stochastic fluctuations in the value of α. Stability analysis and the identification of maxima and minima are commonly carried out when the behavior of economic systems is studied as the solutions to ordinary differential equations.[4,28] The relative optimality (be it higher profits, lower costs, increased utility or raised consumption) of various possible equilibrium solutions is determined by the shape of functions such as Eq. (28). To the extent that stochastic fluctuations affect whether one or another equilibrium state is occupied, a description of economic dynamics is incomplete when it ignores noise.

REFERENCES

1. Barsky, R., and J.B. De Long. "Why Does the Stock Market Fluctuate?" *Qtr. J. Econ.* **CVIII** (1993): 291–311.
2. Berg, H. *Random Walks in Biology.* Princeton University Press, 1993.
3. Billingsley, P. *Probability and Measure.* John Wiley & Sons, 1986.
4. Blanchard, O. J., and S. Fischer. *Lectures on Macroeconomics.* Cambridge, MA: MIT Press, 1992.
5. Blume, L. "The Statistical Mechanics of Strategic Interaction." *Games & Economic Behavior* **5** (1993): 367–424.
6. Brockwell, P., and R. Davis. *Time-Series: Theory and Methods.* Berlin: Springer-Verlag, 1987.
7. Davis, S., and J. Haltinwanger. "Gross Job Creation, Gross Job Destruction and Employment Reallocation." *Qtr. J. Econ.* **CVII** (1992): 819–863.
8. DeGroot, M. *Probability and Statistics.* Reading, MA: Addison-Wesley, 1989.
9. Doering, C. "Modelling Complex Systems: Stochastic Processes, Stochastic Differential Equations and Fokker-Planck Equations." In *1990 Lectures in Complex Systems*, edited by L. Nadel and D. Stein, 3–52. Santa fe Institute Studies in the Sciences of Complexity, Lect. Vol. III. Reading, MA: Addison-Wesley, 1991.
10. Federal Reserve Bank of Chicago. *Stochastic Trends and Economic Fluctuations*, Working paper series WP-91-4, Research Department, Federal Reserve Bank of Chicago, IL, February, 1991.
11. Gard, T. C. *Introduction to Stochastic Differential Equations.* Marcel Dekker, 1988.
12. Glasser, E., H. Kallal, J. Scheinkman, and A. Shleifer. "Growth in Cities." *J. Pol. Econ.* **100** (1992): 1126–1152.
13. Granger, C. W. D., and P. Newbold. *Forecasting and Economic Time Series.* Academic Press, 1986.
14. Haken, H. *Synergetics.* Berlin: Springer-Verlag, 1978.
15. Horsthemke, W., and R. Lefever. *Noise-Induced Transitions.* Berlin: Springer-Verlag, 1984.
16. Isard, W., and P. Liossatos. *Spatial Dynamics and Optimal Space-Time Development.* North-Holland, 1979.
17. Johnston, J. *Econometric Methods.* McGraw-Hill, 1984.
18. Kauffman, S. *The Origins of Order.* Oxford University Press, 1993.
19. Landauer, R. "Motion Out of Noisy States." *J. Stat. Phys.* **53** (1988): 233–248.
20. Landauer, R. "Statistical Physics of Machinery: Forgotten Middle-Ground." *Physica A* **194** (1993): 551–562.
21. Lillien, D. M. "Sectoral Shifts and Cyclical Unemployment." *J. Pol. Econ.* **90** (1982): 770–793.

22. Long Jr., J. B., and C. I. Plosser. "Real Business Cycles." *J. Pol. Econ.* **91** (1983): 39–69.
23. Mackey, M. *Time's Arrow: The Origins of Thermodynamic Behavior.* Berlin: Springer-Verlag, 1992.
24. Nicolis, G., and I. Prigogine. *Self-Organization in Non-Equilibrium Systems.* John-Wiley & Sons, 1977.
25. Reichl, L. E. *A Modern Course in Statistical Physics.* University of Texas Press, 1980.
26. Rensig, L., ed. *Oscillations and Morphogenesis.* Marcel Dekker, 1993.
27. Risken, H. *The Fokker-Planck Equation.* Springer-Verlag, 1989.
28. Varian, H. R. *Microeconomic Analysis.* W. W. Norton, 1992.
29. Zarnowitz, V. *Business Cycles: Theory, History, Indicators and Forecasting.* University of Chicago Press, 1992.

Brett McDonnel
Department of Economics, Standford University, Standford, CA 94305
e-mail: brettmac@leland.stanford.edu

An Introductory Note on the Krugman Spatial Model

Economic geography seems a promising field for developing insights in the emerging (emergent?) Santa Fe approach to economics. This approach concentrates on how external economies (unpriced ways in which the activities of some agents affect the actions of others) and increasing returns (where doing more of an activity makes doing yet more increasingly attractive) lead to multiple equilibria, path dependence, and self-organization. At the Summer School a group of students interested in economics used a recent model by Paul Krugman[3,4] to explore these questions. Three contributions to this volume present various extensions and modifications of the Krugman model.[2,5] This note introduces a bare outline of Krugman's model as background for those chapters.

Krugman presents a simple model for how the concentration of population and production in one or a few cities might occur. He creates a mixture of increasing and decreasing returns. He assumes increasing returns to scale within the production of individual nonagricultural industries, and transportion costs create agglomeration effects because producers want to be near both consumers and suppliers. As the population becomes more concentrated in one place, the desire to be near these markets leads more producers to locate in that place, leading to a positive feedback cycle. On the other hand, the presence of an agricultural sector whose workers are assumed immobile creates a counter-incentive to spread out manufacturing to be near such people.

Formally, using the setup of Krugman,[3] agents have the utility functions

$$U - C_M^{\mu} C_A^{1-\mu} \tag{1}$$

where C_A is the amount of the agricultural good consumed, μ is the share of manufactured good in expenditure, and C_M is a composite index of the amount consumed of a number of manufactured goods, defined by

$$C_M = [\Sigma_i c_i^{(\sigma-l)/\sigma}]^{\sigma/(\sigma-l)} . \tag{2}$$

The amount produced of a particular good using a particular amount of labor is

$$L_{Mij} = a + \beta Q_{Mij} \tag{3}$$

where $L-Mij$ is the amount of labor used to make manufacturing good i at location j, Q_{Mij} is the corresponding amount of good i produced at location j, and α and β are parameters (a represents increasing returns to scale). Transportation costs are given by

$$z_{ijk} = e^{-\tau D(jk)} X_i jk \tag{4}$$

where X_{ijk} is the amount of good i shipped from j to k, and Z_{ijk} is the amount that arrives. The distance between j and k is $D(jk)$, and is the transportation cost parameter. Let l_j be the fraction of manufacturing workers who are at location j, and Ω_j be the real wage at that location, then the economy-wide average real wage is

$$\underline{\Omega} = \Sigma_j l_j, \Omega_j . \tag{5}$$

Krugman then calculates equilibrium price and quantity conditions.[3] The crucial conditions he arrives at, and that the following chapters use or modify, are as follows.

$$\frac{n_j}{n} = l_j . \tag{6}$$

This is because each product i is produced only in one location (due to increasing returns).

$$Y_j = (1-\mu)\phi_j + \mu l_j w_j , \tag{7}$$

where Y_j is total output in location j, ϕ_j is the share of the total farm labor force located at j, and w_j is the nominal manufacturing wage at j.

$$T_j = [\Sigma_k 1_k (w_k e^{\tau D(jk)})^{1-\sigma}]^{1/(1-\sigma)} . \tag{8}$$

T is a price index.

$$w_j = [\Sigma_k Y_k (T_k e^{-\tau D(jk)})^{\sigma-1}]^{1/\sigma} . \tag{9}$$

In Eq. (9), we give the instantaneous equilibrium market nominal wage. The real wage is then

$$\Omega_j = w_j . T_j^{-\mu} . \tag{10}$$

Equations (7) through (10) give the instanteous equilibrium wages that cause supply to equal demand in each product and labor market, given the current location of workers. Krugman assumes these markets clear, and then over a longer period of time workers move to higher real-wage locations according to the main equation of motion:

$$dl_j/overdt = rl_j(\Omega_j - \underline{\Omega}). \tag{11}$$

After such movements, product and labor markets clear again, and the problem interates until a steady state is achieved.

For the two location case, Krugman calculates parameter values for which multiple equilibria occur, and those for which concentration at one location occur. For the case of 12 locations in a circle, he searches for such parameters via computer simulations. Some important extensions clearly present themselves. One can vary the number of cities and/or the geometry of their location, test the sensitivity of Krugman's parameter values, and describe the transient patterns of movement towards equilibrium. Anderson discusses these and other issues.

A major concern of the Santa Fe group was how stochastic dynamics might affect both the quantitative and qualitative outcomes of such models.[1] Lobo explores this issue. One ultimate goal of this line of questioning is to explicitly model the individual agent adaptive behavior that underlies the equilibrium behavior of Eqs. (7) through (10) and the dynamic behavior of Eq. (11). Auerswald and Kim present work along these lines.

ACKNOWLEDGMENTS

The participants in the Summer School discussions were Eric Anderson, Philip Auerswald, Ann Bell, Jan Kim, James Kittock, Jose Lobo, David McCormack, and Arjendu Pattanayak. This note gives a little background to the work of that group. Mark Millonas was a major inspiration, and answered many questions about stochastic differential equations (see footnote 1) feedback cycle. On the other hand, the presence of an agricultural sector whose workers are assumed immobile creates a counterincentive to spread out manufacturing to be near such people.

[1]The work of Mark Millonas was a major inspiration for this question.[6]

REFERENCES

1. Anderson, Eric. "Numerical Investigations of the Krugman Population Model." This volume.
2. Auerswald, Philip, and Jan Kim. "Self-Organization in a Dynamic Spatial Model." This volume.
3. Krugman, Paul. "A Dynamic Spatial Model." Working paper, National Bureau of Economic Research, 1992.
4. Krugman, Paul. "Increasing Returns and Economic Geography." *J. Pol. Econ.* **99(3)** (1991): 483–499.
5. Lobo, Jose. "Does Noise Matter for Economic Systems?" This volume.
6. Millonas, Mark. "Swarms, Phase Transitions, and Collective Intelligence." In *Artificial Life III*, edited by Chris Langton. Santa Fe Institute Studies of the Sciences of Complexity, Proc. Vol. XVII, 417–446. Reading, MA: Addison-Wesley, 1993.

Arjendu K. Pattanayak* and Alfred Hübler**
*Prigogine Center for Statistical Mechanics and Complex Systems, Department of Physics, University of Texas, Austin, TX 78712
**Center for Complex Systems Research, Beckman Institute, and Physics Department, University of Illinois, Urbana, IL 61801

Cellular Automata with Changing Radii: A Model for Intraspecific Computation in Plant Populations

1. INTRODUCTION

Cellular Automata (CAs) have been used in various disciplines as discrete versions (in space *and* time) of nonlinear partial differential equations (see[12,2,14,5] and references therein). In particular, CAs have been used in modeling plant populations to study (a) intraspecific competition,[11] (b) the effect of the fire on forest mosaics,[4] (c) competition between annual and perennial plants,[1] and other such phenomena (see 5[11] for more references). With the usual caveats about possibly excessive simplification, CAs provide a powerful mode for numerical experimentation; they are especially useful in studying the effect of spatial patterns and varied initial conditions on the process and outcome of competition. In this preliminary study, we use CAs to model intraspecific competition in plants.

In a standard CA, there is a regular lattice of cells, and a defined range of states that each cell could have. In one dimension, these cells may be simply labeled by the integers. At a given integer time t, the nth cell has a well-defined state $S_n(t)$.

1993 Lectures in Complex Systems, Eds. L. Nadel and D. L. Stein,
SFI Studies in the Sciences of Complexity, Lect. Vol. VI, Addison-Wesley, 1995

In general, at time $(t+1)$, the new state $S_n(t+1)$ is a function of $S_n(t)$ and of $S_{n\pm i}(t)$ where the range of i defines the neighborhood for the nth cell. (The size of the neighborhood is commonly termed the radius of the CA.) To study plant populations, we may identify $S_n(t)$ with the height (size) of the nth plant.

In general CAs, the radius is predefined as part of the model. We believe that this inadequately models the asymmetry inherent in plant competition: bigger plants compete over larger neighborhoods. Therefore we introduce, in this study, the notion of a varying radius for CAs. The size of the neighborhood of the nth cell at time t is now determined by $S_n(t)$.

This, of course, opens up a wide range of dynamics to be explored, and effects to be studied, all of which are quite interesting. We restrict this chapter, however, to the study of a simple model and one well-defined problem. We consider the issue from the perspective of crop cultivation and explore the following question: given intraspecific competition, does one get a higher crop yield with regular or with random seeding (the difference to be defined more precisely later)?

The model is motivated and discussed in Section 2, the computer experiments are covered in Section 3, and we conclude by presenting the results with a short discussion in Section 4.

2. THE MODEL

There is an ongoing debate in the plant ecology literature (see[6,3,8,10,7] and references therein) about the so-called neighborhood models of competition: the various factors that influence the size of the individual plant over and above the overall density of cultivation. The model that we use translates into cellular automaton from some of these ideas, in particular: (a) that plants compete for resources or areas proportional to their size and (b) that the "competitive pressure" of neighborhood plants is due both to their density and size.

The size of the region over which the plant competes is taken to be linearly proportional to the height—in fact we take a symmetric neighborhood twice the height of the plant ($i = S_n(t)$ in the notation of the previous section). Further, the competition is modeled as a mean field effect: the mean size of the plants in the neighborhood is taken as a simplest measure of density *and* size. Assuming, further, a constant growth rate G in the absence of competition, we have the following equations (modulo boundary and initial conditions) for our CA:

$$S_n(t+1) = S_n(t) + G - \frac{c}{S_n(t)} \sum_{t=1}^{S(n)} [S_{n+i} + S_{n-i}] \tag{1}$$

where c is the competition parameter.

This model may be made more sophisticated in a variety of ways, of course notably by (a) modeling the growth rate as a nonlinear function of S and (b) assuming a distance-dependent function for the competition (for instance an inverse square dependence). There would thence be an infinite neighborhood.[13] A realistic model might also have statistically distributed values for G and c for different plants. This is material for future studies; we believe, however, that this model is qualitatively quite valid without these modifications.

We note here that a general mathematical analysis of Eq. (1) is quite difficult. It can be shown that since

$$f(x + j\Delta x) = \sum_{n}^{\infty} \frac{f^{(n)}(x)}{n!} j^n (\Delta x)^n , \tag{2}$$

where $f^{(n)}(x) = \partial^n f / \partial x^n$, we have that

$$\sum_{j=1}^{r} [f(x + j\Delta x) + f(x - j\Delta x) = \sum_{n}^{\infty} \sum_{j=1}^{r} 2j^{2n} (\Delta x)^{2n} \frac{f^{(2n)}(x)}{(2n)!} . \tag{3}$$

Since r is a function of $f(x)$ in Eq. (1), the coefficients of the various higher-order derivatives in the partial equation underlying this CA are functions of f. This is very like the results obtained by Schaffer and Leigh[9] in deriving the density of a given tree species in a forest while including the effects of spatial heterogeneity. Truncation at the second term, for instance, gives a generalized diffusion equation which itself is difficult to understand, in general.

We also note that this model permits "growth" to be negative; i.e., plants can be exposed to competition enough to decrease in (effective) size. We consider small fluctuations (both positive and negative) to be realistic and do not bar them in the dynamics. We do, however, construct "absorbing barrier" boundary conditions of the form:

a. if $S_n(t) = 0$, then $S_n(t+1) = 0$;

b. if $S_n(t+1) \leq 0$, then $S_n(t+1) = 0$;

c. if $S_n(t+1) \geq S_{\max}$, then $S_n(t+1) = S_{\max}$.

This provides that there are no "negative"-sized plants, that no dead plants are resuscitated and that they have a maximum size. (The last condition is often rendered unnecessary by the competitive dynamics, as we shall see.) We also wish to avoid large and rapid fluctuations in size as being unphysical (or unbiological!). To that end, we first need to establish the parameter range G and c. We also have to choose between periodic or finite-sized lattices. We shall see all this in the next section.

3. EXPERIMENTS

The first system we study is a periodic lattice with homogenous initial conditions. This ensures that the spatial behavior of the system is trivial. Some of the results are summarized in Table 1. We use the notation that

i. Spacing 0 means that all cells were seeded, i.e., had initial conditions $S_n(0) = 1$, spacing 1 means that alternate cells were seeded, i.e., we had initial conditions $S_{2n} = 0$ and $S_{2n+1}(0) = 0$, etc.

ii. $S = a, b, c$ is a temporal listing for any one of the seeded lattice sites (and hence valid for all of them). The unseeded lattice sites remain at 0, of course.

iii. Brackets (a, b) indicate a periodic orbit $S = a, b, a, b \ldots$.

TABLE 1 Summary of the results for a periodic lattice with homogenous seeding. See text for notation.

	Spacing	c value	G value	S summary
1	0	1	G	$S = (G)$
2	0	2	G	$S = (1, G-1)$
3	0	3	$G = 3$	$S = (1)$
4	0	3	G	$S = G-2, (0)$
5	0	4	$G = 1$	$S = (1)$
6	0	4	G	$S = G-3, (0)$
7	1	1	G	$S = (2G)$
8	1	2	G even	$S = (G+1)$
9	1	2	G odd	$S = (G+1, G+2)$
10	1	3	$G = 1$	$S = (0)$
11	1	3	$G = 2$	$S = (1, 3, 2)$
12	1	3	$G = 3$	$S = (1, 4)$
13	1	3	$G = 4$	$S = (2, 3, 4)$
14	1	3	$G = 5$	$S = (3, 5, 4)$
15	1	3	$G = 6$	$S = (4)$
16	1	3	$G = 7$	$S = (4, 5, 6)$
17	1	3	$G = 8$	$S = (5, 7, 6)$
18	1	3	$G = 9$	$S = (7)$
19	1	4	G even	$S = (0)$
20	1	4	G odd	$S = (1, G-1)$

iv. A value such as $2G$ is valid only when it is less than S_{max}, of course, else it is equal to S_{max}.

v. We do not indicate transient behavior in general.

We see first of all that large values of c, irrespective of G, lead to large fluctuations. For the rest of our study, therefore, we restrict considerations to $c \leq 3$. Smaller c values lead to fixed point or periodic behavior. We point out the obvious fact that given a discrete state space (quite apart from the discreteness in space and time) periodic behavior may indicate a fixed point in the physical system that the CA cannot access. This is especially clear if we look at the spacing 1 results.

We now consider the lattice without periodic boundary conditions to see the importance of edge effects. Typical runs are illustrated in Figures 1-3. The initial conditions are still homogenous; the importance of the nonlinear competition can be seen in the way that edge effects propagate (with a "speed" proportional to c).

```
6 0 6 0 6 0 6 0 6 0 6 0 6 0 6 0 6 0 6 0 6 0 6 0 6 0 6 0 6 0 6 0 6 0 6 0 6 0 6 0
8 0 5 0 5 0 2 0 2 0 2 0 2 0 2 0 2 0 2 0 2 0 2 0 2 0 2 0 2 0 2 0 5 0 5 0 8 0
10 0 7 0 7 0 4 0 4 0 4 0 4 0 4 0 4 0 4 0 4 0 4 0 4 0 4 0 4 0 4 0 7 0 7 0 10 0
10 0 9 0 6 0 3 0 3 0 3 0 3 0 3 0 3 0 3 0 3 0 3 0 3 0 3 0 3 0 3 0 6 0 9 0 10 0
10 0 10 0 5 0 5 0 5 0 5 0 5 0 5 0 5 0 5 0 5 0 5 0 5 0 5 0 5 0 5 0 5 0 5 0 10 0 10 0
10 0 10 0 1 0 4 0 4 0 4 0 4 0 4 0 4 0 4 0 4 0 4 0 4 0 4 0 4 0 4 0 4 0 1 0 10 0 10 0
10 0 10 0 6 0 3 0 6 0 3 0 3 0 3 0 3 0 3 0 3 0 3 0 3 0 3 0 3 0 6 0 3 0 6 0 10 0 10 0
10 0 10 0 5 0 2 0 5 0 5 0 5 0 5 0 5 0 5 0 5 0 5 0 5 0 5 0 5 0 5 0 2 0 5 0 10 0 10 0
10 0 10 0 4 0 1 0 7 0 7 0 4 0 4 0 4 0 4 0 4 0 4 0 4 0 4 0 7 0 7 0 1 0 4 0 10 0 10 0
10 0 10 0 0 0 6 0 6 0 9 0 3 0 3 0 3 0 3 0 3 0 3 0 3 0 3 0 9 0 6 0 6 0 0 0 10 0 10 0
10 0 10 0 0 0 2 0 5 0 10 0 2 0 5 0 5 0 5 0 5 0 5 0 5 0 2 0 10 0 5 0 2 0 0 0 10 0 10 0
10 0 10 0 0 0 4 0 7 0 9 0 0 0 4 0 7 0 4 0 4 0 7 0 4 0 0 0 9 0 7 0 4 0 0 0 10 0 10 0
10 0 10 0 0 0 0 0 9 0 8 0 0 0 3 0 6 0 3 0 3 0 6 0 3 0 0 0 8 0 9 0 0 0 0 0 10 0 10 0
10 0 10 0 0 0 0 0 8 0 10 0 0 0 5 0 8 0 5 0 5 0 8 0 5 0 0 0 10 0 8 0 0 0 0 0 0 10 0 10 0
10 0 10 0 0 0 0 0 7 0 9 0 0 0 4 0 7 0 4 0 4 0 7 0 4 0 0 0 9 0 7 0 0 0 0 0 10 0 10 0
10 0 10 0 0 0 0 0 9 0 10 0 0 0 3 0 6 0 3 0 3 0 6 0 3 0 0 0 10 0 9 0 0 0 0 0 0 10 0 10 0
10 0 10 0 0 0 0 0 8 0 9 0 0 0 5 0 5 0 5 0 5 0 5 0 5 0 0 0 9 0 8 0 0 0 0 0 10 0 10 0
10 0 10 0 0 0 0 0 7 0 10 0 0 0 7 0 7 0 4 0 4 0 7 0 7 0 0 0 10 0 7 0 0 0 0 0 0 10 0 10 0
10 0 10 0 0 0 0 0 9 0 9 0 0 0 6 0 6 0 0 0 0 0 6 0 6 0 0 0 9 0 9 0 0 0 0 0 10 0 10 0
10 0 10 0 0 0 0 0 8 0 10 0 0 0 5 0 8 0 0 0 0 0 8 0 5 0 0 0 10 0 8 0 0 0 0 0 0 10 0 10 0
10 0 10 0 0 0 0 0 7 0 9 0 0 0 7 0 7 0 0 0 0 0 7 0 7 0 0 0 9 0 7 0 0 0 0 0 10 0 10 0
10 0 10 0 0 0 0 0 9 0 10 0 0 0 9 0 9 0 0 0 0 0 9 0 9 0 0 0 10 0 9 0 0 0 0 0 0 10 0 10 0
10 0 10 0 0 0 0 0 8 0 9 0 0 0 8 0 8 0 0 0 0 0 8 0 8 0 0 0 9 0 8 0 0 0 0 0 10 0 10 0
10 0 10 0 0 0 0 0 7 0 10 0 0 0 7 0 7 0 0 0 0 0 7 0 7 0 0 0 10 0 7 0 0 0 0 0 0 10 0 10 0
10 0 10 0 0 0 0 0 9 0 9 0 0 0 9 0 9 0 0 0 0 0 9 0 9 0 0 0 9 0 9 0 0 0 0 0 10 0 10 0
10 0 10 0 0 0 0 0 8 0 8 0 0 0 8 0 8 0 0 0 0 0 8 0 8 0 0 0 8 0 8 0 0 0 0 0 10 0 10 0
10 0 10 0 0 0 0 0 7 0 7 0 0 0 7 0 7 0 0 0 0 0 7 0 7 0 0 0 7 0 7 0 0 0 0 0 10 0 10 0
10 0 10 0 0 0 0 0 9 0 9 0 0 0 9 0 9 0 0 0 0 0 9 0 9 0 0 0 9 0 9 0 0 0 0 0 10 0 10 0
10 0 10 0 0 0 0 0 8 0 8 0 0 0 8 0 8 0 0 0 0 0 8 0 8 0 0 0 8 0 8 0 0 0 0 0 10 0 10 0
10 0 10 0 0 0 0 0 7 0 7 0 0 0 7 0 7 0 0 0 0 0 7 0 7 0 0 0 7 0 7 0 0 0 0 0 10 0 10 0
10 0 10 0 0 0 0 0 9 0 9 0 0 0 9 0 9 0 0 0 0 0 9 0 9 0 0 0 9 0 9 0 0 0 0 0 10 0 10 0
```

FIGURE 1 Portion of a run with a finite lattice with homogenous seeding. $G = 5, c = 3$, Spacing $= 1$.

```
1 1 1 1 1 1 1 1 1 1 1 1 1 1 1 1 1 1 1 1 1 1 1 1 1 1 1 1 1 1 1 1 1 1 1 1 1 1 1 1


6 5 5 5 5 5 5 5 5 5 5 5 5 5 5 5 5 5 5 5 5 5 5 5 5 5 5 5 5 5 5 5 5 5 5 5 5 5 5 6
9 7 7 6 6 5 5 5 5 5 5 5 5 5 5 5 5 5 5 5 5 5 5 5 5 5 5 5 5 5 5 5 5 5 5 6 6 7 7 9
10 9 9 7 7 4 5 5 5 5 5 5 5 5 5 5 5 5 5 5 5 5 5 5 5 5 5 5 5 5 5 5 5 5 4 7 7 9 9 10
10 10 10 8 8 3 4 5 5 5 6 5 5 5 5 5 5 5 5 5 5 5 5 5 5 5 5 5 5 6 5 5 5 4 3 8 8 10 10 10
10 10 10 9 9 2 3 5 5 5 6 5 5 5 5 5 5 5 5 5 5 5 5 5 5 5 5 5 5 6 5 5 5 3 2 9 9 10 10 10
10 10 10 10 10 1 3 5 5 5 7 6 5 5 5 5 5 5 5 5 5 5 5 5 5 5 5 5 6 7 5 5 5 3 1 10 10 10 10 10
10 10 10 10 10 0 2 4 5 5 7 7 5 5 5 5 5 5 5 5 5 5 5 5 5 5 5 5 7 7 5 5 4 2 0 10 10 10 10 10
10 10 10 10 10 0 3 4 5 5 7 8 5 5 5 5 5 5 5 5 5 5 5 5 5 5 5 5 8 7 5 5 4 3 0 10 10 10 10 10
10 10 10 10 10 0 3 3 5 5 7 8 5 5 5 5 5 5 5 5 5 5 5 5 5 5 5 5 8 7 5 5 3 3 0 10 10 10 10 10
10 10 10 10 10 0 3 3 5 5 7 8 5 5 5 5 5 5 5 5 5 5 5 5 5 5 5 5 8 7 5 5 3 3 0 10 10 10 10 10
```

FIGURE 2 Portion of a run with a finite lattice with homogenous seeding. $G = 5, c = 3$, Spacing $= 0$.

```
1 1 1 1 1 1 1 1' 1 1 1 1 1 1 1 1 1 1 1 1 1 1 1 1 1 1 1 1 1 1 1 1 1 1 1 1 1 1 1 1


5 3 3 3 3 3 3 3 3 3 3 3 3 3 3 3 3 3 3 3 3 3 3 3 3 3 3 3 3 3 3 3 3 3 3 3 3 3 3 5
7 3 3 1 1 1 1 1 1 1 1 1 1 1 1 1 1 1 1 1 1 1 1 1 1 1 1 1 1 1 1 1 1 1 1 1 1 3 3 7
10 3 3 1 3 3 3 3 3 3 3 3 3 3 3 3 3 3 3 3 3 3 3 3 3 3 3 3 3 3 3 3 3 3 3 3 1 3 3 10
10 3 1 0 3 3 3 1 1 1 1 1 1 1 1 1 1 1 1 1 1 1 1 1 1 1 1 1 1 1 1 1 1 3 3 3 0 1 3 10
10 3 3 0 5 5 5 1 3 3 3 3 3 3 3 3 3 3 3 3 3 3 3 3 3 3 3 3 3 3 3 3 1 5 5 5 0 3 3 10
10 1 1 0 3 3 5 0 1 1 3 1 1 1 1 1 1 1 1 1 1 1 1 1 1 1 1 1 1 3 1 1 0 5 3 3 0 1 1 10
10 0 5 0 5 5 7 0 5 1 7 1 3 3 3 3 3 3 3 3 3 3 3 3 3 3 3 3 1 7 1 5 0 7 5 5 0 5 0 10
10 0 5 0 3 1 5 0 3 0 5 0 1 1 3 1 1 1 1 1 1 1 1 1 1 3 1 1 0 5 0 3 0 5 1 3 0 5 0 10
10 0 7 0 5 0 7 0 5 0 7 0 5 1 7 1 3 3 3 3 3 3 3 3 1 7 1 5 0 7 0 5 0 7 0 5 0 7 0 10
10 0 9 0 5 0 7 0 5 0 7 0 5 0 7 0 1 1 3 1 1 3 1 1 0 7 0 5 0 7 0 5 0 7 0 5 0 9 0 10
10 0 10 0 3 0 7 0 5 0 7 0 5 0 9 0 5 1 5 1 1 5 1 5 0 9 0 5 0 7 0 5 0 7 0 3 0 10 0 10
10 0 10 0 3 0 7 0 5 0 7 0 5 0 9 0 5 0 5 0 0 5 0 5 0 9 0 5 0 7 0 5 0 7 0 3 0 10 0 10
10 0 10 0 3 0 7 0 5 0 7 0 5 0 9 0 5 0 5 0 0 5 0 5 0 9 0 5 0 7 0 5 0 7 0 3 0 10 0 10
```

FIGURE 3 Portion of a run with a finite lattice with homogenous seeding. $G = 4, c = 2$, Spacing $= 0$.

At this point we introduce some randomness into the initial conditions. Each cell is now assigned an initial condition of 1 or 0 (i.e., is randomly seeded) with a probability p ($0 \leq p \leq 1$). In practice, we use a pseudo-random generator with a range $(0, 1)$ to generate a value ξ for each cell; if $\xi \leq p$ the cell is seeded, else it is left unseeded. This gives us some rather interesting space-time patterns as can be seen in Figures 4 and 5.

Finally, with this as our definition of random seeding, we put all the pieces together to study the yield of the system as a function of p. We do this for the periodic lattice in the main; edge effects are too strong for us to be able to interpret the finite-size system with confidence.

```
0 1 0 1 0 0 0 0 1 1 0 1 1 0 1 0 0 1 0 0 1 1 1 1 0 0 0 0 0 1 0 1 1 0 0 1 0 0 0 0

0 5 0 5 0 0 0 0 5 5 0 5 5 0 5 0 0 5 0 0 5 3 3 5 0 0 0 0 0 5 0 5 5 0 0 5 0 0 0 0
0 9 0 7 0 0 0 0 5 5 0 5 5 0 5 0 0 5 0 0 7 3 3 7 0 0 0 0 0 7 0 7 7 0 0 7 0 0 0 0
0 10 0 9 0 0 0 0 5 5 0 5 5 0 5 0 0 5 0 0 9 3 3 9 0 0 0 0 0 7 0 9 9 0 0 9 0 0 0 0
0 10 0 10 0 0 0 0 5 5 0 5 5 0 5 0 0 5 0 0 9 1 1 9 0 0 0 0 0 7 0 9 10 0 0 10 0 0 0 0
0 10 0 10 0 0 0 0 5 5 0 5 5 0 5 0 0 5 0 0 9 0 0 9 0 0 0 0 0 7 0 9 10 0 0 10 0 0 0 0
0 10 0 10 0 0 0 0 5 5 0 5 5 0 5 0 0 7 0 0 9 0 0 9 0 0 0 0 0 7 0 9 10 0 0 10 0 0 0 0
0 10 0 10 0 0 0 0 5 5 0 5 5 0 5 0 0 7 0 0 9 0 0 9 0 0 0 0 0 7 0 9 10 0 0 10 0 0 0 0
```

FIGURE 4 Portion of a run with a finite lattice and random seeding. $G = 4, c = 2, p = 0.5$.

```
0 1 0 0 0 1 1 0 1 0 1 0 1 1 1 1 1 0 1 1 1 0 1 1 0 1 1 1 0 1 0 1 1 0 1 1 0 1 0 1

0 6 0 0 0 6 6 0 6 0 6 0 6 4 4 4 6 0 6 4 6 0 6 6 0 6 4 6 0 6 0 6 6 0 6 6 0 6 0 6
0 9 0 0 0 7 7 0 7 0 5 0 5 3 1 3 5 0 5 1 5 0 5 5 0 5 3 5 0 5 0 5 5 0 5 5 0 7 0 9
0 10 0 0 0 8 8 0 8 0 4 0 6 2 0 2 6 0 6 0 6 0 6 6 0 6 2 4 0 6 0 6 4 0 4 4 0 8 0 10
0 10 0 0 0 9 9 0 9 0 3 0 5 3 0 3 7 0 7 0 7 0 5 5 0 5 3 3 0 5 0 5 5 0 5 1 0 9 0 10
0 10 0 0 0 10 10 0 10 0 4 0 6 2 0 2 8 0 8 0 8 0 4 4 0 6 2 4 0 6 0 6 6 0 4 2 0 10 0 10
0 10 0 0 0 10 10 0 9 0 3 0 5 3 0 3 7 0 9 0 9 0 3 3 0 7 3 3 0 7 0 5 7 0 3 1 0 10 0 10
0 10 0 0 0 10 10 0 10 0 4 0 6 2 0 0 8 0 10 0 10 0 2 2 0 6 2 4 0 8 0 6 6 0 2 4 0 10 0 10
0 10 0 0 0 10 10 0 9 0 3 0 5 5 0 0 9 0 10 0 10 0 1 3 0 7 3 3 0 9 0 5 5 0 3 1 0 10 0 10
0 10 0 0 0 10 10 0 10 0 2 0 6 4 0 0 10 0 10 0 10 0 4 2 0 6 2 2 0 10 0 6 4 0 2 4 0 10 0 10
0 10 0 0 0 10 10 0 9 0 0 0 5 5 0 0 9 0 10 0 10 0 3 3 0 5 3 0 0 9 0 5 5 0 3 1 0 10 0 10
0 10 0 0 0 10 10 0 10 0 0 0 6 4 0 0 10 0 10 0 10 0 2 2 0 6 4 0 0 10 0 6 6 0 2 4 0 10 0 10
0 10 0 0 0 10 10 0 9 0 0 0 5 5 0 0 9 0 10 0 10 0 1 3 0 7 5 0 0 9 0 5 5 0 3 1 0 10 0 10
0 10 0 0 0 10 10 0 10 0 0 0 6 4 0 0 10 0 10 0 10 0 4 2 0 6 6 0 0 10 0 6 6 0 2 4 0 10 0 10
```

FIGURE 5 Portion of a run with a finite lattice and random seeding. $G = 5, c = 2, p = 0.5$.

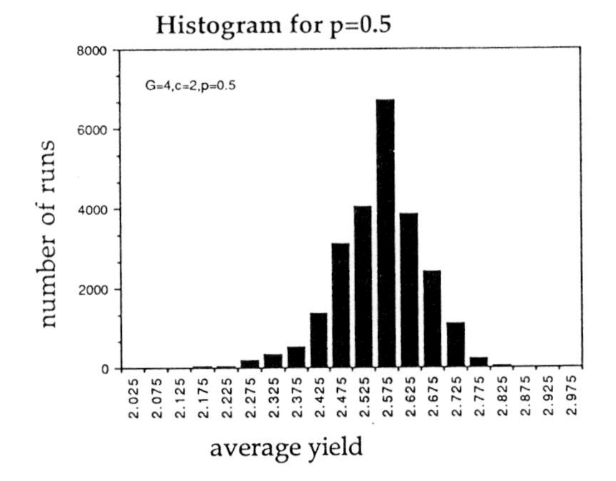

FIGURE 6 Probability distribution of average yield. $G = 4, c = 2, p = 0.5$.

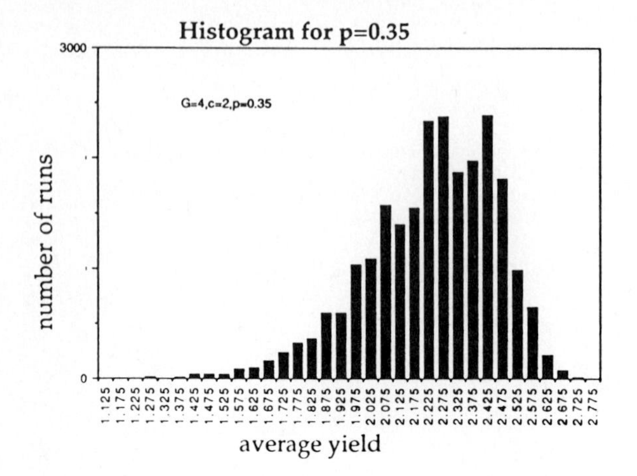

FIGURE 7 Probability distribution of average yield. $G = 4, c = 2, p = 0.35$.

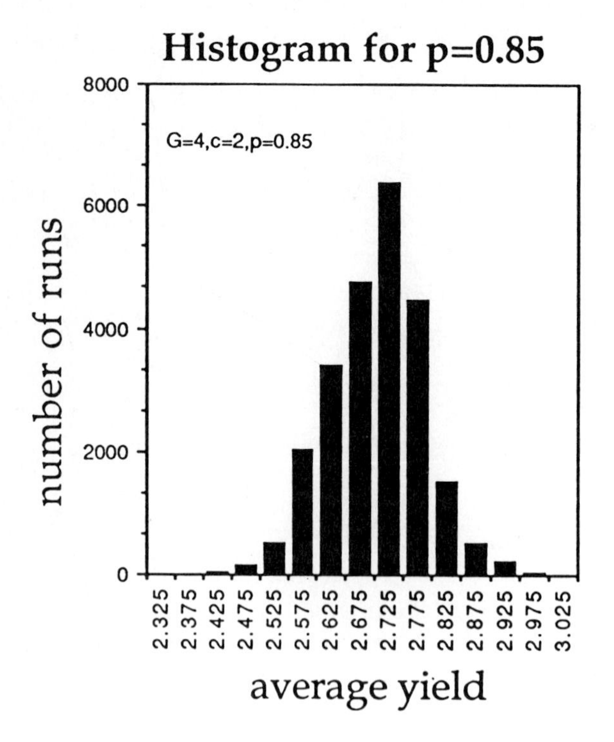

FIGURE 8 Probability distribution of average yield. $G = 4, c = 2, p = 0.85$.

We assume that the yield of a plant is linearly proportional to the size of the plant, and without loss of generality, the constant of proportionality is set to unity. Since the orbits rarely display fixed-point behavior, an estimate of the yield is derived by averaging over the last four time steps in a run of 80 time steps. Some of these results are displayed as histograms (Figures 6–8). These are for the representative case of $G = 4, c = 2$. The mean value for the yield (averaged over the histogram) has been obtained, then converted to the per-plant average by dividing the above number(s) by the density d. For the homogenous seeding case $d =$ (spacing $+1)^{-1}$ and for the probabilistic case $d = p$. This has been plotted as a function of p in Figure 9. The average yield corresponding to the homogenous seeding is shown, as is the maximum per-plant yield for a run at a given density.

DISCUSSION OF RESULTS

The lines in Figure 9 are drawn to guide the eye. Even though the lines for the mean yield with random and with homogeneous seeding are very similar, they do indicate a transition from favoring homogenous seeding to favoring random seeding. At low densities, evenly spaced plants have a higher yield than randomly scattered ones. This is intuitively satisfying; at low enough densities, there is essentially no competitive pressure. Random seeding can only hurt under these circumstances by bringing plants into competition. At higher densities, however, the plants that are randomly seeded have a higher yield. This may be attributed to the presence of effective boundaries generated by the randomness.

More important, however, is the third curve; this and the histograms indicate the presence of exceptional initial conditions amongst the "random" ones that greatly increase the per-plant yield. A careful analysis is required to isolate and characterize these initial conditions that exploit the asymmetry of competition.

Therefore, with the linear yield model, these studies suggest an advantage to "random" (nonhomogenous) seeding when the density is high enough to generate a significant competition. This is particularly true of certain initial configurations. This is not without practical significance: it challenges our standard image of crops cultivated in neat rows to maximize yield.

There are a lot of considerations that may modify this picture. Firstly, the study needs to be extended to two dimensions. Next, the yield might have to be modeled as a nonlinear function of plant size. Modifications in modeling competitive pressure (as mentioned in Section 2) have to be examined. Finally, of course, we have to check results carefully against experimental data from the field. We note, however, that though this particular question was not considered by them, our results are consistent with those of Miller and Weiner[6] and Firbank and Watkinson.[3]

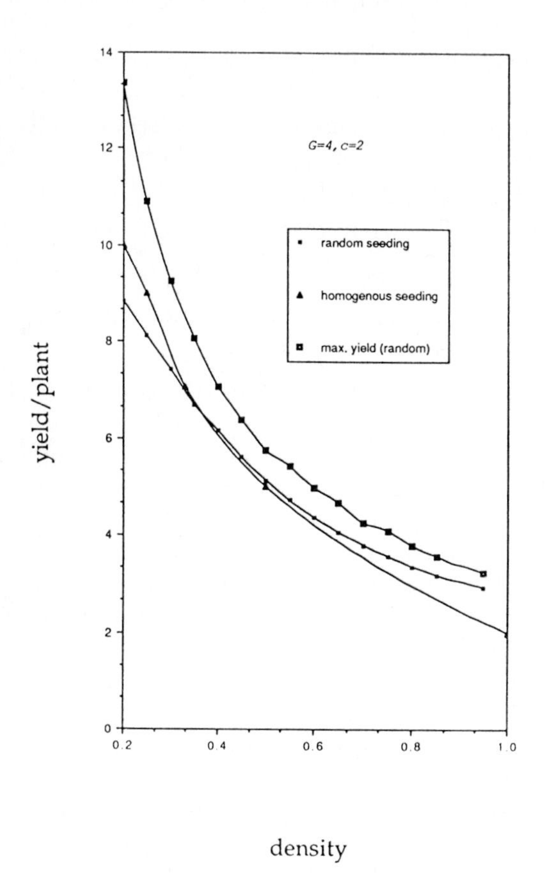

FIGURE 9 Per-plant yield as a function of density. We show the results for homogenous seeding, the mean yield over the histograms for random seeding, and the highest yield possible for random seeding.

In summary, therefore, the pattern of seeding, whether random, trivially periodic, or some more complicated structure, is very important in determining the yield from plants, independent of the density of cultivation. This is in part, at least, due to the asymmetry in intraspecific competition in plants: larger individuals compete over larger areas, and may usurp more of the available resources. We have introduced cellular automata with changing radii to model this behavior. Our preliminary studies, conducted in one dimension, would seem to indicate that the effect of nonhomogenous seeding is to increase yield, on the average. A more thorough investigation is required and constitutes work in progress.

ACKNOWLEDGMENTS

This work was begun and largely completed during the Sixth Annual Summer School on Complex Systems conducted by the Santa Fe Institute (May 30 – June 25, 1993). AKP would like to thank the SFI for the opportunity to attend the School, which led to this collaboration with Prof. A. Hübler, and the Robert A. Welch Foundation (Grant No. F-0365) for partial support during this work. He would also like to thank Clay Williams, Rajarshi Das, and Prof. W. C. Schieve for various forms of help related to this work.

REFERENCES

1. Crawley, M. J., and R. M. May. "Population Dynamics and Plant Community Structure: Competition Between Annuals and Perennials." *J. Theor. Biol.* **125** (1987): 475–489.
2. Farmer, D., T. Toffoli, and S. Wolfram, eds. *Cellular Automata: Proceedings of an Interdisciplinary Workshop*, Vol. 1, 2. Also in *Physica D* **10** (1984).
3. Firbank, L. G., and A. R. Watkinson. "On the Analysis of Competition at the Level of the Individual." *Oecologia* **71** (1987): 308–317.
4. Green, D. G. "Simulated Effects of Fire, Dispersal ans Spatial Pattern on Competition Within Forest Mosaics." *Vegetatio* **82** (1989): 138–153.
5. Gutowitz, H., ed. "Cellular Automata: Theory and Experiment." *Physica D* **45(1-3)** (1990). Proceedings of a workshop sponsored by CNLS, Los Alamos National Laboratory, Los Alamos, NM 87545.
6. Miller, T. E., and J. Weiner. "Local Density Variation May Mimic Effects of Asymmetric Competition in Plant Size Variability." *Ecology* **70** (1989): 1188–1191.
7. Mithen, R., J. L. Harper, and J. Weiner. "Growth and Mortality of Individual Plants as a Function of 'Available Area.'" *Oecologia* **62** (1984): 57–60.
8. Pacala, S. W., and J. A. Silander. "Neighborhood Models of Plant Population Dynamics I. Single-Species of Annuals." *Am. Nat.* **125** (1985): 385–411.
9. Schaffer, W. M., and E. G. Leigh. "The Prospective Role of Mathematical Theory in Plant Ecology." *Systematic Botany* **1** (1976): 209–232.
10. Silander, J. A., and S. W. Pacala. "Neighborhood Predictors of Plant Performance." *Oecologia* **66** (1985): 256–263.
11. Silvertown, J., S. Holtier, J. Johnson, and P. Dale. "Cellular Automata Models of Interspecific Competition for Space—The Effect of Pattern on Process." *J. Ecol.* **80** (1992): 527–534.
12. Wolfram, S. *Theory and Applications of Cellular Automata.* Singapore: World Scientific, 1986.

13. See Silander and Pacala[10] for a discussion that concludes that distance-dependent models are not necessarily superior to simple neighborhood density models.
14. *Cellular Automata and the Modeling of Complex Physical Systems.* Proceedings of the 1989 Les Houches Workshop. Berlin: Springer-Verlag, 1989.

S. L. Pepke
Department of Physics, University of California, Santa Barbara, CA 93106

Dynamical Models of Earthquake Faults and Forecasting in Complex Systems

Prediction in complex systems is a problem with many obvious and important applications: economic trends, weather patterns, species populations, and earthquakes are just a few examples. These systems have several properties in common which tend to make prediction a difficult task. Firstly, they appear highly nonlinear, and, in fact, chaotic, hence exhibiting sensitive dependence on the initial conditions. Secondly, they involve many interacting degrees of freedom which makes it difficult to analyze such problems using dynamical systems techniques to find attractors, calculate Lyapunov exponents, etc. Thirdly, they are far from equilibrium which limits the applicability of the methods of statistical mechanics—that can be used effectively in many systems that display the two features above—since these often rely upon equilibrium concepts such as temperature and free energy. The situation is made yet more difficult by the lack of controlled experiments which may be done to elucidate even the empirical features of the different problems. Thus one is often left with little idea of how inherently predictable a given system is and only precedent or intuition as guides for finding the best prediction methods.

Under such circumstances dynamical models of these systems can begin to play a new and pragmatic role in complex systems theory. It is important to understand that here the term modeling does not refer to data fitting or other statistical time series extrapolation techniques. Rather it refers to construction of a dynamical

1993 Lectures in Complex Systems, Eds. L. Nadel and D. L. Stein,
SFI Studies in the Sciences of Complexity, Lect. Vol. VI, Addison-Wesley, 1995

system based upon possible specific physical (biological, economic, etc. depending upon the system at hand) mechanisms or interactions at work within the system. Furthermore, the models we examine are *spatially extended* and therefore generate spatial as well as temporal information. Numerical simulations of the models can generate the equivalent of thousands of years of perfect data. This allows one to separate the effects of limited statistics from the actual physical limitations on predictability. Particular prediction methods can be tested under controlled circumstances. The long error-free catalogs allow for the critical study of algorithm optimization techniques. In addition, understanding of the physical properties of specific models may lead to innovative and more effective predictive methods.

As a specific example consider a Burridge-Knopoff model of an earthquake fault.[2,3] This model consists of a set of N massive blocks resting on a surface and connected to one another via coupling springs that represent linear elastic compressional forces along the earthquake fault. In addition, each block is coupled to a fixed point on an upper surface via leaf springs that represent linear elastic shear forces in the earth. The blocks are constrained to move along the lower surface and obey a stick-slip friction law. The dynamic friction is velocity-weakening which means that once the force due to both coupling and pulling springs on a block exceeds the static friction threshold it becomes unstuck and begins to slide, such that the rate of energy dissipation due to frictional sliding along the surface decreases with increasing block speed. The system is driven by pulling the lower plate very slowly relative to the top plate, thus loading the system via the fixed end leaf springs. Each block satisfies the following equation of motion:

$$m\ddot{X}_j = k_c(X_{j+1} - 2X_j + X_{j-1}) - k_pX_j - F(\nu + X_j)$$

where X_j is the displacement of the jth block from its equilibrium position, k_c and k_p are the spring constants for the coupling and pulling (leaf) springs, respectively, ν is the pulling speed, and F denotes the frictional force. The geometry of a one-dimensional system is illustrated in Figure 1. Once a block begins to slide the resultant change in force on neighbor blocks through the coupling springs may cause those blocks to become unstuck as well. Thus a single block's motion may trigger an event that ruptures large portions of the model fault. In addition, the pulling speed is taken to be extremely small so that events do not overlap in time.

In the earth, convective forces below the surface drive tectonic plates into and past one another. The resulting strain rate appears constant in magnitude on a human time scale so that one expects the series of dramatic rupture events observed in the crust to be the result of inter-plate frictional forces. It is for this reason that the Burridge-Knopoff (BK) model is interpreted as a physical model of the fault dynamics. While the effects of local properties such as geometry or rock porosity may ultimately play important roles as well, a sufficient understanding of the basic stick-slip dynamics and, in particular, the consequences for intermediate or

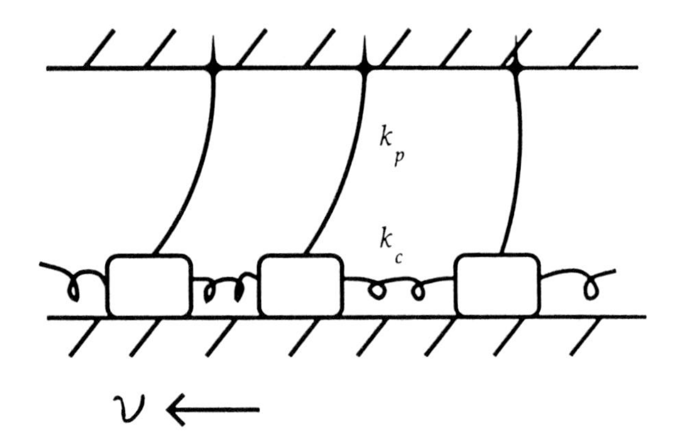

FIGURE 1 Schematic drawing of a Burridge-Knopoff model of an earthquake fault. Massive blocks interact with one another via coupling springs, representing linear elastic compressional forces in the earth, which obey Hooke's law with force constant k_c. Additionally, each block interacts with the pictured upper side of the fault via pulling springs, which represent linear elastic shear forces, characterized by force constant k_p. The blocks rest on one side of the fault, motionless relative to one another, initially. The system is driven by pulling one side of the fault at a very slow speed ν until the combined force on a block due to the coupling and pulling springs overcome a ststic friction threshold. Once a block begins sliding, the frictional force it feels decreases with increasing sliding speed.

long-term prediction is still lacking. In the BK model, the velocity-weakening form of the friction law is essential in order for the model to sustain a complex series of events over long periods of time; hence, it is reasonable to expect that predictability in the earth will be influenced by the physics of dynamic friction as well.

Given any nonuniform initial condition for the block positions, it is found that after an initial transient period (typically tens of thousands of sliding events) the system reaches a statistically stationary state described by $R(\mu) = Ae^{-b}\mu$ where $R(\mu)$ is the rate of occurrence per unit fault length of events of magnitude μ and μ is the logarithm of total displacement along the fault during the slipping time. Over a wide range of model parameters, the BK model yields $b = 1$ for small- and intermediate-sized events while the largest events satisfy the relation with $b = -2/3$ in one dimension. The exponent for small events is the same in two-dimensions; however, the large events exponent is somewhat larger (closer to zero).

Power-law scaling as a function of event size is characteristic of systems that exhibit "self-organized criticality," a conceptual framework proposed by P. Bak, C. Tang, and K. Wiesenfeld while the BK model is not critical (there is a sharp cutoff in the frequency distribution for events larger than a characteristic size), like SOC systems intended to encompass a broad class of driven, dissipative, spatially extended dynamical systems.[1] In such cases the frequency of occurrence of small events is significantly greater than that of larger events, so it would be of obvious benefit to develop prediction schemes for large events by detecting correlations between the statistics of the small events and the subsequent occurrence of large events, i.e., by detecting *precursory phenomena* in the small event statistics.

Algorithms utilizing this idea were applied with some apparent success to real earthquake data by a group led by V. I. Keilis-Borok of the International Institute of Earthquake Prediction Theory and Mathematical Geophysics in Moscow.[6,7] In these algorithms the earth is divided into overlapping spatial regions (circles) and various seismic phenomena are observed within each region and within sliding time windows. These include functions such as the number of aftershocks following small- and intermediate-sized events, rate of change of seismic activity, and spatial clustering. Potential precursory patterns are identified from analysis of catalogs from many different fault regions around the world. The algorithm may be used for forward prediction by continuously monitoring the precursor functions in each region. A precursor function is interpreted as indicating a coming large event when the function value exceeds some threshold that has been specified based upon the patterns observed in the training set of catalogs. If several of the precursor functions independently indicate an impending large event, a "Time of Increased Probability" or TIP alert is declared within the region. If a large earthquake then occurs before the end of some designated time period, a successful prediction is counted.

These algorithms were reported to have correctly predicted eighty percent of the target earthquakes when TIPs occupied about twenty percent of the total observation time, yet they were a subject of much controversy for several reasons. In particular, there were questions as to how rigorously the algorithms had been tested and whether they could perform significantly better than standard long-term prediction schemes based upon estimates of repeat times of large events for local regions in the earth.[4,5] In the earth the case is difficult to decide due to the incompleteness and relatively short length of seismic catalogs available. Because of this, a group at UCSB including Jean Carlson, James Langer, Bruce Shaw, and myself began prediction studies at UCSB by embarking on a careful analysis of the performance of similar algorithms on the BBK model.[11,9] The general problem is this: given the history of an earthquake fault as recorded in seismicity catalogs or as can be deduced from other measurable quantities over some time do there exist algorithms which can discern the relevant precursory phenomena sufficiently accurately to make reliable predictions about the immediate future (the next 1–5 years)? This statement of the problem assumes that precursory phenomena on relatively short time scales actually do exist. This question is not yet answered for the earth, but can be answered in the affirmative for the BK model. Figure 2 illustrates a portion of a catalog (the horizontal axis is space and the vertical axis is time) generated by the one-dimensional model. The horizontal lines represent the areas ruptured during earthquakes while X's mark the epicenters for great events. There is a clear increase in small-scale activity near the future spatial location of a large event epicenter and just prior to it in time. This premonitory activity makes the BK model a particularly appealing one for the studies of intermediate-term prediction methods.

Based upon this, the BK model is clearly somewhat predictable but how particularly well the algorithms do depends upon how one evaluates their performance. Initially, we studied single precursor functions only so that each function was evaluated separately. When this is done (using simple activity-based precursors), one can predict most of the large events while alarms occupy only a tiny fraction of the total observation time but only if one is willing to tolerate a large number of false alarms. When the number of false alarms is used as one of the performance criteria, for most precursor functions it becomes desirable to leave alarms on for longer periods of time, corresponding to decades in the earth. In this case one is

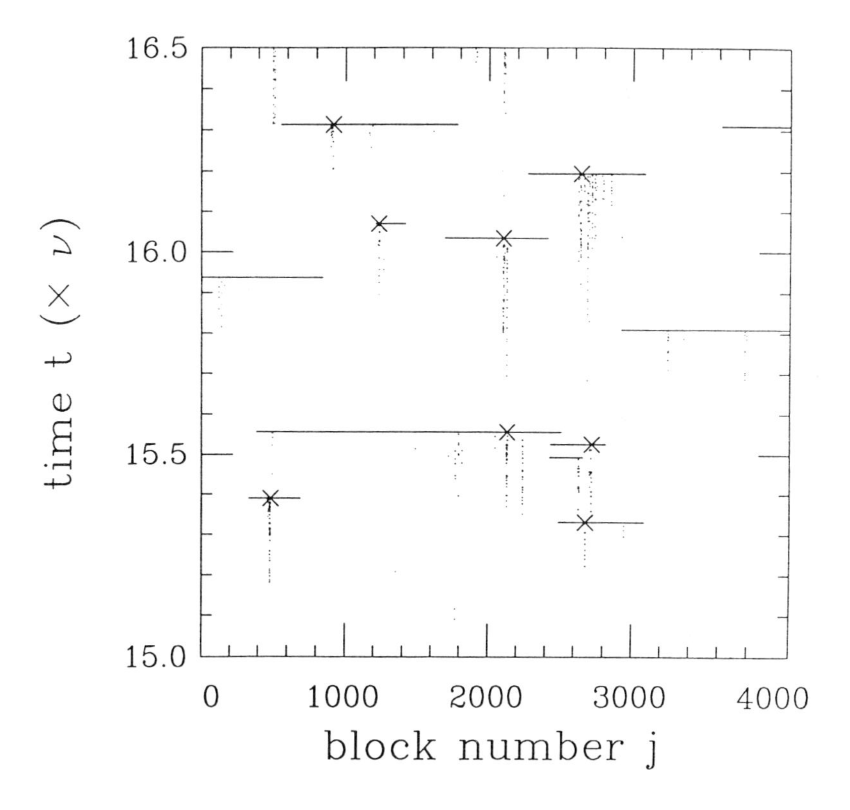

FIGURE 2 Time series of events along a portion of a one-dimensional Burridge-Knopoff model fault. A line is drawn through each block which slipped in a given event and X's mark the positions of large event epicenters. A notable feature here is the precursory activity localized in space and time near the future site of a large event.

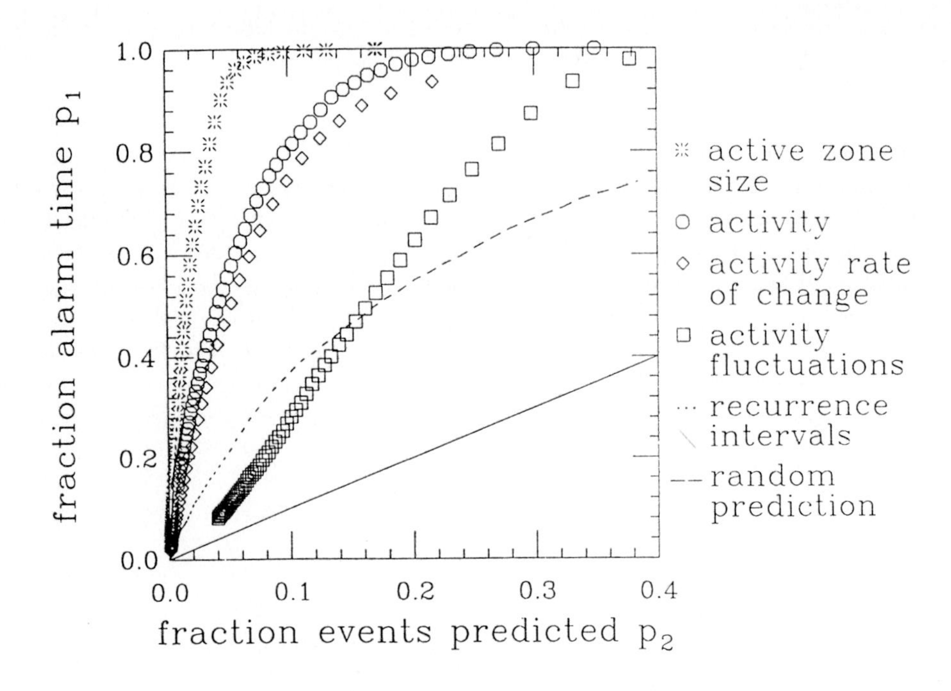

FIGURE 3 Typical predictability results for intermediate-term precursor functions calculated from a Burridge-Knopoff model catalog. All intermediate-term precursors outperform random prediction (solid line) and prediction based on recurrence intervals (dashed line). Active zone size clearly outperforms the other precursors, predicting nearly all of the large events when alarms occupy roughly 5% of the total time.

not doing any better than more conventional long-term techniques based upon the distribution of repeat intervals for the large events. The fraction of large events predicted as a function of the fraction of time occupied by alarms for several different measures is shown in Figure 3. Interestingly, we find that the precursor function which allows the most reliable predictions (fewest false alarms) on an acceptable time scale (corresponding to a few years) is not seismic activity, but a function we call active zone size.

The active zone size function is based upon the physics of the threshold dynamics which govern the BK model behavior. In the BK model small events relieve very little of the local stress. Hence the occurrence of a small event is a direct signal that the force at that location is close to the threshold for slipping. If many of the blocks in some local region are close to threshold, then an event which begins within the region generates a pulse which may "snowball" as it propagates, the potential energy stored in the springs of the blocks near threshold turning into the kinetic energy of the pulse once the blocks begin sliding. If the near-threshold region is long enough, the pulse may acquire sufficient energy to penetrate more stuck regions and a great event is observed. Thus the active zone size is taken to be the number of blocks which have slipped in small events, regardless of how many times each has slipped.

Since active zone size is a measure that performs well on the BK model because it is based on physical knowledge of the dynamics, one might well ask how generalizable the positive predictability results are. Generally, seismic activity within a region may be thought of as measuring temporal correlations, while active zone size is a more direct measure of spatial correlations (as well as temporal). Bak, Tang, and Wiesenfeld's SOC was an attempt to link the development of spatial and temporal correlations in complex dynamical systems, primarily in the simplest sense with larger events corresponding to longer event lifetimes. With the problem of prediction, we examine the role of spatial and temporal scales not only within events themselves but between events as well. Thus one may distinguish between dynamical systems that behave as a superposition of many random fluctuators and those that exhibit dynamical organization in the sense of events on one space or time scale being correlated with the occurrence of events on another scale. In a system of the second type the coupling of events on different scales this way leads to the breaking of time reversal invariance in the system which should be able to be exploited for prediction purposes.[8] Further, the extent to which a complex system such as the earth's fault network is predictable may be determined by the coupling strength between the localized discrete modes represented by individual events. The significance of spatial correlations for predictability in the BK model has led us to study related measures in different fault models and the results look promising.[10] It will be fascinating to explore these issues further in the contexts of other models that exhibit power-law spectra and in the natural systems as well.

Another issue which we have considered in some detail is the ergodicity of the seismic record. For the purpose of prediction algorithm optimization, on real systems it is useful to assume that optimizing over short catalogs from many different spatial regions (fault zones) is equivalent to optimizing over a single zone record for a very long time (much longer than the average great event cycle time). This is the ergodic assumption. To study this problem, we consider an ensemble (borrowing just a little terminology from equilibrium statistical mechanics) of short catalogs on the BK model fault. The prediction algorithm was optimized with respect to the seismic activity precursor function alarm threshold using this ensemble. Here performance is evaluated in terms of a quality function Q taken to be simply $Q = p_1 - p_2 - p_3$ where p_1 is the fraction of large events successfully predicted, while p_2 is the fraction of time occupied by alarms and p_3 is the fraction of alarms issued which turned out to be false. The factor p_1 has a positive coefficient since it represents a benefit while p_2 and p_3 have negative coefficients because they correspond to costs. Optimization with respect to the activity threshold was performed by choosing the threshold which yielded the maximum average Q, denoted $\bar{Q}_{\mathbf{max}}$. Stability of the results was examined by calculating the Q of the algorithm for each of the catalogs in the training set and thence the mean and standard deviation $\sigma^2(Q) = (1/N)\sum_{i=1}^{N}(Q_i - \bar{Q}_{\max})^2$ of this quantity. The catalogs were then lengthened and the procedure was repeated.

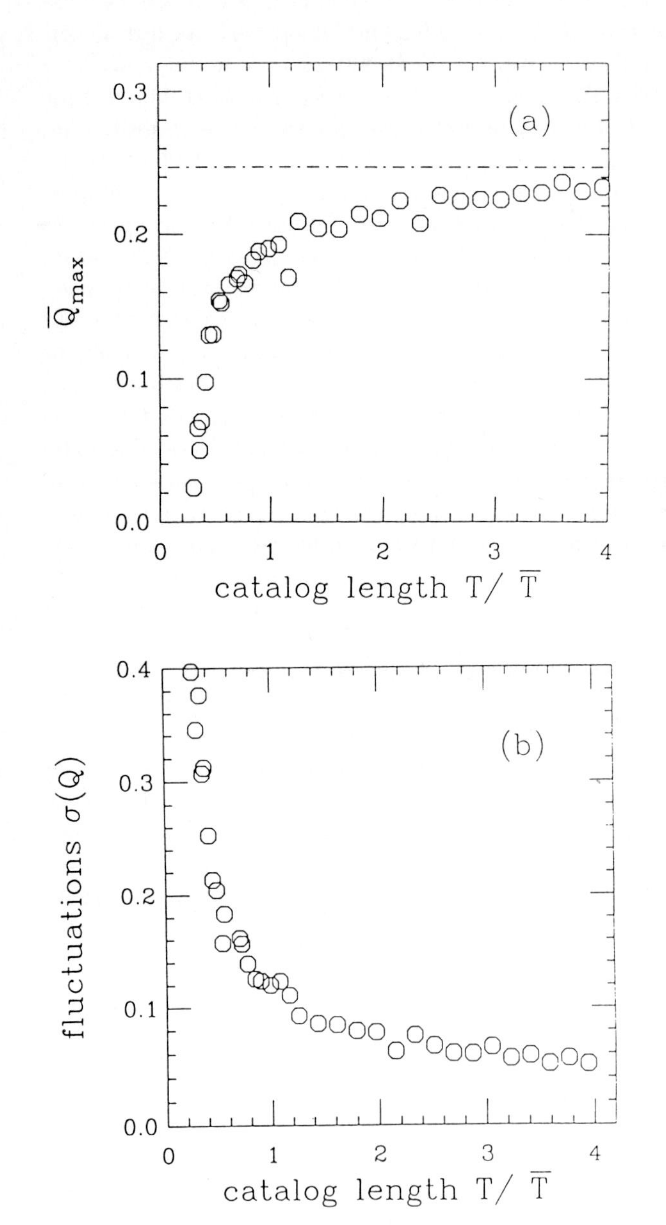

FIGURE 4 The effects of restricted catalog length (space and time windows are fixed). For each catalog length 100 catalogs were combined and optimized with respect to the activity threshold so that the threshold for declaration of an alarm was set by the maximum average $\bar{Q}_{\max}$. Above is plotted this average value as a function of catalog length and also the width of the distribution of Q values (for the independent catalogs with the threshold fixed) as a function of catalog length. The mean converges to its optimal value and the width approaches zero once the catalog length becomes roughly of the order of the mean recurrence interval.

The results are shown in Figure 4(a) where the mean $\bar{Q}_{\max}$ for the ensemble is plotted as a function of catalog length normalized by the mean repeat interval for large events. Figure 4(b) shows the standard deviation as a function of normalized catalog length. It is clear that the mean performance increases and fluctuations in the algorithm performance decrease rapidly until the catalogs are nearly as long as the average time interval between successive great events. This suggests that the stability and reliability of the intermediate-term prediction algorithms may be improved significantly as more data is acquired. This is an important issue since current seismic catalogs contain only about 30 years of data while a typical time interval between great events may be hundreds of years.

In summary, dynamical models provide significant insight into the problem of predictability in complex systems. While no model can be expected to mimic all details of a natural system's behavior, the relative simplicity is beneficial in allowing one to study the effects on predictability of just one or a few types of interactions in a context in which they may be understood. Our studies at UCSB on models of earthquake faults have yielded information about the role of correlations in spatially extended systems that may lead to better and more consistent prediction algorithms. Further studies on a variety of models should increase our knowledge of how to use this to predict well in the most general case. The relative time scale over which temporal correlations are important has been seen to be short enough in the BK mechanical model to allow for substantial increases in algorithm effectiveness as the length of reliable activity catalogs increases. This suggests that while the best algorithms may be insufficient right now, in a matter of a decade or two their performance may be greatly improved. These are just a couple of the interesting results to be had from studies of predictability on dynamical models and yet by themselves demonstrate the merit of such efforts and provide justification for further research in this area.

ACKNOWLEDGMENTS

I would like to thank my collaborators on this work at UCSB, Jean Carlson and Bruce Shaw. In addition, we profited greatly from discussions with V. Keilis-Borok, A. Gabrielov, D. Turcotte, J. Dieterich, G. Swindle, and especially J. S. Langer. The work supported by an INCOR grant from CNLS as Los Alamos National Laboratories, the David and Lucille Packard Foundation, and NSF grant DMR-9212396.

REFERENCES

1. Bak, P., C. Tang, and K. Wiesenfeld. "Self-Organized Criticality: An Explanation of l/f Noise." *Phys. Rev. Lett.* **59** (1987): 381–384.
2. Burridge, R., and L. Knopoff. "Model and Theoretical Seismicity." *Bull. Seismol. Soc. Am.* **57** (1967): 3411–3471.
3. Carlson, J. M., and J. S. Langer. "Mechanical Model of an Earthquake Fault." *Phys. Rev. A* **40** (1989): 6470.
4. Dieterich, J. H. "An Alternate Null Hypothesis." *U.S. Geol. Sur. Open File Rep.* (1992): 88–398.
5. Healy, J. H., V. G. Kossobokov, and J. W. Dewey. "A Test to Evaluate the Earthquake Prediction Algorithm M8." *U. S. Geo. Sur. Open File Rep.* (1992): 92–401.
6. Keilis-Borok, V. I., and I. M. Rotwain. "Diagnosis of Time of Increased Probability of Strong Earthquakes in Different Regions of the World: Algorithm CN." *Phys. Earth Planet. Inter.* **61** (1990): 57–72.
7. Keilis-Borok, V. I., and V. G. Kossobokov. "Premonitory Activation of Earthquake Flow: Algorithm M8." *Phys. Earth Planet. Inter.* **61** (1990): 73–83.
8. O'Brien, K. P, and M. B. Weissman. "Statistical Signatures of Self Organization." *Phys. Rev. A* **46** (1992): R4475–R4478.
9. Pepke, S. L., J. M. Carlson, and B. E. Shaw. "Prediction of Large Events on a Dynamical Model of a Fault." *J. Geophys. Res.* **99** (1994): 6769.
10. Pepke, S. L., and J. M. Carlson. "Predictability of Self-Organizing Systems." *Phys. Rev. E* **40** (1994): 236.
11. Shaw, B. E., J. M. Carlson, and J. S. Langer. "Patterns of Seismic Activity Preceding Large Earthquakes. *J. Geophys. Rev.* **97** (1992): 479.

William Sulis
Departments of Psychiatry* and Psychology,** McMaster University, Hamilton, Ontario, Canada
*Assistant Clinical Professor
**Associate Member

Driven Cellular Automata

The introduction of driving terms into the dynamics of cellular automata results in substantial alterations in the dynamics of the automaton, particularly the introduction of aperiodicity, disruption of attractor basins, and development of synchronization. These ideas are studied through an exhaustive simulation of all two-state, three-neighbor cellular automata on periodic lattices.

1. INTRODUCTION

Cellular automata constitute the prototypical example of a complex system. They have been extensively simulated and have served as exemplars for several fundamental concepts of complex systems theory: self-organization, edge of chaos, emergent computation. Yet cellular automata have never been seriously studied in connection with driving forces or temporal inputs. Virtually every study has observed only autonomous behavior. This is remarkable given the striking differences which have

1993 Lectures in Complex Systems, Eds. L. Nadel and D. L. Stein,
SFI Studies in the Sciences of Complexity, Lect. Vol. VI, Addison-Wesley, 1995

been observed in other complex systems, for example the harmonic oscillator, under autonomous and driven dynamics.

Wolfram[10] recognized the importance of this difference in his list of twenty problems for the study of cellular automata. His Problem 11 states "How are cellular automata affected by noise and other imperfections?" Disordered cellular automata have been studied.[11] In this automata, different lattice sites are permitted the use of differing rules (inhomogeneity). Still, inputs are not considered. It is possible that the lack of attention paid to the effects of inputs and noise is due, in part, to the extensive focus on cellular automata as models of self-organization and emergence, where autonomous function is a defining characteristic of the phenomena under consideration.

Considerable attention has been paid to cellular automata as an example of a complex system capable of emergent computation. Forrest[2] has formally defined emergent computation as follows: there exists (1) a collection of agents, each following explicit instructions; (2) interactions among the agents (according to the instructions) which form implicit global patterns at the macroscopic level, i.e., epiphenomena; and (3) a natural interpretation of the epiphenomena as computations. Langton[4] has presented evidence to suggest that cellular automata are most likely to display emergent computation if they operate at the "edge of chaos" as defined by lambda parameter values around 0.35.

The notion of emergent computation is intriguing for the study of mind, particularly in the context of illness such as Alzheimer's disease. Here the question is to understand the process of psychological dysfunction which develops as neuronal numbers decline. A knowledge of the local conditions required to support emergent computation at the system level may provide insight into methods to sustain psychological function in dementia, or, at the very least, prevent its further degradation through misapplied psychological or pharmacological interventions.

In order to address such a question, it is necessary to consider emergent computation in the context of external inputs since such inputs play a powerful organizing role in psychological processes. Cellular automata represent perhaps the simplest examples of emergent computation and are certainly among the simplest complex systems to simulate. Although they bear little relationship to neuronal structures, cellular automata are simple systems and provide a useful benchmark against which to study more complicated structures.

As presently formulated, the concept of emergent computation rests very much in the eye of the beholder. One observes the presence of complex patterns which are held as presumptive evidence of computational capability. However, such an approach fails when considering cellular automata with input. Even small inputs over time into a cellular automata tend to disrupt and randomize the observed patterns (see Figures 3, 4, 5, and 6). As a result, simply observing pattern complexity fails to inform about the computational ability of the cellular automaton.

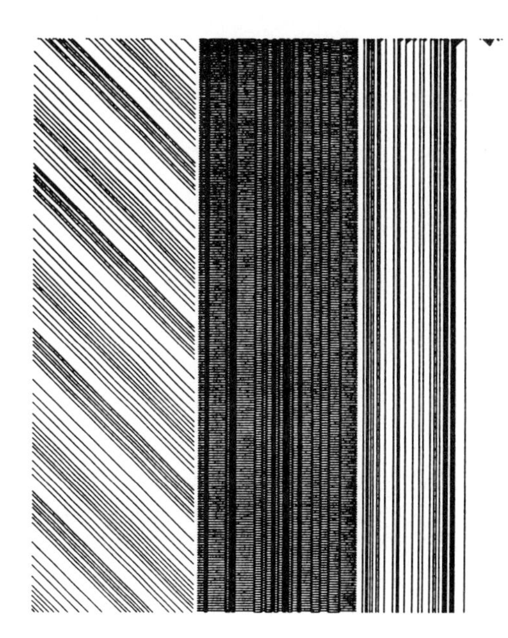

FIGURE 1
Representatives of symmetry classes. Left to Right: Diagonal (Rule 24), Vertical (Rule 123) Fixed (Rule 140), Uniform (Rule 96).

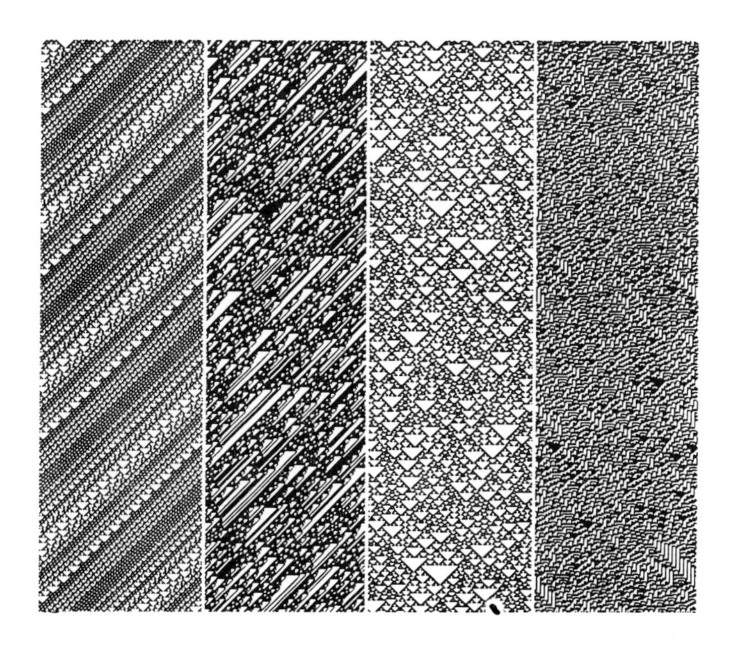

FIGURE 2
Representatives of symmetry classes. Left to Right: Complex (Rule 26), Complex (Rule 106), Chaotic (Rule 22), Chaotic (Rule 45).

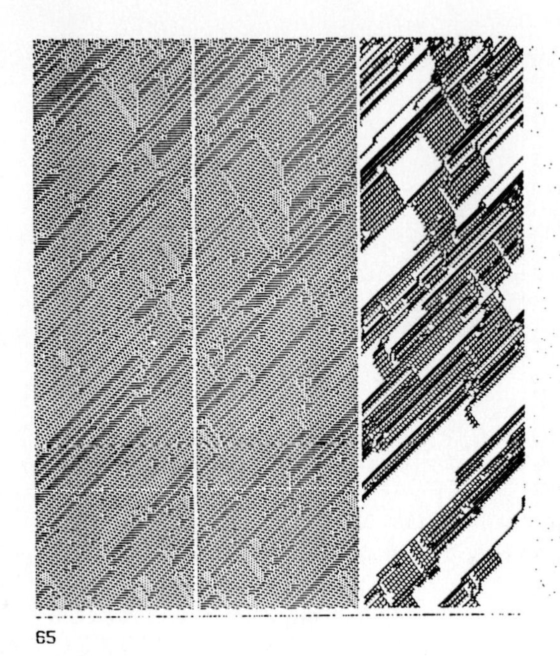

FIGURE 3 Synchronization: Diagonal (Rule 65). Left to Right: Run 1, Run State 2. Discordance Between Runs 1 and 2, Input.

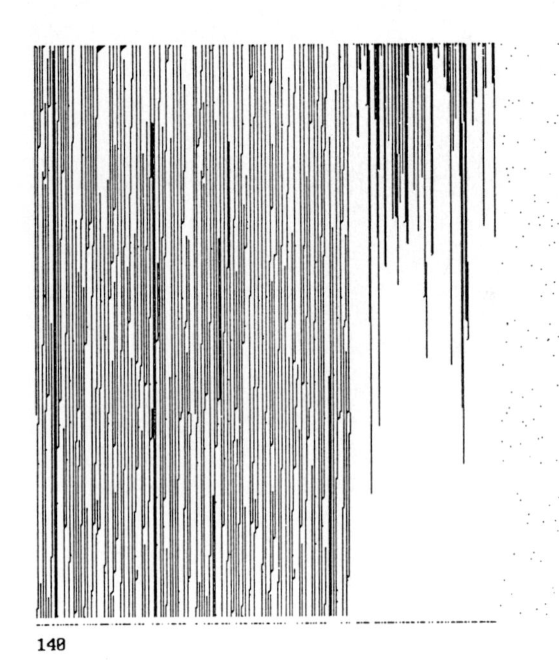

FIGURE 4 Synchronization: Fixed (Rule 140). Left to Right: Run 1, Run 2. Discordance Between Runs 1 and 2, Input.

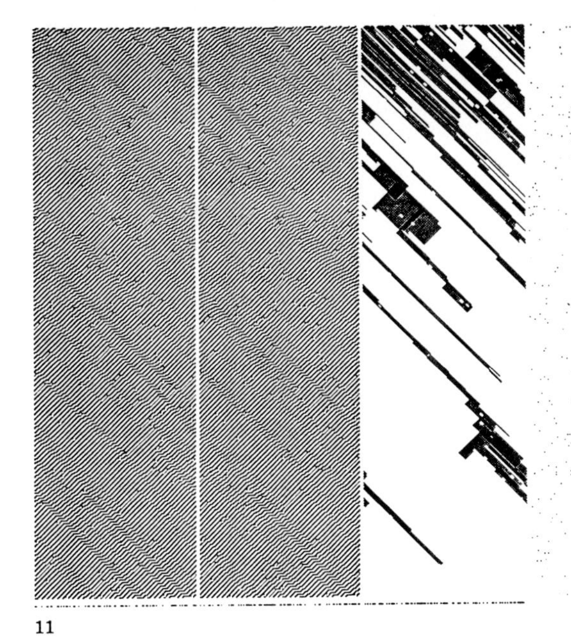

FIGURE 5
Synchronization: Complex (Rule 11). Left to Right: Run 1, Run 2. Discordance Between Runs 1 and 2, Input.

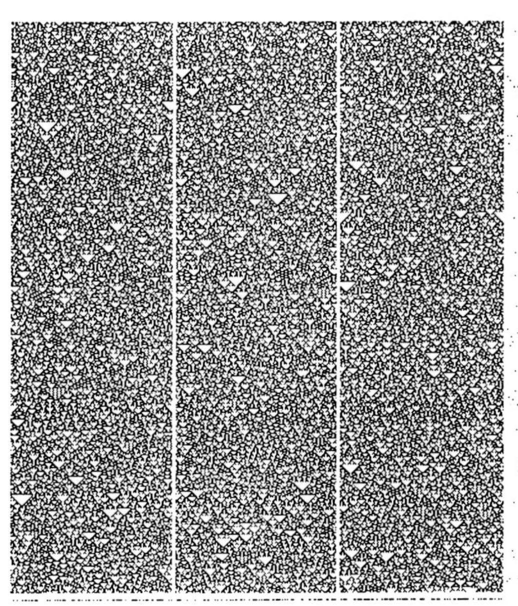

FIGURE 6
Synchronization: Chaotic (Rule 90). Left to Right: Run 1, Run 2. Discordance Between Runs 1 and 2, Input.

For this study a different approach was taken. A more objective marker was sought which could serve as an indicator of the possibility of computational behavior under inputs. That such behavior exists is not guaranteed. However, the presence of the marker indicates a place to search. The particular marker chosen has been described variably in several papers (dynamical memory,[5] dynamical coupling,[6,7,8] synchronization[9]) but the notion of synchronization seems to describe it the best.

Consider a cellular automaton A and two different initial states x, y. Given a time t let $A(x,t)$ denote the state of automaton at time t where $A(x,0) = x$. Clearly $A(x,t+1) = A(A(x,t),1)$. An input is a function from a time set (finite or infinite) to the state space of the automaton. The duration of the input $D(i)$ is just the supremum of the time set. Given an input $i(t)$, the state of the automaton at time t is denoted $A(x,i(t),t)$. Note that $A(x,i(t+1),t+1) = A(A(x,i(t),t),i(t+1),1)$. Let $d(,)$ denote the Hamming distance metric on the cellular automaton state space. The input i synchronizes A at x, y if

$$\lim_{t \to D(i)} d(A(x,i(t),t), \quad A(y,i(t),t)) = 0\,.$$

In particular one is interested in the case where $D(i)$ is finite and $\lim_{t \to D(i)} d(A(x,i(t),t), A(y,i(t),t)) = 0$ since this corresponds to the situation which must pertain in any realistic case.

It should be clear that if $d(A(x,i(t),t), A(y,i(t),t)) = 0$ at time t then $d(A(x,i(t'),t'), A(y,i(t'),t')) = 0$ for all $t' > t$.

Synchronization in cellular automata represents a form of dynamical coupling of responses to a given input. This coupling provides a foundation upon which computation can be based. This is discussed briefly in the next section. It thus serves as a simple marker for potential computational behavior. Its advantage is that it is easy to measure: simply simulate the automaton under an input from two different states and check for matches in states as time progresses.

2. COMPUTATIONAL COMPETENCE AND DYNAMICAL AUTOMATA

In this section I wish to briefly present ideas about computational competence and motivate the assertion that synchronization provides a reasonable marker for competence. This section is sketchy. The interested reader is referred to the original papers[6,7,8] for more detail.

The assertion of emergent computation requires the detection of computational competence, that is, the ability to carry out a specific computation. This requires having a specific computation in mind prior to demanding its performance by the system. This may be impossible given an arbitrary system. However, the nascent study of computational competence provides some general behavioral constraints which help to simplify this task.

Computational competence has not been addressed in the standard literature. Churchland et al.[1] formulated a loose version of this notion but did not proceed any further with it. A more formal definition of computational competence has recently been formulated[7,8] using the language of dynamical automata. Dynamical automata offer a means of discussing complex systems in relevant automata theoretic terms, making it easier to address computational questions. The definition is long but straightforward.

2.1 DYNAMICAL AUTOMATA

DEFINITION **2.1** Let T be a totally ordered group with identity 0. For $a, b \in T, a \leq b$ define $[a, b) = \{x | a \leq x < b\}$ if $a \neq b$ and $[a, a) = \emptyset$. Define $[a, \infty) = \{b | a \leq b\}$. Define $T^* = \{[0, b) | b \in T \cup \{\infty\}\}$. Define an operation on T^* as follows:

We say $[0, b)[0, d) =$

1. $[0, b + d)$ if $b, d \in T$
2. $[0, \infty)$ if $b = \infty$ or $d = \infty$. With this operation T^* is a monoid.

DEFINITION **2.2** Let X be a set and T a totally ordered monoid. We define $T^*(X) = \{f | p \overset{f}{\mapsto} X \text{ for } p \in T^*\}$. That is, $T^*(X)$ consists of functions from elements of T^*, that is, intervals on T, to X. Given $f, g \in T^*(X)$, we may define an operation of concateantion fg as follows: Suppose that $[0, b) \overset{f}{\mapsto} X$ and $[0, d) \overset{g}{\mapsto} X$. Then $[0, b)[0, d) \overset{fg}{\mapsto} X$ and

$$fg(x) = \begin{cases} f(x) & \text{if } x \in [0, b) \\ g(x - b) & \text{if } x \geq b \end{cases}.$$

On $[0, 0)$ we define the empty function and denote it as 0. $T^*(X)$ is a monoid. A monoid is said to be (T)-historical if it is a submonoid of $T^*(X)$.

DEFINITION **2.3** Let M be a (T)-historical monoid. Define a map $M \overset{\mu}{\mapsto} T$ as follows: for any $p \in M$ there exists an element $[0, b) \in T^*$ such that $[0, b) \overset{p}{\mapsto} X$. Define $\mu p = b$. We call μ the duration of p.

DEFINITION **2.4** A dynamical automaton is a 4-tuple (S, E, Δ, T) where T is a totally ordered group, S, E are (T)-historical monoids and Δ is a transition function $S \times \Delta E \overset{\Delta}{\mapsto} S$ satisfying the following conditions:

1. $\Delta(0, \rho) = 0$
2. $\Delta(\psi, \rho) = \psi\psi'$
3. $\mu\psi' = \mu\rho$
4. $\Delta(\psi, 0) = \psi$
5. $\Delta(\psi, \rho\rho') = \Delta(\Delta(\psi, \rho), \rho')$

for all $\psi \in S, \rho, \rho' \in E$. A dynamical automaton with output (S, E, Δ, T, f) is a dynamical automaton (S, E, Δ, T) together with a map $S \overset{f}{\mapsto} E$ such that

1. $\mu f(\psi) = \mu\psi$
2. $f(\psi\psi') = f(\psi)f(\psi')$

2.2 COMPUTATIONAL COMPETENCE

The definition of computational competence is a straightforward generalization of the prototypical psychological experiment. A question is asked and the response measured against an expected standard.

DEFINITION **2.5** Let $\mathcal{A} = (S, E, \Delta, T)$ be a dynamical automaton. Two states $\psi, \psi' \in S$ are dynamically equivalent if for any $\rho \in E$ we have $\Delta(\psi, \rho) = \psi\bar{\psi}$ and $\Delta(\psi', \rho) = \psi'\bar{\psi}$.

DEFINITION **2.6** Let $\mathcal{A} = (S, E, \Delta, T, f)$ denote a dynamical automaton with output. For $\psi \in S$ and $\eta \in E$ we define $\bar{\Delta}(\psi, \eta) = \psi'$ such that $\Delta(\psi, \eta) = \psi\psi'$. If $W \subset S$ and $\bar{E} \subset E$, define $\bar{\Delta}(W, \bar{E}) = \{\psi' | \Delta(\psi, \eta) = \psi\psi' \text{ where } \psi \in W, \eta \in \bar{E}\}$. If ψ, ψ' are dynamically equivalent, then $\bar{\Delta}(\psi, \eta) = \bar{\Delta}(\psi', \eta)$ for all $\eta \in E$. If W is a set of dynamically equivalent states, then $\Delta(W, \bar{E}) = W\bar{\Delta}(W, \bar{E})$.

DEFINITION **2.7** Let $\mathcal{A} = (S, E, \Delta, T, f)$ be a dynamical automaton with output. An information event is a pair (Q, R) where $Q, R \subset E$. Q is the question, R the response. An instance of an information process $\mathcal{P}$ is finite sequence of information events. Given a set $M = \{(Q_i, R_i)\}$ of information events, let $\mathcal{M}$ denote the set of all instances formed from elements of M. $\mathcal{M}$ is just the free monoid over M under concatenation. An information process $\mathcal{P}$ is a subset of $\mathcal{M}$. Therefore $\mathcal{P}$ is a language, an instance is a word and an event is a letter.[3]

DEFINITION **2.8** Let $\mathcal{A} = (S, E, \Delta, T, f)$ be a dynamical automaton with output. Let (Q, R) be an information event. A subset $\bar{S} \subset S$ is said to support (Q, R) if

1. For any $\psi \in \bar{S}, \bar{\Delta}(\psi, Q) \subset f^{-1}(R)$.
2. $\Delta(\bar{S}, Q)$ has nonempty interior.

DEFINITION **2.9** Let $\mathcal{A} = (S, E, \delta, T, f)$ be a dynamical automaton with output, $W = \{(Q_i, R_i) | i \in I\}$ an instance of an information process, and $\bar{E} \subset E$. Then $\mathcal{A}$ supports W in $\bar{E}$ if there exists a subset $\bar{S} \subset S$ such that $\bar{S}$ supports (Q_1, R_1) and for all $j > 1$ we have that $\Delta(\bar{S}, Q_1\bar{E} \ldots Q_{j-1}\bar{E})$ supports (Q_j, R_j).

DEFINITION **2.10** Let $\mathcal{A} = (S, E, \Delta, T, f)$ be a dynamical automaton with output. Let $\mathcal{P}$ be an information process. We say that

1. $\mathcal{A}$ processes $\mathcal{P}$ weakly if for any instance $\{(Q_i, R_i)\}$ there exists a subset $\bar{S} \subset S$ and subsets $\bar{Q}_i \subset Q_i$ such that $\bar{S}$ supports $\{(\bar{Q}_i, R_i)\}$.
2. $\mathcal{A}$ processes $\mathcal{P}$ effectively if there exists a subset $\bar{S} \subset S$ such that $\bar{S}$ supports W for any instance $W \in \mathcal{P}$.
3. $\mathcal{A}$ processes $\mathcal{P}$ strongly if S supports every instance W in $\mathcal{P}$.

DEFINITION **2.11** Let $\mathcal{A} = (S, E, \Delta, T, f)$ be a dynamical automaton with output and (Q, R) an information event. Then $\mathcal{A}$ is said to dynamically couple under (Q, R) if $S_R = f^{-1}(R)$ is a set of dynamically equivalent states and there exists a subset $\bar{S} \subset C$ such that $\bar{S}$ supports (Q, R).

It should be clear that synchronization in cellular automata provides a mechanism for dynamical coupling.

DEFINITION **2.12** Let $\mathcal{A} = (S, E, \Delta, T, f)$ be a dynamical automaton. $\mathcal{A}$ is t-local if $\Delta(\psi\sigma, E) = \psi\Delta(\sigma, E)$ for every $\psi, \sigma \in S$ such that $\mu(\sigma) \geq t$.

It should be clear that standard cellular automata are t-local.

LEMMA. **2.1** Let $\mathcal{A} = (S, E, \Delta, T, f)$ be a dynamical automaton with output. Assume that $\mathcal{A}$ is t-local and that W be a set of dynamically equivalent states such that $\mu(\psi) \geq t$ for all $\psi \in W$. Then for any $H \subset SW$ and any $F \subset E$ we have that

$$\Delta(H, F) = H\bar{\Delta}(H, F).$$

THEOREM. **2.1** Let $\mathcal{A} = (S, E, \Delta, T, f)$ denote a t-local dynamical automaton with output, $\mathcal{M}$ the free monoid over $\{(Q_i, R_i)\}$ and assume (1) that $\mu(\eta) \geq t$ for all $\eta \in \cup_{i\in I} Q_i$, and (2) that $\Sigma_i = f^{-1}(R_i)$ is a set of dynamically equivalent states for each $i \in I$. Let $\bar{E} \subset E$. Then $\mathcal{A}$ effectively processes $\mathcal{M}$ in $\bar{E}$ iff

1. For all $i, j, \Delta(\Sigma_i, \bar{E})$ supports (Q_j, R_j).
2. For all $i, j, W_{ij} = \bar{\Delta}(\Sigma_i, \bar{E}Q_j)$ and Σ_i have nonempty interiors.

This theorem ensures effective computation. It should be clear that the presence of synchronization in a cellular automaton results in the possibility of effective computation. Thus it serves as a useful initial marker.

Synchronization has been detected in cellular automata as well as in coupled map lattices and tempered neural networks.[6,7,8,9]

3. RESULTS

The purpose of this study was to explore the presence of synchronization in cellular automata and its relationship to the level of activation of the input. All 256 two-state, three-neighbor cellular automata rules were examined. The simulation was conducted using a 128-cell lattice with periodic boundary conditions. Two copies of the cellular automaton were simulated starting with different random initial states having 50% of the cells set to 1. Identical random inputs of fixed average activity levels were applied to these cellular automata for a total of 450 time steps. At each time step, the automaton was updated based upon the previous state according to the local rule and then the input was applied. Each cell corresponding to an "on" cell of the input had its state set to 1 and this was carried on to the next updating step. The remaining cells were left unchanged. At each time step, the states of the two automata were compared by Hamming distance. A run was successful if the Hamming distance became 0. Ten trials were carried out for each rule. The input was presented at several activity levels: 0, 0.5, 1, 5, 10, 15, 20, 25, 30, and 35 percent.

The first thing to note from the sample runs, Figures 3, 4, 5, and 6, is that the introduction of an input into the cellular automaton results in a clear shift in the pattern towards randomization. Thus the presence of observable structure is not an adequate marker for computational behavior in the setting of cellular automata with input.

The bottom row of Table 1 presents the overall failure rate, that is the percentage of cellular automata which did not demonstrate synchronization at that particular level of input. Failure of synchronization during the trials does not guarantee that synchronization does not occur. Only 10 trials were run at each input level. Nevertheless the entire experiment has been repeated several times and the results stand suggesting that any synchronization which might occur in these cellular automata is a fairly rare event.

It is interesting to note that synchronization will always occur for any rule for some input at the 35% level or higher. This suggests that a kind of mass effect or perhaps even a kind of crystallization effect is taking place.

The rules demonstrate considerable differences in their ability to be synchronized at each input level. Since all of these runs involved brief transients, the long-term behavior seemed unlikely to be a significant factor governing the presence of this behavior. Thus an examination was made of the actual autonomous patterns generated by these rules to see if any correlations might emerge. From an examination of the 450 time step patterns, it appeared reasonable to divide the rules into six classes: Diagonal, Vertical, Fixed, Uniform, Complex, and Chaotic, based upon symmetry considerations. Examples are given in Figure 1 and 2. Symmetry was considered in the time direction (vertical), in the lattice direction (horizontal), and along the diagonals. For further details, see Sulis.[9]

Two classes show dominant symmetry vertically. The Vertical class shows periodic behavior, usually period 2 with a vertical pattern. The Fixed class corresponds to non-trivial fixed points. The Uniform class shows symmetry in all directions and corresponds to the two trivial fixed points (all 0's and all 1's). Two classes show diagonal symmetry. The Diagonal class shows diagonal symmetry with simple patterns of alternating lines and blanks. The Complex class shows diagonal symmetry but with more complex patterns of lines and periodic structure. The Chaotic pattern shows no characteristic symmetry. Some of the Chaotic rules had runs which sometimes terminated rapidly in one of the two trivial fixed points.

TABLE 1 Synchronization failure rates by symmetry class.

	Input Level (Percentage Activity)									
Symmetry Class	0	0.5	1	5	10	15	20	25	30	35
Diagonal	100	69	35	13	3	1	0	0	0	0
Vertical	100	85	37	15	11	4	0	0	0	0
Fixed	100	63	60	7	0	0	0	0	0	0
Uniform	0	0	0	0	0	0	0	0	0	0
Complex	100	94	79	61	48	36	15	6	0	0
Chaotic	73	100	100	100	100	86	34	17	7	0
Total Rate	88	70	49	27	20	15	6	3	0.8	0

TABLE 2 Classification of rules by symmetry class.

Symmetry Class	Rules
Diagonal	2, 3, 6, 7, 10, 14, 16, 17, 20, 21, 24, 25, 31, 34, 35, 38, 39, 42, 43, 46, 47, 48, 49, 52, 53, 56, 57, 58, 59, 61, 63, 65, 66, 67, 74, 80, 81, 83, 84, 85, 87, 88, 98 99, 103, 111, 112, 113, 114, 115, 116, 117, 119, 125, 130, 134, 138, 139, 143, 144, 148, 152, 155, 158, 159, 162, 163, 170, 171, 174, 175, 176, 177, 184, 185, 186, 187, 188, 189, 190, 191, 194, 208, 209, 211, 214, 215, 226, 227, 229, 230, 231, 240, 242, 243, 244, 245, 246, 247
Vertical	1, 5, 19, 23, 28, 29, 33, 37, 50, 51, 55, 69, 70, 71, 91, 94, 95, 108, 123, 127, 133, 156, 157, 178, 179, 198, 199
Fixed	4, 12, 13, 36, 44, 68, 72, 76, 77, 78, 79, 92, 93, 100, 104, 132, 140, 141, 164, 172, 196, 197, 200, 201, 202 203, 204, 205, 206, 207, 216, 217, 218, 219, 220, 221 222, 223, 228, 232, 233, 236, 237
Uniform	0, 8, 32, 40, 64, 96, 128, 136, 160, 192, 224, 234, 235 238, 239, 248, 249, 250, 251, 252, 253, 254, 255
Complex	9, 11, 15, 26, 27, 41, 62, 73, 82, 97, 106, 107, 109, 118, 120, 121, 131, 142, 145, 154, 166, 167, 169 173, 180, 181, 210, 212, 213, 225
Chaotic	18, 22, 30, 45, 54, 60, 75, 86, 89, 90, 101, 102, 105, 110, 122, 124, 126, 129, 135, 137, 146, 147, 149, 150, 151, 153, 161, 165, 182, 183, 193, 195

In Table 1 are also given the failures rates for these rules at varying input levels relative to their symmetry class. Each class shows a distinct pattern of failure rate versus input level. The most striking result concerns the notion that the edge of chaos is important for emergent computation. Although this may be true following the autonomous paradigm, it does not apply in the situation of external inputs. Synchronization occurs most widely for the Uniform, Diagonal, Vertical, and Fixed rules, which correspond to Wolfram's Class 1 and 2 rules. Both the Complex and Chaoticc rules demonstrate marked resistance to synchronization until fairly high levels of activity are reached. Complex rules lie in Wolfram's Class 2, yet behave more like Chaotic rules which lie in Wolfram's Class 3 and 4. It is clear that rules lying at the edge of chaos or in the chaotic realm are quire resistant to synchronization.

The experiment has been completed for differrent lattice sizes and the classification and results appear to be robust for lattice sizes of greater than 60 cells. In addition, the failure of synchronization for all but the Uniform rules in the autonomous run shows that the synchronization of the cellular automata by the input is not a trivial result. It arises from a restructuring of the basins of attraction of the cellular automaton. Thus the introduciton of an input, and hence noise, can have a significant effect upon the dynamics of the cellular automaton. Further elaboration of these ideas is presented elsewhere.[9]

4. CONCLUSION

Cellular automata have primarily been studied under autonomous conditions. Many of the most popular notions in complex systems have come out of that paradigm. However, the introduction of inputs results in a completely new scenario. The dynamics of the cellular automata change in ways which are no longer predicted by these "classical" notions. The study of cellular automata under input provides a rich and fertile area for the search for novel ideas and paradigms and deserves much more attention than it has received.

REFERENCES

1. Churchland, P., T. Sejnowski, and C. Koch. "What is Computational Neuroscience?" In *Computational Neuroscience*, edited by J. T. Schwartz. Cambridge, MA: MIT Press, 1990.
2. Forrest, S. "Emergent Computation: Self Organizing, Collective and Cooperative Phenomena in Natural and Artificial Computing Networks." *Physica D* **42** (1990): 1–11.
3. Lallement, G. *Semigroups and Combinatorial Applications.* Wiley, 1979.
4. Langton, C. "Computation at The Edge of Chaos: Phase Transitions and Emergent Computation." *Physica D* **42** (1990): 12–37.
5. Sulis, W. "Tempered Neural Networks." In *Proceedings of the International Joint Conference on Neural Networks*, Vol. III, 421–426. Baltimore, 1992.
6. Sulis, W. "Emergent Computation in Tempered Neural Networks 1: Dynamical Automata." In *Proceedings of the World Conference on Neural Networks*, Vol. IV, 448–451. Erlbaum, 1993.

7. Sulis, W. "Emergent Computation in Tempered Neural Networks 2: Computation Theroy." In *Proceedings of the World Conference on Neural Networks*, Vol. IV, 452–455. Erlbaum, 1993.
8. Sulis, W. "Computation in Complex Systems." *Physica D* (submitted).
9. Sulis, W. "Synchronization in Complex Systems. *Phys. Rev. Lett.* (submitted).
10. Wolfram, S. "Twenty Problems in the Theory of Cellular Automata." *Physica Scripta* **T9** (1985): 170–183.
11. Wuensche, A. "The Ghost in the Machine: Basin of Attraction Fields of Disordered Cellular Automata Networks." Working paper #92-04-017, Santa Fe Institute, 1992.

Manfred Tacker* and Peter F. Stadler*†
*Institut für Theoretische Chemie, Universität Wien, Währingerstr. 17, A-1090 Vienna, Austria; e-mail: stadler@tbi.univie.ac.at
†Santa Fe Institute, 1339 Hyde Park Road, Santa Fe, NM 87501

RNA: Genotype and Phenotype

RNA folding is a map assigning (secondary) structures to sequences. The frequency distribution of structures is highly nonuniform: there are few abundant and many rare structures. The distances between sequences folding into the same structure are distributed randomly, but extensive neutral networks of nearest neighbors with common structures percolate the sequence space. Consequently, only a small number of (point) mutations is necessary to obtain a desired structure from an arbitrary initial sequence.

INTRODUCTION

The dichotomy of genotypes and phenotypes is the basis of biological evolution. The information for the unfolding of an organism, i.e., a phenotype, is contained in (DNA) genes and all inheritable variations are, as far as we know today, the effects of changes in the nucleotide sequence of the genome. Selection, on the other hand, acts by enhancing the relative frequency of "fitter" phenotypes, causing the eventual extinction of less fit ones. This is, in a nut shell, the Darwinian principle of evolution.

Fitness is related to the rate of reproduction (and the probability of survival of the off-spring until the age of its own reproduction), which in turn are properties of the phenotype (and the environment, of course). A crucial step in the understanding of evolutionary dynamics on molecular level is hence the development of a theory for the mapping from genotype to phenotype and the subsequent evaluation of the phenotype by the environment.

RNA molecules provide a unique model for the interplay of genotype, phenotype, and fitness. They can be replicated in a special environment containing $Q\beta$ replicase. Spiegelman[18,26] pointed out that such an RNA replication system shows all ingredients of the Darwinian scenario. RNA molecules are replicated template directed, and replication errors occur at the level of the nucleotide sequence (genotype). The replication rate depends on the three-dimensional structure of the molecule which must be recognized by the replicase, and which must be melted during the replication process. The underlying mechanism and its reaction kinetics are known in detail.[1]

It is a lucky coincidence that among all biopolymers RNA is the easiest case for a computational approach. The three-dimensional structure is dominated by its secondary structure, the Watson-Crick and **GU** base pairs. (Usually, triple helical regions and so-called pseudo knots are, by definition, considered part of the tertiary interactions folding the two-dimensional (planar) secondary structure into the full three-dimensional shape of the molecule.) Secondary structure can be efficiently predicted from the sequences by efficient computer programs (see Table 1).

The Hamming distance is defined as the number of positions in which two sequences differ. This metric arranges the sequences space, $\mathcal{C}$, i.e., the set of all sequences of length n, in a highly symmetric graph known as hypercube. In a similar way it is possible to define a quite natural distance between secondary structures. For the details of the definition of distances between secondary structures we refer to Hogeweg and Hesper,[14] Shapiro and Zhang,[24,25] and Fontana et al.[10] In the following we will denote by $d(x, y)$ the distance in sequence space between the configurations (sequences) x and y. For the distance of two structures a and b in shape space we will write $D(a, b)$.

The mapping from genotypes, i.e., sequences, to phenotypes, i.e., secondary structure graphs, and further-on to properties of the phenotype defining its fitness are schematically depicted in Figure 1.

Both the sequence space and the shape space are discrete spaces of combinatorial complexity; mappings between such spaces have been termed *combinatory maps.*[9] The special case of objects being mapped to numerical properties the term (combinatory) *landscape* is widely used. In the RNA case we have landscapes of numerical properties as functions of secondary structures $p : \mathcal{S} \to \mathbf{R}$.

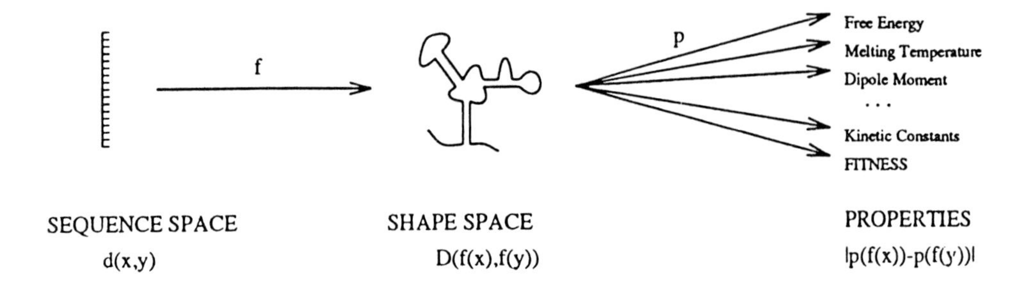

FIGURE 1 Schema of RNA folding.

The composition of the combinatory map $f : \mathcal{C} \to \mathcal{S}$ describing the folding process with these landscapes yields again a family of landscapes $p(f) : \mathcal{C} \to \mathbf{R}$ relating the sequences directly to the physical properties of the unfolded phenotype. In fact, the free energy ΔG of the folding process is obtained as a by-product of the calculation of the secondary structure. RNA free energy landscapes, $\Delta G : \mathcal{C} \to \mathbf{R}$ have been studied in detail.[1,7,8,9,10] Landscapes of kinetic properties as functions of the sequences are described in Tacker et al.[28]

The mathematical object "landscape" occurs in many different contexts. Classical examples are spin glass Hamiltonians and the cost functions of combinatorial optimization problems. (For details see the contribution by Macken and Stadler in this volume.)

It is clear that the dynamics of evolution is closely linked to the structure of the fitness landscape,[3,4,5,6,7] and to the structure of the combinatory map f of phenotype formation in a more general context.[12,13,23,22]

THE COMBINATORY MAP OF RNA FOLDING

A variety of computer programs predicting RNA secondary structures have been published. A very brief overview is given in Table 1. It has been shown recently[29] that the structure of f does not strongly depend on the algorithm or parameter set used to predict the structures. The choice of the distance measure D in shape space has also only a minor influence on the results. We will focus on the data that have been obtained from a minimum free energy approach with a parameter set taken from Freier et al. All calculation reported here have been performed using the `Vienna RNA Package`.[13]

TABLE 1 Folding Algorithms.

Algorithm	ψ	Remark	Reference
deterministic			
Minimum Free Energy	-	fast	Zuker[30]
Kinetic Folding	+	fast	Martinez[16]
5'-3' Folding	+	fast	Tacker[29]
Partition Function	-	ensemble	McCaskill[17]
Maximum Matching	-	unrealistic	Nussinov et al.[21]
stochastic			
Simulated Annealing	+	very slow	Mironov[19,20]

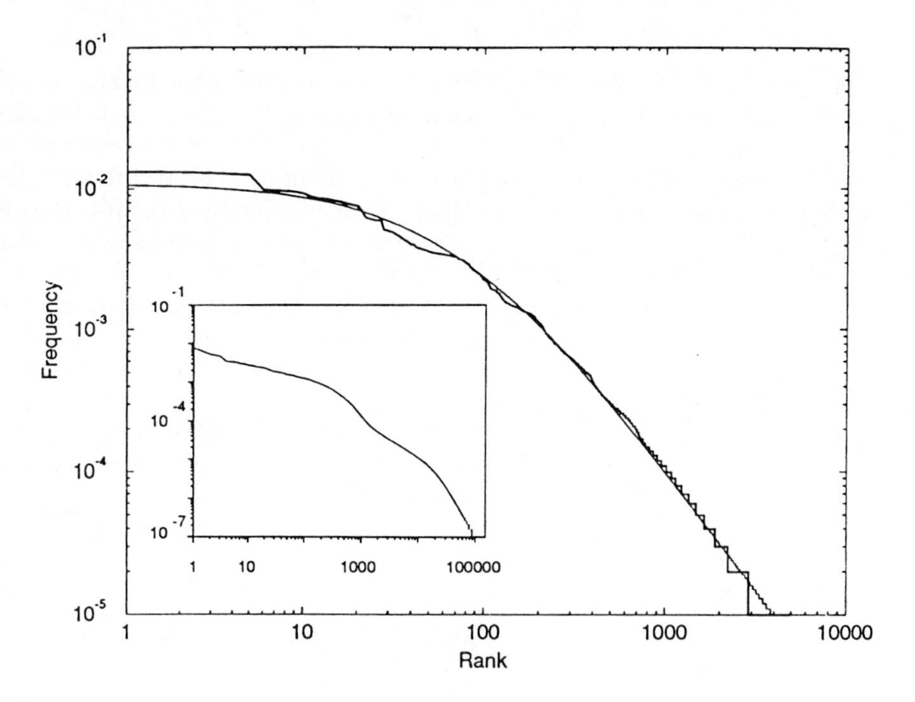

FIGURE 2 Distribution of secondary structures. The large plot has been obtained from "loop-structures" for sequence length $n = 100$, the inset contains data for chain length $n = 30$ at full resolution of the secondary structure representation. Loop structures are obtained by ignoring the size of structural elements. For details see, e.g., Hofacker et al.[12]

There are many more sequences than structures. Counting only planar secondary structures without isolated base pairs and with at least three unpaired bases in hairpin loops one finds that there are approximately

$$S_n \approx 1.4848 \times n^{-3/2} \, (1.8488)^n \tag{1}$$

secondary structures for sequences of length n for the 4^n sequences.[12,27] Hence the question arises how these relatively few structures are distributed over sequence space.

We find that the frequency distribution of secondary structures obtained by folding a random sample of sequences follows roughly a generalized Zipf law,[15]

$$f(r) = a(b + r)^{-c}, \tag{2}$$

where r is the rank (by frequency) of the structure S and $f(r)$ is the fraction of occurrences of S in the sample.[22] The exponent c describes the distribution of rare sequences, the constant b is a rough measure for the number of frequent structures. Our data suggest that there are few frequent structures and many very rare ones. The few thousand most common structures cover already more than 90% of the sequence space for **AUGC** sequences with chain length $n = 30$. For much longer sequences a direct computation of the frequency distribution becomes too costly. One can, however, investigate the distribution of coarse grained structure representations.[23] It turns out that a frequency distribution of the generalized Zipf type occurs at all resolutions of the secondary structure representation.

A basic property of combinatory landscape is *ruggedness*. A landscape is rugged if it has lots of local optima, if adaptive (up-hill) walks are short, and if the correlation between nearest neighbors is small. Adaptation and optimization is harder on more rugged landscapes. While the notions of local optima and adaptive walks do not have counter-parts in general combinatory maps (their definition require the image set to be ordered), we can generalize the definition of pair-correlation to mapping from one metric space into another one[9]:

$$\rho(d) = 1 - \frac{\langle D^2(f(x), f(y)) \rangle_{d(x,y)=d}}{\langle D^2(f(x), f(y)) \rangle_{\text{random}}} \,. \tag{3}$$

The average in the enumerator runs over all pairs of configurations with fixed Hamming distance d while the average in the denominator runs over all pairs of configurations. An empirical correlation length ℓ is obtained from the definition $\rho(\ell) = 1/\mathrm{e}$.

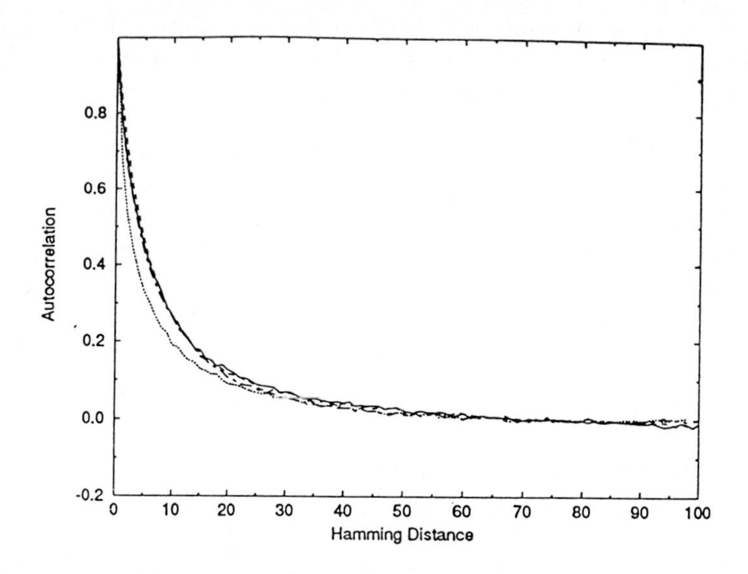

FIGURE 3 Autocorrelation function of the combinatory map of secondary structure formation for the natural **GCAU** alphabet. Data for different algorithms are shown. Distance in shape space is measured by the tree-edit distance.[10]

The *density surface* $\wp(D|d)$ is the conditional probability that two secondary structures have distance D in shape space provided their underlying sequences have distance d in sequence space.[9] Density surfaces contain more information then the autocorrelation functions. For instance, the probability for finding a neutral neighbor, i.e., a sequence in Hamming distance 1 which folds into the same structure is $\wp(0,1)$. In fact, the autocorrelation function can be obtained from the density surface:

$$\rho(d) = 1 - \frac{\sum_D D^2 \wp(D|d)}{\sum_d \sum_D D^2 p(d) \wp(D|d)} \tag{4}$$

where $p(d)$ is the probability for two randomly chosen sequences to have distance d.

The particular form of the density surfaces (see Fontana et al.[9]; data not shown here) suggest that fairly small balls in sequence space already show the global characteristics of the folding map f—in other words (nearly) all structures can be found in a small neighbourhood around *every* sequence. Not all sequences can fold into a particular secondary structure. The set of sequences *compatible* with a secondary structure S can be characterized as follows: On each unpaired position of the structure we can have an arbitrary base, and on the two positions corresponding to a base pair we must have bases that can pair. Our "shape space covering" conjecture states, then, that in a small neighborhood of any sequence compatible with a common structure we will find almost certainly a sequence actually folding into this structure. Small means here of the order of once or twice the correlation length ℓ. Simulations using an inverse folding algorithm[12,13] strongly support this statement.[22] If we start at a random sequence then the majority of mutation steps is necessary to reach a compatible sequence. Once a compatible sequence is reached the next sequence which folds into the desired secondary structure is not very far off.

Furthermore, large samples of sequences folding into the same secondary structure have been generated using the inverse folding algorithm. We find that these sequences cannot be distinguished from a random sample of compatible sequences. The distribution of sequences folding into a common secondary structure is therefore as random as can be.

The definition of a set $C(S)$ of sequences compatible with a secondary structure S suggest to use a modified definition of neighborhood on these sets. We will say that two sequences are neighbors in $C(S)$ if they differ either in a single unpaired base or in the type of a single base pair of S. This definition makes $C(S)$ again to a connected graph.

A *neutral network* is a connected subgraph $\mathcal{N}$ of $C(S)$ such that all sequences in $\mathcal{N}$ fold into the same structure and if $x \in \mathcal{N}$ then so are all its neutral neighbors with respect to the neighborhood defined above for $C(S)$. Another way of specifying a neutral net is the following: A neutral net is a connected component of the set of all sequences folding into a given secondary structure. In the case of RNA secondary structures a neutral net connects all neighboring sequences folding into the same secondary structure.

Computer simulations show a surprising result: neutral networks are very large, in fact in general they are too large to be listed exhaustively. We have performed the following simulation in order to determine how far neutral net reach out in sequence space. Starting with an arbitrary initial sequence we search for neutral neighbors in $C(S)$ subject to the constraint that the distance for the starting point increases after each step. Figure 4 shows the average distance $\bar{d}$ from the starting point that such a walk reaches. The distance is expressed in Hamming distance for easier interpretation. Note that since only a single walk is performed in each network we calculate only a lower bound on the diameter of neutral networks. We find that most neutral networks reach through essentially the entire configuration space.

Neutral networks of rare structures are, as expected, smaller than the neutral networks of the most frequent structures. Even for the rarest structures that we could find in our simulations (which occur with a frequecy of about 10^{-5}) we observe an average distances $\bar{d}$ of well above 24 for chain length 30. This value is to be compared with the average distance of two random structures, $d = 22.5$ and the average distance of two random compatible structures, $d \approx 22$.

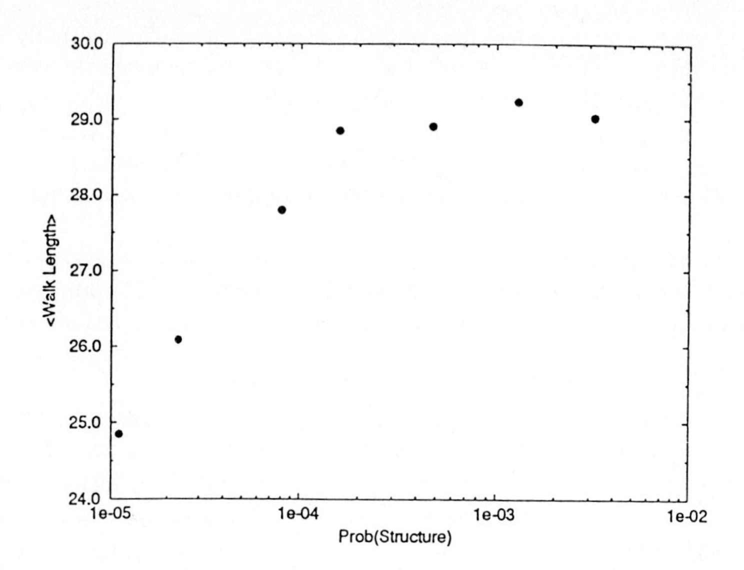

FIGURE 4 Average length of a neutral path as function of the abundancy of the corresponding secondary structure. This is (good) lower bound on the diameter of the neutral network containing the path. Data are for **AUGC**-alphabet and chain length 30.

DISCUSSION

Our data show that optimization of structures by evolutionary trial and error strategies is much simpler than often assumed. In fact, the combinatory map of RNA secondary structures is *optimally suited* for evolutionary adaptation. Exploration is easy because of huge neutral networks that are nearly neutral, and optimization is not a hard task since a molecule with desired secondary structure is just a few mutations away from almost anywhere in sequence space. Populations replicating with sufficiently high error rates will readily spread along these networks and can reach more distant regions in sequence space. A reduced mutation rate causes the population to concentrate in the most favourable part of the neutral network and to adapt locally in this region.

The consequences of our results for natural and artificial selection are immediate. We predict that there is no need to systematically search huge portions of the sequence space, nor does one need specially designed initial conditions. These properties provide further support to the idea of widespread applicability of molecular evolution.[4,5,6]

ACKNOWLEDGMENTS

The work reported here is joint research with Walter Fontana, Ivo L. Hofacker, and Peter Schuster. Financial support by the Austrian *Fonds zur Förderung der wissenschaftlichen Forschung*, proj. nos. S5305-PHY and P8526-MOB, is gratefully acknowledged.

REFERENCES

1. Biebricher C. K., and M. Eigen. "Kinetics of RNA Replication by Qβ Replicase." In *RNA Genetics*, edited by E. Domingo, J. J. Holland and P. Ahlquist, Vol. I, 211–245. Boca Raton, FL: CRC Press, 1988.
2. Bonhoeffer S., J. S. McCaskill, P. F. Stadler, and P. Schuster. "RNA-Multistructure Landscapes." *Eur. Biophys. J.* **22** (1993): 13–24.
3. Bonhoeffer S., and P. F. Stadler. "Errorthreshold on Complex Fitness Landscapes." *J. Theor. Biol.* **164** (1993): 359–372.
4. Eigen M. Self-Organization of Matter and the Evolution of Biological Macromolecules." *Naturwissenschaften* **58** (1971): 465–523.
5. Eigen M., and P. Schuster. *The Hypercycle —A Principle of Natural Self-Organization.* Berlin: Springer-Verlag, 1979.
6. Eigen M., J. S. McCaskill, and P. Schuster. "The Molecular Quasispecies." *Adv. Chem. Phys.* **75** (1989): 149–263.
7. Fontana W., W. Schnabl, and P. Schuster. "Physical Aspects of Evolutionary Optimization and Adaptation." *Phys. Rev. A* **40** (1989): 3301–3321.
8. Fontana W., T. Griesmacher, W. Schnabl, P. F. Stadler, and P. Schuster. "Statistics of Landscapes Based on Free Energies, Replication and Degradation Rate Constants of RNA Secondary Structures." *Mh. Chem.***122** (1991): 795–819.
9. Fontana W., P. F. Stadler, E. G. Bornberg-Bauer, T. Griesmacher, I. L. Hofacker, M. Tacker, P. Tarazona, E. D. Weinberger, and P. Schuster. "RNA Folding Landscapes and Combinatory Landscapes." *Phys. Rev. E* **47** (1993): 2083–2099.
10. Fontana W., D. A. M. Konings, P. F. Stadler, and P. Schuster. "Statistics of RNA Secondary Structures." *Biopolymers* **33** (1993): 1105–1119.
11. Freier, S. M., R. Kierzek, J. A. Jaeger, N. Sugimoto, M. H. Caruthers, T. Nielson, and D. H. Turner. "Improved Free-Energy Parameters for Predictions of RNA Duplex Stability." *Proc. Natl. Acad. Sci., USA* **83** (1986): 9373–9377.

12. Hofacker I. L., P. Schuster, and P. F. Stadler. "Combinatorics of Secondary Structures." Santa Fe Institute Working Paper # 94-04-026. *SIAM J. Disc. Math.*: submitted.
13. Hofacker I. L., W. Fontana, P. F. Stadler, L. S. Bonhoeffer, M. Tacker, and P. Schuster. "Fast Folding and Comparison of RNA Secondary Structures (The Vienna RNA Package)." *Mh. Chem.* **125** (1994): 167–188.
14. Hogeweg, P., and B. Hesper. "Energy Directed Folding of RNA Sequences." *Nucleic Acids Resh.* **12** (1984): 67–74.
15. Mandelbrot, B. B. *The Fractal Geometry of Nature.* New York: Freeman & Co., 1982.
16. Martinez, H. M. "An RNA Folding Rule." *Nucl. Acid. Res.* **12** (1984): 323–335.
17. McCaskill, J. S. "The Equilibrium Partition Function and Base Pair Binding Probabilities for RNA Secondary Structure." *Biopolymers* **20** (1990): 1105–1119.
18. Mills D. R., R. L. Peterson, and S. Spiegelman. "An Extracellular Darwinian Experiment with a Self-Duplicating Nucleic Acid Molecule." *Proc. Natl. Acad. Sci. USA* **58** (1967): 217–224.
19. Mironov A. A., L. P. Dyakonova, and A. E. Kister. "A Kinetic Approach to the Prediction of RNA Secondary Structures." *J. Biomol. Struct. Dyn.* **2** (1985): 953–962.
20. Mironov A. A., and A. E. Kister. "RNA Secondary Structure Formation During Transcription." *J. Biomol. Struct. Dyn.* **4** (1986): 1–9.
21. Nussinov R., G. Pieczenik, J. R. Griggs, and D. J. Kleitman. "Algorithms for Loop Matching." *SIAM J. Appl. Math.* **35** (1978): 68–82.
22. Schuster P. "RNA Based Evolutionary Optimization." *Origins of Life* (1993): 373–391.
23. Schuster P., W. Fontana, P. F. Stadler, and I. L. Hofacker. "From Sequences to Shapes and Back: A Case Study in RNA Secondary Structures." *Proc. Roy. Soc. B* **255** (1994): 279–284.
24. Shapiro B. A. 1988. "An Algorithm for Comparing Multiple RNA Secondary Structures." *CABIOS* **4** (1988): 381–393.
25. Shapiro B. A., and K. Zhang. "Computing Multiple RNA Secondary Structures Using Tree Comparisons." *CABIOS* **6** (1990): 309–318.
26. Spiegelman S. "An Approach to the Experimental Analysis of Precellular Evolution." *Qtr. Rev. Biophys.* **4** (1971): 213–253.
27. Stein, P. R., and M. S. Waterman. "On Some New Sequences Generalizing the Catalan and Motzkin Numbers." *Discrete Math.* **26** (1978): 261–272.
28. Tacker M., W. Fontana, P. F. Stadler, and P. Schuster. "Statistics of RNA Melting Kinetics." *Eur. Biophys. J.* **23** (1994): 29–38.
29. Tacker M., P.F. Stadler, E.G. Bornberg-Bauer, I.L. Hofacker, and P. Schuster. "Robust Properties of RNA Secondary Structure Folding Algorithms." Working Paper, Santa Fe Institute, 1995.

30. Zuker M. 1989. "The Use of Dynamic Programming Algorithm in RNA Secondary Structure Prediction." In *Mathematical Methods for DNA Sequences*, edited by M. S. Waterman, 159–184. Boca Raton, FL: CRC-press, 1989.

Patrick Tufts
Department of Computer Science,Volen Center, Brandeis University, Waltham, MA 02254-9110; e-mail: zippy@cs.brandeis.edu

Parallel Case Evaluation for Genetic Programming

This chapter describes a possible approach to classification problems with many variables using Genetic Programming.connection machine It also outlines an implementation of parallel evaluation for Koza's Genetic Programming Kernel that runs on the CM-5. The system allows training cases to be evaluated in parallel giving a linear speedup with the number of processors.

GENETIC PROGRAMMING PARADIGM

Genetic Programming is an extension of Genetic Algorithms to encompass hierarchical, variable length programs. Programs are created from a fixed set of primitive functions, and are then evaluated with respect to the problem to be solved.

For example, if the task is to find a function that fits a simple curve, a potential set of primitive functions would be: multiplication, division, addition, subtraction, and exponentiation.

Programs are promoted to the next generation in two ways: asexual replication and mating. Asexual reproduction means that the program is copied. Mating means that two programs are selected, and a subtree from each program is exchanged with the other. The two new programs are then made part of the population for the next generation.

The closer the program is to a solution, the higher its fitness. This fitness is measured by an evaluation function supplied by the experimenter. In our curve fitting example, this function might take a set of known points and see how many of them the curve described by the program crosses. The fitness would then be the number of points crossed.

It is the program's fitness that determines whether it gets to mate with another program. Typically, crossover is used on the top 90% of the programs ranked by fitness[2]; the remaining programs are simply copied.

LISP

Genetic Programming requires that programs be created dynamically, and that their tree structure be manipulable (in the case of crossover, for example).

Lisp provides both of these features. The parse tree is an explicit part of the program (parenthesis denote levels in the hierarchy). Furthermore, Lisp is ideal for handling programs that are created on the fly.

*LISP

*Lisp is a data-parallel version of Lisp that runs on the CM-5.[3] I chose *Lisp for the implementation of the parallel evaluation engine because it simplified the interface to Koza's Lisp code.

WHY PUT EFFORT IN TO PARALLEL EVALUATION?

Most of the computer time during a GP experiment is spent on the evaluation of programs.[1] Since the programs do not interact with each other during evaluation, making this step parallel yields linear speedup with the number of processors.

IMPLEMENTING PARALLEL CASE EVALUATION

Rather than evaluating cases one at a time, I load them into a parallel varable (pvar) in *Lisp. Counting up the correct cases looks like this:

```
(*max
    (if!!
     (=!! prediction!! actual!!)
         (enumerate!!)
         (!! 0)))
```

EXPERIMENT

I wanted to apply Genetic Programming to the problem of classification. The general format for this is described in Koza.[1] I wanted to see if GP would scale up to a large classification task.

The problem selected for this experiment is prediction of credit card attrition. Attrition is when a customer chooses not to renew their credit card. Credit card companies are interested in keeping their customers, and even a very small increase in predictive power can translate into great savings for the company.

These companies like to be able to predict which customers will attrite so that they can then offer some sort of incentive (waiving the annual card fee, for example). Systems that do a better job at classification allow a company to prevent the same number of customers from leaving while spending less money on incentives. With several million customers, even small savings per account add up.

The database supplied by a financial institution contained 18 months worth of information broken down by month. The several gigabyte database contained roughly 500 fields of data on several million customers.

I chose to work with only the numeric fields (dollar amounts, number of days since last activity), which meant that this experiment used around 200 of the 500 possible fields.

The fields in the database describe how much the customer spent in the current month, how much interest the customer has paid over the lifetime of their card, whether or not the customer has ever been late on payments, number of days since last purchase, number of months they had carried a balance, etc.

TABLE 1

Objective:	Induce a function that will classify each case in a test database into one of two categories
Terminal Set:	Variables representing 200 fields from the database, RANDOM-INT
Function Set:	IF-LESS-OR-EQ, *, +, /, -
Fitness Cases:	32768 correctly classified cases
Raw Fitness:	Sum of cases correctly classified by program
Standardized Fitness:	Same as raw fitness
Hits:	Same as raw fitness
Wrapper:	If program result is less than or equal to 0, the classify as attrite. If result is greater than 0, than non-attrite
Parameters:	M = 1000, G = 50
Success predicate:	As S-expression scores 100%

This prediction task is very difficult because fewer than 1% of the customers in any given month actually attrite. This means that there are plenty of negative cases (people who do not attrite), and very few positive ones.

PRELIMINARY RESULTS

My work so far has indicated that the Genetic Programming Paradigm cannot find an adequate classifier for this problem given the table above. One problem is that there is no concept of a confidence value in this experiment, and so the system cannot restrict its predictions to an easy subset of the problem.

Solutions with confidence values allow you to ignore answers below a certain threshold, which means if a solution is good at classifying part of the search space very well, it can point this out.

If the program turned out to be extremely good at classifying people with Brandeis addresses, it could then attach a higher confidence to these solutions. If, at the same time, it was terrible at classifying people with Santa Fe addresses, then it could give a low confidence for these predictions.

By hedging its answers, the program is able to point out what it is good at classifying. This makes it possible to measure the effectiveness of classification above a certain confidence level, rather than effectiveness over the entire dataset.

To implement genetic programming with confidence values, I plan on creating two distinct populations. One population consisting of classifiers, the other of functions that estimate a classifier's confidence level for a prediction. The two populations would evolve in parallel.

With prediction and confidence in hand, the programs can then be given variable rewards. One scheme is to use the confidence value as a reward multiplier. Assuming confidence is a real number between 0 and 1, the payoff could be

```
confidence * 1
```

for a correct prediction, and

```
confidence * -1
```

for an incorrect one.

Another way I am interested in extending the Genetic Programming Paradigm is by constraining the genetic operators. Tufts and Hughes[5] describes an experiment in which a nondeterministic tree search of the space of possible solutions outperforms a GA when searching a large solution space.

ACKNOWLEDGMENTS

Thanks to Dave Waltz, Steve Smith, Mario Bourgoin, Brij Masand, Gary Drescher, and Kurt Thearling for helpful comments and discussions. Thanks also to Thinking Machines for the use of their computers, and to Ann Bell, Dan Stein, and the Santa Fe Insitute for making the proceedings of the 1993 Complex Systems Summer School possible.

REFERENCES

1. Koza, John. "Concept Formation and Decision Tree Induction Using the Genetic Programming Paradigm." In *Parallel Problem Solving from Nature*, edited by Schwefel and Maenner. Berlin: Springer-Verlag, 1990.
2. Koza, John. *Genetic Programming.* Cambridge: MIT Press, 1992
3. **Lisp Reference Manual.* Thinking Machines, 1990.
4. *Guide to Porting CM-5 *Lisp.* Thinking Machines, 1991
5. Tufts, Patrick, and Hugues Juille. "Evolving Non-Deterministic Algorithms for Efficient Sorting Networks." Poster presentation at Artificial Life IV Conference, 1994.

Ken Umeno
Department of Physics, University of Tokyo 7-3-1 Hongo Bunkyo-ku, Tokyo 113, Japan

Singular Point Analysis and Nonintegrable Hamiltonian Systems

Analysis of dynamical systems in the complex time plane is useful. As an example, we can constructively prove the existence of Hamiltonian systems with an infinite number of degrees of freedom which have no additional global analytic integrals besides the Hamiltonian itself. Here "complex" is not the adjective of "complexity" but the inverse of "real."

If an integrable Hamiltonian system $H_0(q, p)$ with n degrees of freedom is perturbed to H as follows:

$$H = H_0 + \epsilon H_1 + \epsilon^2 H_2 + \epsilon^3 H_3 \dots , \tag{1}$$

then *in general* the system H does not have an additional first integral $\Phi(q, p)$ where

$$\Phi = \Phi_0 + \epsilon\Phi_1 + \epsilon^2\Phi_2 + \epsilon^3\Phi_3 \dots , \tag{1}$$

and Φ_0 is an additional first integral of the system H_0. This is Poincaré's theorem which shows generic nonintegrability of Hamiltonian systems. However, this does not mean that distinction between integrable systems and nonintegrable systems for given Hamiltonian systems is easy, although this distinction makes a serious effect on the real computation of hamiltonian flows.[9,10,11]

The main purpose here is to prove the following theorem using S. Kovalevskya's singular point analysis of differential equations in the complex time plane.[5,8]

THEOREM 1. There exist Hamiltonian systems with an infinite number of degrees of freedom which have no additional integrals besides the Hamiltonian itself.

More than 100 years ago (she was born in 1850 in Russia),[7] Kovalevskya had a powerful intuition that only movable singularities with solutions in the complex plane would be ordinary poles if the underlying dynamical system is integrable. In essence, this is one strong evidence of the conjecture at present that singularities with the solutions in the complex time plane is deeply related to the fundamental question of integrability or nonintegrability. We consider the following Hamiltonian system

$$H_n = \sum_{i=1}^{n} \left(\frac{1}{2} p_i^2 + q_i^4 \right) + \sum_{\langle i,j \rangle} c_{i,j} q_i^2 q_j^2 \tag{3}$$

with n degrees of freedom where $\langle i,j \rangle$ is a summation of all the different pairs $i \neq j$ and the equations of motion are given by

$$\frac{dq_i}{dt} = \frac{\partial H}{\partial p_i}, \frac{dp_i}{dt} = -\frac{\partial H}{\partial q_i}. \tag{4}$$

In this case of the systems with an homogeneous potential function, we can always get a straight-line solution as follows:

$$\begin{aligned} q_1^{(0)}(t) = C\phi(t), q_i^{(0)} = 0 (2 \leq i \leq n) \\ p_1^{(0)}(t) = C\dot{\phi}(t), p_i^{(0)} = 0 (2 \leq i \leq n) \end{aligned} \tag{5}$$

where

$$\frac{d^2}{dt^2}\phi(t) + \phi(t)^3 = 0 \tag{6}$$

and C is derived to be equal to $\sqrt{-1/2}$. Because C is a complex number, we can regard the solution (5) as a complex analytic function with a complex variable t. We expand the solution as follows:

$$q_i(t) = q_i^0(t) + \xi_i(t), \quad p_i(t) = p_i^0(t) + \eta_i(t). \tag{7}$$

Then we get the following linear variational equation

$$\frac{d\vec{\xi}}{dt} = \vec{\eta}, \quad \frac{d\vec{\eta}}{dt} = -\phi^2(t) V_{CC\vec{\eta}} \tag{8}$$

where the Hessian matrix V_{CC} of the potential function with the solution (5) is given by

$$V_{CC} = \begin{bmatrix} 3 & 0 & \cdots & \cdots & 0 \\ 0 & \frac{c_{1,2}}{2} & 0 & \cdots & 0 \\ 0 & 0 & \frac{c_{1,3}}{2} & 0 & \vdots \\ \vdots & \vdots & \vdots & \vdots & \vdots \\ 0 & \cdots & \cdots & \cdots & \frac{c_{1,n}}{2} \end{bmatrix}. \tag{9}$$

This vector Eq. (9) has already been shown by Ziglin[1,2] and Yoshida[3] to be essential for proving nonintegrability using singular point analysis. The first eigenvalue λ_1 of V_{CC} is trivially 3 because λ_1 is always $l-1$ in the case that the degree of the potential function is l.[4] Thus, the essential eigenvalues of the Hessian matrix V_{CC} are $(c_{1,2})/2, (c_{1,3})/2, (c_{1,n})/2$. Therefore, the monodromy matrix M of the essential part of the vector variational equation (which is called as NVE[l, 2]) is expressed in terms of the $n-1$ block-diagonals form

$$M = \text{diag}[M(\lambda_2), M(\lambda_3), \ldots, M(\lambda_n)] \tag{10}$$

where $\lambda_i = (c_{1,i})/2$ in this case and 2×2 matrices $M(\lambda_i)$ satisfy the following relation

$$\det M(\lambda_i) = 1 \tag{11}$$

from the symplectic property of the variational equation. A monodromy matrix is called nonresonant when the eigenvalues

$$(\sigma_2, \sigma_2^{-1}, \sigma_3, \sigma_3^{-1}, \ldots, \sigma_n, \sigma_n^{-1}) \tag{12}$$

satisfy the following relation:

$$\sigma_2^{l_2} \sigma_3^{l_3} \ldots \sigma_n^{l_n} = 1 \Rightarrow \quad l_2 = l_3 = \ldots = l_n = 0 \tag{13}$$

for integers $l_2, l_3, \ldots l_n$. Ziglin and Yoshida's theorems say that if there exist two different monodromy matrices M_1, M_2 associated with two different loops on the Riemann surface which are nonresonant and if none of the pairs, $M_1(\lambda_i)$ and $M_2(\lambda_i)$ commute, then the corresponding Hamiltonian system has no additional integrals besides the Hamiltonian itself.[1,3] Here we use Yoshida's choice of loops of the monodromy matrix on the Riemann surface.[4] Then the eigenvalues of the different monodromy matrices M_1, M_2 are given explicitly as follows:

$$\sigma_i = \exp\left(2\pi i \sqrt{4 + 32\lambda_i}\right) \tag{14}$$

where λ_i are the eigenvalues of the Hessian V_{CC}.[4] Now we obtained the eigenvalues

$$(\exp(4\pi i\sqrt{1+4c_{1,2}}), \exp(4\pi i\sqrt{1+4c_{1,3}}), \ldots, \exp(4\pi i\sqrt{1+4c_{1,n}})) \tag{15}$$

of the monodromy matrix from the above formula. These eigenvalues clearly satisfy the nonresonant condition if we choose parameters $c_{i,j}$ such that:

$$c_{1,i} = (m_{2i-2})/(m_{2i-3}) - \frac{1}{4} \quad (i > 1) \tag{16}$$

where m_i is the ith prime in the natural numbers $1, 2, 3, 4 \ldots$:

$$m_1 = 2, m_2 = 3, m_3 = 5, m_4 = 7, m_5 = 11 \ldots . \tag{17}$$

We remark here that n real numbers

$$\left(\sqrt{\frac{m_2}{m_1}}, \sqrt{\frac{m_4}{m_3}}, \ldots \sqrt{\frac{m_{2n}}{m_{2n-1}}}\right) \tag{18}$$

are rationally independent because there are no integer solutions of $l_1, \ldots, l_n$ that satisfy the following relation:

$$l_1\sqrt{\frac{m_2}{m_1}} + l_2\sqrt{\frac{m_4}{m_3}} + \ldots l_n\sqrt{\frac{m_{2n}}{m_{2n-1}}} = 0 \tag{19}$$

except the case

$$l_1 = l_2 = \ldots = l_n = 0\,. \tag{20}$$

Since $M_1(\lambda_i)$ and $M_2(\lambda_i)$commute only when

$$\text{trace } M_{1,2}(\lambda_i) = 2\cos(2\pi\sqrt{4 + 32\lambda_i}) = \pm 2 \tag{21}$$

they[4] cannot commute each other. Moreover from the following fact (Chebyshev's theorem in the classical number theory):

$$1 < \frac{m_{2i}}{m_{2i-1}} < 2 \text{ for any } i\,, \tag{22}$$

we have now the restrictions on the coefficients $c_{1,i}$ as follows:

$$0 < c_{1,i} < \frac{1}{4} \text{ for any } i\,. \tag{23}$$

This Eq. (23) also shows that even in the infinite dimensional limit ($n \to \infty$), the coefficients $c_{1,i}$ do not vanish and they have the upper bound of 1/4:

$$0 < \lim_{n\to\infty} c_{1,n} < \frac{1}{4}\,. \tag{24}$$

Finally, we get the following explicit and infinite-dimensional potential function:

$$V = q_1^4 + \frac{1}{9}q_1^2q_2^2 + \frac{1}{10}a_1^2q_3^2 + \frac{1}{22}q_1^2q_3^2 + \frac{1}{34}q_1^2q_5^2 + \frac{3}{46}q_1^2q_6^2 + \ldots \tag{25}$$

because there are an infinite number of prime numbers in the set of natural numbers. Thus, we can construct a Hamiltonian system with an arbitrary number (up to an infinite number) of degrees of freedom which has no additional integrals besides the Hamiltonian itself. The existence of infinite-dimensional nonintegrability here suggests the existence of "infinite-dimensional chaos."[12]

ACKNOWLEDGMENTS

The present author thanks Profs M. Suzuki and K. Kaneko for continual encouragement and thanks also Prof. H. Yoshida for useful discussions. He also thanks L. Mahadevan, X. Lu, X. Tang, and T. Iwamoto for stimulating comments on the reality of nonintegrable Hamiltonian systems during this summer seminar at Santa Fe.

REFERENCES

1. Ziglin, S. L. "Branching of Solutions and Non-Existence of First Integrals in Hamiltonian Mechanics. I." *Funct. Anal. Appl.* **16** (1983): 181–189.
2. Ziglin, S. L. "Branching of Solutions and Non-Existence of First Integrals in Hamiltonian Mechanics. II." *Funct. Anal. Appl.* **17** (1983): 6–17.
3. Yoshida, H. "A Criterion for the Non-Existence of an Additional Integral in Hamiltonian Systems with a Homogeneous Potential." *Physica* **29D** (1987): 128–142.
4. Yoshida, H. "A Criterion for the Non-Existence of an Additional Integral in Hamiltonian Systems with n Degrees of Freedom." *Phys. Lett.* **141A** (1989): 108–112.
5. Kovalevskaya, S. "Sur le probléme de la rotation d'un corps solide autour d'un point fixe." *Acta. Math.* **12** (1889): 177.
6. Kovalevskaya, S. "Sur une propriété du systéme d'équations différéntielles qui définit la rotation d'un corps solide autour d'un point fixe." *Acta. Math.* **14** (1890): 81.
7. Kovalevskaya, S. *A Russian Childhood*, edited by B. Stillman. New York: Springer, 1978.
8. Tabor, M. "Modern Dynamics and Classical Analysis." *Nature* **310** (1984): 277.
9. Ge, Z., and J. E. Marsden. "Lie-Poisson Hamilton-Jacobi Theory and Lie-Poisson Integrators." *Phys. Lett.* **133A** (1988): 134–139.
10. Umeno, K., and M. Suzuki. "Symplectic and Intermittent Behavior of Hamiltonian Flow." *Phys. Lett. A* **181A** (1993): 387–392.
11. Suzuki, M., and K. Umeno. "Higher-Order Decomposition Theory of Exponential Operators and Its Applications to QMC and Nonlinear Dynamics." *Computer Simulations in Condensed Matter Phys. VI*, edited by D. P. Landau, K. K. Mon, and H. B. Shuttler. Berlin: Springer-Verlag, 1993.
12. Umeno, K. "Nonintegrable Character of Hamiltonian Systems with Global and Symmetric Couplings." Preprint

Andreas Wagner
Yale University, Department of Biology, OML#327, P. O. Box 6666, New Haven, CT 06511; e-mail: waganda@doliolum.biology.yale.edu

Reductionism in Evolutionary Biology: A Perceptional Artifact?

THE PROBLEM

What is a genotype-phenotype map? Gene loci produce gene products, gene products interact and these "epigenetic" interactions create and maintain living things, organisms with phenotypes: in some convoluted way, genetic information is mapped onto phenotypes. To characterize this map is to characterize living things and different subdisciplines in biology focus on its different aspects. In evolutionary biology, one unsolved problem is central: how deep can natural selection penetrate this web of epigenetic interactions? Do the effects of natural selection always reach the individual gene, the basis of the epigenetic system, or do they act on a higher level of epigenetic organization? Whatever the answer is, it will determine the level of organization that should be the focus of our research efforts. Maybe, however, no one-for-all solution exists. Different case studies might yield different answers, some favoring genes as the "unit of selection," some favoring higher level entities. This, in and by itself, is not a problem. A problem is that limitations in available methods seem to create a perceptional artifact, the assumption—more implicit than pronounced—is that the proper level of focus is the individual gene. A brief overview of a debate related to the problem will be given; reasons for the persistence of the problem will be discussed.

THE DEBATE, ITS CONCEPTS . . .

There are defenders of the idea that higher levels of genetic organization than the individual gene, in the limit, the whole genome, are the relevant players in evolution. For example, Lewontin[12] concludes the review of a formidable amount of experimental data and theory with the statement:

> The fitness at a single locus, ripped from its interactive context, is about as relevant to real problems of evolutionary genetics as the study of the psychology of individuals isolated from their social context is to an understanding of man's sociopolitical evolution. In both cases context and interaction are not simply second-order effects to be superimposed on a primary monadic analysis. Context and interaction are of the essence.

From the other side of the ideological fence, Williams[15] writes:

> No matter how functionally dependent a gene may be, and no matter how complicated its interactions with other genes and environmental factors, it must always be true that a given gene substitution will have an arithmetic mean effect on fitness in any population. One allele can always be regarded as having a certain selection coefficient relative to another at the same locus at any given point in time. Such coefficients are numbers that can be treated algebraically, and conclusions inferred from one locus can be iterated over all loci. Adaptation can thus be attributed to the effect of selection acting independently at each locus.

WHOM ARE WE TO BELIEVE?

The debate gained widespread public attention with Dawkins'[4,5,6] advocacy of gene selectionism. If nothing else, his contributions have led to conceptual advances: he introduced the concept of the "replicator" to the debate, a concept that was subsequently used and extended by others to a canonical terminological framework. In this terminology, natural selection takes place in a world of "replicators" and "interactors," implying the distinction between "units of selection" and "levels of selection."

According to Dawkins[4] a "replicator" is "any entity in the universe which interacts with its world, including other replicators, in such a way that copies of itself are made." DNA molecules, dividing cells, and asexually reproducing multicellular organisms, along with their genomes, may qualify as replicators, but sexually reproducing organisms and their genomes may not. Sexual reproduction is quite another matter, since genetic recombination is involved.

Hull introduces the concept of the "interactor" and defines it[8] as "an entity that directly interacts as a cohesive whole with its environment in such a way that replication is differential." The individual organism is the prototypic example of an interactor, but there may well be interactors on different levels of organization, such as cell lineages within multicellular organisms. Within such lineages, selective processes may occur, favoring cells that divide more rapidly than others. Here, the dividing cell is the interactor within the organism as its environment. The interplay between different levels of interaction may profoundly affect the evolutionary potential of each level, as emphasized by Buss.[3]

Given these terms, the process of natural selection is defined as "a process in which the differential extinction and proliferation of interactors cause the differential perpetuation of the replicators that produced them."[8] In these terms, a "unit of selection" is any level of organization that qualifies as a replicator, and a "level of selection"[2] is any level of organization on which interaction can occur. Note that interactors and replicators may, but need not necessarily, designate different things. In the above example of cell lineages, cells are interactors as well as replicators. Their genome is a replicator, but not an interactor. And the multicellular organism in which they proliferate, provided that it is sexually reproducing, is an interactor but not a replicator.

. . .AND SOME SHORTCOMINGS

The definition of the replicator, as given above, harbors a conceptual problem. If replicators were always replicated with perfect accuracy, we would not be here to think about them. Changes in genetic replicators are what allows evolution. Since the definition of replicators is purely structural, every change, be it through mutation or recombination, creates a new replicator. And those replicators that are most abundant in a population because they convey higher fitness onto their carriers are most likely to be hit by mutations and, thus, extinguished. Considering its transience, Dawkins'[6] standpoint—that the beneficiary of all adaptation is the replicator—seems somewhat contrived. The strictly structural replicator concept may be too Platonic an idea to be perfectly fit for evolution. Here is a further argument for its inadequacy: if we do not want to seem prejudiced, we have at least to admit the possibility that phenomena typical for other "many body" systems occur in organisms. For example, the information necessary to produce a character may be "distributed" over a set of genes. If this is this case, the outcome may not be determined predominantly by the activity of individual gene products, but rather by patterns and strengths of epigenetic interactions. In other words, the organization of the epigenetic system might be the relevant carrier of information, analogous to what we know from different systems, e.g., neural networks. Changes in individual genes may very well have reproducible effects, but from them alone we may not learn much about the causal relationships that constitute this production system. It may have so many degrees of freedom that a multitude of allelic combinations may produce the same trait. All possible alleles on any contributing locus may be allowed, provided that the population contains appropriate alleles on other loci. No individual replicator might confer selective advantage to its carrier: "context and interaction are of the essence," as Lewontin says.

Should we look for a modification of the available concepts, or must the debate be redefined radically? There may be more appropriate ways to think about evolution than in the replicator-interactor framework. If so, they have not yet been found.

In sum, the adequate level of description for evolutionary phenomena has been the source of an ideologically charged debate for some decades. Standard concepts have partly been introduced by gene selectionists, a fact that may account for their inadequacy with respect to phenomena involving many interacting genes. Polygenic phenomena may require more elaborate concepts. But given the available experimental methods and the body of quantitative theory that characterize the state of the art in the life sciences, can we reasonably expect anything but concepts supporting reductionism to arise? This question lies at the heart of the problem.

THE HIDDEN, CRUCIAL ISSUES

While many students of evolution would agree to statements similar to Lewontin's, such agreement is often mere lipservice, as suggested by the current prevalence of a, sometimes simple-minded, molecular viewpoint of evolution and development. What might be the reasons?

The canonical way to study epigenetic interactions is by constructing mutants and in order to find out about how gene products interact, mutants in more than one gene are needed. However, the combinatorial explosion of possible mutant genotypes sets a very low limit to the number of genes whose interactions can be analyzed: to determine qualitative order relationships in a biochemical pathway, one usually constructs mutants in two genes. Formal genetic analysis of mutants in three genes often requires great sophistication. Moreover, formal genetics is most often not useful for an understanding of epigenetic interactions that is more than qualitative. This is where the toolbox of molecular biology has led to significant advances. But still, organisms are sources of noise, distorting most molecular signals heavily. Quantitative data on concentrations of gene products, on their activity, or on the strength of their interactions are sometimes hard and often impossible to obtain. These and other difficulties associated with molecular methods have not significantly increased the number of epigenetic interactions that can be analyzed simultaneously. And regardless of the method of analysis, one can only hope that all the factors uncontrollable by standard precautions such as, say, utilization of isogenic strains, do not contribute more than random noise.

Of the many phenomena we observe, we are likely to filter those that teach us causal relationships that are "simple," and available methods do not even allow us to *analyze* those that are complex: students of the epigenetic system are not blessed with multivariate data sets, let alone accurate ones. Thus, one must not be surprised by the prevalence of a reductionist viewpoint which is mostly supported by the lack of methods to investigate its antithesis.

Given that empirical methods have insufficient resolving power, is there any candidate body of theory, that (i) claims to be an accurate representation of epigenetic interactions and (ii) supports the possibility of higher order units of selection? The answer has two parts. The first regards models, the second regards concepts.

In biology, experiment and quantitative theory have a long history of interacting poorly. Next to no quantitative population genetical models on epigenetic interactions exist that are acceptable to the experimentalist, a deplorable situation that has its reasons in the methodological problems just outlined. But even browsing the literature for potentially relevant models, regardless of their empirical appeal, results in disappointment: the bulk of population genetical literature contains models in which a phenotypic trait is formed by *additive* interaction of some hypothetical, underlying genetic variables. Since vanishingly little data speaks in support of the assumption of additivity,[18] why use it? One reason is the formidable degree of complexity displayed by models involving even only a small number of loci. Lewontin[12] provides an example in which a simple dynamical system modeling five segregating loci in a population of diploid, sexually reproducing organism has approximately 4×10^9 equilibria, many of which might be stable. Such models have the potential to provide even the seasoned mathematician with a sense of despair. And if analytical approaches to many unrealistic additive models with few genes are not feasible, insight into realistic nonlinear systems with many genes seems even more hopeless. This is also reflected by the observation that the few available nonlinear models involve mostly small numbers of loci as well as types of nonlinearities tailored towards analytical tractability rather than biological realism. The distinction between additive and nonlinear models is emphasized here, because additive epigenetic interactions are bad candidates for higher order units of selection: changes at any locus are transmitted linearly onto the phenotypic level. Effects of allelic variation are therefore independent of the genetic context in which alleles occur.

In sum, available quantitative models are often biologically unrealistic or mathematically intractable, and frequently both. Moreover, their simplicity makes them inadequate tools to investigate the central conceptual question, which is: what level of complexity of epigenetic interactions is required to make selection affect a group of genes as a cohesive whole? What is a cohesive whole, anyway? In other words, if we look for higher order units of selection, what are we looking for? Note that the debate outlined above is only concerned with those biological entities that can *in principle*, i.e., because of their ontological status, be units of selection. It does not provide us with any operational criterion that would identify units of selection irreducible to lower levels of organization. Our current depth of understanding of such irreducible phenomena is best summarized by Mike Simmons, vice president of the Santa Fe Institute, who characterizes "complexity," a closely related and similarily ill-understood term: "It's a lot like the Supreme Court's definition of pornography: it's very hard to define, but you know it when you see it."

Despite the absence of any criteria other than heuristic ones, some distinctions can help to localize the problem. On a very elementary level of understanding one might say that also in the unrealistic case of linear genotype-phenotype maps, genetic variables produce the phenotype "jointly" and respond therefore as a unit to selective forces on the phenotype. However, as discussed above, this is unlikely to be satisfactory. Caution is also appropriate when the existence of correlations between alleles at different loci is taken as a hint towards the presence of higher order units of selection. Such correlations may exist for reasons unrelated to selective forces, e.g., genetic drift.[13] Their existence does not necessarily mean that the respective genes are functionally related in any way.

Wimsatt[17] proposed an operational criterion for higher order units of selection that makes use of Lewontin's notion of context dependence. It is based on the fact that on a nonlinear fitness landscape (i.e., a genotype-phenotype map where the phenotype is fitness itself), one's position on the landscape determines what amount of variation in the units of genetic variation is translated into variation in fitness. Details of a somewhat technical discussion can be found in Wimsatt[17] as well as in Lloyd.[10] According to this criterion, most nonlinear epigenetic interactions imply the existence of higher order units of selection. One might, however, feel ill at ease with such an abundance of higher order units of selection and it might be argued that more than simple nonlinearity is required. Consider, for the purpose of the argument, three fundamentally different types of hypothetical genotype-phenotype maps from N units of genetic variation, $\{G_1, \ldots, G_N\}$, to some phenotype (fitness), P. First, the linear map $P = G_1 + \ldots + G_N$; second, a "simple" nonlinear map, such as the quadratic form $P = \sum_{i,j} G_i G_j$; third, some map that can not be written down in closed form and that transforms genotype space by folding, contracting and stretching it in some complicated way, as in many well known examples from the theory of nonlinear dynamics. The persistence of biology as a field of active research indicates that the latter type is the one closest to reality. According to Wimsatt's criterion, no map of type one, but many maps of type two will imply the existence of higher order units of selection. However, common sense suggests that maps one and two are very similar with regard to one basic property. Statistical regression analysis, linear or nonlinear, can be used to predict P from $\vec{G}$. We feel that in both types we can quantitatively "understand" the production of the phenotype in terms of each underlying variable, since we can describe the map analytically; we can precisely map the "causal" relationship between each of the variables and the phenotype. Therefore, we might be tempted to call the phenotype not emergent, but reducible. The gene and no higher level of organization would be the unit of selection. According to this viewpoint, only those maps that drive even Laplace's demon close to capitulation, might pose a problem. The theory of dynamical systems teaches us that our failure to analytically understand some type three maps may be an intrinsic property of the map, and not of our incompetence. In these cases, only a phenomenological description of the system as a whole may be possible. Currently, however, merely heuristic criteria are available for identification of the kind of nonlinearities that make maps "complex."

At issue is our notion of causality, as the above examples demonstrate. When can we speak of "distributed causality?" Is causality distributed in every system with more than one independent variable? Is any form of nonlinear dependence sufficient? Do we need genuinely complex maps? It seems that we have not developed notions of causality adequate to the analysis of complex systems, a shortcoming that biology shares with the physical sciences.

THE FUTURE

Three imperfections—experimental, theoretical, and conceptual—mutually consolidating each other's persistence, form a stable ecology that prevents any major shift in our perception of evolutionary processes, a shift towards entities more inclusive than the individual gene. The virtual absence of quantitative experimental data on complex epigenetic interactions, together with the sparse representation of such interactions in our mathematical models are especially fatal: experimentalists deride theorists, but, equally deplorable, the bulk of available experimental data is unfit for theory development. Our simplistic notion of causality further stabilizes the deadlock. These imperfections, taken together, prevent us from identifying order on higher levels of organization. Our perceptional filter is not fit to detect it.

The sociological peculiarities of the life sciences indicate that only experiment, if anything, will be able to destabilize this ecology. Do we have reasons to hope for advances? "Cross-talking" in signal transduction processes, "genetic redundancy" in development and, "networks" of genetic regulators are catchwords that have appeared in the jargon of molecular biology in recent years.[7,9,14] Their usage is still anecdotal but already widespread. The underlying empirical observations are reminiscent of phenomena observed in completely different systems, systems with distributed representations of information, systems that are among the best understood examples of complex behavior, systems that will not be described efficiently on the level of their parts.[1] The paradigm shift may be on its way.

ACKNOWLEDGMENTS

I am grateful to Maria J. Blanco, Leo Buss, Janet Stites, Günter P. Wagner, and Martha Woodruff for critical comments and helpful suggestions.

REFERENCES

1. Amit, D. J. *Modeling Brain Function.* Cambridge University Press, 1989.
2. Brandon, R. *The Levels of Selection.* PSA 1982, Vol. 1, 315–322. East Lansing, MI: Philosophy of Science Association, 1982.
3. Buss, L. W. *The Evolution of Individuality.* Princeton University Press, 1987.
4. Dawkins, R. *The Selfish Gene.* New York: Oxford University Press, 1976.
5. Dawkins, R. "Replicator Selection and the Extended Phenotype." *Z. Tierpsychol.* **47**, (1978): 61–76.
6. Dawkins, R. *The Extended Phenotype.* San Francisco: W. H. Freeman, 1982.
7. Hoffmann, M. F. "*Drosophila* and Genetic Redundancy in Signal Transduction." *Trends in Genetics* **7** (1991): 351–355.
8. Hull, D. L. "Individuality and Selection." *Annu. Rev. Ecol. Syst.* **11** (1980): 311–332.
9. Ingham, P. W. "The Molecular Genetics of Embryonic Pattern Formation in *Drosophila.*" *Nature* **335** (1988): 25–32.
10. Lloyd, E. A. *The Structure and Confirmation of Evolutionary Theory.* Westport, CT: Greenwood Press, 1988.
11. Lewontin, R. C. "The Units of Selection." *Annu. Rev. Ecol. Syst.* **1** (1970): 1–18.
12. Lewontin, R. C. *The Genetic Basis of Evolutionary Change.* New York, NY: Columbia University Press, 1974.
13. Ohta, T., and M. Kimura. "Linkage Disequilibrium Due to Random Genetic Drift." *Genet. Res.* **13** (1969): 47–55.
14. Schüle, R., and R. E. Evans. "Cross-coupling of Signal Transduction Pathways: Zinc Finger Meets Leucine Zipper." *Trends in Genetics* **7** (1991): 377–381.
15. Williams, G. C. *Adaptation and Natural Selection.* Princeton University Press, 1966
16. Williams, G. C. *Natural Selection: Domains, Levels and Challenges.* New York: Oxford University Press, 1992
17. Wimsatt, W. C. *Units of Selection and the Structure of the Multi-Level Genome.* PSA 1980, Vol. 2, 122–183. East Lansing, MI: Philosophy of Science Association, 1980.

18. Wright, S. *Evolution and the Genetics of Populations*, Vol. 1. Chicago, IL: University of Chicago Press, 1968.

Index

D

E

F

G

H

I

J

K

L

M

Q

R

S

T

U

V

W

Z